Problems in Classical Electromagnetism

Andrea Macchi · Giovanni Moruzzi ·
Francesco Pegoraro

Problems in Classical Electromagnetism

203 Exercises with Solutions

Second Edition

Andrea Macchi
Research unit "Adriano Gozzini"
CNR/INO
Pisa, Italy

Giovanni Moruzzi
Department of Physics "Enrico Fermi"
Universita di Pisa
Pisa, Italy

Francesco Pegoraro
Department of Physics "Enrico Fermi"
University of Pisa
Pisa, Italy

ISBN 978-3-031-22237-5   ISBN 978-3-031-22235-1   (eBook)
https://doi.org/10.1007/978-3-031-22235-1

This Springer imprint is published by the registered company Springer Nature Switzerland AG
The registered company address is: Gewerbestrasse 11, 6330 Cham, Switzerland

Preface to the Second Edition

The second edition of this book contains 46 new problems which, as the preceding 157 ones, originate from our teaching experience. The lines along which the problems are presented are the same as in the first edition. In particular, we kept trying to relate these problems to real applications.

While most problems have been created for final examinations, a few of them are more appropriate to be discussed in a class. These latter problems may require a bigger mathematical effort; however, they still do not require preresiquites beyond the typical level of a second-year undergraduate course. A teacher willing to use these more challenging problems for an exam may simplify them and provide a "guided" text with additional hints.

Some of the 157 problems of the first edition have been revised for additional clarity or insight. For these improvements, a precious and continuous input has been given by our students whom we warmly acknowledge.

Pisa, Tuscany, Italy
September 2022

Andrea Macchi
Giovanni Moruzzi
Francesco Pegoraro

Preface to the First Edition

This book comprises 157 problems in classical electromagnetism, originating from the second-year course given by the authors to the undergraduate students of Physics at the University of Pisa in the years from 2002 to 2017. Our course covers the basics of classical electromagnetism in a fairly complete way. In the first part we present electrostatics and magnetostatics, electric currents and magnetic induction, introducing the complete set of Maxwell's equations. The second part is devoted to the conservation properties of Maxwell's equations, the classical theory of radiation, the relativistic transformation of the fields, and the propagation of electromagnetic waves in matter or along transmission lines and waveguides. Typically, the total amount of lectures and exercise classes is about 90 and 45 hours, respectively. Most of the problems of this book were prepared for the intermediate and final examinations. In an examination test, a student is requested to solve two or three problems in three hours. The more complex problems are presented and discussed in detail during the classes.

The prerequisite for tackling these problems is having successfully passed the first year of undergraduate studies in Physics, Mathematics or Engineering, acquiring a good knowledge of elementary classical mechanics, linear algebra, differential calculus for functions of one variable. Obviously, classical electromagnetism requires differential calculus involving functions of more than one variable. This, in our undergraduate programme, is taught in parallel courses in the second year. Typically, however, the basic concepts needed to write down the Maxwell equations in differential form are introduced and discussed in our electromagnetism course, in the simplest possible way. Actually, while we do not require higher mathematical methods as a prerequisite, the electromagnetism course is probably the place where the students will encounter for the first time topics such as Fourier series and transform, at least in a heuristic way.

In our approach to teaching we are convinced that checking the ability to solve a problem is the best way, or perhaps the only way, to verify the understanding of the theory. At the same time, the problems offer examples of the application of the theory to the real world. For this reason, we present each problem with a title that often highlights its connection to different areas of Physics or technology, so that the book is also a survey of historical discoveries and applications of classical electro-magnetism. We tried in particular to pick examples from different contexts, such as, e.g., Astrophysics or Geophysics, and to include topics that, for some reason, seem not to be considered in several important textbooks, such as, e.g., radiation pressure or homopolar/unipolar motors and generators. We also included a few examples inspired by recent and modern research areas, including, e.g., optical metamaterials, plasmonics, superintense lasers. These latter topics show that nowadays, more than 150 years after Maxwell's equations, classical electromagnetism is still a vital area, which continuously needs to be understood and revisited in its deeper aspects. These certainly cannot be covered in detail in a second year course, but a selection of examples (with the removal of unnecessary mathematical complexity) can serve as a useful introduction to them. In our problems, the students can have a first glance at "advanced" topics such as, e.g, the angular momentum of light, longitudinal waves and surface plasmons, the principles of laser cooling and of optomechanics, or the longstanding issue of radiation friction. At the same time, they can find the essential notions on, e.g., how an optical fiber works, where a plasma display gets its name from, or the principles of funny homemade electrical motors seen on YouTube.

The organization of our book is inspired by at least two sources, the book *Selected Problems in Theoretical Physics* (ETS Pisa, 1992, in Italian; World Scientific, 1994, in English) by our former teachers and colleagues A. Di Giacomo, G. Paffuti and P. Rossi, and the great archive of *Physics Examples and other Pedagogic Diversions* by Prof. K. McDonald (http://puhep1.princeton.edu/%7Emcdonald/examples/) which includes probably the widest source of advanced problems and examples in classical electromagnetism. Both these collections are aimed at graduate and postgraduate students, while our aim is to present a set of problems and examples with valuable physical contents, but accessible at the undergraduate level, although hopefully also a useful reference for the graduate student as well.

Because of our scientific background, our inspirations mostly come from the physics of condensed matter, materials, and plasmas as well as from optics, atomic physics, and laser-matter interactions. It can be argued that most of these subjects essentially require the knowledge of quantum mechanics. However, many phenomena and applications can be introduced within a classical framework, at least in a phenomenological way. In addition, since classical electromagnetism is the first field theory met by the students, the detailed study of its properties (with particular regard to conservation laws, symmetry relations, and relativistic covariance) provides an important training for the study of wave mechanics and quantum field theories, that the students will encounter in their further years of Physics study.

In our book (and in the preparation of tests and examinations as well), we tried to introduce as many original problems as possible, so that we believe that we have reached a substantial degree of novelty with respect to previous textbooks. Of course, the book also contains problems and examples which can be found in existing literature: this is unavoidable since many classical electromagnetism problems are, indeed, classics! In any case, the solutions constitute the most important part of the book. We did our best to make the solutions as complete and detailed as possible, taking typical questions, doubts, and possible mistakes by the students into account. When appropriate, alternative paths to the solutions are presented. To some extent, we tried not to bypass tricky concepts and ostensible ambiguities or "paradoxes" which, in classical electromagnetism, may appear more often than one would expect.

The sequence of Chaps. 1–12 follows the typical order in which the contents are presented during the course, each Chapter focusing on a well-defined topic. Chapter 13 contains a set of problems where concepts from different chapters are used, and may serve for a general review. To our knowledge, in some undergraduate programs the second-year Physics may be "lighter" than at our Department, i.e., mostly limited to the contents presented in the first six Chapters of our book (i.e., up to Maxwell's equations) plus some preliminary coverage of radiation (Chap. 10) and wave propagation (Chapter 11). Probably this would be the choice also for Physics courses in the Mathematics or Engineering programs. In a Physics program, most of the contents of our Chaps. 7–12 might be possibly presented in a more advanced course at the 3rd year, for which we believe our book can still be an appropriate tool.

Of course, this book of problems must be accompanied by a good textbook explaining the theory of the electromagnetic field in detail. In our course, in addition to lecture notes (unpublished so far) we mostly recommend the volume II of the celebrated *Feynman Lectures on Physics* and the volume 2 of the *Berkeley Physics Course* by E. M. Purcell. For some advanced topics the famous *Classical Electrodynamics* by J. D. Jackson is also recommended, although most of this book is adequate for an higher course. The formulas and brief descriptions given at the beginning of the chapter are not meant at all to provide a complete survey of theoretical concepts, and should serve mostly as a quick reference for most important equations and to clarify the notation we use as well.

In the first Chaps. 1–6, we use both the SI and Gaussian c.g.s. system of units. This choice was made because, while we are aware of the wide use of SI units, still we believe the Gaussian system to be the most appropriate for electromagnetism because of fundamental reasons, such as the appearance of a single fundamental constant (the speed of light c) or the same physical dimensions for the electric and magnetic fields, which seems very appropriate when one realizes that such fields are parts of the same object, the electromagnetic field. As a compromise we used both units in that part of the book which would serve for a "lighter" and more general course as defined above, and switched definitely (except for a few problems) to Gaussian units in the "advanced" part of the book, i.e., Chaps. 7–13. This choice is similar to what was made in the 3rd Edition of the famous *Classical Electrodynamics* book by J. D. Jackson.

Problem-solving can be one of the most difficult tasks for the young physicist, but also one of the most rewarding and entertaining ones. This is even truer for the older physicist who tries to *create* a new problem, and admittedly we learned a lot from this activity which we pursued for 15 years (some say that the only person who certainly learns something in a course is the teacher!). Over this long time, occasionally we shared this effort and amusement with colleagues including in particular Francesco Ceccherini, Fulvio Cornolti, Vanni Ghimenti and Pietro Menotti, whom we wish to

warmly acknowledge. Our final thanks go to the students who did their best to solve
these problems, contributing to an essential extent to improve them.

Pisa, Tuscany, Italy   Andrea Macchi
May 2017   Giovanni Moruzzi
                                                                                      Francesco Pegoraro

Contents

Contents

Chapter 1
Basics of Electrostatics

Topics. The electric charge. The electric field. The superposition principle. Gauss's law. Symmetry considerations. The electric field of simple charge distributions (plane layer, straight wire, sphere). Point charges and Coulomb's law. The equations of electrostatics. Potential energy and electric potential. The equations of Poisson and Laplace. Electrostatic energy. Multipole expansions. The field of an electric dipole.

Units. An aim of this book is to provide formulas compatible with both SI (French: *Système International d'Unités*) 3 units in this chapter and Chaps. 2–6, while only Gaussian units will be used in Chaps. 7–13. This is achieved by introducing some system-of-units-dependent constants.

The first constant we need is *Coulomb's constant*, k_e, which for instance appears in the expression for the force between two electric point charges q_1 and q_2 in vacuum, with position vectors $\mathbf{r}_1$ and $\mathbf{r}_2$, respectively. The Coulomb force acting, for instance, on q_1 is

$$\mathbf{f}_1 = k_e \frac{q_1 q_2}{|\mathbf{r}_1 - \mathbf{r}_2|^2} \, \hat{\mathbf{r}}_{12}, \qquad (1.1)$$

where k_e is Coulomb's constant, dependent on the units used for force, electric charge, and length. The vector $\mathbf{r}_{12} = \mathbf{r}_1 - \mathbf{r}_2$ is the distance from q_2 to q_1, pointing towards q_1, and $\hat{\mathbf{r}}_{12}$ the corresponding unit vector. Coulomb's constant is

$$k_e = \begin{cases} \dfrac{1}{4\pi\varepsilon_0} = 8.987\cdots \times 10^9 \ \text{N}\cdot\text{m}^2\cdot\text{C}^{-2} \simeq 9 \times 10^9 \ \text{m/F} & \text{SI} \\ 1 & \text{Gaussian.} \end{cases} \qquad (1.2)$$

© The Author(s), under exclusive license to Springer Nature Switzerland AG 2023
A. Macchi et al., *Problems in Classical Electromagnetism*,
https://doi.org/10.1007/978-3-031-22235-1_1

Constant $\varepsilon_0 \simeq 8.854\,187\,817\,620\cdots \times 10^{-12}$ F/m is the so-called "dielectric permittivity of free space", and is defined by the formula

$$\varepsilon_0 = \frac{1}{\mu_0 c^2} , \tag{1.3}$$

where $\mu_0 = 4\pi \times 10^{-7}$ H/m (by definition) is the vacuum magnetic permeability, and c is the speed of light in vacuum, $c = 299\,792\,458$ m/s (this is a precise value, since the length of the meter is defined from this constant and the international time standard).

Basic equations The two basic equations of this Chapter are, in differential and integral form,

$$\nabla \cdot \mathbf{E} = 4\pi k_e \varrho , \qquad\qquad \oint_S \mathbf{E} \cdot d\mathbf{S} = 4\pi k_e \int_V \varrho \, d^3 r \tag{1.4}$$

$$\nabla \times \mathbf{E} = 0 , \qquad\qquad \oint_C \mathbf{E} \cdot d\boldsymbol{\ell} = 0 . \tag{1.5}$$

where $\mathbf{E}(\mathbf{r}, t)$ is the electric field, and $\varrho(\mathbf{r}, t)$ is the volume charge density, at a point of location vector $\mathbf{r}$ at time t. The infinitesimal volume element is $d^3 r = dx\,dy\,dz$. In (1.4) the functions to be integrated are evaluated over an arbitrary volume V, or over the surface S enclosing the volume V. The function to be integrated in (1.5) is evaluated over an arbitrary closed path C. Since $\nabla \times \mathbf{E} = 0$, it is possible to define an electric potential $\varphi = \varphi(\mathbf{r})$ such that

$$\mathbf{E} = -\nabla\varphi . \tag{1.6}$$

The general expression of the potential generated by a given charge distribution $\varrho(\mathbf{r})$ is

$$\varphi(\mathbf{r}) = k_e \int_V \frac{\varrho(\mathbf{r}')}{|\mathbf{r} - \mathbf{r}'|} \, d^3 r' . \tag{1.7}$$

The force acting on a volume charge distribution $\varrho(\mathbf{r})$ is

$$\mathbf{f} = \int_V \varrho(\mathbf{r}')\, \mathbf{E}(\mathbf{r}')\, d^3 r' . \tag{1.8}$$

As a consequence, the force acting on a point charge q located at $\mathbf{r}$ (which corresponds to a charge distribution $\varrho(\mathbf{r}') = q\delta(\mathbf{r} - \mathbf{r}')$, with $\delta(\mathbf{r})$ the Dirac delta function) is

$$\mathbf{f} = q\,\mathbf{E}(\mathbf{r}) \,. \tag{1.9}$$

The electrostatic energy U_{es} associated with a given distribution of electric charges and fields is given by the following expressions

$$U_{es} = \int_V \frac{\mathbf{E}^2}{8\pi k_e}\, d^3r \,. \tag{1.10}$$

$$U_{es} = \frac{1}{2} \int_V \varrho\,\varphi\, d^3r \,, \tag{1.11}$$

Equations (1.10)–(1.11) are valid provided that the volume integrals are finite and that all involved quantities are well defined.

The multipole expansion allows us to obtain simple expressions for the leading terms of the potential and field generated by a charge distribution at a distance much larger than its extension. In the following we will need only the expansion up to the dipole term,

$$\varphi(\mathbf{r}) \simeq k_e \left(\frac{Q}{r} + \frac{\mathbf{p} \cdot \mathbf{r}}{r^3} + \cdots \right) \,, \tag{1.12}$$

where Q is the total charge of the distribution and the electric dipole moment is

$$\mathbf{p} \equiv \int_V \mathbf{r}'\rho(\mathbf{r}')\, d^3r' \,. \tag{1.13}$$

If $Q = 0$, then $\mathbf{p}$ is independent on the choice of the origin of the reference frame. The field generated by a dipolar distribution centered at $\mathbf{r} = 0$ is

$$\mathbf{E} = k_e \frac{3\hat{\mathbf{r}}\,(\mathbf{p} \cdot \hat{\mathbf{r}}) - \mathbf{p}}{r^3} \,. \tag{1.14}$$

We will briefly refer to a localized charge distribution having a dipole moment as "an electric dipole" (the simplest case being two opposite point charges $\pm q$ with a spatial separation $\boldsymbol{\delta}$, so that $\mathbf{p} = q\boldsymbol{\delta}$). A dipole placed in an external field $\mathbf{E}_{ext}$ has a potential energy

$$U_p = -\mathbf{p} \cdot \mathbf{E}_{ext} \,. \tag{1.15}$$

1.1   Overlapping Charged Spheres

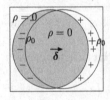

Fig. 1.1

We assume that a neutral sphere of radius R can be regarded as the superposition of two "rigid" spheres: one of uniform positive charge density $+\varrho_0$, comprising the nuclei of the atoms, and a second sphere of the same radius, but of negative uniform charge density $-\varrho_0$, comprising the electrons. We further assume that its is possible to shift the two spheres relative to each other by a quantity $\boldsymbol{\delta}$, as shown in Fig. 1.1, without perturbing the internal structure of either sphere.

Find the electrostatic field generated by the global charge distribution

(a) in the "inner" region, where the two spheres overlap,

(b) in the "outer" region, i.e., outside both spheres, discussing the limit of small displacements $\delta \ll R$.

1.2   Charged Sphere with Internal Spherical Cavity

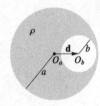

Fig. 1.2

A sphere of radius a has uniform charge density ϱ over all its volume, excluding a spherical cavity of radius $b < a$, where $\varrho = 0$. The center of the cavity, O_b is located at a distance $\mathbf{d}$, with $|\mathbf{d}| < (a - b)$, from the center of the sphere, O_a (Fig. 1.2). The mass distribution of the sphere is proportional to its charge distribution.

(a) Find the electric field inside the cavity.

Now we apply an external, uniform electric field $\mathbf{E}_0$. Find

(b) the force on the sphere,

(c) the torque with respect to the center of the sphere, and the torque with respect to the center of mass.

1.3   Energy of a Charged Sphere

A total charge Q is distributed uniformly over the volume of a sphere of radius R. Evaluate the electrostatic energy of this charge configuration in the following three alternative ways:

(a) Evaluate the work needed to assemble the charged sphere by moving successive infinitesimals shells of charge from infinity to their final location.

(b) Evaluate the volume integral of $u_E = |\mathbf{E}|^2/(8\pi k_e)$ where $\mathbf{E}$ is the electric field (Eq.(1.10)).

(c) Evaluate the volume integral of $\varrho\,\phi/2$ where ϱ is the charge density and ϕ is the electrostatic potential (Eq.(1.11)). Discuss the differences with the calculation made in (b).

1.4 Plasma Oscillations

A square metal slab of side L has thickness h, with $h \ll L$. The conduction-electron and ion densities in the slab are n_e and $n_i = n_e/Z$, respectively, Z being the ion charge.

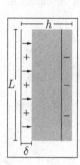

An external electric field shifts all conduction electrons by the same amount δ, such that $|\delta| \ll h$, perpendicularly to the base of the slab (Fig. 1.3). We assume that both n_e and n_i are constant, that the ion lattice is unperturbed by the external field, and that boundary effects are negligible.

(a) Evaluate the electrostatic field generated by the displacement of the electrons.

Fig. 1.3

(b) Evaluate the electrostatic energy of the system.

Now the external field is removed, and the "electron slab" starts oscillating around its equilibrium position.

(c) Find the oscillation frequency, at the small displacement limit ($\delta \ll h$).

1.5 Mie Oscillations

Now, instead of a the metal slab of Problem 1.4, consider a metal sphere of radius R. Initially, all the conduction electrons (n_e per unit volume) are displaced by $-\delta$ (with $\delta \ll R$) by an external electric field, analogously to Problem 1.1.

(a) At time $t = 0$ the external field is suddenly removed. Describe the subsequent motion of the conduction electrons under the action of the self-consistent electrostatic field, neglecting the boundary effects on the electrons close to the surface of the sphere.

(b) At the limit $\delta \to 0$ (but assuming $en_e\delta = \sigma_0$ to remain finite, i.e., the charge distribution is a surface density), find the electrostatic energy of the sphere as a function of δ and use the result to discuss the electron motion as in point (a).

1.6   Coulomb Explosions

At $t = 0$ we have a spherical cloud of radius R and total charge Q, comprising N point-like particles. Each particle has charge $q = Q/N$ and mass m. The particle density is uniform, and all particles are at rest.

(a) Evaluate the electrostatic potential energy of a charge located at a distance $r < R$ from the center at $t = 0$.

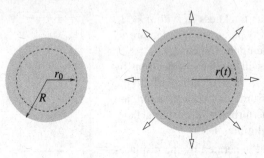

(b) Due to the Coulomb repulsion, the cloud begins to expand radially, keeping its spherical symmetry (Fig. 1.4). Assume that the particles do not overtake one another, i.e., that if two particles were initially located at $r_1(0)$ and $r_2(0)$, with $r_2(0) > r_1(0)$, then $r_2(t) > r_1(t)$ at any subsequent time $t > 0$. Consider the particles located in the infinitesimal spherical shell

Fig. 1.4

$r_0 < r_s < r_0 + dr$, with $r_0 + dr < R$, at $t = 0$. Show that the equation of motion of the layer is

$$m \frac{d^2 r_s}{dt^2} = k_e \frac{q Q}{r_s^2} \left(\frac{r_0}{R} \right)^3 \tag{1.16}$$

(c) Find the initial position of the particles that acquire the maximum kinetic energy during the cloud expansion, and determinate the value of such maximum energy.

(d) Find the energy spectrum, i.e., the distribution of the particles as a function of their final kinetic energy. Compare the total kinetic energy with the potential energy initially stored in the electrostatic field.

(e) Show that the particle density remains spatially uniform during the expansion.

1.7   Plane and Cylindrical Coulomb Explosions

Particles of identical mass m and charge q are distributed with zero initial velocity and uniform density n_0 in the infinite slab $|x| < a/2$ at $t = 0$. For $t > 0$ the slab expands because of the electrostatic repulsion between the pairs of particles.

(a) Find the equation of motion for the particles, its solution, and the kinetic energy acquired by the particles.

(b) Consider the analogous problem of the explosion of a uniform distribution having cylindrical symmetry.

1.8 Collision of Two Charged Spheres

Two rigid spheres have the same radius R and the same mass M, and opposite charges $\pm Q$. Both charges are uniformly and rigidly distributed over the volumes of the two spheres. The two spheres are initially at rest, at a distance $x_0 \gg R$ between their centers, such that their interaction energy is negligible compared to the sum of their "internal" (construction) energies.

(a) Evaluate the initial energy of the system.

The two spheres, having opposite charges, attract each other, and start moving at $t = 0$.

(b) Evaluate the velocity of the spheres when they touch each other (i.e. when the distance between their centers is $x = 2R$).

(c) Assume that, after touching, the two spheres penetrate each other without friction. Evaluate the velocity of the spheres when the two centers overlap ($x = 0$).

1.9 Oscillations in a Positively Charged Conducting Sphere

An electrically neutral metal sphere of radius a contains N conduction electrons. A fraction f of the conduction electrons ($0 < f < 1$) is removed from the sphere, and the remaining $(1 - f)N$ conduction electrons redistribute themselves to an equilibrium configurations, while the N lattice ions remain fixed.

(a) Evaluate the conduction-electron density and the radius of their distribution in the sphere.

Now the conduction-electron sphere is rigidly displaced by δ relatively to the ion lattice, with $|\delta|$ small enough for the conduction-electron sphere to remain inside the ion sphere.

(b) Evaluate the electric field inside the conduction-electron sphere.

(c) Evaluate the oscillation frequency of the conduction-electron sphere when it is released.

1.10 Interaction Between a Point Charge and an Electric Dipole

An electric dipole **p** is located at a distance **r** from a point charge q, as in Fig. 1.5. The angle between **p** and **r** is α.

(a) Evaluate the electrostatic force on the dipole.

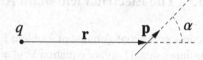

Fig. 1.5

(b) Evaluate the torque acting on the dipole with respect to the dipole center.

1.11 Electric Field of a Charged Hemispherical Surface

A hemispherical surface of radius R is uniformly charged with
surface charge density σ. Evaluate the electric field and potential at the center of curvature (hint: start from the electric field
of a uniformly charged ring along its axis) (Fig. 1.6).

Fig. 1.6

1.12 The Electric Field of an Array of Charged Wires (1)

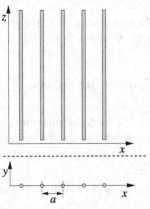

Fig. 1.7

Consider a uniformly spaced infinite array of parallel
charged wires in the $y = 0$ plane of a Cartesian reference system. The wires have infinite lengths, negligible diameters, are spaced by a and are parallel to
the z axis, as shown in Fig. 1.7. Each wire carries a
uniform linear electric charge density λ. Using suitable approximations evaluate the electric field and the
electrostatic potential
(a) far away from the array, i.e., at $|y| \gg a$;
(b) close to the wires, i.e., at distances $r \ll a$ from
each wire.
(c) Now show that the exact solution for the electric
potential is the following expression

$$\varphi(x, y) = -C \ln \left[2 \left(\cosh ky - \cos kx \right) \right] , \qquad (1.17)$$

where C and k are constants to be determined. Hint:

   i. Show that (1.17) is a solution of the two-dimensional Laplace equation $\nabla^2 \varphi = 0$;
  ii. determine k from the symmetry of the problem;
 iii. determine C from the boundary condition at great distances from the array;
 iv. check that the values obtained for k and C lead to the correct electrostatic
potential also close to the wires.

1.13 The Electric Field of an Array of Charged Wires (2)

Reconsider the configuration of Problem 1.12 and verify that (1.17) is a solution of the
two-dimensional Laplace equation $\nabla^2 \varphi = 0$ by exploiting the properties of the functions of complex variables in the representation of two dimensional electrostatic fields
[1]. More specifically introduce the complex conjugate variables $\zeta = (x + iy)/2$
and $\bar{\zeta} = (x - iy)/2$, (where a bar denotes complex conjugation) and

(a) Show that the two-dimensional Laplacian operator

$$\frac{\partial^2}{\partial x^2} + \frac{\partial^2}{\partial y^2} \tag{1.18}$$

can be written as

$$\frac{\partial^2}{\partial x^2} + \frac{\partial^2}{\partial y^2} = \frac{\partial}{\partial \zeta}\frac{\partial}{\partial \bar{\zeta}} = \frac{\partial}{\partial \bar{\zeta}}\frac{\partial}{\partial \zeta} \tag{1.19}$$

where ζ and $\bar{\zeta}$ are treated as independent variables.

Hint: use the chain rule for the x, y derivatives of a function of $\zeta = \zeta(x, y)$ and $\bar{\zeta} = \bar{\zeta}(x, y)$ [2]

(b) Show that the real function $\ln [2 (\cosh ky - \cos kx)]$ is the real part of a function of complex variable $f(\zeta)$ and thus by definition is a solution of the two-dimensional Laplace equation.

Hint use the following definitions of the trigonometric functions of complex variable:
$\cos (x + iy) = \cos (x) \cosh (y) - i \sin (x) \sinh (y), \quad \sin (x + iy) = \sin (x) \cosh (y)$
$+ i \cos (x) \sinh (y)$, *and the product-to-sum relationship*

$$\cos (\theta) - \cos (\phi) = -2 \sin [(\theta + \phi)/2)] \sin [(\theta - \phi)/2)]$$

where ϕ and θ are complex variables [3].

(c) Compute the imaginary part of the complex function corresponding to

$$\ln [2 (\cosh ky - \cos kx)].$$

Show that the curves of constant real and imaginary parts respectively are orthogonal to each other. Relate the curves of constant imaginary part to the electric field lines if the curves of constant real part correspond to constant values of the electric potential.

Hint: The orthogonality is a general property in the x-y plane of the curves of constant real and imaginary parts respectively and does not depend on the specific solution described above. Provide a general proof.

1.14 Mean Value Property and the Force on a Spherical Charge

Assume that in a given compact spatial region U there are no electric charges, so that the electric potential $\varphi(\mathbf{r})$ satisfies satisfies Laplace's equation $\nabla^2 \varphi = 0$ over U, i.e., $\varphi(\mathbf{r})$ is harmonic over U.

(a) Prove that the value of φ at the center C of a sphere of radius r contained in U equals the average value of φ on the surface S of the sphere

$$\varphi(C) = \overline{\varphi}(r) = \frac{1}{4\pi r^2} \oint_S \varphi(r, \theta, \phi) \, dS , \qquad (1.20)$$

where (r, θ, ϕ) is a spherical coordinate system with the origin a the center of the sphere. This is the *mean value property* of the harmonic functions. (Hint: show that $\overline{\varphi}(r)$ does not depend on r).

(b) A sphere of radius a carrying a spherically symmetric charge distribution is located in a space region where the electrostatic potential $\varphi(\mathbf{r})$, prior to the introduction of the sphere, was a solution of Laplace's equation. Show that the force exerted on the sphere by the external field is the same as if its whole charge were concentrated at its center. (Hint: calculate the potential energy of the sphere in the external potential $\varphi(\mathbf{r})$ using the mean value property of the harmonic functions).

References

1. R.P. Feynman, R.B. Leighton, M. Sands, *The Feynman Lectures in Physics*, Vol. II, 7–2 (Addison-Wesley Publishing Company, Reading-MA, London, Amsterdam, Don Mills-ON, Sydney, 2006)
2. https://en.wikipedia.org/wiki/Chain~rule
3. M. Abramovitz, I.A. Stegun, *Handbook of Mathematical Functions, Applied Mathematical Series*, vol. 55 (National Bureau of Standards, Washington, 1964)

Chapter 2
Electrostatics of Conductors

Topics. The electrostatic potential in vacuum. The uniqueness theorem for Poisson's equation. Laplace's equation, harmonic functions and their properties. Boundary conditions at the surfaces of conductors: Dirichlet, Neumann and mixed boundary conditions. The capacity of a conductor. Plane, cylindrical and spherical capacitors. Electrostatic field and electrostatic pressure at the surface of a conductor. The method of image charges: point charges in front of plane and spherical conductors.

Basic equations. Poisson's equation is

$$\nabla^2 \varphi(\mathbf{r}) = -4\pi k_e \, \varrho(\mathbf{r}) , \qquad (2.1)$$

where $\varphi(\mathbf{r})$ is the electrostatic potential, and $\varrho(\mathbf{r})$ is the electric charge density, at the point of vector position $\mathbf{r}$. The solution of Poisson's equation is unique if one of the following boundary conditions is true

1. **Dirichlet boundary condition**: φ is known and well defined on all of the boundary surfaces.
2. **Neumann boundary condition**: $\mathbf{E} = -\nabla\varphi$ is known and well defined on all of the boundary surfaces.
3. **Modified Neumann boundary condition** (also called Robin boundary condition): conditions where boundaries are specified as conductors with known charges.
4. **Mixed boundary conditions**: a combination of Dirichlet, Neumann, and modified Neumann boundary conditions:

Laplace's equation is the special case of Poisson's equation

$$\nabla^2 \varphi(\mathbf{r}) = 0 , \qquad (2.2)$$

which is valid in vacuum.

© The Author(s), under exclusive license to Springer Nature Switzerland AG 2023
A. Macchi et al., *Problems in Classical Electromagnetism*,
https://doi.org/10.1007/978-3-031-22235-1_2

2.1 Metal Sphere in an External Field

A metal sphere of radius R consists of a "rigid" lattice of ions, each of charge $+Ze$, and valence electrons each of charge $-e$. We denote by n_i the ion density, and by n_e the electron density. The net charge of the sphere is zero, therefore $n_e = Zn_i$. The sphere is located in an external, constant, and uniform electric field $\mathbf{E}_0$. The field causes a displacement $\boldsymbol{\delta}$ of the "electron sea" with respect to the ion lattice, so that the total field inside the sphere, $\mathbf{E}$, is zero. Using Problem 1.1 as a model, evaluate

(a) the displacement $\boldsymbol{\delta}$, giving a numerical estimate for $E_0 = 10^3$ V/m;

(b) the field generated by the sphere at its exterior, as a function of $\mathbf{E}_0$;

(c) the surface charge density on the sphere.

2.2 Electrostatic Energy with Image Charges

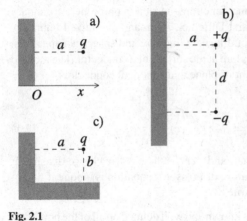

Fig. 2.1

Consider the configurations of point charges in the presence of conducting planes shown in Fig. 2.1. For each case, find the solution for the electrostatic potential over the whole space and evaluate the electrostatic energy of the system. Use the method of image charges.

(a) A charge q is located at a distance a from an infinite conducting plane.

(b) Two opposite charges $+q$ and $-q$ are at a distance d from each other, both at the same distance a from an infinite conducting plane.

(c) A charge q is at distances a and b, respectively, from two infinite conducting half planes forming a right dihedral angle.

2.3 Fields Generated by Surface Charge Densities

Consider the case **(a)** of Problem 2.2: we have a point charge q at a distance a from an infinite conducting plane.

(a) Evaluate the surface charge density σ, and the total induced charge q_{ind}, on the plane.

(b) Now assume to have a *nonconducting* plane with the same surface charge distribution as in point **(a)**. Find the electric field in the whole space.

(c) A non conducting spherical surface of radius a has the same charge distribution as the conducting sphere of Problem 2.4. Evaluate the electric field in the whole space.

2.4 A Point Charge in Front of a Conducting Sphere

A point charge q is located at a distance d from the center of a conducting grounded sphere of radius $a < d$ (Fig. 2.2). Evaluate

(a) the electric potential φ over the whole space;

(b) the force on the point charge;

(c) the electrostatic energy of the system.

Answer the above questions also in the case of an isolated, uncharged conducting sphere.

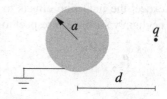

Fig. 2.2

2.5 Dipoles and Spheres

An electric dipole $\mathbf{p}$ is located at a distance d from the center of a conducting sphere of radius a. Evaluate the electrostatic potential φ over the whole space assuming that (Fig. 2.3)

(a) $\mathbf{p}$ is perpendicular to the direction from $\mathbf{p}$ to the center of the sphere,

(b) $\mathbf{p}$ is directed towards the center of the sphere.

(c) $\mathbf{p}$ forms an arbitrary angle θ with respect to the straight line passing through the center of the sphere and the dipole location.

In all three cases consider the two possibilities of (i) a grounded sphere, and (ii) an electrically uncharged isolated sphere.

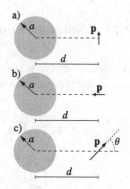

Fig. 2.3

2.6 Coulomb's Experiment

Coulomb, in his original experiment, measured the force between two charged metal spheres, rather than the force between two "point charges". We know that the field of a sphere whose surface is uniformly charged equals the field of a point charge, and that the force between two charge distributions, each of spherical symmetry, equals the force between two point charges

$$F = k_e \frac{q_1 q_2}{r^2} \, \hat{r}, \tag{2.3}$$

where q_1 and q_2 are the charges on the spheres, and $\mathbf{r} = r\,\hat{r}$ is the distance between the two centers of symmetry. But we also know that electric induction modifies the surface charge densities of conductors, so that a correction to (2.3) is needed. We expect the induction effects to be important if the radius a of the spheres is not negligibly small with respect to r (Fig. 2.4).

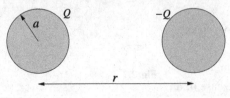

(a) Using the method of image charges, find the solution for the electrical potential outside the spheres as a series expansion, and identify the expansion parameter. For simplicity, assume the spheres to be identical and to have the same charge Q, as in the figure.

Fig. 2.4

(b) Evaluate the lowest order correction to the force between the spheres with respect to Coulomb's law (2.3).

2.7   A Solution Looking for a Problem

An electric dipole $\mathbf{p}$ is located at the origin of a Cartesian frame, parallel to the z axis, in the presence of a uniform electric field $\mathbf{E}$, also parallel to the z axis (Fig. 2.5).

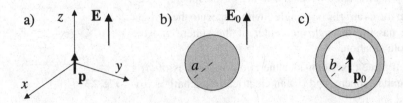

Fig. 2.5

(a) Find the total electrostatic potential $\varphi = \varphi(\mathbf{r})$, with the condition $\varphi = 0$ on the xy plane. Show that, in addition to the xy plane, there is another equipotential surface with $\varphi = 0$, that this surface is spherical, and calculate its radius R.

Now use the result from point (a) to find the electric potential in the whole space for the following problems:

(b) A conducting sphere of radius a is placed in a uniform electric field $\mathbf{E}_0$;

(c) a dipole $\mathbf{p}_0$ is placed in the center of a conducting spherical shell of radius b.

(d) Find the solution to problem **(c)** using the method of image charges.

2.8 Electrically Connected Spheres

Two conducting spheres of radii a and $b < a$, respectively, are connected by a thin metal wire of negligible capacitance. The centers of the two spheres are at a distance $d \gg a > b$ from each othe (Fig. 2.6). A total net charge Q is located on the system.

Evaluate to zeroth order approximation, neglecting the induction effects on the surfaces of the two spheres,

(a) how the charge Q is partitioned between the two spheres.

(b) the value V of the electrostatic potential of the system (assuming zero potential at infinity) and the capacitance $C = Q/V$,

(c) the electric field at the surface of each sphere, comparing the intensities and discussing the limit $b \to 0$.

(d) Now take the electrostatic induction effects into account and improve the preceding results to the first order in a/d and b/d.

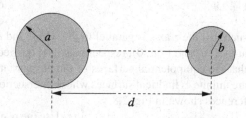

Fig. 2.6

2.9 A Charge Inside a Conducting Shell

A point charge q is located at a distance d from the center of a spherical conducting shell of internal radius $R > d$, and external radius $R' > R$. The shell is grounded, so that its electric potential is zero (Fig. 2.7).

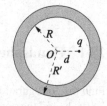

(a) Find the electric potential and the electric field in the whole space.

(b) Evaluate the force acting on the charge.

Fig. 2.7

(c) Show that the total charge induced on the surface of the internal sphere is $-q$.

(d) How does the answer to **(a)** change if the shell is not grounded, but electrically isolated with a total charge equal to zero?

2.10   A Charged Wire in Front of a Cylindrical Conductor

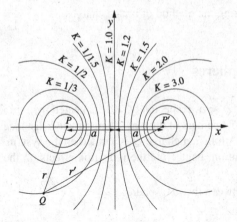

Fig. 2.8

We have two fixed points $P \equiv (-a, 0)$ and $P' \equiv (+a.0)$ on the xy plane, and a third, generic point $Q \equiv (x, y)$. Let $r = \overline{QP}$ and $r' = \overline{QP'}$ be the distances of Q from P and P', respectively.

(a) Show that the family of curves defined by the equation $r/r' = K$, with $K > 0$ a constant, is the family of circumferences drawn in Fig. 2.8.

(b) Now consider the electrostatic field generated by two straight infinite, parallel wires of linear charge densities λ and $-\lambda$, respectively. We choose a Cartesian reference frame such that the z axis is parallel to the wires, and the two wires intersect the xy plane at $(-a, 0)$ and $(+a, 0)$, respectively. Use the geometrical result of point **(a)** to show that the equipotential surfaces of the electrostatic field generated by the two wires are infinite cylindrical surfaces whose intersections with the xy plane are the circumferences shown in Fig. 2.8.

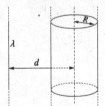

Fig. 2.9

(c) Use the results of points **(a)** and **(b)** to solve the following problem by the method of image charges. An infinite straight wire of linear charge density λ is located in front of an infinite conducting cylindrical surface of radius R. The wire is parallel to the axis of the cylinder, and the distance between the wire and axis of the cylinder is d, with $d > R$, as shown in Fig. 2.9. Find the electrostatic potential in the whole space.

2.11   Hemispherical Conducting Surfaces

Find the configurations of image charges that solve the problems represented in Fig. 2.10a, b, and the corresponding induced-charge distributions, remembering that the electric potential of an infinite conductor is zero.

(a) The plane infinite surface of a conductor has a hemispherical boss of radius R, with curvature center in O. A point charge q is located at a distance $a > R$ from O, the line segment from O to q forms an angle θ with the symmetry axis of the problem.

(b) An infinite conductor has a hemispherical cavity of radius R. A point charge q is located inside the cavity, at a distance $b < R$ from O. Again, the line segment from O to q forms an angle θ with the symmetry axis of the problem.

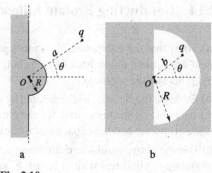

a   b

Fig. 2.10

2.12 The Force Between the Plates of a Capacitor

The plates of a flat, parallel-plate capacitor have surface S and separation $h \ll \sqrt{S}$. Find the force between the plates, both for an isolated capacitor (as a function of the constant charge Q), and for a capacitor connected to an ideal voltage source (as a function of the constant voltage V). In both cases, use *two* different methods, i.e., calculate the force

(a) from the electrostatic pressure on the surface of the plates,

(b) from the expression of the energy as a function of the distance between the plates.

2.13 Electrostatic Pressure on a Conducting Sphere

A conducting sphere of radius a has a net charge Q and it is electrically isolated. Find the electrostatic pressure at the surface of the sphere

(a) directly, from the surface charge density and the electric field on the sphere,

(b) by evaluating variation of the electrostatic energy with respect to a.

(c) Now calculate again the pressure on the sphere, assuming that the sphere is not isolated, but connected to an ideal voltage source, keeping the sphere at the constant potential V with respect to infinity.

2.14 Conducting Prolate Ellipsoid

(a) Show that the equipotential surfaces generated by a uniformly charged line segment are prolate ellipsoids of revolution, with the focal points coinciding with the end points of the segment.

(b) Evaluate the electric field generated by a conducting prolate ellipsoid of revolution of major axis $2a$ and minor axis $2b$, carrying a charge Q. Evaluate the electric capacity of the ellipsoid, and the capacity of a confocal ellipsoidal capacitor.

(c) Use the above results to evaluate an approximation for the capacity of a straight conducting cylindrical wire of length h, and diameter $2b$.

2.15 A Non-coaxial Cylindrical Capacitor

A non-coaxial cylindrical capacitor comprises two conducting cylindrical surfaces of radii R_1 and $R_2 > R_1$, respectively, both of height $h \gg R_2$, disposed as in Fig. 2.11. The axes of the two cylinders are parallel, separated by a distance $c < (R_2 - R_1)$, as shown in the figure. Evaluate the capacity of the system disregarding the border effects.

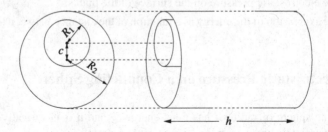

Fig. 2.11 .

Hint: use the equipotential cylindrical surfaces discussed in Problem 2.10.

2.16 Induced Charge Density on a Conducting Plane

A point charge q is located at a distance h from an infinite conducting plane at zero potential.

(a) Show, using Gauss's law, that the total induced charge on the plane is $-q$.

(b) Show that the induced charge on any area of the plane is proportional to the solid angle Ω that the area subtends from the point charge.

2.17 Charge Density on a Metal Sphere in Front of a Point Charge

A point charge q is located at a distance h from the center of a grounded conducting sphere of radius $a < h$, as shown in Fig. 2.12.

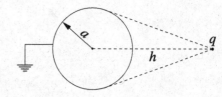

Fig. 2.12

(a) Evaluate the density of induced charge on the sphere surface.

(b) Evaluate the ratio of the induced charge on the part of the sphere visible from q to the induced charge on the rest of the sphere.

(c) Evaluate the density of induced charge on the sphere surface if the sphere is not grounded, but isolated and uncharged.

Chapter 3
Electrostatics of Dielectric Media

Topics. Polarization charges. Dielectrics. Permanent and induced polarization. The auxiliary vector **D**. Boundary conditions at the surface of dielectrics. Relative dielectric permittivity ε_r.

Basic equations. Let **P** denote the *electric polarization* (electric dipole moment per unit volume). Some special materials have a permanent non-zero value of **P**, but in many more cases a polarization appears in the presence of an electric field **E**. We consider linear dielectric materials fow which **P** and **E** are parallel and proportional, i.e.

$$\mathbf{P} = \begin{cases} \varepsilon_0 \chi \, \mathbf{E} \,, & \text{where } \chi = \varepsilon_r - 1 \,, \quad \text{SI} \\ \chi \, \mathbf{E} \,, & \text{where } \chi = \dfrac{\varepsilon_r - 1}{4\pi} \,, \quad \text{Gaussian,} \end{cases} \tag{3.1}$$

where χ is called the *electric susceptibility* and ε_r the *relative permittivity* of the material.[1] Notice that ε_r is a dimensionless quantity with the same numerical value both in SI and Gaussian units.

We shall denote by ϱ_b and ϱ_f the volume densities of bound electric charge and of free electric charge, respectively, and by σ_b and σ_f the surface densities of bound and free charge. Quantities ϱ_b and σ_b are related to the electric polarization **P** by

$$\varrho_b = -\nabla \cdot \mathbf{P} \,, \quad \text{and} \quad \sigma_b = \mathbf{P} \cdot \hat{\mathbf{n}} \,, \tag{3.2}$$

---

[1] In anisotropic media (such as non-cubic crystals) **P** and **E** may not be parallel and thus χ and ε_r are actually second rank tensors, but here we are interested only in isotropic and homogeneous media, for which χ and ε_r are scalar quantities.

© The Author(s), under exclusive license to Springer Nature Switzerland AG 2023
A. Macchi et al., *Problems in Classical Electromagnetism*,
https://doi.org/10.1007/978-3-031-22235-1_3

where $\hat{\mathbf{n}}$ is the unit vector pointing outwards from the boundary surface of the polarized material. We may thus write Eq. (1.4) as

$$\nabla \cdot \mathbf{E} = \begin{cases} \dfrac{\varrho_f + \varrho_b}{\varepsilon_0} = \dfrac{\varrho_f}{\varepsilon_0} - \dfrac{1}{\varepsilon_0} \nabla \cdot \mathbf{P} , & \text{SI} \\ 4\pi(\varrho_f + \varrho_b) = 4\pi \varrho_f - 4\pi \nabla \cdot \mathbf{P} , & \text{Gaussian.} \end{cases} \tag{3.3}$$

We may also introduce the auxiliary vector $\mathbf{D}$ (also called *electric displacement*) defined as

$$\mathbf{D} = \begin{cases} \varepsilon_0 \mathbf{E} + \mathbf{P} , & \text{SI,} \\ \mathbf{E} + 4\pi \mathbf{P} , & \text{Gaussian,} \end{cases} \tag{3.4}$$

so that

$$\nabla \cdot \mathbf{D} = \begin{cases} \varrho_f , & \text{SI,} \\ 4\pi \varrho_f , & \text{Gaussian.} \end{cases} \tag{3.5}$$

In addition, $\nabla \times \mathbf{E} = 0$ holds in static conditions. Thus, at the interface between two different dielectric materials, the component of $\mathbf{E}$ parallel to the interface surface, and the perpendicular component of $\mathbf{D}$ are continuous. In a material of electric permittivity ε_r

$$\mathbf{D} = \begin{cases} \varepsilon_0 \varepsilon_r \mathbf{E} , & \text{SI} \\ \varepsilon_r \mathbf{E} , & \text{Gaussian.} \end{cases} \tag{3.6}$$

To facilitate the use of the basic equations in this chapter also with the system independent units, we summarize some of them in the following Table 3.1.

Table 3.1

Quantity	SI	Gaussian	System independent
Polarization $\mathbf{P}$ of an isotropic dielectric medium of relative permittivity ε_r	$\varepsilon_0(\varepsilon_r - 1)\,\mathbf{E}$	$\frac{\varepsilon_r - 1}{4\pi}\,\mathbf{E}$	$\frac{\varepsilon_r - 1}{4\pi k_e}\,\mathbf{E}$
$\nabla \cdot \mathbf{E}$	$\frac{\varrho_f + \varrho_b}{\varepsilon_0}$	$4\pi(\varrho_f + \varrho_b)$	$4\pi k_e\,(\varrho_f + \varrho_b)$
$\nabla \cdot (\varepsilon_r \mathbf{E})$	$\frac{\varrho_f}{\varepsilon_0}$	$4\pi \varrho_f$	$4\pi k_e\,\varrho_f$
$\nabla \times \mathbf{E}$	0	0	0

3.1 An Artificial Dielectric

We have a tenuous suspension of conducting spheres, each of radius a, in a liquid
dielectric material of relative dielectric permittivity $\varepsilon_r = 1$. The number of spheres
per unit volume is n.
(a) Evaluate the dielectric susceptibility χ of the system as a function of the fraction
of the volume filled by the conducting spheres. Use the mean field approximation
(MFA), according to which the electric field may be assumed to be uniform throughout the medium.
(b) The MFA requires the field generated by a single sphere on its nearest neighbor
to be much smaller than the mean field due to the collective contribution of all the
spheres. Derive a condition on n and a for the validity of the MFA.

3.2 Charge in Front of a Dielectric Half-Space

A plane divides the whole space into two halves, one of
which is empty and the other filled by a dielectric medium
of relative permittivity ε_r. A point charge q is located in
vacuum at a distance d from the medium as shown in
Fig. 3.1.
(a) Find the electric potential and electric field in the
whole space, using the method of image charges.
(b) Evaluate the surface polarization charge density on
the interface plane, and the total polarization charge of
the plane.
(c) Find the field generated by the polarization charge in
the whole space.

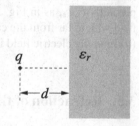

Fig. 3.1

3.3 An Electrically Polarized Sphere

Ferroelectricity is the property of some materials like Rochelle salt, carnauba wax,
barium titanate, lead titanate, ..., that possess a spontaneous electric polarization in
the absence of external fields.
(a) Consider a ferroelectric sphere of radius a and uniform polarization $\mathbf{P}$, in the
absence of external fields, and evaluate the electric field and the displacement field
in the whole space.
(b) Now consider again a ferroelectric sphere of radius a and uniform polarization
$\mathbf{P}$, but with a concentrical spherical hole of radius $b < a$. Evaluate the electric field
and the displacement field in the whole space.

3.4   Dielectric Sphere in an External Field

A dielectric sphere of relative permittivity ε_r and radius a is placed in vacuum, in an initially uniform external electric field $\mathbf{E}_0$, as shown in Fig. 3.2.
(a) Find the electric field in the whole space (hint: use the results of Problem 3.3 and the superposition principle).

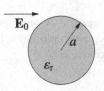

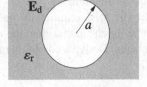

Fig. 3.2   **Fig. 3.3**

A spherical cavity of radius a is located inside an infinite dielectric medium of relative permittivity ε_r, as in Fig. 3.3. The system is in the presence of an external electric field which, far from the cavity (i.e. at a distance $\gg a$), is uniform and equal to $\mathbf{E}_d$.
(b) Find the electric field in the whole space.

3.5   Refraction of the Electric Field at a Dielectric Boundary

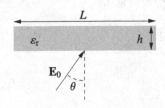

Fig. 3.4

A dielectric slab of thickness h, length $L \gg h$, and dielectric permittivity ε_r, is placed in an external uniform electric field $\mathbf{E}_0$. The angle between $\mathbf{E}_0$ and the normal to the slab surface is θ, as in Fig. 3.4.
(a) Find the electric field $\mathbf{E}'$ inside the slab and the angle θ' between $\mathbf{E}'$ and the normal to the slab surface.
(b) Find the polarization charge densities in the dielectric medium.
(c) Evaluate the torque exerted by the external field on the slab, if any. Neglect all boundary effects.

3.6 Contact Force Between a Conductor and a Dielectric

A conducting square slab of surface $S = a^2$ and thickness $h \ll a$ is in contact with a dielectric medium of relative permittivity ε_r. The dielectric medium is much larger than the slab, we can consider it a hemisphere of radius $R \gg a$, with the slab in contact with its base, as shown in Fig. 3.5a. Part (b) of Fig. 3.5 is an enlargement of the area enclosed in the dashed rectangle of part (a). With this assumption, we can assume the slab to be in contact with a semi-infinite medium filling the half-space $x > 0$, while we have vacuum

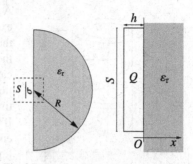

Fig. 3.5

in the half space $x < 0$. The conducting slab carries a total charge Q, and we assume that the boundary effects at its edges are negligible.

(a) Considering both the cases in which the slab is in contact with the dielectric, and in which it is displaced by an amount $\xi \ll a$ to the left, find the free charge densities on the left (σ_1) and right (σ_2) surfaces of the slab, the polarization charge density (σ_b) at the surface of the dielectric, and the electric field in the whole space.

(b) Calculate the electrostatic force acting on the slab.

(c) How do these results change if the dielectric medium is assumed to be an infinite (in the y and z directions) layer of *finite* thickness w in the x direction?

3.7 A Conducting Sphere Between Two Dielectrics

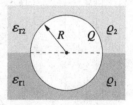

A conducting sphere of mass density ϱ and radius R floats in a liquid of density $\varrho_1 > 2\varrho$ and relative dielectric permittivity ε_{r1} in the presence of the gravitational field. Above the liquid there is a gaseous medium of mass density $\varrho_2 \ll \varrho$ and relative dielectric permittivity $\varepsilon_{r2} < \varepsilon_{r1}$.

Fig. 3.6

The sphere is given a charge Q such that exactly one half of its volume is submerged (Fig. 3.6). Evaluate

(a) the electric field in the whole space, the surface free charge densities on the sphere, and the surface polarization charge densities of the two dielectrics, as functions of R, ε_{r1}, ε_{r2} and Q;

(b) the value of Q.

3.8   Measuring the Dielectric Constant of a Liquid

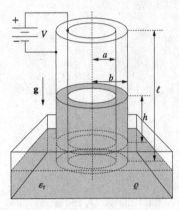

A cylindrical capacitor has internal radius a, external radius $b > a$, and length $\ell \gg b$, so that the boundary effects are negligible. The axis of the capacitor is vertical, and the bottom of the capacitor is immersed in a vessel containing a liquid of mass density ϱ and dielectric permittivity ε_r, in the presence of the gravitational field. If a voltage source maintains a potential difference V between the two cylindrical plates, the liquid rises for a height h in the cylindrical shell between the plates. Show how one can evaluate the value of ε_r from the measurement of h (Fig. 3.7).

Fig. 3.7   (This is a problem from Ref. [1]).

3.9   A Conducting Cylinder in a Dielectric Liquid

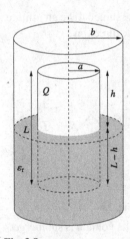

A conducting cylinder of mass M, radius a and height $L \gg a$ is immersed for a depth $L - h$ (with $h \gg a$) in a dielectric liquid having relative permittivity ε_r. The liquid is contained in a cylindrical vessel of radius $b > a$, with conducting lateral surface. A free charge Q is located on the internal cylinder. Boundary effects are assumed to be negligible. The cylinder is free to move vertically preserving its axis (Fig. 3.8). Find

(a) the electric field $\mathbf{E}(a)$ at the surface of the internal cylinder, and the surface charge densities;

(b) the electric field in the region between the lateral surface of the internal cylinder and the container of the liquid $(a < r < b)$;

(c) the electrostatic force on the internal cylinder.

Fig. 3.8   **(d)** Assume that the internal cylinder has mass M, and the liquid has mass density $\varrho > M/(\pi a^2 L)$. Discuss the

equilibrium conditions.

3.10   A Dielectric Slab in Contact with a Charged Conductor

A dielectric slab of relative permeability ε_r, thickness h and surface $S \gg h$ is in contact with a plane conducting surface, carrying a uniform surface charge density σ, as in Fig. 3.9. Boundary effects are negligible.
(a) Evaluate the electric field in the whole space.
(b) Evaluate the polarization surface-
charge densities on the dielectric surfaces.
(c) How do the answers to points (a) and
(b) change if the slab is moved at a distance $s < h$ from the conducting plane?

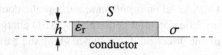

Fig. 3.9

How does the electrostatic energy of the system depend on s? Is there an interaction force between slab and conductor?

3.11   A Transversally Polarized Cylinder

An infinite cylinder of radius a has an internal uniform electric polarization **P**, perpendicular to its axis,
as shown in Fig. 3.10. Evaluate the electric charge density on the lateral surface of the cylinder, the electric
potential and the electric field in the whole space.

Hint: see Problem 1.1.

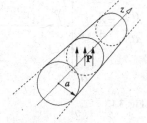

Fig. 3.10

3.12   Force Between a Parallel-Plate Capacitor and a Dielectric Slab

An isolated parallel-plate
capacitor consists of two
rectangular metal plates of
length a and width b, separated by distance $h \ll a, b$,
and has a charge Q. A
dielectric slab of dielectric
permittivity ε_r and edges of

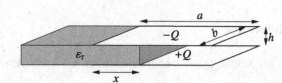

Fig. 3.11

lengths a, b and h, respectively, is partially inserted a depth $x \gg h$ into the capacitor,
as shown in Fig. 3.11. Assume that boundary effects can be neglected.
(a) Evaluate the force exerted by the capacitor on the slab.

(b) Now evaluate the force exerted on the slab assuming that the capacitor is not isolated, but connected to a battery that keeps it at a constant potential difference V.
(c) Explain qualitatively *where* the force is exerted.

3.13   A Cylindrical Quadrupole

Cylindrical multipole moments are the coefficients in a series expansion of the electrostatic potential generated by a 3D charge distribution that is translation invariant in one direction, for instance along the z axis of a Cartesian coordinate system. The expansion is in terms of r'/r, where r' and r are the distances from the z axis of the element of charge distribution and of the point where we evaluate the potential, respectively, under the assumption $r' \ll r$. Problem 3.11 provides a first example of a *cylindrical dipole* moment.

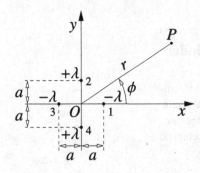

Fig. 3.12

As a special example of a *cylindrical quadrupole* consider four infinite straight, uniformly charged wires, parallel to one another and parallel to the z axis of a Cartesian coordinate system. Wires 1 and 3 carry linear charge densities $-\lambda$ and intersect the $z = 0$ plane at $(x = \pm a, y = 0)$, respectively, wires 2 and 4 carry linear charge densities $+\lambda$ and intersect the $z = 0$ plane at $(x = 0, y = \pm a)$, as shown in Fig. 3.12. This system has zero cylindrical monopole and dipole moments.

(a) Evaluate the electrostatic potential and the electric field generated by the four wires at a point P distant r from the z axis, with $r \gg a$. Hint: use a cylindrical coordinate system (r, ϕ, z) with the z axis in common with our Cartesian coordinate system, as shown in Fig. 3.12.

(b) Evaluate the electrostatic potential and the electric field generated by the four wires at a point P distant r from the z axis, with $r \ll a$.

Reference

1. J.D. Jackson, *Classical Electrodynamics* (Wiley, New York, 1975). Problem 4.13

Chapter 4
Electric Currents

Topics. Electric current density. Continuity equation. Stationary electric currents. Drude model for a conductor. Ohm's law. Joule heating.

Basic equations. The electric current density $\mathbf{J} = \mathbf{J}(\mathbf{r}, t)$ is the local flow of charge per unit area and surface, which appears in the continuity equation

$$\partial_t \varrho + \nabla \cdot \mathbf{J} = 0, \tag{4.1}$$

that states the conservation of the total electric charge. In integral form

$$\frac{dQ}{dt} \equiv \int_V \partial_t \varrho \, d^3 r = \int_S \mathbf{J} \cdot d\mathbf{S} \equiv I . \tag{4.2}$$

where Q is the total charge contained into the volume V bounded by the closed surface S. Usually the flux (or electric current) I is defined also for an open surface, as the total charge crossing the surface per unit time.

The quantity

$$w = \mathbf{J} \cdot \mathbf{E} \tag{4.3}$$

is interpreted as the work per unit time and volume done by the EM fields on a distribution of currents.

In a model of matter where there are several species of charged particles (labeled with the index s) each having a charge q_s, a density of particles $n_s = n_s(\mathbf{r}, t)$ and flowing with velocity $\mathbf{v}_s = \mathbf{v}_s(\mathbf{r}, t)$, the current density is given by

$$\mathbf{J} = \sum_s q_s n_s \mathbf{v}_s . \tag{4.4}$$

In a metal where electrons are the only charge carrier, $\mathbf{J} = -e n_e \mathbf{v}_e$.

© The Author(s), under exclusive license to Springer Nature Switzerland AG 2023
A. Macchi et al., *Problems in Classical Electromagnetism*,
https://doi.org/10.1007/978-3-031-22235-1_4

Drude's model for electrons in a metal assumes the classical equation of motion

$$m_e \frac{dv_e}{dt} = -e\mathbf{E} - m_e \nu v_e \,, \tag{4.5}$$

where ν is a phenomenological friction coefficient. In a steady state $(dv_e/dt = 0)$ this leads to Ohm's law for a conductor

$$\mathbf{J} = \frac{n_e e^2}{m_e \nu}\mathbf{E} \equiv \sigma \mathbf{E} \equiv \frac{\mathbf{E}}{\rho} \,, \tag{4.6}$$

where σ is the conductivity and $\rho = 1/\sigma$ the resistivity of the material.[1]

In a material satisfying (4.6), the latter implies that the current I flowing between two points (or layers) at different values of the electric potential, the potential drop V is proportional to I, leading to the definition of the resistance R:

$$V = RI \,. \tag{4.7}$$

In the common (but particular) example of a straight conductor of length ℓ and cross-section area A, such that the electric field is uniform inside the conductor, one obtains $R = \ell/(\sigma A) = \rho\ell/A$. The equations (Kirchoff's laws) describing DC electric circuits, i.e., networks of interconnected conductors each satisfying (4.7), can be found in any textbook and will not be repeated here.

Equation (4.7) is known as Ohm's law, but it is appropriate to use this name also for the underlying and more general Eq. (4.6) due to G. Kirchoff. An Ohmic conductor is defined as any material which satisfies (4.6). For such materials, Eq. (4.3) gives the power per unit volume dissipated into the material as a consequence of the friction term,

$$\mathbf{J} \cdot \mathbf{E} = \sigma E^2 = \frac{E^2}{\rho} \,, \tag{4.8}$$

which causes the heating of the material (Joule effect). For the above mentioned example, this is equivalent to state that the power dissipated into the whole conductor is $W = RI^2$.

Notice that all the above equations have the same form both in the SI and in the Gaussian system. However, the units of measure are different. For example, the current I is measure in C/s or Ampère (1 A = 1 C/S) in SI, and in statCoulomb/s or "statAmpère" in Gaussian units, while the resistance is measured in Ohms (Ω) in SI and in s/cm in Gaussian units. For the latter, σ has the dimensions of the inverse of a time, and is thus measured in s^{-1}, while ρ can be measured in s.

---

[1] Unfortunately the commonly used symbols for the resistivity and the conductivity are almost the same as those used for the volume and surface densities of charge. However, throughout the book the meaning of the symbols used should be clear from the context.

4.1   The Tolman-Stewart Experiment

The experiment of Tolman and Stewart [1] was conceived in
order to show that conduction in metals is due to electrons.
A metallic torus (ring) of major radius a and minor radius
b is spun at a very high angular velocity ω around its axis
(Fig. 4.1). We assume that $b \ll a$, so that the radial motion of
the charge carriers can be neglected. The cross section of the
ring is $S = \pi b^2$.

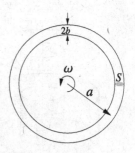

   At time $t = 0$ the rotation of the ring is suddenly stopped.
A current $I = I(t)$ flowing in the ring and decaying in time is
observed for $t > 0$.

Fig. 4.1

(a) Using the Drude model for conduction in metals, find $I =
I(t)$ and its characteristic decay time τ for a ring of copper (electrical conductivity
$\sigma \simeq 10^7 \ \Omega^{-1}\mathrm{m}^{-1}$ and electron density $n_e = 8.5 \times 10^{28} \ \mathrm{m}^{-3}$).
(b) Evaluate the charge that flows in the ring from $t = 0$ to $t = \infty$ as a function of σ.

4.2   Charge Relaxation in a Conducting Sphere

A conducting sphere of radius a and conductivity σ has a net charge Q. At time
$t = 0$ the charge is uniformly distributed over the volume of the sphere, with a
volume charge density $\varrho_0 = Q (3/4\pi a^3)$. Since in static conditions the charge in
an isolated conductor can only be located on the conductor's surface, for $t > 0$ the
charge progressively migrates to the surface of the sphere.
(a) Evaluate the time evolution of the charge distribution on the sphere, and of the
electric field everywhere in space. Give a numerical value for the time constant τ in
the case of a good conductor (e.g., copper).
(b) Evaluate the time evolution of the electrostatic energy of the sphere during the
charge redistribution.
(c) Show that the energy dissipated into Joule heat equals the loss of electrostatic
energy.

4.3 A Coaxial Resistor

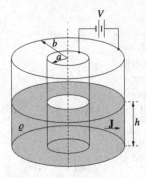

Two coaxial cylindrical plates of very low resistivity ρ_0 have radii a and b, respectively, with $a < b$. The space between the cylindrical plates is filled up to a height h with a medium of resistivity $\rho \gg \rho_0$, as in Fig. 4.2. A voltage source maintains a constant potential difference V between the plates.

(a) Evaluate the resistance R of the system.

(b) Discuss the relation between R and the capacitance of a cylindrical capacitor of radii a and b.

Fig. 4.2

4.4 Electrical Resistance Between Two Submerged Spheres (1)

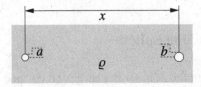

Fig. 4.3

(a) Two highly conducting spheres of radii a and b, respectively, are deeply submerged in the water of a lake, at a distance x from each other, with $x \gg a$ and $x \gg b$. The water of the lake has resistivity ρ. Evaluate the approximate resistance between the two spheres, using the results of the answer to point (b) of Problem 4.3 (Fig. 4.3).

(b) Now suppose that the two spheres are not completely submerged, but just sunk so that their centers are exactly at the level of the surface of the lake, as shown in Fig. 4.4. Evaluate the resistance between them.

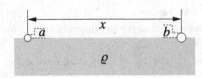

Fig. 4.4

4.5 Electrical Resistance Between Two Submerged Spheres (2)

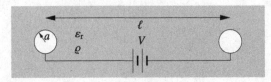

Fig. 4.5

Two identical, perfectly conducting spheres of radius a are immersed in a fluid of resistivity ρ and relative electric permittivity ε_r. The distance between the centers of the two spheres is $\ell \gg a$. A constant potential difference difference V is maintained between the spheres by a suitable voltage source (Fig. 4.5).

As a first approximation, assume the charge to be uniformly distributed over the surface of each sphere, neglecting electrostatic induction effects. Evaluate
(a) the charge on each sphere,
(b) the resistance R and the current I flowing between the spheres.
(c) Find the temporal law and the decay time for the discharge of the spheres when the voltage source is disconnected.
(d) Discuss how electrostatic induction modifies the previous answers, to the lowest order in a/ℓ.

4.6 Effects of Non-uniform Resistivity

Two geometrically identical cylindrical conductors have both height h and radius a, but different resistivities ρ_1 and ρ_2. The two cylinders are connected in series as in Fig. 4.6, forming a single conducting cylinder of height $2h$ and cross section $S = \pi a^2$. The two opposite bases are connected to a voltage source maintaining a potential difference V through the system, as shown in the figure.

(a) Evaluate the electric fields, the electric current and the current densities flowing in the two cylinders in stationary conditions.
(b) Evaluate the surface charge densities at the surface separating the two materials, and at the base surfaces connected to the voltage source.

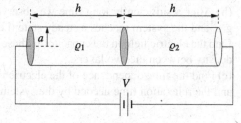

Fig. 4.6

4.7 Charge Decay in a Lossy Spherical Capacitor

A spherical capacitor has internal radius a and external radius b. The spherical shell $a < r < b$ is filled by a lossy dielectric medium of relative dielectric permittivity ε_r and conductivity σ. At time $t = 0$, the charge of the capacitor is Q_0 (Fig. 4.7).
(a) Evaluate the time constant for the discharge of the capacitor.
(b) Evaluate the power dissipated by Joule heating inside the capacitor, and compare it with the temporal variation of the electrostatic energy.

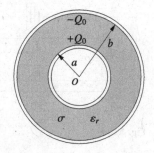

Fig. 4.7

4.8   Dielectric-Barrier Discharge

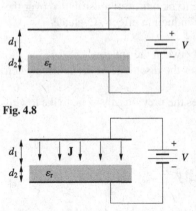

Fig. 4.8

Fig. 4.9

The plates of a parallel-plate capacitor have surface S and separation d. The space between the plates is divided into two layers, parallel to the plates, of thickness d_1 and d_2, respectively, with $d_1 + d_2 = d$, as in Fig. 4.8. The layer of thickness d_1 is filled with a gas of negligible dielectric susceptibility ($\chi = 0$, $\varepsilon_r \simeq 1$), while the layer of thickness d_2 is filled with a dielectric material of dielectric permittivity $\varepsilon_r > 1$. The electric potential difference between the plates, V, is kept constant by a voltage source. Boundary effects can be neglected.

(a) Find the electric field inside the capacitor.

An ionization discharge is started in the gaseous layer at $t = 0$, and the gas instantaneously becomes conducting (Fig. 4.9). We assume that, for $t > 0$, the ionized gas can be considered as an Ohmic conductor of constant and uniform resistivity ρ.

(b) After a sufficiently long time we observe that the current stops flowing in the gas, and the system reaches a steady state (i.e., all physical quantities are constant). Find the electric field in the capacitor in these conditions, and the surface free charge density between the two layers.

(c) Find the time dependence of the electric field during the transient phase ($t > 0$), and the relaxation time needed by the system to reach the steady state condition.

4.9   Charge Distribution in a Long Cylindrical Conductor

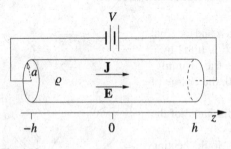

Fig. 4.10

Consider a conducting homogeneous cylindrical wire of radius a and length $2h$, with $a \ll h$, and resistivity ρ. The wire is connected to a voltage source that keeps a constant potential difference V across its ends. We know that the electric field $\mathbf{E}$ and, consequently, the current density $\mathbf{J} = \mathbf{E}/\rho$ must be uniform inside the wire, see Problem 4.6. This implies the presence of charge distributions generating the uniform field. Only surface charge distributions are allowed in a conductor in steady conditions. The charge distributions on the bases of the cylinder are not sufficient for generating an even approximately uniform field in our case of $a \ll 2h$. Thus, a charge density σ_L must be present also on the lateral surface. Verify that a surface charge density $\sigma_L = \gamma z$, where γ is a constant and z is the coordinate along the cylinder axis, leads to a good approximation for the field inside the conductor far from the ends [2] (Fig. 4.10).

4.10   An Infinite Resistor Ladder

An infinite resistor ladder consists of an infinite number of resistors, all of resistance R, arranged as in Fig. 4.11. Evaluate the resistance measured between the terminals A and B. Hint: use an invariance property of the ladder.

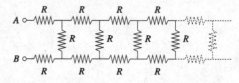

Fig. 4.11

References

1. R.C. Tolman, T.D. Stewart, The electromotive force produced by the acceleration of metals. Phys. Rev. **8**, 97–116 (1916)
2. C.A. Coombes, H. Laue, Electric fields and charge distributions associated with steady currents. Am. J. Phys. **49**, 450–451 (1981)

Chapter 5
Magnetostatics

Topics. Stationary magnetic field in vacuum. Lorentz force. Motion of an electric point charge in a magnetic field. The magnetic force on a current. The magnetic field of steady currents. "Mechanical" energy of a circuit in a magnetic field. Biot-Savart law. Ampères' circuital law. The magnetism of matter. Volume and surface magnetization current densities (bound currents). Magnetic susceptibility. The "auxiliary" vector **H**. Magnetic field boundary conditions. Equivalent magnetic charge method.

Units. In order to write formulas compatible with both SI and Gaussian units, we introduce two new "system dependent" constants, k_m and b_m, defined as

$$k_\mathrm{m} = \begin{cases} \dfrac{\mu_0}{4\pi}, & \text{SI,} \\ \dfrac{1}{c}, & \text{Gaussian,} \end{cases} \qquad b_\mathrm{m} = \begin{cases} 1, & \text{SI,} \\ \dfrac{1}{c}, & \text{Gaussian,} \end{cases} \tag{5.1}$$

where, again, $\mu_0 = 4\pi \times 10^{-7}$ T·m/A is the "magnetic permeability of vacuum", and $c = 29\,979\,245\,800$ cm/s is the light speed in vacuum.

Basic equations. The two Maxwell equations for the magnetic field **B** relevant to this chapter are

$$\nabla \cdot \mathbf{B} = 0 , \tag{5.2}$$
$$\nabla \times \mathbf{B} = 4\pi k_\mathrm{m} \mathbf{J} . \tag{5.3}$$

Equation (5.2) is always valid (in the absence of *magnetic monopoles*), while (5.3) is valid in the absence of time-dependent electric fields. It is thus possible to introduce a *vector potential* **A**, such that

$$\mathbf{B}(\mathbf{r}) = \nabla \times \mathbf{A}(\mathbf{r}) , \tag{5.4}$$

A. Macchi et al., *Problems in Classical Electromagnetism*,
https://doi.org/10.1007/978-3-031-22235-1_5

Imposing the gauge condition $\nabla \cdot \mathbf{A} = 0$, the vector potential satisfies

$$\nabla^2 \mathbf{A}(\mathbf{r}) = -4\pi k_m \mathbf{J}(\mathbf{r}) , \tag{5.5}$$

which is the vector analogous of Poisson's equation (2.1). Thus,

$$\mathbf{A}(\mathbf{r}) = k_m \int_V \frac{\mathbf{J}(\mathbf{r}')}{|\mathbf{r} - \mathbf{r}'|} d^3 r'. \tag{5.6}$$

A particular and typical case is that of closed "line" currents, e.g. flowing in a circuit having wires of negligible thickness. In such case one may replace $\mathbf{J}(\mathbf{r}') d^3 r'$ by $I(\mathbf{r}') d\boldsymbol{\ell}$ and calculate the field via the Biot-Savart formula

$$\mathbf{B}(\mathbf{r}) = k_m \oint \frac{I(\mathbf{r} - \mathbf{r}') d\boldsymbol{\ell} \times (\mathbf{r} - \mathbf{r}')}{|\mathbf{r} - \mathbf{r}'|^3}, \tag{5.7}$$

where the integral is extended to the closed path of the current.

The force exerted by a magnetic field over a distribution of currents is

$$\mathbf{f} = b_m \int_v \mathbf{J}(\mathbf{r}') \times \mathbf{B}(\mathbf{r}') d^3 r'. \tag{5.8}$$

A single point charge q located at $\mathbf{r}$ and moving with velocity $\boldsymbol{v}$ is equivalent to a current density $\mathbf{j}(\mathbf{r}') = q\delta(\mathbf{r} - \mathbf{r}')\boldsymbol{v}$, so that the magnetic force on the point charge is

$$\mathbf{f} = b_m q \boldsymbol{v} \times \mathbf{B} . \tag{5.9}$$

The energy associated to a magnetic field distribution is given by the expression

$$U_m = \int_V \frac{b_m B^2}{8\pi k_m \mu_r} d^3 r. \tag{5.10}$$

In the absence of magnetic monopoles, the first non-vanishing term of the multipole expansion is the magnetic dipole $\mathbf{m}$

$$\mathbf{m} = \frac{1}{2} \int_V \mathbf{r}' \times \mathbf{J}(\mathbf{r}') d^3 r'. \tag{5.11}$$

In the simple case of a small plane coil of area A and electric current I this reduces to the line integral over the coil path C

$$\mathbf{m} = \frac{I}{2} \oint_C \mathbf{r}' \times d\boldsymbol{\ell} = AI\hat{\mathbf{n}} , \tag{5.12}$$

where $\hat{\mathbf{n}}$ is perpendicular to the coil surface. A magnetic dipole located at $\mathbf{r} = 0$ generates a magnetic field

$$\mathbf{B}(\mathbf{r}) = k_m \frac{3\hat{\mathbf{r}} (\mathbf{m} \cdot \hat{\mathbf{r}}) - \mathbf{m}}{r^3} . \tag{5.13}$$

In an external magnetic field $\mathbf{B}_{\text{ext}}$, the magnetic force on a magnetic dipole is

$$\mathbf{f} = (\mathbf{m} \cdot \nabla) \mathbf{B}_{\text{ext}}. \tag{5.14}$$

The magnetization density $\mathbf{M}$ of a material is defined as the dipole moment per unit volue,

$$\mathbf{M} = \frac{d\mathbf{m}}{dV} . \tag{5.15}$$

Ampère's equivalence theorem states that a magnetization density $\mathbf{M} = \mathbf{M}(\mathbf{r})$ is always equivalent to a distribution of volume current density $\mathbf{J}_m$ and surface current density $\mathbf{K}_m$ *bound* to the material, and given by

$$\mathbf{J}_m = \frac{1}{b_m} \nabla \times \mathbf{M} , \tag{5.16}$$

$$\mathbf{K}_m = \frac{1}{b_m} \mathbf{M} \cdot \hat{\mathbf{n}} , \tag{5.17}$$

where $\hat{\mathbf{n}}$ is the unit normal vector pointing outwards from the boundary surface of the material. The total volume and surface current densities are thus

$$\mathbf{J} = \mathbf{J}_f + \mathbf{J}_m , \qquad \mathbf{K} = \mathbf{K}_f + \mathbf{K}_m , \tag{5.18}$$

the subscript f denoting the *free* (e.g., conduction) current densities.

The auxiliary field $\mathbf{H}$ is defined as

$$\mathbf{H} = \begin{cases} \dfrac{\mathbf{B}}{\mu_0} - \mathbf{M} , & \text{SI,} \\ \mathbf{B} - 4\pi\mathbf{M} , & \text{Gaussian,} \end{cases} \tag{5.19}$$

so that Eq. (5.3) becomes

$$\nabla \times \mathbf{H} = 4\pi k_m \mathbf{J}_f , \tag{5.20}$$

A material may have either a permanent magnetization, or a magnetization induced by a magnetic field. In linear, isotropic diamagnetic and paramagnetic materials $\mathbf{M}$ is parallel and proportional to $\mathbf{H}$,

$$\mathbf{M} = \chi_m \mathbf{H} , \tag{5.21}$$

where χ_m is the *magnetic susceptibility* of the material, with $\chi_m < 0$ for diamagnetic materials and $\chi_m > 0$ for paramagnetic materials.[1] The (*relative*) *magnetic permeability* μ_r is defined as

$$\mu_r = \begin{cases} 1 + \chi_m , & \text{SI,} \\ 1 + 4\pi\chi_m , & \text{Gaussian.} \end{cases} \tag{5.22}$$

We have $\mu_r < 1$ for diamagnetic materials and $\mu_r > 1$ for paramagnetic materials. Inserting (5.21) and (5.22) into (5.19) we obtain

$$\mathbf{B} = \begin{cases} \mu_0 \mu_r \, \mathbf{H} , & \text{SI,} \\ \mu_r \, \mathbf{H} , & \text{Gaussian,} \end{cases} \tag{5.23}$$

valid for isotropic, non-ferromagnetic, materials.

To facilitate the use of the basic equations in this chapter also with the system independent units, we summarize some of them in the following Table 5.1.

Table 5.1

Quantity	SI	Gaussian	System independent
$\nabla \times \mathbf{B}$	$\mu_0 \mathbf{J}$	$\frac{4\pi}{c}\mathbf{J}$	$4\pi k_m \mathbf{J}$
Magnetic force on a point charge q moving with velocity $\boldsymbol{v}$ in a magnetic field $\mathbf{B}$	$q\,\boldsymbol{v}\times\mathbf{B}$	$q\,\frac{\boldsymbol{v}}{c}\times\mathbf{B}$	$b_m\,q\,\boldsymbol{v}\times\mathbf{B}$
Magnetic field $\mathbf{dB}$ generated by a wire element $d\boldsymbol{\ell}$ carrying a current I at a distance $\mathbf{r}$ (Biot-Savart's law)	$\frac{\mu_0}{4\pi}\frac{I d\boldsymbol{\ell}\times\hat{\mathbf{r}}}{r^2}$	$\frac{1}{c}\frac{I d\boldsymbol{\ell}\times\hat{\mathbf{r}}}{r^2}$	$k_m\frac{I d\boldsymbol{\ell}\times\hat{\mathbf{r}}}{r^2}$
Magnetic moment $\mathbf{m}$ of a ring circuit carrying an electric current I, and enclosing a surface $\mathbf{S}$	$I\,\mathbf{S}$	$\frac{1}{c}\,I\,\mathbf{S}$	$b_m\,I\,\mathbf{S}$
Volumetric magnetic energy density u_m	$\frac{B^2}{2\mu_0\mu_r}$	$\frac{B^2}{8\pi\mu_r}$	$\frac{b_m B^2}{8\pi k_m \mu_r}$
Self-inductance L of a solenoid of cross-section S, length ℓ and n turns per unit length, filled with a medium of relative permittivity μ_r	$\mu_0\,\mu_r\,n^2\ell\,S$	$\frac{4\pi\,\mu_r\,n^2\ell\,S}{c^2}$	$4\pi k_m b_m\,\mu_r\,n^2\ell\,S$

[1] The magnetization is expressed in terms of the auxiliary field $\mathbf{H}$, rather than in terms of the magnetic field $\mathbf{B}$, for historical reasons. In ferromagnetic materials there is no one-to-one correspondence between $\mathbf{M}$ and $\mathbf{H}$ (between $\mathbf{M}$ and $\mathbf{B}$) because of *magnetic hysteresis*.

5.1 The Rowland Experiment

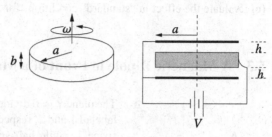

This experiment by Henry A. Rowland (1876) aimed at showing that moving charges generate magnetic fields. A metallic disk or radius a and thickness $b \ll a$ is electrically ·charged and kept in rotation with a constant angular velocity ω.

Fig. 5.1

(a) The disk rotates between two conducting plates, one at a distance $h \simeq 0.5$ cm above its upper surface, and the other at h below its lower surface, as in Fig. 5.1. The two plates are connected to the same terminal of a voltage source maintaining a potential difference $V_0 = 10^4$ V, while the other terminal is connected to the disk by a sliding contact. Evaluate the surface charge density on the disk surfaces.

(b) Calculate the magnetic field $\mathbf{B}_c$ near the center of the disk and the magnetic field component B_r parallel and close to the disk surfaces, as a function of the distance r from the axis. Typical experimental values were $a = 10$ cm, and $\omega \simeq 2\pi \times 10^2$ rad/s (period $T = 2\pi/\omega = 10^{-2}$ s).

(c) The field component B_r generated by the disk at $r = a$ can be measured by orienting the apparatus so that $\hat{\mathbf{r}}$ is perpendicular to the Earth's magnetic field $\mathbf{B}_\oplus$, of strength $B_\oplus \simeq 5 \times 10^{-5}$ T, and measuring the deviation of a magnetic needle when the disk rotates. Find the deviation angle of the needle.

5.2 Pinch Effect in a Cylindrical Wire

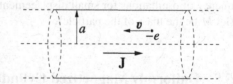

A uniform current density $\mathbf{J}$ flows in an infinite cylindrical conductor of radius a. The current carriers are electrons (charge $-e$) of number volume density n_e and drift velocity $\mathbf{v}$, parallel to the axis of the cylinder. Ions can be considered as fixed in space, with uniform

Fig. 5.2

number density n_i and charge Ze. The system is globally neutral (Fig. 5.2).

(a) Evaluate the magnetic field generated by the current, and the resulting magnetic force on the electrons.

The magnetic force modifies the volume distribution of the electrons and this, in turn, gives origin to a static electric field. At equilibrium the magnetic force on the electrons is compensated by the electrostatic force.

(b) Evaluate the electric field that compensates the magnetic force on the electrons, and the corresponding charge distribution.

(c) Evaluate the effect in "standard" conditions for a good Ohmic conductor.

5.3 A Magnetic Dipole in Front of a Magnetic Half-Space

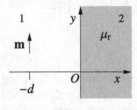

Fig. 5.3

The plane $x = 0$ divides the space into two half-spaces, labeled 1 and 2, respectively. We have vacuum in half-space 1, while half-space 2 is filled by a medium of magnetic permittivity μ_r. A magnetic dipole **m**, parallel to the y axis, is located in vacuum at position $x = -d$ (Fig. 5.3). Find

(a) the magnetic field **B** in the whole space,

(b) the force acting on the magnetic dipole.

5.4 Magnetic Levitation

In a given region of space we have a static magnetic field, which, in a cylindrical reference frame (r, ϕ, z), is symmetric around the z axis, i.e., is independent of ϕ, and can be written $\mathbf{B} = \mathbf{B}(r, z)$. The field component along z is $B_z(z) = B_0 z / L$, where B_0 and L are constant parameters.

(a) Find the radial component B_r close to the z axis.

A particle of magnetic polarizability α (such that it acquires an induced magnetic dipole moment $\mathbf{m} = \alpha \mathbf{B}$ in a magnetic field $\mathbf{B}$), is located close to the z axis.

(b) Find the potential energy of the particle in the magnetic field.

(c) Discuss the existence of equilibrium positions for the particle, and find the frequency of oscillations for small displacements from equilibrium either along z or r (let M be the mass of the particle).

5.5 Uniformly Magnetized Cylinder

A magnetically "hard" cylinder of radius R and height h, with $R \ll h$, carries a uniform magnetization **M** parallel to its axis.

(a) Show that the volume magnetization current density $\mathbf{J}_m$ is zero inside the cylinder, while the lateral surface of the cylinder carries a surface magnetization current density $\mathbf{K}_m$, with $|\mathbf{K}_m| = |\mathbf{M}|$.

(b) Find the magnetic field **B** inside and outside the cylinder, at the limit $h \to \infty$.

(c) Now consider the opposite case of a "flat" cylinder, i.e., $h \ll R$, and evaluate the magnetic field $\mathbf{B}_0$ at the center of the cylinder.

(d) According to the result of **(c)**, $\lim_{R/h\to\infty} \mathbf{B}_0 = 0$. Obtain the same result using the equivalent magnetic charge method.

5.6 Charged Particle in Crossed Electric and Magnetic Fields

A particle of electric charge q and mass m is initially at rest in the presence of a uniform electric field $\mathbf{E}$ and a uniform magnetic field $\mathbf{B}$, perpendicular to $\mathbf{E}$.
(a) Describe the subsequent motion of the particle.

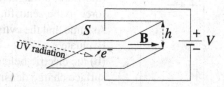

Fig. 5.4

(b) What is the time dependence of the kinetic energy of the particle? Which field exerts the force responsible for the variation of the kinetic energy in time?
(c) Use the result of point **(a)** to discuss the following problem. We have a parallel-plate capacitor with surface S, plate separation h and voltage V, as in Fig. 5.4. A uniform magnetic field $\mathbf{B}$ is applied to the capacitor, perpendicular to the capacitor electric field, i.e., parallel to the plates. Ultraviolet radiation causes the negative plate to emit electrons with zero initial velocity. Evaluate the minimum value of $\mathbf{B}$ for which the electrons cannot reach the positive plate.

5.7 Cylindrical Conductor with an Off-Centered Cavity

An infinite cylindrical conductor of radius a has a cylindrical hole of radius b bored parallel to, and centered at a distance h from, the cylinder axis, as in Fig. 5.5. We have $h + b < a$. The conductor carries a total current I, due to a current density $\mathbf{J}$ uniform throughout the metal part of the cylinder. The current density $\mathbf{J}$ is parallel to the axis of the conductor. The figure shows a section of the conductor. Evaluate the magnetic field $\mathbf{B}$ everywhere in space.

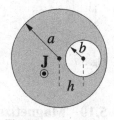

Fig. 5.5

5.8   Conducting Cylinder in a Magnetic Field

A conducting cylinder of radius a and height $h \gg a$ rotates around its axis at constant angular velocity ω in a uniform magnetic field $\mathbf{B}_0$, parallel to the cylinder axis (Fig. 5.6).

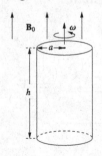

(a) Evaluate the magnetic force acting on the conduction electrons, assuming $\omega = 2\pi \times 10^2\,\mathrm{s}^{-1}$ and $B = 5 \times 10^{-5}\,\mathrm{T}$ (the Earth's magnetic field), and the ratio of the magnetic force to the centrifugal force.

Assume that the cylinder is rotating in stationary conditions. Evaluate

(b) the electric field inside the cylinder, and the volume and surface charge densities;

(c) the magnetic field $\mathbf{B}_1$ generated by the rotation currents inside the cylinder, and the order of magnitude of B_1/B_0 (assume $a \approx 0.1$ m).

Fig. 5.6

5.9   Rotating Cylindrical Capacitor

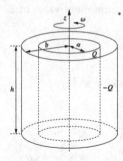

Fig. 5.7

The concentric cylindrical shells of a cylindrical capacitor have radii a and $b > a$, respectively, and height $h \gg b$. The capacitor charge is Q, with $+Q$ on the inner shell of radius a, and $-Q$ on the outer shell of radius b, as in Fig. 5.7. The whole capacitor rotates about its axis with angular velocity $\omega = 2\pi/T$. Boundary effects are negligible.

(a) Evaluate the magnetic field $\mathbf{B}$ generated by the rotating capacitor over the whole space.

(b) Evaluate the magnetic forces on the charges of the two rotating cylindrical shells, and compare them to the electrostatic forces.

5.10   Magnetized Spheres

(a) A sphere of radius R has a uniform and permanent magnetization $\mathbf{M}$. Calculate the magnetic field inside and outside the sphere. (Hint: see Problem 3.3.)

(b) A sphere of radius R has a magnetic permeability μ_r and is located in an external, uniform magnetic field $\mathbf{B}_0$. Calculate the total magnetic field inside and outside the sphere, discussing the limit of a perfectly diamagnetic material ($\mu_r = 0$), as a superconductor.

5.11 A Transversally Magnetized Cylinder

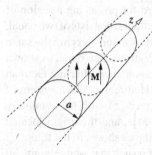

(a) An infinite cylinder of radius a has a permanent, uniform magnetization **M** perpendicular to its axis, as shown in Fig. 5.8. Determine the magnetic field inside the cylinder, showing that it is uniform.

(b) A portion of a particle accelerator consists of a long, thin conducting cylindrical shell. An electric current is distributed over the cylinder surface in order to generate a uniform magnetic field B_0 inside, perpendicular to the cylinder axis. The magnetic field deflects charged particles traveling along the accelerator. Determine the current distribution.

Fig. 5.8

5.12 A Rotating Charged Spherical Shell

A spherical shell of radius a, uniformly charged with surface charge density σ, is rotating about an axis passing through its center with constant angular velocity ω.

(a) Evaluate the magnetic field inside and outside the spherical shell by comparison to the results of point **(a)** of Problem 5.10.

(b) Now evaluate directly the magnetic vector potential **A** and the magnetic field **B** inside and outside rotating the shell using (5.6) and (5.4).

5.13 Magnetic Field of a Polygonal Loop

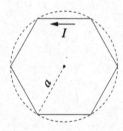

A loop carrying a current I has the form of a regular polygon of n sides inscribed in a circle of radius a. Fig. 5.9 shows the case of a hexagon.

(a) Evaluate the magnetic field at the center of the loop.

(b) Show that the limit for $n \to \infty$ is the magnetic field at the center of a circular loop of radius a.

Fig. 5.9

5.14   Helmholtz Coils

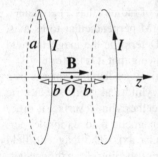

A device commonly used for producing a region of nearly uniform magnetic field **B** consists of two equal, coaxial coils, both of radius a and carrying the same current I, at a distance $2b$ from each other, as shown in Fig. 5.10. The device is named after the German physicist Hermann von Helmholtz. Use a system of cylindrical coordinates (r, ϕ, z) with the symmetry axis of the system as z axis, and the origin located halfway between the coils, as shown in the figure.

Fig. 5.10

(a) Evaluate the value of b such that, at the origin, all derivatives of **B** with respect to z vanish up to, and including, the third order. This ensures a high uniformity of the magnetic field.

(b) Evaluate the radial component of the magnetic field close to the symmetry axis.

Chapter 6
Magnetic Induction and Time-Varying Fields

Topics. Magnetic induction. Faraday's law. Electromotive force. The slowly varying current approximation. Mutual inductance and self-inductance. Energy stored in an inductor. Magnetically coupled circuits. Magnetic energy. Displacement current and the complete Maxwell's equations.

Basic equations. In the presence of a time-varying magnetic field, Eq. (1.5) is modified into the exact equation

$$\nabla \times \mathbf{E} = -b_\mathrm{m} \partial_t \mathbf{B} , \qquad (6.1)$$

so that the line integral of $\nabla \times \mathbf{E}$ around a closed path C is

$$\oint_C \mathbf{E} \cdot \mathrm{d}\boldsymbol{\ell} = -b_\mathrm{m} \int_S \partial_t \mathbf{B} \cdot \mathrm{d}\mathbf{S} \qquad (6.2)$$

Thus, for a *fixed* path, the line integral of $\mathbf{E}$ equals the time derivative of the flux of the time-varying field $\mathbf{B}$ through a surface delimited by the contour C.

The *electromotive force* (emf) $\mathcal{E}$ in a *real* circuit having *moving* parts is the work done by the Lorentz force on a unit charge over the circuit path,

$$\mathcal{E} = \oint_\mathrm{circ} (\mathbf{E} + b_\mathrm{m} \mathbf{V} \times \mathbf{B}) \cdot \mathrm{d}\boldsymbol{\ell} \equiv -b_\mathrm{m} \frac{\mathrm{d}}{\mathrm{d}t} \Phi_\mathrm{circ}(\mathbf{B}) , \qquad (6.3)$$

where $\mathbf{V}$ is the velocity of the circuit element. In (6.3) the flux $\Phi_\mathrm{circ}(\mathbf{B})$ of $\mathbf{B}$ through the circuit may vary because of *both* the temporal variations of $\mathbf{B}$ and of the circuit geometry. Equation (6.3) is the general Faraday's law of induction.

For a system of two electric circuits, the magnetic flux through each circuit can be written as a function of the currents flowing in each circuit,

$$\Phi_1 = L_1 I_1 + M_{21} I_2 , \qquad \Phi_1 = L_2 I_2 + M_{12} I_1 , \qquad (6.4)$$

A. Macchi et al., *Problems in Classical Electromagnetism*,
https://doi.org/10.1007/978-3-031-22235-1_6

where the terms containing the (self-)inductance coefficients L_i are the contribution to flux generated by the circuit itself, and the terms containing the mutual inductance coefficients $M_{21} = M_{12}$ give the flux generated by one circuit over the other.

Finally, for time-varying fields the complete Maxwell's equation replacing (5.3) is

$$\mathbf{\nabla} \times \mathbf{B} = 4\pi k_m \mathbf{J} + \frac{k_m}{k_e}\, \partial_t \mathbf{E} = \begin{cases} \dfrac{4\pi}{c}\mathbf{J} + \dfrac{1}{c}\partial_t \mathbf{E} & \text{(Gaussian)}, \\[2mm] \mu_0 \mathbf{J} + \mu_0 \varepsilon_0 \partial_t \mathbf{E} & \text{(SI.)} \end{cases} \qquad (6.5)$$

6.1 A Square Wave Generator

We have a uniform magnetic field $\mathbf{B} = B\,\hat{\mathbf{z}}$ in the half space $x < 0$ of a Cartesian coordinate system, while the field is zero for $x > 0$. A semicircular loop of radius a and resistance R lyes in the xy plane, with the center of the full circumference at the origin O of our coordinate system, as in Fig. 6.1. The loop rotates around the z axis at constant angular velocity ω.

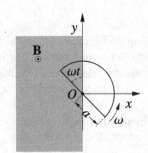

First, assume that the self-inductance of the coil is negligible and evaluate
(a) the current circulating in the coil;
(b) the torque exerted by the magnetic forces on the coil, and the mechanical power needed to keep the coil in rotation. Compare this to the electric power dissipated in the coil.

Fig. 6.1

(c) Now consider the presence of the self-inductance of the coil, and discuss how it changes the answer to point (a).

6.2 A Coil Moving in an Inhomogeneous Magnetic Field

A magnetic field has rotational symmetry around a straight line, that we choose as longitudinal axis, z, of a cylindrical reference frame (r, ϕ, z). The z component of the field on the z axis, $B_z(0, z)$, is known and equals $B_z(0, z) = B_0\, z/L$, where L is a constant. A circular coil has radius a, resistance R, and axis coinciding with the z axis of our reference frame. The coil performs a translational motion at constant velocity $\mathbf{v} = v\,\hat{\mathbf{z}}$, and its radius a is assumed to be small enough that the magnetic field is always approximately uniform over the surface limited by the coil (Fig. 6.2).
(a) Find the current $\mathbf{I}$ flowing in the coil.

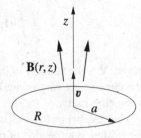

Fig. 6.2

(b) Find the power P dissipated by the coil due to Joule heating, and the corresponding frictional force **f** on the coil.
(c) Calculate **f** as the resultant magnetic force on the loop carrying the current I.

6.3   A Circuit with "Free-Falling" Parts

In the presence of the Earth's gravitational field **g**, two high-conducting bars are located vertically, at a distance a from each other. A uniform, horizontal magnetic field **B** is perpendicular to the plane defined by the vertical bars. Two horizontal bars, both of mass m, resistance $R/2$ and length a, are constrained to move, without friction, with their ends steadily in contact with the two vertical bars (Fig. 6.3). The resistance of the two fixed vertical bars is assumed to be much smaller than $R/2$, so that the net resistance of the resulting rectangular circuit is, with very good approximation, always R, independently of the positions of the two horizontal bars.

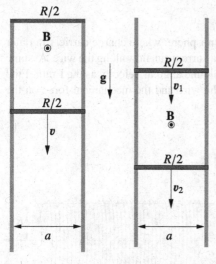

Fig. 6.3

First, assume that the upper horizontal bar is fixed, while the lower bar starts a "free" fall at $t = 0$. Let's denote by $v = v(t)$ the velocity of the falling bar at time t, with $v(0) = 0$.

(a) Write the equation of motion for the falling bar, find the solution for $v(t)$ and show that, asymptotically, the bar approaches a terminal velocity v_t.

(b) Evaluate the power dissipated in the circuit by Joule heating when $v(t) = v_t$, and the mechanical work done per unit time by gravity in these conditions. Now consider the case in which, at $t = 0$, the upper bar already has a velocity $v_0 \neq 0$ directed downwards, while the lower bar starts a "free" fall.

(c) Write the equations of motion for both falling bars, and discuss the asymptotic behavior of their velocities $v_1(t)$ and $v_2(t)$, and of the current in the circuit $I(t)$.

6.4   The Tethered Satellite

The Earth's magnetic field at the Earth's
surface roughly approximates the field of a
magnetic dipole placed at the Earth's center. Its magnitude ranges from 2.5×10^{-5}
to 6.5×10^{-5} T (0.25 to 0.65 G in Gaussian
units), with a value $B_{eq} \simeq 3.2 \times 10^{-5}$ T at
the equator.

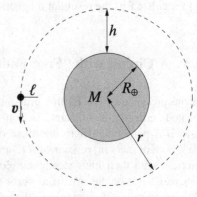

   A satellite moves on the magnetic equatorial plane with a velocity $v \simeq 8$ km/s
at a constant height $h \simeq 100$ km over the
Earth's surface, as shown in the figure (not
to scale!). A tether (leash, or lead line), consisting in a metal cable of length $\ell = 1$ km,   **Fig. 6.4**
hangs from the satellite, pointing to the
Earth's center (Fig. 6.4).

(a) Find the electromotive force on the wire.

(b) The satellite is traveling through the ionosphere, where charge carriers in outer
space are available to close the circuit, thus a current can flow along the wire. Assume
that the ionosphere is rigidly rotating at the same angular velocity as the Erath. Find
the power dissipated by Joule heating in the wire and the mechanical force on the
wire as a function of its resistance R.

6.5   Eddy Currents in a Solenoid

A long solenoid consists of a helical
coil of n turns per unit length wound
around a soft ferromagnetic cylinder of
radius R and length $\ell \gg R$. The ferromagnetic material has a relative magnetic permittivity μ_r, and an electrical conductivity σ. An AC current $I =
I_0 \cos \omega t$ flows in the coil (Fig. 6.5).

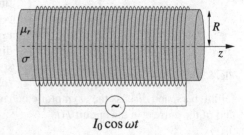

(a) Find the *electric* field induced in
the solenoid.   **Fig. 6.5**

(b) Explain why the cylinder warms up
and evaluate the dissipated power.

(c) Evaluate how the induced currents affect the magnetic field in the solenoid.
(Boundary effects and the displacement current are assumed to be negligible).

6.6   Feynman's "Paradox"

A non-conducting ring of radius R is at rest on the xy plane, with its center at the origin of the coordinate system. The ring has mass m, negligible thickness, and an electric charge Q distributed uniformly on it, so that the ring has a linear charge density $\lambda = Q/(2\pi a)$. The ring is free to rotate around its axis without friction.

A superconducting circular ring of radius $a \ll R$, coaxial to the charged ring and carrying an electric current I_0, also lies on the xy plane, as in Fig. 6.6. At time $t = 0$ the superconducting loop is heated above its critical temperature, and switches to normal conductivity. Consequently, its current decays to zero according to a law $I = I(t)$.
(a) Neglecting self-induction effects, evaluate the angular velocity $\omega = \omega(t)$ of the charged ring as a function of the current $I(t)$ in the smaller ring. Evalu-

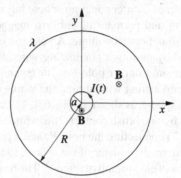

Fig. 6.6

ate the final angular velocity ω_f, and the final angular momentum L_f, of the charged ring.
(b) Evaluate the magnetic field at the ring center, $\mathbf{B}_c$, generated by the rotation of the ring.
(c) Discuss how the results of (a) are modified by taking the "self-inductance" $\mathcal{L}$ of the charged ring into account.

This is one of the possible versions of the so-called *Feynman's disc paradox* [2], presented in Vol. II, Sect. 17–4, of *The Feynman's Lectures on Physics*. The apparent paradox arises because the initial total *mechanical* angular momentum of the system is zero, no external torque is applied, and one could (wrongly) expect the final total angular momentum to be zero, i.e. no rotation of the ring. This conclusion is wrong, of course, for reasons further discussed in Problem 8.8.

6.7   Induced Electric Currents in the Ocean

A fluid flows with uniform velocity v in the presence of a constant and uniform magnetic field $\mathbf{B}$ perpendicular to v. The fluid has an electrical conductivity σ and volumetric mass density ϱ.
(a) Evaluate the electric current density $\mathbf{J}$ induced in the fluid.
(b) Give a numerical estimate of $|\mathbf{J}|$ for the terrestrial oceans, knowing that the Earth's magnetic field has an average value $B \simeq 0.5\,\mathrm{G} = 5 \times 10^{-5}\,\mathrm{T}$, the conductivity of sea water is $\sigma \simeq 4\,\Omega^{-1}\mathrm{m}^{-1}$ ($\sigma \simeq 3.6 \times 10^{10}\,\mathrm{s}^{-1}\mathrm{cm}^{-1}$ in Gaussian units), and a typical value of the flow velocity is $v = 1\,\mathrm{m/s}$.

(c) Due to the presence of the induced current, the magnetic force tends to slow down the fluid. Estimate the order of magnitude for the time constant of this effect.

6.8   A Magnetized Sphere as Unipolar Motor

A magnetized, non-conducting sphere has radius a, mass m and permanent, uniform magnetization $\mathbf{M}$ throughout its volume. An electric circuit is formed by pasting a conducting wire along a half meridian, from the pole P to the equator, and another conducting wire around the whole equator of the sphere, as shown in Fig. 6.7. The circuit is closed by two brush contacts (the white arrows in Fig. 6.7) connecting the pole P, and a point A of the wire on the equator of the sphere, to a voltage source of electromotive force V. The resulting circuit has resistance R.

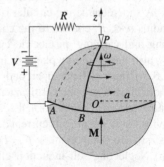

Fig. 6.7

(a) Evaluate the torque on the sphere when a current I flows in the circuit.
(b) If the sphere is free to rotate without friction around the z axis of a cylindrical coordinate system, parallel to $\mathbf{M}$ and passing through the center O of the sphere, it reaches asymptotically a terminal angular velocity ω_t. Evaluated ω_t and the characteristic time of the system.

6.9   Induction Heating

Consider a homogeneous material of electrical conductivity σ and relative magnetic permeability μ_r, both real, positive and independent of frequency. The electric permittivity is $\varepsilon_r = 1$.
(a) Show that, if the displacement current density $\partial_t \mathbf{E}/(4\pi k_e)$ can be neglected, the magnetic field $\mathbf{B}$ inside the material obeys the equation

$$\partial_t \mathbf{B} = \alpha \nabla^2 \mathbf{B} , \tag{6.6}$$

and determine the value of the real constant α.
The material fills the half-space $x > 0$ in the presence of a uniform oscillating magnetic field $\mathbf{B}_0 = \hat{\mathbf{y}} B_0 \cos(\omega t) = \hat{\mathbf{y}} \operatorname{Re}\left(B_0 e^{-i\omega t}\right)$ in the half-space $x < 0$.
(b) Evaluate the magnetic field $\mathbf{B}(x, t)$ for $x > 0$, assuming that the displacement current is negligible. Discuss under what conditions the result is a good approximation for the case of a finite slab of the material.
(c) Evaluate the power dissipated in the medium by Joule heating.

6.10 A Magnetized Cylinder as DC Generator

A long hard-iron cylinder has height h, radius $a \ll h$, and permanent, uniform magnetization $\mathbf{M}$ throughout its volume. The magnetization is parallel to the cylinder axis, which we choose as the z axis of a cylindrical coordinate system (r, ϕ, z).

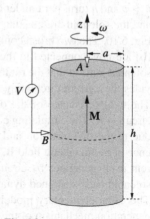

(a) Show that the magnetic field inside the cylinder, far from the two bases, is $\mathbf{B}_0 \simeq 4\pi (k_m/b_m) \mathbf{M}$, or $\mathbf{B}_0 \simeq \mu_0 \mathbf{M}$ in SI units, $\mathbf{B}_0 = 4\pi \mathbf{M}$ in Gaussian units. Show that the magnitude of the z component of the field at the two bases is $B_z \simeq B_0/2$.

(b) Two brush contacts (the white arrows in Fig. 6.8) connect the center of the upper base of the cylinder, A, and a point on the equator of the cylinder, B, to a voltmeter. The cylinder is kept in rotation around the z axis with constant angular velocity ω. Evaluate the electromotive force measured by the voltmeter.

Fig. 6.8

This problem is taken from an example of [1], Sect. 88, p. 379.

6.11 The Faraday Disk and a Self-Sustained Dynamo

A perfectly conducting disk, of radius a and thickness $h \ll a$, rotates at constant angular velocity ω (parallel to the disk axis), in the presence of a uniform and constant magnetic field $\mathbf{B}$ parallel to ω.

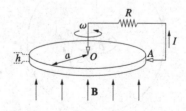

(a) Evaluate the electric field $\mathbf{E}$ in the disk in steady state conditions, and the corresponding potential drop between the center and the boundary of the disk (hint: the total force on charge carriers must be zero at equilibrium).

Fig. 6.9

(b) We now form a closed circuit by connecting the center of the disk to a point of the circumference by brush contacts (white arrows in the figure), as in Fig. 6.9. Let R be the total resistance of the resulting circuit. Calculate the external torque needed to keep the disk in rotation at constant angular speed.

(c) Finally, we place the rotating disk at the center of a long solenoid of radius $b > a$ and n turns per unit length. The disk and the solenoid are coaxial, as shown in Fig. 6.10. The two brush contacts of point (b) are now connected to the ends of the solenoid coil, so that the rotating disk provides the current circulating in the turns. The total resistance of the disk-solenoid circuit is R. The circulating current is thus due to the disk rotation and to the presence of the magnetic field **B**, that the current itself generates by circulating in the

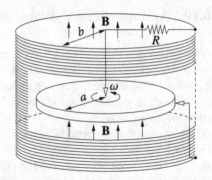

Fig. 6.10

solenoid (self-sustained dynamo). Find a the value of ω for steady-state conditions. This is an elementary model for a dynamo self-sustained by rotation, such as the generation mechanism of the Earth's magnetic field [3].

6.12   Mutual Induction Between Circular Loops

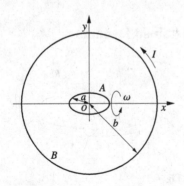

Fig. 6.11

The centers of two circular conducting loops A and B, of radii a and $b \gg a$, respectively, are located at the origin O of a Cartesian reference frame. At time $t = 0$ both loops lie on the xy plane. While the larger loop remains at rest, the smaller loop, of resistance R, rotates about one of its diameters, lying on the x axis, with angular velocity ω, as shown in Fig. 6.11. A constant current I circulates in the larger loop.

(a) Evaluate the current I_A induced in loop A, neglecting self-inductance effects.

(b) Evaluate the power dissipated in loop A due to Joule heating.

(c) Evaluated the torque needed to keep loop A in rotation, and the associated mechanical power.

(d) Now consider the case when loop A is at rest on the xy plane, with a constant current I circulating in it, while loop B rotates around the x axis with constant angular velocity ω. Evaluate the electromotive force induced in B, neglecting self-inductance effects.

6.13 Mutual Induction Between a Solenoid and an Internal Loop

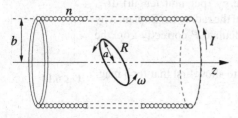

Fig. 6.12

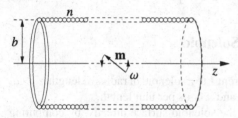

Fig. 6.13

A conducting loop of radius a and resistance R is located with its center at the center of solenoid of radius $b > a$ and n turns per unit length, as in Fig. 6.12. The loop rotates at constant angular velocity ω around a diameter perpendicular to the solenoid axis, while a steady current I flows in the solenoid.

(a) Evaluate the flux of the magnetic field through the rotating coil as a function of time.

(b) Evaluate the torque exerted by the external forces on the loop in order to keep it rotating at constant angular velocity.

Now assume that the solenoid is disconnected from the current source, and that the rotating loop is replaced by a magnetic dipole **m**, still rotating at constant angular velocity ω, as in Fig. 6.13.

(c) Evaluate the electromotive force induced in the solenoid.

6.14 Skin Effect and Eddy Inductance in an Ohmic Wire

A long, straight cylindrical wire of radius r_0 and conductivity σ (which we assume to be real and constant in the frequency range considered) carries an alternating current of angular frequency ω. The impedance per unit length of the wire, Z_ℓ, can be defined as the ratio of the electric field at the wire surface to the total current through the wire cross section. Evaluate Z_ℓ as a function of ω.

6.15   Magnetic Pressure and Pinch Effect for a Surface Current

A current I flows on the surface of a conducting cylinder of radius a and infinite length, in the direction parallel to the axis $\hat{\mathbf{z}}$. The current layer has negligible thickness, so that we can write $I = 2\pi a K$, with $\mathbf{K} = K\hat{\mathbf{z}}$ the surface current density (Fig. 6.14). Calculate
(a) the magnetic field $\mathbf{B}$ in the whole space,
(b) the force per unit surface P on the cylinder surface
(c) the variation of magnetic field energy (per unit length) $\mathrm{d}U_{\mathrm{m}}$ associated to an infinitesimal variation of the radius $\mathrm{d}a$. Explain why $P \neq -(2\pi a)^{-1}\mathrm{d}U_{\mathrm{m}}/\mathrm{d}a$ and how to calculate P correctly from the energy variation.

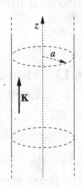

Notice: for point **(b)** it might be useful to show first that for a magnetostatic field we have

Fig. 6.14

$$\mathbf{J} \times \mathbf{B} = \frac{1}{4\pi k_{\mathrm{m}}}\left[(\mathbf{B} \cdot \nabla)\,\mathbf{B} - \tfrac{1}{2}\nabla B^2\right]. \tag{6.7}$$

6.16   Magnetic Pressure on a Solenoid

A current source supplies a constant current I to a solenoid of radius a, length $h \gg a$, so that boundary effects are negligible, and n coils per unit length.
(a) Evaluate the magnetic pressure on the solenoid surface directly, by computing the magnetic force on the coils.
(b) Now evaluate the magnetic pressure on the solenoid surface by evaluating the variation of the magnetic energy of the system for an infinitesimal increase $\mathrm{d}a$ of the radius of the solenoid, and the corresponding work done by the current source in order to keep I constant.

6.17   A Homopolar Motor

A *homopolar motor* is a direct current electric motor consisting of a circuit carrying a direct current I in the presence of a static magnetic field. The circuit is free to rotate around a fixed axis, so that the angle between the current and the magnetic field remains constant in time in each part of the circuit. The resulting electromotive force is continuous, and the homopolar motor needs no device, like a commutator, to switch the current flow. But it still requires slip rings (or brush contacts) to operate. "Homopolar" means that the electri-

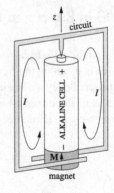

Fig. 6.15

cal polarity of the conductor (the direction of the current flow at each point of the circuit) and the magnetic field do not change in time, and the motor does not require commutation. A simple practical realization of a homopolar motor is shown in Figs. 6.15 and 6.16, based on a Wikipedia entry.

The idea is the following: an electrochemical cell drives a DC current into the double circuit shown in the figures, while a magnetic field is generated generated by the permanent magnet cylindrical located at the bottom of the cell, in electrical contact with its negative pole, as shown in Fig. 6.16. The magnetic field has rotational symmetry around the z axis and is constant in time, in spite of the magnet rotation, and the circuit is free to rotate around the z axis. The magnetic forces on the current-carrying circuit exert a torque, and the circuit starts to rotate.

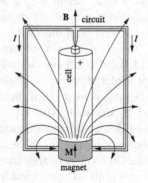

Fig. 6.16

The dimensions, mass and resistance of the circuit (the mass includes battery and magnet), the voltage of the battery and the magnetic field strength generated by the magnet at each point of the circuit are known. Find the torque acting on the circuit, and the angular velocity of the system as a function of time,

6.18 A Magnetic Cylinder Sliding inside a Solenoid

A solenoid of length a, radius $b \ll a$ and n turns per unit length is connected to a current source that supplies a constant electric current I_0. A cylinder of relative magnetic permeability $\mu_r \neq 1$ and radius b is introduced into the solenoid for a depth $Z < a$, as shown in Fig. 6.17, with $Z \gg b$ and $(a - Z) \gg b$.

(**a**) The system may be considered as a series of two inductors. Using the expression of the inductance for a long solenoid, calculate the total inductance as a function of Z. Use this result to evaluate the magnetic energy of the system and the force acting on the cylinder.

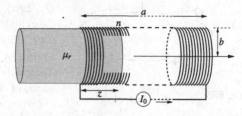

(**b**) Show that, because of the boundary conditions at the interface

Fig. 6.17

of two materials, the magnetic field **B** cannot be uniform inside the partially filled solenoid. Discuss how this affects the assumptions underlying the approach followed in **a**), and explain why, however, these assumptions lead to an acceptable result for the force. Also discuss *where* the force is exerted. (Hint: compare this discussion to that of Problem 3.12.)

6.19   Conducting Cylindrical Shell in a Magnetic Field

A conducting cylindrical shell has radius a, thickness $h \ll a$, height $b \gg a$ and conductivity σ, as
shown in Fig. 6.18. We choose a cylindrical coordinate system with the shell axis as longitudinal axis
z, The shell is in a uniform oscillating magnetic
field $\mathbf{B}_0(t) = \hat{\mathbf{z}} B \cos(\omega t) = \mathrm{Re}\left(\hat{\mathbf{z}} B\,\mathrm{e}^{-\mathrm{i}\omega t}\right)$. Assume
that border effects are negligible.
(a) Evaluate the magnetic field in the region surrounded by the shell. Use the slowly-varying currents
and fields approximation, and neglect the displacement current. Discuss the validity of the approximations. Give a numerical estimate for the case of copper
($\sigma \simeq 6 \times 10^7 \ \Omega^{-1}\mathrm{m}^{-1}$) at a frequency of 50 Hz.

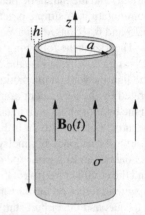

Fig. 6.18

(b) Evaluate the entering and exiting flux of electro-
magnetic energy through the internal and external surfaces of the shell. Show that their
difference equals the power dissipated by Joule heating in the conducting material.
(c) Now evaluate the magnetic field in the region delimited by the cylindrical shell
assuming it to be uniform and the displacement current to be negligible, but that the
slowly-varying fields approximation is not valid. Compare the result to the result of
question (a).

6.20   Electromagnetic Inertia of a Rotating Charged Ring

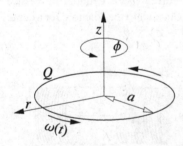

Fig. 6.19

An electric charge Q is uniformly distributed
over a thin *non*-conducting ring of radius
a, resulting in a linear charge density $\lambda =
Q/(2\pi a)$. The ring lies on the $z = 0$ plane of
the cylindrical coordinate system shown in Fig.
6.19. The ring rotates at a given time-dependent
angular velocity $\omega(t)$, parallel to the ring axis
(the z axis of our coordinate system) under the
action of an appropriate external torque $\tau_{\mathrm{ext}}(t)$.
For what follows it is useful to remember that a
conducting loop of radius a has inductance $L = \alpha a/c^2 \ \mathrm{s}^2/\mathrm{cm}$, where α is a numerical
coefficient of the order of unity. Evaluate as functions of $\omega(t)$ and its derivatives
(a) the azimuthal electric field $\mathbf{E}_i(a) = E_i(a)\,\hat{\boldsymbol{\phi}}$ induced on the ring;
(b) the torque τ_E exerted by $\mathbf{E}_i(a)$ on the ring, showing that its effect is equivalent
to an increase of the moment of inertia $\mathcal{I}$ of the ring;

(c) the mechanical power P_E provided by τ_E, comparing it to the rate of change of the magnetic energy.

6.21 A Bar Sliding in a Magnetic Field

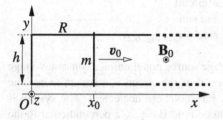

Fig. 6.20

Two straight horizontal, semi-infinite, and highly conducting wires are parallel to the x axis of a Cartesian coordinate system at a distance h from each other along the y axis, in the presence of a uniform magnetic field $\mathbf{B}_0$ parallel to the z axis, as shown in Fig. 6.20. The initial points of the wires, both located at $x = 0$, are joined by a fixed bar of length h and resistance $R/2$. At time $t = 0$ a second, movable bar of mass m and resistance $R/2$ is located at $x = x_0$ with an initial velocity v_0, as shown in the figure. The movable bar slides without friction over the wires keeping electrical contact, so that the resistance of the circuit is R independently of the bar position.

(a) Disregarding the self-inductance of the circuit, evaluate the position of the movable bar as a function of time and the maximum distance it will reach.

(b) Now assume that the circuit has a self-inductance L. How does this affect the motion?

6.22 Magnetic Levitation of a Superconducting Ring

A superconducting ring of mass m and radius r lies on a support at a small height z_0 above a circular coil of radius $R >> r$, in the presence of the Earth gravitational field $\mathbf{g}$. The superconducting ring is constrained to move vertically maintaining its axis vertical and coinciding with the axis of the coil. The self-inductances of the ring and of the coil are L_r and L_c, respectively.

At time $t = 0$ the coil is connected to a current source providing a constant electric current I_c.

(a) Evaluate the mutual inductance $M(z)$ between coil and superconducting ring in the approximation $r \ll R$.

(b) Evaluate the maximum height z_{max} reached by the superconducting ring in the approximation $M(z_{max}) \ll M(z_0)$. Why do we need the initial position of the ring to be above the coil by a finite amount?

(c) Describe the subsequent motion.

6.23 Electromagnetic Brake with Energy Recovery

Figure 6.21 shows an elementary model of an electromagnetic brake, where magnetic induction is
used to stop the rotation of a wheel. A rotor has
moment of inertia $\mathcal{I}$ and four conducting arms of
length a connected to an external circuit through
sliding contacts. The geometry is such that circuit
is always closed by a single arm of the rotor. For simplicity we assume the absence of external mechanical torques, friction and self-induction.

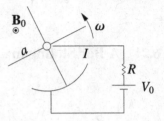

Fig. 6.21

   The circuit contains a battery, i.e., a voltage source maintaining a constant voltage
drop V_0 across its terminals. The resistance R of the circuit is assumed to be time
independent (the effects of circuit length variations are neglected). The system is
immersed in a uniform and constant magnetic field $\mathbf{B}_0 = B_0\,\hat{\mathbf{z}}$ perpendicular to the
plane of the rotor.
(**a**) Evaluate the electromotive force induced on the circuit.
(**b**) Evaluate the torque exerted by the magnetic forces on the rotor.
(**c**) Show that the time variation of the kinetic energy of the rotor equals the sum of
the power dissipated in the resistance and the power either *supplied* or *absorbed* by
the battery, discussing the conditions on V_0 for each case.
(**d**) Find the time dependence of $\omega(t)$ assuming that the circuit is closed at $t = 0$
with the initial angular velocity $\omega(0) = \omega_0 > 0$ such that $\mathcal{E}(\omega_0) + V_0 > 0$. Discuss
the cases $V_0 = 0$ (no battery), $V_0 > 0$ and $V_0 < 0$.
(**e**) Using the solution for $\omega(t)$ calculate the total energy that can be absorbed by the
battery, and discuss the conditions on the system parameters for which the highest
percentage of the initial kinetic energy of the rotor is eventually stored into the battery.

6.24 A High Frequency Capacitor

A plane capacitor consists of two circular, coaxial
and parallel metallic plates at a distance $h \ll a$ from
each other. All boundary effects are assumed to be
negligible (Fig. 6.22).
   Assume the capacitor to be in series with an ideal
current source that supplies the current $I = I(t) =$
$I_0 e^{-i\omega t}$ in complex notation.

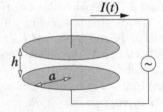

Fig. 6.22

Using the slowly varying currents approximation
(SVCA), calculate
(**a**) the electric field $\mathbf{E}_0 = \mathbf{E}_0(t)$ inside the capacitor and the surface charge density
$\pm\sigma_0 = \pm\sigma_0(t)$ on the two plates to the lowest order in the SVCA, for which both $\mathbf{E}_0$
and σ_0 are considered as uniform;

(b) the magnetic field $\mathbf{B}_1 = \mathbf{B}_1(r, t)$ inside the capacitor, to the lowest non-vanishing order (r being the distance from the capacitor axis);

(c) the first-order correction to the electric field $\mathbf{E}_1 = \mathbf{E}_1(r, t)$ and to the surface density $\sigma_1 = \sigma_1(r, t)$; to fix any integration constant, remember that the current source maintains a fixed current in the circuit.

(d) Now assume that the capacitor is connected to a voltage source that maintains a voltage drop $E_0 h e^{-i\omega t}$ *across the axis of the capacitor* (z axis for reference), so that the electric field on axis is $\mathbf{E}(r = 0, t) = E_0 e^{-i\omega t} \hat{\mathbf{z}}$. Show that the SVCA iteration leads to the following expression for the electric field as an infinite power series,

$$\mathbf{E}(r, t) = e^{-i\omega t} \sum_{n=0}^{\infty} \mathbf{E}_n = \mathbf{E}_0 e^{-i\omega t} \sum_{n=0}^{\infty} \frac{(-1)^n}{(n!)^2} \left(\frac{\omega r}{2c}\right)^{2n} \equiv \mathbf{E}(0) e^{-i\omega t} J_0\left(\frac{\omega r}{2c}\right), \quad (6.8)$$

with the convergent series defining the function $J_0(x)$, known as the Bessel function of zero order. (Hint: use the principle of *mathematical* induction.)

This problem is taken from Feynman's lectures ([2], Volume II, Sect. 23–2) with minor additions mostly to discuss a different choice of boundary conditions and how the series solution can be checked.

References

1. R. Becker, *Electromagnetic Fields and Interactions, Vol. I, Electromagnetic Theory and Relativity* (Blackie, London and Glasgow, 1964)
2. R.P. Feynman, R.B. Leighton, M. Sands, *The Feynman Lectures on Physics* (Addison-Wesley Publishing Company, Reading, MA, 2006)
3. R.T. Merrill, *Our Magnetic Earth: The Science of Geomagnetism*, Chap. 3 (Chicago University Press, 2010)
4. J.D. Jackson, *Classical Electrodyamics*, 3rd edn. (Wiley, New York, 1999)

Chapter 7
Electromagnetic Oscillators and Wave Propagation

Topics. Harmonic oscillators. Resonances. Coupled oscillators, normal modes and eigenfrequencies. Basics of the Fourier transform. Electric circuits: impedances, simple LC and RLC circuits. Waves. The wave equation. Monochromatic waves. Dispersion. Wavepackets. Transmission lines.

Useful formulas for this chapter:

Fourier transform of the Gaussian function

$$\int_{-\infty}^{+\infty} e^{-(\alpha k)^2} e^{ikx} dk = \frac{\sqrt{\pi}}{\alpha} e^{-x^2/4\alpha^2} , \tag{7.1}$$

where in general α is a complex number with $\mathrm{Re}(\alpha) > 0$.

7.1 Coupled RLC Oscillators (1)

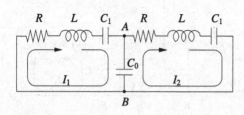

Fig. 7.1

Consider an electrical circuit consisting of two identical resistors R, two identical inductors L, two identical capacitors C_1, and a capacitor C_0, all arranged in two meshes as in Fig. 7.1. Let I_1 and I_2 be the current intensities flowing in the left and right mesh of the circuit, respectively, as shown in the figure. Initially, assume that I_1 and I_2 are flowing in the absence of voltage sources, and assume $R = 0$.

(a) Find the equations for the time evolution of I_1 and I_2. Describe the normal modes of the system, i.e., look for steady-state solutions of the form

$$I_1(t) = A_1 e^{-i\omega t} , \qquad I_2(t) = A_2 e^{-i\omega t} , \qquad (7.2)$$

determining the possible values for ω. Find a mechanical equivalent of the circuit.
(b) Now consider the effect of the nonzero resistances R in series with each of the two inductances L. Find the solutions for I_1 and I_2 in this case.
(c) Evaluate I_1 and I_2 as functions of ω if a voltage source $V = V_0 e^{-i\omega t}$ is inserted into the left mesh of the circuit.

7.2 Coupled RLC Oscillators (2)

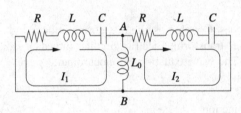

Fig. 7.2

An electrical circuit consists of two identical resistors R, two identical inductors L, two identical capacitors C, and an inductor L_0, all arranged in two meshes as in Fig. 7.2. Let I_1 and I_2 be the currents flowing in the left and right mesh of the circuit, respectively, as shown in the figure.

(a) Initially, assume that the currents are flowing in the absence of sources, and assume $R = 0$. Find the equations for the time evolution of $I_1 = I_1(t)$ and $I_2 = I_2(t)$. Determine the normal modes of the circuit.
(b) Now assume $R \neq 0$. Show that now the modes of the system are damped, and determine the damping rates.

7.3 Coupled RLC Oscillators (3)

An electrical circuit consists of three identical resistors R, three identical inductors L, and two identical
capacitors C, arranged in
three meshes as in Fig. 7.3.
Let I_1, I_2, and I_3 be the currents flowing in the three
meshes, as in the figure. Initially, assume $R = 0$.

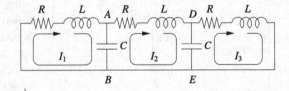

Fig. 7.3

(a) Write the equations for the time evolution of $I_n(t)$. Find a mechanical system with three degrees of freedom and the same equations of motion as those for $I_n(t)$.
(b) Determine the normal oscillation modes of the system and their frequencies.
(c) Now assume $R \neq 0$, and determine the decay rate of the normal modes.

7.4 The *LC* Ladder Network

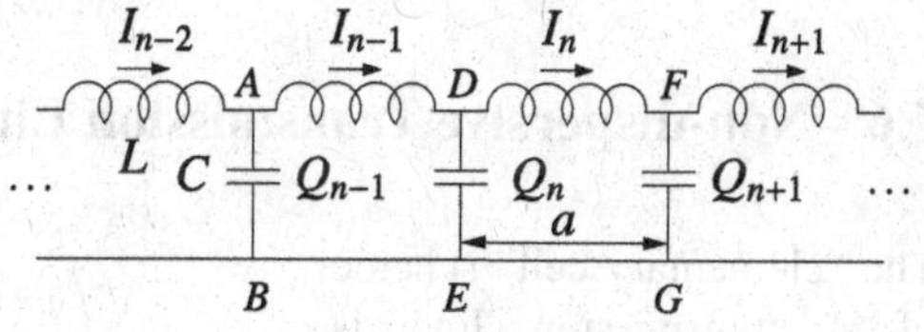

An *LC* ladder network is formed by N inductors L, and N capacitors C, arranged as shown in Fig. 7.4. We denote by $I_n = I_n(t)$ the current in the nth inductor. Resistance effects are assumed to be negligible. The distance between two neighboring nodes is a.

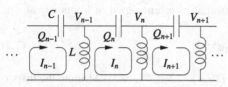

Fig. 7.4

(a) Find the equations for the time evolution of I_n. Which is a mechanical equivalent of the system?
(b) Show that solutions exist in the form of propagating monochromatic waves

$$I_n = C\, e^{i(kna - \omega t)} \tag{7.3}$$

and find the dispersion relation between k and ω.
(c) For a given value of ω, find the allowed values of k with the boundary conditions $I_0 = I_N = 0$.
(d) Discuss the limit to a continuum system, i.e., $N \to \infty$, $n \to \infty$, $a \to 0$, with $na \to x$, where inductance and capacity are continuously distributed, i.e., defined per unit length.

7.5 The *CL* Ladder Network

Fig. 7.5

Consider an infinite ladder network of identical capacitors C and inductors L, arranged as shown in Fig. 7.5. Let $Q_n = Q_n(t)$ be the charge on the nth capacitor, $V_n = V_n(t)$ the voltage drop on the nth inductor, and $I_n = I_n(t) = dQ_n/dt$ is the current flowing in the nth mesh, across the nth capacitor, i.e., between the network nodes at V_{n-1} and V_n.

(a) Show that the currents I_n satisfy the coupled equations

$$L \frac{d^2}{dt^2}(I_{n+1} - 2I_n + I_{n-1}) = \frac{I_n}{C} .$$
(7.4)

(b) Show that the solutions of (7.4) have the form

$$I_n = A\, e^{i(kna - \omega t)} ,$$
(7.5)

with a the distance between two adjacent network elements, and determine the dispersion relation $\omega = \omega(k)$.

7.6   Non-dispersive Transmission Line

The "elementary cell" scheme of a transmission line is sketched in (Fig. 7.6). In addition to the inductance L and capacitance C typical of the ideal "LC" transmission line, there is a resistance R in series with L, which accounts for the finite resistivity of the two con-

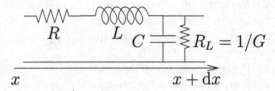

Fig. 7.6

ductors which form the line. In addition, we assume a finite leakage of current between the two conductors (i.e., in the direction "transverse" to the propagation) which is modeled by a second resistance R_L in parallel to C. The corresponding *conductance* is $G = 1/R_L$. In the limit of a continuous system with homogeneous, distributed properties, we define all quantities per unit length by replacing R with $R_\ell dx$, L with $L_\ell dx$, C with $C_\ell dx$ and G with $G_\ell dx$ (it is proper to use G as a quantity defined per unit length instead of R_L because the latter is proportional to the *inverse* of the length of the line).

(a) Show that the current intensity $I = I(x,t)$ satisfies the equation

$$(\partial_x^2 - L_\ell C_\ell \partial_t^2)I = (R_\ell C_\ell + L_\ell G_\ell)\partial_t I + R_\ell G_\ell I = 0 .$$
(7.6)

(b) Study the propagation of a monochromatic current signal of frequency ω, i.e., search for solutions

$$I = I_0 e^{ikx - i\omega t} ,$$
(7.7)

for $x > 0$ with the boundary condition $I(0,t) = I_0 e^{-i\omega t}$, and determine the dispersion relation $k = k(\omega)$.

(c) Find the condition on the line parameters for which a wavepacket traveling along the lines undergoes attenuation of the amplitude but *no* dispersion. This condition corresponds to solutions having the general form

$$I(x,t) = e^{-\kappa x} f(x - vt) ,$$
(7.8)

where $f(x)$ is an arbitrary differentiable function. Find the expression for v and κ.

7.7 An "Alternate" LC Ladder Network

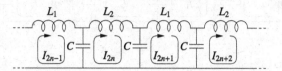

Fig. 7.7

Consider an "alternate" LC ladder network formed by identical capacitors C and inductors of value alternatively L_1 and L_2, as shown in Fig. 7.7. Let I_{2n} be the current flowing in the mesh of the nth inductor of value L_2, and I_{2n+1} the current flowing in the n-th inductor of value L_1.

(a) Show that the currents satisfy the equations

$$L_2 \frac{d^2 I_{2n}}{dt^2} = \frac{1}{C}(I_{2n-1} - 2I_{2n} + I_{2n+1}) , \quad L_1 \frac{d^2 I_{2n+1}}{dt^2} = \frac{1}{C}(I_{2n} - 2I_{2n+1} + I_{2n+2}) .$$
(7.9)

What is a mechanical equivalent of this network?

(b) Search for solutions of (7.9) of the form

$$I_{2n} = I_e \, e^{i[2nka - \omega t]} , \quad I_{2n+1} = I_o \, e^{i[(2n+1)ka - \omega t]} ,$$
(7.10)

where I_e and I_o (the subscripts "e" and "o" stand for *even* and *odd*, respectively) are two constants, and determine the dispersion relation $\omega = \omega(k)$. Determine the allowed frequency range for wave propagation (for simplicity, assume $L_2 \ll L_1$).

7.8 Resonances in an LC Ladder Network

Consider the semi-infinite LC ladder network shown in Fig. 7.8. Let $I_n = I_n(t)$ be the current flowing in the n-th mesh of the circuit. An ideal current source provides the input current

$$I(t) = I_s e^{-i\omega t},$$
(7.11)

where

(a) Assuming $\omega < 2\omega_0$, evaluate $I_n(t)$ as a function of I_s and ω.

(b) Now find $I_n(t)$ assuming $\omega > 2\omega_0$. Hint: search for a solution of the form

$$I_n(t) = A\alpha^n e^{-i\omega t},$$
(7.12)

determining the dependence of α on ω and ω_0.

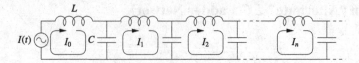

Fig. 7.8

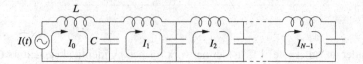

Fig. 7.9

Now assume that our LC ladder is finite, comprising N meshes numbered from 0 to $N-1$, as in Fig. 7.9. Evaluate $I_n(t)$ both for the case $\omega > 2\omega_0$ and for the case $\omega < 2\omega_0$, determining for which values of ω resonances are observed.

7.9 Cyclotron Resonances (1)

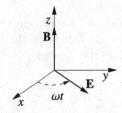

Fig. 7.10

Consider a particle of charge q and mass m in the presence of a constant, uniform magnetic field $\mathbf{B} = B_0\,\hat{\mathbf{z}}$, and of a uniform electric field of amplitude E_0, *rotating* with frequency ω in the (x, y) plane, either in clockwise or in counterclockwise direction (Fig. 7.10 shows the counterclockwise case).

(a) Describe the motion of the particle as a function of B, E, and ω, and show that, given B, a resonance is observed for the appropriate sign and value of ω.

(b) Evaluate the solution of the equations of motion at resonance in the absence of friction.

(c) Now assume the presence of a frictional force $\mathbf{f} = -m\gamma\,\mathbf{v}$, where $\mathbf{v}$ is the velocity of the particle. Find the steady-state solution of the equations of motion, and calculate the power dissipated by friction as a function of ω.

7.10 Cyclotron Resonances (2)

Consider a particle of charge q and mass m in the presence of a constant uniform magnetic field $\mathbf{B} = B_0\hat{\mathbf{z}}$, and of an oscillating uniform electric field $\mathbf{E} = E_0\hat{\mathbf{x}}\cos\omega t$.

(a) Write the equations of motion (assuming no friction) and determine the resonance frequency of the system (hint: show that the equations for the velocity components

v_x and v_y can be separated into two uncoupled equations of the forced harmonic oscillator type)

(b) Now assume the presence of a frictional force $\mathbf{f} = -m\gamma\mathbf{v}$ where $\gamma \ll \omega$ and $\gamma \ll q B_0/m$. Find the steady state solution of the equations of motion and the spectrum of the absorbed power (hint: the equations for v_x and v_y cannot be separated in this case, but seeking a solution in the form $\mathbf{v} = \mathbf{v}_0 e^{-i\omega t}$, with $\mathbf{v}_0$ a complex vector, will work).

7.11   A Quasi-Gaussian Wave Packet

Let us consider a wave packet of Gaussian profile propagating with velocity v along the x axis in a non-dispersive medium, with dispersion relation $\omega(k) = kv$ (Fig. 7.11). In these conditions, the wave packet's profile remains constant, and the packet is described by the function $g(x - vt)$

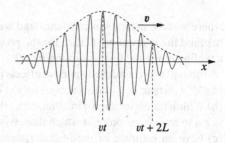

Fig. 7.11

$$g(x - vt) = \sqrt{\pi}\,\frac{A}{L}\,e^{ik_0(x-vt)}e^{-(x-vt)^2/4L^2}$$

$$= \int_{-\infty}^{+\infty} \tilde{g}(k)\,e^{ik(x-vt)}dk,$$

$$(7.13)$$

where L, A and k_0 are constant parameters, and $\tilde{g}(k) = Ae^{-(k-k_0)^2 L^2}$ is the Fourier transform of g. Now consider a second wave packet described by a function f, whose Fourier transform is

$$\tilde{f}(k) = \tilde{g}(k)\,e^{i\phi(k)} = Ae^{-(k-k_0)^2 L^2}e^{i\phi(k)},\qquad (7.14)$$

where the "phase perturbation" $\phi(k)$ is a smooth function, that can be approximated by its Taylor polynomial expansion of degree 2 around $k = k_0$,

$$\phi(k) \simeq \phi(k_0) + \phi'(k_0)(k - k_0) + \frac{1}{2}\phi''(k_0)(k - k_0)^2,\qquad (7.15)$$

where ϕ' and ϕ'' are the first and second derivatives of ϕ. The second wave packet can be considered as an "attempt" to build up a Gaussian wave packet from its spectral components, but with some error on the relative phases of the components themselves. Find the width of the wave packet and discuss its shape in order to show its deviations from the Gaussian profile.

7.12   A Wave Packet Along a Weakly Dispersive Line

A transmission line extends from $x = 0$ to $x = +\infty$. A generator at $x = 0$ inputs a signal

$$f(t) = Ae^{-i\omega_0 t}e^{-t^2/\tau^2}, \tag{7.16}$$

where A and τ are constant and $\omega_0\tau \gg 1$, i.e., the signal is "quasi-monochromatic". The dispersion relation of the transmission line can be written

$$\omega = \omega(k) = kv\,(1 + bk)\,, \tag{7.17}$$

where v and b are known constants, and we assume $k > 0$.
(a) Find the expression $f(x, t)$ for the propagating signal, i.e., for the wave packet traveling along the line, assuming $b = 0$.
   From now on, assume dispersive effects to be small but not negligible, i.e., assume $bk_0 \ll 1$, where $k_0 = k(\omega_0)$ according to (7.17).
(b) Within the above approximation, write the phase and group velocities as functions of ω_0 to the lowest order at which dispersive effects are present.
(c) Give an estimate of the instant t_x when the "peak" of the signal reaches the position x, and of the corresponding length of the wave packet.
(d) Now find the expression of the wave-packet shape as a function of (x, t), by calculating the integral

$$f(x, t) = \int e^{ik(\omega)x - i\omega t}\,\tilde{f}(\omega)\,d\omega\,, \tag{7.18}$$

where $\tilde{f}(\omega)$ is the Fourier transform of the wave packet. As a reasonable approximation, keep only factors up to the second order in $(k - k_0)^2$, for instance use

$$k(\omega) \simeq k(\omega_0) + k'(\omega_0)(\omega - \omega_0) + \frac{1}{2}k''(\omega_0)(\omega - \omega_0)^2. \tag{7.19}$$

Chapter 8
Maxwell Equations and Conservation Laws

Topics. Maxwell's equations. Conservation laws: energy, momentum and angular momentum of the electromagnetic field. Poynting's theorem. Radiation pressure.

Basic equations.
(Note: Gaussian c.g.s. units are used in this chapter unless otherwise specified.)
Maxwell's equations

$$\nabla \cdot \mathbf{E} = 4\pi \rho \, , \tag{8.1}$$

$$\nabla \cdot \mathbf{B} = 0 \, , \tag{8.2}$$

$$\nabla \times \mathbf{E} = -\frac{1}{c}\partial_t \mathbf{B} \, , \tag{8.3}$$

$$\nabla \times \mathbf{B} = \frac{4\pi}{c}\mathbf{J} + \frac{1}{c}\partial_t \mathbf{E} \, . \tag{8.4}$$

Energy conservation (Poynting's) theorem

$$\partial_t u + \nabla \cdot \mathbf{S} = -\mathbf{J} \cdot \mathbf{E} \, , \tag{8.5}$$

where

$$u = \frac{1}{8\pi}\left(\mathbf{E}^2 + \mathbf{B}^2\right) \tag{8.6}$$

is the energy density of the EM field, and

$$\mathbf{S} = \frac{c}{4\pi}\mathbf{E} \times \mathbf{B} \tag{8.7}$$

is the Poynting (also named Poynting-Umov) vector.
Momentum conservation theorem:

$$\partial_t \mathbf{g} + \nabla \cdot \mathsf{T} = -\left(\rho \mathbf{E} + \frac{1}{c}\mathbf{J} \times \mathbf{B}\right) \, , \tag{8.8}$$

© The Author(s), under exclusive license to Springer Nature Switzerland AG 2023
A. Macchi et al., *Problems in Classical Electromagnetism*,
https://doi.org/10.1007/978-3-031-22235-1_8

where

$$\mathbf{g} = \frac{1}{4\pi c}(\mathbf{E} \times \mathbf{B}) = \frac{\mathbf{S}}{c^2} \tag{8.9}$$

is the momentum density of the EM field and T is Maxwell's stress tensor with components

$$T_{ij} = \frac{1}{4\pi}\left[\frac{1}{2}(\mathbf{E}^2 + \mathbf{B}^2)\delta_{ij} - E_i E_j - B_i B_j\right]. \tag{8.10}$$

Thus, $\boldsymbol{\nabla} \cdot \mathsf{T}$ is a vector with components

$$(\boldsymbol{\nabla} \cdot \mathsf{T})_i = \sum_{j=1}^{j=3} \partial_j T_{ij}. \tag{8.11}$$

Angular momentum density of the EM field

$$\boldsymbol{\ell} = \mathbf{r} \times \mathbf{g} = \mathbf{r} \times \frac{\mathbf{S}}{c}. \tag{8.12}$$

8.1   Poynting Vector(s) in an Ohmic Wire

A constant and uniformly distributed current density $\mathbf{J} = \sigma \mathbf{E}$ flows inside an infinite straight wire of radius a and conductivity σ.
(a) Calculate the Poynting vector $\mathbf{S} = (c/4\pi)\mathbf{E} \times \mathbf{B}$ and discuss the energy conservation in the wire.
(b) The Poynting vector occurs in Poynting's theorem only through its divergence, since the theorem only requires that the flux of the Poynting vector through any a closed surface describes the net flow of electromagnetic energy. Show that, consequently, $\mathbf{S}' = \varphi\mathbf{J}$, where φ is the electrostatic potential, is also a suitable choice for $\mathbf{S}$ (hint: substitute $\mathbf{E} = -\boldsymbol{\nabla}\varphi$ into (8.7) and manipulate the result).

8.2   Poynting Vector(s) in a Capacitor

A plane capacitor consists of two parallel circular plates of radius a, at a distance $h \ll a$ from each other. The electric field inside the capacitor is slowly varying in time, $\mathbf{E} = E(t)\hat{\mathbf{z}}$, for instance, assume $E = E_0 t/\tau$. Boundary effects are negligible (Fig. 8.1).
(a) Evaluate the magnetic field $\mathbf{B}$ inside the capacitor.

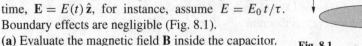

Fig. 8.1

(b) Calculate the Poynting vector $\mathbf{S} = (c/4\pi)\,\mathbf{E} \times \mathbf{B}$, and show that the flux of $\mathbf{S}$ though any surface enclosing the capacitor equals the time variation of the energy associated to the electromagnetic field.

(c) Show that an alternative Poynting vector is

$$\mathbf{S}' = \frac{1}{4\pi}\,\varphi\,\partial_t \mathbf{E}\,, \tag{8.13}$$

where φ is the electric potential ($\mathbf{E} = -\nabla\varphi$). Verify that also the flux of $\mathbf{S}'$ through the closed surface of point **(b)** equals the variation of the energy in the volume inside the surface (hint: proceed as in point **(b)** of Problem 8.1).

8.3 Poynting's Theorem in a Solenoid

A time-dependent current, $I = I(t) = I_0\,t/\tau$, flows through the coils of an infinitely long, cylindrical solenoid. The solenoid has radius a and n turns per unit length.
(a) Find the magnetic and electric fields, $\mathbf{B}$ and $\mathbf{E}$, inside the solenoid.
(b) Verify the law of energy conservation (Poynting's theorem), for a closed internal cylindrical surface, coaxial to the solenoid.
(c) Now verify Poynting's theorem for an *external*, coaxial cylindrical surface (remember that $\mathbf{B} = 0$ outside an infinite solenoid).

8.4 Poynting Vector in a Capacitor with Moving Plates

A plane capacitor consists of two circular metallic plates of radius a, parallel to each other. One plate is kept at rest while the other moves at constant velocity v, so that the distance between the plates is $h = h(t) = h_0 + vt$ (Fig. 8.2). In the following we consider only the case in which $h \ll a$ at any time t, so that boundary effects are negligible. We also assume that v is small enough to ensure the validity of the slowly varying current approximation.

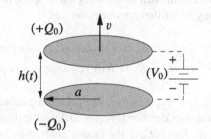

Fig. 8.2

Considering both the case of electrically isolated plates having opposite charges $\pm Q_0$, and the case of plates connected through a voltage source keeping a constant electric potential drop V_0 between them, calculate
(a) the force F needed to keep v constant,
(b) the rate of change of the electrostatic energy U,

(c) the magnetic field between the plates,
(d) the Poynting vector **S** and its flux through a cylindrical surface enclosing, and coaxial with, the capacitor; use this last result to discuss energy conservation in the system.

8.5 Radiation Pressure on a Perfect Mirror

A perfect mirror is defined as a medium inside which $\mathbf{E} = 0$ and $\mathbf{B} = 0$. Thus, an EM wave cannot penetrate the mirror surface and will be reflected by it.

Find the radiation pressure P_{rad}, i.e., the cycle–averaged force per unit surface exerted by a plane wave incident on the surface of a perfect plane mirror, as a function of the intensity I of the wave by each of the following three methods:
(a) Consider the reflection of a square wave packet of arbitrary, but finite, duration. Determine P_{rad} from the difference between the total momentum of the incident wave packet and the momentum of the reflected wave packet.
(b) Calculate the force on the mirror directly, from the knowledge of the EM fields and of the charge and current densities on the mirror surface.
(c) Determine P_{rad} from Maxwell's stress tensor.

8.6 A Light Beam

A beam of monochromatic light with frequency ω, linearly polarized, propagates along the z direction. The beam has a finite width in the plane perpendicular to z. We assume that in some region of space the *transverse* components of the EM fields may be written approximately as

$$
\begin{aligned}
E_x &= E_0(r)\cos(k_0 z - \omega t) = E_0 e^{-r^2/r_0^2}\cos(k_0 z - \omega t), \\
B_y &= B_0(r)\cos(k_0 z - \omega t) = B_0 e^{-r^2/r_0^2}\cos(k_0 z - \omega t),
\end{aligned}
\tag{8.14}
$$

where $r = \sqrt{x^2 + y^2}$. The parameter r_0 is called the *waist* of the beam. The assumption of a Gaussian function to describe the trasverse profile is very common in optics, since for instance the beams emitted by most laser sources have profiles close to a Gaussian shape.
(a) Show that, in addition to the transverse components (8.14), *longitudinal* components E_z and B_z must exist because of the divergence equations, and use the latter to give an expression for E_z and B_z and to estimate their amplitude with respect to E_0 and B_0.
(b) Show that, by imposing that $\partial_t E_z$ and $\partial_t B_z$ satisfy the relevant Maxwell's equations, a relation between E_0, B_0 and k_0 is obtained.
(c) Compute the Poynting vector of the beam **S** and its average over a period $\langle \mathbf{S} \rangle$, showing which components are vanishing.

(d) Verify that the fields (8.14) do *not* satisfy the wave equation in vacuum, hence they are only an approximate expression, as mentioned above. Explain in which range of z, depending on the value of $k_0 r_0$, the approximate expressions are accurate.

8.7 Intensity and Angular Momentum of a Light Beam

A circularly polarized monochromatic light beam of frequency ω propagates along the z direction. The beam has a finite width in the plane perpendicular to z. We assume that in a region of space, close to the "waist" (i.e. to the plane where the beam has minimal width) the *transverse* components of the EM fields can be written approximately as

$$E_x = +E_0(r)\cos(kz - \omega t), \qquad E_y = -E_0(r)\sin(kz - \omega t),$$
$$B_x = E_0(r)\sin(kz - \omega t), \qquad B_y = E_0(r)\cos(kz - \omega t), \qquad (8.15)$$

where $r = \sqrt{x^2 + y^2}$, $k = \omega/c$, and $E_0(r)$ is a known real function.
(a) Write the intensity $I = I(r)$, defined as the "energy flow along z", i.e., $I(r) = S_z = \mathbf{S} \cdot \hat{\mathbf{z}}$ where $\mathbf{S}$ is the Poynting vector.
(b) Show that, in addition to the transverse components of the fields, also *longitudinal* components (E_z, B_z) must exist, and give their expression.
(c) Evaluate the S_x and S_y component of $\mathbf{S}$, and discuss the result.
(d) Show that the density of angular momentum (Eq. (8.12)) of the beam can be written as

$$\ell_z = \ell_z(r) = -\frac{r}{2c\omega}\, \mathrm{d} r\, I, \qquad (8.16)$$

and compute the quantity

$$L_z = \int_0^\infty \ell_z(r)\, 2\pi r\, \mathrm{d} r \qquad (8.17)$$

as a function of the total power of the beam $W = \int_0^\infty I(r)\, 2\pi r\, \mathrm{d} r$.

8.8   Feynman's Paradox solved

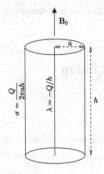

Fig. 8.3

The system in Fig. 8.3 is composed by a non-conducting cylindrical surface of height h and radius a, over which there is a net charge Q uniformly distributed with surface density $\sigma = Q/(2\pi ah)$, and a wire of same length oriented along the cylinder axis and having charge $-Q$ distributed with uniform linear density $\lambda = -Q/h$, so that the system is globally neutral. The cylindrical surface is free to rotate around its axis without friction, and has moment of inertia $\mathcal{I}$ per unit length. The system is at rest in the presence of an external uniform magnetic field $\mathbf{B}_{\text{ext}}$, parallel to the system axis. Assume that boundary effects can be neglected.

Starting at time $t = 0$, the external magnetic field is reduced from its initial value $\mathbf{B}_{\text{ext}} = \mathbf{B}_0$ to zero at a time $t_f \gg a/c$, according to some temporal law $\mathbf{B}_{\text{ext}} = \mathbf{B}_{\text{ext}}(t)$.
(a) Initially assuming that the field generated by the motion of the charges on the cylinder is negligible, evaluate the angular velocity $\omega = \omega(t)$ of the cylinder as a function of time during the decay of $\mathbf{B}_{\text{ext}}$, and the corresponding mechanical angular momentum $\mathbf{L}_c$ of the cylinder.
(b) Now take the field generated by rotating charges into account, and evaluate how the results of **(a)** change.
(c) Consistently with Eqs. (8.8)–(8.9), we introduce the *angular momentum* of a given distribution of electromagnetic fields as

$$\mathbf{L}_{\text{EM}} = \int \mathbf{r} \times \mathbf{g}\, d^3 r \,, \tag{8.18}$$

where $\mathbf{g} = \mathbf{E} \times \mathbf{B}/4\pi$ is the electromagnetic momentum density. Use Eq. (8.18) to check the conservation of the *total* angular momentum for the system (thus solving the "paradox" as outlined in Problem 6.6).

8.9   Magnetic Monopoles

Assume that an experiment gives evidence of the existence of "magnetic monopoles", i.e., of point-like particles with a *net magnetic charge* q_{m}, such that the magnetic field $\mathbf{B}_m$ generated by such charge is

$$\mathbf{B}_m = \alpha \frac{q_{\text{m}}}{r^2} \hat{\mathbf{r}} \,, \tag{8.19}$$

while in the presence of an "external" magnetic field $\mathbf{B}_{ext}$ the force on the particle is $\mathbf{f} = q_m\mathbf{B}_{ext}$. Thus, for example, the interaction force between two particles with magnetic charges q_{m1} and q_{m2} is given by

$$\mathbf{f}_{1\to2} = \alpha\,\frac{q_{m1}q_{m2}}{r_{12}^2}\,\hat{\mathbf{r}}_{12}\,, \qquad \mathbf{f}_{2\to1} = -\mathbf{f}_{1\to2}\,. \tag{8.20}$$

where $\mathbf{r}_{12}$ is the distance vector directed from charge 1 to charge 2. We also assume that conservation of the total magnetic charge holds.

(a) Determine, both in SI and Gaussian units, the expressions for the coefficient α and the dimensions of the magnetic charge q_m with respect to the electric charge q_e. (Hint: we may assume that the field generated by two magnetic charges $+q_m$ and $-q_m$, separated by a distance $\mathbf{h}$, is equivalent to the field of a magnetic dipole $\mathbf{m} = q_m\mathbf{h}$ at distances $r \gg |\mathbf{h}|$.)

(b) Complete Maxwell's equations in order to take the presence of magnetic monopoles into account.

(c) Now consider a beam of magnetic monopoles of radius a, of uniform density and infinite length. The number density of the particles of the beam is n, and all particles have the same magnetic charge q_m and the same velocity v. Find the electric and magnetic fields generated by the beam.

8.10 An Interstellar Light Sail

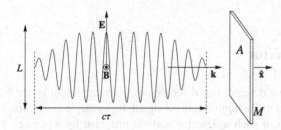

Fig. 8.4

A quasi-monochromatic wave packet of central wavelength λ has total energy U, duration $\tau \gg c/\lambda$ and square cross-section of side $L \gg \lambda$, so that it can be approximated by a plane wave. The wave packet is propagating along the x direction of a Cartesian reference frame.

(a) Show that the wave packet has momentum $\mathbf{p} = \hat{\mathbf{x}}\,U/c$.

(b) The wave packet impinges normally on a perfectly reflecting mirror of area $A > L^2$ and mass M, as shown in Fig. 8.4. The mirror is initially at rest and free to move. Show that its final velocity is $v \simeq 2U/(Mc)$, in the approximation that $v \ll c$.

The aim of the Breakthrough Starshot project is to develop light-sail interstellar probes capable of making the journey to the Alpha Centauri star system, which, at 4.37 light-years from the Sun, is the closest star system to our Solar System. Each "sail" will have a mass $M = 10$ g, a reflecting surface $A = 16$ m^2, and will be accelerated by a laser beam radiating from the Earth.

(c) Assuming $L^2 = A$ evaluate the laser energy needed to accelerate the sail up to a velocity $v = 0.3\,c$. Assuming a laser power of 10^{10} W evaluate the distance ℓ_a traveled by the sail during acceleration. Estimate the laser wavelength required for the Rayleigh length of the laser beam to be larger than ℓ_a. The Rayleigh length ℓ_R is the distance along the propagation direction of a laser beam from the waist to the place where the area of the cross section of the beam is doubled. Approximately we have $\ell_R = S/\lambda$, where S is the cross-section of the beam at the waist.

8.11   Radiation Pressure at Oblique Incidence

Evaluate the radiation pressure, i.e., the force exerted per unit area by an electromagnetic wave of intensity I impinging at an angle θ_i on the planar surface of a medium for which the reflectivity is R, as shown in Fig. 8.5. Show that, in general, the force has both components perpendicular and parallel to the surface.

Hint: consider the momentum balance for a long and wide "squared" quasi-monochromatic wavepacket of duration τ and cross-sectional area A, which is partially reflected by the surface.

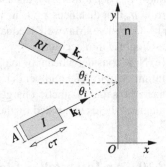

Fig. 8.5

8.12   A Square Law Detector

In electromagnetism a square law detector is a detector that gives an output signal directly proportional to the power of the input electrical signal, for example of a radio signal. A conceptual realization of such a detector may be provided by a mirror, of mass M, kept in place by a spring of Hooke constant κ, as shown in Fig. 8.6. The mirror is subject to the radiation pressure of a reflected beam of light which compresses the spring.

Referring again to Fig. 8.6 let us consider a circularly polarized plane wave with an electric field of amplitude E and frequency ω, being reflected at normal incidence by a mirror with a reflecting surface of area D^2. The mirror-spring system forms an oscillator with oscillation frequency ω_0.

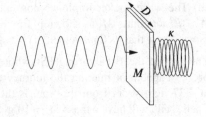

Fig. 8.6

(a) Compute the displacement of the mirror under the effect of the pressure exerted by the wave.

Consider now an additional linearly polarized plane wave impinging normally on the mirror with a small amplitude $E_1 << E$ of frequency ω_1 such that $(\omega - \omega_1)^2 << \omega^2$.

(b) Compute the displacement of the mirror with respect to its rest position and discuss the solution obtained for the case where $(\omega - \omega_1)^2 << \omega_0^2$ and $(\omega + \omega_1)^2 >> \omega_0^2$, i.e., $\omega_1^2 >> \omega_0^2$.

For the sake of simplicity assume that the displacements and the displacement velocities are small so that you can disregard any Doppler effect in the reflection of the two waves.

(c) Compare the result obtained with the case where $E = 0$ while $E_1 \neq 0$.

8.13 Poynting Vector for a Rotating Charged Spherical Shell

Consider the rotating charged spherical shell of Problem 5.12.
(a) Evaluate the Poynting vector in the whole space, due to the simultaneous presence of an electric and a magnetic field.
(b) Evaluate the angular momentum associated to the electromagnetic field.
(c) For a particle of mass m and charge q in the presence of a magnetic field described by a vector potential $\mathbf{A}$ we define the *canonical momentum* as

$$\mathbf{p}^{\mathrm{can}} = m\boldsymbol{v} + \frac{q}{c}\mathbf{A} , \qquad (8.21)$$

see also Problem 13.30. In the case of a medium with volume-distributed mass and electric charge we have a *canonical momentum density* $\mathcal{P}^{\mathrm{can}}$

$$\mathcal{P}^{\mathrm{can}} = \varrho_{\mathrm{m}}\boldsymbol{v} + \frac{\rho}{c}\mathbf{A} , \qquad (8.22)$$

where ϱ_{m} is the mass density and ρ the charge density of the medium. Evaluate accordingly the *canonical angular momentum* of our rotating charged shell, and compare the result to the result of point (b).

Chapter 9
Relativistic Transformations of the Fields

Topics. Relativistic covariance of Maxwell's equations. Four-vectors in electromagnetism: four-current, four-potential. The electromagnetic four-tensor. Lorentz transformations of the fields.

Basic equations. Relation of four-current and four-potential to densities and potentials in three-dimensional space

$$J_\mu \equiv (\rho c, \mathbf{J}) , \qquad A_\mu = (\phi, \mathbf{A}) . \tag{9.1}$$

Lorentz transformations of a four-vector $K_\mu = (K_0, \mathbf{K})$ from the frame S to the frame S' moving with relative velocity $\boldsymbol{v} = \boldsymbol{\beta} c$ with respect to S:

$$K_0' = \gamma(K_0 - \boldsymbol{\beta} \cdot \mathbf{K}) , \qquad K_\parallel' = \gamma(K_\parallel - \beta K_0) , \qquad \mathbf{K}_\perp' = \mathbf{K}_\perp , \tag{9.2}$$

where "$\parallel$" and "$\perp$" denote the directions parallel and perpendicular to $\boldsymbol{\beta}$, respectively, and $\gamma = 1/\sqrt{1 - \beta^2}$.

Compact three-dimensional formulas for the transformation of the EM fields

$$E_\parallel' = E_\parallel , \quad \mathbf{E}_\perp' = \gamma(\mathbf{E}_\perp + \boldsymbol{\beta} \times \mathbf{B}) , \quad B_\parallel' = B_\parallel , \quad \mathbf{B}_\perp' = \gamma(\mathbf{B}_\perp - \boldsymbol{\beta} \times \mathbf{E}) \tag{9.3}$$

or, equivalently

$$\mathbf{E}' = \gamma \, (\mathbf{E} + \boldsymbol{\beta} \times \mathbf{B}) - \frac{\gamma^2}{\gamma + 1} \boldsymbol{\beta}(\boldsymbol{\beta} \cdot \mathbf{E}) , \tag{9.4}$$

$$\mathbf{B}' = \gamma \, (\mathbf{B} - \boldsymbol{\beta} \times \mathbf{E}) - \frac{\gamma^2}{\gamma + 1} \boldsymbol{\beta}(\boldsymbol{\beta} \cdot \mathbf{B}) . \tag{9.5}$$

Three-dimensional transformation of the "Newtonian" force

© The Author(s), under exclusive license to Springer Nature Switzerland AG 2023   81
A. Macchi et al., *Problems in Classical Electromagnetism*,
https://doi.org/10.1007/978-3-031-22235-1_9

$$F'_{\parallel} = F_{\parallel} - \frac{V\mathbf{F} \cdot \mathbf{v}_{\perp}/c^2}{1 - V v_{\parallel}/c^2} \,, \qquad F'_{\perp} = \frac{F_{\perp}}{\gamma(1 - V v_{\parallel}/c^2)} \,, \qquad (9.6)$$

where $\mathbf{F} = d\mathbf{p}/dt$ in the S frame and $\mathbf{F}' = d\mathbf{p}'/dt'$ in the S' frame.

9.1 The Fields of a Current-Carrying Wire

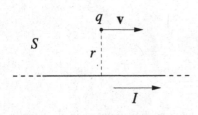

Fig. 9.1 .

In the laboratory frame S, a constant current I flows in an infinitely long wire. The wire has no net charge density. A test charge q moves with a velocity v parallel to the current at a distance r from the wire, as in Fig. 9.1.

(a) Find the force $\mathbf{F}$ acting on q in the laboratory frame. Then evaluate the force $\mathbf{F}'$ acting on the charge in the reference frame S' where the charge is at rest, applying the appropriate Lorentz transformation. What can be inferred on the EM fields in S' from the expression of $\mathbf{F}'$?

(b) Use the Lorentz transformations to obtain the charge and current densities of the wire in S', and the related EM fields. Evaluate the scalar and vector potentials in S'. Compare the results to what obtained from the direct transformation rules for the EM field.

(c) The answers to points (a) and (b) imply that in S' there is a net charge density on the wire. Recover this result by calculating the linear densities of electrons (flowing with velocity v_e in S) and ions (at rest in S) in S' via the Lorentz transformations for velocity and length. (this last point corresponds to the one presented by E. M. Purcell in Ref. [1].)

9.2 The Fields of a Plane Capacitor

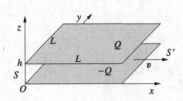

Fig. 9.2 .

In the laboratory frame S, a plane capacitor has parallel square plates of area $A = L^2$, located at a distance $h \ll L$ from each other, so that the boundary effects can be assumed to be negligible. The plates have electric charges $\pm Q$, uniformly distributed over their surfaces, with surface charge density $\pm \sigma = \pm Q/A$, respectively (Fig. 9.2).

Evaluate, in a reference frame S' moving with respect to S with velocity $v = \beta c$ parallel to the capacitor plates,

(a) the electric and magnetic fields in the region between the plates;

(b) the sources of the fields;
(c) the force per unit surface and the total force on each plate, comparing the the results to the corresponding values in S.

9.3   The Fields of a Solenoid

In the laboratory frame S a constant current I flows in an infinite solenoid of radius R and n turns per unit length. At a given instant $t = 0$ a test particle of charge q is located inside the solenoid, with a velocity v perpendicular to the axis of the solenoid, as shown in Fig. 9.3.
(a) Find the electromagnetic fields and the force on the particle both in S, and in the frame S' where the particle is instantaneously at rest ($v' = 0$).
(b) Assuming $v/c \ll 1$, evaluate the sources of the fields in S' up to the first order in v/c.

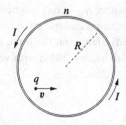

Fig. 9.3

9.4   The Four-Potential of a Plane Wave

Consider a monochromatic plane wave, propagating in vacuum along the x axis of the (Cartesian) laboratory frame S, linearly polarized along $\hat{\mathbf{y}}$, and of frequency ω.
(a) Show that the electric field $\mathbf{E} = \mathbf{E}(x, t)$ and the magnetic field $\mathbf{B} = \mathbf{B}(x, t)$ of the wave can be obtained from a suitable four-potential $A_\mu = (\Phi, \mathbf{A}) = (0, 0, A_y, 0)$. Now consider the same wave observed in a frame S', moving with velocity $\mathbf{v} = v\,\hat{\mathbf{y}}$ with respect to S.
(b) Evaluate the frequency ω' and the wave vector $\mathbf{k}'$ of the wave in S'. Calculate the electric field $\mathbf{E}' = \mathbf{E}'(\mathbf{r}', t')$ and the magnetic field $\mathbf{B}' = \mathbf{B}'(\mathbf{r}', t')$ in S' as functions of $\mathbf{E}$ in the S frame.
(c) Verify that the wave is linearly polarized in S' and show that $\mathbf{E}'$ and $\mathbf{B}'$ can be obtained from a four-potential $A'_\mu = (0, \mathbf{A}')$ where $\mathbf{A}' = \mathbf{A}'(\mathbf{r}', t')$.
(d) Find the four-potential $\bar{A}'_\mu$ obtained from A_μ through a Lorentz transformation. Verify that $\mathbf{E}'$ and $\mathbf{B}'$ can be obtained also from $\bar{A}'_\mu$.
(e) Show that A'_μ and $\bar{A}'_\mu$ are related by a gauge transformation.

9.5   The Force on a Magnetic Monopole

Assume that an experiment has given evidence for the existence of magnetic monopoles, i.e., point-like particles which, in the presence of a magnetic field **B**, are subject to a force

$$\mathbf{F}_m = q_m \mathbf{B} \, , \tag{9.7}$$

where q_m is the magnetic charge of the monopole. We assume that these particles have no electric charge.

(a) Show that the force exerted by an electric field **E** on a monopole moving in the laboratory frame with velocity v is

$$\mathbf{F}_e = -q_m \frac{v}{c} \times \mathbf{E} \, . \tag{9.8}$$

(b) The "Lorentz force" on a magnetic monopole is the sum of (9.7) and (9.8). Use this expression to study the motion of a magnetic monopole of mass m in either an electric field **E** or in perpendicular **E** and **B** fields, where the fields are both constant and uniform, and $E > B$. (For simplicity assume a non-relativistic motion. Compare the results to those of Problem 5.6, point (a)).

9.6   Reflection from a Moving Mirror

An electromagnetic wave of frequency ω and electric field amplitude E_i, linearly polarized along the y axis, is perpendicularly incident on a perfect conductor whose bounding surface lies on the yz plane. The perfect conductor behaves as a perfect mirror, i.e., we have $\mathbf{E} = 0$ and $\mathbf{B} = 0$ inside the material ($x > 0$).

(a) Evaluate the the field of the reflected wave and the total electromagnetic field. The mirror is now set in motion with respect to the laboratory frame S, with a constant velocity $v = \hat{\mathbf{x}} v$ parallel to the x axis.

(b) Find the frequencies and the fields of the incident and reflected waves in the S' frame, where the mirror is at rest.

(c) Find the frequency and the fields of the reflected wave in the S frame.

(d) Discuss the continuity of the fields at the moving mirror surface.

9.7 Oblique Incidence on a Moving Mirror

In the laboratory frame S, a perfectly reflecting mirror moves with constant velocity $\boldsymbol{v}$, perpendicular to its surface. In S, the wave vector $\mathbf{k}_i$ of an incident EM wave makes an angle θ_i with the normal to the mirror surface, as in Fig. 9.4. The incident wave has frequency ω_i. Find
(a) the frequency ω_i' of the incident wave, the incidence angle θ_i', and the reflection angle θ_r' in the in the S' frame, where the mirror is at rest;
(b) the frequency ω_r of the reflected wave, and the reflection angle θ_r, in the S frame. What happens if $\cos\theta_i \geqslant v/c$?

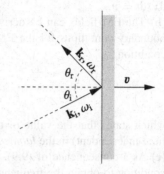

Fig. 9.4

9.8 Pulse Modification by a Moving Mirror

In the laboratory frame S we have an EM square wave packet of amplitude E_i, comprising N complete oscillations of frequency ω_i, therefore of duration $\tau_i = 2\pi N/\omega_i$. Assume that $N \gg 1$, so that the packet is "quasi-monochromatic". The wave packet impinges perpendicularly on a perfect mirror. In the laboratory frame, the mirror itself is moving with

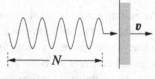

Fig. 9.5

constant velocity $\boldsymbol{v}$ perpendicularly to its surface (Fig. 9.5).
(a) Determine the form, duration and amplitude of the reflected wave packet.
(b) Compare the total energies of the incident and reflected wave packets, and determine the amount of mechanical work W done by the mirror during the reflection stage (consider all quantities per unit surface).
(c) Show that W is equal to the integral over time and volume of $\mathbf{J} \cdot \mathbf{E}$, in agreement with Poynting's theorem.

9.9 Boundary Conditions on a Moving Mirror

The reflecting surface of a perfect mirror is parallel to the yz plane of a laboratory (Cartesian) reference frame S. The mirror is translating with constant velocity $\boldsymbol{v}$ parallel to the x axis, as in Fig. 9.6. A plane monochromatic wave of frequency ω_i, amplitude E_i, and wave vector $\mathbf{k}_i = \hat{\mathbf{x}}\,\omega_i/c$, linearly polarized along the y axis, impinges onto the mirror.

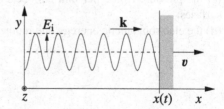

Fig. 9.6

(a) Show that in the laboratory frame S both *total* (i.e., incident + reflected) fields $\mathbf{E}$ and $\mathbf{B}$ are *discontinuous* at the mirror surface, lying on the plane $x = x(t)$, with $dx/dt = v$.

(b) The EM fields can be derived from a vector potential $\mathbf{A}(x, t)$. Show that the boundary conditions for the EM fields at the mirror surface are equivalent to the condition

$$\frac{d}{dt}\mathbf{A}[x(t), t] = 0 , \qquad (9.9)$$

which states that the value of the vector potential at the surface is constant (i.e., time-independent) in the *laboratory* frame.

(c) As a consequence of (9.9), we can assume $\mathbf{A}[x(t), t] \equiv 0$. Use this boundary condition to obtain the frequency (ω_r) and amplitude (E_r) of the reflected wave in the laboratory frame.

9.10 Lorentz Transformations on a Coaxial Cable

An inifinite coaxial cable consists of an inner cylindrical core of radius a and a concentric external cylindrical shield of radius $b > a$. In the laboratory reference frame S the internal conductor has a linear electric charge density λ and carries a global courrent I. Both λ and I are constant in time and uniformly distributed over the cylindrical surface $r = a$. Correspondingly, a linear electric charge density $-\lambda$ and an electric current $-I$, are uniformly distributed over the outer cylindrical conducting surface $r = b$, as in Fig. 9.7.

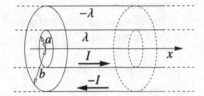

Fig. 9.7 .

(a) Find two reference frames S' and S'', moving with velocities $\mathbf{V}_1$ and $\mathbf{V}_2$, respectively, both parallel to the cable axis, such that $\lambda' = 0$ and $I'' = 0$.
In the three reference frames S, S' and S'' evaluate

(b) the electric and magnetic fields, checking that they satisfy the transformation laws;

(c) the pressure (force per unit surface, with the correct sign) on both conductor surfaces;

(d) the electromagnetic eneregy and momentum densities.

9.11 Lorentz Transformations on a Rectangular Metal Pipe

An infinite rectangular metal pipe is at
rest in the laboratory frame S. Figure 9.8
shows the cross section of the pipe: the
pipe (infinite) length is parallel to the z
axis, the cross section has sides a and b,
parallel to the x and y axes, respectively.
The pipe walls have thickness $h \ll a, b$.
Inside the walls flows a uniform and
constant current density $\mathbf{J}$, as shown in
Fig. 9.8. According to Ampère's circuital law the magnetic field is $\mathbf{B}_0 =$
$\hat{\mathbf{z}} \, 4\pi J h / c$ inside the pipe and zero outside.

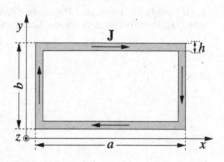

Fig. 9.8

Now consider the pipe as seen from a reference frame S' moving with constant
velocity $\mathbf{V} = V \hat{\mathbf{x}} = \beta c \, \hat{\mathbf{x}}$ with respect to S.
(a) Evaluate the current density, the charge density and the electromagnetic fields
inside and outside the pipe in S'.
(b) Evaluate the energy and electromagnetic momentum densities in S and S'.
(c) Check the boundary conditions for the electromagnetic fields at the pipe walls in
S', at the limit of very thin walls.
(d) Check that the continuity equation (in integral and differential form) holds on
the pipe walls in S'. In particular, check the behavior at the pipe edges.

9.12 Force on an Ohmic Wire

A straight, infinite, cylindrical conducting wire of cross section $A = \pi a^2$ and conductivity σ is located in a constant uniform electric field $\mathbf{E}_0$,
(a) Evaluate the power lost by Joule heating per unit length of the wire and compare
it to the flux of the Poynting vector at the wire surface.
Now consider the wire as seen from a reference frame S' in motion with respect to the
laboratory frame with constant velocity $\mathbf{V}$ perpendicular to the wire and to $\mathbf{E}$. Assume
$V \ll c$ so that terms in V/c of the second order and higher can be disregarded.
(b) Evaluate the electromagnetic fields, the charge densities and the current densities
in S'.
(c) Show that in S' there is a force per unit length acting on the wire, and determine
its value. What is the physical origin of this force?

Reference

1. E.M. Purcell, *Electricity and Magnetism (Berkeley Physics Course - Vol. 2, Section 5.9)*, 2nd edn. (McGraw-Hill Book Company, New York, 1984)

Chapter 10
Radiation Emission and Scattering

Topics. The radiation field. Multipole expansion. Electric dipole radiation. Magnetic dipole radiation.

Basic equations. Fields in the radiation zone of a point-like source at $r = 0$ having an electric dipole moment $\mathbf{p}(t)$:

$$\mathbf{E}(\mathbf{r}, t) = \frac{\left[\ddot{\mathbf{p}}(t_{\text{ret}}) \times \hat{\mathbf{r}}\right] \times \hat{\mathbf{r}}}{rc^2}, \qquad \mathbf{B}(\mathbf{r}, t) = \hat{\mathbf{r}} \times \mathbf{E} \qquad (10.1)$$

where $t_{\text{ret}} = t - r/c$.

Instantaneous radiation power from the electric dipole source and its angular distribution

$$P_{\text{rad}} = \frac{2}{3c^3}|\ddot{\mathbf{p}}|^2, \qquad \frac{dP_{\text{rad}}}{d\Omega} = \frac{3P_{\text{rad}}}{4\pi}\sin^2\theta, \qquad (10.2)$$

where θ is the angle between $\mathbf{p}$ and $\mathbf{r}$, and the infinitesimal solid angle $d\Omega = 2\pi \sin\theta d\theta$.

Analogous formulas for the fields and the power of a magnetic dipole $\mathbf{m}(t)$:

$$\mathbf{E}(\mathbf{r}, t) = -\frac{\left[\ddot{\mathbf{m}}(t_{\text{ret}}) \times \hat{\mathbf{r}}\right]}{rc^2}, \qquad \mathbf{B}(\mathbf{r}, t) = \hat{\mathbf{r}} \times \mathbf{E}, \qquad (10.3)$$

$$P_{\text{rad}} = \frac{2}{3c^3}|\ddot{\mathbf{m}}|^2, \qquad \frac{dP_{\text{rad}}}{d\Omega} = \frac{3P_{\text{rad}}}{4\pi}\sin^2\theta. \qquad (10.4)$$

10.1   Cyclotron Radiation

An electron moves in the xy plane in the presence of a constant and uniform magnetic field $\mathbf{B} = B_0\,\hat{\mathbf{z}}$. The initial velocity is $v_0 \ll c$, so that the motion is non-relativistic and the electron moves on a circular orbit of radius $r_L = v_0/\omega_L$ and frequency $\omega_L = eB_0/m_e c$ (Larmor frequency).

(a) Describe the radiation emitted by the electron in the dipole approximation specifying its frequency, its polarization for radiation observed along the z axis, and along a direction lying in the xy plane, and the total irradiated power P_{rad}. Discuss the validity of the dipole approximation.

(b) The electron gradually loses energy because of the emitted radiation,. Use the equation $P_{\text{rad}} = -dU/dt$, where U is the total energy of the electron, to show that the electron actually spirals toward the "center" of its orbit. Evaluate the time constant τ of the energy loss, assuming $\tau \gg \omega_L^{-1}$, and provide a numerical estimate.

(c) The spiral motion cannot occur if we consider the Lorentz force $\mathbf{f}_L = -(e/c)\,\mathbf{v} \times \mathbf{B}$ as the only force acting on the electron. Show that a spiral motion can be obtained by adding a friction force $\mathbf{f}_{\text{fr}}$ proportional to the electron velocity.

10.2   Atomic Collapse

In the classical model for the hydrogen atom, an electron travels in a circular orbit of radius a_0 around the proton.

(a) Evaluate the frequency ω of the radiation emitted by the orbiting electron, and the emitted radiation power, both as functions of a_0.

(b) Use the results of point (a) to show that, classically, the electron would collapse on the nucleus, and find the decay time assuming $a_0 = 0.53 \times 10^{-8}$ cm (Bohr radius, actually obtained from quantum considerations) .

10.3   Radiative Damping of the Elastically Bound Electron

The motion of a classical, elastically bound electron in the absence of external fields is described by the equation

$$\frac{d^2\mathbf{r}}{dt^2} + \eta\frac{d\mathbf{r}}{dt} + \omega_0^2\mathbf{r} = 0 \,, \tag{10.5}$$

where the vector $\mathbf{r}$ is the distance of the electron from its equilibrium position, η is a friction coefficient, and ω_0 is the undamped angular frequency. We assume that at time $t = 0$ the electron is located at $\mathbf{r}(0) = \mathbf{s}_0$, with zero initial velocity.

(a) As a first step, find the solution of (10.5) assuming $\eta = 0$, and evaluate the cycle-averaged emitted radiation power P_{rad} due to the electron acceleration.

(b) Assuming the oscillation amplitude to decay due to the radiative energy loss, estimated the decay time τ using the result of point (a) for the emitted power P_{rad}. Determine under which conditions τ is much longer than one oscillation period. Now assume $\eta \neq 0$, with $\eta \ll \omega_0$, in Eq. (10.5). In the following, neglect quantities of the order $(\eta/\omega_0)^2$ or higher.

(c) Describe the motion of the electron and determine, *a posteriori*, the value of η that reproduces the radiative damping.

10.4 Radiation Emitted by Orbiting Charges

Two identical point charges q rotate with constant angular velocity ω on the circular orbit $x^2 + y^2 = R^2$ on the $z = 0$ plane of a Cartesian reference frame.

(a) Write the most general trajectory for the charges both in polar coordinates $r_i = r_i(t)$, $\phi_i = \phi_i(t)$ and in Cartesian coordinates $x_i = x_i(t)$, $y_i = y_i(t)$ (where $i = 1, 2$ labels the charge) and calculate the electric dipole moment of the system.

(b) Characterize the dipole radiation emitted by the two-charge system, discussing how the power depends on the initial conditions, and finding the polarization of the radiation emitted along the $\hat{x}$, $\hat{y}$ and $\hat{z}$ directions.

(c) Answer questions (a) and (b) in the case where the charges are orbiting with *opposite* angular velocity.

(d) Now consider a system of *three* identical charges on the circular orbit with the same angular velocity. Find the initial conditions for which the radiation power is either zero or has its maximum.

(e) Determine whether the magnetic dipole moment gives some contribution to the radiation, for each of the above specified cases.

10.5 Spin-Down Rate and Magnetic Field of a Pulsar

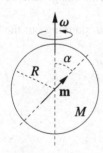

Fig. 10.1

A pulsar is a neutron star with mass $M \approx 1.4 M_\odot \approx 2.8 \times 10^{33}$ g (where $M_\odot$ is the Sun mass), and radius $R \simeq 10\,\text{km} = 10^5$ cm. The star rotates with angular velocity ω and has a magnetic moment $\mathbf{m}$, which is, in general, *not* parallel to the rotation axis [1].

(a) Describe the radiation emitted by the pulsar, and find the total radiated power, assuming that the angle between the magnetic moment and the rotation axis is α, as in Fig. 10.1.

(b) Find the "spindown rate" (decay constant of the rotation) of the pulsar, assuming that energy loss is due to radiation only.

(c) Explain how, from the knowledge of mass, radius, rotation period T, and time derivative dT/dt of the pulsar one can estimate the magnetic field at the pulsar surface. Give a a numerical approximation based on the results of observations [3] which give $T = 7.476551 \pm 3$ s and $\dot{T} = (2.8 \pm 1.4) \times 10^{-11} \simeq 10^{-3}$ s/year (for simplicity assume that $\mathbf{m}$ is perpendicular to $\boldsymbol{\omega}$).

10.6   A Bent Dipole Antenna

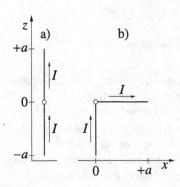

A dipole antenna consists of two identical conductive elements, usually two metal rods, each of length a and resistance R. The driving current is applied between the two halves of the antenna, so that the current flows as shown in Fig. 10.2a. For a "short" antenna ($a \ll \lambda = 2\pi c/\omega$) the current can be approximately specified as [2]

$$I = I(z,t) = \mathrm{Re}\left[I_0\left(1 - \frac{|z|}{a}\right)e^{-i\omega t}\right]. \quad (10.6)$$

Fig. 10.2

The dependence of the current oscillation amplitude on z is shown in Fig. 10.3. Calculate
(a) the cycle-averaged the dissipated power P_{diss};
(b) the linear charge density q_ℓ on the rods of the antenna, and the antenna electric dipole moment $\mathbf{p}$;
(c) the cycle-averaged radiated power P_{rad} and the ratio $P_{\mathrm{rad}}/P_{\mathrm{diss}}$.

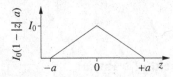

(d) Find the directions along which there no radiation is observed.

Now assume that the upper rod of the dipole antenna is bent by 90°, so that it is parallel to the x axis, as shown in Fig. 10.2b, without perturbing either the current or the charge density anywhere in the two rods.

Fig. 10.3

(e) Answer questions (a), (b) and (c) again for the bent antenna, pointing out the differences with the straight antenna.

10.7 A Receiving Circular Antenna

A receiving circular antenna is a circular coil of radius a and resistance R. The amplitude of the received signal is proportional to the current induced in the antenna by an incoming EM wave (Fig. 10.4).

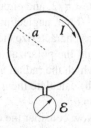

Fig. 10.4

(a) Assume that the incoming signal is a monochromatic, linearly polarized wave of wavelength $\lambda \gg a$, and electric field amplitude E_0. Find how the antenna must be oriented with respect to the wave vector $\mathbf{k}$ and to the polarization in order to detect the maximum signal, and evaluate the signal amplitude.

(b) In a receiving linear antenna the signal is approximately proportional to $E_\parallel \ell$, where $E_\parallel$ is the component of the electric field of the wave parallel to the antenna, and ℓ is the length of the antenna. Old portable TV sets were provided with both a linear and a circular antenna, typical dimensions were $\ell \simeq 50\,\text{cm}$ and $a \simeq \ell/2$. Which antenna is best suited to detect EM waves with λ in the $10^2 - 10^3$ cm range?

(c) Calculate the power P_{rad} scattered by the antenna, and the ratio $P_{\text{rad}}/P_{\text{diss}}$, where P_{diss} is the power dissipated in the antenna by Joule heating.

10.8 Polarization of Scattered Radiation

An EM wave impinges on a particle that acquires an electric dipole moment $\mathbf{p} = \alpha\mathbf{E}$, where $\mathbf{E}$ is the electric field of the wave at the position of the particle. Assume that the size of the particle is much smaller than the wavelength of the incoming wave.

(a) Find the polarization of the scattered radiation as a function of the polarization of the incoming wave, and of the angle between the directions of observation and propagation.

(b) If the incoming radiation is *unpolarized*, what can be said about the polarization of the scattered radiation?

10.9 Polarization Effects on Thomson Scattering

An electron is in the field of an elliptically polarized plane wave of frequency ω propagating along the z axis of a Cartesian reference frame. The electric field of the wave can be written as

$$\mathbf{E} = E_0\left[\hat{\mathbf{x}}\cos\theta\cos(kz - \omega t) + \hat{\mathbf{y}}\sin\theta\sin(kz - \omega t)\right], \qquad (10.7)$$

where θ is a constant real number with $0 \leqslant \theta \leqslant \pi/2$. such that we have linear polarization along the x axis for $\theta = 0$, linear polarization along the y axis for $\theta = \pi/2$, and circular polarization for $\theta = \pi/4$.

First, neglecting the effects of the magnetic force $-e\,\boldsymbol{v} \times \mathbf{B}/c$,

(a) characterize the radiation scattered by the electron by determining the frequency and the polarization observed along each axis (x, y, z), and find a direction along which the radiation is circularly polarized;

(b) calculate the total (cycle-averaged) scattered power and discuss its dependence on θ;

Now consider the effect of the magnetic force on the scattering process.

(c) Evaluate the $-e\boldsymbol{v} \times \mathbf{B}/c$ term by calculating the $\mathbf{B}$ field from (10.7) and using the result of point a) for $\boldsymbol{v}$. Discuss the direction and frequency of the magnetic force and its dependence on θ as well.

(d) Discuss how the scattering of the incident wave is modified by the magnetic force by specifying which new frequencies are observed, in which direction and with which polarization, and the modification of the scattered power.

10.10   Scattering and Interference

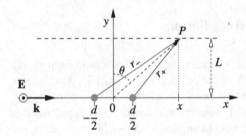

Fig. 10.5

A monochromatic plane wave propagates along the x axis of a Cartesian coordinate system. The wave is linearly polarized in the $\hat{\mathbf{z}}$ direction, and has wavelength λ. Two identical, point-like scatterers are placed on the x axis at $x = \pm d/2$, respectively, as in Fig. 10.5. The dipole moment of each scatterer is $\mathbf{p} = \alpha\mathbf{E}$, where $\mathbf{E}$ is the electric field of the incoming wave at the scatterer position. The intensity I_s of the scattered radiation is measured on the $y = L$ plane, with both $L \gg d$ and $L \gg \lambda$.

(a) Evaluate the phase difference $\Delta\phi$ between the two scattered waves in a generic point $P \equiv (x, L, 0)$, with L a constant, as a function of the observation angle $\theta = \arctan(x/L)$, as shown in Fig. 10.5.

(b) Study the scattered intensity distribution $I_s = I_s(\theta)$ as a function of kd, where $\mathbf{k}$ is the wave vector of the incoming wave. Determine for which values of kd interference fringes appear.

10.11   Optical Beats Generating a "Lighthouse Effect"

Two oscillating dipoles, $\mathbf{p}_-$ and $\mathbf{p}_+$, are located at $(0, -d/2, 0)$ and $(0, +d/2, 0)$, respec-
tively, in a Cartesian reference frame. The
two dipoles are parallel to the z axis and
oscillate, with equal amplitude, at slightly
different frequencies $\omega_\pm = \omega_0 \pm \delta\omega/2$, with
$\delta\omega \ll \omega_0$. In complex representation we have
$\mathbf{p}_\pm = \mathbf{p}_0\, e^{-i\omega_\pm t}$. The distance between the two
dipoles is $d = \lambda_0/2 = \pi c/\omega_0$. The radiation
emitted by the dipoles is observed at a point P
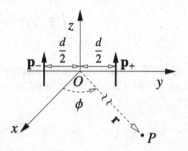
Fig. 10.6

at a distance $\mathbf{r}$ from the origin, with $r \gg \lambda_0$,
on the $z = 0$ plane. Let ϕ be the angle between $\mathbf{r}$ and the x axis, as shown in Fig.
10.6.
(a) Determine the direction of the electric field in P and its dependence on ϕ and
$\omega_\pm$, up to the first order in $\delta\omega/\omega_0$.
The wave intensity in P is measured by two detectors with different temporal reso-
lutions: the first detector measures the "instantaneous" flux averaged over an inter-
val Δt such that $2\pi/\omega_0 \ll \Delta t \ll 2\pi/\delta\omega$, while the second detector averages over
$\Delta t' \gg 2\pi/\delta\omega$.
(b) Determine the dependence on the angle ϕ and the time t of the fluxes measured
with the two detectors.
(c) How do the above results change if the observation point is located in the $x = 0$
plane?

10.12   Radiation Friction Force

An accelerated point charge emits radiation. Considering for definiteness an electron performing a periodic (non-relativistic for simplicity) motion in an oscillating
external field, there is a finite amount of energy leaving the electron as radiation, but
on the average the external field produces no work. Thus, to account self-consistently
for the energy lost as radiation, it is necessary to modify the Newton-Lorentz force
by adding a new "friction" term $\mathbf{F}_{rad}$ so that the mechanical work done by $\mathbf{F}_{rad}$ equals
the radiated energy.[1]

We thus write for the electron

$$m_e \frac{d\boldsymbol{v}}{dt} = -e\left(\mathbf{E} + \frac{\boldsymbol{v}}{c} \times \mathbf{B}\right) + \mathbf{F}_{rad}\,,   \tag{10.8}$$

---

[1] From another viewpoint, $\mathbf{F}_{rad}$ aims to describe the back-action or *reaction* of the self-generated
EM fields on the accelerated charge.

and look for a suitable expression for $\mathbf{F}_{rad}$ starting from the condition

$$\int_{t}^{t+T} \mathbf{F}_{rad}(t) \cdot \boldsymbol{v}(t)\, dt = -\int_{t}^{t+T} P_{rad}(t)\, dt \,, \tag{10.9}$$

where T is the period of the electron motion and $P_{rad}(t)$ is the instantaneous radiated power, which is given by the Larmor formula

$$P_{rad}(t) = \frac{2e^2}{3c^2}\left|\frac{d\boldsymbol{v}}{dt}\right|^2 . \tag{10.10}$$

(a) Show by direct substitution of the expression for $\mathbf{F}_{rad}$

$$\mathbf{F}_{rad} = m_e \tau \, \frac{d^2\boldsymbol{v}}{dt^2} \tag{10.11}$$

into (10.9), that the equation is verified, and find the expression of the constant τ, estimating its numerical value.

(b) Determine the steady state solution of (10.8), where $\mathbf{F}_{rad}$ is given by (10.11), for an electron in a uniform, oscillating electric field

$$\mathbf{E}(t) = \mathrm{Re}\left(-e\,\mathbf{E}_0\, e^{-i\omega t}\right) . \tag{10.12}$$

Compare the result with what obtained using the simple classical model an electron subject to a frictional force

$$m_e \frac{d\boldsymbol{v}}{dt} = \mathbf{F}_{ext} - m_e \eta\, \boldsymbol{v} . \tag{10.13}$$

10.13   Radiation of an Electron Falling on a Proton

An electron, initially at rest at infinity, is attracted by a proton by Coulomb force. Assume the proton mass to be infinite.

(a) Evaluate the electron velocity $\boldsymbol{v}$ as a function of the distance r from the proton assuming a non-relativistic motion. Use the result to estimate the minimum distance r_1 at which the assumption is valid. For instance, you may assume that the motion is non-relativistic for $v < c/\sqrt{10}$.

(b) Evaluate the instantaneous radiated power as a function of r, using the electric dipole approximation.

(c) Evaluate the total energy U_r lost by radiation as a function of r for $r > r_1$, assuming it to be negligible with respect to the kinetic energy of the electron. Use the result to estimate the minimum distance r_2 at which this assumption is valid.

10.14 Scattering by a Perfectly Conducting Sphere (1)

In a Cartesian reference frame (x, y, z), a linearly polarized, plane monochromatic electromagnetic wave has the electric field

$$\mathbf{E}(x, t) = \hat{\mathbf{y}}\, E_0 \cos(kx - \omega t)\,, \quad \text{where} \quad k = \frac{\omega}{c}\,, \qquad (10.14)$$

and impinges on a metallic sphere of radius $a \ll \lambda = 2\pi/k$, located at the origin of the reference frame. We further assume that the sphere is "perfectly conducting" at the frequency of the wave, so that the total electric field can be assumed to be zero over the whole volume of the sphere.

(a) Find the power scattered by the sphere in the electric dipole approximation, and the corresponding scattering cross section.

(b) Assuming that the sphere is also perfectly diamagnetic ($\mathbf{B} = 0$ inside the sphere), find the contribution of the magnetic dipole to the scattering cross section.

10.15 Scattering by a Perfectly Conducting Sphere (2)

A plane, monochromatic, linearly polarized electromagnetic wave of electric field amplitude E_i and wavevector $\mathbf{k} = \hat{\mathbf{x}}\,\omega/c$ impinges on a small conducting sphere of radius $a \ll \lambda = 2\pi c/\omega$, as shown in Fig. 10.7. As in Problem 10.14, assume that the sphere is perfectly conducting and perfectly diamagnetic, so that the total electric and magnetic fields are zero at its inside.

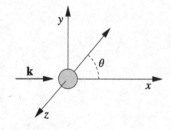

Fig. 10.7

(a) Evaluate the intensity of the radiation emitted by the sphere in the xy plane as a function of the angle θ between $\mathbf{k}$ and the direction of emission, both for y and z polarization of the impinging wave. What are the directions of maximum and minimum emission? *Hint: show that the electric field of the emitted wave depends on the vector* $\mathbf{p}_s \times \mathbf{m}_s$, *where* $\mathbf{p}_s$ *and* $\mathbf{m}_s$ *are the electric and magnetic dipole moments induced by the wave on the sphere, respectively. Consider that only the dependence on the angles is required.*

(b) What is the dependence on θ in the case of unpolarized impinging radiation? Unpolarized radiation means that the radiation can be considered as consisting of wave packets of finite duration whose polarization changes randomly between successive packets, and that the radiation is measured by integrating over times long compared to the duration of a single packet.

References

1. C. Bernardini, C. Guaraldo, *Fisica del Nucleo* (Editori Riuniti, Roma, 1982)
2. J.D. Jackson, *Classical Electrodinamics*, Sects. 9.2 and 9.4, 3rd ed. (Wiley, New York, London, Sidney, 1998)
3. C. Kouveliotou et al., An X-ray pulsar with a superstrong magnetic field in the soft γ-ray repeater SGR1806-20. Nature **393**, 235–237 (1998)

Chapter 11
Electromagnetic Waves in Matter

Topics. Wave equation in continuous media. Classical model of the electron, bound and free electrons. Frequency-dependent conductivity $\sigma(\omega)$ and dielectric permittivity $\varepsilon(\omega)$ for harmonic fields. Relation between $\sigma(\omega)$ and $\varepsilon(\omega)$. Transverse and longitudinal waves. The refraction index. Propagation of monochromatic waves in matter. Dispersion relations. Reflection and transmission at a plane interface: Snell's law, Fresnel's formulas, total reflection, Brewster's angle. Anisotropic media.

Basic equations. Wave equation for the electric field:

$$\nabla^2 \mathbf{E} - \frac{1}{c^2}\partial_t^2 \mathbf{E} - \nabla(\nabla \cdot \mathbf{E}) = \frac{4\pi}{c^2}\partial_t \mathbf{J} = \frac{4\pi}{c^2}\partial_t^2 \mathbf{P} . \tag{11.1}$$

(Note that $\mathbf{J} = \partial_t \mathbf{P}$.)
Definition of $\sigma(\omega)$, $\chi(\omega)$ and $\varepsilon(\omega)$ for harmonic fields $\mathbf{E}(\mathbf{r}, t) = \mathrm{Re}\left[\tilde{\mathbf{E}}(\mathbf{r})\mathrm{e}^{-\mathrm{i}\omega t}\right]$,
$\mathbf{J}(\mathbf{r}, t) = \mathrm{Re}\left[\tilde{\mathbf{J}}(\mathbf{r})\mathrm{e}^{-\mathrm{i}\omega t}\right]$, $\mathbf{P}(\mathbf{r}, t) = \mathrm{Re}\left[\tilde{\mathbf{P}}(\mathbf{r})\mathrm{e}^{-\mathrm{i}\omega t}\right]$:

$$\tilde{\mathbf{J}} = \sigma(\omega)\tilde{\mathbf{E}} , \qquad\qquad \tilde{\mathbf{P}} = \chi(\omega)\tilde{\mathbf{E}} , \tag{11.2}$$

$$\varepsilon(\omega) = 1 + 4\pi\chi(\omega) , \qquad \chi(\omega) = \frac{\mathrm{i}\sigma(\omega)}{\omega} \quad (\omega \neq 0) . \tag{11.3}$$

Dispersion relation in a medium and refraction index $\mathsf{n}(\omega)$:

$$\frac{k^2 c^2}{\omega^2} = \varepsilon(\omega) = \mathsf{n}^2(\omega) . \tag{11.4}$$

© The Author(s), under exclusive license to Springer Nature Switzerland AG 2023
A. Macchi et al., *Problems in Classical Electromagnetism*,
https://doi.org/10.1007/978-3-031-22235-1_11

Let us consider the reflection and refraction of a plane electromagnetic wave at the boundary between two media of refractive indices n_1 and n_2, respectively, both real. We also assume that both media have magnetic permittivity $\mu_r = 1$. The incident wave has wave vector $\mathbf{k}_i$ lying in the xy plane of Fig. 11.1 and forming an angle θ_i with the x axis, as shown in the figure. Also the wave vectors of the reflected and transmitted wave, $\mathbf{k}_r$ and $\mathbf{k}_t$, respectively, lie in the xy plane forming angles θ_r and θ_t with the x axis, as shown in Fig. 11.1. We have $\theta_r = \theta_i$, while

$$n_2 \sin \theta_t = n_1 \sin \theta_i \qquad (11.5)$$

according to Snell's law. If the electric field $\mathbf{E}_i$ of the incident wave is parallel to the z axis, thus perpendicular to the xy incidence plane (S polarization, S for *senkrecht*, "perpendicular" in German), also the electric fields $\mathbf{E}_r$ and $\mathbf{E}_t$ of the reflected and transmitted waves are perpendicular to the incidence plane. According to the Fresnel equations we have for the field amplitudes E_r and E_t

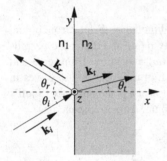

Fig. 11.1 .

$$\frac{E_r}{E_i} = \frac{n_1 \cos \theta_i - n_2 \cos \theta_t}{n_1 \cos \theta_i + n_2 \cos \theta_t} , \qquad (11.6)$$

$$\frac{E_t}{E_i} = \frac{2n_1 \cos \theta_i}{n_1 \cos \theta_i + n_2 \cos \theta_t} . \qquad (11.7)$$

If $\mathbf{E}_i$ lies on the incidence plane so do also $\mathbf{E}_r$ and $\mathbf{E}_t$, and we have P polarization (P for *parallel* to the incidence plane). In this case the Fresnel equations are

$$\frac{E_r}{E_i} = \frac{n_2 \cos \theta_i - n_1 \cos \theta_t}{n_2 \cos \theta_i + n_1 \cos \theta_t} , \quad \text{and} \quad \frac{E_t}{E_i} = \frac{2n_1 \cos \theta_i}{n_2 \cos \theta_i + n_1 \cos \theta_t} . \qquad (11.8)$$

Other polarizations can be treated as superpositions of S and P polarizations.

According to the Fresnel equations P-polarized light is not reflected if the angle of incidence is

$$\theta_i = \theta_B = \arctan\left(\frac{n_2}{n_1}\right) , \qquad (11.9)$$

which is called *Brewster's angle*.

11.1   Wave Propagation in a Conductor at High and Low Frequencies

In a classical treatment, a metal has n_e conduction electrons per unit volume, whose equations of motion in the presence of an external electric field $\mathbf{E}(\mathbf{r}, t)$ are

$$m_e \frac{dv}{dt} = -e\mathbf{E}(\mathbf{r}, t) - m_e \eta \, v \, , \qquad v = \frac{d\mathbf{r}}{dt} \, , \qquad (11.10)$$

where $-e$ and m_e are electron charge and mass, respectively, and η is a constant describing friction.

(a) Determine the complex conductivity of the metal, $\sigma = \sigma(\omega)$, as a function of the angular frequency ω of the electric field, and the values of ω for which σ is either purely real or purely imaginary. Discuss these limits for a good conductor, whose DC conductivity (i.e., its conductivity for static fields) has values of the order of $\sigma_{DC} \sim 5 \times 10^{17} \, s^{-1}$.

Now consider a monochromatic, plane EM wave, linearly polarized along the y axis and traveling in the positive direction along the x axis of a Cartesian coordinate system. The wave is incident on a conductor filling the $x > 0$ half-space, while we have vacuum in the $x < 0$ half-space.

(b) Consider both cases of σ purely real and purely imaginary, and determine the frequency ranges in which the wave is evanescent inside the metal.

(c) Find the time-averaged EM energy flux through the metal surface and show that it is equal to the amount of energy dissipated inside the metal.

11.2   Energy Densities in a Free Electron Gas

A plane, monochromatic, transverse electromagnetic wave propagates in a medium containing n_e free electrons per unit volume. The electrons move with negligible friction. Calculate

(a) the dispersion relation of the wave, the phase (v_φ) and group (v_g) velocities, and the relation between the amplitudes of the electric (E_0) and magnetic (B_0) fields;

(b) the EM energy density u_{EM} (averaged over an oscillation period) as a function of E_0;

(c) the kinetic energy density u_K (averaged over an oscillation period), defined as $u_K = n_e m_e \langle v^2 \rangle / 2$, where v is the electron oscillation velocity, and the *total* energy density $u = u_{EM} + u_K$.

(d) Assume that the medium fills the half-space $x > 0$, while we have vacuum in the half-space $x < 0$. An EM wave, propagating along the x axis, enters the medium. Assume that both v_g and v_φ are real quantities. Use the above results to verify the conservation of the energy flux, expressed by the relation

$$c(u_i - u_r) = v_g u_t \, , \qquad (11.11)$$

where u_i, u_r and u_t are the total energy densities for the incident, reflected and transmitted waves, respectively.

11.3 Longitudinal Waves

Consider a *longitudinal* monochromatic plane wave, propagating in a medium along the x axis of a Cartesian reference frame. "Longitudinal" means that the electric field $\mathbf{E}$ of the wave is parallel to the wavevector $\mathbf{k}$. Assume that the electric and magnetic fields of the wave are

$$\mathbf{E} = \mathbf{E}(x, t) = \hat{\mathbf{x}}\, E_0\, e^{ikx - i\omega t}, \qquad \mathbf{B} \equiv 0, \qquad (11.12)$$

respectively, and that the optical properties of the medium are described by a given frequency-dependent dielectric permittivity $\varepsilon_{\mathrm{r}}(\omega)$.

(a) Show that the possible frequencies for the wave (11.12) correspond to zeros of the dielectric permittivity, $\varepsilon_{\mathrm{r}}(\omega) = 0$.

(b) Find the charge and current densities in the medium associated to the presence of the wave fields (11.12).

(c) Assuming that the optical properties of the medium are determined by n_{e} classical electrons per unit volume, bound to atoms by an elastic force $-m_{\mathrm{e}}\omega_0^2 \mathbf{r}$, determine $\varepsilon_{\mathrm{r}}(\omega)$ and the dispersion relation for the longitudinal wave.

11.4 Transmission and Reflection by a Thin Conducting Foil

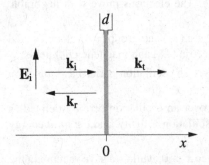

Fig. 11.2 .

A plane wave of frequency $\omega = 2\pi c/\lambda$ strikes at normal incidence a thin metal foil of thickness $d \ll \lambda$. At the limit of an infinitely thin foil (Fig. 11.2), the volume electron density in space can be approximated as $n_{\mathrm{v}}(x) = n_{\mathrm{e}}d\,\delta(x)$, where n_{e} is the volume electron density in the conductor, so that $n_{\mathrm{e}}d$ is the surface electron density on the foil, and $\delta(x)$ is the Dirac delta function. Analogously, the volume current density in space can be approximated as $\mathbf{J}(x, t) = \mathbf{K}(t)\,\delta(x)$, where $\mathbf{K}(t)$ is the surface current density on the foil.

(a) Prove the following relations for the field components parallel to the foil surface

$$E_{\|}(0^+) - E_{\|}(0^-) = 0, \qquad B_{\|}(0^+) - B_{\|}(0^-) = \frac{4\pi}{c}K. \qquad (11.13)$$

(b) Evaluate the EM field in the whole space as a function of the foil conductivity σ, with σ, in general, a complex scalar quantity. Assume a linear dependence of the current density $\mathbf{J}$ on the electric field $\mathbf{E}$, using the complex notation $\mathbf{J} = \mathrm{Re}\left(\tilde{\mathbf{J}}\,e^{-i\omega t}\right)$.

(c) Now use the classical equation of motion for the electrons in the metal

$$m_e \frac{d\mathbf{v}}{dt} = -e\mathbf{E} - m_e \nu \mathbf{v}, \tag{11.14}$$

where ν is a damping constant, to obtain an expression for σ, and evaluate the cycle-averaged absorbed power at the limits $\nu \gg \omega$ and $\nu \ll \omega$, respectively.

(d) Verify the conservation of energy for the system by showing that the flux of EM energy into the foil equals the absorbed power.

11.5   Anti-reflection Coating

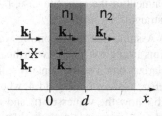

A monochromatic plane EM wave of angular frequency ω travels in vacuum ($x < 0$) along the x direction of a Cartesian coordinate system. On the plane $x = 0$ the wave strikes normally a semi-infinite composite medium. The medium comprises a first layer, between the planes $x = 0$ and $x = d$, of real refractive index n_1, followed by a semi-infinite layer filling the half-space $x > d$, of real refractive index n_2, as shown in Fig. 11.3.

Fig. 11.3

We want to determine the conditions on n_1 and d in order to have a *total* transmission of the incident wave, so that there is *no* reflected wave in the vacuum region. Proceed as follows:

(a) write the general solution for the EM wave in each region of space;

(b) write the relations between the amplitudes of the EM fields in each region due to matching conditions at the two interfaces;

(c) having determined from point b) the relation between n_1, n_2 and d necessary to the absence of reflection, find the values of n_1 and d for which a solution exists in the $n_2 = 1$ case.

(d) How does the answer to point c) change if $n_2 \neq 1$?

11.6   Birefringence and Waveplates

The refractive index of *anisotropic* crystals depends on both the propagation direction and the polarization of the incoming EM wave. We choose a Cartesian reference frame such that the $x = 0$ plane separates the investigated medium from vacuum.

The wave vector of the incident wave, $\mathbf{k}_i$, lies in the xy plane and forms an angle θ_i with the x axis, as shown in Fig. 11.4. In this context we consider a material whose refractive index has the values n_s for a wave polarized perpendicularly to the incidence xy plane (S polarization, from German *senkrecht*, perpendicular), and n_p for waves whose electric field lies in the xy plane (P polarization, from *parallel*). Here, both n_s and n_p are assumed to be real and positive, with $n_p > n_s$. The treatment of the opposite case, $n_s > n_p$, is straightforward.

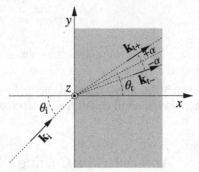

Fig. 11.4 .

(a) Assume that the incoming wave is linearly polarized, and that its electric field forms an angle $\psi = \pi/4$ with the z axis, so that its polarization is a mixture of S and P polarizations. The incident ray splits into two refracted rays at different angles, $\theta_{t\pm} = \theta_t \pm \alpha$, as shown in Fig. 11.4, where $\mathbf{k}_{t+}$ corresponds to S, and $\mathbf{k}_{t-}$ to P polarization. Show how the values of n_s and n_p can be obtained from the measurements of θ_t and α. Assume that $n_p = \bar{n} + \delta n$, and $n_- = \bar{n} - \delta n$, with $\delta n / \bar{n} \ll 1$, and keep only first-order therms in $\delta n / \bar{n}$.

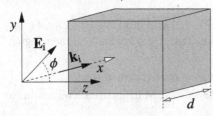

Fig. 11.5 .

(b) Now assume normal incidence ($\theta_i = 0$), and that the electric field of the linearly polarized incoming wave, $\mathbf{E}_i$, still forms an angle $\psi = \pi/4$ with the $\hat{\mathbf{z}}$ axis, as in Fig. 11.5. The crystal has a thickness $d \gg \lambda$. Find the values of d such that the light exiting the crystal is either *circularly* polarized, or linearly polarized, but rotated by $\pi/2$ with respect to the polarization of the incident light. Neglect the difference between the reflection coefficients for S and P polarizations.

11.7   Magnetic Birefringence and Faraday Effect

An EM plane wave of frequency ω travels in a medium in the presence of a static uniform magnetic field $\mathbf{B}_0 = B_0\,\hat{\mathbf{z}}$, where $\hat{\mathbf{z}}$ is the z unit vector of a Cartesian reference frame. B_0 is much stronger than the magnetic field of the wave. The direction of the wave propagation is also parallel to $\hat{\mathbf{z}}$. The medium contains n_e bound electrons per unit volume, obeying the classical equations of motion

$$
m_e \frac{d^2\mathbf{r}}{dt^2} = -e\left(\mathbf{E} + \frac{v}{c}\times\mathbf{B}\right) - m_e\omega_0^2\,\mathbf{r}\,, \qquad v = \frac{d\mathbf{r}}{dt}\,. \tag{11.15}
$$

where m_e and $-e$ are the electron mass and charge, respectively.

(a) Show that the propagation of the wave depends on its polarization by evaluating the refractive index for *circular* polarization, either left-handed or right-handed.

(b) Now consider the propagation of a *linearly* polarized wave. Assume the electric field at $z = 0$ to be given by $\mathbf{E}_i(z = 0, t) = \hat{\mathbf{x}}\,E_i\,e^{-i\omega t}$, and a relatively weak magnetic field so that $\omega \gg \omega_c$ and terms of order higher than ω_c/ω may be neglected. Find the electric field at the position $z = \ell$, showing that the polarization has *rotated* (Faraday effect).

11.8   Whistler Waves

Lightnings excite transverse EM signals which propagate in the ionosphere, mostly in the direction parallel to the Earth's magnetic field lines.

(a) Show that, in a frequency range to be determined, and depending on the wave polarization, the dispersion relation for such signals has the form

$$
\omega = \alpha k^2\,, \tag{11.16}
$$

with α a constant depending on the free electron density n_e and the magnetic field B_0 (both assumed to be uniform for simplicity). Give a numerical estimate for the frequency range, knowing that typical values are $n_e \approx 10^5$ cm^{-3}, and $B_0 \approx 0.5$ G.

(b) Determine the group and phase velocities following from (11.16) as functions of ω, and compare them to c.

(c) Suppose that a lightning locally excites a pulse having a frequency spectrum extending from a value ω_1 to $\omega_2 = 2\omega_1$, within the frequency range determined at point (a). Assuming the pulse to be "short" (in a sense to be clarified *a posteriori*), estimate the pulse length after propagation over a distance $L \simeq 10^4$ km. Try to explain why these signals are called *whistlers*.

(Refer to [1], Sect. 7.6, and to Problem 11.7 for the propagation of EM waves along a magnetic field).

11.9   Wave Propagation in a "Pair" Plasma

A "pair" plasma is composed by electrons and positrons with equal density n_0 (pair annihilation is neglected).

(a) In the absence of external fields, find the dispersion relation for transverse EM waves, determining cut-off and/or resonance frequencies, if any.

(b) Find and discuss the dispersion relation as in **(a)**, but for waves propagating along the direction of an external, static magnetic field $\mathbf{B}_0$ (see also Problem 11.7).

11.10   Surface Waves

A homogeneous medium fills the $x > 0$ half-space of a Cartesian reference frame, while we have vacuum for $x < 0$. The dielectric permittivity of the medium, $\varepsilon = \varepsilon(\omega)$, assumes real values in the frequency range of interest. A monochromatic EM wave propagates along the y-direction, parallel to the interface between the medium and vacuum. Inside the medium, the magnetic field of the wave has the z-component only, given by

$$B_z = B_0\,\mathrm{e}^{-qx}\cos(ky - \omega t) = \mathrm{Re}\left(B_0\,\mathrm{e}^{-qx}\mathrm{e}^{iky-i\omega t}\right) \qquad (x > 0)\,, \quad (11.17)$$

where q is a real and positive quantity.

(a) Using the wave equation for $\mathbf{B}$ inside the dielectric medium, find a relation between q, k and ω.

(b) Write the expression for the electric field $\mathbf{E}$ inside the medium.

(c) Calculate the Poynting vector $\mathbf{S}$ and specify the direction of the time-averaged EM energy flow.

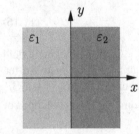

Now consider two different homogeneous media of dielectric permittivities ε_1 and ε_2, respectively, filling the $x < 0$ and $x > 0$ half-spaces (Fig. 11.6). A linearly-polarized EM wave propagates along the y-axis on the $x = 0$ interface, with the magnetic field given by

$$\mathbf{B} = \mathrm{Re}\left[\hat{\mathbf{z}}\,B_z(x)\,\mathrm{e}^{iky-i\omega t}\right]\,, \quad (11.18)$$

Fig. 11.6

where

$$B_z(x) = \begin{cases} B_1\,\mathrm{e}^{+q_1 x}\,, & x < 0 \\ B_2\,\mathrm{e}^{-q_2 x}\,, & x > 0 \end{cases}\,. \quad (11.19)$$

(d) Using the boundary conditions for B_z at the $x = 0$ surface, find the relation between B_1 and B_2.

(e) Using the continuity of E_y at the $x = 0$ surface, find the relation between q_1 and q_2. Show that ε_1 and ε_2 must have *opposite* sign in order to have $q_{1,2} > 0$, i.e., vanishing fields for $|x| \to \infty$.

(f) From the results of points (a) and (e) find the dispersion relation $\omega = \omega(k)$ as a function of ε_1 and ε_2, showing that wave propagation requires $\varepsilon_1 + \varepsilon_2 < 0$.

(g) If medium 1 is vacuum ($\varepsilon_1 = 1$), how should medium 2 and the wave frequency be chosen in order to fulfill the condition found at point (f)?

11.11 Mie Resonance and a "Plasmonic Metamaterial"

A plane, monochromatic wave of frequency ω impinges on a a small sphere of radius $a \ll \lambda = 2\pi c/\omega$. The sphere is made of a material whose dielectric function $\varepsilon = \varepsilon(\omega)$ can be written as

$$\varepsilon = 1 - \frac{\omega_p^2}{\omega^2 - \omega_0^2 + i\omega\eta} \tag{11.20}$$

where, according to the model of the elastically bound electron, ω_p is the plasma frequency, ω_0 is the resonance frequency of bound electrons, and η is a damping constant.

(a) Find the induced field and polarization inside the sphere, and discuss any resonant behavior. (Hint: have a look back at Problem 3.4.)

(b) Assume that the EM wave is propagating inside a material where there are n_s metallic ($\omega_0 = 0$) nanospheres per unit volume, with $n_s\lambda^3 \gg 1 \gg \lambda/a$ (Fig. 11.7). Find the *macroscopic* polarization of the material and discuss the propagation of the wave as a function of the frequency ω.

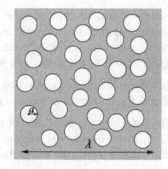

Fig. 11.7 .

11.12 Wave Incident at Brewster's Angle

A medium of real refractive index $n = n(\omega) > 0$ fills the half-space $x > 0$. A monochromatic plane wave of angular frequency ω and electric field amplitude $\mathbf{E}_0$ is incident on the $x = 0$ plane at an angle θ_i, as shown in Fig. 11.8. The wave is linearly polarized in the plane of incidence, and the incidence angle θ_i equals Brewster's angle $\theta_B = \arctan(n)$, so that there is no reflected wave.

(a) Evaluate the amplitudes of the electric and magnetic fields of the transmitted wave.

(b) Show that the flux (averaged over one period) of the Poynting vector **S** is continuous through the vacuum-medium interface.

(c) Show that there is a surface charge density on the $x = 0$ plane, and evaluate its value.

(d) Evaluate the electromagnetic pressure (averaged over one period) on the surface.

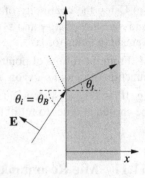

Now assume that the medium is not infinite in the x direction, but has a thickness d.

Fig. 11.8

(e) Evaluate the electromagnetic fields in the whole space.

11.13   Electromagnetic Wave in a Conducting Medium

An electromagnetic plane wave of angular frequency ω is incident normally on a conducting half-space limited by the yz plane of a Cartesian reference frame. We have vacuum for $x < 0$ and conducting medium for $x > 0$. In the conducting medium Ohm's law $\mathbf{J} = \sigma \mathbf{E}$ holds, with σ real and $\sigma \gg \omega$.

(a) Write the dielectric function and the refractive index of the medium, showing that at the limit $\sigma \gg \omega$ we have $\mathsf{n} \simeq \mathsf{n}_0(1 + i)$, with n_0 real.

(b) Let E_t be the amplitude of the electric field of the transmitted wave at $x = 0$. Evaluate the energy flux, averaged over one period, through the medium surface, and show that it equals the power dissipated by Joule heating per unit volume.

11.14   Wave Reflection at the Ionosphere

The ionosphere is a layer of ionized gas, located from about $h \simeq 48\,000$ m altitude to about $965\,000$ m. As a very rough model we assume the ionized gas as a medium with free electrons without the effects of either dissipation or the Earth's magnetic field, so that it can be described by the dielectric permittivity $\varepsilon_r = \varepsilon_r(\omega) = 1 - \omega_p^2/\omega^2$; we further assume the free electron density n_e (and thus the plasma frequency ω_p) to be uniform (a typical value is $n_e \simeq 10^{12}$ m^{-3}).

The antenna of a radio broadcaster (S in Fig. 11.9) is located at sea level and can be modeled as an electric dipole oscillating in the direction perpendicular to the Earth surface. The radio is transmitting signals to a receiver (R in Fig. 11.9) located at a distance L, via reflection from the lower boundary of the ionosphere (assumed to be of infinite extension above h).

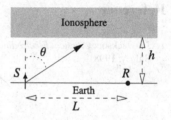

Fig. 11.9 .

(a) Determine the conditions on the broadcaster frequency ω and the distance L for total reflection from the ionosphere of the signal sent from S to R.

Let θ be the angle between the wave vector of the emitted signal and the dipole, as in Fig. 11.9, determine what values of θ correspond to total reflection of the wave at the ionosphere.

(b) Evaluate the intensity of the signal reaching R as a function of the total power emitted by the antenna (P_a).

(c) The Earth's magnetic field has an intensity of approximately 0.5 G. Is this intensity strong enough to affect the propagation of radio waves in the ionosphere? Wavelengths used for radio broadcasting range from approximately 0.1 m to approximately 100 m.

11.15 Waves in a Dipole Chain

Consider a one-dimensional chain of small spheres of radius a, located at the positions $x_n = nd$ (with n an integer) along the x-axis, with a spacing $d \gg a$, as shown in Fig. 11.10. The spheres are made of

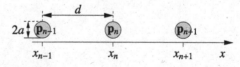

Fig. 11.10 .

a simple metal with dielectric permittivity $\varepsilon_r = 1 - \omega_p^2/\omega^2$, where ω_p is the plasma frequency of the metal, and negligible dissipation.

By considering only the interactions between neighboring dipoles, discuss the propagation along the chain of monochromatic waves of frequency ω for which the n-th dipole is given by

$$\mathbf{p}_n = \bar{\mathbf{p}}\, e^{ikx_n - i\omega t} \ . \tag{11.21}$$

with $\bar{\mathbf{p}}$ a constant amplitude, for the cases in which $\bar{\mathbf{p}}$ is either parallel or transverse to the chain axis. Assume $\lambda = 2\pi c/\omega \gg d$.

Reference

1. J.D. Jackson, *Classical Electrodinamics*, Sects. 9.2 and 9.4, 3rd edn. (Wiley, New York, London, Sidney, 1998)

Chapter 12
Transmission Lines, Waveguides, Resonant Cavities

Topics. Guided propagation of EM waves. Transmission lines, TEM mode. Waveguides, TE and TM modes. Resonant cavities and discretization of frequencies.

12.1 The Coaxial Cable

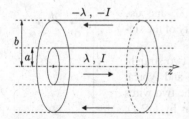

Fig. 12.1

A coaxial cable consists of two coaxial, infinitely long conductors: an inner cylinder of radius a, and an outer cylindrical shell of internal radius $b > a$. In general, if there is a charge per unit length λ on the inner conductor, there is an opposite charge $-\lambda$ on the outer conductor. Similarly, if a total current I flows through the inner conductor, an opposite "return" current $-I$ flows in the outer one (Fig. 12.1).

We use a cylindrical coordinate system (r, ϕ, z) with the cable axis as z axis, and, at first, we assume that the region $a < r < b$ is filled by an insulating medium of dielectric permittivity $\varepsilon = 1$ and magnetic permeability $\mu = 1$.

(a) Evaluate the capacitance and inductance per unit length of the cable.

(b) Describe the propagation of a current signal $I(z, t)$ and of an associated linear charge signal $\lambda(z, t)$ along the cable, remembering the results of Sect. 7.4. How are $I(z, t)$ and $\lambda(z, t)$ related to each other?

(c) For given $I(z, t)$ and $\lambda(z, t)$, find the electric field **E** and the magnetic field **B** in the space between the conductors, assuming that both **E** and **B** are *transverse*, i.e. perpendicular to the direction of propagation (such configuration is called TEM mode).

© The Author(s), under exclusive license to Springer Nature Switzerland AG 2023
A. Macchi et al., *Problems in Classical Electromagnetism*,
https://doi.org/10.1007/978-3-031-22235-1_12

(d) Now consider a semi-infinite cable with an ideal source imposing the voltage $V(t)$ between the inner and outer conductors at the end of the cable. Show that the work done by the generator equals the flux of the Poynting vector through the cable (far enough from the end, so that we may neglect boundary effects).
(e) How do the preceding answers change if the medium between the internal and external conductors has real and positive values for ε and μ, but different from unity?

12.2   Electric Power Transmission Line

Consider a thin, infinite straight wire along the z axis of a cylindrical coordinate system (r, ϕ, z). The wire is located in a medium of relative electric permittivity $\varepsilon_r = 1$ and relative magnetic permeability $\mu_r = 1$. Assume a current $I = I(z, t)$ to flow in the wire, with

$$I = I(z,t) = I_0\, e^{ikz - i\omega t}. \tag{12.1}$$

(a) Calculate the linear charge density $\lambda = \lambda(z, t)$ on the wire.
(b) Assume that the electric and magnetic fields have only their radial and azimuthal components, respectively,

$$E_\phi = E_z = 0, \quad E_r = E_r(r)\, e^{ikz - i\omega t}, \quad B_r = B_z = 0, \quad B_\varphi = B_\varphi(r)\, e^{ikz - i\omega t}. \tag{12.2}$$

Calculate E_r and B_φ as functions of I_0 and ω, and use Maxwell's equations to evaluate the phase velocity of the signal $v_\varphi = \omega/k$.
(c) A high voltage transmission line comprises two straight parallel wires, at a constant distance $d = 5$ m and typical height over the ground $h = 30$ m (Fig. 12.2). The two wires have opposite current intensities $\pm I(z, t)$ given by (12.1), where typically $I_0 = 10^3$ A and $\omega = 2\pi \times 50$ s^{-1}. Calculate the electric and magnetic fields on the symmetry plane between the two wires, and evaluate their magnitude on the ground.

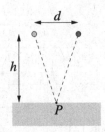

Fig. 12.2

12.3   TEM and TM Modes in an "Open" Waveguide

An "open" waveguide comprises two parallel, perfectly conducting planes, between which the waves propagate. Let us choose a Cartesian coordinate system (x, y, z) such that the two conducting planes are at $y = \pm a/2$, respectively, as in Fig. 12.3. An EM wave of frequency ω propagates in the waveguide along $\hat{\mathbf{x}}$. The magnetic field of the

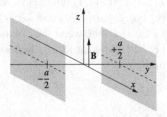

Fig. 12.3

wave is directed along $\hat{\mathbf{z}}$ and has the form

$$B_z(x, y, t) = B_0 \cos(k_y y) \, e^{ik_x x - i\omega t} \ . \tag{12.3}$$

(a) Find the relations between ω, k_x and k_y.
(b) Find the expression for the electric field $\mathbf{E} = \mathbf{E}(x, y, t)$ of the EM wave.
(c) Find how the possible values for k_y are determined by the boundary conditions on $\mathbf{E}$, and discuss the existence of cut-off frequencies.
(d) Find the flux of energy along the direction of propagation $\hat{\mathbf{x}}$, showing that it is proportional to the group velocity of the wave.

12.4 Square and Triangular Waveguides

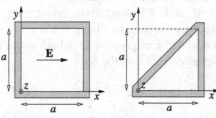

Fig. 12.4   Fig. 12.5

A waveguide has perfectly conducting walls and a square section of side a, as shown in Fig. 12.4. We choose a Cartesian coordinate system (x, y, z) where the interior of the waveguide is delimited by the four planes $x = 0$, $x = a$, $y = 0$ and $y = a$. Consider the propagation along $\hat{\mathbf{z}}$ of a wave of frequency ω, whose electric field $\mathbf{E}(x, y, z, t)$ is perpendicular to $\hat{\mathbf{z}}$ (a TE mode). Assume that the electric field can be written as

$$\mathbf{E}(x, y, z, t) = \tilde{\mathbf{E}}(x, y) \, e^{ik_z z - i\omega t} \ , \tag{12.4}$$

where $\tilde{\mathbf{E}}(x, y)$ depends on x and y only.
(a) Assume that $\mathbf{E}$ is parallel to $\hat{\mathbf{x}}$, i.e. $\mathbf{E} = \hat{\mathbf{x}} \, E_x$, and determine the lowest value of ω for which the TE mode can propagate in the waveguide, and the corresponding expressions for the electric and magnetic fields.
(b) Determine the lowest frequency and the EM fields for a waveguide delimited by the conducting planes $x = 0$, $y = 0$, and $y = x$, whose cross section is the right isosceles triangle shown in Fig. 12.5.

12.5   Waveguide Modes as an Interference Effect

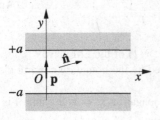

Fig. 12.6

An electric dipole $\mathbf{p} = p\,\hat{\mathbf{y}}$ is located at the origin of a Cartesian coordinate system (x, y, z), between two infinite, perfectly conducting planes located at $y = \pm a$, respectively, as shown in Fig. 12.6.

(a) Find the the electrostatic potential between the two conducting planes, using the method of images.

Now assume that the dipole is oscillating, in complex notation $\mathbf{p} = \mathbf{p}_0\,e^{-i\omega t}$, and consider the emitted radiation in the region between the two conducting planes, at large distances from the dipole, i.e., with both $|x| \gg \lambda$ and $|x| \gg a$.

(b) Find in which directions $\hat{\mathbf{n}}$, lying in the $z = 0$ plane, we observe constructive interference between the waves emitted by the dipole and its images, and the corresponding constraints on the possible values of the oscillation frequency ω.

Now consider two types of waves, labeled "0" and "1", respectively, propagating between the two conducting planes with their wavevectors $\mathbf{k}_{0,1}$ lying in the $z = 0$ plane. Assume that the only nonzero component of the magnetic field of both waves is parallel to $\hat{\mathbf{z}}$ (TM waves), and that the magnetic fields have the form

$$\mathbf{B}_0 = \hat{\mathbf{z}}\,B_0\,e^{ik_{0x}x - i\omega t}\ ,\qquad \mathbf{B}_1 = \hat{\mathbf{z}}\,B_1\,\sin(k_{1y}y)\,e^{ik_{1x}x - i\omega t}\ . \qquad (12.5)$$

(c) Find the relation between the components of the wavevectors and ω for both waves.

(d) Find the expressions for the electric fields $\mathbf{E}_{0,1}$ of the waves corresponding to the magnetic fields (12.5).

(e) Verify (or impose when appropriate) that for the expressions found in **(d)** the component of $\mathbf{E}$ parallel to the planes vanishes at their surface, and the related constraints on $\mathbf{k} = (k_x, k_y)$. What is the relation with the orders of interference found at point **(b)**?

12.6   Propagation in an Optical Fiber

Figure 12.7 represents a simple model for an optical fiber. In a Cartesian reference frame (x, y, z) the space between the planes $y = \pm a/2$ is filled by a material of a real and positive refractive index $n > 1$ (in the frequency range of interest), while we have vacuum ($n = 1$) in the regions $y > a/2$ and $y < -a/2$. A monochromatic electromagnetic wave of frequency ω propagates parallel to $\hat{\mathbf{x}}$ inside the

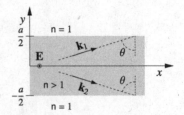

Fig. 12.7

fiber. We assume that the only nonzero component of the electric field $\mathbf{E}$ of the wave is parallel to z (i.e. perpendicular to the plane of the figure). Further, we assume that the wave is the superposition of two plane waves with wavevectors $\mathbf{k}_1 \equiv (k_x, k_y, 0) \equiv k(\sin\theta, \cos\theta, 0)$, and $k_2 \equiv (k_x, -k_y, 0) \equiv k(\sin\theta, -\cos\theta, 0)$, where θ is the angle of incidence shown in the figure. We have, in complex notation,

$$
\begin{aligned}
\mathbf{E} = \hat{\mathbf{y}}\, E_z(x,y,t) &= \hat{\mathbf{y}}\left(E_1\, e^{i\mathbf{k}_1\cdot\mathbf{r}-i\omega t} + E_2\, e^{i\mathbf{k}_2\cdot\mathbf{r}-i\omega t}\right) \\
&= \hat{\mathbf{y}}\left(E_1\, e^{ik_x x + ik_y y - i\omega t} + E_2\, e^{ik_x x - ik_y y - i\omega t}\right).
\end{aligned} \tag{12.6}
$$

(a) Find the relation between k and ω, and the range of θ for which the wave propagates without energy loss through the boundary surfaces at $y = \pm a/2$.
(b) The *amplitude reflection coefficient* $\mathsf{r} = E_\mathrm{r}/E_\mathrm{i}$ is the ratio of the complex amplitude of the reflected wave to the amplitude of the incident wave, at the surface separating two media. In the case of total reflection we have $\mathsf{r} = e^{i\delta}$, with δ a real number. Show that, in our case, we have

$$
k_y a + \delta = m\pi, \quad \text{with } m \in \mathbb{N}, \tag{12.7}
$$

and write the equation for the cut-off frequencies of the fiber. Find the values of k_y explicitly at the $n\sin\theta \gg 1$, $\theta \to \pi/2$ limit.
(c) How do the results change if $\mathbf{E}$ lies in the xy plane?

12.7 Wave Propagation in a Filled Waveguide

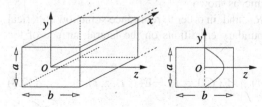

Fig. 12.8

A waveguide has rectangular cross section and perfectly conducting walls. We choose a Cartesian reference frame where the waves propagate parallel to the x axis, and the conducting walls lie on the $y = \pm a/2$ and $z = \pm a/2$ planes, as in Fig. 12.8. The waveguide is uniformly filled with a medium having refractive index $n = n(\omega)$.

(a) Consider the propagation of a TE mode of frequency ω, for which the electric field is $\mathbf{E} = \hat{\mathbf{z}}\, E_z(y)\, e^{ikx - i\omega t}$. Find the general expression for $E_z(y)$ and the dispersion relation $\omega = \omega(k)$. Determine the cut-off frequencies for the particular case in which the filling medium is a gas of free electrons, i.e., a plasma, with plasma frequency ω_p. In this case we have for the refractive index $n^2(\omega) = 1 - \omega_\mathrm{p}^2/\omega^2$.
(b) Now assume that the medium fills only the $x > 0$ region of the waveguide. A monochromatic wave of the lowest frequency that can propagate in both regions

($x < 0$ and $x > 0$) travels in the guide from $x = -\infty$. Find the amplitudes of the reflected and transmitted waves at the $x = 0$ interface.

12.8   Schumann Resonances

The system formed by the Earth and the ionosphere can be considered as a resonant cavity. The cavity is delimited by two conducting, concentric spherical surfaces: the Earth's surface (radius $R_\oplus \simeq 6400$ km) and to the lower border of the ionosphere, located at an altitude $h \simeq 100$ km above, as shown in Fig 12.9, obviously out of scale. Inside this "cavity" there are standing electromagnetic waves of particular frequencies, called *Schumann resonances*.

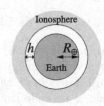

Fig. 12.9

We want to estimate the typical frequency ω of these resonances, assuming that both the Earth and the ionosphere are perfect conductors, and thus completely reflect the electromagnetic waves in the resonant frequency range.

In order to avoid mathematical complications due to the spherical geometry of the problem, we choose a simplified, flat model consisting in a rectangular parallelepiped with two square, conducting bases of side L, and height h. In a Cartesian reference frame, the base standing for the Earth surface lies on the $z = 0$ plane, while the base standing for the surface at the bottom of the ionosphere lies on the $z = h$ plane, as shown

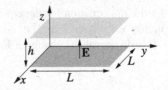

Fig. 12.10

in Fig. 12.10. We choose $L = 2\pi R_\oplus$, and, in order to reproduce somehow spherical geometry, we impose periodic boundary conditions on the lateral surface of the parallelepiped, namely

$$\mathbf{E}(0, y, z, t) = \mathbf{E}(L, y, z, t) , \quad \mathbf{E}(x, 0, z, t) = \mathbf{E}(x, L, z, t) , \qquad (12.8)$$

where $\mathbf{E}$ is the field of the wave, the same conditions are assumed for the magnetic field of the wave. We assume $\varepsilon_r = 1$ and $\mu_r = 1$ in the interior of our parallelepiped. Further, we assume a TE mode with an electric field of the form

$$\mathbf{E} = \hat{\mathbf{z}} \, E_0 \, e^{ik_x x + ik_y y - i\omega t} . \qquad (12.9)$$

(a) Find the possible values of k_x, k_y, ω and give a numerical estimate of ω and the corresponding wavelength for the lowest frequency mode.
(b) The low-frequency conductivity of sea water is $\sigma \simeq 4.4 \ \Omega^{-1}\mathrm{m}^{-1}$. Discuss if approximating the surface of the oceans as a perfect conductor is reasonable at the frequency of the Schumann resonances.

12.9   A One-Dimensional Cavity Fed by an Antenna (1)

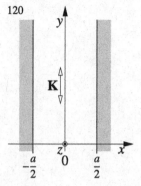

Fig. 12.11

As a toy model for a resonant cavity fed by an antenna, consider two infinite conducting planes placed at $x = \pm a/2$ in a Cartesian reference frame, and a surface current $\mathbf{K}(t) = K(t)\,\hat{\mathbf{y}}$ flowing on the $x = 0$ plane, as shown in Fig. 12.11. This surface current is equivalent to a volume current density $\mathbf{J}(x, y, z, t) = K(t)\,\delta(x)\,\hat{\mathbf{y}}$, where $\delta(x)$ is the Dirac delta function. The current is driven by an external generator at frequency ω, so that $K(t) = K_0\,e^{-i\omega t}$.

Assuming that the confining walls are perfectly conducting, determine

(a) the electromagnetic fields inside the cavity,

(b) the instantaneous and average power provided by the generator. In particular, discuss the behavior at $\omega = \omega_n = (2n + 1)\,\pi\,c/a$, with $n = 0, 1, 2, \ldots$, corresponding to the normal modes of the cavity.

12.10   A One-Dimensional Cavity Fed by an Antenna (2)

Reconsider Problem 12.9 with the assumption that the walls are not perfectly conducting, but are made of a metal with resistive skin depth δ at the frequency ω.

(a) Determine the electromagnetic fields inside the cavity and the power provided by the antenna in the case of the normal modes of the cavity, $ka = \pi(2n + 1)$

(b) Extend the results of **(a)** to the general case of an arbitrary value of ω.

Chapter 13
Further Problems

13.1 Electrically and Magnetically Polarized Cylinders

Let us consider a cylinder of relative magnetic perme-
abilty μ_r, located in a uniform magnetic field $\mathbf{B}_0$ parallel
to the cylinder axis, and the analogous problem of a cylinder of relative electric permittivity ε_r located in a uniform
electric field $\mathbf{E}_0$ parallel to the cylinder axis. In both cases
the cylinder has radius a an height h (Fig. 13.1).
(a) First, consider "long" cylinders, with $a \ll h$. Evaluate
the magnetic field $\mathbf{B}_i$, and, respectively, the electric field
$\mathbf{E}_i$, inside the cylinders, neglecting boundary effects.
(b) Now evaluate the internal magnetic and electric fields
in the case of "flat" cylinders, $a \gg h$, again neglecting
boundary effects.
(c) Evaluate the fields of point **(a)** at the next order of accu-
racy, taking the boundary effects at the lowest nonzero
order in a/h into account.
(d) Evaluate the fields of point **(b)** at the lowest nonzero order in h/a.

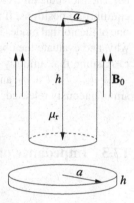

Fig. 13.1

13.2 Oscillations of a Triatomic Molecule

A triatomic symmetric linear molecule,
like CO_2, can be schematized as a central
point mass M, of charge $2q$, and two identical point masses m, each of charge $-q$,
which, when the molecule is at rest, are
located symmetrically around M, as shown
in Fig. 13.2. In the case of longitudinal

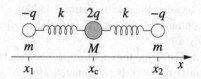

Fig. 13.2

© The Author(s), under exclusive license to Springer Nature Switzerland AG 2023
A. Macchi et al., *Problems in Classical Electromagnetism*,
https://doi.org/10.1007/978-3-031-22235-1_13

small-amplitude vibrations, the interactions between the three masses (including the electrostatic forces) can be described as two identical springs, each of rest length ℓ and elastic constant k, located as in the figure.

Let x_1 and x_2 be the positions of the two lateral masses, and x_c the position of the central mass. We want to study the longitudinal vibrations of the molecule in its center-of-mass reference frame, defined by the condition

$$x_{cm} = \frac{mx_1 + mx_2 + Mx_c}{2m + M} = 0 .$$ (13.1)

When the molecule is at rest we have thus

$$x_1 = -\ell , \quad x_c = 0 , \quad x_2 = \ell .$$ (13.2)

(a) Find the normal longitudinal oscillation modes of the molecule, and their frequencies.
(b) The molecule emits radiation because the charged masses oscillate around their equilibrium positions. If the electric dipole term is dominant, the frequency of only one of the normal modes is observed in the spectrum of the emitted radiation. Explain why, and evaluate the observed frequency.
c) Assume that, initially, the molecule is "excited" by locating the masses at $x_1 = -\ell + d_1, x_2 = \ell + d_2$, and x_c such that $x_{cm} = 0$. Then, at $t = 0$, the three masses are simultaneously released. Find the power radiated at $t > 0$.

13.3 Impedance of an Infinite Ladder Network

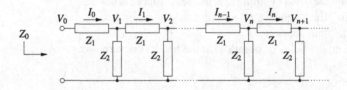

Fig. 13.3

Consider the (semi-)infinite ladder of Fig. 13.3, where each (identical) section contains a "horizontal" impedance $Z_1 = Z_1(\omega)$, and a "vertical" impedance $Z_2 = Z_2(\omega)$.
(a) Calculate the input impedance Z_0 of the semi-infinite network. How can a real, finite network be terminated after N sections, so that its impedance has also the value Z_0?

(b) Let V_n be the voltage at the nth node. Find the relation between V_n and V_{n+1} and, from this, the dependence of V_n on n and on the input voltage V_0. Discuss the result for the case of a purely resistive network ($Z_1 = R_1$, $Z_2 = R_2$).

Now consider the case of an LC network, with $Z_1 = -i\omega L$ and $Z_2 = -1/i\omega C$, in the presence of an input signal $V_0(t) = V_0 e^{-i\omega t}$.

(c) Find the frequency range in which signals can propagate in the network. Show that there is a cut-off frequency ω_{co}, such that the signal is damped if $\omega > \omega_{co}$, and evaluate the damping factor.

(d) Discuss the case of a CL network, with $Z_1 = -1/i\omega C$ and $Z_2 = -i\omega L$.

13.4 Discharge of a Cylindrical Capacitor

A cylindrical capacitor has internal radius a, external radius $b > a$, and height $h \gg b$. For $t < 0$, the two cylindrical plates have charges $\pm Q_0$, respectively, and are disconnected. At $t = 0$ the plates are connected through a resistor R as in Fig. 13.4. We assume that during the discharge (i) the slowly-varying current approximation holds, (ii) the surface charge density on the plates remains uniform, (iii) we can neglect the effects of the external circuit and the resistance of the plates, (iv) other boundary effects are negligible. We use a cylindrical coordinate system (r, ϕ, z) with the capacitor axis as z axis, and the origin at the center of the capacitor.

(a) Calculate the current $I = I(z, t)$ flowing on the plates, and the magnetic field $\mathbf{B} = \hat{\boldsymbol{\phi}} \, B_\phi(r, z, t)$ inside the capacitor, for $|z| \ll (h/2)$.

(b) Calculate the Poynting vector $\mathbf{S}$ for $|z| \ll (h/2)$, and verify that its flux through a cylindrical surface coaxial to the capacitor equals the time variation of the electrostatic energy inside the surface.

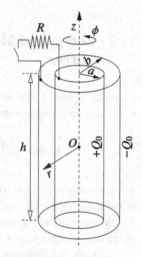

Fig. 13.4

(c) Discuss the validity of the slowly varying current approximation and of the assumption of uniform charge distribution over the plates.

13.5 Fields Generated by Spatially Periodic Surface Sources

Evaluate the electromagnetic fields and potentials generated by the following three surface charge and/or current densities, located on the $y = 0$ plane of a Cartesian coordinate system,

(a) a static surface charge density $\sigma = \sigma_0 \cos kx$;
(b) a static surface current density $\mathbf{K} = \hat{\mathbf{z}} K_0 \cos kx$;
(c) a time-dependent surface current density $\mathbf{K} = \hat{\mathbf{z}} K_0 e^{-i\omega t} \cos kx$, discussing for which values of ω (for fixed k) the fields are propagating.
(d) In case (c), calculate the time- and space-averaged power dissipated per unit surface on the $y = 0$ plane

$$W = \frac{k}{2\pi} \int\limits_{-\pi/k}^{+\pi/k} \langle \mathbf{K}(x, t) \cdot \mathbf{E}(x, y = 0, t) \rangle \, dx, \qquad (13.3)$$

and find for which values of ω we have $W = 0$. Discuss the result with respect to the findings of point (c).

13.6 Energy and Momentum Flow Close to a Perfect Mirror

Consider a plane EM wave, propagating along the x axis of a Cartesian coordinate system, of frequency ω and *elliptical* polarization, with electric field

$$\mathbf{E}_i = \frac{E_0}{\sqrt{1 + \epsilon^2}} \left[\hat{\mathbf{y}} \cos(kx - \omega t) - \hat{\mathbf{z}} \epsilon \sin(kx - \omega t) \right], \qquad (13.4)$$

where $k = \omega/c$, and ϵ is a real parameter, $0 < \epsilon < 1$, characterizing the eccentricity of the polarization ellipse. The normalization factor $1/\sqrt{1 + \epsilon^2}$ has been chosen so that the intensity of the wave is $I = cE_0^2/8\pi$ for any value of ϵ. The wave is incident on a plane, perfect mirror located at $x = 0$.
(a) Evaluate the Poynting vector $\mathbf{S} = \mathbf{S}(x, t)$ in front of the mirror, including the contribution of the reflected wave. Find the value of ϵ for which $\mathbf{S} = 0$ everywhere, and the corresponding angle between the total electric ($\mathbf{E}$) and magnetic ($\mathbf{B}$) fields.
(b) Find the force per unit surface on the mirror $F_x = T_{xx}$, where $T_{xx} = T_{11}(x = 0^-)$ is the $(1, 1)$ component of the stress tensor at $x = 0^-$. Show that, in general, F_x has both a steady and an oscillating component, and find the frequency of the latter. For which value of ϵ the oscillating component is missing?

13.7 Laser Cooling of a Mirror

A plane mirror has surface area A, finite thickness, mass M, and its two opposite surfaces are perfectly reflecting. At $t = 0$ the mirror lies on the $x = 0$ plane of a Cartesian coordinate system, as in Fig. 13.5. Two plane EM waves of intensities I_1 and I_2, respectively, are impinging at normal incidence on the two surfaces.

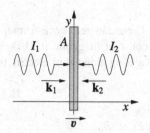

(a) Find the total force on the mirror and the direction of its acceleration if $I_1 > I_2$.

Fig. 13.5

Now assume that the two waves have equal intensities, $I_1 = I_2 = I$, and that the mirror is moving with velocity $v = \hat{x} v$.

(b) Evaluate the force on the mirror, in the system where the mirror is at rest.

(c) Discuss the motion of the mirror under the action of the force found at point (b), at the limit $v \ll c$.

13.8 Radiation Pressure on a Thin Foil

An EM wave of frequency ω, traveling along the x axis of a Cartesian coordinate system, is impinging normally on a very thin foil of thickness d and surface A (see Problem 11.4, in particular Fig. 11.2). The foil is perfectly conducting in a frequency range containing ω. As shown in the solution of Problem 11.4, the (complex) transmission and reflection coefficients of the foil are

$$t = \frac{1}{1 + \eta} ; \quad r = -\frac{\eta}{1 + \eta} , \quad \text{where} \quad \eta = i \frac{\omega_p^2 d}{2\omega c} , \tag{13.5}$$

and ω_p is the plasma frequency of the foil.

(a) Show that the radiation pressure on the thin foil is

$$P_{\text{rad}} = \frac{2RI}{c} , \tag{13.6}$$

where I is the intensity of the wave and $R \equiv |r|^2$.

(b) Now assume that the foil is moving with velocity $v = \beta c \hat{x}$ in the laboratory frame. Assuming $R = 1$ (a perfectly reflecting foil), evaluate the force on the foil.

(c) How does the answer to (b) change if $R = R(\omega) < 1$?

13.9   Thomson Scattering in the Presence of a Magnetic Field

In a Cartesian reference frame, an electron moves in the presence of a uniform and constant magnetic field $\mathbf{B}_0 = \hat{\mathbf{z}} B_0$ and of a monochromatic plane EM wave, propagating along $\hat{\mathbf{z}}$, of electric field

$$\mathbf{E}(z, t) = \hat{\mathbf{y}} \, E_i \, e^{ikz - i\omega t} , \tag{13.7}$$

with $E_i \ll B_0$.
(a) Describe the motion of the electron in steady state conditions, neglecting friction and the effect of the magnetic field of the wave.
(b) Calculate the power radiated by the electron. Discuss the dependence of the emitted spectrum on ω, and the angular distribution of the emitted radiation at the limits $\omega \ll \omega_c$ and $\omega \gg \omega_c$, where $\omega_c = e B_0 / m_e$ is the cyclotron frequency of the electron in the presence of $\mathbf{B}_0$.

13.10   Undulator Radiation

In a Cartesian laboratory reference frame $S \equiv (x, y, z)$, we have a static magnetic field $\mathbf{B}$. In a certain region of space, free of charges and currents, the magnetic field is independent of z, and its y and z components can be written as

$$B_y = b(y) \cos(kx) , \qquad B_z \equiv 0 , \tag{13.8}$$

where $b(y)$ is an even function of y. The field is generated by sources located outside the region of interest, at finite values of $|y|$.
(a) Show that, in the region of interest, we must have

$$B_y = B_0 \cos(kx) \cosh(ky) , \tag{13.9}$$

and determine the expression for B_x.
    Now assume that, in the laboratory frame S, an electron enters our magnetic field region with initial velocity $\mathbf{v} = \hat{\mathbf{x}} \, v$.
(b) Describe the electron motion in the frame S', moving at the velocity $\mathbf{v}$ relative to the laboratory frame S, and discuss the emitted radiation. (Assume the electron motion to be non-relativistic and keep only linear terms in the equation of motion.)
(c) Determine the frequency of the radiation emitted in the directions both parallel and antiparallel to $\mathbf{v}$, as observed in S. In which directions the radiation intensity is zero in S'?

13.11   Electromagnetic Torque on a Conducting Sphere

A plane, monochromatic, circularly polarized electromagnetic wave, of wavelength $\lambda = 2\pi c/\omega$ and amplitude E_0, impinges on a small metallic sphere of radius $a \ll \lambda$. We assume that ω is low enough so that the metal can be considered as an Ohmic conductor, of conductivity σ independent of frequency.
(a) Evaluate the dipole moment induced on the sphere.
(b) Show that the EM wave exerts a torque on the sphere.

13.12   Surface Waves in a Thin Foil

A very thin conductive foil is located between the $x = -\ell$ and $x = +\ell$ planes of a Cartesian coordinate system. A *surface wave* propagates along both sides of the foil, in the y-direction. The fields of the surface wave have only the following non-zero components : E_x, E_y, and B_z, all of them independent of z. We know that the electric field component *parallel* to the propagation direction is

$$E_y(x, y, t) = E_0\, e^{-q|x|} e^{i(ky-\omega t)} , \qquad (13.10)$$

where the frequency ω is such that $\lambda = 2\pi c/\omega \gg \ell$. In these conditions, the foil can be treated, with good approximation, as the superposition of a surface charge $\sigma(y, z, t)$, and a surface current $\mathbf{K}(y, z, t)$, all lying on the $x = 0$ plane.

Starting from (13.10) and Maxwell's equations in vacuum, evaluate
(a) the field components E_x and B_z, specifying their parity with respect to $\hat{\mathbf{x}}$, and the surface current $\mathbf{K}(y, z, t)$,
(b) the Poynting vector and the time-averaged flux of electromagnetic energy associated to the surface wave,
(c) the relations between q, k, and ω.

Now assume that, in the relevant frequency range, the relation between the current density $\mathbf{J}$ and the electric field $\mathbf{E}$ in the foil can be written (for harmonic fields) as

$$\mathbf{J} = 4\pi i \frac{\omega_p^2}{\omega}\, \mathbf{E} , \qquad (13.11)$$

where ω_p is the plasma frequency of the foil. Equation (13.11) characterizes of an ideal conductor in the high-frequency regime (Problem 11.1).
(d) Using (13.11) and the boundary conditions for the fields of a thin foil discussed in Problem 11.4, obtain an additional relation between q, k, and ω.
(e) By combining the results of points (d) and (e) find the dispersion relation $\omega = \omega(k)$ and discuss its limits of validity.

13.13   The Fizeau Effect

A plane electromagnetic wave of frequency ω and wavevector $\mathbf{k} = \hat{\mathbf{x}}k$ propagates in a homogenous medium, while the medium itself is moving with velocity $\mathbf{u} = \hat{\mathbf{x}}u$ (thus parallel to $\mathbf{k}$) in the laboratory frame. The refractive index of the medium is real and positive, $n > 0$, *in the rest frame of the medium*. Assume $u \ll c$, and answer the following questions evaluating all results up to the first order in $\beta = u/c$.

First, assume a non-dispersive medium, with n *independent* of frequency.
(a) Evaluate the phase velocity of the wave, v_φ, in the *laboratory frame*.
Now assume that the medium is dispersive, with n depending on frequency according to a known law $n = n(\omega)$, defined *in the rest frame of the medium*.
(b) Evaluate the phase velocity in the laboratory frame in this case, showing that now

$$v_\varphi \simeq \frac{c}{n(\omega)} + \beta c \left(1 - \frac{1}{n^2(\omega)} + \frac{\omega}{n(\omega)} \partial_\omega n(\omega) \right) + \mathcal{O}(\beta^2). \qquad (13.12)$$

Hint: use the first-order Doppler effect for evaluating the relation between the frequency in the laboratory frame and the frequency observed in the rest frame of the medium.
(c) Use (13.12) to show that, in a medium containing free electrons moving with negligible friction (a simple metal or an ideal plasma), the phase velocity does *not* depend on β up to to first order [1].

13.14   Lorentz Transformations for Longitudinal Waves

Consider a longitudinal wave with fields

$$\mathbf{E} = \mathbf{E}(x, t) = \hat{\mathbf{x}} E_0 \, e^{i(kx - \omega t)}, \qquad \mathbf{B} \equiv 0, \qquad (13.13)$$

in the (Cartesian) laboratory frame S. We have shown in Problem 11.3 that the phase velocity of this wave, $v_\varphi = \omega_L/k_L$ is undetermined, and can have arbitrary values.

Find the frequency, wavevector and fields of the wave in a frame S', moving with velocity $\mathbf{v}$ with respect to S, for the three following cases:
(a) $\mathbf{v} = v_\varphi \, \hat{\mathbf{x}}$, with $v_\varphi < c$;
(b) $\mathbf{v} = (c^2/v_\varphi)\hat{\mathbf{x}}$, with $v_\varphi > c$;
(c) $\mathbf{v} = V\hat{\mathbf{y}}$ with $V < c$.

13.15 Lorentz Transformations for a Transmission Cable

A transmission cable can be schematized as an infinite straight conducting wire.
We choose a cylindrical coordinate system (r, ϕ, z) with the z axis along the wire.
A monochromatic charge and current signal, of frequency ω, propagates along the
cable, with total current I and linear charge density λ given by, in complex notation,

$$I = I(z, t) = I_0 \, e^{ikz - i\omega t} , \qquad \lambda = \lambda(z, t) = \lambda_0 \, e^{ikz - i\omega t} . \qquad (13.14)$$

The cable is located in a uniform medium of real dielectric permittivity $\varepsilon > 1$, and
magnetic permeability $\mu = 1$, in the frequency region of interest.
(a) Find the relation between I_0 and λ_0.
(b) Evaluate the electric and magnetic fields in the medium, $\mathbf{E}(r, z, t)$ and $\mathbf{B}(r, z, t)$,
assuming that they are in a TEM mode, i.e.,

$$\mathbf{E}(r, z, t) = \mathbf{E}(r) e^{ikz - i\omega t} , \qquad \mathbf{B}(r, z, t) = \mathbf{B}(r) e^{ikz - i\omega t} , \qquad (13.15)$$

with $\mathbf{E} \cdot \hat{\mathbf{z}} = 0$ and $\mathbf{B} \cdot \hat{\mathbf{z}} = 0$. Evaluate the dispersion relation between the frequency
ω and the wave vector k.

(c) Show that the fields and their sources are independent of time in a reference frame
S', moving at the phase velocity $\hat{\mathbf{z}} \, v_\varphi = \hat{\mathbf{z}} \, (\omega/k)$ relative to the laboratory frame S
where the wire is at rest. Show that, in S', we have $\mathbf{E}' = 0$ and $I' = 0$, while $\lambda' \neq 0$
and $\mathbf{B}' \neq 0$. Explain this apparently surprising result.

13.16 A Waveguide with a Moving End

Two perfectly conducting plane surfaces located at $y = \pm a/2$, respectively, form a waveguide. The waveguide is terminated at $x = 0$ by a perfectly conducting wall, as shown in
Fig. 13.6. Consider the propagation of
a monochromatic TE$_{10}$ wave along the
x axis. The electric field of the wave
has only the z component. In complex notation we have $E_z(x, y, t) =$
$E_z(y) \, e^{ikx - i\omega t}$, where ω and k are related by the dispersion relation of the TE$_{10}$
mode.

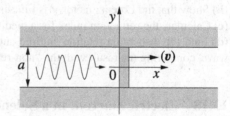

Fig. 13.6

(a) Find the total electric and magnetic fields inside the waveguide.

Now assume that the end of the waveguide moves with constant velocity $\boldsymbol{v} = v \, \hat{\mathbf{x}}$.

(b) Assuming $v < kc^2/\omega$, determine the frequency ω_r and the wavevector k_r of the reflected wave. Verify that ω_r and k_r are related by the dispersion relation of the TE_{10} mode.
(c) What happens in the $v > kc^2/\omega$ case?

13.17   A "Relativistically" Strong Electromagnetic Wave

We consider a circularly polarized, plane electromagnetic wave propagating parallel to the z axis of a Cartesian reference frame. The wave fields are

$$\mathbf{E} = E_0 \left[\hat{\mathbf{x}} \cos(kz - \omega t) - \hat{\mathbf{y}} \sin(kz - \omega t) \right] , \qquad (13.16)$$

$$\mathbf{B} = E_0 \left[\hat{\mathbf{x}} \sin(kz - \omega t) + \hat{\mathbf{y}} \cos(kz - \omega t) \right] . \qquad (13.17)$$

We assume that the field is strong enough that electrons oscillate at *relativistic* velocities *not* much smaller than c. The relativistically correct equation of motion for an electron in the presence of the intense wave is

$$\frac{d\mathbf{p}}{dt} = -e \left(\mathbf{E} + \frac{v}{c} \times \mathbf{B} \right) , \qquad (13.18)$$

where $\mathbf{p} = m_e \gamma v$, and $\gamma = 1/\sqrt{1 - v^2/c^2} = \sqrt{1 + \mathbf{p}^2/(m_e c)^2}$.
We want to study the propagation of such a "relativistic" wave in a medium with free electrons (ions are considered at rest).
(a) Show that it is self-consistent to assume that the electron motion occurs on the xy plane. Do this in two steps. First we assume that, at $t = 0$, the z component of the momentum of the electrons is zero, $p_z = 0$. Then solve the equations of motion in steady state conditions and verify the consistency of the assumption *a posteriori*.
(b) Show that the Lorentz factor γ is time-independent, and give its expression.
(c) Calculate the refraction index for a medium with free electron density n_e.
(d) Find the dispersion relation and the cut-off frequency for the electromagnetic wave, comparing the result with the "non-relativistic" case of low field amplitudes.

13.18   Electric Current in a Solenoid

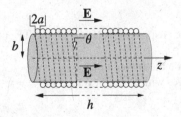

A solenoid is made by winding a thin conducting wire of radius a and conductivity σ around a non-conducting cylinder of radius $b \gg a$ and height $h \gg b$. Thus the solenoid coil has a pitch

Fig. 13.7

$$\theta = \arctan\left(\frac{a}{\pi b}\right) \ll 1 , \tag{13.19}$$

since the wire moves in the $\hat{z}$ direction by a step of length $2a$ at every turn of length $2\pi b$. The solenoid is located in an external uniform electric field $\mathbf{E} = E\hat{z}$ (Fig. 13.7).
(a) Evaluate the magnetic field $\mathbf{B}$ both *inside* $(r < b)$ and *outside* $(r > b)$ the solenoid, neglecting boundary effects.
(b) Calculate the flux of the Poynting vector $\mathbf{S} = c\mathbf{E} \times \mathbf{B}/4\pi$ through a cylindrical surface external and coaxial to the solenoid, and compare its value with the power dissipated by Joule heating.

13.19 An Optomechanical Cavity

We have a one-dimensional cavity limited by two perfectly conducting plane surfaces located at $x = \pm d/2$,
respectively, as in Fig. 13.8. The electromagnetic field
inside the cavity has frequency ω, peak amplitude of
the electric field E_0, and is linearly polarized parallel
to the walls.
(a) Find the possible values for ω and write the most
general form of the electromagnetic field.
(b) Calculate the electromagnetic energy per unit surface U inside the cavity.
(c) Calculate the radiation pressure P on the walls as
a function of E_0, and the ratio U/P.
(d) Now assume that the two cavity walls are finite
squares, each of mass M and surface $S \gg d^2$. Each
cavity wall is connected to an external fixed wall
by a spring of Hooke's constant K, as shown in
Fig. 13.9. Neglecting boundary effects, evaluate the
relation between frequency and amplitude of the electromagnetic modes of the cavity.

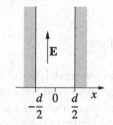

Fig. 13.8

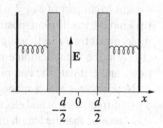

Fig. 13.9

13.20 Radiation Pressure on an Absorbing Medium

In an appropriate Cartesian reference frame we have vacuum in the $x < 0$ half-space,
while the $x > 0$ half-space is filled with a medium of *complex* refractive index

$$n = n_1 + in_2 , \tag{13.20}$$

with $n_1 > 1 \gg n_2$. A monochromatic plane wave of frequency ω and intensity I_i, propagating in the positive x direction, is incident on the $x = 0$ plane. Calculate
(a) the power absorbed by the medium, W_{abs}, showing that $W_{abs} = (1 - R) I_i$, where R is the reflection coefficient ($R = |r|^2$ where r is the usual amplitiude coefficient for the reflected wave as defined in the Fresnel formulas);
(b) the pressure on the medium, P_{rad}, showing that $P_{rad} = (1 + R) I_i/c$.

13.21   Radiation and Scattering from a Linear Molecule

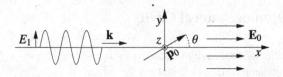

Fig. 13.10

A simple model for a polar linear molecule, neglecting vibrations, is a one-dimensional rigid rotor associated to an electric dipole moment $\mathbf{p}_0$. The molecule has moment of inertia $\mathcal{I}$ about any rotational axis passing through its barycenter and perpendicular to the molecule. Let us consider a polar linear molecule located in a uniform and constant electric field $\mathbf{E}_0$, parallel to the x axis of a Cartesian coordinate system (right part of Fig. 13.10).
(a) Find the equilibrium positions of the molecule, and discuss the motion when the molecule at time $t = 0$ is slightly displaced from its stable equilibrium position.
(b) Describe the radiation emitted by the molecule during small amplitude oscillations, and estimate the damping time of such oscillations.

Now assume that a monochromatic plane wave, linearly polarized along the y axis, of frequency ω and electric field amplitude E_1, is propagating along the x axis. Also assume that the length of the molecule, d, is much smaller than the wavelength, $d \ll \lambda = 2\pi c/\omega = 2\pi/k$.
(c) Describe the motion of the molecule in these conditions.
(d) Calculate the power scattered by the molecule and its scattering cross section.

13.22   Radiation Drag Force

The classical motion of a particle of charge q and mass m, under the simultaneous action of an electric field $\mathbf{E}$ and a magnetic field $\mathbf{B}$, is described by the equation

$$\frac{d^2\mathbf{r}}{dt} = \frac{d\mathbf{v}}{dt} = \frac{q}{m}\left(\mathbf{E} + \frac{\mathbf{v}}{c} \times \mathbf{B}\right) - \nu\mathbf{v}, \qquad (13.21)$$

where ν is a damping coefficient.

Given a Cartesian reference frame (x, y, z), consider the motion of the particle in the presence of a plane, monochromatic electromagnetic wave propagating along the x axis. The wave is linearly polarized along y with electric field amplitude E_0, has frequency ω and wave vector $\mathbf{k} = \hat{\mathbf{x}}\,\omega/c$. Assume the velocity of the particle to be much smaller than c, so that, as a first order approximation, we can neglect the $\boldsymbol{v} \times \mathbf{B}/c$ term.

(a) Solve (13.21) in steady-state conditions.

(b) Calculate the cycle-averaged power P_{abs} absorbed by the particle, i.e., the work made by the electromagnetic force over an oscillation period.

(c) Calculate the cycle-averaged power P_{rad} radiated by the particle, and obtain an expression for the damping coefficient ν assuming that all the absorbed power is re-emitted as radiation, $P_{\text{rad}} = P_{\text{abs}}$.

(d) Now use the result of point b) to evaluate the effect of the term $\boldsymbol{v} \times \mathbf{B}/c$. The cycle-averaged force along x, which accelerates the particle in the wave propagation direction, is

$$F_x = \left\langle q\left(\frac{\boldsymbol{v}}{c} \times \mathbf{B}\right)_x \right\rangle. \qquad (13.22)$$

Calculate F_x and the P_{abs}/F_x ratio.

(e) Assume that instead of a point particle we have a small sphere of radius a, such that $ka \ll 1$, containing $N \gg 1$ particles (plus a neutralizing background). Find the force on the sphere and the related acceleration as a function of N (neglect any collective effect such as screening of the electromagnetic field inside the sphere).

13.23 A Coaxial Cable with a Faraday Generator

A conducting toroid has rectangular cross section, inner radius a, outer radius $b > a$ and height $h \ll a$, as shown in Fig. 13.11. The toroid terminates a coaxial cable whose inner conductor has radius a and the outer shield has radius b.

A constant current I is flowing in the cable as shown in Fig. 13.11. The current is uniformly distributed in the inner conductor and over the outer shield. Assume that the current flows from the inner conductor of the coaxial cable to the outer shield through the toroidal termination, and that, inside the termination, the current density $\mathbf{J}$ is radial in the cylindrical coordinate system (r, ϕ, z) shown in Fig. 13.11, with $\mathbf{J} = J_r(r)\,\hat{\mathbf{r}}$, independent of z and ϕ. Assume that the border effects are negligible.

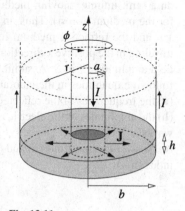

Fig. 13.11

(a) Write $\mathbf{J}$ as a function of I and the coordinate r.

(b) Assume that the toroid is made of an Ohmic metal where the law $\mathbf{E} = \eta\mathbf{J}$ holds, η being the resistivity of the metal. Evaluate the resistance R of the termination.
(c) Evaluate the surface charge densities at the contact surfaces at $r = a$ and $r = b$, and show that the volumic charge density ρ is zero inside the toroid.

Now assume that the conducting toroid is rotating with constant angular velocity ω around the z axis, in the presence of an external, constant and uniform magnetic field $\mathbf{B}$ parallel to the z axis, as shown in Fig. 13.12, thus acting as a Faraday disk generator (see Problem 6.11). Assume a radial current density in the toroid. Assume that the resistivity of the cable is negligible with respect to η.

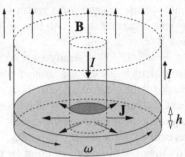

Fig. 13.12

(d) Evaluate the electric field $\mathbf{E}$ inside the toroid as a function of $I, \mathbf{B}, \omega$ and r. Evaluate the surface and volume charge densities in the toroid.
(e) Evaluate the torque exerted by the magnetic field on the rotating toroid and the mechanical power P_m needed to maintain the rotation at constant angular velocity.
(f) Assuming that P_m equals the power P_d dissipated by Joule heating in the toroid, evaluate a relation between I and B. Assume that friction is negligible.

13.24   Reflection and Transmission from a Moving Transparent Medium

A plane monochromatic electromagnetic wave is incident normally from vacuum on a semi-infinite, moving medium of refractive index n, real and positive (valid for the medium at rest). Thus, in the laboratory frame S we have vacuum for $x < Vt$, and the refractive medium for $x \geq Vt$, with V constant in time. The wave is traveling in the positive x direction. The incident wave has frequency ω_i and electric-field amplitude E_i in S. Assuming that n is independent of frequency, evaluate the following quantities in the laboratory frame S
(a) the frequencies of the reflected and of the transmitted waves;
(b) the amplitudes of the electric and magnetic fields of the reflected and transmitted waves.
(c) Answer the questions (a) and (b) in the case of a frequency-dependent refractive index n(ω).

13.25 The Electromotive Force in Two Different Frames

In a Cartesian reference frame S fixed with the laboratory we have a static non-uniform magnetic field $\mathbf{B} = \hat{\mathbf{z}} B_z(x) = \hat{\mathbf{z}} B_0 x/L$, where B_0 and L are constants. A square coil with sides of length $a \ll L$ parallel to the x and y axes is kept in motion by a force $\mathbf{F}$ so that it has constant velocity $\mathbf{V} = V\hat{\mathbf{x}}$, with $V \ll c$, parallel to the x axis,

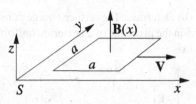

Fig. 13.13

as shown in Fig. 13.13. Assuming that the self-inductance of the coil is negligible, calculate
(a) the electromotive force in the coil, both from Faraday's law and from its definition as a line integral of the electric field around the coil;
(b) the external mechanical force $\mathbf{F}$ exerted on the coil to keep its velocity constant, and the work done per unit time by $\mathbf{F}$.
(c) Now answer to questions (a) and (b) in the reference frame where the coil is at rest (use Galilean transformations, consistently with $V \ll c$).

13.26 Electron Orbits in a Magnetic Field

Consider the motion of an electron in a given stationary magnetic field of the form $\mathbf{B} = B_z(x)\hat{\mathbf{z}}$.
(a) Write the equation of motion of the electron and show that there are three constants of motion: (i) the energy, (ii) the momentum along z and (iii) a generalized momentum along y which involves the y component of the vector potential.
(b) Reduce the solution of the electron motion to a 1-D problem using these invariants and defining an effective potential.
(c) Consider the case $B_z(x) = B_0 (x/L)$ and show, on the basis of the shape of the effective potential, that one can distinguish three classes of orbits, i.e., those (class a) that remain in the half-plane $x < 0$, those (class b) that remain in the half-plane $x > 0$ and those (class c) that can cross the magnetic null line at $x = 0$.

13.27 A Parallel-Wire Metamaterial

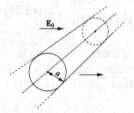

(a) A dielectric cylinder of relative permittivity ε_r, radius a and infinite length is in the presence of an external uniform electric field $\mathbf{E}_0$ perpendicular to the cylinder axis, as shown in Fig. 13.14. Using the results of Problem 3.11 show that the cylinder acquires the polarization density

Fig. 13.14

$$P = \frac{\varepsilon_r - 1}{2\pi k_e(\varepsilon_r + 1)} E_0 . \tag{13.23}$$

(b) Determine the surface charge density of an inifinite conducting cylinder of radius a in the presence of an external uniform electric field E_0 perpendicular to the cylinder axis

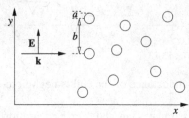

(c) A metamaterial is made of parallel infinite dielectric cylinders of permittivity ε_r and sadius a, as described in point (a), with an average distance $b \gg a$ between the cylinder axes, as shown in Fig. 13.15. A plane linearly-polarized monochromatic electromagnetic wave of electric field

Fig. 13.15

$$E(x, t) = E_0 \, e^{i(kx - \omega t)} \tag{13.24}$$

is incident on the metamaterial. The wave is propagating perpendicularly to the cylinder axes and is polarized perpendicularly to the cylinder axes, and has wavelength $\lambda = 2\pi/k \gg b$.

Evaluate the macroscopic (i.e., *averaged* over a volume containing a large number of cylinders) dielectric permittivity $\varepsilon_r^{\mathrm{met}\perp}$ of the metamaterial.

(d) Evaluate $\varepsilon_r^{\mathrm{met}\perp} = \varepsilon_r^{\mathrm{met}\perp}(\omega)$ for the case of cylinders made of a metal with dielectric permittivity $\varepsilon_r = \varepsilon_r(\omega) = 1 - \omega_p^2/\omega^2$, where ω_p is the plasma frequency of the metal and ω the frequency of the incident wave. Determine the spectral regions where the wave is propagating or evanescent. Determine the cutoff and resonance frequencies, if any.

(e) Answer to points (c) and (d) in the case of a wave polarized parallel to the cylinder axes.

13.28   Motion in an Inhomogeneous Magnetic Field

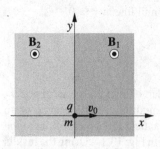

In a given region of space of the laboratory frame S, a constant (time-independent) magnetic field **B** has the following spatial profile

$$\mathbf{B} = \mathbf{B}(x) = \begin{cases} B_2\hat{\mathbf{z}} & (x < 0) \\ B_1\hat{\mathbf{z}} & (x > 0) \end{cases} \tag{13.25}$$

Fig. 13.16

as shown in Fig. 13.16. At time $t = 0$ a particle of mass m and charge q is located at $(x = 0, y = 0)$ with initial velocity $v(0) = v_0 \hat{\mathbf{x}}$, with $v_0 > 0$.

(a) Study the motion and the trajectory of the particle, showing that it drifts in the $\hat{\mathbf{y}}$ direction with an average velocity $\mathbf{v}_d = v_d\hat{\mathbf{y}}$ (to be determined).
(b) Study the motion and the trajectory of the particle in a reference frame S' moving with velocity $\mathbf{v}_d$ with respect to S (assume $v_d \ll c$ and non-relativistic motion).

13.29 Solar Sail

The Sun emits most of its light in the visible region of the spectrum, i.e., at wavelegths between approximately 4×10^{-7} m and 7.5×10^{-7} m. In this spectral range, the intensity of sunlight impinging on the upper atmosphere of the Earth is about $I_E = 1.6 \times 10^3$ W/m^2. The average distance between the Earth and the Sun is $r_E \simeq 1.5 \times 10^{11}$ m. The Sun mass is $M_\odot \simeq 2 \times 10^{30}$ kg. The gravitational constant is $G = 6.67 \times 10^{-11}$ in SI units. Using this information only, show that
(a) a spherical grain of dust having a sufficiently small radius a_g may be swept away from the solar system (assume a_g to be larger than the wavelength of visible light, and the density of the grain ρ_g to be approximately that of water);
(b) a thin, perfectly reflecting sail of surface a_s^2 and sufficiently small thickness $d_s \ll a_S$ can be accelerated outwards from the Sun (assume the density of the sail material to be approximately 2.3 times that of water).
(c) If the sail is made of Graphene with area density $\sigma_s = 7.6 \times 10^{-7}$ kg m^{-2} estimate its final velocity when launched in space from the neighborhood of the Earth.

13.30 Canonical Momentum

Consider the motion of a particle of mass m and charge q on the xy plane of a Cartesian coordinate system (x, y, z) in the presence of a magnetic field $\mathbf{B}(x) = \hat{\mathbf{z}} B_z(x)$, constant in time, but depending on the x coordinate. The magnetic field can be obtained from a vector potential of the form $\mathbf{A}(x) = \hat{\mathbf{y}} A_y(x)$.
(a) Show that the quantity $P_y = mv_y + (q/c) A_y(x)$, where v_y is the y component of the particle velocity $\mathbf{v}$, is a constant of motion. The quantity $\mathbf{P} = m\mathbf{v} + q\mathbf{A}/c$ is called the canonical momentum of the particle in the presence of a magnetic field.
   Now we consider the special case of a uniform field $\mathbf{B} = \hat{\mathbf{z}} B_0$, which can be obtained from a vector potential of the form $\mathbf{A}(x) = \hat{\mathbf{y}} B_0 x$.
(b) Show that the conserved quantity P_y is proportional to the x coordinate of the center of the circular orbit described by the particle, called the center of gyration.
(c) In the presence of the same uniform magnetic field, how can we choose the vector potential so that the conserved quantity corresponds to the y coordinate of the center of gyration?

13.31 Classical Zeeman Effect

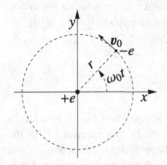

In the classical model of a hydrogen atom the electron follows a circular orbit around the nucleus, initially in the absence of external fields. The orbit has radius r, and the electron has angular velocity ω_0 and peripheral velocity $v_0 = \omega_0 \times r$, as shown in Fig. 13.17. At time $t = 0$ a uniform magnetic field $\mathbf{B}_0$, perpendicular to the orbit plane, and out of page in the figure, is switched on. For simplicity, assume that all fields have axial symmetry around the orbit axis.

Fig. 13.17

(a) Assume (i) arbitrarily for the moment, that the orbit remains a circle of radius r, and (ii) that the new angular frequency ω is such that $|\omega - \omega_0| = |\delta\omega| \ll \omega_0$. Evaluate an approximation for $\delta\omega$. What would $\delta\omega$ be if $\mathbf{B}_0$ were antiparallel, rather than parallel, to the orbital angular momentum?

(b) Now assume that the applied magnetic field changes gradually from 0 to $\mathbf{B}_0$ in a finite time interval Δt. Show that assumption (i) of question a) is justified within the limits of assumption (ii).

(c) What is the change of the electron orbital angular momentum after applying the magnetic field? In Problem 13.30 we defined as $\mathbf{P} = m\mathbf{v} + q\mathbf{A}/c$ the canonical momentum of a particle of mass m, velocity $\mathbf{v}$ and charge q in the presence of a magnetic field $\mathbf{B} = \nabla \times \mathbf{A}$. Evaluate the *canonical angular momentum* $\mathbf{L}^c = \mathbf{r} \times \mathbf{P}$ of the electron before and after applying the magnetic field. Following Problem 13.30 and with our assumed axial symmetry we can write the vector potential as

$$\mathbf{A} = \frac{B_0}{2}\left(-\hat{\mathbf{x}}\,y + \hat{\mathbf{y}}\,x\right) .$$

13.32 Force-Free Magnetic Field

A force-free magnetic field is a magnetic field in which the Lorentz force is equal to zero in spite of the presence of an often very large magnetic energy density $B^2/(8\pi\mu_r)$ [2]. This occurs rather frequently, for instance, in astrophysics. As an example, the magnetic field in the Sun's corona is often approximated as a force-free field.

Here, as a simple model, we consider a cylindrical conducting medium of radius R where the field lines of the current density $\mathbf{J}$ are helices wound around the cylinder axis, as shown in Fig. 13.18. We shall use a cylindrical coordinate system (r, ϕ, z) with the z axis coinciding with the cylinder axis. The current density $\mathbf{J}(r)$ has components $J_z(r)$ and $J_\phi(r)$ depending on r only, with $0 \leq r \leq R$, and the conditions $J_\phi(0) = 0$ and $J_z(0) = J_0$.

(a) What is the condition on the directions of the current density $\mathbf{J}(r)$ and of the magnetic field $\mathbf{B}(r)$, which satisfies the relation $\nabla \times \mathbf{B} = (4\pi/c)\,\mathbf{J}$, for the total force acting on the moving charges to be zero? Consider the special case where the ratio $|\mathbf{J}|/|\mathbf{B}|$ is independent of r.
(b) Write the relation between the components of the magnetic field $\mathbf{B}$ and their derivatives with respect to r implied by the condition on $|\mathbf{J}|/|\mathbf{B}|$ of point **a)**.
(c) Show that the equation for B_z can be decoupled from the equation for B_ϕ and written in terms of the Laplace operator in cylindrical coordinates.
(d) Using the properties of Bessel's functions write the expressions for B_z and B_ϕ and determine how the ratio $|\mathbf{B}|/|\mathbf{J}|$ must be chosen in order to have $B_z = 0$ at $r = R$.

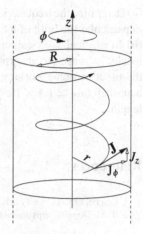

Fig. 13.18

13.33 Field Inside a Spherical Shell

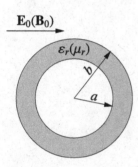

$\mathbf{E_0(B_0)}$

$\varepsilon_r(\mu_r)$

b

a

Fig. 13.19

(a) A spherical shell has inner radius a, outer radius $b > a$, and is made of a dielectric material of relative permittivity ε_r. The shell is located in a uniform external electric field $\mathbf{E}_0$ (Fig. 13.19). Calculate the electric field (assumed to be uniform) inside the shell. (Hint: make the following educated guess for the field,

$$
\mathbf{E} = \begin{cases}
\mathbf{E}_0 + \dfrac{(3\mathbf{p}_0 \cdot \hat{\mathbf{r}})\,\hat{\mathbf{r}} - \mathbf{p}_0}{r^3} & (r > b) \\[2mm]
\mathbf{E}_1 + \dfrac{(3\mathbf{p}_1 \cdot \hat{\mathbf{r}})\,\hat{\mathbf{r}} - \mathbf{p}_1}{r^3} & (b > r > a) \\[2mm]
\mathbf{E}_2 & (r < a)
\end{cases} \tag{13.26}
$$

and determine the values of $\mathbf{p}_0$, $\mathbf{p}_1$, $\mathbf{E}_1$ and $\mathbf{E}_2$ from the boundary conditions.)
        (b) Consider the analogous problem of a shell of magnetic permeability μ_r located in an external magnetic field $\mathbf{B}_0$.

13.34 Resonances in a Spherical Shell

A metallic shell of inner radius a and outer radius $b > a$ is placed in an uniform, oscillating electric field $\mathbf{E}(t) = \mathbf{E}_0 e^{-i\omega t}$. In the relevant frequency range, the permittivity of the material can be described by the usual expression $\varepsilon = \varepsilon(\omega) = 1 - \omega_p^2/\omega^2$, assumed to be real for simplicity.

(a) Determine the frequencies for which the response of the shell is resonant. Consider in particular the limit of a thin shell having thickness $d = b - a \ll a$.

(b) In medical laser surgery, metallic nanoshells are inserted in organic tissue to locally increase the absorption of electromagnetic waves at frequencies for which the tissue is non-absorbing. Determine suitable fabrication parameters for a Gold nanoshell ($\omega_p \simeq 1.4 \times 10^{16}$ Hz) and a laser source operating at $\lambda_L = 800$ nm wavelength.

References

1. I. Lerche, On a curiosity arising from Fizeau's experiment. Am. J. Phys. **43**, 910 (1975)
2. K.T. McDonald, https://www.hep.princeton.edu/~mcdonald/examples/-forcefree.pdf

Chapter S-1
Solutions for Chapter 1

S-1.1 Overlapping Charged Spheres

(a) The electrostatic field at any point in space is the sum of the fields generated by each charged sphere (superposition principle). The field generated by a single uniformly charged sphere at its interior is $\mathbf{E}(\mathbf{r}) = 4\pi k_e \varrho_0 \mathbf{r}/3$, where ϱ_0 is the charge density and $\mathbf{r}$ is the position vector relative to the center of the sphere. Thus, the two spheres generate at their interiors the fields $\mathbf{E}_\pm = \pm 4\pi k_e \varrho_0 \mathbf{r}_\pm/3$, respectively, $\mathbf{r}_\pm$ being the position vectors relative to the two centers. We assume that the centers are located on the x

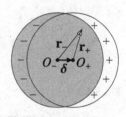

Fig. S-1.1

axis at points $O_+ \equiv (+\delta/2, 0, 0)$ and $O_- \equiv (-\delta/2, 0, 0)$ (Fig. S-1.1). We thus have $\mathbf{r}_\pm = \mathbf{r} \pm \delta/2$, where $\mathbf{r}$ is the position vector relative to the origin $O \equiv (0, 0, 0)$. The total field in the overlap region is

$$\mathbf{E}_{\text{in}} = +\frac{4\pi k_e}{3}\varrho_0\left(\mathbf{r} - \frac{\delta}{2}\right) - \frac{4\pi k_e}{3}\varrho_0\left(\mathbf{r} + \frac{\delta}{2}\right) = -\frac{4\pi k_e}{3}\varrho_0\,\delta . \quad \text{(S-1.1)}$$

The internal field $\mathbf{E}_{\text{in}}$ is thus *uniform* and proportional to $-\delta$.

(b) The electrostatic field generated by a uniformly charged sphere, with volume charge density ϱ_0, *outside* its volume equals is the field of a point charge $Q = 4\pi R^3 \varrho_0/3$ located at its center. Thus, the electrostatic field in the outer region (outside both spheres) is the sum of the fields of two point charges $\pm Q$ located at O_+ and O_-, respectively. If $R \gg \delta$, this is equivalent to the field of an electric dipole of moment

$$\mathbf{p} = Q\delta = \frac{4\pi R^3}{3}\varrho_0\,\delta \quad \text{(S-1.2)}$$

© The Editor(s) (if applicable) and The Author(s), under exclusive license to Springer   139
Nature Switzerland AG 2023
A. Macchi et al., *Problems in Classical Electromagnetism*,
https://doi.org/10.1007/978-3-031-22235-1

located at the origin and lying on the x axis. The external field is thus

$$\mathbf{E}_{\text{ext}}(\mathbf{r}) = k_{\text{e}} \frac{3\hat{\mathbf{r}}\,(\mathbf{p}\cdot\hat{\mathbf{r}}) - \mathbf{p}}{r^3}\,,\qquad\qquad\text{(S-1.3)}$$

where $\mathbf{r} = r\,\hat{\mathbf{r}}$ is the vector position relative to the origin.

In the two transition "shell" regions, of net charge densities $\pm\varrho_0$, the field is the sum of the inner field of one sphere and of the outer field of the other. We omit to write down the expression for brevity.

S-1.2 Charged Sphere with Internal Spherical Cavity

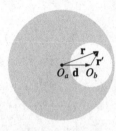

Fig. S-1.2

(a) Once again we use the superposition principle. Our charged sphere with an internal spherical cavity can be thought of as the superposition of two uniformly charged spheres: a sphere of radius a centered in O_a, with charge density ϱ, and a smaller sphere of radius b centered in O_b, with charge density $-\varrho$. The electric field everywhere in space is the sum of the fields generated by the two spheres. The field generated by a uniformly charged sphere at its inside is $\mathbf{E} = (4\pi k_{\text{e}}/3)\,\varrho\,\mathbf{r}$, where $\mathbf{r}$ is the distance from the center of the sphere. The total field inside the cavity at a point of vector position $\mathbf{r}$ relative to O_a, and vector position $\mathbf{r}'$ relative to O_b (Fig. S-1.2), is thus

$$\mathbf{E}_{\text{cav}} = \frac{4\pi k_{\text{e}}}{3}\,\varrho\,(\mathbf{r} - \mathbf{r}') = \frac{4\pi k_{\text{e}}}{3}\,\varrho\,\mathbf{d}\,,\qquad\qquad\text{(S-1.4)}$$

uniform and parallel to the straight line passing through O_a and O_b. If $\mathbf{d} = 0$ we obtain $\mathbf{E} = 0$, as expected from Gauss's law and symmetry considerations.

(b) In an external field $\mathbf{E}_0$ the total force on the system is the sum of the forces that $\mathbf{E}_0$ would exert on the two point charges $Q_a = 4\pi a^3\varrho/3$ and $Q_b = -4\pi b^3\varrho/3$, located in O_a and O_b, respectively, so that

$$\mathbf{F} = \frac{4\pi}{3}\varrho\,(a^3 - b^3)\,\mathbf{E}_0\,.\qquad\qquad\text{(S-1.5)}$$

(c) Since the vector sum of the forces is different from zero, the torque depends on our choice of the origin. The torque about the center of the sphere O_a is

$$\boldsymbol{\tau} = \mathbf{d}\times\mathbf{F} = -\frac{4\pi}{3}\,\varrho b^3 \mathbf{d}\times\mathbf{E}_0\,.\qquad\qquad\text{(S-1.6)}$$

Let us introduce a reference system with the x axis passing through O_a and O_b, and the origin in O_a. We denote by $\varrho_m = \alpha \varrho$ the mass density, with α some constant value. Our coordinate origin is thus the location of the center of mass of the cavity-less sphere of radius a and mass $M_{tot} = 4\pi a^3 \varrho_m/3$, while $x = d$ is the location of the center of mass of a sphere of radius b and mass $M_b = 4\pi b^3 \varrho_m/3$. Let us denote by x_c the center of mass of the sphere with cavity, of mass $M_c = 4\pi(a^3 - b^3)\varrho_m/3$. We have

$$0 = \frac{M_c x_c + M_b d}{M_{tot}}, \quad \text{thus} \quad x_c = -d \frac{M_b}{M_c} = -d \frac{b^3}{a^3 - b^3}. \tag{S-1.7}$$

The torque about the center of mass x_c is thus

$$\tau_c = \frac{b^3 d}{a^3 - b^3} \frac{4\pi}{3} \varrho a^3 \, \hat{\mathbf{x}} \times \mathbf{E}_0 - \left(d + \frac{b^3 d}{a^3 - b^3}\right) \frac{4\pi}{3} \varrho b^3 \, \hat{\mathbf{x}} \times \mathbf{E}_0 = 0, \tag{S-1.8}$$

as was to be expected, since each charged volume element $\mathrm{d}^3 r$ is subject to the force $\varrho \mathbf{E}_0 \, \mathrm{d}^3 r$, and acquires an acceleration

$$\mathbf{a} = \frac{\varrho \, \mathbf{E}_0 \, \mathrm{d}^3 r}{\varrho_m \mathrm{d}^3 r} = \frac{\mathbf{E}_0}{\alpha}, \tag{S-1.9}$$

equal for each charged volume element.

S-1.3   Energy of a Charged Sphere

(a) We can assemble the sphere by moving successive infinitesimal shells of charge from infinity to their final location. Let us assume that we have already assembled a sphere of charge density ϱ and radius $r < R$, and that we are adding a further shell of thickness $\mathrm{d}r$. The assembled sphere has charge $q(r) = \varrho(4\pi r^3/3)$, and its potential $\varphi(r, r')$ at any point at distance $r' \geq r$ from the center of the sphere is

$$\varphi(r, r') = k_e \frac{q(r)}{r'} = k_e \, \varrho \frac{4\pi r^3}{3} \cdot \frac{1}{r'}. \tag{S-1.10}$$

The work needed to move the new shell of charge $\mathrm{d}q = \varrho \, 4\pi r^2 \mathrm{d}r$ from infinity to r is

$$\mathrm{d}W = \varphi(r, r') \, \mathrm{d}q = k_e \, \varrho \frac{4\pi r^3}{3} \frac{1}{r} \varrho 4\pi r^2 \mathrm{d}r = k_e \frac{(4\pi \varrho)^2}{3} r^4 \mathrm{d}r. \tag{S-1.11}$$

The total work needed to assemble the sphere is obtained by integrating dW from $r = 0$ (no sphere) up to the final radius R

$$U_0 = k_e \frac{(4\pi\varrho)^2}{3} \int\limits_0^R r^4 dr = k_e \frac{(4\pi\varrho)^2 R^5}{15} = \frac{3k_e}{5} \frac{Q^2}{R}, \qquad \text{(S-1.12)}$$

where $Q = (\varrho 4\pi R^3)/3$ is the total charge of the sphere.

(b) The electric field everywhere in space is, according to Gauss's law,

$$E(r) = k_e Q \times \begin{cases} \dfrac{r}{R^3}, & r \leq R \\ \dfrac{1}{r^2}, & r \geq R, \end{cases} \qquad \text{(S-1.13)}$$

and the integral of the corresponding energy density $u_E = E^2/(8\pi k_e)$ over the whole space is

$$U_0 = \int\limits_0^\infty \frac{E^2(r)}{8\pi k_e} 4\pi r^2 \, dr = \frac{k_e^2 Q^2}{2k_e} \left[\int\limits_0^R \left(\frac{r}{R^3} \right)^2 r^2 dr + \int\limits_R^\infty \left(\frac{1}{r^2} \right)^2 r^2 dr \right]$$

$$= k_e \frac{Q^2}{2} \left(\frac{1}{5R} + \frac{1}{R} \right) = \frac{3k_e}{5} \frac{Q^2}{R}. \qquad \text{(S-1.14)}$$

(c) The electrostatic potential generated by the sphere everywhere in space is

$$\varphi(r) = k_e Q \times \begin{cases} -\dfrac{r^2}{2R^3} + \dfrac{3}{2R}, & r \leq R \\ \dfrac{1}{r}, & r > R \end{cases} \qquad \text{(S-1.15)}$$

where the constant $3k_e Q/(2R)$ appearing for $r \leq R$ is needed for $\varphi(r)$ to be continuous at $r = R$. Since $\varrho = 0$ for $r > R$, we need only the integral of $\varrho\varphi/2$ inside the sphere

$$U_0 = \frac{1}{2} \int\limits_0^R \varrho k_e Q \left(-\frac{r^2}{2R^3} + \frac{3}{2R} \right) 4\pi r^2 \, dr = k_e \frac{Q}{4} \frac{Q}{R^3/3} \left(-\frac{R^2}{5} + R^2 \right)$$

$$= \frac{3k_e}{4} \frac{Q^2}{R^3} \frac{4R^2}{5} = \frac{3k_e}{5} \frac{Q^2}{R}. \qquad \text{(S-1.16)}$$

All methods, including **(b)** and **(c)**, lead to the correct result, as expected. However, a comparison between methods **(b)** and **(c)** shows that it is incorrect to interpret the "energy density" of the electric field as the "energy stored in a given region of space per unit volume". If we give this meaning to quantity $E^2/(8\pi k_e)$, as in **(b)**, we conclude that the energy is spread over the whole space. If, on the other hand, we assume the energy density to be $\frac{1}{2}\varrho\varphi$, as in **(c)**, the energy is "stored" only inside the volume of the sphere, i.e., "where the charge is". Thus, the concept of energy density is ambiguous, while the total electrostatic energy of the system is a well defined quantity, at least in the absence of point charges.

S-1.4 Plasma Oscillations

(a) Assuming $\delta > 0$, the collective rigid displacement of the conduction electrons due to the external field gives origin to the charge density

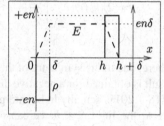

$$\varrho(x) = \begin{cases} 0, & x < 0, \\ +en, & 0 < x < \delta, \\ 0, & \delta < x < h, \\ -en, & h < x < h+\delta, \\ 0, & x > h+\delta. \end{cases} \quad \text{(S-1.17)}$$

Fig. S-1.3

The electrostatic field $\mathbf{E}(x)$ generated by this charge distribution is obtained by integrating the equation $\nabla \cdot \mathbf{E} = \partial_x E_x = 4\pi k_e \varrho$ with the boundary condition $\mathbf{E}(-\infty) = 0$:

$$E_x(x) = 4\pi enk_e \begin{cases} 0, & x < 0, \\ x, & 0 < x < \delta, \\ \delta, & \delta < x < h, \\ h+\delta-x, & h < x < h+\delta, \\ 0, & x > h+\delta. \end{cases}$$
$$\text{(S-1.18)}$$

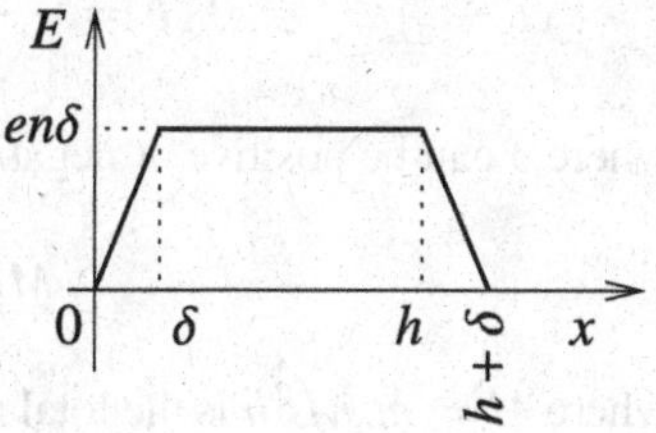

If we assume a negative displacement $-\delta$ (with $\delta > 0$) the charge density and the electric field are

Fig. S-1.4

$$\varrho(x) = \begin{cases} 0, & x < -\delta, \\ -en, & -\delta < x < 0, \\ 0, & 0 < x < h-\delta, \\ +en, & h-\delta < x < h, \\ 0, & x > h. \end{cases} \quad E_x(x) = 4\pi enk_e \begin{cases} 0, & x < -\delta, \\ -x-\delta, & -\delta < x < 0, \\ -\delta, & 0 < x < h-\delta, \\ x-h-\delta, & h-\delta < x < h, \\ 0, & x > h. \end{cases}$$
$$\text{(S-1.19)}$$

The plots are obtained from Figs. S-1.3 and S-1.4, respectively, by flipping around the x axis and translating by δ towards the negative x values.

(b) The electrostatic energy of the system, in the case of a positive displacement, can be evaluated by integrating the "energy density" $u = E_x^2/(8\pi k_e)$ over the whole space:

$$
U_{es} = \int \frac{E_x^2}{8\pi k_e}\, d^3r = \frac{L^2}{8\pi k_e} \int\limits_0^{h+\delta} E_x^2\, dx
$$

$$
= \frac{L^2}{8\pi k_e} (4\pi en)^2 \left[\int\limits_0^{\delta} x^2\, dx + \int\limits_{\delta}^{h} \delta^2\, dx + \int\limits_{h}^{h+\delta} (h+\delta-x)^2\, dx \right]
$$

$$
= 2\pi k_e (enL)^2 \left[\frac{\delta^3}{3} + \delta^2(h-\delta) + \frac{\delta^3}{3} \right] = 2\pi k_e (enL)^2 \left(h\delta^2 - \frac{\delta^3}{3} \right),
$$

$$\text{(S-1.20)}$$

where, due to the symmetry of the problem, we used $d^3r = L^2\, dx$. Exactly the same result is obtained for a negative displacement by $-\delta$. The δ appearing in the last line of (S-1.20) is actually to be interpreted as the absolute value $|\delta|$.

(c) At the limit $\delta \ll h$ we can neglect the third-order term in δ of (S-1.20), and approximate $U_{es} \simeq 2\pi k_e (enL)^2 h\delta^2$, which is the potential energy of a harmonic oscillator. The force on the "electron slab" is thus

$$
F = -\frac{\partial U_{es}}{\partial \delta} = -4\pi k_e (enL)^2 h\delta, \qquad\qquad \text{(S-1.21)}
$$

where δ can be positive or negative. The equation of motion for the electrons is

$$
M\ddot{\delta} = F \equiv -M\omega^2 \delta, \qquad\qquad \text{(S-1.22)}
$$

where $M = m_e nL^2 h$ is the total mass of the conduction-electron slab. We thus have

$$
\omega^2 = \frac{4\pi k_e ne^2}{m_e} \equiv \omega_p^2, \qquad\qquad \text{(S-1.23)}
$$

where ω_p is called the *plasma frequency*, and is an intrinsic property of the given conductor, dependent only on the density of free electrons.

S-1.5 Mie Oscillations

(a) In Problem 1.1 we showed that the electric field is
uniform and equal to $-4\pi k_e \, \varrho_0 \, \delta/3$, with $\varrho_0 = en_e$, in
the region where the conduction-electrons sphere and
the ion-lattice sphere overlap,

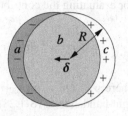

Fig. S-1.5

$$\mathbf{E}_{int} = -\frac{4\pi k_e}{3} en_e \delta \ . \qquad (S\text{-}1.24)$$

We assume that the displacement δ is sufficiently small
for the volumes a (only conduction electrons) and c
(only ion lattice) of Fig. S-1.5 to be negligible com-
pared to the overlap volume b, an order of magnitude for δ is found in (S-2.4) of
Solution S-2.1. Assuming further that conduction electrons behave like a "rigid"
body, oscillating in phase with the same displacement $\delta = \delta(t)$ from their rest positions, the equation of motion for the single electron is

$$m_e \frac{d^2\delta}{dt^2} = -e\mathbf{E}_{int} = -e\frac{4\pi k_e}{3} en_e \delta = -m_e \frac{\omega_p^2}{3} \delta \ , \qquad (S\text{-}1.25)$$

where ω_p is the plasma frequency (S-1.23). Thus, the displacement of each conduction
electron from its rest position is

$$\delta(t) = \delta(0) \cos\left(\frac{\omega_p}{\sqrt{3}} t\right) , \qquad (S\text{-}1.26)$$

where $\delta(0)$ is a constant and $\omega_p/\sqrt{3}$ is called the Mie frequency. This type of motion
is known as Mie oscillation (or surface plasmon of the sphere).
(b) The electrostatic energy of the system is given by the integral

$$U_{es} = \int \frac{E^2}{8\pi k_e} \, d^3r \ . \qquad (S\text{-}1.27)$$

For δ approaching 0, the electric field is given by (S-1.24) inside the sphere ($r < R$),
and by (S-1.3), i.e., an electric dipole field, outside the sphere ($r > R$). Thus we
can split the integral of (S-1.27) into the sum of two terms, corresponding to the
integration domains $r < R$ and $r > R$, respectively

$$U_{es} = U_{es}^{in} + U_{es}^{out} = \frac{1}{8\pi k_e} \left(\int\limits_{r<R} E^2 \, d^3r + \int\limits_{r<R} E^2 \, d^3r \right) . \qquad (S\text{-}1.28)$$

For $r < R$ the field is uniform and we immediately find

$$U_{\text{es}}^{\text{in}} = \frac{1}{8\pi k_e} \left(k_e \frac{4\pi}{3} e n_e \delta \right)^2 \frac{4\pi}{3} R^3 = k_e \frac{8\pi^2}{27} (e n_e \delta)^2 R^3 . \quad \text{(S-1.29)}$$

For evaluating the contribution of the outer region, we substitute

$$d^3 r = r^2 \sin\theta \, dr \, d\theta \, d\phi , \quad \text{and} \quad E^2 = \left(\frac{k_e p}{r^3} \right)^2 (3\cos^2\theta + 1) , \quad \text{(S-1.30)}$$

where $\mathbf{p} = Q\delta = \delta \, e n_e \, 4\pi R^3/3$ (see Problem 1.1), into the second integral at the right-hand side of (S-1.28)

$$U_{\text{es}}^{\text{out}} = 2\pi \frac{1}{8\pi k_e} k_e^2 p^2 \int\limits_R^{+\infty} r^2 \, dr \int\limits_0^\pi \sin\theta \, d\theta \, \frac{3\cos^2\theta + 1}{r^6} , \quad \text{(S-1.31)}$$

and obtain

$$U_{\text{es}}^{\text{out}} = k_e \frac{p^2}{3R^3} = k_e \frac{4\pi}{9} (e n_e \delta)^2 \frac{4\pi}{3} R^3 = k_e \frac{16\pi^2}{27} (e n_e \delta)^2 R^3 , \quad \text{(S-1.32)}$$

thus $U_{\text{es}}^{\text{out}} = 2 U_{\text{es}}^{\text{in}}$. For the total energy we have finally

$$U_{\text{es}} = U_{\text{es}}^{\text{in}} + U_{\text{es}}^{\text{out}} = k_e \frac{8\pi^2}{9} (e n_e \delta)^2 R^3 . \quad \text{(S-1.33)}$$

The derivative of U_{es} with respect to δ gives the force associated to the displacement of the electron sphere:

$$F = -\frac{\partial U_{\text{es}}}{\partial \delta} = -k_e \frac{16\pi^2}{9} R^3 (e n_e)^2 \delta . \quad \text{(S-1.34)}$$

The equation of motion for the rigid sphere of electrons is $M \, d^2\delta/(dt^2) = F$, where $M = m_e n_e 4\pi R^3/3$ is the total mass of electrons. Thus

$$\frac{d^2\delta}{dt^2} = -k_e \frac{4\pi n_e e^2}{3 m_e} \delta \equiv -\frac{\omega_p^2}{3} \delta , \quad \text{(S-1.35)}$$

and we are back to the oscillations at the Mie frequency of (S-1.26).

S-1.6 Coulomb Explosions

(a) The electric field has radial symmetry, $\mathbf{E} = E(r)\,\hat{\mathbf{r}}$. According to Gauss's law we have $4\pi r^2 E(r) = 4\pi k_e Q_{\text{int}}(r)$, where $Q_{\text{int}}(r)$ is the charge inside the sphere of radius r. At $t = 0$ we have $Q_{\text{int}} = Q\,(r/R)^3$ for $r < R$, and $Q_{\text{int}} = Q$ for $r > R$, thus the electric field inside and outside the cloud is, respectively,

$$E(r) = k_e Q \times \begin{cases} \dfrac{r}{R^3}, & r \leq R, \\[2mm] \dfrac{1}{r^2}, & r \geq R. \end{cases} \tag{S-1.36}$$

Due to the spherical symmetry of the problem, we have $E = -\partial\varphi/\partial r$, where φ is the electric potential, and the potential at $t = 0$ can be obtained by a simple integration:

$$\varphi(r) = k_e Q \times \begin{cases} -\dfrac{r^2}{2R^3} + \dfrac{3}{2R}, & r \leq R, \\[2mm] \dfrac{1}{r}, & r \geq R. \end{cases} \tag{S-1.37}$$

As in Eq. (S-1.15) of Problem 1.3, the integration constants have been chosen so that $\varphi(\infty) = 0$ and $\varphi(r)$ is continuous at $r = R$. The potential energy of a *test* charge q_{t} located at distance r from the center is thus $q_{\text{t}}\,\varphi(r)$.

(b) Under the action of the electric field, the test charge would move and convert all its potential energy into kinetic energy *if the field remained stationary during the charge motion*, i.e., *if all the source charges of the field remained fixed*. At $t = 0$ the electric field inside the spherical cloud increases with r. Thus, the "outer" particles, located at larger r, have a higher acceleration than the "inner" particles, located at smaller r. After an infinitesimal time interval the "outer" particles will acquire a higher velocity (we are assuming that all particles are at rest at $t = 0$), and will not be overtaken by the "inner" ones. Moreover, also the acceleration has radial symmetry, and thus any spherical shell preserves its shape in time. These arguments can be iterated for any following time, proving the validity of our assumptions that the particles do not overtake one another, and that the spherical symmetry is preserved.

Let us denote by $r_s(r_0, t)$ the position of a particle initially located at r_0. Since the particles do not overtake one another, the charge inside a sphere of radius $r_s(r_0, t)$ is constant. The electric field intensity at $r_s(r_0, t)$ can be evaluated by applying Gauss's law: from

$$4\pi r_s^2(r_0, t)\, E\,[r_s(r_0, t)] = 4\pi k_e Q \left(\frac{r_0}{R}\right)^3 \tag{S-1.38}$$

we obtain

$$E\left[r_s(r_0, t)\right] = k_e \frac{Q}{r_s^2(r_0, t)} \left(\frac{r_0}{R}\right)^3,\qquad\text{(S-1.39)}$$

from which Eq. (1.16) can be derived. The forces on the particles, and thus their accelerations, increase with increasing r_0, in agreement with our "non-overtaking" result. Note that the electric field, and thus the force, at $t = 0$ is proportional to r_0, not to r_0^3, because we have $r_s^2(r_0, 0) = r_0^2$ at the denominator.

(c) Each infinitesimal spherical shell expands from its initial radius r_0 to its final radius $r_s(r_0, \infty) = \infty$ under the action of the force (1.16). The final kinetic energy of a particle belonging to the shell, $K_{\text{fin}}(r_0)$, equals the work done by the force on the particle

$$K_{\text{fin}}(r_0) = \int\limits_{r_{i0}}^{+\infty} k_e \frac{qQ}{r_i^2} \left(\frac{r_0}{R}\right)^3 dr = k_e \frac{qQ}{r_0} \left(\frac{r_0}{R}\right)^3 = k_e q \, Q \frac{r_0^2}{R^3}. \qquad\text{(S-1.40)}$$

Quantity $K_{\text{fin}}(r_0)$ is a monotonically increasing function of r_0, thus its maximum value $K_{\max}$ is observed for $r_0 = R$

$$K_{\max} = K_{\text{fin}}(R) = k_e \frac{qQ}{R}. \qquad\text{(S-1.41)}$$

This means that the particles initially located at $r_0 = R$, i.e., at the cloud surface, acquire the maximum final kinetic energy.

(d) The energy distribution, or energy spectrum, function $f(K)$ is defined so that the number dN of particles with kinetic energy in the interval $(K, K + dK)$ equals $f(K)\,dK$, therefore $f(K) = dN/dK$. A particle belonging to the shell $r_0 < r_s < r_0 + dr_0$ at $t = 0$ has a kinetic energy in the interval $(K_{\text{fin}}(r_0), K_{\text{fin}}(r_0) + dK_{\text{fin}})$ at $t = \infty$, where

$$dK_{\text{fin}} = \frac{2k_e \, qQ \, r_0}{R^3} \, dr_0. \qquad\text{(S-1.42)}$$

On the other hand, at $t = 0$ the number of particles in the shell $(r_0, r_0 + dr_0)$ is

$$dN = N \frac{3}{4\pi R^3} 4\pi r_0^2 \, dr_0 = N \frac{3r_0^2}{R^3} \, dr_0, \qquad\text{(S-1.43)}$$

and the number of particles in a given shell is constant during the motion, since the particles do not overtake one another. Thus, inserting (S-1.40) and (S-1.41), we obtain

$$f(K_{\text{fin}}) = \frac{dN}{dK_{\text{fin}}} = N \frac{3r_0^2}{R^3} \frac{R^3}{2k_e q \, Qr_0} = N \frac{3r_0}{2k_e q \, Q} = \frac{3N}{2K_{\max}^{3/2}} \sqrt{K_{\text{fin}}}, \qquad\text{(S-1.44)}$$

valid for $K_{\text{fin}} \le K_{\max}$.

The final total kinetic energy is

$$K_{\text{tot}} = \int_0^{K_{\max}} K f(K) \, dK = \frac{3N}{2K_{\max}^{3/2}} \int_0^{K_{\max}} K^{3/2} dK = \frac{3N}{5} K_{\max}$$

$$= \frac{3k_e}{5} \frac{NqQ}{R} = \frac{3k_e}{5} \frac{Q^2}{R}, \tag{S-1.45}$$

which equals the total electrostatic energy stored in the charged sphere at $t = 0$ (Problem 1.3). Here we have substituted $Nq = Q$. Thus, all the electrostatic energy stored in the initial configuration is eventually converted into kinetic energy.

It is a relatively common error to assume that the final kinetic energy of a particle initially in the shell $r_0 < r_s < r_0 + dr$ is equal to the potential energy of the same particle at $t = 0$, i.e., that $K_{\text{fin}} = q \varphi(r_0)$, where φ is given by (S-1.37). This is obviously wrong, because a particle initially at $r_0 = 0$ has the highest possible initial potential energy, $\varphi(0) = 3k_e Q/(2R)$, while it undergoes the lowest possible gain in kinetic energy (zero)! Moreover, this behavior would not preserve the total energy of the system, because the initial potential energy of the sphere is (see Problem 1.3)

$$U(0) = \frac{1}{2} \sum_i q \, \varphi[r_i(0)], \quad \text{not} \quad U(0) = \sum_i q \, \varphi[r_i(0)].$$

The point is that while the field is electrostatic ($\nabla \times \mathbf{E} = 0$) at any time, it is *time dependent*. Thus, φ can be defined for any value of t, but it *cannot* be used to evaluate the final kinetic energy, because φ changes as the particles move.

The gain in kinetic energy equals the initial potential energy $q\varphi(R, t = 0)$ for the particles initially at $r_i = R$, i.e., for the most external ones. Only these particles are accelerated by a field that can be treated as static, being simply equal to the field of a point charge Q located at $r = 0$ at any time.

(e) If we introduce the new variable $x(t) = r_s(r_0, t)/r_0$, (1.16) becomes

$$m \frac{d^2x}{dt^2} = k_e \frac{qQ}{R^3 x^2}, \tag{S-1.46}$$

which is independent of r_0. The solution of (S-1.46), $x = x(t)$, with the initial condition $x(0) = 1$, is thus valid for all the particles of the cloud. Thus, if two shells, labeled 1 and 2, have initial radii r_{10} and r_{20}, with $r_{20} > r_{10}$, their subsequent radii will be $r_1(t) = r_s(r_{10}, t) = r_{10} x(t)$ and $r_2(t) = r_s(r_{20}, t) = r_{20}x(t)$. It will always be $r_2(t) > r_1(t)$, and the internal shell cannot overtake the external one. The number of particles contained between the layers 1 and 2 is constant and equal to

$$\delta N_{12} = \frac{N}{R^3} \left(r_{20}^3 - r_{10}^3 \right). \tag{S-1.47}$$

Thus the particle density between the two layers at time t is

$$n(t) = \frac{3}{4\pi \left[r_2^3(t) - r_1^3(t)\right]} \, \delta N_{12} = \frac{3N}{4\pi R^3} \, \frac{r_{20}^3 - r_{10}^3}{(r_{20}^3 - r_{10}^3) \, x^3(t)} = \frac{3N}{4\pi R^3} \, x^{-3}(t) \, . \quad \text{(S-1.48)}$$

This result does not depend on the particular choice of the two layers, and the particle density is uniform at any time t, and decreases with increasing time as $x^{-3}(t)$ (Fig. S-1.6).

S-1.7   Plane and Cylindrical Coulomb Explosions

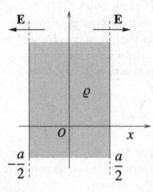

Fig. S-1.6

(a) The electric field is parallel to the x axis and independent of the y and z coordinates for symmetry reasons, thus we have $\mathbf{E}(x, y, z) = E(x)\,\hat{\mathbf{x}}$. Again for symmetry reasons, the electric field is antisymmetric with respect to the $x = 0$ plane, so that $E(-x) = -E(x)$. Thus it is sufficient to consider the field for $x \geq 0$. The charge density at $t = 0$ is $\varrho(x) = qn_0\Theta(a/2 - |x|)$, where $\Theta(x)$ is the Heaviside step function, defined as $\Theta(x) = 1$ for $x > 0$, and $\Theta(x) = 0$ for $x < 0$. The electric field at $t = 0$ can be evaluated by integrating the equation $\nabla \cdot \mathbf{E} = \partial_x E_x = 4\pi k_e \varrho(x)$, with the boundary condition $\mathbf{E}(0) = 0$, obtaining

$$E(x) = 4\pi k_e \begin{cases} qn_0 x \, , & x < \dfrac{a}{2} \, , \\[2mm] qn_0\dfrac{a}{2} \, , & x > \dfrac{a}{2} \, . \end{cases} \quad \text{(S-1.49)}$$

Since the particles are at rest at t = 0, and the electric field increases with increasing x, the particles cannot overtake one another. The motion of a particle initially at x_0 is described by an equation $x_s = x_s(x_0, t)$. Let us consider a parallelepiped of base S lying on the $x = 0$ plane and height $x_s(x_0, t)$. The charge inside our parallelepiped is constant in time since no particle can cross the moving base. We can apply Gauss's law to evaluate the electric field on the particle located at $x_s(x_0, t)$

$$E[x_s(x_0, t), t] \, S = 4\pi k_e Q_{\text{in}}(t) = 4\pi k_e Q_{\text{in}}(0) = 4\pi k_e qn_0 x_0 S \, . \quad \text{(S-1.50)}$$

Thus, the field accelerating each particle is constant in time, and equals

$$E(x_0) = 4\pi k_e qn_0 x_0 \, , \quad \text{(S-1.51)}$$

where x_0 is the particle position at $t = 0$. The equation of motion is thus

$$m \frac{d^2 x_s(x_0, t)}{dt^2} = q E(x_0) = 4\pi k_e q^2 n_0 x_0 \,, \tag{S-1.52}$$

with the initial conditions $x_s(x_0, 0) = x_0$ and $\dot{x}_s(x_0, 0) = 0$. The solution is

$$x_s(x_0, t) = x_0 + 2\pi k_e \frac{q^2 n_0 x_0}{m} t^2 = x_0 \left(1 + \frac{\omega_p^2 t^2}{2} \right) , \tag{S-1.53}$$

where $\omega_p = \sqrt{4\pi k_e q^2 n_0 / m}$ is the "plasma frequency" of the infinite charged layer at $t = 0$ (see Problem 1.4). Thus, the acceleration of an infinitesimal plane layer of thickness dx is proportional to its initial x coordinate, and more external layers are faster than more internal ones. The velocity, and the kinetic energy, of each layer grow indefinitely with time, which is not surprising since the system has an infinite initial potential energy.

If we introduce the dimensionless variable $\xi = x_s/x_0$, its equation of motion

$$m \frac{d^2 \xi}{dt^2} = q E(x_0) = 4\pi k_e q^2 n_0, \quad \xi(0) = 1 \,, \quad \frac{d\xi(0)}{dt} = 0 \,, \tag{S-1.54}$$

is independent of x_0. Thus the position of any particle can be written in the form

$$x_s(x_0, t) = x_0 \, \xi(t) \,, \tag{S-1.55}$$

which shows that the particle density, and the charge density, remain uniform during the explosion.

(b) The case of the Coulomb explosion of a system of charged particles initially confined, at rest, inside an infinite cylinder of radius a, is similar. We use cylindrical coordinates (r, ϕ, z), and assume that the initial particle density is uniform and equal to n_0 for $r < a$, and zero for $r > a$. All particles have mass m and charge q. According to Gauss's law, at $t = 0$ the field at position (r_0, ϕ, z), with $r_0 < a$, is

$$E(r_0) = 2\pi k_e n_0 q \, r_0 \,. \tag{S-1.56}$$

Again, the particles cannot overtake one another because the electric field increases with increasing r_0. A particle initially at r_0 will move along the r coordinate according to a law $r_s = r_s(r_0, t)$, with $r_s(r_0, 0) = r_0$. The field acting on the particle at time t is

$$E(r_0, t) = \frac{2\pi k_e n_0 q \, r_0^2}{r_s(r_0, t)} \,, \tag{S-1.57}$$

and its equation of motion is

$$m \frac{d^2 r_s(r_0, t)}{dt^2} = q \frac{2\pi k_e n_0 q\, r_0^2}{r_s(r_0, t)} . \tag{S-1.58}$$

It is not possible to solve (S-1.58) for $r_s(r_0, t)$ in a simple way, however, we can multiply both sides by $dr_s(r_0, t)/dt$, obtaining

$$m \frac{dr_s(r_0, t)}{dt} \frac{d^2 r_s(r_0, t)}{dt^2} = 2\pi k_e n_0 q^2 r_0^2 \frac{1}{r_s(r_0, t)} \frac{dr_s(r_0, t)}{dt} . \tag{S-1.59}$$

Equation (S-1.59) can be rewritten

$$\frac{m}{2} \frac{d}{dt} \left[\frac{dr_s(r_0, t)}{dt} \right]^2 = 2\pi k_e n_0 q^2 r_0^2 \frac{d}{dt} \ln[r_s(r_0, t)] , \tag{S-1.60}$$

which can be integrated with respect to time, leading to

$$\frac{m}{2} \left[\frac{dr_s(r_0, t)}{dt} \right]^2 = 2\pi k_e n_0 q^2 r_0^2 \ln[r_s(r_0, t)] + C$$

$$= 2\pi k_e n_0 q^2 r_0^2 \ln\left(\frac{r_s(r_0, t)}{r_0} \right) , \tag{S-1.61}$$

where the integration constant C has been determined by the condition that the kinetic energy of the particle must be zero at $t = 0$, when $r_s(r_0, t) = r_0$. The first side of (S-1.61) is the kinetic energy $K(r_s)$ at time t, when the particle is located at $r_s(r_0, t)$, which we can simply denote by r_s. Thus we have the following, seemingly time-independent equation for the kinetic energy of a particle initially located at r_0

$$K(r_s) = 2\pi k_e n_0 q^2 r_0^2 \ln\left(\frac{r_s}{r_0} \right) . \tag{S-1.62}$$

At the limit $r_s \to \infty$, $t \to \infty$, the integral diverges logarithmically. Again, this is due to the infinite potential energy initially stored in the system.

S-1.8 Collision of Two Charged Spheres

(a) The electrostatic energy of a uniformly charged sphere of radius R and total charge Q is, according to the result of Problem 1.3,

$$U_0 = \frac{3}{5} k_e \frac{Q^2}{R} , \tag{S-1.63}$$

so that the initial energy of our system of two spheres is

$$U_{\text{tot}} = 2U_0 = \frac{6}{5} k_e \frac{Q^2}{R}. \tag{S-1.64}$$

(b) Let us denote by x the distance between the centers of the two spheres. When x is such that the interaction energy $U_{\text{int}}(x)$ is no longer negligible with respect to U_0, but still larger than $2R$, the total potential energy $U_{\text{pot}}(x)$ of the system is

$$U_{\text{pot}}(x) = 2U_0 + U_{\text{int}}(x). \tag{S-1.65}$$

As long as $x \geqslant 2R$ the force between the spheres is identical to the force between two point charges $\pm Q$ located at the centers of the spheres, and

$$U_{\text{int}}(x) = -k_e \frac{Q^2}{x}. \tag{S-1.66}$$

Both the total momentum and the total energy of the system are conserved. Thus, the velocities of the two sphere are always equal and opposite. As long as $x \geqslant 2R$ the total energy of the system $U_{\text{tot}} = 2U_0$ equals the sum of the potential and kinetic energies of the system

$$U_{\text{tot}} = 2\frac{1}{2} M v^2(x) + 2U_0 + U_{\text{int}}(x), \tag{S-1.67}$$

where M is the mass of each sphere, and $\pm v(x)$ are the velocities of the two spheres. Thus

$$\frac{6}{5} k_e \frac{Q^2}{R} = M v^2(x) + \frac{6k_e}{5} \frac{Q^2}{R} - k_e \frac{Q^2}{x}, \quad \text{and} \quad v(x) = \sqrt{k_e \frac{Q^2}{Mx}}. \tag{S-1.68}$$

When $x = 2R$, the velocity is

$$v(2R) = \sqrt{k_e \frac{Q^2}{2MR}}. \tag{S-1.69}$$

(c) When the two spheres overlap completely, the charge density and the electrostatic field are zero over the whole space, so that also the electrostatic energy is zero. This means that all the initial energy has been converted into kinetic energy, i.e.,

$$2\frac{1}{2} M v^2(0) = 2U_0, \tag{S-1.70}$$

from which we obtain

$$v(0) = \sqrt{\frac{6}{5} k_e \frac{Q^2}{MR}}. \tag{S-1.71}$$

S-1.9 Oscillations in a Positively Charged Conducting Sphere

(a) At equilibrium, the remaining $(1 - f)N$ conduction electrons must be subject to zero electric field. For symmetry reasons, this is possible only if they occupy a spherical volume of radius $b < a$ concentric with the conducting sphere, where the where e is the elementary electric charge, and n_e and n_i are the conduction-electron density and the ion density, respectively (Fig. S-1.7). Thus, we must have $n_e = n_i$, with total charge density ϱ is zero. We thus have

$$\varrho(r) = e(n_i - n_e) = 0 \quad \text{for} \quad r < b, \tag{S-1.72}$$

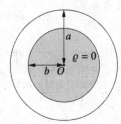

where

$$n_i = \frac{3N}{4\pi a^3} \quad \text{and} \quad n_e = \frac{3(1 - f)N}{4\pi b^3}, \tag{S-1.73}$$

and we get

$$b = a\sqrt[3]{1 - f}, \tag{S-1.74}$$

Fig. S-1.7   and

$$\varrho(r) = \frac{3Ne}{4\pi a^3} \quad \text{for} \quad b < r < a. \tag{S-1.75}$$

Note that the electric field is nonzero in the spherical shell $b < r < a$. However, this region is not conducting, since the conduction electrons are confined in the inner region $r < b$.

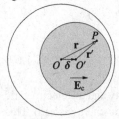

(b) Now the conduction electron sphere is rigidly displaced by an amount δ relative to the metal sphere centered in O, so that its center is in O', as in Fig. S-1.8. The electric field in any point of space can be evaluated by superposition, adding the field generated by the ion lattice, of charge density $\varrho_i = en_i$ and the field generated by the conduction electrons of charge density $\varrho_e = -en_e$. The electric field $\mathbf{E}_c$ in a point P inside the conduction-electron sphere, of vector position $\mathbf{r}$ relative to O, and $\mathbf{r}'$ relative to O', is

Fig. S-1.8

$$\mathbf{E}_c = \frac{4\pi k_e}{3}(\varrho_i\mathbf{r} - \varrho_e\mathbf{r}') = \frac{4\pi k_e}{3}\frac{3Ne}{4\pi a^3}(\mathbf{r} - \mathbf{r}') = \frac{k_e Ne}{a^3}\boldsymbol{\delta}, \tag{S-1.76}$$

spatially uniform uniform and parallel to $\boldsymbol{\delta}$.

(c) Each conduction electron is subject to the force

$$\mathbf{F}_c = -e\mathbf{E}_c = -\frac{k_e N e^2}{a^3}\,\boldsymbol{\delta}\,,$$

(S-1.77)

proportional to the displacement from its equilibrium position. The equation of motion for each electron is thus

$$m_e\frac{d^2\boldsymbol{\delta}}{dt^2} = \mathbf{F}_c = -\frac{k_e N e^2}{a^3}\,\boldsymbol{\delta} = -m_e\,\omega_M^2\,\boldsymbol{\delta}\,,$$

(S-1.78)

where m_e is the electron mass, and ω_M the oscillation frequency for the resulting harmonic motion. The oscillation frequency is thus

$$\omega_M = \sqrt{\frac{k_e N e^2}{m_e a^3}}\,,$$

(S-1.79)

i.e., the Mie frequency of (S-1.26).

S-1.10 Interaction Between a Point Charge and an Electric Dipole

(a) The simplest way to evaluate the force exerted by the point charge q on the dipole $\mathbf{p}$ is to evaluate the force exerted by $\mathbf{p}$ on q and reversing the sign, according to Newton's third law. We choose a spherical coordinate system (r, β, ϕ) with the origin located at the dipole position, the polar axis parallel to $\mathbf{p}$ and such that $\mathbf{p}$ and q lie on the $\phi = 0$ plane as in Fig. S-1.9. We denote the polar angle by β, and the field generated by $\mathbf{p}$ at the location of q by $\mathbf{E_p}$. We have for the force $\mathbf{f}$ acting on q

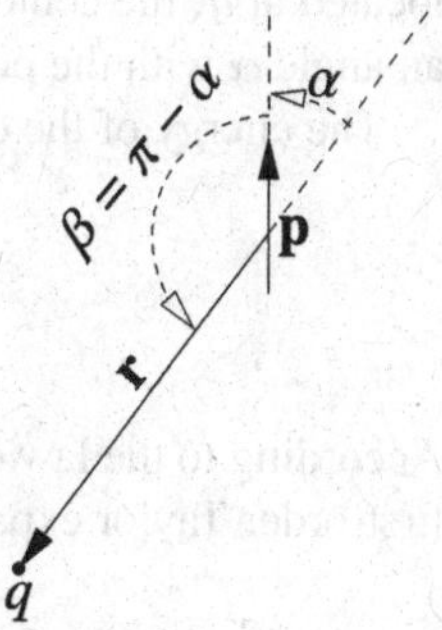

$$\mathbf{f} = q\mathbf{E_p} = k_e\frac{qp}{r^3}\left(2\cos\beta\,\hat{\mathbf{r}} + \sin\beta\,\hat{\boldsymbol{\beta}}\right) = k_e\frac{qp}{r^3}\left(-2\cos\alpha\,\hat{\mathbf{r}} + \sin\alpha\,\hat{\boldsymbol{\beta}}\right)\,,$$

(S-1.80)

Fig. S-1.9

the compoment parallel to $\mathbf{r}$ is attractive if $\cos\alpha > 0$, repulsive if $\cos\alpha < 0$, the component pependicular to $\mathbf{r}$ is directed downwards in Fig. 1.5 if $\sin\alpha > 0$ ($\hat{\boldsymbol{\beta}}$ is directed counterclockwise). Thus the force $\mathbf{F} = -\mathbf{f}$ exerted by q on $\mathbf{p}$ has the components

$$F_r = -k_e\frac{2qp\cos\alpha}{r^3}\,, \qquad F_\perp = k_e\frac{2qp\sin\alpha}{r^3}\,.$$

(S-1.81)

(b) The energy of the dipole in the field of the point charge is

$$U = -\mathbf{p} \cdot \mathbf{E} = -k_e \frac{q}{r^2} \mathbf{p} \cdot \hat{\mathbf{r}} = -k_e \frac{qp \cos \alpha}{r^2} \tag{S-1.82}$$

thus the torque acting on the dipole is

$$\tau = -\frac{\partial U}{\partial \alpha} = -k_e \frac{qp \sin \alpha}{r^2} , \tag{S-1.83}$$

if $q > 0$, the torque tends to align $\mathbf{p}$ to $\mathbf{r}$.

Alternatively we can directly evaluate the force exerted by q on $\mathbf{p}$. For this we model the dipole as a system of two charges $+q'$ and $-q'$ at a distance $2h$ from each other, and consider the limit $h \to 0$,

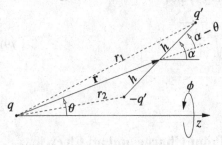

Fig. S-1.10

$q' \to \infty$ with $2hq' = p$ constant. It is important to note that this does not mean that every dipole consists of two opposite charges at a small distance from each other, but simply that this model leads to the correct result. We start by evaluating the energy of the dipole in the field of the point charge q. We choose a spherical coordinate system (r, θ, ϕ) such that $\mathbf{p}$ and q lie on the $\phi = 0$ plane, the origin is located at q, the center of the dipole has coordinates $(r, \theta, 0)$, and the dipole forms an angle α with the polar axis z, as shown in Fig. S-1.10.

The energy of the dipole, neglecting its self-energy, is

$$U = k_e q \left(\frac{q'}{r_1} - \frac{q'}{r_2} \right) . \tag{S-1.84}$$

According to the law of cosines we have $r_1 = \sqrt{r^2 + h^2 + 2rh \cos(\alpha - \theta)}$, thus the first-order Taylor expansion for $1/r_1$ is

$$\frac{1}{r_1} = \frac{1}{\sqrt{r^2 + h^2 + 2rh \cos(\alpha - \theta)}} \simeq \frac{1}{r} - \frac{\cos(\alpha - \theta)}{r^2} h , \tag{S-1.85}$$

and, analogously, we have for $1/r_2$

$$\frac{1}{r_2} \simeq \frac{1}{r} + \frac{\cos(\alpha - \theta)}{r^2} h . \tag{S-1.86}$$

Substituting into (S-1.84) we obtain

$$U \simeq k_e qq' \left(\frac{1}{r} - \frac{\cos(\alpha - \theta)}{r^2} h - \frac{1}{r} - \frac{\cos(\alpha - \theta)}{r^2} h \right) = -k_e qq' \frac{2h\cos(\alpha - \theta)}{r^2}$$

$$= -k_e \frac{qp}{r^2} (\cos\alpha\cos\theta + \sin\alpha\sin\theta) , \qquad (S\text{-}1.87)$$

which is independent of ϕ, thus the force $\mathbf{F}$ acting on the dipole has only the F_r and F_θ components. We are interested in F_r and F_θ evaluated at $\theta = 0$, see Fig. 1.5. The gradient components in spherical coordinates are

$$F_r|_{\theta=0} = -\frac{\partial U}{\partial r}\bigg|_{\theta=0} = -k_e \frac{2qp\cos\alpha}{r^3} , \qquad (S\text{-}1.88)$$

$$F_\theta|_{\theta=0} = -\frac{1}{r}\frac{\partial U}{\partial \theta}\bigg|_{\theta=0} = k_e \frac{qp\sin\alpha}{r^3} , \qquad (S\text{-}1.89)$$

in agreement with (S-1.81).

A further possibility is evaluating directly the forces acting on the charges q' and $-q'$ of our model, see Fig. S-1.11, and their limits as $q' \to \infty$, $h \to 0$ with $2q'h = p$ constant. We can approximate $1/r_1^2$ and $1/r_2^2$ as

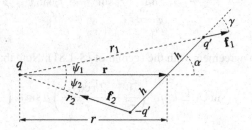

$$\frac{1}{r_1^2} \simeq \frac{1}{r^2} - \frac{2\cos\alpha}{r^3} h , \qquad (S\text{-}1.90)$$

Fig. S-1.11

$$\frac{1}{r_2^2} \simeq \frac{1}{r^2} + \frac{2\cos\alpha}{r^3} h , \qquad (S\text{-}1.91)$$

obtaining for the force $\mathbf{f}_1$ acting on the positive charge q' of the dipole and for $\mathbf{f}_2$ acting on the negative charge $-q'$

$$\mathbf{f}_1 \simeq k_e qq' \left(\frac{1}{r^2} - \frac{2h\cos\alpha}{r^3} \right) \hat{\mathbf{r}}_1 , \quad \mathbf{f}_2 \simeq -k_e qq' \left(\frac{1}{r^2} + \frac{2h\cos\alpha}{r^3} \right) \hat{\mathbf{r}}_2 , \quad (S\text{-}1.92)$$

respectively. As $h \to 0$ both angles ψ_1 and ψ_2 tend to zero, and $\hat{\mathbf{r}}_1$ and $\hat{\mathbf{r}}_2$ tend both to $\hat{\mathbf{r}}$, thus the r component of the total force acting on the dipole is

$$F_r = \lim_{\substack{h\to 0 \\ q'\to\infty \\ 2hq'=p}} (f_1 + f_2) = -k_e qq' \frac{4h\cos\alpha}{r^3} = -2k_e \frac{qp\cos\alpha}{r^3} , \qquad (S\text{-}1.93)$$

in agreement with the first of (S-1.81). Note that

$$\hat{\mathbf{r}} \lim_{\substack{h \to 0 \\ q' \to \infty \\ 2hq'=p}} (f_1 + f_2) \neq \lim_{\substack{h \to 0 \\ q' \to \infty \\ 2hq'=p}} \left(\hat{\mathbf{r}}_1 f_1 + \hat{\mathbf{r}}_2 f_2\right) , \tag{S-1.94}$$

in fact the right-hand side of (S-1.94) has also a component perpendicular to $\mathbf{r}$ that we are going to evaluate. We have

$$\sin \psi_1 = \frac{h \sin \alpha}{r_1} \simeq h \sin \alpha \left(\frac{1}{r} - \frac{h \cos \alpha}{r^2}\right) ,$$

$$\sin \psi_2 = -\frac{h \sin \alpha}{r_2} \simeq -h \sin \alpha \left(\frac{1}{r} + \frac{h \cos \alpha}{r^2}\right) , \tag{S-1.95}$$

thus the component of $\mathbf{F}$ perpendicular to $\mathbf{r}$ is, disregarding the terms in h^2 and higher,

$$F_\perp = \lim_{\substack{h \to 0 \\ q' \to \infty \\ 2hq'=p}} (f_1 \sin \psi_1 + f_2 \sin \psi_2) = k_e \frac{qp \sin \alpha}{r^3} , \tag{S-1.96}$$

in agreement with the second of (S-1.81). Note that

$$f_1 \sin \psi_1 \simeq k_e qq' \left(\frac{1}{r^2} - \frac{2h \cos \alpha}{r^3}\right) h \sin \alpha \left(\frac{1}{r} - \frac{h \cos \alpha}{r^2}\right) \simeq k_e \frac{qp \sin \alpha}{2r^3} . \tag{S-1.97}$$

As $h \to 0$ and $\psi_1 \to 0$, the angle γ approaches α, and the limit of the torque of $\mathbf{f}_1$ on the dipole is

$$\tau_1 = - \lim_{\substack{h \to 0 \\ q' \to \infty \\ 2hq'=p}} h f_1 \sin \gamma = - \lim_{\substack{h \to 0 \\ q' \to \infty \\ 2hq'=p}} k_e \frac{qq'h \sin \gamma}{r^2 + h^2 + 2rh \cos \alpha} = -k_e \frac{qp \sin \alpha}{2r^2} , \tag{S-1.98}$$

analogously, the limit of the torque of $\mathbf{f}_2$ is

$$\tau_2 = -k_e \frac{qp \sin \alpha}{2r^2} , \tag{S-1.99}$$

and the total torque on the dipole is

$$\tau = \tau_1 + \tau_2 = -k_e \frac{qp \sin \alpha}{r^2} , \tag{S-1.100}$$

in agreement with (S-1.83).

S-1.11 Electric Field of a Charged Hemispherical Surface

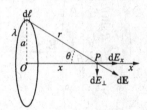

Fig. S-1.12

We start from the electric field generated by a ring of radius a and linear charge density λ in a generic point P on its axis, at a distance x from the center O of the ring. With reference to Fig. S-1.12, the infinitesimal ring arc $d\ell$ of charge $\lambda \, d\ell$ generates a field $d\mathbf{E}$ in P. The magnitude of $d\mathbf{E}$ is

$$dE = k_e \frac{\lambda \, d\ell}{a^2 + x^2}. \tag{S-1.101}$$

The field $d\mathbf{E}$ has a component dE_x parallel to the ring axis, and a component $dE_\perp$ perpendicular to the axis. We need only the parallel component

$$dE_x = \cos\theta \, dE = k_e \frac{\lambda \, d\ell}{a^2 + x^2} \frac{x}{\sqrt{a^2 + x^2}} = k_e \frac{\lambda x \, d\ell}{\left(a^2 + x^2\right)^{3/2}}, \tag{S-1.102}$$

because the perpendicular component cancels out because of symmetry when we integrate over the whole ring. When we integrate, θ and r do not depend on the position of $d\ell$, and the total field in P is

$$E_x = k_e \frac{\lambda x}{\left(a^2 + x^2\right)^{3/2}} \int\limits_0^{2\pi a} d\ell = k_e \frac{2\pi a \lambda x}{\left(a^2 + x^2\right)^{3/2}}$$

$$= \begin{cases} \dfrac{1}{4\pi\varepsilon_0} \dfrac{2\pi a \lambda x}{\left(a^2 + x^2\right)^{3/2}} & \text{SI} \\[3ex] \dfrac{2\pi a \lambda x}{\left(a^2 + x^2\right)^{3/2}} & \text{Gaussian,} \end{cases} \tag{S-1.103}$$

which can be rewritten

$$\mathbf{E} = \hat{\mathbf{x}} \, k_e \frac{Q x}{\left(a^2 + x^2\right)^{3/2}}, \tag{S-1.104}$$

where $Q = 2\pi a \lambda$ is the total charge of the ring.

The charged hemispherical surface can be divided into infinitesimal strips between "parallels" of colatitude θ and $\theta + d\theta$ with respect to the symmetry axis of the hemisphere, as in Fig. S-1.13. Each infinitesimal strip is equivalent to a charged ring of radius $R \sin\theta$ and total charge $dQ = \sigma 2\pi R^2 \sin\theta\, d\theta$. The curvature center of the hemisphere is located on the axis of the rings, at a distance $x = R \cos\theta$ from the center of each ring. Thus, the contribution of each strip to the field at the center is

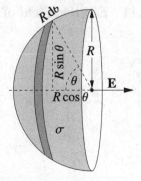

Fig. S-1.13

$$
dE = k_e \frac{\sigma 2\pi R^2 \sin\theta\, d\theta\, R \cos\theta}{\left(R^2 \sin^2\theta + R^2 \cos^2\theta\right)^{3/2}}
$$

$$
= k_e \frac{\sigma 2\pi R^3 \cos\theta \sin\theta\, d\theta}{R^3} = k_e \sigma 2\pi \cos\theta \sin\theta\, d\theta\,, \tag{S-1.105}
$$

and the total field is

$$
E = k_e \sigma 2\pi \int\limits_0^{\pi/2} \cos\theta \sin\theta\, d\theta = k_e \pi\sigma\,, \tag{S-1.106}
$$

independent of the radius R.

S-1.12   The Electric Field of an Array of Charged Wires (1)

(a) At distances $|y| \gg a$ our wire array is equivalent to an infinite charged plane with surface charge density $\sigma = \lambda/a$, thus the electric field is

$$
\mathbf{E} = \pm 2\pi k_e \sigma\, \hat{\mathbf{y}} = \pm 2\pi k_e \frac{\lambda}{a}\, \hat{\mathbf{y}} =
\begin{cases}
\pm \dfrac{\lambda}{2\varepsilon_0 a}\, \hat{\mathbf{y}}\,, & \text{(SI)}, \\[2ex]
\pm 2\pi \dfrac{\lambda}{a}\, \hat{\mathbf{y}}\,, & \text{(Gaussian)},
\end{cases}
\tag{S-1.107}
$$

where we have the plus sign for $y > 0$ and the minus sign for $y < 0$. If we set $\varphi = 0$ on the $y = 0$ plane we have

$$
\varphi(y) = -E\,|y| = -2\pi k_e \frac{\lambda}{a}\, |y| =
\begin{cases}
-\dfrac{\lambda}{2\varepsilon_0 a}\, |y|\,, & \text{(SI)}, \\[2ex]
-2\pi \dfrac{\lambda}{a}\, |y|\,, & \text{(Gaussian)}.
\end{cases}
\tag{S-1.108}
$$

(b) At distances $r \ll a$ from a wire (but $r \geq r_w$, where r_w is the radius of the wire, which we assumed to be negligible) we can neglect the contribution of the other wires to the electric field, thus

$$
\mathbf{E}(r) = 2k_e \frac{\lambda}{r} \hat{\mathbf{r}} = \begin{cases} \dfrac{\lambda}{2\pi\varepsilon_0 r} \hat{\mathbf{r}} \,, & \text{(SI)}, \\[2ex] \dfrac{2\lambda}{r} \hat{\mathbf{r}} \,, & \text{(Gaussian)}, \end{cases}
\tag{S-1.109}
$$

and the potential is

$$
\varphi(r) = -2k_e \lambda \ln\left(\frac{r}{r_0}\right) = \begin{cases} -\dfrac{\lambda}{2\pi\varepsilon_0} \ln\left(\dfrac{r}{r_0}\right) \,, & \text{(SI)}, \\[2ex] -2\lambda \ln\left(\dfrac{r}{r_0}\right) \,, & \text{(Gaussian)}, \end{cases}
\tag{S-1.110}
$$

where r_0 is an arbitrary constant, corresponding to the distance from the wire at which we set $\varphi(r_0) = 0$. If we choose a particular wire as reference, so that it is located at $x = 0$, close to the nth wire we have $r = \sqrt{\xi^2 + y^2}$, with $\xi = x - na$.

(c) Differentiating (1.17) we obtain

$$
\frac{\partial\varphi}{\partial x} = -C \frac{2k \sin kx}{2(\cosh ky - \cos kx)} \,,
$$

$$
\frac{\partial^2\varphi}{\partial x^2} = -C \frac{k^2 \cos kx \cosh ky - k^2 \cos^2 kx - k^2 \sin^2 kx}{(\cosh ky - \cos kx)^2}
$$

$$
= -C \frac{k^2 \cos kx \cosh ky - k^2}{(\cosh ky - \cos kx)^2} \,,
\tag{S-1.111}
$$

$$
\frac{\partial\varphi}{\partial y} = -C \frac{2k \sinh ky}{2(\cosh ky - \cos kx)} \,,
$$

$$
\frac{\partial^2\varphi}{\partial y^2} = -C \frac{k^2 \cosh^2 kx - k^2 \cosh ky \cos 2kx - k^2 \sinh^2 ky}{(\cosh ky - \cos kx)^2}
$$

$$
= C \frac{k^2 \cos kx \cosh ky - k^2}{(\cosh ky - \cos kx)^2} \,,
\tag{S-1.112}
$$

thus $\nabla^2\varphi = \partial^2\varphi/\partial x^2 + \partial^2\varphi/\partial y^2 = 0$. From the periodicity of the wire array we obtain

$$
k = \frac{2\pi}{a} \,.
\tag{S-1.113}
$$

At the limit $|y| \to \infty$ we have $\cosh ky \to e^{k|y|}/2$, which is dominant compared to $\cos kx$. Thus

$$
\lim_{|y|\to\infty} \varphi(x, y) = -Ck|y| \,,
\tag{S-1.114}
$$

comparing to (S-1.110) we obtain

$$C = \frac{2\pi k_e \lambda}{ka} = k_e \lambda = \begin{cases} \dfrac{\lambda}{4\pi\varepsilon_0} \,, & \text{(SI)} \\[2mm] \lambda \,, & \text{(Gaussian).} \end{cases} \tag{S-1.115}$$

Now we investigate the behavior at the limits $x \to 0$ and $y \to 0$. From the Taylor series for the cos and cosh functions we have

$$\lim_{x \to 0} \cos kx = 1 - \frac{k^2 x^2}{2} \,, \qquad \lim_{y \to 0} \cosh ky = 1 + \frac{k^2 y^2}{2} \,, \tag{S-1.116}$$

thus

$$\lim_{x,y \to 0} \varphi(x, y) = -\lim_{x,y \to 0} C \ln[2(\cosh ky - \cos kx)]$$

$$= -C \ln\left(\frac{k^2 y^2}{2} + \frac{k^2 x^2}{2}\right) = -C \ln(k^2 r^2) = -2C \ln(kr)$$

$$= -2C \ln\left(\frac{2\pi}{a} r\right) = \begin{cases} -\dfrac{\lambda}{2\pi\varepsilon_0} \ln\left(r \dfrac{2\pi}{a}\right) \,, & \text{(SI),} \\[3mm] -2\lambda \ln\left(r \dfrac{2\pi}{a}\right) \,, & \text{(Gaussian),} \end{cases}$$

$$\tag{S-1.117}$$

which coincides with (S-1.110) if we set the arbitrary constant $r_0 = 1/k = a/(2\pi)$.

S-1.13   The Electric Field of an Array of Charged Wires (2)

(a) Using the chain rule we find

$$\frac{\partial}{\partial x} = \frac{\partial \zeta}{\partial x} \frac{\partial}{\partial \zeta} + \frac{\partial \bar{\zeta}}{\partial x} \frac{\partial}{\partial \bar{\zeta}} = \frac{1}{2}\left[\frac{\partial}{\partial \zeta} + \frac{\partial}{\partial \bar{\zeta}}\right] \,, \tag{S-1.118}$$

and

$$\frac{\partial}{\partial y} = \frac{\partial \zeta}{\partial y} \frac{\partial}{\partial \zeta} + \frac{\partial \bar{\zeta}}{\partial y} \frac{\partial}{\partial \bar{\zeta}} = \frac{1}{2}\left[\frac{\partial}{\partial \zeta} - \frac{\partial}{\partial \bar{\zeta}}\right] \,. \tag{S-1.119}$$

Thus we have for the two-dimensional Laplace operator

$$\frac{\partial^2}{\partial x^2} + \frac{\partial^2}{\partial y^2} = \frac{1}{4}\left[\frac{\partial}{\partial \zeta} + \frac{\partial}{\partial \bar{\zeta}}\right]^2 - \frac{1}{4}\left[\frac{\partial}{\partial \zeta} - \frac{\partial}{\partial \bar{\zeta}}\right]^2 = \frac{\partial}{\partial \zeta} \frac{\partial}{\partial \bar{\zeta}} \,. \tag{S-1.120}$$

(b) First we see that the general solution of the equation

$$\frac{\partial}{\partial \zeta} \frac{\partial}{\partial \bar{\zeta}} F(\zeta, \bar{\zeta}) = 0 \,, \tag{S-1.121}$$

is

$$F(\zeta, \bar{\zeta}) = g(\zeta) + h(\bar{\zeta}) \tag{S-1.122}$$

with g and h arbitrary functions.[1]

We are interested in real solutions of the Laplacian in the x, y variables which can be expressed in terms of the real and of the imaginary parts of a function $f(\zeta)$ (or equivalently of $f(\bar{\zeta}) = \bar{f}(\zeta)$)

$$f_R = \mathrm{Re}\,[f(\zeta)] = \frac{[f(\zeta) + f(\bar{\zeta})]}{2} \,, \qquad f_I = \mathrm{Im}\,[f(\zeta)] = \frac{[f(\zeta) - f(\bar{\zeta})]}{2\,i} \,. \tag{S-1.123}$$

In mathematical terms f_R and f_I are called conjugate harmonic functions. Any linear combination with real coefficients of f_R and f_I is a solution of (S-1.121).

For the specific case under consideration, using the trigonometric relationships given in the text we find

$$\cosh\,(ky) - \cos\,(kx) = -2\sin\left(k\,\frac{x+iy}{2}\right)\sin\left(-k\,\frac{x-iy}{2}\right)$$

$$= 2\sin\,(k\zeta)\,\sin\,(k\bar{\zeta}) = 2\,\big|\sin\,(k\zeta)\big|^2 \,. \tag{S-1.124}$$

Then

$$\ln\,[2\,(\cosh ky - \cos kx)] = \ln\left[4|\sin\,(k\zeta)|^2\right] = 2\mathrm{Re}\left\{\log[2\sin(k\zeta)]\right\}, \tag{S-1.125}$$

where log denotes the logarithm in the complex plane: $\log\zeta = \ln|\zeta| + i\phi$ where we have used the polar representation $\zeta = |\zeta|\,e^{i\phi}$.

(c) The imaginary part is given by the phase ϕ of $\sin\,(k\zeta)$. Using the definition of $\sin\,(k\zeta)$ and its polar representation we find

$$\phi = \arctan\left[\frac{\sinh(ky/2)\cos\,(kx/2)}{\cosh\,(ky/2)\sin(kx/2)}\right] \,. \tag{S-1.126}$$

The plot of the contour lines of $\ln\left[4|\sin\,(k\zeta/2)|^2\right]$ (dashed lines) and of ϕ (solid lines) in Fig. S-1.14 shows that these lines are orthogonal.

---

[1] It may be interesting to note that a similar procedure is used in the case of a 1–D wave equation to show that the general solutions is of the form $f(x + vt) + g(x - vt)$ with v the wave propagation velocity.

This orthogonality property can be proved by computing the gradients of the the real part $[f(\zeta) + f(\bar{\zeta})]/2$ and of the imaginary part $[f(\zeta) - f(\bar{\zeta})]/(2i)$ of a generic function of complex variable $f(\zeta)$ and showing that their scalar product vanishes. Using (S-1.118) and (S-1.119) we obtain

$$\left\{ \hat{\mathbf{x}} \left[\frac{\partial f(\zeta)}{\partial \zeta} + \frac{\partial f(\bar{\zeta})}{\partial \bar{\zeta}} \right] + i\hat{\mathbf{y}} \left[\frac{\partial f(\zeta)}{\partial \zeta} - \frac{\partial f(\bar{\zeta})}{\partial \bar{\zeta}} \right] \right\} \cdot$$

$$\left\{ -i\hat{\mathbf{x}} \left[\frac{\partial f(\zeta)}{\partial \zeta} - \frac{\partial f(\bar{\zeta})}{\partial \bar{\zeta}} \right] + \hat{\mathbf{y}} \left[\frac{\partial f(\zeta)}{\partial \zeta} + \frac{\partial f(\bar{\zeta})}{\partial \bar{\zeta}} \right] \right\} = 0. \qquad \text{(S-1.127)}$$

The correspondence between the curves of constant imaginary part and the field lines of the electric field $\mathbf{E}$ can be shown directly from (S-1.127) by recognizing that the first term in braces is simply proportional to $\mathbf{E}$ and the second term in braces to the gradient of the imaginary part. Thus (S-1.127) can be read as

$$\mathbf{E} \cdot \nabla \operatorname{Im}[f(\zeta)] = 0, \qquad \text{(S-1.128)}$$

which implies that $\operatorname{Im}[f(\zeta)]$ is constant along the field lines of the electric field.

This property shows that the complex variable approach in 2D electrostatics provides us directly with both the curves of constant potential and the field lines of the electric field.

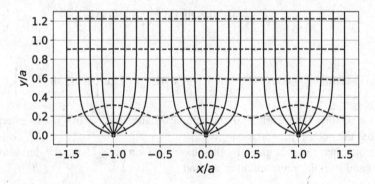

Fig. S-1.14 .

S-1.14 Mean Value Property and the Force on a Spherical Charge

(a) Let $\varphi = \varphi(r, \theta, \phi)$ be the electrostatic potential in spherical coordinates, and $\overline{\varphi}(r)$ its average over a spherical surface S of radius r centered at the origin, defined as

$$\overline{\varphi}(r) = \frac{1}{4\pi r^2} \int_S \varphi(r, \theta, \phi) \, dS = \frac{1}{4\pi r^2} \int_S \varphi(r, \theta, \varphi) r^2 \, d\Omega = \frac{1}{4\pi} \int_S \varphi(r, \theta, \phi) \, d\Omega \, .$$
(S-1.129)

Note that we are allowed to take r out of the integral because the integration is performed at constant r. Differentiating $\overline{\varphi}(r)$ with respect to r leads to

$$\frac{d\overline{\varphi}}{dr} = \frac{1}{4\pi} \int_S \left(\frac{\partial \varphi}{\partial r} \right) d\Omega = \frac{1}{4\pi} \int_S (\nabla \varphi)_r \, d\Omega \, ,$$
(S-1.130)

where we used $(\nabla\varphi)_r = \partial_r\varphi$, valid in spherical coordinates. Multiplying and dividing by r^2, reinserting $dS = r^2 d\Omega$ and introducing the vector $d\mathbf{S} = \hat{\mathbf{r}} \, dS$ we obtain

$$\frac{d\overline{\varphi}}{dr} = \frac{1}{4\pi r^2} \int_S \nabla\varphi \cdot d\mathbf{S} \, .$$
(S-1.131)

Now we apply the divergence theorem in order to convert the surface integral into a volume integral,

$$\int_S \nabla\varphi \cdot d\mathbf{S} = \int_{vol} \nabla \cdot (\nabla\varphi) \, d^3r' = \int_{vol} \nabla^2\varphi \, d^3r' = 0 \, ,$$
(S-1.132)

where the last equality follows immediately from $\nabla^2\varphi = 0$. We have thus shown that $d\overline{\varphi}/dr = 0$, i.e., that $\overline{\varphi}$ does not depend on r. Thus we are free to evaluate its value for $r \to 0$ so that $\overline{\varphi} \to \varphi(0)$. Setting $r = 0$ in the l.h.s. of (S-1.129) yields

$$\varphi(0) = \frac{1}{4\pi} \int_S \varphi(r, \theta, \phi) \, d\Omega \, ,$$
(S-1.133)

which demonstrates the mean value property.

(b) Let **r** be the position vector of the center of the charged sphere, O'', in an arbitrary coordinate system, as shown in Fig. S-1.15. The energy of the sphere in the external electric field is given by the following integral, evaluated over the volume of the sphere,

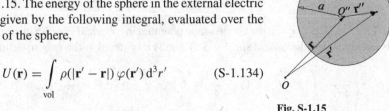

$$U(\mathbf{r}) = \int_{\text{vol}} \rho(|\mathbf{r}' - \mathbf{r}|)\,\varphi(\mathbf{r}')\,d^3r' \qquad (\text{S-1.134})$$

Fig. S-1.15

where d^3r' is the infinitesimal volume element of the sphere, $\mathbf{r}'$ its position vector in our coordinate system, and the charge density ρ depends only on $|\mathbf{r}' - \mathbf{r}| = r'' \equiv |\mathbf{r}''|$ because of the assumed spherical symmetry. Thus, using $\mathbf{r}''$ as integration variable, we can write

$$U(\mathbf{r}) = \int_{\text{vol}} \rho(r'')\,\varphi(\mathbf{r} + \mathbf{r}'')\,d^3r'' . \qquad (\text{S-1.135})$$

Introducing spherical coordinates (r'', θ'', ϕ'') with origin at the center of the sphere (O'') we rewrite the integral as

$$U(\mathbf{r}) = \int_0^a dr''\, r''^2 \rho(r'') \int d\Omega\, \varphi(\mathbf{r} + \mathbf{r}'') . \qquad (\text{S-1.136})$$

The second integral is, apart from a $1/4\pi$ factor, the angular average of φ over the sphere of radius r'', which, according to the mean value property, equals the value in $r'' = 0$:

$$\int d\Omega\, \varphi(\mathbf{r} + \mathbf{r}'') = 4\pi\varphi(\mathbf{r}) . \qquad (\text{S-1.137})$$

Since this is independent of $\mathbf{r}''$ we can take it outside of the first integral of (S-1.136). Thus, it remains to calculate

$$\int_0^a dr''\, r''^2 \rho(r'') = \frac{1}{4\pi} Q , \qquad (\text{S-1.138})$$

where Q is the total charge of the sphere. Thus the potential energy of the sphere in the external field is

$$U(\mathbf{r}) = 4\pi\varphi(\mathbf{r}) \frac{1}{4\pi} Q = Q\,\varphi(\mathbf{r}) , \qquad (\text{S-1.139})$$

and the force acting on the sphere is

$$\mathbf{F} = -\nabla U(\mathbf{r}) = -Q\,\nabla\varphi(\mathbf{r}) = Q\,\mathbf{E}(\mathbf{r}) , \qquad (\text{S-1.140})$$

where $\mathbf{E}(\mathbf{r})$ is the "external" electric field evaluated at the center of the sphere.

This result shows that an extended body, carrying a spherically-symmetric charge distribution and interacting with an external electric field, behaves like a point-like particle located at its center carrying the same total charge. This, combined with the more familiar property that the electric field generated by a spherically-symmetric charge distribution is equivalent (outside the distribution) to the field generated by a point-like particle carrying the same total charge and located at its center, allows to treat rigid charge distributions with spherical symmetry as models for point-like charges, removing the singularities of the field and potential bound to point-like charges.

Chapter S-2
Solutions for Chapter 2

S-2.1 Metal Sphere in an External Field

(a) The total electric field inside a conductor must be zero in static conditions. Thus, in the presence of an external field $\mathbf{E}_0$, the surface charge distribution of our sphere must generate a field $\mathbf{E}_{in} = -\mathbf{E}_0$ at its inside. As we found in Problem 1.1, a rigid displacement $-\boldsymbol{\delta}$ of the electron sphere (or "electron sea") with respect to the ion lattice gives origin to the internal uniform field (S-1.1)

$$\mathbf{E}_{in} = -k_e \frac{4\pi}{3} \varrho_0 \boldsymbol{\delta} , \tag{S-2.1}$$

where $\varrho_0 = e n_e$ is the charge density of the "electron sphere". The magnitude of the displacement δ is thus

$$\delta = \frac{3E_0}{4\pi k_e \varrho_0} . \tag{S-2.2}$$

For a rough numerical estimate for n_e, we can assume that each atom contributes a single conduction electron ($Z = 1$). If M is the atomic mass of our atoms, M grams of metal contain $N_A \simeq 6.0 \times 10^{23}$ atoms (Avogadro constant), and occupy a volume of M/ϱ_m cm^3, where ϱ_m is the mass density. Typical values for a metal are $M \sim 60$ and $\varrho_m \sim 8$ g/cm^3, leading to

$$n_e \sim \frac{N_A \varrho_m}{M} \sim 10^{22} \text{ cm}^{-3}, \quad \text{and} \quad \varrho_0 = e n_e \sim 5 \times 10^{12} \text{ statC/cm}^3 . \tag{S-2.3}$$

In SI units we have $n_e \sim 10^{29}$ m^{-3} and $\varrho_0 \sim 1.6 \times 10^{-10}$ C/m^3. Substituting into (S-2.2) and assuming $E_0 = 1000$ V/m (0.003 statV/cm), we finally obtain

$$\delta \sim 10^{-15} \text{ cm}. \tag{S-2.4}$$

A. Macchi et al., *Problems in Classical Electromagnetism*, https://doi.org/10.1007/978-3-031-22235-1

This value for δ is smaller by orders of magnitude than the spacing between the atoms in a crystalline lattice ($\sim 10^{-8}$ cm), therefore it makes sense to consider the charge as distributed on the surface. Formally, this is equivalent to take the limits $\delta \to 0$ and $\varrho_0 \to \infty$, keeping constant the product

$$\sigma_0 = \varrho_0 \, \delta = \frac{3E_0}{4\pi k_e} \, . \qquad (S\text{-}2.5)$$

(b) According to Problem 1.1, the field generated by the charge distribution of the metal sphere outside its volume equals the field of an electric dipole $\mathbf{p} = Q \, \boldsymbol{\delta}$, where $Q = (4\pi/3)R^3 e n_e$, located at the center of the sphere. Replacing $\boldsymbol{\delta}$ by its value of (S-2.2) we have for the dipole moment

$$\mathbf{p} = \frac{R^3}{k_e}\mathbf{E}_0 \, . \qquad (S\text{-}2.6)$$

The field outside the sphere ($r > R$) is the sum of $\mathbf{E}_0$ and the field generated by $\mathbf{p}$

$$\mathbf{E} = \mathbf{E}_0 + [3 \, (\mathbf{E}_0 \cdot \hat{\mathbf{r}}) \, \hat{\mathbf{r}} - \mathbf{E}_0] \left(\frac{R}{r}\right)^3 \, . \qquad (S\text{-}2.7)$$

(c) The external field at the surface of the sphere is obtained by replacing r by R in (S-2.7)

$$\mathbf{E}_{\text{surf}} = \mathbf{E}_0 + 3 \, (\mathbf{E}_0 \cdot \hat{\mathbf{r}}) \, \hat{\mathbf{r}} - \mathbf{E}_0 = 3 \, (\mathbf{E}_0 \cdot \hat{\mathbf{r}}) \, \hat{\mathbf{r}} \, , \qquad (S\text{-}2.8)$$

which is perpendicular to the surface, as expected. The surface charge density is

$$\sigma = \frac{1}{4\pi k_e}\mathbf{E}_{\text{surf}} \cdot \hat{\mathbf{r}} = k_e \frac{3}{4\pi}E_0 \cos\theta = \sigma_0 \cos\theta \, , \qquad (S\text{-}2.9)$$

where $\sigma_0 = 3k_e E_0/(4\pi)$, and θ is the angle between $\hat{\mathbf{r}}$ and $\mathbf{E}_0$.

S-2.2 Electrostatic Energy with Image Charges

In all cases, the conducting (half-)planes divide the whole space into two regions: one free of charges (A), and one containing electrical charges (B), as shown in Fig. S-2.1. Since the charge distribution is finite, the electric potential φ equals zero at the boundaries of both regions, i.e., on the conducting surfaces and at infinity. We can thus use the uniqueness theorem for Poisson's equation. The potential φ (and therefore the electric field $\mathbf{E}$) is uniformly equal to zero in region A. The potential problem in region B is solved if we find an *image* charge distribution, located in region A, that replicates the boundary conditions of region B. The poten-

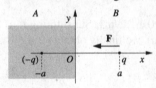

Fig. S-2.1

tial and the electric field (and thus the forces on the real charges) in region B are the same as if the image charges were real.

(a) We introduce a reference frame with the x axis perpendicular to the conducting plane, and passing through the charge. The origin, and the y and z axes, lie on the plane. The charge is thus located at $(x = a, y = 0, z = 0)$, and the potential problem for $x > 0$ is solved by placing an image charge $q' = -q$ at $(x = -a, y = 0, z = 0)$. The force on the real charge is $F = -k_e q^2/(4a^2)$. The electrostatic energy U_{es} of the system equals the work W done by the field when the real charge q moves from $x = a$ to $x = +\infty$. Simultaneously, the image charge will move from $x = -a$ to $x = -\infty$, but no additional work is needed for this, since what actually moves is the surface charge on the conducting plane, which is constantly at zero potential. Thus we have

$$U_{es} = W = \int_a^\infty F \, dx = -k_e \frac{q^2}{4} \int_a^\infty \frac{dx}{x^2} = -k_e \frac{q^2}{4a} . \tag{S-2.10}$$

This is half the electrostatic energy U_{real} of a system comprising two *real* charges, q and $-q$, at a distance $2a$ from each other. The $\frac{1}{2}$ factor is due to the fact that, if two *real* charges move to infinity in opposite directions, the work done by the field is

$$W_{real} = \int_{+a}^{+\infty} F \, dx - \int_{-a}^{-\infty} (-F) \, dx = 2 \int_{+a}^{+\infty} F_x \, dx = -k_e \frac{q^2}{2a} , \tag{S-2.11}$$

since the force acting on $-q$ is the opposite of the force acting on q, and $U_{real} = W_{real}$.

The $\frac{1}{2}$ factor can also be explained by evaluating the electrostatic energies for our system, and for the system of the two real charges. In both cases, because of the cylindrical symmetry around the x axis, the electrostatic field is a function of the longitudinal coordinate x and of the radial distance $r = \sqrt{y^2 + z^2}$ only, i.e., $\mathbf{E} = \mathbf{E}(x, r)$. In the case of the two real charges we have

$$U_{real} = \frac{1}{8\pi k_e} \int d^3 r \, \mathbf{E}^2 = \frac{1}{8\pi k_e} \int_{-\infty}^\infty dx \int_0^\infty 2\pi r \, dr \, \mathbf{E}^2(x, r)$$

$$= 2 \frac{1}{8\pi k_e} \int_0^\infty dx \int_0^\infty 2\pi r \, dr \, \mathbf{E}^2(x, r) , \tag{S-2.12}$$

since $\mathbf{E}(x, r) = -\mathbf{E}(-x, r)$, so that $\mathbf{E}^2(x, r) = \mathbf{E}^2(-x, r)$. In the case of the charge in front of a conducting plane we have

$$U_{es} = \frac{1}{8\pi k_e} \int_0^\infty dx \int_0^\infty 2\pi r \, dr \, \mathbf{E}^2(x, r) \,, \qquad (S\text{-}2.13)$$

because $\mathbf{E} = 0$ for $x < 0$ (in region A), while the field is the same as in the "real" case for $x > 0$. Thus $U_{es} = U_{real}/2$. The electrostatic energy includes both the interaction energy between the charges, U_{int}, and the "self-energy", U_{self}, of each charge. For the "real" system we have

$$U_{real} = U_{self}(q) + U_{self}(-q) + U_{int}(q, -q) = 2U_{self}(q) + U_{int}(q, -q) \,, \quad (S\text{-}2.14)$$

since $U_{self}(-q) = U_{self}(q)$. For the charge in front of the conducting plane we have

$$U_{es} = U_{self}(q) + U_{int}(q, \text{plane}) \,, \qquad (S\text{-}2.15)$$

since there is only one real charge. Actually, the self-energy U_{self} approaches infinity if we let the charge radius approach 0, but this issue is not really relevant here. In any case, the divergence may be treated by assuming an arbitrarily small, but non-zero radius for the charge. Since $U_{es} = U_{real}/2$, we also have $U_{int}(q, \text{plane}) = U_{int}(q, -q)/2$.

(b) Again, we choose a reference frame with the x axis perpendicular to the conducting plane, so that q has coordinates $(a, d/2, 0)$ and $-q$ has coordinates $(a, -d/2, 0)$ (Fig. S-2.2). The potential problem for the $x > 0$ half-space is solved by placing an image charge $-q$ at $(-a, d/2, 0)$, and an image charge q at $(-a, -d/2, 0)$. According to the arguments at the end of point **(a)**, the electrostatic energy U_{es} of our system is one half of the energy U_{real} of a system of four charges, all of them real, at the same locations. We can evaluate U_{real} by inserting the four charges one by one, each interacting only with the previously inserted charges,

$$U_{real} = k_e \left(-2\frac{q^2}{d} - 2\frac{q^2}{2a} + 2\frac{q^2}{\sqrt{d^2 + 4a^2}} \right) \,. \qquad (S\text{-}2.16)$$

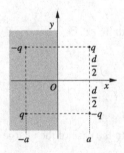

Fig. S-2.2

The same result is obtained by evaluating the work of the electric forces when the two real charges are moved to infinite distance from the plane, and infinite distance from each other. This can be done in two steps. First we move the charge at $(a, d/2, 0)$, then the charge at $(a, -d/2, 0)$. When we move the first charge, three forces are acting on it: $\mathbf{F}_1$, due to its own image, which is simultaneously moving to $-\infty$, and $\mathbf{F}_2$ and $\mathbf{F}_3$, due to the second real charge and to its image, at distances r_2 and r_3, respectively. The total work on the first charge is thus

$$W = W_1 + W_2 + W_3$$

$$= \int\limits_a^\infty \mathbf{F}_1 \cdot d\mathbf{r} + \int\limits_a^\infty \mathbf{F}_2 \cdot d\mathbf{r} + \int\limits_a^\infty \mathbf{F}_3 \cdot d\mathbf{r} , \qquad (S\text{-}2.17)$$

where $\mathbf{r}$ is the position vector of the first charge, and the first integral is the same as the integral of (S-2.10) and equals $-k_e\, q^2/(4a)$. The second integral can be rewritten, in terms of the angle θ of Fig. S-2.3,

$$W_2 = -k_e \int\limits_a^\infty \frac{q^2}{r_2^2} \sin\theta \, dx$$

$$= -k_e\, q^2 \int\limits_0^{\pi/2} \frac{\cos^2\theta}{d^2} \sin\theta \frac{d}{\cos^2\theta}\, d\theta$$

$$= -k_e \frac{q^2}{d} \int\limits_0^{\pi/2} \sin\theta \, d\theta$$

$$= -k_e \frac{q^2}{d} , \qquad (S\text{-}2.18)$$

where we have used the facts that $r_2 = d/\cos\theta$ and $dx = (d/\cos^2\theta)\, d\theta$. The third integral of (S-2.17) can be treated analogously, in terms of the angle ψ of Fig. S-2.3,

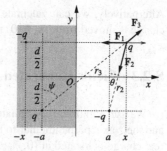

Fig. S-2.3

$$W_3 = k_e \int\limits_a^\infty \frac{q^2}{r_3^2} \sin\psi \, dx$$

$$= k_e \frac{q^2}{d} \int\limits_{\psi_0}^{\pi/2} \sin\psi \, d\psi$$

$$= k_e \frac{q^2}{\sqrt{4a^2 + d^2}} , \qquad (S\text{-}2.19)$$

where ψ_0 is the value of ψ when q is at $x = a$, i.e., $\psi_0 = \arccos(d/\sqrt{4a^2 + d^2})$. Thus, the work done by the electric field when the first charge is moved to infinity is

$$W = k_e \left(-\frac{q^2}{4a} - \frac{q^2}{d} + \frac{q^2}{\sqrt{4a^2 + d^2}} \right) . \qquad (S\text{-}2.20)$$

We must still move the second real charge to infinity, this is done in the presence of its own image charge only, and the work is $-k_e q^2/(4a)$. We finally have

$$U_{es} = W - k_e \frac{q^2}{4a} = k_e \left(-\frac{q^2}{2a} - \frac{q^2}{d} + \frac{q^2}{\sqrt{4a^2 + d^2}} \right), \qquad (S\text{-}2.21)$$

i.e., one half of the value of U_{real} of (S-2.16), as expected.

(c) We choose a reference frame with the half planes $(x = 0, y \geqslant 0)$ and $(y = 0, x \geqslant 0)$ coincid-ing with the two conducting half-planes. Thus, the real charge q is located at $(x = a, y = b, z = 0)$ (Fig. S-2.4). If we add two image charges $q'_1 = q'_2 = -q$ at $(-a, b, 0)$ and $(a, -b, 0)$, respectively, and an image charge $q'_3 = q$ at $(-a, -b, 0)$, the potential is zero on the $x = 0$ and $y = 0$ planes, and at infin-ity. This solves the potential problem in the dihedral

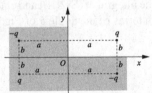

Fig. S-2.4

angle where the real charge is located. Following the discussions of points (a) and (b), the electrostatic energy of this system is one quarter of the energy of a system comprising four charges, all of them real, in the same locations, since the energy density is zero in three quarters of the whole space.

$$U_{es} = \frac{1}{4} k_e \left(\frac{-q^2}{a} - \frac{q^2}{b} + \frac{q^2}{\sqrt{b^2 + a^2}} \right). \qquad (S\text{-}2.22)$$

Alternatively, we can calculate the work done by the electric field when the real charge is moved from $(a, b, 0)$ to $(\infty, \infty, 0)$.

S-2.3   Fields Generated by Surface Charge Densities

(a) We use cylindrical coordinates (r, ϕ, z) with the origin O on the conducting plane, and the z axis perpendicular to the plane and passing through the real charge q. The real charge is located at $(0, \phi, z)$, and the image charge at $(0, \phi, -a)$, ϕ being irrel-evant when $r = 0$. The electric field on the conducting plane is perpendicular to the plane, and depends only on r. At a generic point $P \equiv (r, \phi, 0)$ on the plane the magnitude of the field $\mathbf{E}^{real}$ generated by the real charge is

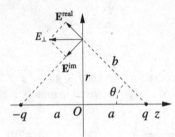

Fig. S-2.5

$$E^{real} = k_e \frac{q}{b^2} = k_e \frac{q}{a^2 + r^2}. \qquad (S\text{-}2.23)$$

The field generated at P by the image charge, $\mathbf{E}^{im}$, has the same magnitude, the same z com-ponent, but opposite r component of $\mathbf{E}^{real}$, as in Fig. S-2.5. The total electric field in P is thus perpendicular to the plane and has magnitude

$$E(r) = 2E_z^{\text{real}}(r) = -2k_e \frac{q}{a^2+r^2} \frac{a}{\sqrt{(a^2+r^2)}} = -2k_e \frac{q\,a}{(a^2+r^2)^{3/2}}. \quad \text{(S-2.24)}$$

The surface charge density is thus

$$\sigma(r) = \frac{1}{4\pi k_e} E(r) = -\frac{1}{2\pi} \frac{q\,a}{(a^2+r^2)^{3/2}}, \quad \text{(S-2.25)}$$

and the annulus between r and $r + dr$ on the conducting plain has a charge

$$dq_{\text{ind}} = \sigma\, 2\pi r\, dr = \frac{q a r\, dr}{(a^2+r^2)^{3/2}}$$

$$= -\frac{1}{2\pi} \frac{q}{a^2} \cos^3\theta\, 2\pi a^2 \tan\theta \frac{d\theta}{\cos^2\theta} = -q \sin\theta\, d\theta, \quad \text{(S-2.26)}$$

since $r = a \tan\theta$. The total induced charge on the conducting plane is

$$q_{\text{ind}} = \int_0^{\pi/2} dq_{\text{ind}} = -q \int_0^{\pi/2} \sin\theta\, d\theta = -q. \quad \text{(S-2.27)}$$

(b) In the problem of a real charge q located on the z axis, at $z = a$, in front of a conducting plane, the only real charges are q and the surface charge distribution σ on the plane. What we observe is no field in the half-space $z < 0$, while in the half-space $z > 0$ we observe a field equivalent to the field of q, plus the field of an image charge $-q$ located on the z axis at $z = -a$ (Fig. S-2.6). The field generated by the surface charge distribution alone is thus equivalent

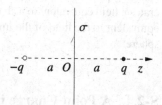

Fig. S-2.6

to the field of a charge $-q$ located at $z = +a$ in the half-space $z - 0$, and to the field of a charge $-q$ located at $z = -a$ in the half-space $z > 0$. In the half space $z < 0$, the field of the surface charge distribution and the field or the real charge cancel each other. The discontinuity of the field at $z = 0$ is due to the presence of a finite surface charge density on the conducting plane, which implies an infinite volume charge density.

(c) Let us introduce a spherical coordinate system (r, θ, ϕ) into Problem 2.4, with the origin O at the center of the conducting sphere and the z axis on the line through O and the real charge q (Fig. S-2.7). The electric potential outside the sphere, $r \geqslant a$, is obtained from (S-2.31) by replacing a by r, and q' and d' by their values of (S-2.37). We have

$$\varphi(r, \theta) = k_e \left[\frac{q}{\sqrt{r^2 + d^2 - 2dr\cos\theta}} - \frac{q\dfrac{a}{d}}{\sqrt{r^2 + \dfrac{a^4}{d^2} - 2r\dfrac{a^2}{d}\cos\theta}} \right] \quad (S\text{-}2.28)$$

independent of ϕ. The electric field at $r = a^+$, on the outer surface of the sphere, is

$$E_\perp(a^+, \theta) = - \partial_r V(r, \theta)\big|_{r=a} , \quad (S\text{-}2.29)$$

and the surface charge density on the sphere is

$$\sigma(\theta) = \frac{1}{4\pi k_e} E_\perp(a^+, \theta) . \quad (S\text{-}2.30)$$

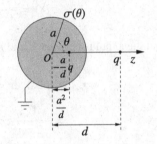

Fig. S-2.7

The actual evaluation does not pose particular difficulties, but is rather involved, and we neglect it here. But we can use the same arguments as in point (b). The only real charges of the problem are the real charge q, and the surface charge distribution of the sphere. There is no net field inside the sphere, and the field for $r > 0$ is equivalent to the field of q, plus the field of an image charge $-q\,a/d$ located at $z = a^2/d$. Thus, the surface charge distribution alone generates a field equivalent to a charge $-q$ located at $z = d$ inside the sphere, and a field equivalent to the field of the image charge $-q\,a/d$, located at $z = a^2/d$, outside the sphere.

S-2.4 A Point Charge in Front of a Conducting Sphere

(a) We have a conducting grounded sphere of radius a, and an electric charge q located at a distance $d > a$ from its center O. Again, the whole space is divided into two regions: the inside (A) and the outside (B) of the sphere. The electrostatic potential is uniformly equal to zero in region A because the sphere is grounded. We try to solve the potential problem in region B by locating an image charge q' inside the sphere, on the line through O and q, at a distance d' from the center O. The problem is solved if we can find values for q' and d' such that the electric potential φ is zero everywhere on the surface of the sphere. This would replicate the boundary conditions for region B, with $\varphi = 0$ both on the surface of the sphere and at infinity, and only the real charge q in between. Let us evaluate the potential $\varphi(P)$ at a generic point P of the sphere surface, such that the line segment OP forms an angle θ with the line segment Oq, as shown in Fig. S-2.8. We must have

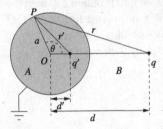

Fig. S-2.8

$$0 = \varphi(P) = k_e \left(\frac{q}{r} + \frac{q'}{r'} \right)$$

$$= k_e \left(\frac{q}{\sqrt{a^2 + d^2 - 2ad\cos\theta}} \right.$$

$$\left. + \frac{q'}{\sqrt{a^2 + d'^2 - 2ad'\cos\theta}} \right) , \quad \text{(S-2.31)}$$

where r is the distance from P to q, r' the distance from P to q', and we have used the cosine rule. We see that the sign of q' must be the opposite of the sign of q. If we take the square of (S-2.31) we have

$$q^2(a^2 + d'^2 - 2ad'\cos\theta) = q'^2(a^2 + d^2 - 2ad\cos\theta) , \quad \text{(S-2.32)}$$

which must hold for any θ. We must thus have separately

$$q^2(a^2 + d'^2) = q'^2(a^2 + d^2) , \quad \text{and} \quad \text{(S-2.33)}$$

$$2q^2 ad' \cos\theta = 2q'^2 ad \cos\theta . \quad \text{(S-2.34)}$$

Equation (S-2.34) leads to

$$q'^2 = q^2 \frac{d'}{d} , \quad q' = -q\sqrt{\frac{d'}{d}} , \quad \text{(S-2.35)}$$

which can be inserted into (S-2.33), leading to

$$dd'2 - (a^2 + d'^2)d' + a^2 d = 0 , \quad \text{(S-2.36)}$$

which has the two solutions $d' = d$ and $d' = a^2/d$. The first solution is not acceptable because it is larger than the radius of the sphere a (it actually corresponds to the trivial solution of superposing a charge $-q$ to the charge q). Thus we are left with $d' = a^2/d$, which can be substituted into (S-2.35), leading to our final solution

$$q' = q \frac{a}{d} , \quad d' = \frac{a^2}{d} . \quad \text{(S-2.37)}$$

If the sphere is isolated and has a net charge Q, the problem in region B is solved by placing an image charge q' at d', as above, and a further point charge $q'' = Q - q'$ in O, so that the potential is uniform over the sphere surface, and the total charge of the sphere is Q. The case $Q = 0$ corresponds to an uncharged, isolated sphere.

(b) The total force $\mathbf{f}$ on q equals the sum of the forces exerted on q by the image charge q' located in d', $q'' = -q'$ and Q, both located in O. Thus $\mathbf{f} = \mathbf{f}' + \mathbf{f}'' + \mathbf{f}'''$, with

$$f' = k_e \frac{qq'}{(d-d')^2} \doteq -k_e \frac{q^2 ad}{(d^2-a^2)^2} \,, \quad f'' = k_e \frac{q^2}{d^3} \,, \quad f''' = k_e \frac{qQ}{d^2} \quad \text{(S-2.38)}$$

with $f'' = f''' = 0$ if the sphere is grounded.

(c) The electrostatic energy U of the system equals the work of the electric field if the real charge q is moved to infinity. When q is at a distance x from O we evaluate the force on it by simply replacing d by x in (S-2.38). The work is thus the sum of the three terms

$$W_1 = \int_d^\infty f' \, dx = k_e \left[\frac{q^2 a}{2(x^2-a^2)} \right]_d^\infty = -k_e \frac{q^2 a}{2(d^2-a^2)} \,,$$

$$W_2 = \int_d^\infty f'' \, dx = -k_e \left[\frac{q^2 a}{2x^2} \right]_d^\infty = k_e \frac{q^2 a}{2d^2} \,,$$

$$W_3 = \int_d^\infty f''' \, dx = k_e \frac{qQ}{d} \,. \qquad\qquad \text{(S-2.39)}$$

Thus we have $U = W_1$ for the grounded sphere, $U = W_1 + W_2$ for the isolated chargeless sphere, and $U = W_1 + W_2 + W_3$ for the isolated charged sphere.

It is interesting to compare this result for the energy of the isolated chargeless sphere with the electrostatic energy U_{real} of a system comprising three real charges q, q', and $-q'$, located in d, d' and O, respectively:

$$U_{\text{real}} = k_e \sum_{i<j} \frac{q_i q_j}{r_{ij}} = k_e \left(\frac{qq'}{d-d'} - \frac{qq'}{d} - \frac{q'^2}{d'} \right)$$

$$= k_e \left(-\frac{q^2 a}{d^2-a^2} + \frac{q^2 a}{d^2} - \frac{q^2}{d} \right) . \qquad \text{(S-2.40)}$$

We see that U is obtained from U_{real} by halving the interaction energies of the real charge with the two image charges, and neglecting the interaction energy between the two image charges.

S-2.5   Dipoles and Spheres

(a) We consider the case of the grounded sphere first, so that its potential is zero. We can treat the dipole as a system of two point charges $\pm q$, separated by a distance $2h$ as in Fig. S-2.9. Eventually, we shall let q approach ∞, and h approach zero, with the product $p = 2hq$ remaining constant. Following Problem 2.4, the two charges induce two images

$$\pm q' = \mp q \, \frac{a}{\sqrt{d^2 + h^2}}, \qquad \text{(S-2.41)}$$

respectively, each at a distance

$$d' = \frac{a^2}{\sqrt{d^2 + h^2}} \qquad \text{(S-2.42)}$$

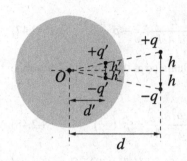

from the center of the sphere O, each lying on the straight line passing through O and the corresponding real charge. Since we are interested in the limit $h \to 0$ (thus, $h \ll d$), we can use the approximations

$$\pm q' \simeq \mp q \, \frac{a}{d}, \quad \text{and} \quad d' = \frac{a^2}{d} \,. \text{(S-2.43)}$$

Fig. S-2.9

The two image charges are separated by a distance

$$2h' = 2h \, \frac{d'}{d} = h \left(\frac{a}{d}\right)^2, \qquad \text{(S-2.44)}$$

so that the moment of the image dipole is

$$\mathbf{p}' = 2q'\mathbf{h}' = -2q\mathbf{h} \left(\frac{a}{d}\right)^3 = -\mathbf{p} \left(\frac{a}{d}\right)^3. \qquad \text{(S-2.45)}$$

The image dipole is antiparallel to the real dipole, i.e., the two dipoles lie on parallel straight lines, but point in opposite directions. The sum of the image charges, which equals the total induced charge on the sphere surface, is zero. Therefore this solution is valid also for an isolated uncharged sphere.

(b) Also in this case, we consider the grounded sphere first. Again, the dipole can be treated as a system of two charges $\pm q$, separated by a distance $h = p/q$. This time the charge $+q$ is at distance d from the center of the sphere O, while $-q$ is at distance $d + h$ (Fig. S-2.10).

Thus, the images q' of $+q$, and q'' of $-q$, have different absolute values, and are located at different distances from O, d' and d'', respectively. We have

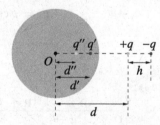

Fig. S-2.10

$$q' = -q\frac{a}{d} = -\frac{p}{h}\frac{a}{d}, \quad d' = \frac{a^2}{d}, \tag{S-2.46}$$

$$q'' = +q\frac{a}{d+h} = +\frac{p}{h}\frac{a}{d+h}, \quad d'' = \frac{a^2}{d+h}.$$

The absolute values of q' and q'' remain different from each other also at the limits $h \to 0, q \to \infty$, so that a net image charge q''' is superposed to the image dipole

$$q''' = \lim_{h\to 0}(q' + q'') = \lim_{h\to 0} -p\frac{a}{h}\frac{h}{d(d+h)} = -p\frac{a}{d^2}. \tag{S-2.47}$$

The moment of the image electric dipole can be calculated as the limit of the absolute value of q' times $(d' - d'')$

$$p' = \lim_{h\to 0}|q'|\,(d' - d'') = \lim_{h\to 0} \frac{p}{h}\frac{a}{d}\frac{a^2 h}{d(d+h)} = p\left(\frac{a}{d}\right)^3, \tag{S-2.48}$$

the same result is obtained by evaluating the limit of $q''(d' - d'')$. Thus the real dipole **p** and the image dipole **p**′ lie on the same straight line and point in the same direction. The image dipole is located at a distance a^2/d from O.

Since a net charge q''' is needed to have zero potential on the surface of the sphere, this solution is valid only in the case of a grounded sphere. The solution for an isolated uncharged sphere requires an image charge $-q''' = +pa/d^2$ at the center of the sphere, so that the total image charge is zero and the surface of the sphere is equipotential.

(c) We start from the case of the grounded sphere, and use a Cartesian reference frame with the origin located at the center of the sphere, O, the x axis passing through the dipole **p**, and the y axis lying in the plane of the dipole. We denote by θ the angle between the electric dipole **p** and the x axis, as in Fig. S-2.11. We can decompose the dipole into the vector sum of its x and y components

$$\mathbf{p}_x = p\cos\theta\,\hat{\mathbf{x}}, \quad \text{and} \quad \mathbf{p}_y = p\sin\theta\,\hat{\mathbf{y}}. \tag{S-2.49}$$

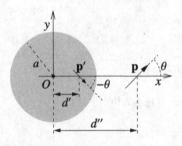

Both components generate images located on the x axis at a distance $d' = a^2/d$ from O. According to (**a**) and (**b**), $\mathbf{p}_y$ and $\mathbf{p}_y$ generate the images

$$\mathbf{p}_y' = -\left(\frac{a}{d}\right)^3 p\sin\theta\,\hat{\mathbf{y}}$$

$$\mathbf{p}_x' = \left(\frac{a}{d}\right)^3 p\cos\theta\,\hat{\mathbf{x}}, \tag{S-2.50}$$

Fig. S-2.11

resulting in an image dipole $\mathbf{p}'$, of modulus $p' = p\,(a/d)^3$, forming an angle $-\theta$ with the x axis, and superposed to a net charge $q''' = +p\cos\theta\,(a/d^2)$, since now it is the "tail" of $\mathbf{p}_x$ which points toward O. In the case of an isolated uncharged sphere, we must add a point charge $-q'''$ in O, so that the net charge of the sphere is zero.

S-2.6   Coulomb's Experiment

(a) The zeroth-order solution is obtained by neglecting the induction effects, considering the charges as uniformly distributed over the surfaces of the two spheres. Thus, at zeroth order, the force between the two spheres equals the force between two point charges, each equal to Q, located at their centers. In order to evaluate higher-order solutions, it is convenient to introduce the dimensionless parameter $\alpha = (a/r) < 1$, where a is the radius of the two spheres, and r the distance between their centers. The solution of order n is obtained by locating inside each sphere a point charge q of the same order of magnitude as Q at its center, plus increasingly smaller point charges q', q'', $\ldots$, $q^{(n)}$ at appropriate positions, with orders of magnitude $|q'| \sim \alpha Q$, $|q''| \sim \alpha^2 Q$, $\ldots$, $|q^{(n)}| \sim \alpha^n Q$. The charges must obey the normalization condition $q + q' + q'' + \cdots + q^{(n)} = Q$.

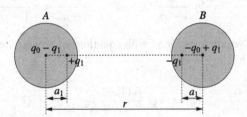

Fig. S-2.12 .

At the first order, the point charge q at the center of each sphere induces an image charge $q' = -\alpha q$ inside the other sphere, located at a distance $a' = a^2/r = r\alpha^2$ from its center (see Problem 2.4), as shown in Fig. S-2.12. Thus, by solving the simultaneous equations $q' = -\alpha q$, and $q + q' = Q$, we obtain for the values of the two charges

$$q = \frac{1}{1-\alpha}\,Q\,, \quad q' = -\frac{\alpha}{1-\alpha}\,Q\,. \tag{S-2.51}$$

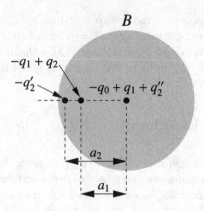

Fig. S-2.13 .

At the second order, the first-order charge q' inside each sphere induces an image charge q'' inside the other sphere, located a distance a'' from its center, as shown in Fig. S-2.13. Since the distance of q' from the center of the other sphere is $r - a' = r(1 - \alpha^2)$, we have

$$q'' = -q' \frac{a}{r - a'} = -q' \frac{\alpha}{1 - \alpha^2}, \qquad a'' = \frac{a^2}{r - a'} = r \frac{\alpha^2}{1 - \alpha^2}. \qquad \text{(S-2.52)}$$

Combining the above equation for q'' with equations $q' = -\alpha q$ and $q + q' + q'' = Q$, we finally obtain

$$q = Q \frac{1 - \alpha^2}{1 - \alpha + \alpha^3}, \quad q' = -Q \frac{\alpha(1 - \alpha^2)}{1 - \alpha + \alpha^3}, \quad q'' = Q \frac{\alpha^2}{1 - \alpha + \alpha^3} \text{(S-2.53)}$$

Higher order approximations are obtained by iterating the procedure. Thus we obtain a sequence of image charges $q, q', q'', q''', \ldots$ inside each sphere. At each iteration, the new image charge is of the order of α times the charge added at the previous iteration. Therefore, the smaller the value of $\alpha = a/r$, the sooner one may truncate the sequence obtaining a good approximation.

(b) We obtain the first order approximation of the force between the two spheres by considering only the charges of (S-2.51) for each sphere. To this approximation, the force between the spheres is the sum of four terms. The first term is the force between the two zeroth-order charges q, at a distance r from each other. The second and third terms are the forces between the zeroth order charge q of one sphere and the first-order charge q' of the other. The distance between these charges is $r - a' = r(1 - \alpha^2)$. The fourth term is the force between the two first-order charges q', at a distance $r - 2a' = r(1 - 2\alpha^2)$ from each other. Summing up all these contributions we obtain

$$F = k_e \frac{Q^2}{r^2} \frac{1}{(1-\alpha)^2} \left[1 - \frac{2\alpha}{(1-\alpha^2)^2} + \left(\frac{\alpha}{1-2\alpha^2} \right)^2 \right]. \qquad \text{(S-2.54)}$$

From the Taylor expansion, valid for $x < 1$,

$$\frac{1}{(1-x)^2} = 1 + 2x + 3x^2 + 4x^3 + O(x^4), \qquad \text{(S-2.55)}$$

we obtain, to the fourth order,

$$\frac{1}{(1-\alpha^2)^2} = 1 + 2\alpha^2 + 3\alpha^4 + O(\alpha^6), \qquad \text{(S-2.56)}$$

and

$$\frac{1}{(1-2\alpha^2)^2} = 1 + 4\alpha^2 + 12\alpha^4 + O(x^6), \qquad \text{(S-2.57)}$$

so that

$$F = k_e \frac{Q^2}{r^2} \left(1 + 2\alpha + 3\alpha^2 + 4\alpha^3 + \cdots \right) \left(1 - 2\alpha + \alpha^2 - 4\alpha^3 + \cdots \right)$$

$$= k_e \frac{Q^2}{r^2} \left[1 - 4\alpha^3 + O(\alpha^4) \right], \qquad \text{(S-2.58)}$$

since all the terms of order α and α^2 vanish. The first non vanishing correction to the "Coulomb" force is thus at the third order in a/r,

$$F = k_e \frac{Q^2}{r^2} \left(1 - 4 \frac{a^3}{r^3} \right). \qquad \text{(S-2.59)}$$

This result can be interpreted in terms of multipole expansions of the charge distributions of the spheres. The first two multipole moments of the charge distribution of each sphere are a monopole equal to the total charge Q, and an electric dipole $\mathbf{p} = -q'a'\hat{\mathbf{r}} = -(\alpha Q)(\alpha^2 r)\hat{\mathbf{r}} = -\alpha^3 Qr\,\hat{\mathbf{r}}$, with $\hat{\mathbf{r}}$ pointing toward the center of the opposite sphere. The contribution of the monopole moments to the total force is $F_{\text{mm}} = k_e\, Q^2/r^2$. Now we need the force exerted by the monopole terms of each sphere on the dipole term of the other. The monopole of, say, the left sphere generates a field $\mathbf{E}^{(0)} = k_e\, Q/r^2$ at the center of the right sphere. We can consider the dipole moment of the right sphere as the limit for $h \to 0$ of two charges, $-q'$ located at $r - h$ from the center of the left sphere, and q' located at r, with $q'h = |p|$. The force between the left monopole and the right dipole is thus

$$F_{\text{md}} = \lim_{h \to 0} k_e \, Qq' \left[-\frac{1}{(r-h)^2} + \frac{1}{r^2} \right] \simeq \lim_{h \to 0} k_e \, Qq' \left(-\frac{1}{r^2} - \frac{2h}{r^3} + \frac{1}{r^2} \right)$$

$$= -\lim_{h \to 0} k_e \, Qq' \frac{2h}{r^3} = -2 \, k_e \frac{Qp}{r^3} = -2 \, k_e \, a^3 \frac{Q^2}{r^3} \,, \tag{S-2.60}$$

where we have used the first-order Taylor expansion of $(r-h)^{-2}$. Adding the force between the right monopole and the left dipole, the total force is thus

$$F = F_{\text{mm}} + 2F_{\text{md}} = k_e \frac{Q^2}{r^2} \left(1 - 4 \frac{a^3}{r^3} \right) \,, \tag{S-2.61}$$

in agreement with (S-2.59). The same result can be obtained by applying the formula for the force between a point charge and an electric dipole, $\mathbf{F} = (\mathbf{p} \cdot \nabla)\mathbf{E}$. See also Problem 1.10 on this subject.

From (S-2.59) we find that a ratio $a/r \simeq 0.13$ is enough to reduce the systematic deviation from the pure inverse-square law below 1%.

S-2.7 A Solution Looking for a Problem

(a) The total electric potential in a point of position vector $\mathbf{r}$ is the sum of the dipole potential and of the potential of the external uniform electric field,

$$\varphi(\mathbf{r}) = k_e \frac{\mathbf{p} \cdot \mathbf{r}}{r^3} - Ez = k_e \frac{p \cos \theta}{r^2} - Er \cos \theta \,, \tag{S-2.62}$$

where θ is the angle between $\mathbf{r}$ and the z axis. Note that it is not possible to take the reference point for the electrostatic potential at infinity, since the potential of our uniform electric field diverges for $z \to \pm\infty$. Thus we have chosen $\varphi = 0$ on the xy plane, which is an equipotential surface both for the dipole and for the uniform electric field. Now we look for a possible further equipotential surface on which $\varphi = 0$. On this surface we must have

$$\varphi = k_e \frac{p \cos \theta}{r^2} - Er \cos \theta = 0 \,, \tag{S-2.63}$$

and, in addition to the solution $\theta = \pi/2$, corresponding to the xy plane, we have the θ-independent solution

$$r = k_e^{1/3} \left(\frac{p}{E} \right)^{1/3} \equiv R \,, \tag{S-2.64}$$

corresponding to a sphere of radius R. Note that the two equipotential surfaces intersect each other on the circumference $x^2 + y^2 = R^2$ on the $z = 0$ plane. This is possible because the electric field of the dipole on the intersection circumference is

$$\mathbf{E}_{\text{dip}} = k_e \frac{3(\mathbf{p} \cdot \hat{\mathbf{r}}) \hat{\mathbf{r}} - \mathbf{p}}{r^3} = -k_e \frac{\mathbf{p}}{R^3} = -\mathbf{E} \,, \tag{S-2.65}$$

so that the total field on the circumference is zero, i.e., the only field that can be perpendicular to both equipotential surfaces.

(b) We must find a solution for the potential φ that satisfies the condition $\varphi = 0$ at the surface of the conducting sphere, i.e. $\varphi(|\mathbf{r}| = a) = 0$, and such that at large distance from the conductor the field is $\mathbf{E}_0$.

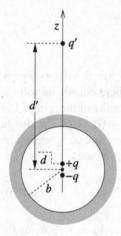

According to point **(a)**, the field outside the sphere must equal $\mathbf{E}_0$ plus the field of an electric dipole $\mathbf{p}_i$, parallel to $\mathbf{E}_0$ and located at the center of the sphere. The moment of the dipole is obtained by substituting $R = a$ into (S-2.64),

$$\mathbf{p}_i = k_e^{-1} a^3 \, \mathbf{E}_0 = \frac{3}{4\pi k_e} V_a \mathbf{E}_0 \,, \tag{S-2.66}$$

where V_a is the volume of the sphere. The potential for $r \geqslant a$ is thus

$$\varphi = k_e \frac{\mathbf{p}_i \cdot \mathbf{r}}{r^3} - E_0 z \,, \tag{S-2.67}$$

while $\varphi = 0$ for $\leqslant a$. The total charge induced on the sphere is zero, so that the solution is the same for a grounded and for an isolated, uncharged sphere. The solution is identical to the one obtained in Problem 2.1 via a different (heuristic) approach.

Fig. S-2.14

(c) For the dipole at the center of a spherical conducting cavity, the boundary condition is $\varphi = 0$ at $r = b$. The polarization charges on the inner surface must generate a uniform field $\mathbf{E}_i$ parallel to $\mathbf{p}_0$ and, according to (S-2.64), of intensity

$$E_i = k_e \frac{p_0}{b^3} = k_e \frac{4\pi p_0}{3 V_b} = k_e \frac{p_0}{b^3} \,. \tag{S-2.68}$$

As in the preceding case, the total induced charge is zero and thus it does not matter whether the shell is grounded, or isolated and uncharged.

(d) We can think of the dipole as a system of two point charges $\pm q$, respectively located at $z = \pm d$, with $p = 2qd$, as in Fig. S-2.14. According to the method of the image charges, the charge $+q$ modifies the charge distribution of the inner surface of the shell, so that it generates a field inside the sphere, equivalent to the field of an image charge $q' = -qb/d$, located at $z = d' = b^2/d$. Also the presence of the charge $-q$ affects the surface charge distribution, so that the total field inside the shell is the sum of the fields of the two real charges, plus the field of two image charges $\mp qb/d$ located at $z = \pm b^2/d$, respectively. Letting $d \to 0$ and $q \to \infty$, keeping the product $2qd = p$ constant, the field of the real charges approaches the field of a dipole $\mathbf{p} = p\hat{\mathbf{z}}$ located at the center of the shell, while the field of the image charges approaches a uniform field. Let us evaluate the field of the image charges at the center of the shell:

$$E_c = 2k_e \frac{qb}{d} \left(\frac{d}{b^2}\right)^2 = k_e \frac{2qd}{b^3} = k_e \frac{p}{b^3} , \qquad (S\text{-}2.69)$$

in agreement with the result of point **(c)**. The method of the image charges can also be used to obtain the result of point **(b)**.

S-2.8 Electrically Connected Spheres

(a) To the zeroth order in a/d and b/d, we assume the surface charges to be uniformly distributed. The electrostatic potential generated by each sphere outside its volume is thus equal to the potential of a point charge located at the center of the sphere. Let us denote by Q_a and Q_b the charges on each sphere, with $Q_a + Q_b = Q$. The charge on the wire is negligible because we have assumed that its capacitance is negligible. The electrostatic potentials of the spheres with respect to infinity are

$$V_a \simeq k_e \frac{Q_a}{a} , \qquad V_b \simeq k_e \frac{Q_b}{b} , \qquad (S\text{-}2.70)$$

respectively. Since the spheres are electrically connected, $V_a = V_b \equiv V$. Solving for the charges we obtain

$$Q_a \simeq Q \frac{a}{a+b} , \qquad Q_b \simeq Q \frac{b}{a+b} , \qquad (S\text{-}2.71)$$

so that $Q_a > Q_b$.

(b) From the results of point **(a)** it follows

$$V \simeq k_e \frac{Q}{a+b} , \qquad C \simeq \frac{a+b}{k_e} . \qquad (S\text{-}2.72)$$

(c) The electric fields at the sphere surfaces are

$$E_a \simeq k_e \frac{Q_a}{a^2} = k_e \frac{Q}{a(a+b)} , \qquad E_b \simeq k_e \frac{Q_b}{b^2} = k_e \frac{Q}{b(a+b)} , \qquad (S\text{-}2.73)$$

with $E_b > E_a$. At the limit $b \to 0$ we have $E_a \to k_e Q/a^2$, while $E_b \to \infty$.

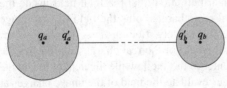

(d) We proceed as in Problem 2.6. To zeroth order, we consider the field of each sphere outside its volume as due to a point charge at the sphere center. We denote by q_a and q_b the values of these point charges.

Fig. S-2.15

To the first orders in a/d and b/d, we consider that each zeroth-order charge induces an image charge inside the other sphere, with values

$$q'_a = -q_b \frac{a}{d}, \qquad q'_b = -q_a \frac{b}{d}, \tag{S-2.74}$$

at distances a^2/d and b^2/d from the centers, respectively (Fig. S-2.15). At each successive order, we add the images of the images added at the previous order. This leads to image charges of higher and higher orders in a/d and b/d.

Up to the first order, we thus have four point charges with the condition $q_a + q_b + q'_a + q'_b = Q$. A further condition is that the potentials at the sphere surfaces are

$$V_a \simeq k_e \frac{q_a}{a}, \qquad V_b \simeq k_e \frac{q_b}{b}, \tag{S-2.75}$$

since, at the surface of each sphere, the potentials due to the external zeroth-order charge and to the internal first-order charge cancel each other. Finally, we must have $V_a = V_b$, because the spheres are connected by the wire, so that

$$q_a \simeq \frac{Q}{1 + b/a - 2b/d}, \qquad q_b \simeq \frac{Q}{1 + a/b - 2a/d}. \tag{S-2.76}$$

S-2.9   A Charge Inside a Conducting Shell

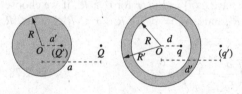

Fig. S-2.16

(a) Let us first recall Problem 2.4, now with a point charge Q at a distance a from the center O of a conducting, grounded sphere, of radius $R < a$. We introduce a spherical coordinate system, with the origin in O. We shall need only the radial coordinate r.

We have seen that the boundary conditions for $r \geqslant R$ are replicated by locating an image charge $Q' = Q\,(a/R)$ inside the sphere, at a distance $a' = R^2/a$ from O, on the line joining O and Q. In the present case we are dealing with the reverse problem, and we can obtain the solution in the region $r \leqslant R$ by reversing the roles of the real and image charges. The real charge q is now inside the cavity of a spherical conducting, grounded shell of internal radius R, at a distance $d < R$ from the center O. The boundary conditions inside the cavity are replicated by locating an *external* image charge $q' = q\,(R/d)$ at a distance $d' = R^2/d$ from O, on the straight line through O and q, as in Fig. S-2.16. Thus, the electric potential inside the cavity equals the sum of the potentials of q and q'. The potential φ in the region $R \leqslant r \leqslant R'$ is constant because here we are inside a conductor in static conditions, and equal to zero because the shell is grounded. We have $\varphi \equiv 0$ also for $r \geqslant R'$, because $\varphi = 0$ both on the spherical surface at $r = R$, and at infinity, and there are no charges in between.

(b) The force between q and the shell equals the Coulomb force between q and its image charge q', and is attractive

$$F = k_e \frac{qq'}{(d' - d)^2} = -k_e \frac{q^2 R d}{(R^2 - d^2)^2}. \tag{S-2.77}$$

(c) Let us consider a spherical surface of radius R'', centered in O, with $R < R'' < R'$. The flux of the electric field through this closed surface is zero, because the field is zero everywhere inside a conductor. The total charge inside the sphere must thus be zero according to Gauss's law. This implies that the charge induced on the inner surface of the shell is $-q$, as may be verified directly by calculating the surface charge and integrating over the whole surface.

(d) The electric potential must still be constant for $R \leqslant r \leqslant R'$, but it is no longer constrained to be zero. The electric potential in the region $r \leqslant R$ is still equivalent to the potential generated by the charges q and q' of point **(a)**, plus a constant quantity φ_0 to be determined. The electric field in the region $R \leqslant r \leqslant R'$ is still zero, so that the potential is constant and equal to φ_0. Since the total charge on the shell must be zero, we must distribute a charge q over its external surface, of radius R', to compensate the charge $-q$ distributed over the internal surface, of radius R. Since the real charge q, and the charge $-q$ distributed over the surface of radius R generate a constant potential for $r \geqslant R$, the charge q must be distributed uniformly over the external surface in order to keep the total potential constant in the region $R \leqslant r \leqslant R'$.

The potential in the region $r \geqslant R'$ is equivalent to the potential generated by a point charge q located in O. Thus we have $\varphi(r) = k_e q/r$ for $r \geqslant R'$, if we choose $\varphi(\infty) = 0$. Thus $\varphi_0 = \varphi(R') = k_e q/R'$, and $\varphi(r) = \varphi_0$ for $R \leqslant r \leqslant R'$. For $r \leqslant R$ we have

$$\varphi(r) = k_e \left(\frac{q}{r_q} + \frac{q'}{r_{q'}} \right) + \varphi_0, \tag{S-2.78}$$

where r_q is the distance of the point from the real charge q, and $r_{q'}$ is the distance of the point from the image charge q'. The field inside the cavity is the same for a grounded or for an isolated shell.

S-2.10 A Charged Wire in Front of a Cylindrical Conductor

(a) We have $r = \sqrt{(x + a)^2 + y^2}$ and $r' = \sqrt{(x - a)^2 + y^2}$, x and y being the coordinates of Q. Thus, squaring the equation $r/r' = K$ we get

$$\frac{(x+a)^2 + y^2}{(x-a)^2 + y^2} = K^2$$

$$x^2 + 2ax + a^2 + y^2 = K^2x^2 - 2K^2ax + K^2a^2 + K^2y^2$$

$$-(x^2 + y^2)(K^2 - 1) + 2ax(K^2 + 1) = a^2(K^2 - 1)$$

$$x^2 + y^2 - 2\frac{K^2 + 1}{K^2 - 1}ax = a^2. \tag{S-2.79}$$

On the other hand, the equation of a circumference centered at $(x_0, 0)$ and radius R is

$$(x - x_0)^2 + y^2 = R^2$$

$$x^2 + y^2 - 2x_0x = R^2 - x_0^2. \tag{S-2.80}$$

Comparing (S-2.80) to (S-2.79) we see that the curves defined by the equation $r/r' = K$ are circumferences centered at

$$x_0(K) = \frac{K^2 + 1}{K^2 - 1}a, \quad y_0 = 0, \tag{S-2.81}$$

of radius

$$R(K) = \frac{2K}{|K^2 - 1|}a. \tag{S-2.82}$$

Note that

$$x_0\left(\frac{1}{K}\right) = -x_0(K), \quad \text{and} \quad R\left(\frac{1}{K}\right) = R(K). \tag{S-2.83}$$

Thus, we may restrict ourselves to $K > 1$, so that $x_0(K) > a > 0$, and omit the absolute-value sign in the expression for $R(K)$. The circumferences corresponding to $0 < K < 1$ are obtained by reflection across the y axis of the circumferences corresponding to $1/K$.

(b) According to Gauss's law, the electrostatic field and potential generated by an infinite straight wire with linear charge density λ are

$$E(r) = 2k_e\frac{\lambda}{r} = \begin{cases} \dfrac{\lambda}{2\pi\varepsilon_0 r}, \\[2mm] \dfrac{2\lambda}{r}, \end{cases} \quad \varphi(r) = -2k_e\lambda\ln\left(\frac{r}{r_0}\right) = \begin{cases} -\dfrac{\lambda}{2\pi\varepsilon_0}\ln\left(\dfrac{r}{r_0}\right), & \text{(SI)}, \\[2mm] -2\lambda\ln\left(\dfrac{r}{r_0}\right), & \text{(Gaussian)}, \end{cases}$$
$$\tag{S-2.84}$$

where r is the distance from the wire and r_0 an arbitrary constant, corresponding to the distance at which we pose $\varphi = 0$. The potential generated by two parallel wires of charge densities λ and $-\lambda$, respectively, is

$$\varphi = -2k_e\lambda\ln\left(\frac{r}{r_0}\right) + 2k_e\lambda\ln\left(\frac{r'}{r_0'}\right) = 2k_e\lambda\ln\left(\frac{r'}{r}\right) + 2k_e\lambda\ln\left(\frac{r_0}{r_0'}\right), \tag{S-2.85}$$

where r_0' is a second arbitrary constant, analogous to r_0. The term $\ln(r_0)/r_0')$ is actually a single arbitrary constant, which we can set equal to zero. With this choice the electrostatic potential is zero on the $x = 0$ plane of a Cartesian reference frame where the two wires lie on the straight lines $(x = -a, y = 0)$ and $(x = +a, y = 0)$. The equation for the equipotential surfaces in this reference frame is

$$2k_e\lambda \ln\left(\frac{r'}{r}\right) = \varphi,$$ (S-2.86)

which leads to

$$\frac{r}{r'} = e^{-\varphi/(2k_e\lambda)} = \begin{cases} e^{-2\pi\varepsilon_0\varphi/\lambda}, & \text{(SI)}, \\ e^{-\varphi/(2\lambda)}. & \text{(Gaussian)}. \end{cases}$$ (S-2.87)

Thus we can substitute $K = e^{-\varphi/(2k_e\lambda)}$ into (S-2.81) and (S-2.82). We see that the equipotential surfaces are infinite cylindrical surfaces whose axes have the equations

$$x_0(\varphi) = \frac{e^{-\varphi/(ke\lambda)} + 1}{e^{-\varphi/k_e\lambda)} - 1}, \quad y_0 = 0,$$ (S-2.88)

and their radii are

$$R(\varphi) = \frac{2\,e^{-\varphi/(2k_e\lambda)}}{|e^{-\varphi/(k_e\lambda)} - 1|}\,a.$$ (S-2.89)

By multiplying the numerators and denominators of the above expressions by $e^{\varphi/(2k_e\lambda)}$ we finally obtain

$$x_0(\varphi) = \frac{e^{-2\varphi/(2k_e\lambda)} + e^{\varphi/(2k_e\lambda)}}{e^{-\varphi(2k_e\lambda)} - e^{\varphi/(2k_e\lambda)}}\,a = -a\,\coth\left(\frac{\varphi}{2k_e\lambda}\right)$$ (S-2.90)

and

$$R(\varphi) = \frac{2}{|e^{-\varphi/(2k_e\lambda)} - e^{\varphi/(2k_e\lambda)}|}\,a = \left|\frac{a}{\sinh(\varphi/(2k_e\lambda))}\right|.$$
                                                                                  (S-2.91)

If the negative wire is located on the $(x = -a, y = 0)$ straight line, the $\varphi > 0$ equipotential cylinders are located in the $x < 0$ half space ($r < r'$ in Fig. 2.8, and the $\varphi < 0$ equipotentials in the $x > 0$ half space.

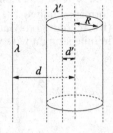

Fig. S-2.17

(c) We can solve the problem by locating an image wire with charge density $\lambda' = -\lambda$ inside the cylinder (Fig. S-2.17). In Fig. 2.8, let the real wire intersect the xy plane at $P \equiv (-a, 0)$, and the image wire at $P' \equiv (a, 0)$. The surface of the conducting cylinder intersects the xy plane on one of the circumferences $r/r' = K$. This is always possible as far as $d > R$. With these locations of the real and image wires the potential of the cylinder surface is constant and equal to a certain value φ_0. Given R and d, we can find the values of a and d' by first defining the dimensionless constant $\varphi' = \varphi_0/(2k_e\lambda)$, and then solving the simultaneous equations

$$2a + d' = d, \quad a + d' = x_0 = a \coth \varphi', \quad \frac{a}{\sinh \varphi'} = R. \tag{S-2.92}$$

From the first equation we obtain $a = (d - d')/2$, which we substitute into the other two equations

$$\frac{d + d'}{2} = \frac{d - d'}{2} \coth \varphi', \quad \frac{d - d'}{2} = R \sinh \varphi', \tag{S-2.93}$$

and the latter equation leads to

$$\sinh \varphi' = \frac{d - d'}{2R}, \tag{S-2.94}$$

independent of λ. From the relations

$$\cosh^2 x - \sinh^2 x = 1, \quad \text{and} \quad \coth x = \frac{\cosh x}{\sinh x},$$

we obtain

$$\coth \varphi' = \frac{\sqrt{4R^2 + (d - d')^2}}{d - d'}, \tag{S-2.95}$$

which, substituted into the first of (S-2.92) leads to

$$\frac{d + d'}{2} = \frac{d - d'}{2} \frac{\sqrt{4R^2 + (d - d')^2}}{d - d'}. \tag{S-2.96}$$

Disregarding the trivial solution $d' = d$ (corresponding to two superposed wires of linear charge density λ and $-\lambda$, generating zero field in the whole space), we have

$$d' = \frac{R^2}{d}, \quad a = \frac{d^2 + R^2}{2d}, \quad \varphi' = \text{arccosh}\left(\frac{d^2 + 3R^2}{d^2 + R^2}\right). \tag{S-2.97}$$

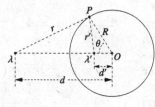

Fig. S-2.18

Alternatively, we may proceed analogously to the well-known problem of the potential of a point charge in front of a grounded, conducting sphere (Problem 2.4).

Figure S-2.18 shows the intersection with the xy plane of the conducting cylinder of radius R, the real charged wire at distance d from the cylinder axis, and the image wire at distance d' from the axis. We have translational symmetry perpendicularly to the figure. The potential φ generated by the real wire of linear charge density λ, and by the image wire of linear of linear charge density λ' must be constant over the cylinder surface. The potential at a generic point P of the surface is

$$\varphi = -2k_e\lambda \ln r - 2k_e\lambda' \ln r' = \text{const} \qquad (S\text{-}2.98)$$

where r is the distance of P from the real wire and r' the distance of P from the image wire. dividing by $-2k_e$ we obtain

$$\lambda \ln r + \lambda \ln r' = \text{const}, \qquad (S\text{-}2.99)$$

which can be rewritten by expressing r and r' in terms of d, d', R and the angle θ between r and the radius joining P to the intersection of the cylinder axis with the xy plane, O, and applying the law of cosines,

$$\lambda \ln \left(\sqrt{d^2 + R^2 - 2Rd\cos\theta}\right) + \lambda' \ln \left(\sqrt{d'^2 + R^2 - 2Rd'\cos\theta}\right) = \text{const}. \qquad (S\text{-}2.100)$$

Differentiating with respect to θ we obtain

$$\frac{\lambda Rd\sin\theta}{d^2 + R^2 - 2Rd\cos\theta} = -\frac{\lambda' Rd'\sin\theta}{d'^2 + R^2 - 2Rd'\cos\theta}, \qquad (S\text{-}2.101)$$

implying that λ and λ' must have opposite signs. Dividing both sides by $R\sin\theta$ we obtain, after some algebra,

$$\lambda d \left(d'^2 + R^2 - 2Rd'\cos\theta\right) = -\lambda'd' \left(d^2 + R^2 - 2Rd\cos\theta\right)$$

$$\lambda \left(dd'^2 + dR^2 - 2Rdd'\cos\theta\right) = -\lambda' \left(d'd^2 + d'R^2 - 2Rdd'\cos\theta\right), \qquad (S\text{-}2.102)$$

which requires $\lambda' = -\lambda$ in order to make the equation independent of θ, and, disregarding the trivial solution $d' = d$, we finally obtain

$$d' = \frac{R^2}{d}. \qquad (S\text{-}2.103)$$

S-2.11 Hemispherical Conducting Surfaces

(a) We choose a cylindrical coordinate system (r, ϕ, z) with the symmetry axis of the problem as z axis, so that the point charge is located in $(a\sin\theta, \phi, a\cos\theta)$, with ϕ a given fixed angle, as in Fig. S-2.19. The conductor surface, comprising the hemispherical boss and the plane part, is equipotential with $\varphi = 0$. If the conductor surface were simply plane, with no boss, the problem would be solved by locating an image charge $q_1 = -q$ in $(a\sin\theta, \phi, -a\cos\theta)$, as in Fig. S-2.19. On the other

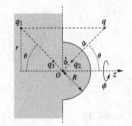

Fig. S-2.19

hand, if the conductor were a grounded spherical surface of radius R, the problem would be solved by locating an image charge $q_2 = -q(R/a)$ in $(a' \sin\theta, \phi, a' \cos\theta)$, with $a' = R^2/a$. The real charge q, with the two image charges q_1 and q_2, gives origin to a potential $\varphi(\mathbf{r}) = \varphi_q(\mathbf{r}) + \varphi_{q_1}(r) + \varphi_{q_2}(r)$ which is different from zero both on the plane surface, where it equals $\varphi_{q_2}(r)$, since $\varphi_q(\mathbf{r}) + \varphi_{q_1}(r) = 0$ on the plane, and on the hemispherical surface, where it equals $\varphi_{q_1}(r)$. The problem is solved by adding a third image charge $q_3 = q(R/a)$ at $(a' \sin\theta, \phi, -a' \cos\theta)$, so that the pairs $\{q, q_1\}$ and $\{q_2, q_3\}$ generate a potential $\varphi = 0$ on the plane surface, and the pairs $\{q, q_2\}$ and $\{q_1, q_3\}$ generate a potential $\varphi = 0$ on the spherical (and hemispherical!) surface. According to Gauss's law, the total charge induced on the conductor equals the sum of the image charges

$$q_{\text{ind}} = q_1 + q_2 + q_3 = -q + \left(-\frac{R}{a}q\right) + \left(\frac{R}{a}q\right) = -q. \qquad \text{(S-2.104)}$$

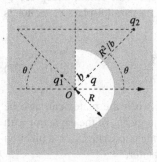

Fig. S-2.20

Note that, since the electric field generated by the real charge plus the three image charges is always perpendicular to the conductor surface, it must be zero on the circumference $(R, \phi, 0)$, here with ϕ any, where the hemisphere joins the plane.

(b) Now the real charge q is located at $(b \sin\theta, \phi, b \cos\theta)$ inside the hemispherical cavity of radius $R > b$ in the conductor, as in Fig. S-2.20. The solution is analogous to the solution of point **(a)**: we locate three image charges in the conductor, outside of the cavity, namely, $q_1 = -q$ in $(b \sin\theta, \phi, -b \cos\theta)$, $q_2 = -(R/b)q$ in $(b' \sin\theta, \phi, b' \cos\theta)$, with $b' = R^2/b > R$, and $q_3 = -q_2 = (R/b)q$ in $(b' \sin\theta, \phi, -b' \cos\theta)$ (Fig. S-2.21).

S-2.12   The Force Between the Plates of a Capacitor

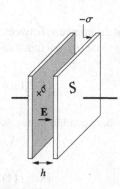

Fig. S-2.21

We present this simple problem in order to point out, and prevent, two typical recurrent errors. The first error regards the electrostatic pressure at the surface of a conductor, the second the derivation of the force from the energy of a system.

(a) Let us consider the electrostatic pressure first. If Q is the charge of the capacitor, and S the surface of its plates, the surface charge density (which is located on the *inner* surfaces only!) is $\pm\sigma = \pm Q/S$. Within our approximations, the electric field is uniform between the two charged surfaces, $E = 4\pi k_e \sigma$, and zero everywhere else. This leads to an electrostatic pressure

$$P = \frac{1}{2}\sigma E = 2\pi k_e \sigma^2 . \qquad (S\text{-}2.105)$$

Here, the typical mistake is to forget the $\frac{1}{2}$ factor and to write $P \overset{w}{=} \sigma E$ (the "w" on the "=" sign stands for wrong!). In fact, only one half of the electric field is due to the charge on the other plate. The force F is attractive because the two plates have opposite charges, and we can write

$$F = -PS = -2\pi k_e \frac{Q^2}{S^2} S = -2\pi k_e \frac{Q^2}{S} . \qquad (S\text{-}2.106)$$

Thus the force depends on Q only, and is independent of the distance h between the plates. Equation (S-2.106) is valid both for an isolated capacitor, and for a capacitor connected to a voltage source maintaining a fixed potential difference V. But, in the latter case, the charge is no longer constant, and it is convenient to replace Q by the product CV, remembering that the capacity of a parallel-plate capacitor is $C = S/(4\pi k_e h)$. Thus

$$F = -2\pi k_e \frac{(CV)^2}{S} = -\frac{V^2 S}{8\pi k_e h^2} . \qquad (S\text{-}2.107)$$

(b) In the case of an isolated capacitor, the force between the plates can also be evaluated as minus the derivative of the electrostatic energy U_{es} of the capacitor with respect to the distance between the plates, h. It is convenient to write U_{es} as a function of the charge Q, which is constant for an isolated capacitor,

$$U_{es} = \frac{Q^2}{2C} = 2\pi k_e \frac{Q^2 h}{2S} , \qquad (S\text{-}2.108)$$

so that the force between the plates is

$$F = -\partial_h U_{es} = -2\pi k_e \frac{Q^2}{S} , \qquad (S\text{-}2.109)$$

in agreement with (S-2.106).

If the capacitor is connected to a voltage source, the potential difference V between the plates is the constant quantity. Thus, it is more convenient to write U_{es} as a function of V

$$U_{es} = \frac{1}{2}CV^2 = \frac{1}{8\pi k_e} \frac{V^2 S}{h} . \qquad (S\text{-}2.110)$$

At this point, it is tempting, but wrong, to evaluate the force between the plates as minus the derivative of U_{es} with respect to h. We would get

$$F \overset{w}{=} -\partial_h U_{es} = +\frac{1}{8\pi k_e} \frac{V^2 S}{h^2} , \qquad (S\text{-}2.111)$$

and, if the "+" sign were correct, now the force would be repulsive, although equal in magnitude to (S-2.107)! Of course, this cannot be true, since the plates have opposite charges and attract each other. The error is that the force equals minus the gradient of the potential energy of an *isolated system*, which now includes also the voltage source. And the voltage source has to do some work to keep the potential difference of the capacitor constant while the capacity is changing. Let us consider an infinitesimal variation of the plate separation, dh which leads to an infinitesimal variation of the capacity, dC. The voltage source must move a charge $dQ = VdC$ across the potential difference V, in order to keep V constant. The source thus does a work

$$dW = VdQ = V^2dC, \tag{S-2.112}$$

and its internal energy (whatever its nature: mechanical, chemical, ...) must change by the amount

$$dU_{source} = -dW = -V^2dC. \tag{S-2.113}$$

Since at the same time the electrostatic energy of the capacitor changes by $\frac{1}{2}V^2dC$, the variation of the *total* energy of the isolated system, dU_{tot}, is

$$dU_{tot} = dU_{source} + dU_{es} = -V^2dC + \frac{V^2}{2}dC = -\frac{V^2}{2}dC = -dU_{es}. \tag{S-2.114}$$

Thus, the force is

$$F = -\partial_h U_{tot} = +\partial_h U_{es} = -\frac{V^2S}{8\pi k_e h^2}, \tag{S-2.115}$$

in agreement with (S-2.107).

S-2.13   Electrostatic Pressure on a Conducting Sphere

(a) The surface charge is $\sigma = Q/S$, where $S = 4\pi a^2$ is the surface of the sphere. The electric field at the surface is $E = 4\pi k_e \sigma$, so that the pressure is

$$P = \frac{1}{2}\sigma E = 2\pi k_e \sigma^2 = k_e \frac{Q^2}{8\pi a^4}. \tag{S-2.116}$$

(b) According to Gauss's law, the electric field of the sphere is

$$E(r) = \begin{cases} 0, & r < a, \\ k_e \dfrac{Q}{r^2}, & r > a, \end{cases} \tag{S-2.117}$$

and thus the electrostatic energy is

$$U_{es} = \int \frac{1}{8\pi k_e} E^2(\mathbf{r})\, d^3\mathbf{r} = \int\limits_a^\infty \frac{k_e}{8\pi} \left(\frac{Q}{r^2}\right)^2 4\pi r^2\, dr = k_e \frac{Q^2}{2a}. \tag{S-2.118}$$

The derivative of U_{es} with respect to a, which has the dimensions of a force, can be interpreted as the integral of the electrostatic pressure over the surface of the sphere. Since the pressure is uniform for symmetry reasons, we can write

$$P = \frac{1}{4\pi a^2}\left(-\frac{dU_{es}}{da}\right) = \frac{1}{4\pi a^2}\frac{k_e Q^2}{2a^2} = k_e \frac{Q^2}{8\pi a^4}, \tag{S-2.119}$$

in agreement with (S-2.116).

(c) This problem is equivalent to locating a charge Q on the sphere, such that the potential difference between the sphere and infinity is V. The problem can also be seen as a spherical capacitor with internal radius a and external radius b, potential difference V, at the limit of b approaching infinity (Fig. S-2.22). The capacity is

$$C = \lim_{b\to\infty} \frac{1}{k_e}\frac{ab}{b-a} = \frac{a}{k_e}, \tag{S-2.120}$$

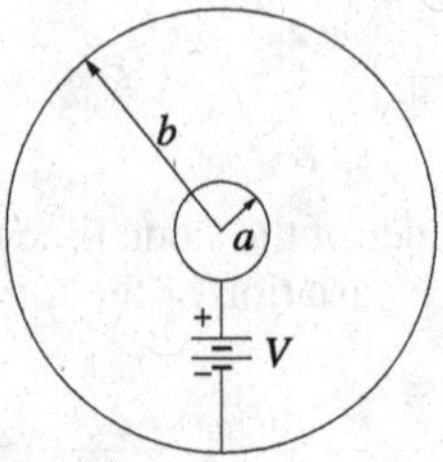

Fig. S-2.22

while the electric potential inside the capacitor is

$$\varphi(r) = \begin{cases} V & (r < a) \\ V\dfrac{a}{r} & (r > a) \end{cases} \tag{S-2.121}$$

so that the charge on the sphere of radius a is $Q = aV/k_e$. By substituting Q in (S-2.116) we obtain

$$P = \frac{V^2}{8\pi k_e a^2}. \tag{S-2.122}$$

Alternatively, we can write the electrostatic energy (S-2.118) as a function of V,

$$U_{es} = \frac{1}{2}CV^2 = \frac{aV^2}{2k_e}, \tag{S-2.123}$$

and remember from Problem 2.12 that, if the radius a is increased by da at constant voltage, the electrostatic energy of our "capacitor" changes by dU_{es}, and, simultaneously, the voltage source does a work $dW = 2dU_{es}$, so that the variation of the "total" energy is

$$dU_{tot} = dU_{es} - dW = -dU_{es} , \qquad (S\text{-}2.124)$$

and the pressure is

$$P = \frac{1}{4\pi a^2} \left(-\frac{dU_{tot}}{da} \right) = \frac{1}{4\pi a^2} \left(\frac{dU_{es}}{da} \right) = \frac{V^2}{8\pi k_e a^2} , \qquad (S\text{-}2.125)$$

in agreement with (S-2.122).

S-2.14   Conducting Prolate Ellipsoid

(a) Let us consider a line segment of length $2c$, of uniform linear electric charge density λ, so that the total charge of the segment is $Q = 2c\lambda$. We start using a system of cylindrical coordinates (r, ϕ, z), such that the end points of the segment have coordinates $(0, \phi, \pm c)$, the value of ϕ being irrelevant when $r = 0$. The electric potential $\varphi(P)$ of a generic point P, of coordinates (r, ϕ, z), is

$$\varphi(P) = k_e \int_{-c}^{+c} \frac{\lambda \, dz'}{s} = k_e \lambda \int_{-c}^{+c} \frac{dz'}{\sqrt{(z - z')^2 + r^2}} , \qquad (S\text{-}2.126)$$

where s is the distance from P to the point of the charged segment of coordinate z', as shown in Fig. S-2.23. The indefinite integral is

$$\int \frac{dz'}{\sqrt{(z - z')^2 + r^2}}$$

$$= -\ln \left[2\sqrt{(z - z')^2 + r^2} + 2z - 2z' \right]$$

$$+ C , \qquad (S\text{-}2.127)$$

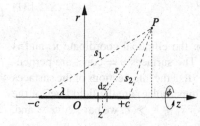

Fig. S-2.23

as can be checked by evaluating the derivative, leading to

$$\varphi(P) = k_e \lambda \ln \left[\frac{\sqrt{(z+c)^2 + r^2} + z + c}{\sqrt{(z-c)^2 + r^2} + z - c} \right] = k_e \frac{Q}{2c} \ln \left(\frac{s_1 + z + c}{s_2 + z - c} \right), \quad \text{(S-2.128)}$$

where $s_1 = \sqrt{(z+c)^2 + r^2}$ and $s_2 = \sqrt{(z-c)^2 + r^2}$ are the distances of P from the end points of the charged line segment, as shown in Fig. S-2.23. We now introduce the elliptic coordinates u and v

$$u = \frac{s_1 + s_2}{2c}, \quad v = \frac{s_1 - s_2}{2c}, \quad \text{(S-2.129)}$$

so that

$$s_1 = c(u + v), \quad s_2 = c(u - v),$$

and

$$uv = \frac{s_1^2 - s_2^2}{4} = \frac{z}{c}. \quad \text{(S-2.130)}$$

Because of (S-2.129), we have $u \geq 1$, and $-1 \leq v \leq 1$. The surfaces $u = \text{const}$ are confocal ellipsoids of revolution, and the surfaces $v = \text{const}$ are confocal hyperboloids of revolution, as shown in Fig. S-2.24. The surface $u = 1$ is the degenerate case of an ellipsoid with major radius $a = c$ and minor radius $b = 0$, coinciding with segment $(-c, c)$. The surface $v = 0$ is the degenerate case of the plane $z = 0$, while $v = \pm 1$ correspond to the degenerate cases of hyperboloids collapsed to the half-lines $(c, +\infty)$ and $(-c, -\infty)$. In terms of u and v, Eq. (S-2.128) becomes

$$\varphi(P) = k_e \frac{Q}{2c} \ln \left[\frac{c(u+v) + cuv + c}{c(u-v) + cuv - c} \right] = k_e \frac{Q}{2c} \ln \left[\frac{(u+1)(v+1)}{(u-1)(v+1)} \right]$$

$$= k_e \frac{Q}{2c} \ln \left(\frac{u+1}{u-1} \right), \quad \text{(S-2.131)}$$

Thus, the electric potential depends only on the elliptical coordinate u, and is constant on the ellipsoidal surfaces $u = \text{const}$. The surfaces $v = \text{const}$ are perpendicular to the equipotential surfaces $u = \text{const}$, so that the intersections of the surfaces $v = \text{const}$ with the planes $\phi = \text{const}$ (confocal hyperbolae) are the field lines of the electric field. If we let u approach infinity, i.e., for $s_1 + s_2 \gg c$, we have $s_1 \simeq s_2$ and

$$\frac{u+1}{u-1} \simeq 1 + \frac{2}{u},$$

$$\ln\left(1 + \frac{2}{u}\right) \simeq \frac{2}{u}, \qquad \text{(S-2.132)}$$

and

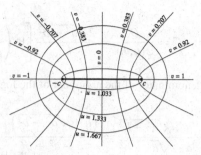

Fig. S-2.24

$$\lim_{u \to \infty} \varphi(P) = k_e \frac{Q}{2c} \frac{2}{u}$$

$$= k_e \frac{Q}{cu} \simeq k_e \frac{Q}{s_1}, \qquad \text{(S-2.133)}$$

since $s_1 \simeq s_2$. This is what expected for a point charge. In other words, the ellipsoidal equipotential surfaces approach spheres as $u \to \infty$.

(b) For a prolate ellipsoid of revolution of major and minor radii a and b, respectively, the distance between the center O and a focal point, c, is (Fig. S-2.25)

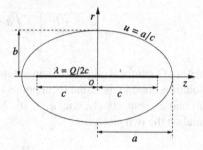

Fig. S-2.25

$$c = \sqrt{a^2 - b^2}. \qquad \text{(S-2.134)}$$

At each point of the surface of the ellipsoid we have $s_1 + s_2 = 2a$, so that the equation of the surface in elliptic coordinates is $u = a/c$. A uniformly charged line segment with end points at $(0, \phi, -c)$ and $(0, \phi, c)$, and linear charge density $\lambda = Q/(2c)$, generates a constant electric potential $\varphi(a, b)$ on the surface of the ellipsoid

$$\varphi(a, b) = k_e \frac{Q}{2c} \ln\left(\frac{u+1}{u-1}\right) = k_e \frac{Q}{2\sqrt{a^2 - b^2}} \ln\left(\frac{a + \sqrt{a^2 - b^2}}{a - \sqrt{a^2 - b^2}}\right). \qquad \text{(S-2.135)}$$

On the other hand, the potential generated by the charged segment at infinity is zero, and there are no charges between the surface of the ellipsoid and infinity. The flux of the electric field through any closed surface containing the ellipsoid is Q. Thus, the potential, and the electric field, generated by the charged segment outside the surface of the ellipsoid equal the potential, and the electric field, generated by the conducting ellipsoid carrying a charge Q, and this solves the problem. The capacity of the ellipsoid is thus

$$C = \frac{Q}{\varphi(a,b)} = \frac{2\sqrt{a^2 - b^2}}{k_e} \left[\ln\left(\frac{a + \sqrt{a^2 - b^2}}{a - \sqrt{a^2 - b^2}} \right) \right]^{-1}. \tag{S-2.136}$$

The denominator of the argument of the logarithm can be rationalized, leading to

$$\frac{a + \sqrt{a^2 - b^2}}{a - \sqrt{a^2 - b^2}} = \frac{\left(a + \sqrt{a^2 - b^2} \right)^2}{a - a^2 + b^2} = \left(\frac{a + \sqrt{a^2 - b^2}}{b}, \right)^2 \tag{S-2.137}$$

and the capacity of the prolate ellipsoid can be rewritten

$$C = \frac{\sqrt{a^2 - b^2}}{k_e} \left[\ln\left(\frac{a + \sqrt{a^2 - b^2}}{b} \right) \right]^{-1}. \tag{S-2.138}$$

The plates of a confocal ellipsoidal capacitor are the surfaces of two prolate ellipsoids of revolution, sharing the same focal points located at $\pm c$ on the z axis, and of major radii a_1 and a_2, respectively, with $a_1 < a_2$. According to (S-2.134) and (S-2.135) the potential on the two plates are

$$\varphi_{1,2} = k_e \frac{Q}{2c} \ln\left(\frac{a_{1,2} + c}{a_{1,2} - c} \right) \tag{S-2.139}$$

so that the capacity is

$$C = \frac{Q}{\varphi_1 - \varphi_2} = \frac{2c}{k_e \ln\left(\dfrac{a_1 + c}{a_1 - c} \dfrac{a_2 - c}{a_2 + c} \right)} = \frac{2c}{k_e \ln\left(\dfrac{a_1 a_2 - c^2 + c(a_2 - a_1)}{a_1 a_2 - c^2 - c(a_2 - a_1)} \right)}. \tag{S-2.140}$$

(c) A straight wire of length h and diameter $2b$, with $h \gg b$, can be approximated by an ellipsoid prolate in the extreme, with major radius $a = h/2$ and minor radius b, with, of course, $b \ll a$. From

$$\sqrt{a^2 - b^2} \simeq a - \frac{b^2}{a} \quad \text{valid for } b \ll a, \tag{S-2.141}$$

and (S-2.138) we have

$$C_{\text{wire}} \simeq \frac{a}{2k_e \ln(2a/b)} = \frac{h}{k_e \ln(h/b)}. \tag{S-2.142}$$

S-2.15 A Non-coaxial Cylindrical Capacitor

Let us assume a charge $+Q$ on the inner cylinder and a charge $-Q$ on the outer cylinder, corresponding to linear charge denisities $\pm\lambda = \pm Q/h$, respectively. The two cylindrical plates are equipotential surfaces with no charge between them and, within our approximations, according to Gauss's theorem the field is zero both inside the smaller cylinder of radius R_1 and outside the larger cylinder of radius R_2. For solving the problem in the region between the two cylindrical shells we can locate an image infinite straight-line wire of linear charge density $+\lambda$ inside the smaller cylinder and an image infinite straight-line wire of linear charge density $-\lambda$ outside the larger cylinder, as shown in Fig. S-2.26, so that the two cylindrical plates correspond to equipotential surfaces. In fact, in Problem 2.10 we saw that two infinite parallel straight wires of linear charge densities $-\lambda$ and λ, respectively, parallel to the z axis of a Cartesian reference frame, and intersecting the $z = 0$ plane at $(x = -a, y = 0)$ and $(x = +a, y = 0)$ generate cylindrical equipotential surfaces of axes

$$x(\varphi) = a \coth\left(\frac{\varphi}{2k_e\lambda}\right), \quad y(\varphi) = 0, \quad \text{(S-2.143)}$$

and radii

$$R(\varphi) = \frac{a}{|\sinh[\varphi/(2k_e\lambda)]|}, \quad \text{(S-2.144)}$$

where φ is the electrostatic potential characterizing the surface. With our choice for the signs of λ the potential is positive for $x > 0$ and negative for $x < 0$. In Fig. S-2.26 we know λ, the radii R_1 and R_2 of the two cylindrical equipotential surfaces and the distance c between their axes, and we must determine the quantities φ_1, φ_2 and a. Then the capacity will be

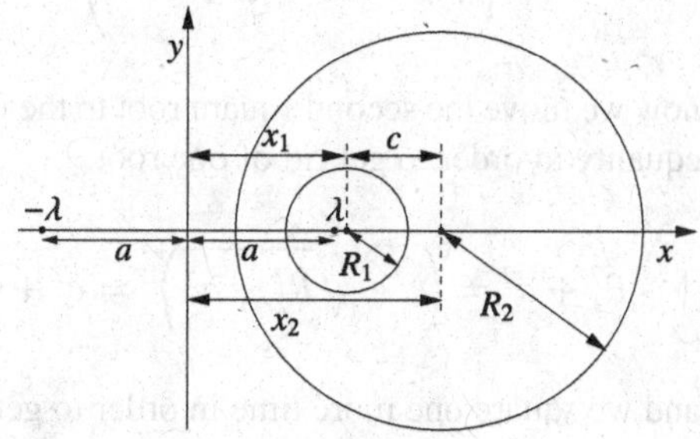

Fig. S-2.26

$$C = \frac{h\lambda}{\varphi_1 - \varphi_2} = \frac{h}{2k_e\left(\varphi_1' - \varphi_2'\right)}, \quad \text{(S-2.145)}$$

where, for convenience, we have introduced the two dimensionless quantities $\varphi_1' = \varphi_1/(2k_e\lambda)$ and $\varphi_2' = \varphi_2/(2k_e\lambda)$. After substituting φ_1' and φ_2' also into (S-2.143) and (S-2.144) we must solve the system of three equations

$$\frac{a}{\sinh \varphi_1'} = R_1 , \qquad \text{(S-2.146)}$$

$$\frac{a}{\sinh \varphi_2'} = R_2 , \qquad \text{(S-2.147)}$$

$$a \left(\coth \varphi_1' - \coth \varphi_2' \right) = c , \qquad \text{(S-2.148)}$$

in the three unknowns φ_1', φ_2' and a. We can solve the system by substitution. From (S-2.146) and Sect. A.3 of the Appendix we have

$$\sinh \varphi_1' = \frac{a}{R_1} , \quad \cosh \varphi_1' = \sqrt{\frac{a^2}{R_1^2} + 1} , \quad \coth \varphi_1' = \sqrt{\frac{R_1^2}{a^2} + 1} , \qquad \text{(S-2.149)}$$

analogously we have from (S-2.147)

$$\sinh \varphi_2' = \frac{a}{R_2} , \quad \cosh \varphi_2' = \sqrt{\frac{a^2}{R_2^2} + 1} , \quad \coth \varphi_2' = \sqrt{\frac{R_2^2}{a^2} + 1} , \qquad \text{(S-2.150)}$$

substituting the hyperbolic cotangents into (S-2.148) we obtain

$$c = a \left(\sqrt{\frac{R_1^2}{a^2} + 1} - \sqrt{\frac{R_2^2}{a^2} + 1} \right) = \sqrt{R_1^2 + a^2} - \sqrt{R_2^2 + a^2} , \qquad \text{(S-2.151)}$$

now we move the second square root to the left side and square both members of the equality in order to get rid of one root

$$R_1^2 + a^2 = \left(c + \sqrt{R_2^2 + a^2} \right)^2 = c^2 + R_2^2 + a^2 + 2c\sqrt{R_2^2 + a^2} , \qquad \text{(S-2.152)}$$

and we square one more time in order to get rid of the last square root

$$\left(R_2^2 - R_1^2 + c^2 \right)^2 = 4c^2 R_2^2 + 4c^2 a^2 ,$$

$$\left(R_2^2 - R_1^2 + c^2 \right)^2 - 4c^2 R_2^2 = 4c^2 a^2 , \qquad \text{(S-2.153)}$$

from which we finally obtain

$$a = \frac{\sqrt{\left(R_2^2 - R_1^2 + c^2 \right)^2 - 4c^2 R_2^2}}{2c} . \qquad \text{(S-2.154)}$$

From (S-2.144) we obtain for φ_1'

$$\varphi_1' = \text{arsinh}\, \frac{a}{R_1} = \text{arsinh}\, \frac{\sqrt{\left(R_2^2 - R_1^2 + c^2\right)^2 - 4c^2 R_2^2}}{2cR_1}$$

$$= \ln\left[\frac{\sqrt{\left(R_2^2 - R_1^2 + c^2\right)^2 - 4c^2 R_2^2}}{2cR_1} + \frac{R_2^2 - R_1^2 - c^2}{2cR_1}\right] \quad \text{(S-2.155)}$$

where we have used (A.21) of the Appendix. Analogously we obtain for φ_2'

$$\varphi_2' = \text{arsinh}\, \frac{a}{R_2} = \text{arsinh}\, \frac{\sqrt{\left(R_2^2 - R_1^2 + c^2\right)^2 - 4c^2 R_2^2}}{2cR_2}$$

$$= \ln\left[\frac{\sqrt{\left(R_2^2 - R_1^2 + c^2\right)^2 - 4c^2 R_2^2}}{2cR_2} + \frac{R_2^2 - R_1^2 + c^2}{2cR_2}\right] \quad \text{(S-2.156)}$$

Thus we have for the potential difference between the plates

$$\varphi_1' - \varphi_2' = \ln\left[\frac{R_2}{R_1}\left(\frac{\sqrt{\left(R_2^2 - R_1^2 + c^2\right)^2 - 4c^2 R_2^2} + R_2^2 - R_1^2 - c^2}{\sqrt{\left(R_2^2 - R_1^2 + c^2\right)^2 - 4c^2 R_2^2} + R_2^2 - R_1^2 + c^2}\right)\right],$$
$$\text{(S-2.157)}$$

we can rationalize the denominator by multiplying numerator and denominator by $\left[\sqrt{\left(R_2^2 - R_1^2 + c^2\right)^2 - 4c^2 R_2^2} - (R_2^2 - R_1^2 + c^2)\right]$, obtaining

$$\varphi_1' - \varphi_2' = \ln\left[\frac{\sqrt{\left(R_2^2 - R_1^2 + c^2\right)^2 - 4c^2 R_2^2} + R_2^2 + R_1^2 - c^2}{2R_1 R_2}\right]. \quad \text{(S-2.158)}$$

The capacity is thus

$$C = \frac{h}{2k_e \ln\left\{\left[\sqrt{\left(R_2^2 - R_1^2 + c^2\right)^2 - 4c^2 R_2^2} + R_2^2 + R_1^2 - c^2\right]/(2R_1 R_2)\right\}}. \quad \text{(S-2.159)}$$

Note that the limit for $c \to 0$ is

$$\lim_{c \to 0} C = \frac{h}{2k_e \ln(R_2/R_1)}, \quad \text{(S-2.160)}$$

as expected for a coaxial cylindrical capacitor. The other limit of interest is when the two cylindrical surfaces tend to be tangent to each other, i.e., $c \to (R_2 - R_1)$. We have

$$\lim_{c \to (R_2-R_1)} a = 0 , \quad \lim_{c \to (R_2-R_1)} \varphi'_1 = \lim_{c \to (R_2-R_1)} \varphi'_2 = 0 , \quad \lim_{c \to (R_2-R_1)} C = \infty ,$$

$$(\text{S-2.161})$$

as we could expect from the analogy with a parallel-plate capacitor, whose capacity, $C = S/(4\pi k_e d)$, where S is the surface of the plates and d their separation, tends to infinity as d tends to zero. Further, the distance between the two image charged wires, $2a$, tends to zero as $c \to (R_2 - R_1)$, while λ and $-\lambda$ are constant, thus the potential tends to zero everywhere because $(r_+ - r_-) \to 0$, r_+ being the distance of the point where we evaluate the potential from the positive wire, and r_- its distance from the negative wire.

S-2.16   Induced Charge Density on a Conducting Plane

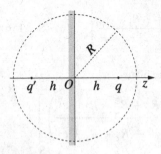

Fig. S-2.27

(a) Let us consider a cylindrical coordinate system (ρ, ϕ, z) with the z axis perpendicular to the conducting plane, the origin on the plane and passing through the point charge q, as shown in Fig. S-2.27. We know that the electrostatic potential in the $z < 0$ half space is uniformly zero, while the problem of the potential in the $z > 0$ half-space is solved by locating an image charge $q' = -q$ on the z axis at $z = -h$. In fact, if we assume to remove the conducting plane and to locate a real charge $-q$ at $z = -h$, the potential is zero on the $z = 0$ plane.

Thus the electric field $\mathbf{E}$ is uniformly zero in the $z < 0$ half space and equal to the vector sum of the fields generated by q and its image q' in the $z > 0$ half-space As a closed surface for applying Gauss's law we consider a sphere of radius R centered at the origin O of our coordinate system at the limit $R \to \infty$. We know that the flux of the electric field through any closed surface S is

$$\Phi(\mathbf{E}) = \oint_S \mathbf{E} \cdot d\mathbf{S} = 4\pi k_e Q , \qquad (\text{S-2.162})$$

where Q is the total electric charge inside the closed surface. In the case of our sphere the contribution to the flux of the surface located in the $z < 0$ half space is zero because the electric field is zero. At the limit $R \to \infty$ on the part of the spherical surface located in the $z > 0$ half space we have an electric dipole field because the distance $2h$ between q and q' is constant, thus

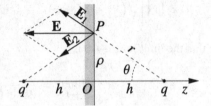

Fig. S-2.28

$2h \ll R$. Thus the electric field on the surface tends to zero as $1/R^3$ while the surface tends to infinity as R^2. As a result $\Phi(\mathbf{E})$ tends to zero. The charges enclosed by the sphere are the real charge q and the charge induced on the conducting plane, which thus must be $-q$.

(b) The electric field $\mathbf{E}$ at a point P located on the $z = 0^+$ plane at a distance ρ from the origin is the vector sum of the field $\mathbf{E}_1$ generated by q and the field $\mathbf{E}_2$ "generated" by its image q', as shown in Fig. S-2.28. For the magnitudes of $\mathbf{E}_1$ and $\mathbf{E}_2$ we have

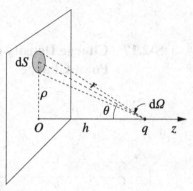

$$E_1 = E_2 = k_e \frac{q}{r^2} = k_e \frac{q \cos^2 \theta}{h^2} , \quad \text{(S-2.163)}$$

where $r^2 = h^2 + \rho^2$ and $\theta = \arctan(\rho/h)$, so that $r = h/\cos\theta$. The ρ components of $\mathbf{E}_1$ and $\mathbf{E}_2$ cancel each other, only thr z components are left and the resulting field is

Fig. S-2.29

$$\mathbf{E}(P) = -\hat{\mathbf{z}} \, 2k_e \frac{q \cos^3 \theta}{h^2} . \quad \text{(S-2.164)}$$

The surface charge density on the conducting plane at P is thus

$$\sigma(P) = \frac{E_z}{4\pi k_e} = -\frac{q \cos^3 \theta}{2\pi h^2} . \quad \text{(S-2.165)}$$

Let us consider a small surface element dS around P (Fig. S-2.29), it carries an induced charge

$$dQ = \sigma(P) \, dS = -\frac{q \, \cos^3 \theta \, dS}{2\pi h^2} . \quad \text{(S-2.166)}$$

On the other hand dS subtends a solid angle $d\Omega$ from q given by

$$d\Omega = \frac{dS}{r^2} \cos\theta = dS \frac{\cos^2 \theta}{h^2} \cos\theta = dS \frac{\cos^3 \theta}{h^2}. \quad \text{(S-2.167)}$$

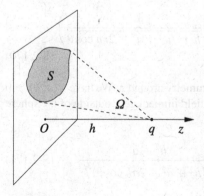

Thus the charge on the surface element dS is

$$dQ = -q \frac{d\Omega}{2\pi} . \quad \text{(S-2.168)}$$

Any area S lying on the conducting plane, whatever its size or shape, can be decomposed into a large number of small surface elements, each of which carries a charge proportional to the solid angle it subtends from q, thus the charge carried by the whole area is $-q \, \Omega/(2\pi)$, where Ω is the total

Fig. S-2.30

solid angle subtended by the area, see Fig. S-2.30. The whole plane subtends a solid angle 2π from q, thus the whole plane carries an induced charge $-q$, in agreement with the result of point (a).

S-2.17   Charge Density on a Metal Sphere in Front of a Point Charge

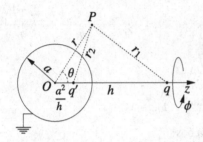

Fig. S-2.31

(a) We know from Problem 2.4 that the problem of the potential outside the conducting sphere is solved by locating an image charge $q' = -qa/h$ inside the sphere at a distance a^2/h from the center O of the sphere, on the line passing through O and the real charge q. We choose the line passing through O and q as the z axis of a spherical coordinate system with the origin at O, as shown in Fig. S-2.31. The electric potential at any point P outside the sphere is thus

$$\varphi(P) = k_e q \left(\frac{1}{r_1} - \frac{a}{h} \frac{1}{r_2} \right) , \qquad (S\text{-}2.169)$$

where r_1 and r_2 are the distances of P from the real charge q and from the image charge q', respectively. Now let us denote by r the distance of P from O, and by θ the angle between r and the z axis, so that r and θ are the radial and polar coordinates, respectively, of P. Using the law of cosines we can rewrite (S-2.169) as

$$\varphi(r, \theta) = k_e q \left(\frac{1}{\sqrt{r^2 + h^2 - 2rh\cos\theta}} - \frac{a}{h} \frac{1}{\sqrt{r^2 + (a^2/h)^2 - 2r(a^2/h)\cos\theta}} \right)$$

$$= k_e q \left(\frac{1}{\sqrt{r^2 + h^2 - 2rh\cos\theta}} - \frac{1}{\sqrt{h^2 r^2/a^2 + a^2 - 2rh\cos\theta}} \right) ,$$

$$(S\text{-}2.170)$$

independent of ϕ because of the rotational symmetry around z. We have $\varphi(a, \theta) = 0$ independently of θ, as expected. The electric field immediately outside of the sphere has the only component $E_r(a^+, \theta)$ given by

$$E_r(a^+, \theta) = - \left. \frac{\partial \varphi(r, \theta)}{\partial r} \right|_{r=a^+} = -k_e q \frac{h^2 - a^2}{a \left(h^2 + a^2 - 2ah\cos\theta \right)^{3/2}} . \qquad (S\text{-}2.171)$$

The surface charge density is

$$\sigma(\theta) = \frac{E_r(a^+, \theta)}{4\pi k_e} = -\frac{q}{4\pi} \frac{h^2 - a^2}{a\left(h^2 + a^2 - 2ah\cos\theta\right)^{3/2}}, \tag{S-2.172}$$

(b) The part of the sphere visible from q ranges from $\theta = 0$ to the points on the sphere surface where the rays starting from q are tangent to the sphere. This corresponds to a maximum value for theta $\theta_v = \arccos(a/h)$, as shown in Fig. S-2.32. The induced charge on the visible part of the sphere is thus

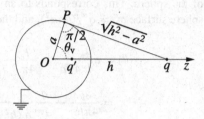

Fig. S-2.32

$$Q_v = \int\limits_0^{\theta_v} \sigma(\theta)\, 2\pi a^2 \sin\theta \,d\theta$$

$$= -\frac{qa\left(h^2 - a^2\right)}{2} \int\limits_0^{\arccos a/h} \frac{\sin\theta \,d\theta}{\left(h^2 + a^2 - 2ah\cos\theta\right)^{3/2}}$$

$$= -\frac{qa\left(h^2 - a^2\right)}{2} \frac{1}{2ah} \int\limits_{h^2+a^2-2ah}^{h^2-a^2} \frac{dx}{x^{3/2}}, \tag{S-2.173}$$

where we have substituted $x = (h^2 + a^2 - 2ah\cos\theta)$, so that $dx = 2ah\sin\theta \,d\theta$.

$$Q_v = \frac{q\left(h^2 - a^2\right)}{4h} \left[-\frac{2}{\sqrt{x}}\right]_{(h-a)^2}^{h^2-a^2} = -q\frac{h^2 - a^2}{2h}\left(\frac{1}{h-a} - \frac{1}{\sqrt{h^2 - a^2}}\right)$$

$$= -\frac{q}{2h}\left(h + a - \sqrt{h^2 - a^2}\right) = -\frac{q}{2h}\sqrt{h+a}\left(\sqrt{h+a} - \sqrt{h-a}\right). \tag{S-2.174}$$

The charge on the rest of the sphere is

$$Q_r = -q\frac{a}{h} - Q_v = -q\frac{a}{h} + \frac{q}{2h}\sqrt{h+a}\left(\sqrt{h+a} - \sqrt{h-a}\right)$$

$$= -\frac{q}{2h}\sqrt{h-a}\left(\sqrt{h+a} - \sqrt{h-a}\right), \tag{S-2.175}$$

and the ratio of the induced charge seen from q to the rest of the induced charge is

$$\frac{Q_v}{Q_r} = \sqrt{\frac{h+a}{h-a}}. \tag{S-2.176}$$

(c) If the sphere is isolated and uncharged the problem of the potential outside the sphere is solved by locating a second image charge $q'' = -q' = qa/h$ at the center of the sphere. This corresponds to an additional uniform charge density over the sphere surface $\sigma'' = q''/(4\pi a^2)$, and the total charge density is

$$\sigma(\theta) = q\frac{a}{h}\frac{1}{4\pi a^2} - \frac{q}{4\pi}\frac{h^2 - a^2}{a\left(h^2 + a^2 - 2ah\cos\theta\right)^{3/2}}$$

$$= \frac{q}{4\pi ha} - \frac{q}{4\pi}\frac{h^2 - a^2}{a\left(h^2 + a^2 - 2ah\cos\theta\right)^{3/2}}. \tag{S-2.177}$$

Chapter S-3
Solutions for Chapter 3

S-3.1 An Artificial Dielectric

(a) According to (S-2.6) of Problem 2.1, a metal sphere in a uniform external field $\mathbf{E}$ acquires a dipole moment

$$\mathbf{p} = \frac{a^3}{k_e} \mathbf{E} = \frac{3}{4\pi k_e} V \mathbf{E}, \qquad (\text{S-3.1})$$

where $V = \frac{4}{3}\pi a^3$ is the volume of the sphere. The polarization of our suspension is

$$\mathbf{P} = n\mathbf{p} = \frac{3n}{4\pi k_e} V \mathbf{E}. \qquad (\text{S-3.2})$$

In SI units we have $\mathbf{P} = \varepsilon_0 \chi \mathbf{E}$, and $\chi = 3f$, while in Gaussian units we have $\mathbf{P} = \chi\mathbf{E}$, and $\chi = 3f/(4\pi)$. In both cases $f = nV$ is the fraction of the volume occupied by the spheres. Since the minimum distance between the centers of two spheres is $2a$, we have

$$f \leq \frac{4\pi a^3}{3} \frac{1}{8a^3} = \frac{\pi}{6}, \qquad (\text{S-3.3})$$

leading to $\chi \leq \pi/2$ in SI units, and $\chi \leq 1/8$ in Gaussian units.

(b) The average distance ℓ between two sphere centers is of the order of $n^{-1/3}$. The electric field of a dipole at a distance ℓ is of the order of

$$E_{\text{dip}} \simeq k_e \frac{p}{\ell^3} \simeq k_e \frac{a^3}{k_e} E\, n = a^3 E\, n. \qquad (\text{S-3.4})$$

Thus, the condition $E_{\text{dip}} \ll E$ requires $n \ll 1/a^3$.

© The Editor(s) (if applicable) and The Author(s), under exclusive license to Springer Nature Switzerland AG 2023
A. Macchi et al., *Problems in Classical Electromagnetism*,
https://doi.org/10.1007/978-3-031-22235-1

S-3.2   Charge in Front of a Dielectric Half-Space

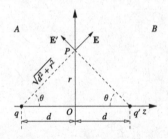

Fig. S-3.1

We denote by A the $z < 0$, vacuum, half-space, containing the real charge q, and by B the $z > 0$, dielectric, half-space, containing no free charge. We shall treat the two half spaces separately, making educated guesses, in order to apply the uniqueness theorem for the Poisson equation. We use cylindrical coordinates (r, ϕ, z), with the real charge located at $(0, \phi, -d)$. All our formulas will be independent of the azimuthal coordinate ϕ, which is not determined, and not relevant, when $r = 0$.

(a) We treat the field in the half-space A assuming vacuum in the whole space, including the half-space B. As ansatz, we locate an image charge q', of value to be determined, at $(0, 0, +d)$, in the half space that we are not considering, as in Fig. S-3.1. Now we evaluate the electric field $\mathbf{E}^{(-)}$ in a generic point $P \equiv (r, \phi, 0^-)$ of the plane $z = 0^-$. The distance between P and q is $\sqrt{d^2 + r^2}$ and forms an angle $\theta = \arccos(d/\sqrt{d^2 + r^2})$ with the z axis. Also the distance between P and q' will be $\sqrt{d^2 + r^2}$. The field at P, $\mathbf{E}^{(-)}$, is the vector sum of the fields $\mathbf{E}$ due to the real charge q, and $\mathbf{E}'$ do to the image charge q'. The components of $\mathbf{E}^{(-)}$, perpendicular and parallel to the $z = 0$ plane are, respectively

$$E_\perp^{(-)} = k_e \frac{q}{d^2 + r^2} \cos\theta - k_e \frac{q'}{d^2 + r^2} \cos\theta = k_e \frac{d}{(d^2 + r^2)^{3/2}} (q - q')$$

$$E_\parallel^{(-)} = k_e \frac{q}{d^2 + r^2} \sin\theta + k_e \frac{q'}{d^2 + r^2} \sin\theta = k_e \frac{r}{(d^2 + r^2)^{3/2}} (q + q'). \quad \text{(S-3.5)}$$

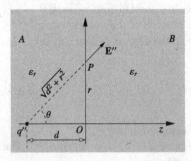

Fig. S-3.2

We treat the half-space B assuming that the whole space, including the half-space A, is filled by a dielectric medium of relative permittivity ε_r. We are not allowed to introduce charges or alter anything in B, but, as an educated guess, we replace the real charge q, located in the half-space A that we are not treating, by a charge q'', of value to be determined (Fig. S-3.2). We evaluate the field $\mathbf{E}^{(+)}$ at the same point P as before, but on the $z = 0^+$ plane. The components of $\mathbf{E}^{(+)}$ perpendicular and parallel to the $z = 0$ plane are

$$E_\perp^{(+)} = \frac{k_e}{\varepsilon_r} \frac{q''}{d^2 + r^2} \cos\theta = \frac{k_e}{\varepsilon_r} \frac{d}{(d^2 + r^2)^{3/2}} q''$$

$$E_\parallel^{(+)} = \frac{k_e}{\varepsilon_r} \frac{q''}{d^2 + r^2} \sin\theta = \frac{k_e}{\varepsilon_r} \frac{r}{(d^2 + r^2)^{3/2}} q'' . \tag{S-3.6}$$

If our educated guesses are correct, the dielectric boundary conditions must hold at $z = 0$. This implies $E_\perp^{(-)} = \varepsilon_r E_\perp^{(+)}$ and $E_\parallel^{(-)} = E_\parallel^{(+)}$, corresponding to the equations

$$q - q' = q'', \quad \text{and} \quad q + q' = \frac{q''}{\varepsilon_r}, \tag{S-3.7}$$

with solutions

$$q' = -\frac{\varepsilon_r - 1}{\varepsilon_r + 1} q, \quad \text{and} \quad q'' = \frac{2\varepsilon_r}{\varepsilon_r + 1} q . \tag{S-3.8}$$

We can easily check that, at the limit $\varepsilon_r \to 1$ (vacuum in the whole space), we have $q' \to 0$ and $q'' \to q$, i.e., in the whole space we have the field of charge q only. At the limit $\varepsilon_r \to \infty$ (dielectric $\to$ conductor limit) we have $q' \to -q$ and $q'' \to 2q$, i.e., the field of the real charge q and its image $-q$ in the half-space A, and zero field in the half space B, as at point (a) of Problem 2.2. The finite value of q'' is irrelevant for the field in the half-space B, because of the infinite value of ε_r.

Note: Alternatively, we can write Eqs. (S-3.6) without ε_r in the denominators, thus including the dielectric bound charge into q''. This leads to the equations

$$q - q' = \varepsilon_r q'', \quad \text{and} \quad q + q' = q'' \tag{S-3.9}$$

with solutions

$$q' = -\frac{\varepsilon_r - 1}{\varepsilon_r + 1} q, \quad \text{and} \quad q'' = \frac{2}{\varepsilon_r + 1} q . \tag{S-3.10}$$

(b) The polarization charge density on the $z = 0$ plane, $\sigma_b(r)$, is

$$\sigma_b(r) = -\frac{1}{4\pi k_e} (E_\perp^{(-)} - E_\perp^{(+)}) = -\frac{1}{4\pi} \frac{d}{(d^2 + r^2)^{3/2}} \left(q - q' - \frac{q''}{\varepsilon_r}\right)$$

$$= -\frac{1}{2\pi} \frac{d}{(d^2 + r^2)^{3/2}} \frac{\varepsilon_r - 1}{\varepsilon_r + 1} q = \frac{1}{2\pi} \frac{d}{(d^2 + r^2)^{3/2}} q' . \tag{S-3.11}$$

The total polarization charge on the $z = 0$ plane is

$$q_p = \int_0^\infty \sigma_b(r) \, 2\pi r \, dr = -\frac{\varepsilon_r - 1}{\varepsilon_r + 1} q \int_0^{\pi/2} \cos\theta \, d\theta = q', \tag{S-3.12}$$

where we have substituted $\cos\theta = d/\sqrt{d^2 + r^2}, r = d/\cos\theta$ and $dr = d \, d\theta/\cos^2\theta$.

(c) The polarization charge of the $z = 0$ plane generates an electric field equal to the field of a charge $q' = -q\,(\varepsilon_r - 1)/(\varepsilon_r + 1)$ located at $(0, 0, +d)$ in the half space $z < 0$, and equal to the field of a charge q' located at $(0, 0, -d)$ in the half space $z > 0$.

S-3.3 An Electrically Polarized Sphere

(a) Since the polarization $\mathbf{P}$ of the sphere is uniform, we have no volume bound-charge density, according to $\varrho_b = \nabla \cdot \mathbf{P}$. If we choose a spherical coordinate system (r, θ, ϕ) with the azimuthal axis parallel to $\mathbf{P}$, as shown in Fig. S-3.3, we see that the surface bound-charge density of the sphere is $\sigma_b = P \cos\theta$, according to $\sigma_b = \mathbf{P} \cdot \hat{\mathbf{n}}$. Thus, in principle, we can evaluate the electric field everywhere in space as the field generated by the bound-charge distribution on the sphere surface. However, it is easier to consider the polarized sphere as the superposition of two uniformly charged sphere, both of radius a, one of volume charge density ϱ, and one of volume charge density $-\varrho$. The centers of the two spheres are separated by a small distance , as in Fig. 1.1 of Problem 1.1. Thus, two initially superposed infinitesimal volume elements d^3r of the two spheres, of charge $\pm\varrho\,d^3r$, respectively, give origin to an infinitesimal electrical dipole moment $d\mathbf{p} = \delta\varrho\,d^3r$ after the displacement.

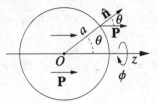

This corresponds to a polarization $d\mathbf{p}/d^3r = \varrho\,\delta$, we must have $\mathbf{P} = \varrho\,\delta$, and are interested in the limit $|\delta| \to 0$, $\rho \to \infty$, with $\varrho\,\delta = \mathbf{P} = $ constant. Now we can follow the solution of Problem 1.1. According to (S-1.1), the electric field inside the sphere is uniform and equals

Fig. S-3.3

$$\mathbf{E}_{in} = -\frac{4\pi k_e}{3}\,\varrho\,\delta = -\frac{4\pi k_e}{3}\,\mathbf{P}\,. \qquad \text{(S-3.13)}$$

The problem of the field outside the sphere is solved at point (b) of Problem 1.1, we have

$$\mathbf{E}_{ext}(\mathbf{r}) = k_e\,\frac{3\hat{\mathbf{r}}\,(\mathbf{p}\cdot\hat{\mathbf{r}}) - \mathbf{p}}{r^3}\,, \qquad \text{(S-3.14)}$$

where $\mathbf{p} = \mathbf{P}(4\pi a^3/3)$ is the total dipole moment of the sphere (Fig. S-3.4).

The external displacement field is simply $\mathbf{D}_{ext} = \mathbf{E}_{ext}$ (Gaussian), and $\mathbf{D}_{ext} = \varepsilon_0\mathbf{E}_{ext}$ (SI). The internal displacement field is

$$\mathbf{D}_{in} = \begin{cases} \varepsilon_0\mathbf{E}_{in} + \mathbf{P} = -\varepsilon_0\dfrac{\mathbf{P}}{3\varepsilon_0} + \mathbf{P} = \dfrac{2}{3}\,\mathbf{P}\,, & \text{SI} \\[4mm] \mathbf{E}_{in} + 4\pi\mathbf{P} = -\dfrac{4\pi}{3}\,\mathbf{P} + 4\pi\mathbf{P} = \dfrac{8\pi}{3}\,\mathbf{P}\,, & \text{Gaussian.} \end{cases} \qquad \text{(S-3.15)}$$

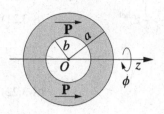

Fig. S-3.4

(b) The problem can be solved by the superposition principle. The hole of radius b can be regarded as a sphere of uniform electrical polarization $-\mathbf{P}$ superposed to the sphere of radius a and polarization $\mathbf{P}$ (Fig. S-3.4). The sphere of radius b generates a field.

$$\mathbf{E}_{in}^{(b)} = \frac{4\pi k_e}{3}\mathbf{P} \qquad (S-3.16)$$

at its interior. Thus, the total field inside the spherical hole is $\mathbf{E}_{in}^{(a+b)} = 0$. The filed inside the spherical shell $b < r < a$ is the sum of the uniform field (S-3.13) and the field generated by an electric dipole of moment $\mathbf{p}^{(b)}$ located at the center O, with

$$\mathbf{p}^{(b)} = -\frac{4\pi b^3}{3}\mathbf{P} . \qquad (S-3.17)$$

Finally, the external field ($r > a$) equals the field generated by a single dipole $\mathbf{p}^{(a+b)}$ located in O with

$$\mathbf{p}^{(a+b)} = \frac{4\pi\left(a^3 - b^3\right)}{3}\mathbf{P} . \qquad (S-3.18)$$

S-3.4   Dielectric Sphere in an External Field

(a) As an educated guess let us assume that the external field induces a uniform electric polarization $\mathbf{P}$ in the sphere. We have seen in Problem 3.3 that a sphere of uniform electric polarization $\mathbf{P}$ generates a uniform electric field $\mathbf{E}^{pol.} = -(4\pi k_e/3)\mathbf{P}$ at its interior. The difference is that in the present case $\mathbf{P}$ is not permanent but it is induced by the local electric field, i.e. it is given by

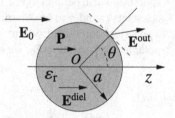

Fig. S-3.5

$$\mathbf{P} = \frac{\varepsilon_r - 1}{4\pi k_e}\mathbf{E}^{diel} , \qquad (S-3.19)$$

where $\mathbf{E}^{diel}$ is the field inside the dielectric sphere, which is the sum of the external and the induced fields:

$$\mathbf{E}^{diel} = \mathbf{E}_0 + \mathbf{E}^{pol} . \qquad (S-3.20)$$

We thus have

$$\mathbf{E}^{\text{diel}} = \mathbf{E}_0 - \frac{4\pi k_e}{3}\mathbf{P} = \mathbf{E}_0 - \frac{\varepsilon_r - 1}{3}\mathbf{E}^{\text{diel}} , \qquad (S\text{-}3.21)$$

which can be solved for $\mathbf{E}^{\text{diel}}$:

$$\mathbf{E}^{\text{diel}} = \frac{3}{\varepsilon_r + 2}\mathbf{E}_0 . \qquad (S\text{-}3.22)$$

Since $\varepsilon_r > 1$, the field inside the dielectric sphere is smaller than $\mathbf{E}_0$.

The electric field outside the sphere $\mathbf{E}^{\text{out}}$ will be given by the sum of $\mathbf{E}_0$ and the field of a dipole

$$\mathbf{p} = \frac{4\pi a^3}{3}\mathbf{P} = \frac{a^3}{k_e}\frac{\varepsilon_r - 1}{\varepsilon_r + 2}\mathbf{E}_0 \qquad (S\text{-}3.23)$$

located at the center of the of the sphere. Thus

$$\mathbf{E}^{\text{out}} = \mathbf{E}_0 + k_e\frac{3(\mathbf{p}\cdot\hat{\mathbf{r}})\hat{\mathbf{r}} - \mathbf{p}}{r^3} = \mathbf{E}_0 + \frac{a^3}{3r^3}\frac{\varepsilon_r - 1}{\varepsilon_r + 2}[3(\mathbf{E}_0\cdot\hat{\mathbf{r}})\hat{\mathbf{r}} - \mathbf{E}_0] . \quad (S\text{-}3.24)$$

It is instructive, and useful for the following, to check that the above solution satisfies the boundary conditions at the surface of the sphere. Let us then restart the problem by assuming that the field $\mathbf{E}^{\text{diel}}$ inside the sphere ($r < a$) is uniform and parallel to the external field $\mathbf{E}_0$, and that the field $\mathbf{E}^{\text{out}}$ outside the sphere ($r > a$) is the sum of the external field and that of a dipole $\mathbf{p}$ located at the center of the sphere and also parallel to $\mathbf{E}_0$. Thus we can write $\mathbf{E}^{\text{diel}} = \alpha\mathbf{E}_0$ and $\mathbf{p} = \eta\mathbf{E}_0$, with the constants α and η to be determined by the boundary conditions at $r = a$. Choosing a spherical coordinate system with the origin O at the center of the sphere, and the polar axis z parallel to $\mathbf{E}_0$, we have

$$
\begin{array}{ll}
E_r^{\text{diel}} = \alpha E_0 \cos\theta & E_r^{\text{out}} = E_0 \cos\theta + k_e\eta E_0\dfrac{2\cos\theta}{r^3} \\[3mm]
E_\theta^{\text{diel}} = \alpha E_0 \sin\theta & E_\theta^{\text{out}} = E_0 \sin\theta - k_e\eta E_0\dfrac{\sin\theta}{r^3} \\[3mm]
E_\phi^{\text{diel}} = 0 & E_\phi^{\text{out}} = 0 . \qquad (S\text{-}3.25)
\end{array}
$$

The boundary conditions at the surface of the sphere are $\varepsilon_r E_\perp^{\text{diel}} = E_\perp^{\text{out}}$ and $E_\parallel^{\text{diel}} = E_\parallel^{\text{out}}$ which yields in spherical coordinates

$$\varepsilon_r E_r^{\text{diel}}(r = a^-) = E_r^{\text{out}}(r = a^+) \qquad E_\theta^{\text{diel}}(r = a^-) = E_\theta^{\text{out}}(r = a^+) . \quad (S\text{-}3.26)$$

Using (S-3.25) and deleting the common factors we obtain

$$\varepsilon_r\alpha = 1 + \frac{2k_e}{a^3}\eta , \qquad \alpha = 1 - \frac{k_e}{a^3}\eta , \qquad (S\text{-}3.27)$$

which are solved for α and η as

$$\alpha = 3/(\varepsilon_r + 2) , \qquad \eta = \frac{a^3}{k_e} \frac{\varepsilon_r - 1}{\varepsilon_r - 2} , \qquad (\text{S-3.28})$$

so that we eventually recover (S-3.22) and (S-3.23).

(b) We make an educated guess analogous to the one of the previous point, i.e., that the field inside the cavity, $\mathbf{E}^{\text{cav}}$, is uniform and parallel to $\mathbf{E}_d$, and that the field in the dielectric medium, $\mathbf{E}^{\text{diel}}$, is the sum of $\mathbf{E}_d$ and the field of an electric dipole $\mathbf{p}_c$, located at the center of the cavity and parallel to $\mathbf{E}_d$. Thus we can write

$$\mathbf{E}^{\text{cav}} = \alpha \mathbf{E}_d , \qquad \mathbf{p}_c = \eta \mathbf{E}_d ; \qquad (\text{S-3.29})$$

where, again, α and η are constants to be determined by the boundary conditions. Using again spherical coordinates with the origin at the center of the spherical cavity and the z axis parallel to $\mathbf{E}_d$, the expressions analogous to (S-3.25) are

$$
\begin{aligned}
E_r^{\text{cav}} &= \alpha E_d \cos \theta & E_r^{\text{med}} &= E_d \cos \theta + k_e \eta E_d \frac{2 \cos \theta}{r^3} \\
E_\theta^{\text{cav}} &= \alpha E_d \sin \theta & E_\theta^{\text{med}} &= E_d \sin \theta - k_e \eta E_d \frac{\sin \theta}{r^3} \\
E_\phi^{\text{cav}} &= 0 & E_\phi^{\text{med}} &= 0 , & (\text{S-3.30})
\end{aligned}
$$

with the boundary conditions

$$E_r^{\text{cav}}(r = a^-) = \varepsilon_r E_r^{\text{med}}(r = a^+) \qquad E_\theta^{\text{cav}}(r = a^-) = E_\theta^{\text{med}}(r = a^+) . \qquad (\text{S-3.31})$$

The values for α and η may thus be easily obtained by solving a linear system of two equations as in point **a)**. However, we can imemdiately obtain the solution by noticing that (S-3.30) and (S-3.31) are identical to (S-3.25) and (S-3.26) but for the replacements $E_d \leftrightarrow E_0$, $\mathbf{E}^{\text{cav}} \leftrightarrow \mathbf{E}^{\text{diel}}$, $E^{\text{med}} \leftrightarrow E^{\text{out}}$, and $\varepsilon_r \leftrightarrow 1/\varepsilon_r$. Thus, the solutions for $\mathbf{E}^{\text{cav}}$ and $\mathbf{p}_c$ are obtained from those for $\mathbf{E}^{\text{diel}}$ and $\mathbf{p}$, (S-3.22) and (S-3.23), by substituting E_d for E_0 and $1/\varepsilon_r$ for ε_r:

$$\mathbf{E}^{\text{cav}} = \frac{3}{1/\varepsilon_r + 2} \mathbf{E}_d = \frac{3\varepsilon_r}{1 + 2\varepsilon_r} \mathbf{E}_d , \qquad (\text{S-3.32})$$

$$\mathbf{p}_c = \frac{a^3}{3k_e} \frac{1/\varepsilon_r - 1}{1/\varepsilon_r + 2} \mathbf{E}_d = \frac{a^3}{3k_e} \frac{1 - \varepsilon_r}{1 + 2\varepsilon_r} \mathbf{E}_d . \qquad (\text{S-3.33})$$

Thus $E^{\text{cav}} > E_d$, i.e. the field inside the cavity is stronger than that outside it, and $\mathbf{p}_c$ is antiparallel to $\mathbf{E}_d$.

S-3.5   Refraction of the Electric Field at a Dielectric Boundary

(a) First, we note that the electric field $\mathbf{E}_0$ outside the dielectric slab equals the field that we would have in vacuum in the absence of the slab. Neglecting the boundary effects, the bound surface charge densities of slab are analogous to the surface charge densities of a parallel-plate capacitor. These generate a uniform electric field inside the capacitor, but no field outside. Thus, the electric field inside the slab is the sum of $\mathbf{E}_0$ and the field generated by the surface polarization charge

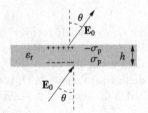

Fig. S-3.6

densities. If we denote by $\mathbf{E}'$ the internal electric field, the boundary conditions at the dielectric surfaces are

$$E_{0\perp} = \varepsilon_r E'_\perp, \quad E_{0\parallel} = E'_\parallel, \tag{S-3.34}$$

where the subscripts $\perp$ and $\parallel$ denote the field components perpendicular and parallel to the interface surface, respectively. In terms of the angles θ and θ' of Fig. S-3.6 we have

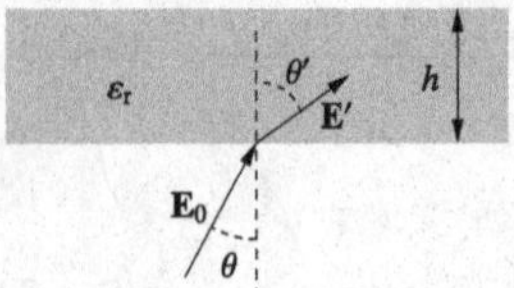

$$E_0 \cos\theta = \varepsilon_r E' \cos\theta'$$
$$E_0 \sin\theta = E' \sin\theta'. \tag{S-3.35}$$

Fig. S-3.7

If we divide the second of (S-3.35) by the first we obtain

$$\frac{1}{\varepsilon_r} \tan\theta' = \tan\theta, \tag{S-3.36}$$

and, since $\varepsilon_r > 1$, we have $\theta' > \theta$ (Fig. S-3.7).

(b) From Gauss's law we obtain

$$\sigma_b = \frac{1}{4\pi k_e}(E_{o,\perp} - E'_\perp) = \frac{1}{4\pi k_e} E_0 \cos\theta \left(1 - \frac{1}{\varepsilon_r}\right). \tag{S-3.37}$$

(c) The electrostatic energy density inside the slab is

$$u_{es} = \frac{\varepsilon_r}{8\pi k_e} \mathbf{E}'^2 = \frac{\varepsilon_r}{8\pi k_e}, (E'^2_\perp + E'^2_\parallel) = \frac{\varepsilon_r}{8\pi k_e} E_0^2 \left(\frac{\cos^2\theta}{\varepsilon_r^2} + \sin^2\theta\right)$$
$$= \frac{1}{8\pi k_e \varepsilon_r} E_0^2 \left[(\varepsilon_r^2 - 1)\sin^2\theta + 1\right], \tag{S-3.38}$$

so that u_{es} increases with increasing θ, and we expect a torque τ tending to rotate the slab toward the angle of minimum energy, i.e., $\theta = 0$. Neglecting boundary effects, the total electrostatic energy of the slab is $U_{es} = V u_{es}$, where V is the volume of the slab, and the torque exerted by the electric field is

$$\tau = -\frac{\partial U_{es}}{\partial \theta} = -\frac{1}{8\pi k_e \varepsilon_r} E_0^2 V \left(\varepsilon_r^2 - 1\right) \sin 2\theta < 0. \qquad \text{(S-3.39)}$$

S-3.6 Contact Force Between a Conductor and a Dielectric

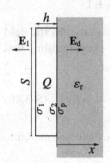

Fig. S-3.8

(a) Neglecting boundary effects at the edges of the slab, the electric field is parallel to the x axis in all the regions of interest because of symmetry reasons. Thus, we can omit the vector notation, and we shall use positive numbers for vectors whose unit vector is $\hat{\mathbf{x}}$, negative numbers otherwise.

According to Gauss's law, a uniformly charged plane with surface charge density σ_a generates uniform fields at both its sides, of intensities $E_a = \pm\sigma_a/2\varepsilon_0$, respectively. In our problem we have three charged parallel plane surfaces: we denote by σ_1 the surface charge density on the left surface of the slab, by σ_2 charge the density on its right surface, and by σ_b the *bound* surface charge density of the dielectric material on its surface, as shown in Fig. S-3.8. Since the total *free* charge on the slab is Q, we have

$$\sigma_1 + \sigma_2 = \frac{Q}{S} = \sigma_{tot}. \qquad \text{(S-3.40)}$$

At any point in space the total electric field is the sum of the fields generated by the three surface charges. Nŏw, the electric field must be zero inside the conducting slab. Thus the sum of all surface charge densities (including *both* free and bound charges) at the left of the slab must equal the sum of all surface charge densities at the right, so that their respective fields cancel out inside the slab. This conclusion holds both when the slab is in contact with the dielectric, and when there is a vacuum gap between them. Thus, we have

$$\sigma_1 = \sigma_2 + \sigma_b. \qquad \text{(S-3.41)}$$

The electric field E_d inside the dielectric medium is $E_d = 4\pi k_e (\sigma_2 + \sigma_b)$. This implies for the dielectric polarization of the medium P

$$P = \frac{\varepsilon_r - 1}{4\pi k_e} E_d = (\varepsilon_r - 1)\sigma_1. \qquad \text{(S-3.42)}$$

Since we also have $\sigma_b = -\mathbf{P} \cdot \hat{\mathbf{x}} = -P$, we obtain the additional relation

$$\sigma_b = -(\varepsilon_r - 1)(\sigma_2 + \sigma_b),$$ (S-3.43)

that leads to

$$\sigma_b = -\frac{\varepsilon_r - 1}{\varepsilon_r}\sigma_2.$$ (S-3.44)

From (S-3.40), (S-3.41) and (S-3.44) we finally obtain

$$\sigma_1 = \frac{1}{\varepsilon_r + 1}\sigma_{tot}, \qquad \sigma_2 = \frac{\varepsilon_r}{\varepsilon_r + 1}\sigma_{tot}, \qquad \sigma_b = -\frac{\varepsilon_r - 1}{\varepsilon_r + 1}\sigma_{tot}.$$ (S-3.45)

The magnitudes of the electric field at the left of the slab E_1, and of the electric field inside the dielectric medium E_d, can be evaluated from Gauss's law, recalling that the field is zero inside the slab. We have

$$E_1 = -4\pi k_e\,\sigma_1 = -\frac{4\pi k_e}{\varepsilon_r + 1}\sigma_{tot} \quad \text{and} \quad E_d = 4\pi k_e\,(\sigma_2 + \sigma_b) = -E_1.$$ (S-3.46)

In the case of a vacuum gap between the conducting slab and the dielectric medium, as shown in Fig. S-3.9, the field E_2 in the gap is

$$E_2 = 4\pi k_e\,\sigma_2 = 4\pi k_e\,\frac{\varepsilon_r}{\varepsilon_r + 1}\sigma_{tot} = -\varepsilon_r E_1.$$ (S-3.47)

The values of E_1 and E_d are not affected by the presence of the vacuum gap.

As an alternative approach we can assume, following Problem 3.2, that the free charge layers σ_1 and σ_2 induce image charge layers σ_1' and σ_2' in the dielectric,

$$\sigma_1' = -\frac{\varepsilon_r - 1}{\varepsilon_r + 1}\sigma_1, \qquad \sigma_2' = -\frac{\varepsilon_r - 1}{\varepsilon_r + 1}\sigma_2,$$ (S-3.48)

with the image planes located in position symmetrical with respect to the dielectric surface. Due to Gauss's law the bound surface charge density is the sum of the image charge densities,

$$\sigma_p = \sigma_1' + \sigma_2' = -\frac{\varepsilon_r - 1}{\varepsilon_r + 1}\frac{Q}{S}.$$ (S-3.49)

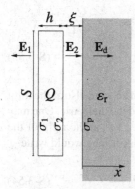

Fig. S-3.9

The free charge densities can be now found by requiring the field to vanish inside the slab: omitting a common multiplying factor we have

$$0 = \sigma_1 - \sigma_2 - \sigma_1' - \sigma_2' = 2\frac{\varepsilon_r \sigma_1 - \sigma_2}{\varepsilon_r + 1}, \qquad \text{(S-3.50)}$$

from which we obtain $\varepsilon_r \sigma_1 = \sigma_2$, and we eventually recover the free and bound surface charge densities of (S-3.45).

(b) In order to evaluate the electrostatic force acting on the conducting slab, we first assume the presence of a small vacuum gap of width ξ between the slab and the dielectric medium, as shown in Fig. S-3.9.

We can evaluate the total electrostatic force F acting on the conducting slab in three equivalent ways:

(i) We can evaluate the variation of the total electrostatic energy U_{es} when the slab is displaced by an infinitesimal amount dx toward the right, thus decreasing the gap. In this case U_{es} increases by $E_1^2 S dx/(8\pi k_e)$ at the left of the slab, because the width of region "filled" by the field E_1 increases by dx, and correspondingly decreases by $E_2^2 S dx/(8\pi k_e)$ at its right. Thus, $dU_{es} = (E_1^2 - E_2^2) S dx/(8\pi k_e)$, from which we obtain the force per unit surface

$$F = -\frac{dU_{es}}{dx} = \frac{S}{8\pi k_e}(E_1^2 - E_2^2) = \frac{S}{8\pi k_e}(\varepsilon_r^2 - 1)\left(\frac{4\pi k_e}{\varepsilon_r + 1}\right)^2\left(\frac{Q}{S}\right)^2$$

$$= 2\pi k_e \frac{\varepsilon_r - 1}{\varepsilon_r + 1}\frac{Q^2}{S}. \qquad \text{(S-3.51)}$$

We have $F > 0$, meaning that the slab is attracted by the dielectric medium.

(ii) We can multiply the charge of the slab Q by the *local* field, i.e., by the field generated by all charges excluding the charges of the slab. In our case the local field is the field E_p generated by the bound surface charge density σ_b. We have

$$E_p = -2\pi k_e \sigma_b = 2\pi k_e \frac{\varepsilon_r - 1}{\varepsilon_r + 1}\sigma_{tot}, \quad \text{and} \quad F = 2\pi k_e \frac{\varepsilon_r - 1}{\varepsilon_r + 1}\frac{Q^2}{S}. \qquad \text{(S-3.52)}$$

(iii) We can evaluate the force on the slab by summing the forces F_1 on its left and F_2 on its right surface. These are obtained by multiplying the respective charges $Q_1 = S\sigma_1$ and $Q_2 = S\sigma_2$ by the average fields at the surfaces

$$F = F_1 + F_2 = Q_1 \frac{E_1}{2} + Q_2 \frac{E_2}{2} = -\frac{Q}{\varepsilon_r + 1} \frac{2\pi k_e}{\varepsilon_r + 1} \sigma_{tot} + \frac{\varepsilon_r Q}{\varepsilon_r + 1} \frac{\varepsilon_r 2\pi k_e}{\varepsilon_r + 1} \sigma_{tot}$$

$$= 2\pi k_e \frac{\varepsilon_r^2 - 1}{(\varepsilon_r + 1)^2} \frac{Q^2}{S} = 2\pi k_e \frac{\varepsilon_r - 1}{\varepsilon_r + 1} \frac{Q^2}{S}. \tag{S-3.53}$$

The force F is independent of ξ, thus the above result should be valid also at the limit $\xi \to 0$, i.e., when there is contact between the metal slab and the dielectric. One may argue, however, that in these conditions the field at $x = 0^+$, i.e., at the right of the slab, is given by $E_d = -E_1$, so that following the approach (iii) one would write

$$F = F_1 + F_2 \overset{?}{=} Q_1 \frac{E_1}{2} + Q_2 \frac{E_d}{2} \neq Q_1 \frac{E_1}{2} + Q_2 \frac{E_2}{2}. \tag{S-3.54}$$

This discrepancy comes out because actually the average field on the *free* charges located on the right surface of the slab is *not* $E_d/2$, which is the average field across the two merging layers of free and bound charges; however, the force on the slab must be calculated by taking the average field on free charges only. To illustrate this issue, let us assume for a moment the free charges at the slab surfaces to be distributed in a layer of small but finite width, so that we can localize exactly where free charges are without merging them with bound charges. In particular, let the free surface charge layer have thickness ℓ_2 and volume charge density $\varrho_2(x)$ such that

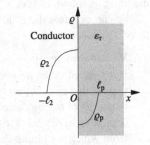

Fig. S-3.10

$$\int_{-\ell_2}^{0} \varrho_2 \, dx = \sigma_2, \tag{S-3.55}$$

as shown in Fig. S-3.10. The electric field is still directed along the x axis for symmetry reasons. Gauss's law in one dimension gives $\partial_x E = 4\pi k_e \varrho$. Since $E(-\ell_2) = 0$ (as deep into the conductor the field should vanish) we have for the electric field in the $-\ell_2 \leqslant x \leqslant 0$ region

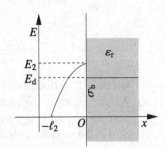

Fig. S-3.11

$$E(x) = 4\pi k_e \int_{-\ell_2}^{x} \varrho_2(x') \, dx'. \tag{S-3.56}$$

The total force *on the free charges only* can thus be evaluated as

$$F_2 = S \int_{-\ell_2}^{0} E(x)\, \varrho_2(x)\, dx = \frac{S}{4\pi k_e} \int_{-\ell_2}^{0} E(x)\, \partial_x E(x)\, dx = \frac{S}{8\pi k_e} \int_{-\ell_2}^{0} \partial_x E^2(x)\, dx$$

$$= \frac{S}{8\pi k_e} E_2^2 = \frac{S(4\pi k_e \sigma_2)^2}{8\pi k_e} = 2\pi k_e\, S\sigma_2^2, \tag{S-3.57}$$

the electric field at $x = 0^-$ is E_2, as shown in Fig. S-3.11, and the resulting electro-static pressure is $p_2 = F_2/S = 2\pi k_e\, \sigma_2^2$, independent of the particular distribution $\varrho_2(x)$, and in agreement with the previous result (S-3.53). However, the electric field at $x = 0^+$ is E_d because of the presence of the surface bound charge.[2]

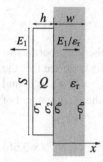

Fig. S-3.12

(c) If the dielectric medium is actually a slab limited at $x = w$, as shown in Fig. S-3.12, a further bound surface charge density $-\sigma_b$, opposite to the density σ_b at $x = 0$, appears at its $x = w$ surface. This charge distribution is iden-tical to that of a plane capacitor, so that the bound charges generate *no* field outside the dielectric slab. As trivial con-sequences the surface charge densities on the conducting slab are $\sigma_1 = -\sigma_2 = Q/2S$, the field inside the dielectric is $E_1/\varepsilon_r = Q/(2\varepsilon_0\varepsilon_r S)$, and there is no force between the slab and the dielectric. Moreover, this result is independent of w, and therefore should be valid also in the limit $w \to \infty$.

The apparent contradiction with the results of points (a) and (b) is that, in two attempts to approximate real conditions by objects of infinite size, we are assuming different boundary conditions at infinity. To discuss this issue let us look again at Fig. 3.5, showing the slab of charge Q located in front of a dielectric hemisphere of radius R. At the limit $R \to \infty$, the field in the dielectric half-space approaches the field that we would have if the dielectric medium filled the whole space, and the surface S had surface charge density $\sigma'' = 2\varepsilon_r\sigma/(\varepsilon_r + 1)$, see Problem 3.2. Thus, the field, the polarization, and the polarization surface charge density all approach zero at the hemispherical surface. Part (b) of Fig. 3.5 is an enlargement of the area enclosed in the dashed rectangle of part (a) of the same figure, and the vanishing charge density on the hemisphere surface does not contribute to the field in this area, according to the result of Problem 1.11. This motivates the boundary condition assumed in points (a) and (b). In contrast, in point (c) the bound surface charge density does not vanish at infinity and generates a uniform field, which in vacuum cancels out the field generated by the dielectric surface at $x = 0$.

---

[2] We might assume that also the polarization charge fills a layer of small, but finite width ℓ_d at the surface of the dielectric. However, this would only imply that the field becomes E_d at $x \geqslant \ell_d$, and would not affect our conclusions on the forces on the conductor.

S-3.7 A Conducting Sphere Between Two Dielectrics

(a) We use a spherical coordinate system
(r, θ, ϕ) with the origin O at the center of the
sphere, and the zenith direction perpendicu-
lar to the plane separating the two dielectric
media, as shown in Fig. S-3.13. The electric
field inside the conducting sphere is zero. The
electric field outside the sphere, $\mathbf{E}(r, \theta, \phi)$, is
independent of ϕ because of the symmetry of
our problem. Since the sphere is conducting,
the electric field $\mathbf{E}(R^+, \theta, \phi)$ must be perpendicular to its surface, and its only nonzero com-

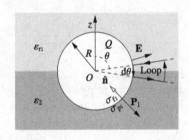

Fig. S-3.13

ponent is E_r. If we write Maxwell's equation $\nabla \times \mathbf{E} = 0$ in spherical coordinates
over the spherical surface $r = R^+$ (see Table A.1 of the Appendix), we see that the
r and θ components of the curl are automatically zero because $E_\phi = 0$, $E_\theta = 0$, and
all derivatives with respect to ϕ are zero. The condition that also the ϕ component of
the curl must be zero is

$$\partial_\theta E_r = \partial_r(r E_\theta) = E_\theta + r \partial_r E_\theta \,. \tag{S-3.58}$$

The right-hand side of (S-3.58) is zero because $E_\theta(R^-) = E_\theta(R^+) = 0$, implying
that also $\partial_r E_\theta(R) = 0$. Thus, $\partial_\theta E_r(R^+) = 0$, and $E_r(R^+)$ does not depend on θ
(and, consequently, on the dielectric medium). If we denote by σ_{tot} the sum of the
free surface charge density σ_f and the bound charge density σ_b, the relation

$$\sigma_{\text{tot}} = \frac{E(R^+, \theta, \phi)}{4\pi k_e} = \frac{E(R^+)}{4\pi k_e} \tag{S-3.59}$$

shows that σ_{tot} is constant over the whole surface of the sphere. Thus the electric field
in the whole space outside the sphere equals the field of a point charge Q_{tot} located
in O, with $Q_{\text{tot}} = Q + Q_b$, where Q_b is the total polarization (bound) charge:

$$\mathbf{E}(r, \theta, \phi) = \mathbf{E}(r) = 4\pi k_e \sigma_{\text{tot}} \frac{R^2}{r^2} \hat{\mathbf{r}}, \quad r > R, \tag{S-3.60}$$

since the field depends on r only. The polarization charge densities on the surfaces
of the two dielectrics in contact with the sphere are, respectively,

$$\sigma_{b1} = \hat{\mathbf{n}} \cdot \mathbf{P}_1 = -\frac{\varepsilon_{r1} - 1}{4\pi k_e} E(R) = -(\varepsilon_{r1} - 1)\sigma_{\text{tot}}$$

$$\sigma_{b2} = \hat{\mathbf{n}} \cdot \mathbf{P}_2 = -\frac{\varepsilon_{r2} - 1}{4\pi k_e} E(R) = -(\varepsilon_{r2} - 1)\sigma_{\text{tot}} \tag{S-3.61}$$

where the unit vector $\hat{\mathbf{n}}$ points toward the center of the sphere. The free surface charge densities in the regions in contact with the two dielectrics, σ_{f_1} and σ_{f_2}, are, respectively,

$$\sigma_{f_1} = \sigma_{tot} - \sigma_{b_1} = \varepsilon_{r_1}\sigma_{tot}$$

$$\sigma_{f_2} = \sigma_{tot} - \sigma_{b_2} = \varepsilon_{r_2}\sigma_{tot}. \tag{S-3.62}$$

Since $2\pi R^2(\sigma_{f_1} + \sigma_{f_2}) = Q$, we finally obtain (Fig. S-3.14).

$$\sigma_{tot} = \frac{Q}{2\pi R^2(\varepsilon_{r_1} + \varepsilon_{r_2})} \qquad \mathbf{E}(r) = 2k_e\frac{Q}{(\varepsilon_{r_1} + \varepsilon_{r_2})\,r^2}\,\hat{\mathbf{r}}$$

$$\sigma_{f_1} = \frac{\varepsilon_{r_1}\,Q}{2\pi R^2(\varepsilon_{r_1} + \varepsilon_{r_2})} \qquad \sigma_{f_2} = \frac{\varepsilon_{r_2}\,Q}{2\pi R^2(\varepsilon_{r_1} + \varepsilon_{r_2})}$$

$$\sigma_{b_1} = -\frac{(\varepsilon_{r_1} - 1)\,Q}{2\pi R^2(\varepsilon_{r_1} + \varepsilon_{r_2})} \qquad \sigma_{b_2} = -\frac{(\varepsilon_{r_2} - 1)\,Q}{2\pi R^2(\varepsilon_{r_1} + \varepsilon_{r_2})}. \tag{S-3.63}$$

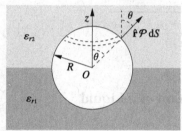

Fig. S-3.14

(b) The electrostatic pressures on the two hemispherical surfaces equal the electrostatic energy densities in the corresponding dielectric media, and are, respectively,

$$\mathcal{P}_{f_i} = \frac{2\pi k_e}{\varepsilon_{r_i}}\sigma_{f_i}^2 = \frac{2\pi k_e}{\varepsilon_{r_i}}\left[\frac{\varepsilon_{r_i}\,Q}{2\pi R^2(\varepsilon_{r_1} + \varepsilon_{r_2})}\right]^2$$

$$= k_e\frac{\varepsilon_{r_i}\,Q^2}{2\pi R^4(\varepsilon_{r_1} + \varepsilon_{r_2})^2} \tag{S-3.64}$$

with $i = 1, 2$. Thus $\mathcal{P}_1 > \mathcal{P}_2$ because $\varepsilon_{r_1} > \varepsilon_{r_2}$, and the pressure pushes the sphere towards the medium of higher permittivity. The force on the sphere surface element $dS = R^2\sin\theta\,d\theta\,d\phi$ is $d\mathbf{F} = \hat{\mathbf{r}}\mathcal{P}_{f_i}\,dS$, with $i = 1$ if $\theta > \pi/2$, and $i = 2$ if $\theta < \pi/2$. The total force acting on the upper hemisphere ($\theta < \pi/2$) is thus

$$\mathbf{F}_2 = \hat{\mathbf{z}}\int_0^{2\pi} d\phi\int_0^{\pi/2} d\theta\,R^2\sin\theta\cos\theta\,k_e\frac{\varepsilon_{r_2}\,Q^2}{2\pi(\varepsilon_{r_1} + \varepsilon_{r_2})^2\,R^4}$$

$$= \hat{\mathbf{z}}\,\pi R^2 k_e\frac{\varepsilon_{r_2}Q^2}{2\pi(\varepsilon_{r_1} + \varepsilon_{r_2})^2\,R^4} = \hat{\mathbf{z}}\,\pi R^2\mathcal{P}_2 = \hat{\mathbf{z}}\,k_e\frac{\varepsilon_{r_2}Q^2}{2\,(\varepsilon_{r_1} + \varepsilon_{r_2})^2\,R^2}, \tag{S-3.65}$$

directed upwards, since the force components perpendicular to the z axis cancel out. Note that $\mathbf{F}_2$ simply equals $\mathcal{P}_{f_2}$ times the section of the sphere πR^2. The total force acting on the lower hemisphere ($\theta > \pi/2$) is, analogously,

$$\mathbf{F}_1 = -\hat{\mathbf{z}}\, k_e\, \frac{\varepsilon_{r1}\, Q^2}{2\,(\varepsilon_{r1} + \varepsilon_{r2})^2 R^2}\,. \tag{S-3.66}$$

The total electrostatic force acting on the conducting sphere is thus

$$\mathbf{F}_{tot} = \mathbf{F}_1 + \mathbf{F}_2 = -\hat{\mathbf{z}}\, k_e\, \frac{(\varepsilon_{r1} - \varepsilon_{r2})\, Q^2}{2\,(\varepsilon_{r1} + \varepsilon_{r2})^2\, R^2}\,. \tag{S-3.67}$$

If the sphere is at equilibrium when half of its volume is submerged, $\mathbf{F}_{tot}$ plus the sphere weight must balance Archimedes' buoyant force

$$k_e\, \frac{(\varepsilon_{r1} - \varepsilon_{r2})\, Q^2}{2\,(\varepsilon_{r1} + \varepsilon_{r2})^2\, R^2} = g\, \frac{2\pi R^3}{3}\, (\varrho_1 + \varrho_2 - 2\,\varrho)\,, \tag{S-3.68}$$

where g is the gravitational acceleration. Thus, at equilibrium, the electric charge on the sphere must be

$$Q = \sqrt{\frac{4\pi R^5\, (\varepsilon_{r1} + \varepsilon_{r2})^2\, (\varrho_1 + \varrho_2 - 2\,\varrho)\, g}{3\, k_e\, (\varepsilon_{r1} - \varepsilon_{r2})}}\,. \tag{S-3.69}$$

S-3.8 Measuring the Dielectric Constant of a Liquid

The partially filled capacitor is equivalent to two capacitors connected in parallel, one with vacuum between the plates, and the other filled by the dielectric liquid. The two capacitors have the same internal and external radii, a and b, but different lengths, $\ell - h$ and h, respectively. The total capacitance is

$$C = \frac{\ell - h}{2k_e\, \ln(b/a)} + \frac{\varepsilon_r h}{2k_e\, \ln(b/a)} = \frac{\ell + (\varepsilon_r - 1)\, h}{2k_e\, \ln(b/a)}\,, \tag{S-3.70}$$

and the electrostatic energy of the capacitor is

$$U_{es} = \tfrac{1}{2}\, C V^2 = \frac{\ell + (\varepsilon_r - 1)\, h}{4k_e\, \ln(b/a)}\, V^2\,. \tag{S-3.71}$$

If the liquid raises by an amount dh the capacity increases by

$$dC = \frac{(\varepsilon_r - 1)\, dh}{2k_e\, \ln(b/a)}\,, \tag{S-3.72}$$

and, if the potential difference V across the capacitor plates is kept constant, the electrostatic energy of the capacitor increases by an amount

$$dU_{es} = \frac{1}{2}V^2 dC = \frac{(\varepsilon_r - 1)\,dh}{4k_e \ln(b/a)}\,V^2\,. \tag{S-3.73}$$

Simultaneously the voltage source does a work

$$dW = V\,dQ = V^2\,dC = \frac{(\varepsilon_r - 1)\,dh}{2k_e \ln(b/a)}\,V^2\,, \tag{S-3.74}$$

because the charge of the capacitor must increase by $dQ = V\,dC$ in order to keep the potential difference across the plates constant, and this implies moving a charge dQ from one plate to the other. The energy of the voltage source changes by

$$dU_{source} = -\frac{(\varepsilon_r - 1)\,dh}{2k_e \ln(b/a)}\,V^2 = -2\,dU_{es}\,. \tag{S-3.75}$$

We must still evaluate the increase in gravitational potential energy of the liquid. When the liquid raises by dh an infinitesimal annular cylinder of mass $dm = \varrho\,\pi(b^2 - a^2)\,dh$ is added at its top, and the gravitational energy increases by

$$dU_g = g\,\varrho\,\pi(b^2 - a^2)\,h\,dh\,. \tag{S-3.76}$$

The *total* energy variation is thus

$$\begin{aligned}
dU_{tot} &= dU_{es} + dU_{source} + dU_g = -dU_{es} + dU_g \\
&= -\frac{(\varepsilon_r - 1)\,dh}{4k_e \ln(b/a)}\,V^2 + g\,\varrho\,\pi(b^2 - a^2)\,h\,dh\,,
\end{aligned} \tag{S-3.77}$$

and the total force is

$$F = -\frac{\partial U_{tot}}{\partial h} = \frac{\varepsilon_r - 1}{4k_e \ln(b/a)}\,V^2 - g\,\varrho\,\pi(b^2 - a^2)\,h\,. \tag{S-3.78}$$

At equilibrium we have $F = 0$, which corresponds to

$$h = \frac{(\varepsilon_r - 1)\,V^2}{4\pi k_e\,g\,\varrho(a^2 - b^2)\ln(b/a)} \quad \text{and} \quad \varepsilon_r = 1 + \frac{4\pi k_e\,g\,\varrho(a^2 - b^2)\ln(b/a)}{V^2}\,h\,. \tag{S-3.79}$$

For the electric susceptibility χ, we have in SI units $\chi = \varepsilon_r - 1$, and

$$\chi = \frac{g\,\varrho\,(a^2 - b^2)\ln(b/a)}{\varepsilon_0 V^2}\,h\,, \tag{S-3.80}$$

while in Gaussian units we have $\chi = (\varepsilon_r - 1)/4\pi$, and

$$\chi = \frac{g\,\varrho\,(a^2 - b^2)\ln(b/a)}{V^2}\,h\,. \tag{S-3.81}$$

S-3.9   A Conducting Cylinder in a Dielectric Liquid

(a) We choose a cylindrical coordinate system (r, ϕ, z) with the longitudinal axis z superposed to the axis of the conducting cylinder, and the origin O at the height of the boundary surface between the dielectric liquid and the vacuum above it (Fig. S-3.15). The azimuthal angle ϕ is irrelevant for the present problem. The electric field $\mathbf{E}(r, \phi, z)$ is perpendicular to the surface of the cylinder, thus we have $\mathbf{E}(r, \phi, z) \equiv [E_r(r, z), 0, 0]$. The field is continuous at the dielectric-vacuum boundary surface, since it is parallel to it. We thus have $E_r(r, z) = E_r(r)$, independently of z. Let us denote by σ_1 and σ_2 the free-charge surface densities on the cylinder lateral surface for $z > 0$ and $z < 0$, respectively. Quantities σ_1 and σ_2 are related to the electric field at the cylinder surface, $E_r(a)$, to ε_r and Q by

$$\sigma_1 = \frac{E_r(a)}{4\pi k_e}, \qquad \sigma_2 = \frac{\varepsilon_r E_r(a)}{4\pi k_e},$$

$$Q = 2\pi a \left[\sigma_1 h + \sigma_2 (L - h)\right]. \qquad \text{(S-3.82)}$$

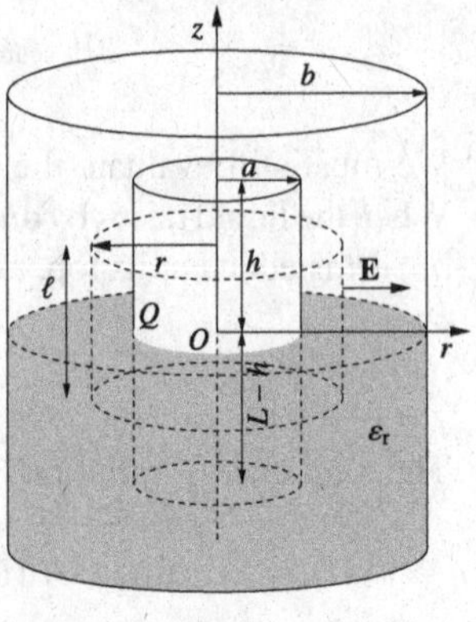

We thus have

$$E_r(a) = Q \frac{2k_e}{a[h + \varepsilon_r(L - h)]}$$

$$= Q \frac{2k_e}{a[\varepsilon_r L - (\varepsilon_r - 1) h]}. \qquad \text{(S-3.83)}$$

Fig. S-3.15

(b) The electric field $E_r(r)$ in the region $a < r < b$ can be evaluated by applying Gauss's law to a closed cylindrical surface of radius r and height $\ell \ll L$, coaxial to the conducting cylinder. Neglecting the boundary effects, the flux of the electric field through the bases of the Gaussian surface is zero, and we have

$$2\pi r \, \ell E_r(r) = 4\pi k_e Q_{\text{int}},$$

$$E_r(r) = \frac{2k_e Q_{\text{int}}}{r \ell}, \qquad \text{(S-3.84)}$$

where Q_{tot} is the *total* charge inside the Gaussian surface, including both free and polarization charges. If we let r approach a keeping ℓ constant, Q_{int} remains constant and we have

$$\lim_{r \to a} E_r = \frac{2k_e Q_{\text{int}}}{a \ell} = E_r(a), \quad \text{so that} \quad E_r(r) = E_r(a)\frac{a}{r} \qquad \text{(S-3.85)}$$

and, inserting (S-3.83),

$$E_r(r) = \frac{2k_e Q}{r \left[\varepsilon_r L - (\varepsilon_r - 1) h\right]}. \qquad \text{(S-3.86)}$$

(c) The electrostatic energy of the system is

$$
U_{es} \simeq \frac{1}{8\pi k_e} \left[\varepsilon_r \int_0^{L-h} dz \int_a^b E_r^2(r) \, 2\pi r \, dr + \int_{L-h}^L dz \int_a^b E_r^2(r) \, 2\pi r \, dr \right]
$$

$$
= \frac{1}{8\pi k_e} \left[\frac{2k_e Q}{[\varepsilon_r L - (\varepsilon_r - 1) h]} \right]^2 \left\{ [\varepsilon_r(L-h) + h] \int_a^b \frac{2\pi}{r} \, dr \right\}
$$

$$
= k_e \left[\frac{Q}{[\varepsilon_r L - (\varepsilon_r - 1) h]} \right]^2 [\varepsilon_r(L-h) + h] \ln\left(\frac{b}{a}\right)
$$

$$
= k_e \frac{Q^2 \ln(b/a)}{\varepsilon_r L - (\varepsilon_r - 1) h}, \tag{S-3.87}
$$

i.e., the electrostatic energy of two cylindrical capacitors connected in parallel, with total charge Q. Both capacitors have internal radius a and external radius b, one has length $L - h$ and is filled with the dielectric material, the other has length h and vacuum between the plates. The electrostatic force, directed along z, is

$$
F_{es} = -\frac{dU_{es}}{dh} = -k_e \frac{(\varepsilon_r - 1) \ln(b/a) \, Q^2}{[\varepsilon_r L - (\varepsilon_r - 1) h]^2} < 0. \tag{S-3.88}
$$

The electrostatic forces tends to decrease h, i.e., to sink the cylinder into the liquid.
(d) The sum of the gravitational and buoyant (due to Archimedes' principle) forces on the cylinder is

$$
F_g = -Mg + \varrho g(L - h) \pi a^2, \tag{S-3.89}
$$

and the cylinder is in equilibrium when $F_{es} + F_g = 0$, i.e., when

$$
\varrho g(L - h) \pi a^2 - Mg = k_e \frac{(\varepsilon_r - 1) \ln(b/a) \, Q^2}{[\varepsilon_r L - (\varepsilon_r - 1) h]^2}. \tag{S-3.90}
$$

Given L, h and ε_r, we have equilibrium for

$$
Q = [\varepsilon_r L - (\varepsilon_r - 1) h] \sqrt{\frac{\varrho g(L - h) \pi a^2 - Mg}{k_e (\varepsilon_r - 1) \ln(b/a)}}. \tag{S-3.91}
$$

S-3.10 A Dielectric Slab in Contact with a Charged Conductor

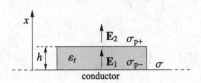

Fig. S-3.16

(a) Within our approximations, the electric fields are perpendicular to the conducting surface. We choose a Cartesian reference frame with the origin on the conductor surface and the x axis perpendicular the surface, as in Fig. S-3.16, so that the only nonzero component of the electric fields is their x component. We denote by E_1 the electric field field inside the dielectric slab, and by E_2 the electric field in vacuum, while the field will is zero inside the conductor.

The fields E_1 and E_2 can be evaluated by applying equation $\nabla \cdot (\varepsilon_r \mathbf{E}) = 4\pi k_e \varrho_f$ to two Gaussian "pillboxes", crossing the $x = 0$ and the $x = h$ surfaces, respectively, as in Fig. S-3.17. We see that $\varepsilon_r E$ is discontinuous at $x = 0$ surface, and continuous at $x = h$:

$$\varepsilon_r E_1 = 4\pi k_e \,\sigma \,,$$
$$E_2 = \varepsilon_r E_1 \,, \qquad\qquad \text{(S-3.92)}$$

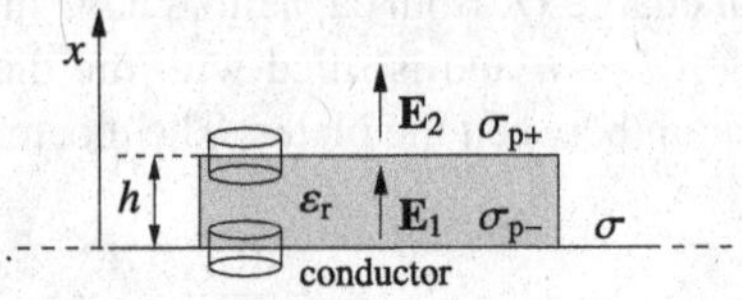

Fig. S-3.17

which lead to

$$E_1 = \frac{4\pi k_e}{\varepsilon_r}\sigma \,, \quad E_2 = 4\pi k_e\,\sigma \,. \qquad\qquad \text{(S-3.93)}$$

(b) We denote by σ_{b-} and σ_{b+} the surface polarization charge densities at $x = 0$ and $x = h$, respectively. These quantities can be calculated by applying Gauss's law $\nabla \cdot \mathbf{E} = 4\pi k_e (\varrho_f + \varrho_b)$ to the two "pillboxes" of Fig. S-3.17, and obtaining

$$E_1 = 4\pi k_e(\sigma + \sigma_{b-})\,, \quad E_2 - E_1 = 4\pi k_e\,\sigma_{b+}\,, \qquad \text{(S-3.94)}$$

introducing (S-3.92) into (S-3.94) we finally have

$$\sigma_{b+} = -\sigma_{b-} = \left(1 - \frac{1}{\varepsilon_r}\right)\sigma = \frac{\varepsilon_r - 1}{\varepsilon_r}\sigma \,. \qquad \text{(S-3.95)}$$

(c) In the vacuum region between the conductor and the dielectric slab the field is $E = 4\pi k_e\,\sigma = E_2$, independent of s. The electric field inside the dielectric slab, and above the slab, are E_1 and E_2, respectively, as in the case of $s = 0$, thus independently of s (Fig. S-3.18).

The net electrostatic force on the slab is
zero, independently of s, since the forces
on the upper and lower surfaces of the slab
are exactly opposite. Further, if we evaluate the electrostatic energy of the system
as the volume integral of $\varepsilon_r E^2/(8\pi k_e)$, we
see that also this quantity is independent
of s, within our approximations.

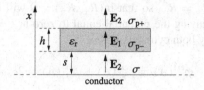

Fig. S-3.18

S-3.11   A Transversally Polarized Cylinder

We choose a cylindrical coordinate system (r, ϕ, z), with the cylinder axis as z axis,
and the reference plane (the plane from which the angle ϕ is measured) parallel to **P**.
We have translational symmetry along z, so that, mathematically, the problem is two-
dimensional. The surface charge polarization density of the cylinder is $\sigma(\phi) = \mathbf{P} \cdot \hat{\mathbf{n}}$,
where $\hat{\mathbf{n}}$ is the outgoing unit vector perpendicular to the cylinder surface, thus

$$\sigma(\phi) = P \cos \phi . \tag{S-3.96}$$

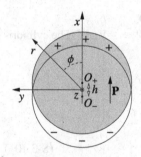

Fig. S-3.19

As suggested in the hint, our transversally polarized cylinder can be considered as the limit for $h \to 0$
and $\varrho \to \infty$ of two partially overlapping cylinders,
of volume charge density $\pm\varrho$, respectively. The two
cylinder axes are the straight lines $x = \pm h/2$, both out
of paper in Fig. S-3.19. The product ϱh is constant,
and equals the polarization P of the original cylinder.

The electrostatic potential $\varphi_\pm^{\text{ext}}(A)$, generated by
each charged cylinder at an external point $A \equiv$
(r, ϕ, z), equals the potential of an infinite line charge
of linear charge density $\lambda_\pm = \pm\pi a^2 \varrho$, located on the
cylinder axis,

$$\varphi_\pm^{\text{ext}}(A) = \mp 2k_e \pi a^2 \varrho \ln \left(\frac{r_\pm}{R_\pm} \right) , \tag{S-3.97}$$

where

$$r_\pm \simeq r \mp \frac{h}{2} \cos \phi \tag{S-3.98}$$

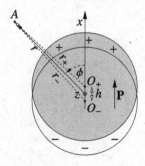

Fig. S-3.20

are the distances of A from the axes of the two cylinders, see Fig. S-3.20. Quantities $R_\pm$ are two arbitrary
constants, such that $\varphi_\pm^{\text{ext}}(r_\pm, \phi, z) = 0$ on the cylindrical surfaces $r_\pm = R_\pm$. It is convenient to choose

$R_+ = R_-$, so that $\ln(R_+/R_-) = 0$ will cancel out in the following computations, leaving the potential equal to zero at $r = \infty$. The electrostatic potential generated by both cylinders is thus

$$\varphi^{\text{ext}}(A) = \varphi_+^{\text{ext}}(A) + \varphi_-^{\text{ext}}(A) = 2k_e\pi a^2 \varrho \ln\left(\frac{r_-}{r_+}\right) + 2k_e\pi a^2 \varrho \ln\left(\frac{R_+}{R_-}\right)$$

$$\simeq 2k_e\pi a^2 \varrho \ln\left[\frac{r + (h/2)\cos\phi}{r - (h/2)\cos\phi}\right] = 2k_e\pi a^2 \varrho \ln\left[\frac{1 + (h/2r)\cos\phi}{1 - (h/2r)\cos\phi}\right]$$

$$= 2k_e\pi a^2 \varrho \left[\ln\left(1 + \frac{h}{2r}\cos\phi\right) - \ln\left(1 - \frac{h}{2r}\cos\phi\right)\right]$$

$$\simeq 2k_e\pi a^2 \varrho\frac{h}{r}\cos\phi = 2k_e\pi a^2 \frac{P\cos\phi}{r} = 2k_e\pi a^2 \frac{\mathbf{P}\cdot\hat{\mathbf{r}}}{r}\,, \qquad (\text{S-3.99})$$

where $\hat{\mathbf{r}}$ is the unit vector of the cylindrical coordinate r. Thus, the potential of our *cylindrical electric dipole* (see Problem 3.13 for a *cylindrical electric quadrupole*) decreases as r^{-1}, while the potential of the ordinary electric dipole decreases as r^{-2}. In Cartesian coordinates we have

$$\varphi^{\text{ext}}(x, y, z) = 2k_e\pi a^2 \frac{Px}{x^2 + y^2}\,, \qquad (\text{S-3.100})$$

where the x and y axes are the ones shown in Fig. S-3.19.

The external electric field is obtained by evaluating $\mathbf{E}^{\text{ext}} = -\nabla\varphi^{\text{ext}}$. The cylindrical components are, from Table A.1 of the Appendix,

$$E_r^{\text{ext}} = -\partial_r\varphi^{\text{ext}} = 2k_e\pi a^2 \frac{P\cos\phi}{r^2}\,,$$

$$E_\phi^{\text{ext}} = -\frac{1}{r}\partial_\phi\varphi^{\text{ext}} = 2k_e\pi a^2 \frac{P\sin\phi}{r^2}\,,$$

$$E_z^{\text{ext}} = -\partial_z\varphi^{\text{ext}} = 0\,, \qquad (\text{S-3.101})$$

the field decreases proportionally to r^{-2}, while the field of the usual electric dipole decreases as r^{-3}. The Cartesian components of the field are

$$E_x^{\text{ext}} = -\partial_x\varphi^{\text{ext}} = 2k_e\pi a^2 P \frac{x^2 - y^2}{(x^2 + y^2)^2}\,,$$

$$E_y^{\text{ext}} = -\partial_y\varphi^{\text{ext}} = 2k_e\pi a^2 P \frac{2xy}{(x^2 + y^2)^2}\,,$$

$$E_z^{\text{ext}} = -\partial_z\varphi^{\text{ext}} = 0\,. \qquad (\text{S-3.102})$$

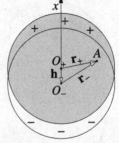

Fig. S-3.21

The electric field generated by each cylinder at its interior is, according to Gauss's law, $\mathbf{E}_{\pm}^{\text{int}} = \pm 2\pi k_e \varrho \, \mathbf{r}_{\pm}$, where $\mathbf{r}_{\pm}$ is the distance from the respective axis, see Fig. S-3.21. The two contributions sum up to a uniform internal field

$$\mathbf{E}^{\text{int}}(A) = 2\pi k_e \varrho \, (\mathbf{r}_+ - \mathbf{r}_-) = -2\pi k_e \varrho \, h \, \hat{\mathbf{x}} = -2\pi k_e \, \mathbf{P} \, . \qquad (\text{S-3.103})$$

The electrostatic potential inside the cylinder, in Cartesian and cylindrical coordinates, is thus

$$\varphi^{\text{int}} = 2\pi k_e \, x + C = 2\pi k_e \, r \cos\phi + C \, , \qquad (\text{S-3.104})$$

where C is an arbitrary constant. Since the potential most be continuous, we must have, in cylindrical coordinates,

$$\varphi^{\text{int}}(a, \phi, z) = \varphi^{\text{ext}}(a, \phi, z) \, , \qquad (\text{S-3.105})$$

which is verified if we choose $C = 0$.

S-3.12 Force Between a Parallel-Plate Capacitor and a Dielectric Slab

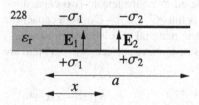

Fig. S-3.22

(a) Since the dielectric-vacuum interface at the x position is far from the boundaries, the electric field inside the capacitor can be assumed to be perpendicular to the plate surfaces, hence parallel to the interface. Thus the electric field is continuous across the interface, and, referring to Fig. S-3.22, we have $\mathbf{E}_1 = \mathbf{E}_2 = \mathbf{E}$. Now, let $\pm\sigma_1$ and $\pm\sigma_2$ be the free electric charge densities on the capacitor plates in front of the dielectric medium and in front of vacuum, respectively. The relation $\sigma_i = D_i = \varepsilon_i E_i = \varepsilon_i E$ (with $i = 1, 2$, and $\varepsilon_1 = \varepsilon_0 \, \varepsilon_r$, $\varepsilon_2 = \varepsilon_0$) leads to $\sigma_1 = \varepsilon_r \, \sigma_2$. In addition, we know that the total charges on the plates are $\pm Q$, respectively, which yields

$$b x \, \sigma_1 + b(a - x) \, \sigma_2 = Q \, , \qquad (\text{S-3.106})$$

since $\pm\sigma_1$ and $\pm\sigma_2$ cover areas equal to bx and $b(a - x)$, respectively. By substituting $\sigma_1 = \varepsilon_r \, \sigma_2$, we obtain

$$\sigma_2 = \frac{Q}{b[(\varepsilon_r - 1) \, x + a]} \, , \qquad (\text{S-3.107})$$

and

$$E = E_1 = E_2 = \frac{4\pi k_e Q}{b[(\varepsilon_r - 1)x + a]} = \begin{cases} \dfrac{Q}{\varepsilon_0 b[(\varepsilon_r - 1)x + a]}, & \text{SI} \\[3mm] \dfrac{4\pi Q}{b[(\varepsilon_r - 1)x + a]}, & \text{Gaussian.} \end{cases}$$

$$\text{(S-3.108)}$$

In order to calculate the energy U_{es} of the system we can integrate over the whole space the generalized "energy density" $u = \varepsilon_r E^2 / 8\pi k_e$. If we neglect the field outside the capacitor, the integral can be split into the two contributions from region 1 of volume xbh filled by the dielectric and the empty region 2 of volume $(a - x)bh$. Thus we obtain

$$U_{es} = \int \frac{\varepsilon_r}{8\pi k_e} E^2 \, d^3r = \frac{E^2}{8\pi k_e} [\varepsilon_r hbx + hb(a - x)]$$

$$= \frac{E^2}{8\pi k_e} hb [(\varepsilon_r - 1)x + a] . \tag{S-3.109}$$

If we now insert (S-3.108) for E we obtain

$$U_{es} = \frac{2\pi k_e h Q^2}{b[(\varepsilon_r - 1) + a]} = \begin{cases} \dfrac{h Q^2}{2\varepsilon_0 b[(\varepsilon_r - 1)x + a]}, & \text{SI} \\[3mm] \dfrac{2\pi h Q^2}{b[(\varepsilon_r - 1) + a]}, & \text{Gaussian.} \end{cases} \tag{S-3.110}$$

Alternatively, the capacitor can be considered as equivalent to two capacitors, C_1 and C_2, connected in parallel. Its capacity is thus $C = C_1 + C_2$. Both capacitors have separation h, C_1 is filled with the dielectric material of relative permittivity ε_r and its plates have surface bx, while the plates of C_2 have surface $b(a - x)$ and are separated by vacuum. Thus we have

$$C_1 = \frac{\varepsilon_r}{4\pi k_e} \frac{bx}{h} , \quad C_2 = \frac{1}{4\pi k_e} \frac{b(a - x)}{h} , \quad C = C_1 + C_2 = \frac{b[(\varepsilon_r - 1)x + a]}{4\pi k_e h} .$$

$$\text{(S-3.111)}$$

The electrostatic energy of the capacitor can be written as

$$U_{es} = \frac{Q^2}{2C} = \frac{2\pi k_e h Q^2}{b[(\varepsilon_r - 1)x + a]} , \tag{S-3.112}$$

in agreement with (S-3.110). Note that U_{es} decreases with increasing penetration depth x. Since the capacitor and the slab constitute a closed system, the force acting on the dielectric slab can be evaluated as

$$F = -\frac{\partial U_{es}}{\partial x} = \frac{2\pi k_e h Q^2 (\varepsilon_r - 1)}{b\left[(\varepsilon_r - 1)x + a\right]^2} = \begin{cases} \dfrac{h Q^2 (\varepsilon_r - 1)}{2\varepsilon_0 b\left[(\varepsilon_r - 1)x + a\right]^2}, & \text{SI} \\[3mm] \dfrac{2\pi h Q^2 (\varepsilon_r - 1)}{b\left[(\varepsilon_r - 1)x + a\right]^2}, & \text{Gaussian,} \end{cases}$$

$$\text{(S-3.113)}$$

the slab is pulled into the capacitor since $F > 0$.

(b) In the case of the capacitor connected to a battery forcing a constant potential difference V between the plates (Fig. S-3.23), the electric field is $E = V/h$ and does not depend on x. Thus, if the slab penetrates by a further amount dx,

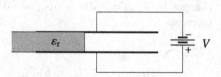

Fig. S-3.23

the dielectric medium fills the further volume $d\tau_1 = hb\,dx$ of Fig. S-3.24, whose contribution to the integral giving the total energy increases from $E^2/(8\pi k_e)\,d\tau_1$ to $\varepsilon_r E^2/(8\pi k_e)\,d\tau_1$. Simultaneously, the dielectric medium frees the volume $d\tau_2$ (of identical value as $d\tau_1$) at its external end, where we assume the field to be negligible. Thus $d\tau_2$ does not contribute to the variation of electrostatic energy. In all the remaining space the field is unchanged, thus the electrostatic energy variation is

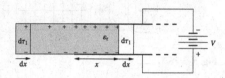

Fig. S-3.24

$$dU_{es} = \frac{(\varepsilon_r - 1) E^2}{8\pi k_e}\,d\tau_1 = (\varepsilon_r - 1)\frac{V^2 b}{8\pi k_e h}\,dx\,.$$

$$\text{(S-3.114)}$$

Of course the same result can be obtained by writing the electrostatic energy as

$$U_{es} = \frac{1}{2}\,C V^2 = \frac{b\left[(\varepsilon_r - 1)x + a\right] V^2}{8\pi k_e\,h}\,,\qquad \text{(S-3.115)}$$

since now V, not Q, is constant. Differentiation with respect to x leads immediately to Eq. (S-3.114), which can also be written as $dU_{es} = V^2\,dC$, where

$$dC = \frac{b(\varepsilon_r - 1)}{4\pi k_e\,h}\,dx\,.\qquad \text{(S-3.116)}$$

Note that now U_{es} increases with increasing penetration depth x. However, the capacitor and the slab no longer constitute a closed system, due to the presence of the battery. The closed system is thus slab + capacitor + battery. When the capacity increases by an amount dC, the charge on the plates increases by $dQ = V\,dC$ in order to keep V constant. The battery must perform a work

$$dW = V\,dQ = \frac{bV^2(\varepsilon_r - 1)}{4\pi k_e\,h}\,dx = 2\,dU_{es} \qquad (S\text{-}3.117)$$

at the expense of its internal energy in order to carry the charge dQ across a voltage drop V. Thus, when the slab penetrates by a further amount dx, the electrostatic energy of the capacitor increases by dU_{es}, but the total energy of the closed system changes by $dU_{tot} = -dW + dU_{es} = -dU_{es}$. Hence the force acting on the slab is given by

$$F = -\frac{\partial U_{tot}}{\partial x} = +\frac{\partial U_{es}}{\partial x} = \frac{bV^2(\varepsilon_r - 1)}{8\pi k_e\,h} = \begin{cases} \dfrac{\varepsilon_0 bV^2(\varepsilon_r - 1)}{2h}, & \text{SI} \\[2mm] \dfrac{bV^2(\varepsilon_r - 1)}{8\pi\,h}. & \text{Gaussian,} \end{cases}$$

$$(S\text{-}3.118)$$

We have again $F > 0$, and the slab is still pulled into the capacitor.

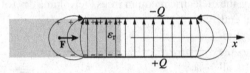

Fig. S-3.25

(c) According to our assumptions the electric field lines in the central part of the capacitor (far from the borders) are straight line segments perpendicular to the capacitor plates. The field lines are thus perpendicular also to the x axis, as shown in Fig. S-3.25. The x component of the electric field being zero, there cannot be a local pulling force on the dielectric slab in this central region. But close to the capacitor borders the field lines start to bend, and outside the capacitor they are curved as approximately shown in Fig. S-3.25. Thus they have x components which exert forces on the surface polarization charges of the slab. The resultant of all these forces pulls the slab into the capacitor.

Alternatively, the force can be calculated by observing that the electric field induces a volume polarization $\mathbf{P}$ in the dielectric slab, and the induced dipoles at the border and outside the capacitor are subject to a non-uniform electric field. This leads to a force density $\mathbf{f}_v = (\mathbf{P} \cdot \nabla)\,\mathbf{E}$, to be integrated over the slab volume. The two computations lead to the same result.

It is interesting to note that while neglecting the border effects leads to a correct evaluation of the pulling force, it masks the location where the force is applied. However, our assumption on the border effects affects also our evaluation of the energy variation, because we are assuming that the field is negligible at the far end of the dielectric slab outside the capacitor.

S-3.13   A Cylindrical Quadrupole

Let us denote by r_n, $n = 1, \ldots, 4$, the distance
of P from wire n. Wire n has the linear charge
density $(-1)^n \lambda$ and generates in P the electrostatic potential

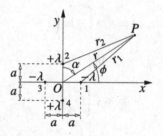

$$\varphi_n(r_n) = (-1)^n 2k_e\lambda \ln\left(\frac{r_n}{R_n}\right) , \quad \text{(S-3.119)}$$

where the R_n are arbitrary constants such that   **Fig. S-3.26**
$\varphi_n(R_n) = 0$. It is convenient to choose $R_1 =$
$R_2 = R_3 = R_4 = R_0$ so that the R_n cancel out
in our final expressions. With this choice the total potential $\varphi = \sum_{n=1}^{4} \varphi_n$ is zero at
infinity and on the two $y = \pm x$ planes. The potential in P is thus

$$\varphi(r, \phi) = 2k_e\lambda \sum_{n=1}^{4}(-1)^n \ln\left(\frac{r_n}{R_0}\right) = 2k_e\lambda \sum_{n=1}^{4}(-1)^n (\ln r_n - \ln R_0)$$

$$= 2k_e\lambda \sum_{n=1}^{4}(-1)^n \ln r_n = k_e\lambda \sum_{n=1}^{4}(-1)^n \ln r_n^2 , \quad \text{(S-3.120)}$$

where we have used $2\ln x = \ln x^2$, since, in the following, it is more convenient to
use r_n^2 than r_n. For the sake of clarity Fig. S-3.26 shows only the distances r_1 and r_2.
According to the cosine law we have

$$r_1^2 = r^2 + a^2 - 2ar\cos\phi , \quad r_2^2 = r^2 + a^2 - 2ar\cos\alpha = r^2 + a^2 - 2ar\sin\phi ,$$
$$\text{(S-3.121)}$$

analogously we have

$$r_3^2 = r^2 + a^2 + 2ar\cos\phi , \quad r_4^2 = r^2 + a^2 + 2ar\sin\phi . \quad \text{(S-3.122)}$$

(a) For the far field, $r \gg a$, it is convenient to rewrite (S-3.121) and (S-3.122) in the
form

$$r_1^2 = r^2\left(1 + \frac{a^2}{r^2} - 2\frac{a}{r}\cos\phi\right) , \quad r_2^2 = r^2\left(1 + \frac{a^2}{r^2} - 2\frac{a}{r}\sin\phi\right) ,$$
$$r_3^2 = r^2\left(1 + \frac{a^2}{r^2} + 2\frac{a}{r}\cos\phi\right) , \quad r_4^2 = r^2\left(1 + \frac{a^2}{r^2} + 2\frac{a}{r}\sin\phi\right) . \quad \text{(S-3.123)}$$

Equation (S-3.120) thus becomes, remembering that $\ln(xy) = \ln x + \ln y$, so that all terms $\ln r^2$ cancel out,

$$
\begin{aligned}
\varphi(r, \phi) = k_e \lambda \Bigg[& -\ln\left(1 + \frac{a^2}{r^2} - 2\frac{a}{r}\cos\phi\right) + \ln\left(1 + \frac{a^2}{r^2} - 2\frac{a}{r}\sin\phi\right) \\
& -\ln\left(1 + \frac{a^2}{r^2} + 2\frac{a}{r}\cos\phi\right) + \ln\left(1 + \frac{a^2}{r^2} + 2\frac{a}{r}\sin\phi\right) \Bigg] \\
= k_e \lambda \Bigg\{ \ln & \left[\left(1 + \frac{a^2}{r^2} - 2\frac{a}{r}\sin\phi\right)\left(1 + \frac{a^2}{r^2} + 2\frac{a}{r}\sin\phi\right)\right] \\
& -\ln\left[\left(1 + \frac{a^2}{r^2} - 2\frac{a}{r}\cos\phi\right)\left(1 + \frac{a^2}{r^2} + 2\frac{a}{r}\cos\phi\right)\right] \Bigg\} . \quad \text{(S-3.124)}
\end{aligned}
$$

Now we expand the products inside the square brackets, disregarding the terms in $(a/r)^4$, obtaining

$$
\varphi(r, \phi) = k_e \lambda \left[\ln\left(1 + 2\frac{a^2}{r^2} - 4\frac{a^2}{r^2}\sin^2\phi\right) - \ln\left(1 + 2\frac{a^2}{r^2} - 4\frac{a^2}{r^2}\cos^2\phi\right) \right] . \quad \text{(S-3.125)}
$$

Now, since $a \ll r$, we can use the approximation $\ln x \simeq 1 + x$, which leads to

$$
\varphi(r, \phi) = 4k_e \lambda \frac{a^2}{r^2}\left(\cos^2\phi - \sin^2\phi\right) = k_e \frac{4a^2\lambda\cos(2\phi)}{r^2} . \quad \text{(S-3.126)}
$$

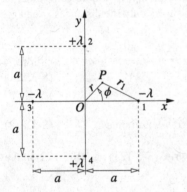

For the electric field in $P \equiv (r, \phi)$ we obtain

$$
E_r = -\frac{\partial\phi}{\partial r} = k_e \frac{8a^2\lambda\cos(2\phi)}{r^3} ,
$$

$$
\text{(S-3.127)}
$$

$$
E_\phi = -\frac{1}{r}\frac{\partial\varphi}{\partial\phi} = k_e \frac{8a^2\lambda\sin(2\phi)}{r^3} .
$$

$$
\text{(S-3.128)}
$$

Fig. S-3.27

(b) The electrostatic potential generated by the four wires close to the z axis ($r \ll a$) is still given by (S-3.120). Fig. S-3.27 shows the distance r_1 of P from wire 1. Quantities $r_1^2, \ldots, r_4^2$ are still given by (S-3.121) and (S-3.122), but now, since we are considering $r \ll a$, it is convenient to rewrite them as

$$r_1^2 = a^2 \left(1 + \frac{r^2}{a^2} - 2\frac{r}{a}\cos\phi \right) , \quad r_2^2 = a^2 \left(1 + \frac{r^2}{a^2} - 2\frac{r}{a}\sin\phi \right) ,$$

$$r_3^2 = a^2 \left(1 + \frac{r^2}{a^2} + 2\frac{r}{a}\cos\phi \right) , \quad r_4^2 = a^2 \left(1 + \frac{r^2}{a^2} + 2\frac{r}{a}\sin\phi \right) . \quad \text{(S-3.129)}$$

Inserting (S-3.129) into (S-3.120), and proceeding analogously to point **(a)**, we obtain

$$\varphi(r, \phi) \simeq k_e \frac{4\lambda r^2 \cos(2\phi)}{a^2} . \tag{S-3.130}$$

Formally, with respect to point **(a)** variables a and r are simply swapped in all formulae. For the electric field at $r \ll a$ we have thus

$$E_r = -\frac{\partial \phi}{\partial r} = k_e \frac{8\lambda r \cos(2\phi)}{a^2} , \quad E_\phi = -\frac{1}{r}\frac{\partial \varphi}{\partial \phi} = k_e \frac{8\lambda r \sin(2\phi)}{a^2} . \quad \text{(S-3.131)}$$

Figure S-3.28 shows the field lines of the cylindrical quadrupole, while Fig. S-3.29 shows the intersections of the equipotential surfaces with the $z = 0$ plane.

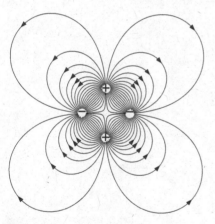

Fig. S-3.28 .   **Fig. S-3.29** .

Chapter S-4
Solutions for Chapter 4

S-4.1 The Tolman-Stewart Experiment

(a) The equation of motion for the "free" (conduction) electrons in a metal is, according to the Drude model,

$$m \frac{d\langle v \rangle}{dt} = \mathbf{F} - m\eta \, \langle v \rangle \, , \tag{S-4.1}$$

where $\langle v \rangle$ is the "average" electron velocity, $\mathbf{F}$ is the external force on the electrons, and $m\eta \, \langle v \rangle$ is a phenomenological friction force. In a steady state $(d \langle v \rangle /dt = 0)$ in the presence of an external electric field $\mathbf{E}$, so that $\mathbf{F} = -e\mathbf{E}$, the electrons have a constant average velocity

$$\langle v \rangle = -\frac{e}{m\eta} \mathbf{E} \, . \tag{S-4.2}$$

The current density is $\mathbf{J} = -e \, n_e \langle v \rangle$, where n_e is the volume density of free electrons. From this we obtain the microscopic form of Ohm's law

$$\mathbf{J} = \frac{n_e e^2}{m\eta} \mathbf{E} \equiv \sigma \mathbf{E} \, . \tag{S-4.3}$$

The value of the damping frequency η for copper is

$$\eta = \frac{n_e e^2}{m\sigma} = \frac{8.5 \times 10^{28}(1.6 \times 10^{-19})^2}{9.1 \times 10^{-31} \times 10^7} \simeq 2.4 \times 10^{14} \, \text{s}^{-1} \, , \tag{S-4.4}$$

$(m = m_e = 9.1 \times 10^{-31} \text{ kg})$.

At $t = 0$ the electron tangential velocity is $v_0 = a\omega$. For $t > 0$, due to the absence of external forces the solution of Eq. (S-4.1) is

© The Editor(s) (if applicable) and The Author(s), under exclusive license to Springer 239
Nature Switzerland AG 2023
A. Macchi et al., *Problems in Classical Electromagnetism*,
https://doi.org/10.1007/978-3-031-22235-1

$$v = v_0 \, e^{-\eta t}. \tag{S-4.5}$$

The total current is thus given by

$$I = I_0 \, e^{-\eta t}, \qquad I_0 = -(e n_e v_0) \, S, \tag{S-4.6}$$

and the decay time is $\tau = 1/\eta \simeq 4 \times 10^{-15}$ s.

(b) The total charge flown in the ring is

$$Q = \int_0^\infty I(t) \, dt = \frac{I_0}{\eta} = -\frac{m}{e} \sigma S v_0. \tag{S-4.7}$$

Thus, measuring σ, S, v_0 and Q the value of e/m can be obtained. In the original experiment, Tolman and Stewart were able to measure Q using a ballistic galvanometer in a circuit coupled with a rotating coil.

S-4.2 Charge Relaxation in a Conducting Sphere

(a) For symmetry reasons the electric field is radial, and it is convenient to use a spherical coordinate system (r, θ, ϕ) with the origin located at the center of the sphere. Coordinates θ and ϕ are irrelevant for this problem. Let us denote by $q(r, t)$ the electric charge contained inside the sphere $r < a$, at time $t \geqslant 0$. If we apply Gauss's law to the surface of our sphere we obtain

$$E(r, t) = k_e \frac{q(r, t)}{r^2}. \tag{S-4.8}$$

According to the continuity equation, the flux of the current density $\mathbf{J} = \sigma \mathbf{E}$ through our spherical surface equals the time derivative of $q(r, t)$:

$$\oint \mathbf{J} \cdot d\mathbf{S} = 4\pi r^2 J(r, t) = 4\pi r^2 \sigma E(r, t) = -\partial_t q(r, t). \tag{S-4.9}$$

By substituting (S-4.8) into (S-4.9) we obtain

$$\partial_t q(r, t) = -4\pi k_e \, \sigma \, q(r, t), \tag{S-4.10}$$

with solution

$$q(r, t) = q(r, 0) \, e^{-t/\tau}, \quad \text{where} \quad \tau = \frac{1}{4\pi k_e \, \sigma}. \tag{S-4.11}$$

Since at $t = 0$ the charge density $\varrho(r, t)$ is uniform all over the volume of the sphere of radius a, we have

$$\varrho(r, 0) = \varrho_0 = Q\,\frac{3}{4\pi a^3}\,, \quad r < a\,, \quad \text{so that} \quad q(r, 0) = Q\,\frac{r^3}{a^3}\,. \tag{S-4.12}$$

Thus, according to (S-4.11), the density $\varrho(r, t)$ remains uniform over the sphere volume (independent of r) at any time $t > 0$

$$\varrho(r, t) = \varrho(t) = \varrho_0\, \mathrm{e}^{-t/\tau}\,. \tag{S-4.13}$$

The *surface* charge density $q_s(t)$ (we have already used the Greek letter σ for the conductivity) can also be evaluated from the continuity equation, since

$$\partial_t q_s(t) = +J(a, t) = \sigma E(a, t) = k_e\, \sigma\frac{Q}{a^2}\, \mathrm{e}^{-t/\tau} = \frac{Q}{4\pi a^2 \tau}\, \mathrm{e}^{-t/\tau}\,, \tag{S-4.14}$$

so that, asymptotically,

$$q_s(\infty) = \int_0^\infty \partial_t q_s\, \mathrm{d}t = \frac{Q}{4\pi a^2 \tau} \int_0^\infty \mathrm{e}^{-t/\tau}\mathrm{d}t = \frac{Q}{4\pi a^2}\,. \tag{S-4.15}$$

The equation for the time evolution of the electric field inside the sphere ($r < a$) is

$$E(r, t) = k_e\,\frac{q(r, t)}{r^2} = k_e\, Q\,\frac{r}{a^3}\, \mathrm{e}^{-t/\tau}\,, \quad r < a\,, \tag{S-4.16}$$

while the electric field is independent of time outside the sphere

$$E(r, t) = E(r) = k_e\,\frac{Q}{r^2}\,, \quad r > a\,. \tag{S-4.17}$$

The time constant $\tau = 1/(4\pi k_e\sigma)$ is extremely short in a good conductor. For copper we have (in SI units) $\sigma \simeq 6 \times 10^7\ \Omega^{-1}\mathrm{m}^{-1}$ at room temperature, thus

$$\tau = \frac{\varepsilon_0}{\sigma} \simeq \frac{8.854 \times 10^{-12}}{6 \times 10^7}\ \mathrm{s} \sim 1.4 \times 10^{-19}\ \mathrm{s} = 0.14\ \mathrm{as} \tag{S-4.18}$$

(1 as = 1 *attosecond* = 10^{-18} s: *atten* means *eighteen* in Danish). This extremely short value should be not surprising, since there is no need for the electrons to travel distances even of the order of the atomic spacing within the relaxation time; a very small collective displacement of the electrons is sufficient to reach equilibrium conditions (see also Problem 2.1). This very short relaxation time shows that, experimentally, it is very difficult to achieve the initial conditions of the problem.

(b) We can easily evaluate the variation of electrostatic energy ΔU_{es} during the charge relaxation by noticing that the electric field $E(r, t)$ is constant outside the conducting sphere $(r > a)$. The electric field inside the sphere decays from the initial profile $E(r, 0) = (Q/4\pi\varepsilon_\theta)(r/a^3)$ to $E(r, \infty) = 0$. Thus, using the "energy density" $u_{es} = E^2/(8\pi k_e)$ we can write

$$\Delta U_{es} = -\frac{1}{8\pi k_e} \int_{\text{sphere}} E^2(r, 0)\, d^3x = -\frac{1}{8\pi k_e} \left(\frac{k_e^2 Q^2}{a^3}\right) \int_0^a r^2\, 4\pi r^2\, dr$$

$$= -\frac{k_e}{10} \frac{Q^2}{a}. \tag{S-4.19}$$

(c) The time derivative of the electrostatic energy can be written as

$$\partial_t U_{es} = \frac{1}{8\pi k_e} \partial_t \int_0^\infty E^2(r, t)\, 4\pi r^2\, dr = \frac{1}{8\pi k_e} \int_0^a \partial_t E^2(r, t)\, 4\pi r^2 dr$$

$$= \frac{1}{8\pi k_e} \int_0^a \left(-\frac{2}{\tau}\right) E^2(r, t)\, 4\pi r^2 dr = -\frac{1}{4\pi k_e \tau} \int_0^a k_e^2 Q^2 \frac{r^2}{a^6}\, e^{-2t/\tau} 4\pi r^2 dr$$

$$= -\frac{k_e}{5\tau} \frac{Q^2}{a}\, e^{-2t/\tau} = -\frac{4\pi k_e^2 \sigma Q^2}{5a}\, e^{-2t/\tau}, \tag{S-4.20}$$

where we used (S-4.16) and (S-4.17). The power loss due to Joule heating is

$$P_d = \int_0^\infty \mathbf{J} \cdot \mathbf{E}\, 4\pi r^2 dr = \int_0^a \sigma E^2(r, t)\, 4\pi r^2 dr$$

$$= \frac{4\pi k_e^2 \sigma Q^2}{5a}\, e^{-2t/\tau}, \tag{S-4.21}$$

since $\mathbf{J} = \sigma \mathbf{E}$ for $r < a$, and $\mathbf{J} = 0$ for $r > a$. Thus $P_d = -\partial_t U_{es}$, and all the electrostatic energy lost by the sphere during the relaxation process is turned into Joule heat.

S-4.3 A Coaxial Resistor

(a) We use a cylindrical coordinate system (r, ϕ, z), with the z axis coinciding with the common axis of the cylindrical plates. The material between the plates can be considered as a series of infinitesimal cylindrical-shell resistors, each of internal and external radii r and $r + dr$ and of height h. The resistance of the cylindrical shell between r and $r + dr$ is

$$dR = \rho \frac{dr}{S(r)} = \rho \frac{dr}{2\pi r h}, \tag{S-4.22}$$

since dr is the "length" of our resistor, and $S(r) = 2\pi r$ its "cross-sectional area". The resistance of the material is thus

$$R = \frac{\rho}{2\pi h} \int_a^b \frac{dr}{r} = \rho \frac{\ln(b/a)}{2\pi h}. \tag{S-4.23}$$

(b) The capacity of a cylindrical capacitor of radii a and b, and length h, is

$$C_0 = \frac{h}{2k_e \ln(b/a)}, \tag{S-4.24}$$

assuming that the space between the plates is filled with vacuum. Thus, in our case we have

$$R = \frac{\rho}{4\pi k_e C_0}. \tag{S-4.25}$$

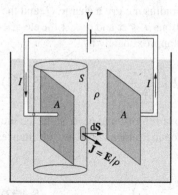

Fig. S-4.1

Equation (S-4.25) is actually of much more general validity, and is a very good approximation for evaluating the resistance between two electrodes of high conductivity and calculable capacity immersed in a medium of known resistivity. As an example, consider two highly-conducting square plates immersed in an ohmic medium, and connected to a voltage source by insulated cables, as in Fig. S-4.1. The current that flows, for instance, from the left plate, can be written

$$I = \int \mathbf{J} \cdot d\mathbf{S} = \frac{1}{\rho} \int \mathbf{E} \cdot d\mathbf{S}, \tag{S-4.26}$$

where the flux is calculated through a surface enclosing the electrode, except for the area through which the current enters it, like the cylindrical closed surface of Fig. S-4.1. In most cases the contribution of the excluded area to the flux of $\mathbf{E}$ is negligible in an electrostatic problem, while, according to Gauss's law, we have

$$\oint \mathbf{E} \cdot d\mathbf{S} = 4\pi k_e Q \tag{S-4.27}$$

where Q is the charge on the electrode that would produce the field $\mathbf{E}$. Within the approximation of considering the last integral of (S-4.26) as equal to the integral through the whole closed surface, we have

$$I = \frac{4\pi k_e}{\rho} Q . \tag{S-4.28}$$

On the other hand, if we consider the two electrodes as the plates of a capacitor of capacitance C_0 we have

$$Q = C_0 V , \tag{S-4.29}$$

where V is the potential difference between them. We thus have

$$I = \frac{4\pi k_e}{\rho} C_0 V = \frac{V}{R} , \tag{S-4.30}$$

from which (S-4.25) follows.

S-4.4   Electrical Resistance Between Two Submerged Spheres (1)

(a) We start by evaluating the capacitance C_0 of the two spheres in vacuum, with the same geometry of the problem. Let the sphere of radius a carry a charge Q, and the sphere of radius b a charge $-Q$. With our assumptions $a \ll x$ and $b \ll x$ the electric potentials φ_a and φ_b of the two spheres are given approximately by

$$\varphi_a \simeq k_e Q \left(\frac{1}{a} - \frac{1}{x} \right) \quad \text{and} \quad \varphi_b \simeq k_e Q \left(-\frac{1}{b} + \frac{1}{x} \right) , \tag{S-4.31}$$

where we have assumed the potential φ to be zero at infinity, and have neglected the induction effects between the two spheres, discussed in Problem 2.6. The capacitance of the two spheres can thus be approximated as

$$C_0 = \frac{Q}{\varphi_a - \varphi_b} \simeq \frac{1}{k_e} \left(\frac{1}{a} + \frac{1}{b} - \frac{2}{x} \right)^{-1} , \tag{S-4.32}$$

and, according to (S-4.25), the resistance between them is

$$R = \frac{\rho}{4\pi k_e C_0} \simeq \frac{\rho}{4\pi} \left(\frac{1}{a} + \frac{1}{b} - \frac{2}{x} \right) , \tag{S-4.33}$$

which can be further approximated to

$$R \simeq \frac{\rho}{4\pi} \left(\frac{1}{a} + \frac{1}{b} \right) , \tag{S-4.34}$$

independent of the distance between x between the centers of the spheres.

(b) In this case the resistance between the spheres is twice the value found at point **(a)**, since at point **(a)** we can introduce a horizontal plane passing through the centers of the spheres, which divides the fluid into two equivalent halves, each of resistance $2R$, so that, in parallel, they are equivalent to a resistance R. In the present case the upper half is replaced by vacuum, so that only the resistance $2R$ of the lower half remains. This problem is of interest in connection with electrical circuits that use the ground as a return path. In this case ρ is the resistivity of the earth (of course, the assumption that ρ is uniform is a very rough approximation). In practical applications, the resistivity of the earth in the neighborhood of the electrodes can be decreased by moistening the ground around them.

S-4.5 Electrical Resistance Between Two Submerged Spheres (2)

(a) According to (S-4.32) of Problem 4.4, and remembering that now the medium has a relative dielectric permittivity ε_r, the charge of each sphere is

$$Q \simeq \varepsilon_r C_0 V = \frac{\varepsilon_r}{k_e} \left(\frac{2}{a} - \frac{2}{\ell} \right)^{-1} V = \frac{\varepsilon_r}{k_e} \frac{\ell}{\ell - a} \frac{a}{2} V$$
$$\simeq \frac{\varepsilon_r a}{2k_e} \left(1 + \frac{a}{\ell} \right) V \simeq \frac{\varepsilon_r a}{2k_e} V \,, \tag{S-4.35}$$

where the last two terms are the first and the zeroth order approximations in a/ℓ.
(b) According to (S-4.33) and (S-4.34) we have

$$R \simeq \frac{\rho}{4\pi} \left(\frac{2}{a} - \frac{2}{\ell} \right) \simeq \frac{\rho}{2\pi a} \,, \tag{S-4.36}$$

again to the zeroth order in a/ℓ. The current I is thus

$$I = \frac{V}{R} = \frac{2\pi a}{\rho} V \,. \tag{S-4.37}$$

This result can be checked by introducing a cylindrical coordinate system (r, ϕ, z) with the z axis through the centers of the two spheres and the origin O so that the sphere centers are at $(0, \phi, -\ell/2)$ and $(0, \phi, +\ell/2)$, respectively, and evaluating the flux of the current density $\mathbf{J}$ through the plane $z = 0$ (Fig. S-4.2).

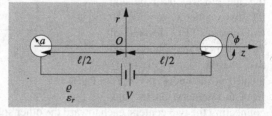

Fig. S-4.2

$$I = \int \mathbf{J} \cdot d\mathbf{S}$$

$$= \frac{1}{\rho} \int_0^\infty E_z(r)\, 2\pi r\, dr \,, \tag{S-4.38}$$

where

$$E_z(r) = \frac{2k_e Q}{\varepsilon_r}\, \frac{\ell/2}{[(\ell/2)^2 + r^2]^{3/2}} \,, \tag{S-4.39}$$

so that

$$I = \frac{k_e Q}{\rho \varepsilon_r} \int_0^\infty \frac{\ell r}{[(\ell/2)^2 + r^2]^{3/2}}\, dr = \frac{4\pi k_e}{\rho \varepsilon_r} Q = \frac{2\pi a}{\rho} V. \tag{S-4.40}$$

(c) Having a system equivalent to a capacitor in parallel with a resistor, we expect an exponential decay of the charge Q, with a time constant $\tau = RC$. The time constant is independent of the geometry of the problem because the capacitance C of the system is $\varepsilon_r C_0$, where C_0 is the capacitance when the medium is replaced by vacuum, while, according to (S-4.25), the resistance is $R = \rho/(4\pi k_e C_0)$, so that

$$\tau = RC = \frac{\varepsilon_r \rho}{4\pi k_e}. \tag{S-4.41}$$

This relations holds for any "leaky capacitor", if the discharge occurs only through leakage. In the present case we obtain from the continuity equation.

$$\frac{dQ}{dt} = -I = -\frac{4\pi k_e}{\varepsilon_r \rho} Q, \quad Q(t) = Q(0)\, e^{-t/\tau}, \quad \tau = \frac{\varepsilon_r \rho}{4\pi k_e}. \tag{S-4.42}$$

(d) To the first order in a/ℓ the electrostatic induction effects can be described by regarding the electric field outside the two spheres as due to two charges $\pm Q$ located at the centers of the spheres, and two charges $\pm q = \pm(a/\ell)\, Q$ located at distances $d = a^2/\ell$ from the centers, on the line connecting the two centers, each toward the other sphere, as in the figure (Fig. S-4.3).

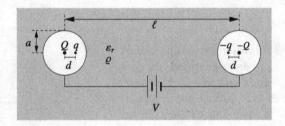

Fig. S-4.3

Thus, to the first order, the potential of each sphere is $\simeq \pm k_e\, Q/(\varepsilon_r\, a)$, since the contribution of the charge $\mp Q$ on the other sphere is canceled by the image charge $\pm q$ present in the sphere. We have $Q \simeq \varepsilon_r\, a V/(2k_e)$, while the absolute value of the total charge on each sphere is $Q + q = Q\,(1 + a/\ell)$. The capacitance of the system is thus

$$C = \frac{Q}{V} = \frac{\varepsilon_r\, a}{2k_e}\left(1 + \frac{a}{\ell}\right) . \tag{S-4.43}$$

The same result is obtained from (S-4.32) of Problem 4.4

$$C = \varepsilon_r\, C_0 = \frac{\varepsilon_r}{k_e}\left(\frac{2}{a} - \frac{2}{\ell}\right)^{-1} = \frac{\varepsilon_r\, a}{2k_e}\frac{\ell}{\ell - a} \simeq \frac{\varepsilon_r\, a}{2k_e}\left(1 + \frac{a}{\ell}\right) , \tag{S-4.44}$$

where the image charges have been disregarded, but the effect of the charge on each sphere on the potential of the other has been taken into account.

According to (S-4.33) the resistance now becomes

$$R = \frac{\rho}{4\pi k_e C_0} = \frac{\rho}{2\pi a}\frac{l - a}{a} , \tag{S-4.45}$$

so that the time constant $\tau = RC = \varepsilon_r\, \rho/(4\pi k_e)$ is unchanged.

S-4.6   Effects of Non-uniform Resistivity

(a) We use a cylindrical coordinate system (r, ϕ, z), with the z axis along the common axis of the two cylinders, and the origin O on the surface separating the two cylinders as in Fig. S-4.4. We denote the volume charge density by q_v, since the Greek letter ρ is already used to denote the resistivities. In a steady state we must have $\partial_t q_v = 0$ everywhere, otherwise the volume charge density would increase, or decrease, indefinitely. Thus, according to the continuity equation, we have also

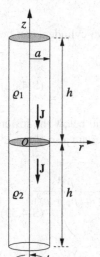

Fig. S-4.4

$$\mathbf{V} \cdot \mathbf{J} = -\partial_t q_v = 0 . \tag{S-4.46}$$

On the other hand, from $\mathbf{V} \cdot \mathbf{E} = 4\pi k_e\, q_v$ and $\mathbf{J} = \mathbf{E}/\rho$ we obtain

$$0 = \mathbf{V} \cdot \mathbf{J} = \frac{1}{\rho}\mathbf{V} \cdot \mathbf{E} = \frac{4\pi k_e}{\rho}q_v , \tag{S-4.47}$$

showing that also the volume charge density q_v must be zero everywhere inside a conductor in stationary conditions. This does not exclude the presence of surface charge densities on the surfaces delimiting a conductor.

If we assume that $h \gg a$, it follows from $\nabla \cdot \mathbf{J} = 0$ and $\nabla \times \mathbf{E} = 0$ that $\mathbf{J}$ is uniform inside the cylinders, pointing downwards along the z direction. Since $\mathbf{E}$ and $\mathbf{J}$ are proportional to each other inside each cylinder, it follows that also $\mathbf{E}$ is uniform inside each cylinder. The current density $\mathbf{J}$ must be continuous through the surface separating the two cylinders, otherwise charge would accumulate indefinitely on the surface. Thus, $\mathbf{J}$ is uniform throughout the whole conductor, and the current is $I = J\pi a^2$.

The resistances $R_{1,2}$ of the two cylinders are, respectively,

$$R_{1,2} = \rho_{1,2} \frac{h}{\pi a^2}, \tag{S-4.48}$$

leading to a total resistance R of the system

$$R = R_1 + R_2 = (\rho_1 + \rho_2) \frac{h}{\pi a^2}. \tag{S-4.49}$$

The current, and the current density, flowing in the system are

$$I = \frac{V}{R} = \frac{\pi a^2 V}{h(\rho_1 + \rho_2)}, \quad J = \frac{V}{h(\rho_1 + \rho_2)}. \tag{S-4.50}$$

Since we have the same current density in two conductors of different resistivities, and $\mathbf{E} = \rho \mathbf{J}$, the electric fields in the two conductors must be different, namely

$$E_1 = \rho_1 J = \frac{\rho_1 V}{h(\rho_1 + \rho_2)}, \quad E_2 = \rho_2 J = \frac{\rho_2 V}{h(\rho_1 + \rho_2)}. \tag{S-4.51}$$

(b) The surface charge density on the surface separating the two cylinders can be evaluated from Gauss's law

$$\sigma = \frac{1}{4\pi k_e} (E_2 - E_1) = \frac{1}{4\pi k_e} \frac{(\rho_2 - \rho_1) V}{h(\rho_1 + \rho_2)}. \tag{S-4.52}$$

Assuming that the electric field is zero above the upper base and below the lower base of the conductor, the surface charge densities at the two bases are also obtained from Gauss's law as

$$\sigma_1 = \frac{E_1}{4\pi k_e} = \frac{1}{4\pi k_e} \frac{\rho_1 V}{h(\rho_1 + \rho_2)}, \quad \sigma_2 = -\frac{E_2}{4\pi k_e} = -\frac{1}{4\pi k_e} \frac{\rho_2 V}{h(\rho_1 + \rho_2)}. \tag{S-4.53}$$

S-4.7 Charge Decay in a Lossy Spherical Capacitor

(a) We use a spherical coordinate system (r, θ, ϕ), with the origin O at the center of the capacitor. We have $\mathbf{E} = 0$ for $r < a$ and $r > b$. For symmetry reasons, the electric field $\mathbf{E}$ is radial and depends on r and t only in the spherical shell $a < r < b$. The flux of $\varepsilon_r \mathbf{E}$ through a spherical Gaussian surface centered in O and of radius r is independent of r and equals

$$\varepsilon_r \oint \mathbf{E}(r, t) \cdot d\mathbf{S} = 4\pi k_e\, Q(t) , \qquad (S\text{-}4.54)$$

where $Q(t)$ is the *free* charge contained in the Gaussian surface, i.e, the free surface charge of the conducting sphere of radius a. Thus we have

$$\mathbf{E}(r, t) = \frac{k_e}{\varepsilon_r} \frac{Q(t)}{r^2} \mathbf{r}_c . \qquad (S\text{-}4.55)$$

In addition to the free charge, our system contains surface polarization charges at $r = a$ and $r = b$, of values $\mp Q (\varepsilon_r - 1)/\varepsilon_r$, respectively. No volume polarization charge is present, because

$$\nabla \cdot \mathbf{P} = \frac{\varepsilon_r - 1}{4\pi k_e} \nabla \cdot \mathbf{E}(r, t) = 0 . \qquad (S\text{-}4.56)$$

The electric field $\mathbf{E}(r, t)$, in the presence of an electrical conductivity σ, gives origin to a current density $\mathbf{J}$

$$\mathbf{J} = \sigma \mathbf{E} = \sigma \frac{k_e}{\varepsilon_r} \frac{Q(t)}{r^2} \hat{\mathbf{r}} , \qquad (S\text{-}4.57)$$

so that we have a total charge flux rate (electric current) through our Gaussian surface

$$I = \frac{dq}{dt} = \oint \mathbf{J} \cdot d\mathbf{S} = \frac{4\pi \sigma k_e}{\varepsilon_r} Q(t) . \qquad (S\text{-}4.58)$$

The charge crossing the Gaussian surface is subtracted from the free charge on the internal conducting sphere, so that

$$\frac{dQ(t)}{dt} = -\frac{4\pi \sigma k_e}{\varepsilon_r} Q(t) , \qquad (S\text{-}4.59)$$

leading to

$$Q(t) = Q_0\, e^{-t/\tau} , \quad \text{with} \quad \tau = \frac{\varepsilon_r}{4\pi \sigma k_e} , \qquad (S\text{-}4.60)$$

and the decay constant is independent of the sizes of the capacitor, in agreement with (S-4.41).

(b) The power dissipated over the volume of the capacitor is

$$P_d = \int \mathbf{J} \cdot \mathbf{E} \, d^3x = \sigma \int \mathbf{E}^2 \, d^3x = \sigma \int_a^b \left[\frac{k_e}{\varepsilon_r} \frac{Q(t)}{r^2} \right]^2 4\pi r^2 \, dr$$

$$= \frac{4\pi \sigma k_e^2}{\varepsilon_r^2} Q_0^2 \, e^{-2t/\tau} \int_a^b \frac{dr}{r^2} = \frac{4\pi \sigma k_e^2 \, (b-a)}{\varepsilon_r^2 \, ab} Q_0^2 \, e^{-2t/\tau}. \qquad \text{(S-4.61)}$$

The electrostatic energy of the capacitor is

$$U_{es} = \frac{1}{2} \frac{Q^2(t)}{C} = \frac{k_e \, (b-a)}{2\varepsilon_r \, ab} Q_0^2 \, e^{-2t/\tau}, \qquad \text{(S-4.62)}$$

so that

$$\frac{dU_{es}}{dt} = -\frac{k_e(b-a)}{\tau \, \varepsilon_r \, ab} Q_0^2 \, e^{-2t/\tau} = -\frac{4\pi \sigma k_e^2(b-a)}{\varepsilon_r^2 \, ab} Q_0^2 \, e^{-2t/\tau} = -P_d. \quad \text{(S-4.63)}$$

Thus, the electrostatic energy of the capacitor is dissipated into Joule heating.

S-4.8 Dielectric-Barrier Discharge

(a) We denote by E_1 and E_2 the electric fields in the gas and in the dielectric layers, respectively. Since the voltage drop between the plates is V, we must have

$$E_1 d_1 + E_2 d_2 = V. \qquad \text{(S-4.64)}$$

In the absence of free surface charges the normal component of $\varepsilon_r \mathbf{E}$ is continuous through the surface separating the two layers, so that

$$E_1 = \varepsilon_r E_2. \qquad \text{(S-4.65)}$$

Combining (S-4.64) and (S-4.65) we obtain

$$E_1 = \frac{\varepsilon_r V}{\varepsilon_r d_1 + d_2}, \qquad E_2 = \frac{V}{\varepsilon_r d_1 + d_2}. \qquad \text{(S-4.66)}$$

(b) In steady-state conditions the current density in the gas, $\mathbf{J}$, must be zero, otherwise the free charge on the surface separating the gas and the dielectric material would increase steadily. Since the current density is $J = E_1/\rho$, we must have $E_1 = 0$. On the other hand (S-4.64) still holds, so that $E_2 = V/d_2$.

The free charge density on the surface separating the layers is

$$\sigma = \frac{1}{4\pi k_e}(\varepsilon_r E_2 - E_1) = \frac{\varepsilon_r}{4\pi k_e}E_2 = \frac{\varepsilon_r}{4\pi k_e}\frac{V}{d_2}. \tag{S-4.67}$$

(c) The continuity equation for σ and J is

$$\partial_t \sigma = J = \frac{E_1}{\rho}. \tag{S-4.68}$$

From (S-4.67), now with $E_1 \neq 0$ (discharge conditions), we have

$$E_1 = \varepsilon_r E_2 - 4\pi k_e \sigma, \tag{S-4.69}$$

which, combined with (S-4.64), leads to

$$E_1 = \varepsilon_r \left(\frac{V}{d_2} - \frac{d_1}{d_2}E_1\right) - 4\pi k_e \sigma, \quad \text{i.e.,} \quad E_1 = \frac{\varepsilon_r V}{\varepsilon_r d_1 + d_2} - \frac{4\pi k_e d_2}{\varepsilon_r d_1 + d_2}\sigma. \tag{S-4.70}$$

Equation (S-4.70), substituted into (S-4.68), gives

$$\partial_t \sigma = -\frac{4\pi k_e d_2}{\rho(\varepsilon_r d_1 + d_2)}\sigma + \frac{\varepsilon_r V}{\rho(\varepsilon_r d_1 + d_2)}. \tag{S-4.71}$$

with solution

$$\sigma = \frac{\varepsilon_r V}{4\pi k_e d_2}\left(1 - e^{-t/\tau}\right), \quad \text{where} \quad \tau = \frac{\rho(\varepsilon_r d_1 + d_2)}{4\pi k_e d_2}. \tag{S-4.72}$$

This problem shows the concept of the *"dielectric-barrier discharge"* (DBD). This scheme, where the dielectric layer acts as a current limiter, is used in various electrical discharge devices, for example in plasma TV displays, where the discharge acts as an ultraviolet micro-source to activate the phosphors in each pixel of the screen.

S-4.9   Charge Distribution in a Long Cylindrical Conductor

(a) As we saw in point (a) of Problem 4.6, the volume charge density q_v is zero everywhere inside our conducting cylinder, while $\mathbf{E}$ and $\mathbf{J}$ are uniform. The presence of an electric field requires the presence of a charge distribution generating it, and, since there cannot be volume charge densities inside a conductor in steady conditions, the charges generating the fields must be distributed on the conductor surfaces. Consider the thin cylindrical conductor shown in Fig. S-4.5, of radius a and length $2h$, with $h \ll a$, connected to a voltage source V_0. In this case, neglect-

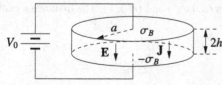

Fig. S-4.5

ing boundary effects, the surface charge densities σ_B and $-\sigma_B$ on the two bases are sufficient to generate the uniform electric field $\mathbf{E}$ inside the conductor. This leads also to a uniform current density $\mathbf{J} = \mathbf{E}/\rho$. Neglecting the boundary effects we have

$$E = \frac{V_0}{2h}, \quad \sigma_B = \frac{E}{4\pi k_e} = \frac{V_0}{8\pi k_e h}. \tag{S-4.73}$$

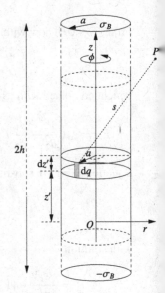

But here we are dealing with the opposite case, when the potential difference V_0 is applied to the bases of a very elongated cylinder, with $h \gg a$. Without loss of generality, we assume the potential to be $+V_0/2$ at the upper base, and $-V_0/2$ at the lower base. With this geometry, the surface charge densities $\pm\sigma_B$ on the two bases alone cannot generate a uniform electric field inside the whole conductor. We need another charge density σ_L, not necessarily uniform, distributed on the lateral surface of the cylinder. In order to treat the problem, we introduce a cylindrical coordinate system (r, ϕ, z), with the z axis coinciding with the axis of the cylinder, the origin O being located so that the upper and lower bases are at $z = \pm h$, respectively (this is not apparent from Fig. S-4.6 for practical reasons).

Because of symmetry reasons, σ_L cannot depend on ϕ. And it cannot be constant along the lateral surface, otherwise, neglecting the boundary effects, it would generate no field inside the conductor. Thus, σ_L must be a function of z, and z only. As an educated guess, we assume that σ_L is proportional to z, so that we have

Fig. S-4.6

$$\sigma_L(z) = \gamma z, \tag{S-4.74}$$

with γ a constant. This choice leads to $\sigma_L(0) = 0$ at $z = 0$, and $|\sigma_L|$ increasing, with opposite signs toward the upper and lower bases. Let us evaluate the electric potential in a point $P \equiv (r, 0, z)$, with $r \ll h$, not necessarily inside the conductor. The choice of $\phi = 0$ does not affect the generality of the approach because of the rotational symmetry around the z axis. The contribution of the charge element $dq = \gamma z' a \, d\phi \, dz'$, located on the lateral surface of the conductor at (a, ϕ, z'), to the potential $V(r, 0, z)$ is

$$dV = k_e \frac{dq}{s},$$

where s is the distance between the points (a, ϕ, z') and $(r, 0, z)$. The distance s can be evaluated by the cosine formula,

$$s = \sqrt{(z'-z)^2 + a^2 + r^2 - 2ar\cos\phi}$$
$$= \sqrt{(z'-z)^2 + a^2\left(1 + \frac{r^2}{a^2} - 2\frac{r}{a}\cos\phi\right)}$$
$$= \sqrt{(z'-z^2) + a^2 f(r,\phi)}, \qquad (S\text{-}4.75)$$

where we have defined

$$f(r,\phi) = \left(1 + \frac{r^2}{a^2} - 2\frac{r}{a}\cos\phi\right). \qquad (S\text{-}4.76)$$

We thus have

$$dV = k_e a\gamma z' \frac{d\phi\, dz'}{\sqrt{(z'-z)^2 + a^2 f(r,\phi)}}, \qquad (S\text{-}4.77)$$

and the electric potential in P is

$$V(P) = k_e a\gamma \int_0^{2\pi} d\phi \int_{-h}^{h} \frac{z'\, dz'}{\sqrt{(z'-z)^2 + a^2 f(r,\phi)}}. \qquad (S\text{-}4.78)$$

In order to evaluate the integral we introduce a new variable $\zeta = z' - z$, so that

$$V(P) = k_e a\gamma \int_0^{2\pi} d\phi \int_{-h-z}^{h-z} \frac{(z+\zeta)\, d\zeta}{\sqrt{\zeta^2 + a^2 f(r,\phi)}}. \qquad (S\text{-}4.79)$$

The indefinite integrals needed in the formula are

$$\int \frac{d\zeta}{\sqrt{\zeta^2 + b}} = \ln\left(2\zeta + 2\sqrt{\zeta^2 + b}\right), \quad \text{and} \quad \int \frac{\zeta\, d\zeta}{\sqrt{\zeta^2 + b}} = \sqrt{\zeta^2 + b}. \quad (S\text{-}4.80)$$

We can split $V(P)$ into the sum of two terms $V(P) = V_1(P) + V_2(P)$, where

$$V_1(P) = k_e a\gamma z \int_0^{2\pi} d\phi \int_{-h-z}^{h-z} \frac{d\zeta}{\sqrt{\zeta^2 + a^2 f(r,\phi)}}$$

$$= k_e a\gamma z \int_0^{2\pi} d\phi \ln\left[\frac{h - z + \sqrt{(h-z)^2 + a^2 f(r,\phi)}}{-h - z + \sqrt{(h+z)^2 + a^2 f(r,\phi)}}\right] \qquad (S\text{-}4.81)$$

and

$$V_2(P) = k_e \, a\gamma \int\limits_0^{2\pi} d\phi \int\limits_{-h-z}^{h-z} \frac{\zeta \, d\zeta}{\sqrt{\zeta^2 + a^2 f(r,\phi)}}$$

$$= k_e \, a\gamma \int\limits_0^{2\pi} d\phi \left[\sqrt{(h-z)^2 + a^2 f(r,\phi)} - \sqrt{(h+z)^2 + a^2 f(r,\phi)} \right].$$

$$(S\text{-}4.82)$$

The square roots appearing in the integrals can be approximated as

$$\sqrt{(h \pm z)^2 + a^2 f(r,\phi)} \simeq h \pm z + \frac{a^2}{2\,(h \pm z)} \, f(r,\phi) \qquad (S\text{-}4.83)$$

up to the second order in a/h and r/h. The second order is needed only in the denominator of the argument of the logarithm appearing in (S-4.81), where the first order cancels out with $-h-z$. Thus, $V_1(P)$ can be approximated as

$$V_1(P) \simeq k_e \, a\gamma z \int\limits_0^{2\pi} d\phi \, \ln \left\{ \frac{2(h-z)}{a^2 f(r,\phi)/[2(h+z)]} \right\},$$

$$= 2\pi k_e \, a\gamma z \int\limits_0^{2\pi} d\phi \, \ln \left[\frac{4(h^2 - z^2)}{a^2 f(r,\phi)} \right] \simeq 2\pi k_e \, acz \int\limits_0^{2\pi} d\phi \, \ln \left[\frac{4h^2}{a^2 f(r,\phi)} \right],$$

$$(S\text{-}4.84)$$

while the approximation for $V_2(P)$ is

$$V_2(P) \simeq -k_e \, a\gamma \int\limits_0^{2\pi} 2z \, d\phi = -4\pi k_e \, acz \,. \qquad (S\text{-}4.85)$$

The two contributions sum up to

$$V(P) = V_1(P) + V_2(P) \simeq 2\pi k_e \, a\gamma z \left\{ \int_0^{2\pi} d\phi \, \ln\left[\frac{4h^2}{a^2 f(r,\phi)}\right] - 2 \right\}$$

$$= 2\pi k_e \, a\gamma z \left\{ 2\pi \ln\left(\frac{h^2}{a^2}\right) + \int_0^{2\pi} d\phi \, \ln\left[\frac{4}{f(r,\phi)}\right] - 2 \right\}$$

$$= 2\pi k_e \, a\gamma z \left\{ 4\pi \ln\left(\frac{h}{a}\right) + \int_0^{2\pi} d\phi \, \ln\left[\frac{4}{f(r,\phi)}\right] - 2 \right\}. \qquad \text{(S-4.86)}$$

If h is sufficiently large, the first terms in braces is dominant, and we have

$$V(r,z) \simeq 8\pi^2 k_e \, a\gamma \ln\left(\frac{h}{a}\right) z, \qquad \text{(S-4.87)}$$

thus independent of r, within our approximations, as expected. Since we have assumed $V(r,h) = V_0/2$, we must have

$$\frac{V_0}{2} = 8\pi^2 k_e \, a\gamma \ln\left(\frac{h}{a}\right) h, \qquad \text{(S-4.88)}$$

which leads to

$$\gamma = \frac{V_0}{16\pi^2 k_e a h \ln(h/a)} \quad \text{and} \quad \sigma_L(z) = \frac{V_0}{16\pi^2 k_e a h \ln(h/a)} z. \qquad \text{(S-4.89)}$$

S-4.10 An Infinite Resistor Ladder

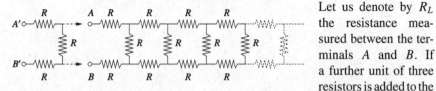

Fig. S-4.7

Let us denote by R_L the resistance measured between the terminals A and B. If a further unit of three resistors is added to the left of the ladder, as in Fig. S-4.7, the "new" resistance measured between terminals A' and B' must equal the "old" resistance R_L. The "old" resistor ladder at the right of terminals A and B can be replaced by

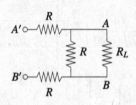

Fig. S-4.8

the equivalent resistance R_L, leading to the configuration of Fig. S-4.8. We see that the resistance between terminals A' and B' is the solution of

$$R_L = 2R + \frac{R R_L}{R + R_L}$$

$$R R_L + R_L^2 = 2R^2 + 2R R_L + R R_L$$

$$R_L^2 - 2R R_L - 2R^2 = 0, \qquad\qquad \text{(S-4.90)}$$

and, disregarding the negative solution, we have

$$R_L = R\left(1 + \sqrt{3}\right). \qquad\qquad \text{(S-4.91)}$$

Chapter S-5
Solutions for Chapter 5

S-5.1 The Rowland Experiment

(a) Neglecting the boundary effects, the electric field $\mathbf{E}_0$ in the regions between the disk and the plates is uniform, perpendicular to the disk surfaces, and its magnitude is $E_0 = V_0/h$ in both regions. In both regions, the field directed outwards form the disk, according to the polarity of the source shown in Fig. 5.1. The charge densities of the lower and upper surfaces of the disk, σ, are equal in modulus and sign, because the field must be zero inside the disk. Thus we have $\sigma = E_0/(4\pi k_e) = V_0/(4\pi k_e h)$. In SI units we have $\sigma = \varepsilon_0 V_0/h$, with $\varepsilon_0 = 8.85 \times 10^{-12}$, $V_0 = 10^4$ V, $h = 5 \times 10^{-3}$ m, resulting in $\sigma = 1.77 \times 10^{-5}$ C/m^2. In Gaussian units we have $\sigma = V_0/(4\pi h)$, with $V_0 = 33.3$ statV and $h = 0.5$ cm, resulting in $\sigma = 5.3$ statC/cm^2.

(b) We evaluate the magnetic field $\mathbf{B}_c$ at the center of the disk by dividing its upper and lower surfaces into annuli of radius r (with $0 < r < a$) and width dr. On each syrface, each annulus carries a charge $dq = \sigma dS = 2\pi \sigma r\, dr$. Due to the rotation of the disk, each annulus is equivalent to a coil with a current intensity

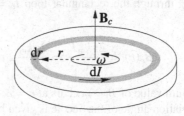

Fig. S-5.1

$dI = \omega\, dq/(2\pi)$, that generates at its center a magnetic field $d\mathbf{B}_c = (2\pi k_m\, dI/r)\,\hat{\omega}$, perpendicular to the disk plane (Fig. S-5.1). The total field at the center of the disk is thus given by the integral

A. Macchi et al., *Problems in Classical Electromagnetism*,
https://doi.org/10.1007/978-3-031-22235-1

$$B_c = 2 \int_0^a 2\pi k_m \frac{dI}{r} = 4\pi k_m \sigma \omega \int_0^a \frac{r \, dr}{r}$$

$$= 4\pi k_m \omega \sigma a = \begin{cases} \mu_0 \omega \sigma a \simeq 1.4 \times 10^{-9} \text{ T} & \text{SI} \\[2ex] \dfrac{4\pi}{c} \omega \sigma a \simeq 1.4 \times 10^{-5} \text{ G} & \text{Gaussian,} \end{cases} \qquad \text{(S-5.1)}$$

where the factor of 2 in front of the first integral is due to contribution of both the upper and the lower surfaces of the disk to the magnetic field.

The magnetic field component B_r, parallel to the disk surface and close to it, can be evaluated by applying Ampère's law to the closed rectangular path C shown in Fig. S-5.2. The path is placed at a distance r from the rotation axis, with the sides parallel to the disk surfaces having length $\ell \ll r$, so that $\mathbf{B}_r$ is approximately constant along the sides. The contribution of the vertical paths to the line integral cancel each other, thus

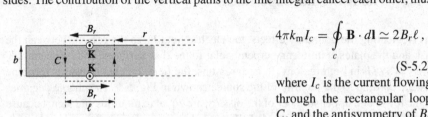

$$4\pi k_m I_c = \oint_c \mathbf{B} \cdot d\mathbf{l} \simeq 2 B_r \ell ,$$

$$\text{(S-5.2)}$$

Fig. S-5.2

where I_c is the current flowing through the rectangular loop C, and the antisymmetry of B_r with respect to the midplane has been used. The rotation of the disk leads to a surface current density $\mathbf{K} = \sigma \mathbf{v} = \sigma \omega r \, \hat{\phi}$, resulting in a total current flowing through the rectangular loop $I_c = 2K\ell = 2\sigma \omega r \ell$. Thus, according to (S-5.2),

$$B_r(r) = \frac{2\pi k_m I_c}{\ell} = \frac{2\pi k_m}{\ell} 2\sigma \omega r \ell = 4\pi k_m \sigma \omega r . \qquad \text{(S-5.3)}$$

The maximum value of $B_r(r)$ occurs at $r = a$, where $B_r(a) = B_c$ (Fig. S-5.1).

(c) The deviation angle of the needle is given by $\tan \theta = B/B_\oplus$, hence

$$\theta \simeq \frac{B}{B_\oplus} = 2.8 \times 10^{-5} \text{ rad} = 1.6 \times 10^{-3} \text{ deg} . \qquad \text{(S-5.4)}$$

The expected angle is very small, and its measurement requires exceptional care.

S-5.2   Pinch Effect in a Cylindrical Wire

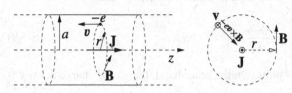

(a) We use a cylindri cal coordinate system (r, ϕ, z) with the z axis along the axis of the cylinder (Fig. S-5.3). The vectors **J** and **v** are along

Fig. S-5.3

z. If we assume $J > 0$ we have $v < 0$ since $J = -n_e\, e\, \boldsymbol{v}$. The magnetic field **B** is azimuthal for symmetry reasons. Its only component $B_\phi(r)$ can easily be evaluated by applying Ampère's circuital law to a circular closed path coaxial with the cylinder axis, as shown in figure. We have

$$2\pi r\, B_\phi = 4\pi^2 r^2 k_{\rm m} J = \begin{cases} \mu_0\, \pi r^2\, J\,, & \text{SI} \\[2mm] \dfrac{4\pi^2 r^2}{c}\, J\,, & \text{Gaussian,} \end{cases} \tag{S-5.5}$$

so that

$$B_\phi = 2\pi k_{\rm m} J\, r = -2\pi k_{\rm m}\, n_e e v\, r\,; \tag{S-5.6}$$

and $B_\phi > 0$, since $v < 0$. Thus the field lines of **B** are oriented counterclockwise with respect to the z axis.

The magnetic force $\mathbf{F}_{\rm m} = -e b_{\rm m}(\boldsymbol{v} \times \mathbf{B})$ is radial and directed towards the z axis

$$\mathbf{F}_{\rm m} = -2\pi k_{\rm m} b_{\rm m} n_e e^2 v^2 = \begin{cases} -\dfrac{\mu_0\, n_e e^2 v^2}{2}\, \mathbf{r}\,, & \text{SI} \\[3mm] -\dfrac{2\pi n_e e^2 v^2}{c^2}\, \mathbf{r}\,, & \text{Gaussian.} \end{cases} \tag{S-5.7}$$

Thus the magnetic force pulls the charge carriers toward the axis of the wire, independently of their sign. A beam of charged particles always gives origin to a magnetic field that tends to "pinch" the beam, i.e., to shrink it toward its axis. However, if the beam is propagating in vacuum, the Coulomb repulsion between the charged particles is dominant. In our case, or in the case of a plasma, the medium is globally neutral, and, initially, the positive and negative charge densities are uniform over the medium, so that the pinch effect can be observed, at least in principle.

(b) The Lorentz force is $\mathbf{F}_{\rm L} = -e\, (\mathbf{E} + b_{\rm m}\, \boldsymbol{v} \times \mathbf{B})$. At equilibrium the r component of $\mathbf{F}_{\rm L}$ must be zero in the presence of conduction electrons (see Problem 1.9), so that the electrons flow only along the z axis. Thus the r component of the electric field, E_r, must be

$$E_r = -b_{\rm m} v B = 2\pi\, k_{\rm m} b_{\rm m} n_e e\, v^2 r\,, \tag{S-5.8}$$

while $E_z = J/\sigma$, where σ is the conductivity of the material. According to Gauss's law, a charge density ϱ, uniform over the cylinder volume, generates a field $\mathbf{E} = 2\pi k_e \varrho \mathbf{r}$, and the required field E_r is generated by the charge density

$$\varrho = \frac{k_m b_m}{k_e} n_e e\, v^2 = n_e e\, \frac{v^2}{c^2}\,, \tag{S-5.9}$$

independent of the system of units. On the other hand, the global charge density is $\varrho = e\,(Zn_i - n_e)$, so that

$$n_e = \frac{Zn_i}{1 - v^2/c^2}\,. \tag{S-5.10}$$

Thus, the electron density is uniform over the wire volume, but it exceeds the value $n_{e0} = Zn_i$, corresponding to $\varrho = 0$. This means that the number density of the electrons is increased by a factor $(1 - v^2/c^2)^{-1}$, and ϱ is negative, *inside* the wire. The "missing" positive charge is uniformly distributed over the surface of the conductor.
(c) For electrons in a usual Ohmic conductor we have $v \simeq 1$ cm/s $= 10^{-2}$ m/s, corresponding to $(v/c)^2 \simeq 10^{-21}$, and the resulting "pinch" effect is so small that it cannot be observed. On the other hand, the effect may be strong in high density particle beams or plasma columns, where v is not negligible with respect to c.

In order to get further insight into the size of the effect, let us consider an Ohmic cylindrical conductor (wire) of radius a. We assume that the electron density is increased in a central cylindrical region of radius $a - d$, where $n_e^{\text{pinch}} = Zn_i/(1 - v^2/c^2)$, and the volume charge density is

$$\varrho^{\text{pinch}} = e\,(Zn_i - n_e^{\text{pinch}}) = -e\,Zn_i\,\frac{v^2}{c^2}\,\frac{1}{1 - v^2/c^2} < 0\,, \tag{S-5.11}$$

while the cylindrical shell between $r = a - d$ and $r = a$ is depleted of conduction electrons, so that its charge density ϱ^{surf} is $\varrho^{\text{surf}} = eZn_i$. The thickness d of the depleted cylindrical shell can be estimated by the constraint of charge conservation. A slice of wire of length ℓ must be globally neutral, thus, assuming $d \ll a$, we must have

$$\pi(a - d)^2 \ell\, \varrho^{\text{pinch}} = -2\pi a d\ell\, \varrho^{\text{surf}}$$

$$\pi(a - d^2)\, e\, Zn_i\,\frac{v^2}{c^2}\,\frac{1}{1 - v^2/c^2} = 2\pi a d e Z\, n_i$$

$$(a^2 - 2ad + d^2)\,\frac{v^2}{c^2}\,\frac{1}{1 - v^2/c^2} = 2ad\,, \tag{S-5.12}$$

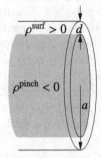

Fig. S-5.4

and, since $v \ll c$ and $d \ll a$, we can approximate

$$a^2\,\frac{v^2}{c^2} \simeq 2ad\,, \quad \text{so that} \quad d \simeq \frac{a\,v^2}{2\,c^2}\,. \tag{S-5.13}$$

Remembering that v^2/c^2 is of the order of 10^{-21}, we see that a value of d of the order of the crystal lattice spacing ($\simeq 10^{-10}$ m) would require a wire of radius $a \simeq 10^{11}$ m, a remarkably large radius! (Fig. S-5.4).

S-5.3 A Magnetic Dipole in Front of a Magnetic Half-Space

(a) Analogously to Problem (3.2), we treat the vacuum half-space and the medium-filled half-space separately, with separate educated guesses for the magnetic field in each half-space. This in order to exploit the uniqueness theorem for the Poisson

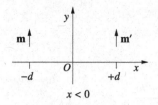
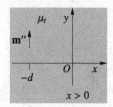

Fig. S-5.5

equation (5.5). Our guess for half-space 1 ($x < 0$) is that the field is same as if the magnetic medium were removed from half-space 2 (thus, vacuum in the whole space), and replaced by with respect to the $x = 0$ plane, at $x =$ d. Our guess for half-space 2 ($x > 0$) is that the field is the same as if the magnetic medium filled the whole space, and the magnetic dipole **m** were replaced by a different magnetic dipole **m″**, placed at the same location (Fig. S-5.5). Thus we look for values of **m′** and **m″** originating a magnetic field $\mathbf{B}_1$ in half-space 1, and a magnetic field $\mathbf{B}_2$ in half-space 2, satisfying the interface conditions at $x = 0$

$$B_{1\perp}(x = 0^-) = B_{2\perp}(x = 0^+) , \qquad B_{1\parallel}(x = 0^-) = \frac{1}{\mu_r} B_{2\parallel}(x = 0^+) \text{ (S-5.14)}$$

The subscripts $\parallel$ and $\perp$ stand for parallel and perpendicular to the $x = 0$ plane, respectively. Thus, at a generic point $P \equiv (0, y, z)$ of the $x = 0$ plane, we must have

$$B_x(0^-, y, z) = B_x(0^+, y, z)$$
$$B_y(0^-, y, z) = \frac{1}{\mu_r} B_y(0^+, y, z)$$
$$B_z(0^-, y, z) = \frac{1}{\mu_r} B_z(0^+, y, z) . \qquad \text{(S-5.15)}$$

The field generated by a magnetic dipole **m** in a medium of relative magnetic permittivity μ_r is

$$\mathbf{B}(\mathbf{r}) = \frac{k_m}{b_m} \mu_r \frac{3(\mathbf{m} \cdot \hat{\mathbf{r}})\hat{\mathbf{r}} - \mathbf{m}}{r^3} \qquad \text{(S-5.16)}$$

where **r** is the distance vector directed from **m** to the point where we evaluate the field, and $\hat{\mathbf{r}} = \mathbf{r}/r$ is the unit vector along **r**. Note that, differently from Problem (3.2), here we do not have cylindrical symmetry around the x axis, because the real magnetic dipole **m** is not lying on x. It is convenient to introduce the angles $\theta = \arcsin(d/r)$ and $\phi = \arctan(z/y)$ (Fig. S-5.6), and write the Cartesian components of **B** separately

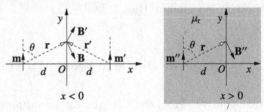

Fig. S-5.6

$$B_x = \frac{k_m}{b_m} \mu_r \frac{3m \cos\theta \sin\theta}{r^3}$$

$$B_y = \frac{k_m}{b_m} \mu_r \frac{3m \cos^2\theta \cos\phi - m}{r^3}$$

$$B_z = \frac{k_m}{b_m} \mu_r \frac{3m \cos^2\theta \sin\phi}{r^3}, \tag{S-5.17}$$

where $r = \sqrt{d^2 + y^2 + z^2}$. If we replace (S-5.17) into (S-5.15) and divide by (k_m/b_m) the boundary conditions become

$$\frac{3m \cos\theta \sin\theta}{r^3} - \frac{3m' \cos\theta \sin\theta}{r^3} = \mu_r \frac{3m'' \cos\theta \sin\theta}{r^3}$$

$$\frac{3m \cos^2\theta \cos\phi - m}{r^3} + \frac{3m' \cos^2\theta \cos\phi - m'}{r^3} = \frac{3m'' \cos^2\theta \cos\phi - m''}{r^3}$$

$$\frac{3m \cos^2\theta \sin\phi}{r^3} + \frac{3m' \cos^2\theta \sin\phi}{r^3} = \frac{3m'' \cos^2\theta \sin\phi}{r^3}, \tag{S-5.18}$$

which can be further simplified into

$$\mathbf{m} - \mathbf{m}' = \mu_r \mathbf{m}''$$

$$\mathbf{m} + \mathbf{m}' = \mathbf{m}'' \tag{S-5.19}$$

leading to the solution

$$\mathbf{m}' = -\frac{\mu_r - 1}{\mu_r + 1} \mathbf{m}, \qquad \mathbf{m}'' = \frac{2}{\mu_r + 1} \mathbf{m}. \tag{S-5.20}$$

What is actually happening is that the magnetization $\mathbf{M}$ of the medium in the half-space 2 generates a surface magnetization current $\mathbf{K}_m(0, y, z) = -\mathbf{M}(0, y, z) \cdot \hat{\mathbf{x}}/b_m$ on the interface plane $x = 0$, according to (5.17). This magnetization-current distribution generates, in each half-space, the same magnetic field that would be generated by a magnetic dipole $\mathbf{m}'$ located on the x axis in the other half-space, at a distance d from the interface. The field due to $\mathbf{K}_m$ adds to the field of the real dipole $\mathbf{m}$, so that an effective dipole $\mathbf{m}'' = \mathbf{m} + \mathbf{m}'$ is "seen" in half-space 2, and two dipoles, $\mathbf{m}$ and $\mathbf{m}'$, are "seen" in half-space 1.

(b) The force exerted by the magnetic half-space on $\mathbf{m}$ equals the force that would be exerted on $\mathbf{m}$ by a real magnetic dipole $\mathbf{m}'$ located at $x = +d$. The force between two magnetic dipoles at a distance $\mathbf{r}$ from each other is

$$\mathbf{f} = -\frac{k_m}{b_m} \nabla \left[\frac{\mathbf{m} \cdot \mathbf{m}' - 3(\mathbf{m} \cdot \hat{\mathbf{r}})(\mathbf{m}' \cdot \hat{\mathbf{r}})}{r^3} \right], \tag{S-5.21}$$

with, in our case, $\hat{\mathbf{r}} = \hat{\mathbf{x}}$, $r = 2d$, and $\mathbf{m} \cdot \hat{\mathbf{r}} = \mathbf{m}'\hat{\mathbf{r}} = 0$, so that the force on $\mathbf{m}$ is

$$\mathbf{f} = -\frac{k_m}{b_m} \frac{3m^2}{r^4} \left(\frac{\mu_r - 1}{\mu_r + 1} \right) \hat{\mathbf{x}} = \frac{k_m}{b_m} \frac{3m^2}{16 d^4} \left(\frac{\mu_r - 1}{\mu_r + 1} \right) \hat{\mathbf{x}}. \tag{S-5.22}$$

The force is repulsive (antiparallel to $\hat{\mathbf{x}}$) for $\mu_r < 1$ (diamagnetic material), and attractive (parallel to $\hat{\mathbf{x}}$) for $\mu_r > 1$ (paramagnetic material). At the limit $\mu_r \to 0$ we have a perfect diamagnetic material (superconductor), and $\mathbf{m}' \to \mathbf{m}$, the two dipoles are *parallel* and the force is repulsive, as expected. In this case $\mathbf{m}'' \neq 0$, so that $\mathbf{H} \neq 0$ in the half-space 2, where, however, $\mu_r = 0$ so that $\mathbf{B} = \mu_0\mu_r\mathbf{H} = 0$. The situation is *opposite* to that of a perfect conductor in electrostatics, where an electric dipole would induce an *opposite* image dipole, and the force would be *attractive*.

At the limit of $\mu_r \to \infty$ (perfect ferromagnetic material), we have $\mathbf{m}' \to -\mathbf{m}$, corresponding to an *attractive* force, while $\mathbf{m}'' \to 0$ and $\mathbf{H} \to 0$ inside the material. This situation is analogous to the case of a conductor in electrostatics. Notice that $\mathbf{B}$ is finite inside the material (since $\mu_r\mathbf{m}'' \to 2\mathbf{m}$) and given by

$$\mathbf{B} = \frac{k_m}{b_m} \mu_r \frac{3\hat{\mathbf{r}}(\hat{\mathbf{r}} \cdot \mathbf{m}'') - \mathbf{m}''}{4\pi r^3} \to 2 \frac{3\hat{\mathbf{r}}(\hat{\mathbf{r}} \cdot \mathbf{m}) - \mathbf{m}}{4\pi r^3}, \qquad (x > 0), \tag{S-5.23}$$

so that the paramagnetic material doubles the value of the magnetic field in vacuum in the limit $\mu_r \to \infty$.

S-5.4   Magnetic Levitation

(a) The radial component B_r of the magnetic field close to the z axis can be evaluated by applying Gauss's law $\nabla \cdot \mathbf{B} = 0$ to a small closed cylinder of radius r, coaxial with the z axis, and with the bases at z and $z + \Delta z$, as shown in Fig. S-5.7. The flux of $\mathbf{B}$ through the total surface of the cylinder must be zero, thus we have

$$0 = \oint_{\substack{\text{closed} \\ \text{cylinder}}} \mathbf{B} \cdot d\mathbf{S} \qquad \qquad \text{(S-5.24)}$$

$$= 2\pi r \Delta z B_r(r) + \pi r^2 \left[B_z(z + \Delta z) - B_z(z) \right] ,$$

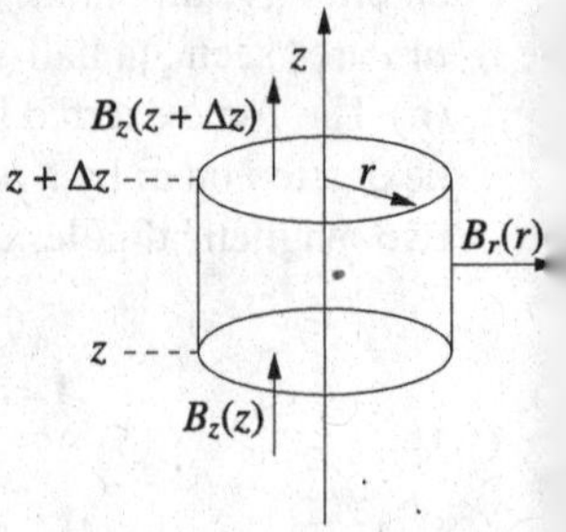

leading to

$$B_r = -\frac{r \left[B_z(z + \Delta z) - B_z(z) \right]}{2\Delta z}$$

Fig. S-5.7

$$\simeq -\frac{B_0}{2L} r \qquad \qquad \text{(S-5.25)}$$

(b) According to Table 5.1, the force exerted by an external magnetic field $\mathbf{B}$ on a magnetic dipole $\mathbf{m}$ is $\mathbf{f} = (\mathbf{m} \cdot \nabla)\mathbf{B}$. If we assume that the dipole is moving in a region free of electric current densities, so that $\nabla \times \mathbf{B} = 0$, the work done on the dipole when it performs an infinitesimal displacement $d\mathbf{r} \equiv (dx, dy, dz)$ is[3]

$$dW = \mathbf{f} \cdot d\mathbf{r} = \left[(\mathbf{m} \cdot \nabla)\mathbf{B} \right] \cdot d\mathbf{r} = \mathbf{m} \cdot d\mathbf{B} . \qquad \text{(S-5.27)}$$

For a permanent magnetic dipole this leads to the well known expression for the potential energy of a dipole located at $\mathbf{r}$

$$U(\mathbf{r}) = -\mathbf{m} \cdot \mathbf{B}(\mathbf{r}) . \qquad \qquad \text{(S-5.28)}$$

Here, however, the magnetic dipole is not permanent. Rather, we have an induced dipole $\mathbf{m} = \alpha \mathbf{B}$. Thus we have

---

[3] We have

$$dW = \left[(\mathbf{m} \cdot \nabla)\mathbf{B} \right] \cdot d\mathbf{r} = \sum_{i,j} m_i \, \partial_i B_j \, dx_j = \sum_{i,j} m_i \, \partial_j B_i \, dx_j = \sum_i m_i \, dB_i = \mathbf{m} \cdot d\mathbf{B} ,$$

$$\text{(S-5.26)}$$

where, as usual, $x_{1,2,3} = x, y, z$, and $\partial_{1,2,3} = \partial_x, \partial_y, \partial_z$. We have used the property $\partial_i B_j = \partial_j B_i$, trivial for $i = j$, while the condition $\nabla \times \mathbf{B} = 0$ implies $\partial_i B_j - \partial_j B_i = 0$ also for $i \neq j$.

$$U(\mathbf{r}_2) - U(\mathbf{r}_1) = -\int_{\mathbf{r}_1}^{\mathbf{r}_2} \mathbf{m} \cdot d\mathbf{B} = -\alpha \int_{\mathbf{r}_1}^{\mathbf{r}_2} \mathbf{B} \cdot d\mathbf{B} = -\frac{\alpha}{2} \int_{\mathbf{r}_1}^{\mathbf{r}_2} dB^2 \qquad (S-5.29)$$

$$= \frac{\alpha}{2} \left[B^2(\mathbf{r}_1) - B^2(\mathbf{r}_2) \right] = \frac{1}{2} \left[\mathbf{m}(\mathbf{r}_1) \cdot \mathbf{B}(\mathbf{r}_1) - \mathbf{m}(\mathbf{r}_2) \cdot \mathbf{B}(\mathbf{r}_2) \right].$$

and the potential energy for the induced dipole at $\mathbf{r}$ is written

$$U(\mathbf{r}) = -\frac{1}{2} \mathbf{m}(\mathbf{r}) \cdot \mathbf{B}(\mathbf{r}). \qquad (S-5.30)$$

For the present problem, this leads to

$$U(\mathbf{r}) = -\frac{1}{2} \mathbf{m}(\mathbf{r}) \cdot \mathbf{B}(\mathbf{r}) = -\frac{1}{2} \alpha B^2(\mathbf{r}) = -\frac{1}{2} \frac{\alpha B_0^2}{L^2} \left(z^2 + \frac{r^2}{4} \right). \qquad (S-5.31)$$

(c) The potential energy U has a minimum in the origin ($r = 0, z = 0$) if $\alpha < 0$ (diamagnetic particle). The force is

$$\mathbf{f} = -\nabla U = -\frac{1}{2} |\alpha| \frac{2B_0^2}{L^2} \left(z\hat{\mathbf{z}} + \frac{r}{4} \mathbf{r} \right). \qquad (S-5.32)$$

Thus, we have a harmonic force both for radial and axial displacements, with corresponding oscillation frequencies

$$\omega_z = \sqrt{\frac{|\alpha| B_0^2}{ML^2}}, \qquad \omega_r = \frac{1}{2} \sqrt{\frac{|\alpha| B_0^2}{ML^2}}. \qquad (S-5.33)$$

S-5.5 Uniformly Magnetized Cylinder

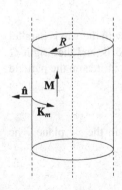

Fig. S-5.8

(a) The volume magnetization current (bound current) density $\mathbf{J}_m$ is zero all over the cylinder volume because the cylinder magnetization $\mathbf{M}$ is uniform, and $\mathbf{J}_m = \nabla \times \mathbf{M}/b_m$. For the surface magnetization current density $\mathbf{K}_m$ we have $\mathbf{K}_m = \mathbf{M} \times \hat{\mathbf{n}}/b_m$, where $\hat{\mathbf{n}}$ is the unit vector perpendicular to the cylinder lateral surface, and b_m is the system dependent constant defined in (5.1). Since $\mathbf{M}$ and $\hat{\mathbf{n}}$ are perpendicular to each other, we have $K_m = |\mathbf{K}_m| = |\mathbf{M}|/b_m$ (Fig. S-5.8).

(b) The magnetized cylinder is equivalent to a solenoid with $nI = K_m$, where n is the number of coils per unit length, and I is the electric current circulating in each coil. Thus, at the $h \gg R$ limit, the magnetic field is uniformly zero outside the cylinder, and it is uniform and equal to

$$B = \frac{4\pi k_m}{b_m} n I = \frac{4\pi k_m}{b_m} K_m = \frac{4\pi k_m}{b_m} M = \begin{cases} \mu_0 M, & \text{SI} \\ 4\pi M, & \text{Gaussian,} \end{cases} \tag{S-5.34}$$

inside. The auxiliary magnetic field $\mathbf{H}$ is zero both inside and outside the cylinder because

$$\mathbf{H}_{in} = \begin{cases} \dfrac{\mathbf{B}_{in}}{\mu_0} - \mathbf{M} = 0, & \text{SI} \\ \mathbf{B}_{in} - 4\pi\mathbf{M} = 0, & \text{Gaussian.} \end{cases} \qquad \mathbf{H}_{out} = \begin{cases} \dfrac{\mathbf{B}_{out}}{\mu_0} = 0, & \text{SI} \\ \mathbf{B}_{out} = 0, & \text{Gaussian.} \end{cases}$$
$$\tag{S-5.35}$$

(c) At the "flat cylinder" limit, $R \gg h$, the cylinder is equivalent to a single coil of radius R carrying a current $I = h K_m = h M / b_m$. Thus we have for the field at its center

$$B_0 = 2\pi k_m \frac{I}{R} = \begin{cases} \dfrac{\mu_0 I}{2R} = \dfrac{\mu_0 M h}{2R}, & \text{SI} \\[2mm] \dfrac{2\pi I}{cR} = \dfrac{2\pi M h}{cR}, & \text{Gaussian,} \end{cases} \tag{S-5.36}$$

and $\mathbf{B}_0$ approaches zero as $h/r \to 0$.

(d) The equivalent magnetic charge density is defined as $\varrho_m = -\nabla \cdot \mathbf{M}$, thus $\varrho_m \equiv 0$ inside the cylinder volume, while the two bases of the cylinder carry surface magnetic charge densities $\sigma_m = \mathbf{M} \cdot \hat{\mathbf{n}} = \pm M$. Therefore our flat magnetized cylinder is the "magnetostatic" equivalent of an electrostatic parallel-plate capacitor. The equivalent magnetic charge "generates" the auxiliary magnetic field $\mathbf{H}$, which is uniform, and equal to $H = -\sigma_m = -M$, inside the volume of the flat cylinder, and zero outside. Thus $B = \mu_0(H + M)$ is zero everywhere, more realistically zero far from the boundaries.

S-5.6   Charged Particle in Crossed Electric and Magnetic Fields

(a) We choose a Cartesian laboratory frame of reference xyz with the y axis parallel to the electric field $\mathbf{E}$, the z axis parallel to the magnetic field $\mathbf{B}$, and the origin O located so that the particle is initially at rest in O. The Lorentz force on the particle

$$\mathbf{f} = q \left[\mathbf{E} + b_m \mathbf{v} \times \mathbf{B} \right]$$

has no z component, and the motion of the particle occurs in the xy plane. The equations of motion are thus

$$m\ddot{x} = b_{\mathrm{m}}qB\,\dot{y}\,,$$
$$m\ddot{y} = -b_{\mathrm{m}}qB\,\dot{x} + qE\,. \qquad (\text{S-5.37})$$

It is convenient to introduce two new variables x', and y', such that

$$x = x' + v_0\,t\,,\quad y = y'\,, \qquad (\text{S-5.38})$$

where v_0 is a constant velocity, which we shall determine in order to simplify the equations of motion. The initial conditions for the primed variables are

$$x'(0) = 0\,, \qquad\qquad\qquad \dot{x}'(0) = -v_0\,,$$
$$y'(0) = 0\,, \qquad\qquad\qquad \dot{y}'(0) = -0\,. \qquad (\text{S-5.39})$$

Differentiating (S-5.38) with respect to time we obtain

$$\dot{x} = v_0 + \dot{x}'\,, \qquad\qquad\qquad \ddot{x} = \ddot{x}'\,,$$
$$\dot{y} = \dot{y}'\,, \qquad\qquad\qquad\qquad \ddot{y} = \ddot{y}'\,, \qquad (\text{S-5.40})$$

which we substitute into (S-5.37), thus obtaining the following equations for the time evolution of the primed variables

$$m\,\ddot{x}' = b_{\mathrm{m}}qB\,\dot{y}'\,,$$
$$m\,\ddot{y}' = -b_{\mathrm{m}}qB\,v_0 - b_{\mathrm{m}}qB\,\dot{x}' + qE\,. \qquad (\text{S-5.41})$$

Now we choose the constant velocity v_0 to be

$$v_0 = \frac{E}{b_{\mathrm{m}}B} = \begin{cases} \dfrac{E}{B}\,, & \text{SI,} \\[2ex] \dfrac{E}{B}\,c\,, & \text{Gaussian}\,, \end{cases} \qquad (\text{S-5.42})$$

independently of the charge and mass of the particle, so that the terms qE and $-b_{\mathrm{m}}qB\,v_0$ cancel each other in the second of (S-5.41). The equations reduce to

$$\ddot{x}' = b_{\mathrm{m}}\frac{qB}{m}\,\dot{y}'\,,$$
$$\ddot{y}' = -b_{\mathrm{m}}\frac{qB}{m}\,\dot{x}'\,, \qquad (\text{S-5.43})$$

which are the equations of a uniform circular motion with angular velocity $\omega = -b_{\mathrm{m}}qB/m = -\omega_{\mathrm{c}}$, the cyclotron frequency. The rotation is clockwise if $q > 0$, counterclockwise if $q < 0$. Since, according to (S-5.39), $\dot{x}'(0) = -v_0$ and $\dot{y}'(0) = 0$, the radius of the circular path is

$$r = \frac{mv_0}{b_m q B} = \frac{mE}{b_m^2 q B^2} . \tag{S-5.44}$$

The time evolution of the primed variables is thus

$$x' = x_0' + r\cos(\omega t + \phi) = r\sin(\omega t) = -\frac{mE}{b_m^2 q B^2}\sin\left(\frac{b_m q B}{m} t\right) , \tag{S-5.45}$$

$$y' = y_0' + r\sin(\omega t + \phi) = r - r\cos(\omega t) = \frac{mE}{b_m^2 q B^2}\left[1 - \cos\left(\frac{b_m q B}{m} t\right)\right] ,$$

where we have chosen the constants $\phi = -\pi/2$, $x_0' = 0$, and $y_0' = -r$, in order to reproduce the initial conditions. The time evolution of the unprimed variables is

$$x = \frac{E}{b_m B} t - \frac{mE}{b_m^2 q B^2}\sin\left(\frac{b_m q B}{m} t\right) ,$$

$$y = \frac{mE}{b_m^2 q B^2}\left[1 - \cos\left(\frac{b_m q B}{m} t\right)\right] , \tag{S-5.46}$$

and the observed motion is a cycloid, as shown in Fig. S-5.9 for a positive charge.

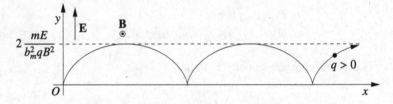

Fig. S-5.9 .

(b) The components of the particle velocity are

$$\dot{x} = \frac{E}{b_m B} - \frac{E}{b_m B}\cos\left(\frac{b_m q B}{m} t\right) , \quad \dot{y} = \frac{E}{b_m B}\sin\left(\frac{b_m q B}{m} t\right) , \tag{S-5.47}$$

thus the kinetic energy is

$$K = \frac{m}{2}\left(\dot{x}^2 + \dot{y}^2\right) = m\left(\frac{E}{b_m B}\right)^2\left[1 - \cos\left(\frac{b_m q B}{m} t\right)\right] , \tag{S-5.48}$$

and oscillates between 0 and $K_{max} = 2mE^2/(b_m B)^2$ at the cyclotron frequency. The magnetic field exerts a force perpendicular to the particle velocity and does no work, while the electric field does a work

$$W_\uparrow = 2rqE = 2\frac{mE}{b_m^2 q B^2} qE = 2m\left(\frac{E}{q_m B}\right)^2 = K_{\text{max}} \qquad \text{(S-5.49)}$$

when the y coordinate of the particle moves from 0 to $2r$, and a work $W_\downarrow = -W_\uparrow'$ when the particle moves from $y = 2r$ to $y = 0$.

(c) From the results of point (a), we know that the motion of the electron will be a cycloid starting from the negative plate, and reaching a maximum distance $2r = 2mE/(b_m^2 q B^2)$ from it, where $E = V/h$. The condition for the electron not reaching the positive plate is thus

$$\frac{2mE}{b_m^2 q B^2} = \frac{2mV}{b_m^2 hq B^2} < h, \quad \text{corresponding to} \quad B > \frac{1}{b_m h}\sqrt{\frac{2m_e V}{e}}, \qquad \text{(S-5.50)}$$

where m_e is the electron mass, and e the absolute value of the electron charge.

S-5.7 Cylindrical Conductor with an Off-Centered Cavity

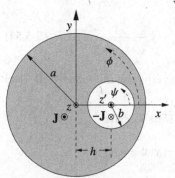

Fig. S-5.10

We shall use three different coordinate systems (Fig. S-5.10):

1. a Cartesian coordinate system with the z axis coinciding with the axis of the cylindrical conductor, and the x passing through the axis of the cavity;

2. a cylindrical coordinate system (r, ϕ, z) sharing the z axis with the Cartesian system, and $\phi = 0$ on the xz plane;

3. a cylindrical coordinate system (ρ, ψ, z') with the z' axis coinciding with the axis of the cavity, and $\psi = 0$ on the xz plane.

According to the superposition principle, a current density $\mathbf{J}$ flowing uniformly through the conducting cross section of the wire in the positive z direction is equivalent to a uniform current density $\mathbf{J}$, flowing through the whole circular section of radius a, superposed to a current density $-\mathbf{J}$, flowing in the negative z direction through the the cavity. Given the current I, we have

$$J = \frac{I}{\pi(a^2 - b^2)}, \qquad \text{(S-5.51)}$$

since there is no net current density flowing through the cavity. In the coordinate system 2, the current flowing in the positive z direction $(+\mathbf{J})$ through the whole section generates an azimuthal magnetic field

$$\mathbf{B}^+ = 2\pi k_{\mathrm{m}} J r \,\hat{\boldsymbol{\phi}} \quad \text{for } r \le a, \quad \mathbf{B}^+ = k_{\mathrm{m}}\frac{2}{r}\pi a^2 J \,\hat{\boldsymbol{\phi}} \quad \text{for } r \ge a, \qquad \text{(S-5.52)}$$

note that the field $\mathbf{B}^+$ outside the wire is equivalent to the field generated by an infinite straight wire superposed to the z axis, carrying a current $I' = I a^2/(a^2 - b^2) > I$. In reference frame 3 the current $-\mathbf{J}$ flowing in the cavity generates an azimuthal magnetic field

$$\mathbf{B}^- = -2\pi k_{\mathrm{m}} J \rho \,\hat{\boldsymbol{\psi}} \quad \text{for } \rho \le b, \quad \mathbf{B}^- = -k_{\mathrm{m}}\frac{2}{\rho}\pi b^2 J \,\hat{\boldsymbol{\psi}} \quad \text{for } \rho \ge b, \qquad \text{(S-5.53)}$$

again, the field $\mathbf{B}^-$ outside the cavity is equivalent to the field generated by an infinite wire superposed to the z' axis, carrying a current $I'' = -I b^2/(a^2 - b^2)$. The cylindrical coordinates of references [2] and [3] can be transformed into the Cartesian coordinates of reference through the relations

$$r = \sqrt{x^2 + y^2}$$
$$\hat{\boldsymbol{\phi}} = -\hat{\mathbf{x}} \sin \phi + \hat{\mathbf{y}} \cos \phi = -\hat{\mathbf{x}}\,\frac{y}{\sqrt{x^2 + y^2}} + \hat{\mathbf{y}}\,\frac{x}{\sqrt{x^2 + y^2}}$$
$$\rho = \sqrt{(x - h)^2 + y^2}$$
$$\hat{\boldsymbol{\psi}} = -\hat{\mathbf{x}} \sin \psi + \hat{\mathbf{y}} \cos \psi = -\hat{\mathbf{x}}\,\frac{y}{\sqrt{(x - h)^2 + y^2}} + \hat{\mathbf{y}}\,\frac{x - h}{\sqrt{(x - h)^2 + y^2}} \qquad \text{(S-5.54)}$$

The field everywhere in the space is obtained by superposing the values of $\mathbf{B}^+$ and $\mathbf{B}^-$. We have three regions

1. $r < a$, $\rho < b$, i.e., the region inside the cavity. Here we have

$$\mathbf{B}(x, y) = 2\pi k_{\mathrm{m}} J [-\hat{\mathbf{x}} y + \hat{\mathbf{y}} x + \hat{\mathbf{x}} y - \hat{\mathbf{y}}(x - h)]$$

$$= 2\pi k_{\mathrm{m}} J h \,\hat{\mathbf{y}} = \begin{cases} \dfrac{\mu_0}{2} J h \,\hat{\mathbf{y}}, & \text{SI} \\[2mm] \dfrac{2\pi}{c} J h \,\hat{\mathbf{y}}, & \text{Gaussian,} \end{cases} \qquad \text{(S-5.55)}$$

the field in the cavity is uniform and directed perpendicular to the plain containing the z and z' axes.
2. $r < a$, $\rho > b$, i.e., inside the conductor, excluding the cavity. Here we have the field $\mathbf{B}^+ = 2\pi k_{\mathrm{m}} J (-\hat{\mathbf{x}} y + \hat{\mathbf{y}} x)$ plus the field of a straight wire coinciding with the z' axis carrying a current I''.
3. $r > a$, $\rho > b$, i.e., the region outside the wire. Here we have the superposition generated by a wire along the z axis, carrying a current I', and a wire along the z' axis, carrying a current I''.

S-5.8   Conducting Cylinder in a Magnetic Field

(a) We us a cylindrical coordinate system (r, ϕ, z), with the z axis along the cylinder axis. The centrifugal force, $\mathbf{F}_c$, and the magnetic force, $\mathbf{F}_m$, are both directed along $\hat{\mathbf{r}}$ and depend on r only:

$$\mathbf{F}_c = m_e \omega^2 \mathbf{r}, \quad \mathbf{F}_m = -e\boldsymbol{v} \times \mathbf{B}_0 = -e\omega B_0 \mathbf{r}, \quad \frac{|\mathbf{F}_c|}{|\mathbf{F}_m|} = \frac{m_e \omega}{e B_0} \simeq 7.2 \times 10^{-5}. \tag{S-5.56}$$

The magnetic force is dominant, and we shall neglect the centrifugal force in the following.

(b) In static conditions the magnetic force must be compensated by an electric field $\mathbf{E}$

$$\mathbf{E} = -\boldsymbol{v} \times \mathbf{B}_0 = -\omega B_0 \mathbf{r}. \tag{S-5.57}$$

The existence of this electric field implies a uniform charge density

$$\rho = \frac{1}{4\pi k_e} \nabla \cdot \mathbf{E} = \frac{E(r)}{2\pi k_e r} = -\frac{\omega B_0}{2\pi k_e}. \tag{S-5.58}$$

Since the cylinder carries no net charge, its lateral surface must have a charge density.

$$\sigma = -\frac{\pi a^2 h \rho}{2\pi a h} = -\frac{a\rho}{2} = \frac{\omega a B_0}{4\pi k_e}. \tag{S-5.59}$$

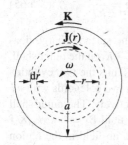

Fig. S-5.11

(c) The volume charge density ρ is associated to a volume rotational current density $\mathbf{J}(r)$ due to the cylinder rotation

$$\mathbf{J}(r) = \rho \omega r \hat{\phi} = -\frac{\omega^2 r B_0}{2\pi k_e} \hat{\phi}. \tag{S-5.60}$$

The contribution of $\mathbf{J}(r)$ to the magnetic field on the cylinder axis, B_J, can be evaluated by dividing the cylinder into infinitesimal coaxial cylindrical shells between r and $r + dr$ (Fig. S-5.11). Each shell is equivalent to a solenoid of radius r and product $nI = J(r)\,dr$, contributing $dB_J = 4\pi k_m J(r)\,dr$ to the field at its inside. The total contribution of $J(r)$ at a distance r from the axis is thus

$$B_J(r) = 4\pi k_m \int_r^a J(r')\,dr' = -4\pi k_m \frac{\omega^2 B_0}{2\pi k_e} \int_r^a r'\,dr'$$

$$= -4\pi k_m \frac{\omega^2 B_0}{2\pi k_e} \left[\frac{r'^2}{2} \right]_r^a = -\frac{k_m}{k_e} \omega^2 B_0 (a^2 - r^2). \tag{S-5.61}$$

Now we must add the contribution B_K of the surface current density $K = \sigma \omega a$

$$B_K = 4\pi k_m \sigma \omega a = 4\pi k_m \omega a \frac{\omega a B_0}{4\pi k_e} = \frac{k_m}{k_e} \omega^2 a^2 B_0 \qquad \text{(S-5.62)}$$

and the total magnetic field $B_1(r)$ due to the rotational currents is

$$B_1(r) = B_J(r) + B_K = \frac{k_m}{k_e} \omega^2 B_0 r^2 , \qquad \text{(S-5.63)}$$

which is zero on the axis and reaches its maximum value at $r = a^-$. We thus have

$$\frac{B_1(a^-)}{B_0} = \frac{k_m}{k_e}\omega^2 a^2 = \frac{\omega^2 a^2}{c^2} \simeq (2.1 \times 10^{-7})^2 \ll 1 . \qquad \text{(S-5.64)}$$

S-5.9 Rotating Cylindrical Capacitor

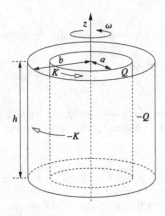

Fig. S-5.12

(a) We use cylindrical coordinates (r, ϕ, z) with the z axis coinciding with capacitor axis (Fig. S-5.12). We assume $\boldsymbol{\omega} = \omega \hat{\mathbf{z}}$, with $\omega = 2\pi/T > 0$. The surface currents due to the capacitor rotation are thus

$$K = \sigma v = \frac{Q}{2\pi a h} \omega a = \frac{Q}{hT} , \qquad \text{(S-5.65)}$$

where $\sigma = Q/(2\pi a h)$ is the surface charge density, on the inner shell, and $-K$ on the outer shell, independently of a and b. Thus the two cylindrical shells are equivalent to two solenoids with nI products $nI = \pm K$, respectively. The outer shell gives origin to a magnetic field $\mathbf{B}_b = -4\pi k_m K \hat{\mathbf{z}}$ in the region $r < b$, and to no field in the region $r > b$. The inner shell gives origin to a field $\mathbf{B}_a = -\mathbf{B}_b$ in the region $r < a$, and to no field in the region $r > b$. The total field $\mathbf{B} = \mathbf{B}_a + \mathbf{B}_b$ is thus

$$\mathbf{B} = \begin{cases} -4\pi k_m K \hat{\mathbf{z}} = -4\pi k_m \dfrac{Q}{hT} \hat{\mathbf{z}}, & a < r < b , \\[2mm] 0, & r < a , r > b. \end{cases} \qquad \text{(S-5.66)}$$

(b) The electric field is zero for $r < a$ and $r > b$, while it is $\mathbf{E}(r) = \hat{\mathbf{r}} \, 2k_e Q/(hr)$ for $a < r < b$, and the force between the two shells is attractive. The electrostatic force per unit area on, for instance, the external shell is thus

$$\mathbf{f}_s^{(e)} = \sigma_b \frac{E(b)}{2} = -\hat{\mathbf{r}} \frac{Q}{2\pi bh} \frac{2k_e Q}{hb} = -\hat{\mathbf{r}} \frac{k_e Q^2}{\pi b^2 h^2},$$ (S-5.67)

where σ_b is the surface charge density on the shell. The magnetic force per unit area on the same shell is

$$\mathbf{f}_s^{(m)} = \frac{1}{2}\sigma_b \, \boldsymbol{v} \times \mathbf{B} = \hat{\mathbf{r}} \frac{1}{2} \frac{Q}{2\pi bh} \frac{2\pi b}{T} 4\pi k_m \frac{Q}{hT} = \hat{\mathbf{r}} \, 2\pi k_m \frac{Q^2}{h^2 T^2},$$ (S-5.68)

directed opposite to the electrostatic force. The ratio $f_s^{(m)}/f_s^{(e)}$ on the outer shell is

$$\frac{f_s^{(m)}}{f_s^{(e)}} = 2\pi k_m \frac{Q^2}{h^2 T^2} \frac{\pi b^2 h^2}{k_e Q^2} = \frac{1}{2} \frac{k_m}{k_e} \left(\frac{2\pi b}{T}\right)^2 = \frac{1}{2} \frac{v_b^2}{c^2}$$ (S-5.69)

where $v_b = 2\pi b/T$ is the tangential velocity of the outer shell. The ratio (S-5.69) is thus negligibly small in all practical cases.

S-5.10 Magnetized Spheres

(a) The quickest way to obtain the solution is to exploit the analogy of the magnetostatic equations $\nabla \times \mathbf{H} = 0$, $\nabla \cdot \mathbf{B} = 0$ with the electrostatic ones $\nabla \times \mathbf{E} = 0$, $\nabla \cdot \mathbf{D} = 0$ (see also Problem 5.3), along with the definitions (3.4) and (5.19). The spatial distribution of $\mathbf{M}$ is the same as that of $\mathbf{P}$ in Problem (3.3), and the boundary conditions for $\mathbf{H}$ are the same as for $\mathbf{E}$. Thus from (S-3.13) we immediately obtain that inside the sphere ($r < R$) the field $\mathbf{H}$ is uniform with constant value $\mathbf{H}^{(\text{int})}$, given by

$$\mathbf{H}^{(\text{int})} = \begin{cases} -\dfrac{\mathbf{M}}{3}, & \text{SI}, \\[2mm] -\dfrac{4\pi\mathbf{M}}{3}, & \text{Gaussian}. \end{cases}$$ (S-5.70)

Using (5.19) we obtain for the magnetic field inside the sphere

$$\mathbf{B}^{(\text{int})} = \frac{8\pi}{3}\frac{k_m}{b_m}\mathbf{M} = \begin{cases} \dfrac{2\mu_0\mathbf{M}}{3}, & \text{SI}, \\[2mm] \dfrac{8\pi\mathbf{M}}{3}, & \text{Gaussian}. \end{cases}$$ (S-5.71)

Outside the sphere the field is that of a magnetic dipole $\mathbf{m} = \mathbf{M}(4\pi R^3/3)$ located at the center of the sphere ($r = 0$).

(b) Analogously to Problem 3.4 for a dielectric sphere in an external electric field, we assume that the induced magnetization $\mathbf{M} = \chi_m \mathbf{H}$ is uniform and parallel to $\mathbf{B}_0$.

The total field will be the sum of the external field $\mathbf{B}_0 \equiv [\mu_0] \, \mathbf{H}_0$ (with $[\mu_0]$ replaced by unity for Gaussian units) and of that generated by the magnetization. Thus, inside the sphere $\mathbf{H}$ is uniform and has the value $\mathbf{H}^{(\mathrm{int})}$ given by

$$
\mathbf{H}^{(\mathrm{int})} =
\begin{cases}
\mathbf{H}_0 - \dfrac{\mathbf{M}}{3} = \mathbf{H}_0 - \dfrac{\chi_{\mathrm{m}}}{3} \mathbf{H}^{(\mathrm{int})} \,, & \text{SI} \,, \\[2mm]
\mathbf{H}_0 - \dfrac{4\pi \mathbf{M}}{3} = \mathbf{H}_0 - \dfrac{4\pi \chi_{\mathrm{m}}}{3} \mathbf{H}^{(\mathrm{int})} \,, & \text{Gaussian} \,.
\end{cases}
\tag{S-5.72}
$$

Solving for $\mathbf{H}^{(\mathrm{int})}$ and finally using $\mathbf{B}^{(\mathrm{int})} = [\mu_0] \, \mu_r \mathbf{H}^{(\mathrm{int})}$ we obtain

$$
\mathbf{H}^{(\mathrm{int})} = \frac{3}{\mu_r + 2} \mathbf{H}_0 \,, \qquad
\mathbf{B}^{(\mathrm{int})} = \frac{3\mu_r}{\mu_r + 2} \mathbf{B}_0 \,,
\tag{S-5.73}
$$

independently of the system of units; it may be interesting to compare the result with (S-3.22) for the dielectric sphere. The magnetization is given by $\mathbf{M} = \chi_{\mathrm{m}} \mathbf{H}^{(\mathrm{int})}$.

In the case of a perfectly diamagnetic sphere (a superconducting sphere) we have $\mu_r = 0$ and $\mathbf{B}^{(\mathrm{int})} = 0$, and the magnetization is

$$
\mathbf{M} = \frac{3\chi_{\mathrm{m}}}{2} \mathbf{H}_0 = -\frac{3}{8\pi} \frac{b_{\mathrm{m}}}{k_{\mathrm{m}}} \mathbf{B}_0 \,.
\tag{S-5.74}
$$

Actually, inside the sphere the external field is completely screened by the surface magnetization currents $\mathbf{K}_{\mathrm{m}}$ corresponding to the magnetization (S-5.74), namely

$$
\mathbf{K}_{\mathrm{m}} = \frac{1}{b_{\mathrm{m}}} \mathbf{M} \times \hat{\mathbf{r}} = \frac{1}{b_{\mathrm{m}}} M \sin\theta \, \hat{\boldsymbol{\phi}} \,,
\tag{S-5.75}
$$

in spherical coordinates (r, θ, ϕ) with the polar axis parallel to $\mathbf{M}$.

It is instructive to double check the above solution by verifying the boundary conditions at the surface of the sphere, analogously to the dielecric case of Solution S-3.4. We choose a spherical coordinate system (r, θ, ϕ) with the zenith direction z parallel to the external magnetic field $\mathbf{B}_0$, and the origin at the center of the sphere

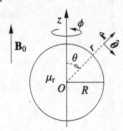

Fig. S-5.13

O, as shown in Fig. S-5.13. As an educated guess, we look for a solution where (i) the magnetic field inside the sphere, $\mathbf{B}^{(\mathrm{int})}$, is uniform and proportional to $\mathbf{B}_0$, and, accordingly, (ii) the magnetization $\mathbf{M}$ of the sphere, proportional to $\mathbf{B}^{(\mathrm{int})}$, is uniform, and (iii) the total external field, $\mathbf{B}^{(\mathrm{ext})}$, is the superposition of the applied external field $\mathbf{B}_0$ and of the field $\mathbf{B}^{(\mathrm{mag})}$, generated by the sphere magnetization. Thus, $\mathbf{B}^{(\mathrm{mag})}$ will be the field generated by a magnetic dipole $\mathbf{m} = \alpha \mathbf{B}_0$ located at the center of the sphere, with α a constant to be determined. Summing up, we are looking for a solution

$$\mathbf{B}^{(\mathrm{int})} = \psi \mathbf{B}_0 ,$$
$$\mathbf{B}^{(\mathrm{ext})} = \mathbf{B}_0 + \mathbf{B}^{(\mathrm{mag})} , \qquad (\text{S-5.76})$$

with ψ a further constant to be determined. $\mathbf{B}^{(\mathrm{mag})}$ and its spherical components are

$$\mathbf{B}^{(\mathrm{mag})} = \alpha B_0 \frac{k_m}{b_m} \left[\left(3 \frac{\hat{\mathbf{z}} \cdot \mathbf{r}}{r^5} \right) \mathbf{r} - \frac{\hat{\mathbf{z}}}{r^3} \right] , \qquad (\text{S-5.77})$$

$$B_r^{(\mathrm{mag})} = \alpha B_0 \frac{k_m}{b_m} \frac{2 \cos \theta}{r^3} , \qquad (\text{S-5.78})$$

$$B_\theta^{(\mathrm{mag})} = \alpha B_0 \frac{k_m}{b_m} \frac{\sin \theta}{r^3} , \qquad (\text{S-5.79})$$

$$B_\phi^{(\mathrm{mag})} = 0 , \qquad (\text{S-5.80})$$

where $k_m/b_m = \mu_0/(4\pi)$ in SI units, and $k_m/b_m = 1$ in Gaussian units. The constants α and ψ are determined from the boundary conditions on $\mathbf{B}$ and $\mathbf{B}/\mu_r$ at the surface of the sphere

$$B_\perp^{(\mathrm{int})}(R, \theta) = B_\perp^{(\mathrm{ext})}(R, \theta) , \qquad \frac{B_\parallel^{(\mathrm{int})}(R, \theta)}{\mu_r} = B_\parallel^{(\mathrm{ext})}(R, \theta) , \qquad (\text{S-5.81})$$

which lead to

$$\psi B_0 \cos \theta = B_0 \cos \theta + B_r^{(\mathrm{mag})}(R, \theta) = B_0 \cos \theta \left(1 + \alpha \frac{k_m}{b_m} \frac{2}{R^3} \right) , \qquad (\text{S-5.82})$$

$$\psi \frac{B_0}{\mu_r} \sin \theta = B_0 \sin \theta - B_\theta^{(\mathrm{mag})}(R, \theta) = B_0 \sin \theta \left(1 - \alpha \frac{k_m}{b_m} \frac{1}{R^3} \right) . \qquad (\text{S-5.83})$$

Dividing (S-5.82) by $B_0 \cos \theta$, and (S-5.83) by $B_0 \sin \theta$, we obtain

$$\psi = 1 + \alpha \frac{k_m}{b_m} \frac{2}{R^3} , \qquad \frac{\psi}{\mu_r} = 1 - \alpha \frac{k_m}{b_m} \frac{1}{R^3} , \qquad (\text{S-5.84})$$

with solutions

$$\psi = \frac{3\mu_r}{\mu_r + 2} , \qquad \alpha = R^3 \frac{b_m}{k_m} \left(\frac{\mu_r - 1}{\mu_r + 2} \right) = \begin{cases} \dfrac{4\pi R^3}{\mu_0} \left(\dfrac{\mu_r - 1}{\mu_r + 2} \right) , & \text{SI,} \\[3mm] R^3 \left(\dfrac{\mu_r - 1}{\mu_r + 2} \right) , & \text{Gaussian,} \end{cases} \qquad (\text{S-5.85})$$

which eventually lead to

$$\mathbf{B}^{(\mathrm{int})} = \frac{3\mu_r}{\mu_r + 2} \mathbf{B}_0 , \qquad \mathbf{m} = \frac{4\pi R^3}{3} \mathbf{M} = R^3 \frac{b_m}{k_m} \left(\frac{\mu_r - 1}{\mu_r + 2} \right) \mathbf{B}_0 . \qquad (\text{S-5.86})$$

S-5.11   A Transversally Magnetized Cylinder

(a) This problem can be solved either by using the fictitious density of magnetic charge $\varrho_m = \nabla \cdot \mathbf{M}$, or, equivalently, the analogy with an electrically polarized cylinder (Problem 3.11). This approach consists in modeling the magnetized cylinder as the superposition of two cylinders of equal radius a and opposite charge densities $\pm \varrho_e$ (or $\pm \varrho_m$), with parallel axes displaced by δ in the y direction, so that $\mathbf{P} = \varrho_e \, \delta$ (or $\mathbf{M} = \varrho_m \, \delta$). Then we consider the limit $\delta \to 0$ and $\varrho_e \to \infty$ (or $\varrho_m \to \infty$), with the product $\varrho_e \, \delta = \mathbf{P}$ (or $\varrho_m \, \delta = \mathbf{M}$) constant. The field can thus be calculated by the superposition of the fields of the two charged cylinders, obtaining for the two cases

$$
\mathbf{E} = -2\pi k_e \mathbf{P} = \begin{cases} -\dfrac{\mathbf{P}}{2\varepsilon_0}, \\ -2\pi \mathbf{P}, \end{cases} \qquad \mathbf{H} = \begin{cases} -\dfrac{\mathbf{M}}{2}, & \text{(SI)}, \\ -2\pi \mathbf{M}, & \text{(Gaussian)}. \end{cases} \qquad \text{(S-5.87)}
$$

The field $\mathbf{B}$ is then obtained as

$$
\mathbf{B} = \begin{cases} \mu_0 (\mathbf{H} + \mathbf{M}) = +\mu_0 \dfrac{\mathbf{M}}{2}, & (SI), \\ \mathbf{H} + 4\pi \mathbf{M} = +2\pi \mathbf{M}, & \text{(Gaussian.)} \end{cases} \qquad \text{(S-5.88)}
$$

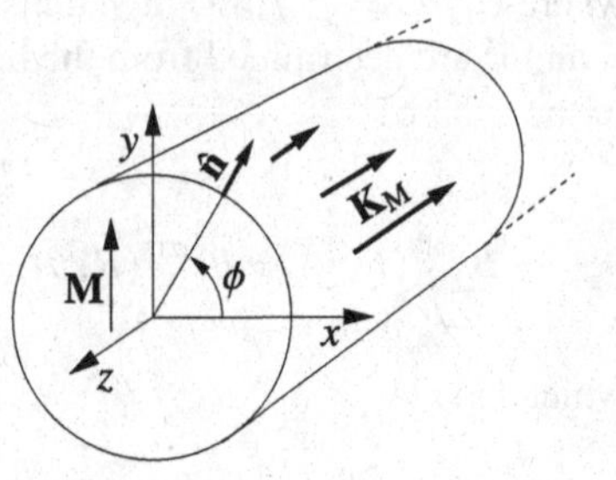

Fig. S-5.14

(b) The magnetization of a material is equivalent to a distribution of volume current density $\mathbf{J}_M = \nabla \times \mathbf{M}/b_m$ and, at the surfaces of discontinuity, of surface current density $\mathbf{K}_M = \mathbf{M} \times \hat{\mathbf{n}}/b_m$, where $\hat{\mathbf{n}}$ is the unit vector perpendicular the surface of discontinuity, the cylinder lateral surface in our case, see Fig. S-5.14. Since in our case the magnetization is uniform we have $\nabla \times \mathbf{M} = 0$, and the transversally magnetized cylinder is equivalent to a surface current distribution $\mathbf{K}_M = -\hat{\mathbf{z}} \, K_0 \cos \phi$, with $K_0 = M/b_m$, as shown in Fig. S-5.14. Thus, in the absence of the magnetized material, one can produce a field $\mathbf{B}_0 = B_0 \, \hat{\mathbf{y}}$ in the transverse (y) direction by distributing a current $\mathbf{K}(\phi) = -\hat{\mathbf{z}} \, K_0 \cos \phi$ over the cylinder lateral surface. Note: this surface current distribution is equivalent to the superposition of two cylinders of radius a, with parallel axes displaced by a small quantity $\mathbf{h}$ in the x direction, carrying uniform volume current densities of equal magnitude but opposite sign $\mathbf{J}_\pm = \pm \hat{\mathbf{z}} \, J_0$, such that $J_0 h = K_0$, as shown in Fig. S-5.15a. We then take the limit $h \to 0$, $J_0 \to \infty$ with the product $J_0 \, h = K_0$ constant. The cylinders carrying the positive and negative current densities give origin to the following magnetic fields at their insides

$$
\mathbf{B}_+ = 2\pi k_m \mathbf{r}_+ \times \mathbf{J}, \qquad \mathbf{B}_- = -2\pi k_m \mathbf{r}_- \times \mathbf{J} = -2\pi k_m (\mathbf{r}_+ - \mathbf{h}) \times \mathbf{J}, \qquad \text{(S-5.89)}
$$

where $\mathbf{r}_+$ and $\mathbf{r}_-$ are the vector distances of the field point from the axes of the cylinders carrying the currents along $\hat{\mathbf{z}}$ and $-\hat{\mathbf{z}}$, respectively. We have used $\mathbf{r}_- = \mathbf{r}_+ - \mathbf{h}$, see Fig. S-5.15b. The total internal field is thus

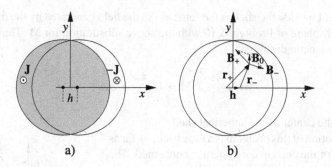

Fig. S-5.15 .

$$\mathbf{B}_0 = \mathbf{B}_+ + \mathbf{B}_- = 2\pi k_m \mathbf{h} \times \mathbf{J} = \hat{\mathbf{y}} \, 2\pi k_m K_0$$

$$= \hat{\mathbf{y}} \, 2\pi k_m \frac{M}{b_m} = \begin{cases} \mu_0 \dfrac{\mathbf{M}}{2} \, , & \text{(SI)}, \\[2mm] 2\pi \mathbf{M} \, , & \text{(Gaussian)}. \end{cases} \tag{S-5.90}$$

S-5.12   A Rotating Charged Spherical Shell

(a) We use a spherical coordinate system (r, θ, ϕ) with ω along the polar axis z and the origin at the center of the sphere. The rotation of the sphere with uniform surface charge density σ generates an azimuthal surface current density

$$\mathbf{K} = \sigma \boldsymbol{v} = \sigma a \omega \sin \theta \, \hat{\boldsymbol{\phi}} \, . \tag{S-5.91}$$

This surface current distribution current is analogous to that of the magnetization current distribution (5.17) for the magnetized sphere of point **(a)** of Problem 5.10,

$$\mathbf{K}_m = \frac{1}{b_m} \mathbf{M} \times \hat{\mathbf{r}} = \frac{1}{b_m} M \sin \theta \, \hat{\boldsymbol{\phi}} \, . \tag{S-5.92}$$

Thus, the magnetic field generated by **K** of the rotating charged sphere is the same as that generated by $\mathbf{K}_m$ of the magnetized sphere, given by (S-5.71), with the replacement $\mathbf{M} = b_m \sigma a \omega$, namely

$$\mathbf{B} = \frac{8\pi}{3} k_m \sigma a \omega = \begin{cases} \dfrac{2}{3} \mu_0 \sigma a \omega \, , & \text{(SI)}, \\[3mm] \dfrac{8\pi}{3c} \sigma a \omega \, , & \text{(Gaussian)}, \end{cases} \tag{S-5.93}$$

Also the field outside the shell is the same as the the field generated by the uniformly magnetized sphere of Problem 5.10 with the above substitution for **M**. Thus it is the field of a magnetic dipole

$$\mathbf{m} = \frac{4\pi}{3} a^3 \mathbf{M} = b_{\mathrm{m}} \frac{4\pi}{3} a^4 \sigma \boldsymbol{\omega} \,, \tag{S-5.94}$$

located at the center of the spherical shell.

(b) The solution of this problem is a little tricky as far as the choice of the coordinate system is concerned. The "natural" choice of a spherical coordinate system with the polar axis z along $\boldsymbol{\omega}$ would lead to both the point where we evaluate the vector potential and the point of the shell surface over which we integrate to have varying θ and ϕ coordinates. If we want to evaluate the vector potential **A** in a generic point P, no matter if inside or outside the spherical shell, it is more convenient to choose a spherical coordinate system (r, θ, ϕ)

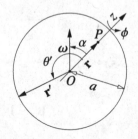

Fig. S-5.16

with the polar axis z passing through the center of the rotating shell O and through P, with the origin in O, as shown in Fig. S-5.16. Simultaneously we use a Cartesian coordinate system with the origin and the z axis in common with the spherical system, and the angular velocity $\boldsymbol{\omega}$ lying in the xz plane. The angular velocity forms an angle α with the z axis and has Cartesian coordinates

$$\boldsymbol{\omega} \equiv (\omega \sin \alpha, 0, \omega \cos \alpha) \,. \tag{S-5.95}$$

The position vector of P has spherical coordinates $\mathbf{r} \equiv (r, 0, 0)$, where the value of ϕ is, obviously, arbitrary. Equation (5.6) for the vector potential can be rewritten

$$\mathbf{A}(\mathbf{r}) = k_{\mathrm{m}} \int_S \frac{\mathbf{K}(\mathbf{r}')}{|\mathbf{r} - \mathbf{r}'|} \, \mathrm{d}S' = k_{\mathrm{m}} \int_0^\pi \mathrm{d}\theta' \sin \theta' \int_0^{2\pi} \mathrm{d}\phi' \frac{\mathbf{K}(\mathbf{r}')}{|\mathbf{r} - \mathbf{r}'|} a^2 \,, \tag{S-5.96}$$

integrated over the shell surface, **K** being the surface current density. The position vector of the infinitesimal surface element $\mathrm{d}S' \equiv a^2 \sin \theta' \mathrm{d}\theta' \mathrm{d}\phi'$ is $\mathbf{r}'$ and has spherical coordinates $\mathbf{r}' \equiv (a, \theta', \phi')$ and Cartesian coordinates

$$\mathbf{r}' \equiv (a \sin \theta' \cos \phi', a \sin \theta' \sin \phi', a \cos \theta') \,. \tag{S-5.97}$$

According to the cosine law we have

$$|\mathbf{r} - \mathbf{r}'| = \sqrt{a^2 + r^2 - 2ra \cos \theta'} \,, \tag{S-5.98}$$

further we have for the surface current density in Cartesian coordinates

$$\mathbf{K}(\mathbf{r}') = \sigma\boldsymbol{\omega} \times \mathbf{r}'$$

$$\equiv \sigma a\omega \left(-\cos\alpha\sin\theta'\cos\phi', \cos\alpha\sin\theta'\cos\phi' - \sin\alpha\cos\theta', \sin\alpha\sin\theta'\sin\phi'\right) .$$

$$(S\text{-}5.99)$$

Thus the integral (S-5.96) is split into three Cartesian components of which both the x and the z components, as well as the first term of the y component, are zero because, with our choice of the coordinates, $|\mathbf{r} - \mathbf{r}'|$ is independent of ϕ' and

$$\int_0^{2\pi} \sin\phi' d\phi' = \int_0^{2\pi} \cos\phi' d\phi' = 0 . \qquad (S\text{-}5.100)$$

Thus we are left with

$$\mathbf{A}(\mathbf{r}) = -\hat{\mathbf{y}}\, k_{\mathrm{m}} 2\pi\sigma a^3 \omega \sin\alpha \int_0^{\pi} \frac{\cos\theta'\sin\theta'}{\sqrt{a^2 + r^2 - 2ar\cos\theta'}}\, d\theta' . \qquad (S\text{-}5.101)$$

We can perform the substitution $\xi = \cos\theta'$, $d\xi = -\sin\theta' d\theta'$, and, remembering that

$$\int \frac{\xi\, d\xi}{\sqrt{a + b\,\xi}} = -\frac{2(2a - b\,\xi)}{3b^2}\sqrt{a + b\,\xi} \qquad (S\text{-}5.102)$$

we have

$$\int_{-1}^{1} \frac{\xi\, d\xi}{\sqrt{a^2 + r^2 - 2ar\xi}} = \left[\frac{4a^2 + 4r^2 + 4ar\xi}{12\, a^2 r^2}\sqrt{a^2 + r^2 - 2ar\xi}\right]_{-1}^{+1}$$

$$= \begin{cases} -\dfrac{2}{3}\dfrac{r}{a^2} , & \text{if } r < a , \\[2mm] -\dfrac{2}{3}\dfrac{a}{r^2} , & \text{if } r > a . \end{cases} \qquad (S\text{-}5.103)$$

Note that $\left|\sqrt{a^2 + r^2 - 2ar}\right|$ equals $(a - r)$ if $r < a$ and $(r - a)$ if $r > a$. Noting further that

$$-\hat{\mathbf{y}}\,\omega r\sin\alpha = \boldsymbol{\omega} \times \mathbf{r} , \qquad (S\text{-}5.104)$$

Equation (S-5.101) for the vector potential finally turns into

$$\mathbf{A}(\mathbf{r}) = \begin{cases} k_{\mathrm{m}}\dfrac{4\pi}{3}\sigma a\,\boldsymbol{\omega} \times \mathbf{r} , & \text{if } r \le a , \\[3mm] k_{\mathrm{m}}\dfrac{4\pi}{3r^3}\sigma a^4\boldsymbol{\omega} \times \mathbf{r} , & \text{if } r \ge a , \end{cases} \qquad (S\text{-}5.105)$$

which are independent of the choice of the reference frame. Note that **A** is continuous at $r = a$. The magnetic field inside the sphere is uniform and equals

$$\mathbf{B} = \nabla \times \mathbf{A} = k_\mathrm{m}\frac{4\pi}{3}\sigma a\, \nabla \times (\boldsymbol{\omega} \times \mathbf{r}) = k_\mathrm{m}\frac{8\pi}{3}\sigma a\,\boldsymbol{\omega} = \begin{cases} \dfrac{2}{3}\mu_0\sigma a\omega\,, & \text{(SI)}, \\[2mm] \dfrac{8\pi}{3c}\sigma a\omega\,, & \text{(Gaussian)}, \end{cases}$$

(S-5.106)

in agreement with (S-5.93). Remembering that the vector potential of a magnetic dipole **m** is

$$\mathbf{A} = \frac{k_\mathrm{m}}{b_\mathrm{m}}\frac{\mathbf{m} \times \mathbf{r}}{r^3} = \begin{cases} \dfrac{\mu_0}{4\pi}\dfrac{\mathbf{m} \times \mathbf{r}}{r^3}\,, & \text{(SI)}, \\[2mm] \dfrac{\mathbf{m} \times \mathbf{r}}{r^3}\,, & \text{(Gaussian)}, \end{cases}$$

(S-5.107)

we see that the magnetic field outside the sphere is the field of a magnetic dipole of value

$$\mathbf{m} = b_\mathrm{m}\frac{4\pi}{3}a^4\sigma\boldsymbol{\omega}\,,$$

(S-5.108)

located at the center of the sphere, in agreement with (S-5.94).

S-5.13   Magnetic Field of a Polygonal Loop

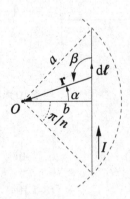

Fig. S-5.17

(a) According to the Biot-Savart law given in Table 5.1 the contribution d**B** to the magnetic field of an infinitesimal vector element d$\boldsymbol{\ell}$ of a circuit carrying an electric current I in a point at a vector distance **r** from d$\boldsymbol{\ell}$ is

$$d\mathbf{B} = k_\mathrm{m}I\,\frac{d\boldsymbol{\ell} \times \mathbf{r}}{r^3}\,.$$

(S-5.109)

All the sides of our polygon lie on the same plane, as does the distance **r** of the center O of the polygon from any point of the loop. Thus all the contributions d**B** are perpendicular to the plane and parallel to one another. Each side of the polygon contributes by the same amount to the magnetic field **B** at O, thus we can evaluate the contribution of a single side and multiply the result by n. Figure S-5.17 shows the special case of one side of a square loop, but the procedure is valid for any n. With the notation of Fig. S-5.17 the numerator of (S-5.109) is

$$d\boldsymbol{\ell} \times \mathbf{r} = \hat{\mathbf{z}}r\sin\beta\,d\ell\,,$$

(S-5.110)

where $\hat{z}$ is the unit vector out of page. The magnitude of $\mathbf{r}$ is $r = b/\cos\alpha$, where b is the apothem of the polygon $b = a\cos(\pi/n)$. The magnitude of $d\boldsymbol{\ell}$ is

$$d\ell = d(b\tan\alpha) = a\cos(\pi/n)\frac{d\alpha}{\cos^2\alpha}. \tag{S-5.111}$$

Thus, noting from the figure that $\sin\beta$ appearing in (S-5.110) equals $\cos\alpha$, we have for the magnitude of the total magnetic field B at O

$$B = k_{\mathrm{m}} nI \int_{-\pi/n}^{\pi/n} \frac{\cos^3\alpha}{a^3\cos^3(\pi/n)} \frac{a\cos(\pi/n)}{\cos\alpha} a\cos(\pi/n)\frac{d\alpha}{\cos^2\alpha}\cos\alpha$$

$$= k_{\mathrm{m}} nI \frac{1}{a\cos(\pi/n)} \int_{-\pi/n}^{\pi/n} \cos\alpha\, d\alpha = k_{\mathrm{m}}\frac{2nI}{a}\tan(\pi/n), \tag{S-5.112}$$

valid for $n \geq 3$, since the triangle is the simplest polygon which can exist in the Euclidean plane.

(b) At the limit $n \to \infty$ we have

$$\lim_{n\to\infty} k_{\mathrm{m}}\frac{2nI}{a}\tan(\pi/n) = \lim_{n\to\infty} k_{\mathrm{m}}\frac{2nI}{a}\frac{\pi}{n} = k_{\mathrm{m}}\frac{2\pi I}{a} = \begin{cases} \dfrac{\mu_0 I}{2a}, & \text{(SI)}, \\[2mm] \dfrac{2I}{ca}, & \text{(Gaussian)}, \end{cases} \tag{S-5.113}$$

which is the magnetic field at the center of a circular loop or radius a carrying a current I.

S-5.14 Helmholtz Coils

(a) In our coordinate system the centers of the two coils are located at $z = -b$ and $z = +b$, respectively. The magnetic field at a generic point z on the axis is thus

$$\mathbf{B}(z) = \hat{z}\, 2\pi k_{\mathrm{m}} I \left\{ \frac{a^2}{\left[a^2 + (z+b)^2\right]^{3/2}} + \frac{a^2}{\left[a^2 + (z-b)^2\right]^{3/2}} \right\}, \tag{S-5.114}$$

where $2\pi k_{\mathrm{m}} = \mu_0/2$ in SI units and $2\pi k_{\mathrm{m}} = 2\pi/c$ in Gaussian units. We introduce the dimensionless variables $\zeta = z/a$ and $\beta = b/a$, obtaining for the magnitude of $\mathbf{B}$

$$B(\zeta) = \frac{2\pi k_{\mathrm{m}} I}{a} \left\{ \frac{1}{\left[1 + (\zeta+\beta)^2\right]^{3/2}} + \frac{1}{\left[1 + (\zeta-\beta)^2\right]^{3/2}} \right\}, \tag{S-5.115}$$

which is an even function of ζ, thus all its odd order derivatives at $\zeta = 0$ are zero. The function itself, at $\zeta = 0$, is

$$B(0) = \frac{4\pi \, k_m I}{a} \frac{1}{\left(1 + \beta^2\right)^{3/2}} . \tag{S-5.116}$$

The first derivative of (S-5.115) with respect to ζ is

$$\frac{\partial B}{\partial \zeta} = -\frac{6\pi \, k_m I}{a} \left\{ \frac{\zeta + \beta}{\left[1 + (\zeta + \beta)^2\right]^{5/2}} + \frac{\zeta - \beta}{\left[1 + (\zeta - \beta)^2\right]^{5/2}} \right\} , \tag{S-5.117}$$

which, as expected, is zero for $\zeta = 0$. The second derivative is

$$\frac{\partial^2 B}{\partial \zeta^2} = -\frac{6\pi \, k_m I}{a} \left\{ \frac{1 - 4(\zeta + \beta)^2}{\left[1 + (\zeta + \beta)^2\right]^{7/2}} + \frac{1 - 4(\zeta - \beta)^2}{\left[1 + (\zeta - \beta)^2\right]^{7/2}} \right\} \tag{S-5.118}$$

and, for $\zeta = 0$, we have

$$\left. \frac{\partial^2 B}{\partial \zeta^2} \right|_{\zeta=0} = \frac{6\pi \, k_m I}{a} \frac{\left(8 \beta^2 - 2\right)}{\left(1 + \beta^2\right)^{7/2}} . \tag{S-5.119}$$

Thus, if we choose $\beta = 1/2$, corresponding to $b = a/2$, the second derivative of B is zero at $\zeta = 0$, while also the first and third derivatives are zero because they are odd-order derivatives of an even function of ζ. If we expand the magnetic field into a Taylor series around $\zeta = 0$ we obtain

$$B(\zeta) = \frac{4\pi \, k_m I}{a} \left(\frac{4}{5}\right)^{3/2} + \frac{1}{4!} \frac{\partial^4 B}{\partial \zeta^4} \zeta^4 + \cdots . \tag{S-5.120}$$

A plot of B versus ζ is shown in Fig. S-5.18, where the two vertical bars at $\zeta = \pm 0.5$ represente the positions of the two coils.

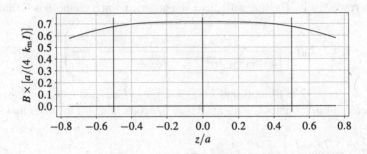

Fig. S-5.18 .

Helmholtz coils are commonly used for generating nearly uniform magnetic fields or to compensate the Earth's magnetic field in experiments where the absence of magnetic fields is required.

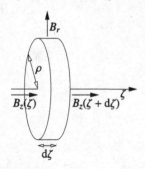

Fig. S-5.19

(b) Because of symmetry our magnetic field has only the longitudinal component B_z and the radial component B_r, both independent of ϕ. We introduce the further dimensionless variable $\rho = r/a$. We consider a small closed cylindrical surface of radius ρ and height $d\zeta$, coaxial with the system, as shown in Fig. S-5.19. The flux Φ of the magnetic field through the surface must be zero because of Maxwell's equation $\nabla \cdot \mathbf{B} = 0$. This corresponds to

$$0 \simeq 2\pi\rho\,d\zeta\,B_r + \pi\rho^2 B_z(\zeta + d\zeta) - \pi\rho^2 B_z(\zeta)$$

$$\simeq 2\pi\rho\,d\zeta\,B_r + \pi\rho^2 \frac{\partial B_z}{\partial \zeta} d\zeta\,, \qquad (S\text{-}5.121)$$

which leads to

$$B_r(\rho, \zeta) \simeq -\frac{\rho}{2}\frac{\partial B_z}{\partial \zeta} = \frac{3\pi\,k_m\,\rho\,I}{a}\left\{\frac{\zeta + \beta}{\left[1 + (\zeta + \beta)^2\right]^{5/2}} + \frac{\zeta - \beta}{\left[1 + (\zeta - \beta)^2\right]^{5/2}}\right\},$$
$$(S\text{-}5.122)$$

where we have substituted (S-5.117) for $\partial B_z/\partial \zeta$, (S-5.122) is valid for $\rho \ll 1$. If we expand $B_r(\rho, \zeta)$ into a power series of ζ around $\zeta = 0$ we note that (S-5.122) is an odd function of ζ, thus only odd powers of ζ will appear in the expansion. The first derivative of (S-5.122) is proportional to (S-5.118) which, with our choice of β, vanishes at $\zeta = 0$. Thus we have

$$B_r(\rho, \zeta) \simeq \frac{1}{3!}\left.\frac{\partial^3 B_r(\rho, \zeta)}{\partial \zeta^3}\right|_{\rho=0,\ \zeta=0} \rho\zeta^3\,. \qquad (S\text{-}5.123)$$

Chapter S-6
Solutions for Chapter 6

S-6.1 A Square Wave Generator

(a) The motion is periodic, and we choose the origin of time, $t = 0$, at an instant when the coil surface is completely in the $x \geqslant 0$ half of the xy plane. With this choice, the flux of the magnetic field through the coil, $\Phi(t)$, increases with time when $2n\pi < \omega t < (2n + 1)\pi$, with n any integer, and equals $\Phi(t) = B(\omega t \bmod 2\pi) a^2/2$. Here, $x \bmod y$ stands for the remainder of the division of x by y with an integer quotient. When $(2n + 1)\pi < \omega t <$ $(2n + 2)\pi$, the flux decreases with time and equals $\Phi(t) = B[2\pi - (\omega t \bmod 2\pi)] a^2/2$. The electromotive force in the coil, $\mathcal{E}(t)$, is thus

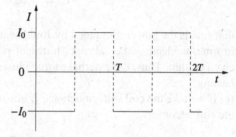

Fig. S-6.1

$$\mathcal{E}(t) = -b_m \frac{d\Phi(t)}{dt} \quad (S\text{-}6.1)$$

$$= -b_m \frac{Ba^2\omega}{2} \text{sign}[\pi - (\omega t \bmod 2\pi)],$$

where $\text{sign}(x) = x/|x|$ is the sign function. Thus, $\mathcal{E}$ reverses its sign whenever $\omega t = n\pi$, with n any integer. The current circulating in the coil is $I = \mathcal{E}/R$, and, as shown in Fig. S-6.1, both $\mathcal{E}$ and I are square waves of period $T = 2\pi/\omega$, and amplitudes

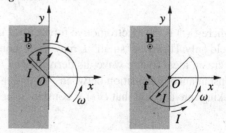

Fig. S-6.2

© The Editor(s) (if applicable) and The Author(s), under exclusive license to Springer Nature Switzerland AG 2023
A. Macchi et al., *Problems in Classical Electromagnetism*,
https://doi.org/10.1007/978-3-031-22235-1

$$\mathcal{E}_0 = b_{\mathrm{m}} \frac{B\omega a^2}{2}, \quad I_0 = b_{\mathrm{m}} \frac{B\omega a^2}{2R}. \tag{S-6.2}$$

(b) The external torque applied to the coil in order to keep its angular velocity constant must balance the torque exerted by the magnetic forces. The magnetic force on a current-carrying circuit element $d\boldsymbol{\ell}$ is $d\mathbf{f} = b_{\mathrm{m}} I\, d\boldsymbol{\ell} \times \mathbf{B}$, and is different from zero only in the $x < 0$ half plane. The corresponding torque $d\boldsymbol{\tau} = \mathbf{r} \times d\mathbf{f} = b_{\mathrm{m}}^2 I \mathbf{r} \times (d\boldsymbol{\ell} \times \mathbf{B})$, where $\mathbf{r}$ is the distance of the coil element $d\boldsymbol{\ell}$ from the z axis, is always equal to zero on the circumference arc of the coil because the three vectors of the triple product are mutually perpendicular here. Thus, $d\boldsymbol{\tau}$ is different from zero only on the half of the straight part of the coil inside the magnetic field, where $d\boldsymbol{\ell} = d\mathbf{r}$. Here we have $d\boldsymbol{\tau} = -b_{\mathrm{m}}^2 I_0 r B\, dr\, \hat{\mathbf{z}}$, as shown in Fig. S-6.2, and the total torque on the coil, $\boldsymbol{\tau}$, is

$$\boldsymbol{\tau} = \int d\boldsymbol{\tau} = -b_{\mathrm{m}}^2\, \omega\, \frac{B^2 a^2}{2R}\, \hat{\mathbf{z}} \int\limits_0^a r\, dr = -b_{\mathrm{m}}^2\, \omega\, \frac{B^2 a^4}{4R}\, \hat{\mathbf{z}}, \tag{S-6.3}$$

corresponding to a power dissipation

$$P_{\mathrm{diss}} = -\boldsymbol{\tau} \cdot \boldsymbol{\omega} = b_{\mathrm{m}}^2\, \omega^2\, \frac{B^2 a^4}{4R} = R I_0^2, \tag{S-6.4}$$

that equals the power dissipated by Joule heating. The power dissipation is constant in time, neglecting the "abrupt" transient phases at $t = n\pi/\omega$, where I instantly changes sign. Thus, the external torque must provide the power dissipated by Joule heating.

(c) If we take the coil self-inductance L into account, the equation for the current in the coil becomes

$$\mathcal{E}(t) - L\frac{dI}{dt} = RI, \tag{S-6.5}$$

where $\mathcal{E}(t)$ is the electromotive force (S-6.1), due to the flux change of the external field only. However "small" L may be, its contribution is not negligible because, if I were an ideal square wave, its derivative dI/dt would diverge whenever $t = n\pi/\omega$ (instantaneous transition between $-I_0$ and $+I_0$). The general solution of (S-6.5) is, taking into account that $\mathcal{E}(t)$ is constant over each half-period,

$$I = \frac{\mathcal{E}}{R} + K e^{-t/t_0}, \tag{S-6.6}$$

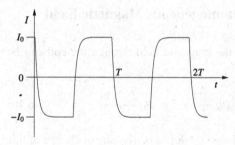

Fig. S-6.3

for $0 < t < \pi/\omega$,

where $t_0 = L/R$ is the characteristic time of the loop, and K is a constant to be determined from the initial conditions. If L is small enough, we can assume that at time $t = 0^-$ we have $\mathcal{E}(t) = +\mathcal{E}_0$ and $I(t) = +I_0$. At time $t = 0$, $\mathcal{E}(t)$ switches instantaneously from $+\mathcal{E}_0$ to $-\mathcal{E}_0$, and the constant K is determined by the initial condition $I(0) = I(0^-) = I_0 = \mathcal{E}_0/R$, leading to $K = 2\mathcal{E}_0/R$. Thus,

$$I(t) = I_0 \left(2e^{-t/t_0} - 1\right). \tag{S-6.7}$$

At $t = (\pi/\omega)^-$ we have $\mathcal{E}(t) = -\mathcal{E}_0$ and we can assume that $I(t) = -I_0$. At $t = \pi/\omega$ $\mathcal{E}(t)$ switches instantaneously from $-\mathcal{E}_0$ to $+\mathcal{E}_0$, and, for $\pi/\omega < t < 2\pi/\omega$, we have

$$I(t) = -I_0 \left(2e^{-(t-\pi/\omega)/t_0} - 1\right), \tag{S-6.8}$$

and so on for the successive periods.

The self-inductance of the coil prevents the current from switching instantaneously between $+I_0$ and $-I_0$: the change occurs following an exponential with characteristic time $t_0 = L/R$. The behavior described by (S-6.7) and (S-6.8) is valid only if $t_0 \ll T = 2\pi/\omega$, as in Fig. S-6.3, representing the case of $t_0 = 0.04\,T$. If t_0 is not negligible with respect to T, the current oscillates between two values

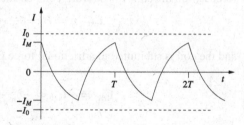

Fig. S-6.4

$+I_M$ and $-I_M$, with $I_M < I_0$. Let us consider the time interval $0 \le t \le \pi/\omega$. We must have $I(0) = I_M$ and $I(\pi/\omega) = -I_M$. Replacing I by (S-6.6), we obtain

$$I_M = I_0 \frac{1 - e^{-T/2t_0}}{1 + e^{-T/2t_0}}. \tag{S-6.9}$$

The plot of $I(t)$ can no longer be approximated by a square wave, as shown in Fig. S-6.4 for the case $t_0 = 0.25\,T$.

S-6.2 A Coil Moving in an Inhomogeneous Magnetic Field

(a) With our assumptions, the flux of the magnetic field through the coil can be approximated as

$$\Phi_{\mathbf{B}}(t) = \Phi_{\mathbf{B}}[z(t)] \simeq \pi a^2 B_0 \frac{z(t)}{L} = \pi a^2 B_0 \frac{z_0 + vt}{L} , \qquad (S\text{-}6.\dot{1}0)$$

where z_0 is the position of the center of the coil at $t = 0$. The rate of change of this magnetic flux is associated to an electromotive force $\mathcal{E}$, and to a current $I = \mathcal{E}/R$ circulating in the coil

$$\mathcal{E} = RI = -b_{\mathrm{m}} \frac{d\Phi}{dt} = -b_{\mathrm{m}} \pi a^2 B_0 \frac{v}{L} . \qquad (S\text{-}6.11)$$

(b) The power dissipated by Joule heating is

$$P = RI^2 = \frac{\mathcal{E}^2}{R} = b_{\mathrm{m}}^2 \frac{(\pi a^2 B_0 v)^2}{L^2 R} . \qquad (S\text{-}6.12)$$

Thus, in order to keep the coil in motion at constant speed, one must exert an external force $\mathbf{f}_{\mathrm{ext}}$ on the coil, whose work compensates the dissipated power. We have

$$\mathbf{f}_{\mathrm{ext}} \cdot \boldsymbol{v} = P = b_{\mathrm{m}}^2 \frac{(\pi a^2 B_0 v)^2}{L^2 R} , \qquad (S\text{-}6.13)$$

and the coil is submitted to a frictional force proportional to its velocity

$$\mathbf{f}_{\mathrm{frict}} = -\mathbf{f}_{\mathrm{ext}} = -b_{\mathrm{m}}^2 \frac{(\pi a^2 B_0)^2}{L^2 R} \boldsymbol{v} . \qquad (S\text{-}6.14)$$

(c) The force $\mathbf{f}_{\mathrm{frict}}$ is actually the net force obtained by integrating the force $d\mathbf{f}_{\mathrm{frict}} = b_{\mathrm{m}} I \, d\boldsymbol{\ell} \times \mathbf{B}$ acting on each coil element $d\boldsymbol{\ell}$:

$$\mathbf{f}_{\mathrm{frict}} = b_{\mathrm{m}} I \oint_{\mathrm{coil}} d\boldsymbol{\ell} \times \mathbf{B} . \qquad (S\text{-}6.15)$$

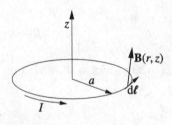

The contribution of the z component of $\mathbf{B}$ is a radial force tending to shrink the coil if $\partial_t \Phi > 0$, or to widen it if $\partial_t \Phi < 0$, according to Lenz's law; the case represented in Fig. S-6.5 corresponds to the latter case. Thus $\mathbf{f}_{\mathrm{ext}}$, directed along z, is due only to the radial component B_r of $\mathbf{B}$. The component B_r is not given by the problem, but, as we saw at

Fig. S-6.5   answer **(a)** of Problem 5.4, it can be evaluated by applying Gauss's law to a closed cylindrical surface of radius r and height Δz. According to (S-5.25)

$$B_r \simeq -\frac{B_0}{2L}\, r\,, \quad \text{thus} \quad \mathrm{df_{frict}} = b_\mathrm{m}\, I\, \mathrm{d}\ell\, \frac{B_0\, a}{2L}\,\hat{\mathbf{z}}\,, \tag{S-6.16}$$

and by substituting (S-6.11) and integrating over the coil we obtain

$$\mathbf{f}_\mathrm{frict} = \hat{\mathbf{z}}\, b_\mathrm{m} I \oint_\mathrm{coil} \mathrm{d}\ell\, \frac{B_0\, a}{2L} = -\hat{\mathbf{z}}\, b_\mathrm{m}\left(b_\mathrm{m}\pi a^2 B_0\,\frac{v}{L}\right)\left(2\pi a\,\frac{B_0\, a}{2L}\right) = -\hat{\mathbf{z}}\, b_\mathrm{m}^2\,\frac{(\pi a^2 B_0)^2}{L^2 R}\, v\,,$$

$$\tag{S-6.17}$$

in agreement with (S-6.14).

S-6.3 A Circuit with "Free-Falling" Parts

(a) We choose the x axis oriented downwards, with the origin at the location of the upper horizontal bar, as in Fig. S-6.6. The current I in the rectangular circuit is

$$I = \frac{\mathcal{E}}{R} = -b_\mathrm{m}\,\frac{1}{R}\,\frac{\mathrm{d}\Phi(\mathbf{B})}{\mathrm{d}t} = -b_\mathrm{m}\,\frac{Ba}{R}\,\frac{\mathrm{d}x}{\mathrm{d}t} = -b_\mathrm{m}\,\frac{Bav}{R}\,, \tag{S-6.18}$$

where x is the position of the falling bar, and $v = \dot{x}$ its velocity. The velocity is positive, and the current I is negative, i.e., it circulates clockwise, in agreement with Lenz's law. The magnetic force on the falling bar is $\mathbf{f}_B = b_\mathrm{m} BaI\,\hat{\mathbf{x}}$, antiparallel to the gravitational force $m\mathbf{g}$, and the equation of motion is

$$m\frac{\mathrm{d}v}{\mathrm{d}t} = mg + b_\mathrm{m} BaI = mg - b_\mathrm{m}^2\,\frac{(Ba)^2}{R}\, v. \tag{S-6.19}$$

The solution of (S-6.19), with the initial condition $v(0) = 0$, is

$$v(t) = v_t\left(1 - \mathrm{e}^{-t/\tau}\right) \tag{S-6.20}$$

where

$$\tau = \frac{mR}{(b_\mathrm{m} Ba)^2} \quad \text{and} \quad v_t = g\tau = \frac{mRg}{(b_\mathrm{m} Ba)^2}. \tag{S-6.21}$$

As $t \to \infty$, the falling bar approaches the terminal velocity v_t.

(b) When $v = v_t$, the power dissipated in the circuit by Joule heating is

$$P_J = RI_t^2 = \frac{(b_\mathrm{m} Bav_t)^2}{R} = \left(\frac{mg}{b_\mathrm{m} Ba}\right)^2 R\,, \tag{S-6.22}$$

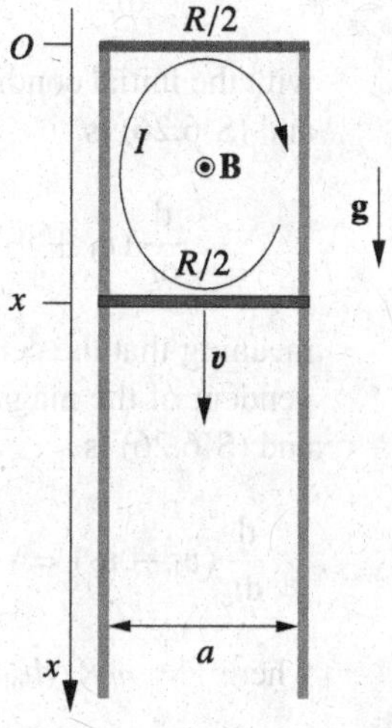

Fig. S-6.6

where $I_t = -b_{\mathrm{m}} B a v_t / R$ is the "terminal current". On the other hand, the work done by the force of gravity per unit time is

$$P_G = m\mathbf{g} \cdot \mathbf{v}_t = mg \frac{mgR}{(b_{\mathrm{m}} B a)^2} = P_J, \qquad (\text{S-6.23})$$

in agreement with energy conservation for the bar moving at constant velocity.

(c) When both horizontal bars are falling, we denote by x_1 the position of the upper bar, and by x_2 the position of the lower bar, as in Fig S-6.7, with $v_1 = \dot{x}_1$ and $v_2 = \dot{x}_2$. The current I circulating in the circuit is

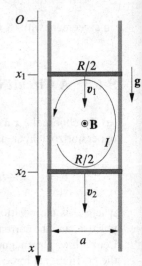

$$I = \frac{\mathcal{E}}{R} = -b_{\mathrm{m}} \frac{1}{R} \frac{\mathrm{d}\Phi(\mathbf{B})}{\mathrm{d}t} = -b_{\mathrm{m}} \frac{Ba}{R} \frac{\mathrm{d}}{\mathrm{d}t}(x_2 - x_1)$$

$$= -b_{\mathrm{m}} \frac{Ba}{R}(v_2 - v_1), \qquad (\text{S-6.24})$$

circulating counterclockwise ($I > 0$) if $v_1 > v_2$, and clockwise ($I < 0$) if $v_1 < v_2$. The magnetic forces acting on the two falling bars are $\mathbf{f}_{B1} = -b_{\mathrm{m}} Ba I\, \hat{\mathbf{x}}$ and $\mathbf{f}_{B2} = b_{\mathrm{m}} Ba I\, \hat{\mathbf{x}}$, respectively. Independently of the sign of I, we have $\mathbf{f}_{B1} = -\mathbf{f}_{B_2}$, so that the net magnetic force on the system comprising the two falling bars is zero. The equations of motion are thus

$$m \frac{\mathrm{d}v_1}{\mathrm{d}t} = mg + b_{\mathrm{m}}^2 \frac{(Ba)^2}{R}(v_2 - v_1) \qquad (\text{S-6.25})$$

$$m \frac{\mathrm{d}v_2}{\mathrm{d}t} = mg - b_{\mathrm{m}}^2 \frac{(Ba)^2}{R}(v_2 - v_1), \qquad (\text{S-6.26})$$

Fig. S-6.7

with the initial conditions $v_1(0) = v_0$ and $v_2(0) = 0$. The sum of equations (S-6.25) and (S-6.26) is

$$\frac{\mathrm{d}}{\mathrm{d}t}(v_1 + v_2) = 2g, \quad \text{with solution} \quad \frac{v_1 + v_2}{2} = \frac{v_0}{2} + gt, \qquad (\text{S-6.27})$$

meaning that the center of mass of the two horizontal bars follows a free fall, independent of the magnetic field $\mathbf{B}$. On the other hand, the difference of Eqs. (S-6.25) and (S-6.26) is

$$\frac{\mathrm{d}}{\mathrm{d}t}(v_1 - v_2) = -\frac{2}{\tau}(v_1 - v_2), \quad \text{with solution} \quad v_1 - v_2 = v_0\, \mathrm{e}^{-2t/\tau}, \qquad (\text{S-6.28})$$

where $\tau = mR/(b_{\mathrm{m}} Ba)^2$. For the velocities of the two horizontal bars we obtain

$$v_1 = \frac{v_0}{2}\left(1 + e^{-2t/\tau}\right) + gt\,, \quad v_2 = \frac{v_0}{2}\left(1 - e^{-2t/\tau}\right) + gt\,. \qquad \text{(S-6.29)}$$

At the steady state limit ($t \gg \tau$) we have

$$\lim_{t\to\infty} v_1 = \lim_{t\to\infty} v_2 = \frac{v_0}{2} + gt \quad \text{and} \quad \lim_{t\to\infty} I = 0\,, \qquad \text{(S-6.30)}$$

since, for $v_1 = v_2$, the flux of **B** through the loop is constant.

S-6.4   The Tethered Satellite

(a) To within our approximations, we can assume that the magnetic field is constant over the satellite orbit, and equal to the field at the Earth's equator, $B_{eq} \simeq 3.2 \times 10^{-5}$ T. The field is parallel to the axis of the satellite orbit, and constant over the tether length. The electromotive force $\mathcal{E}$ on the tether equals the line integral of the magnetic force along the wire,

$$\mathcal{E} = b_{m} \int_{\text{tether}} d\boldsymbol{\ell} \cdot \mathbf{v}(r) \times \mathbf{B}_{eq} = b_{m} \int_{R_{\oplus}+h}^{R_{\oplus}+h-\ell} dr\, \omega r\, B_{eq}\,, \qquad \text{(S-6.31)}$$

where $\omega = v/r$ is the angular velocity of the satellite. To within our approximations we can also assume that also $v(r) \simeq v(R_{\oplus}) \simeq 8\,000$ m/s is constant over the wire length, and obtain

$$\mathcal{E} \simeq b_{m} v \ell B_{eq} = \begin{cases} 8\,000 \times 1\,000 \times 3.2 \times 10^{-5} \simeq 250\,\text{V}\,, & \text{SI,} \\[2mm] \dfrac{1}{c} \times 8 \times 10^{5} \times 10^{5} \times 0.32 \simeq 0.85\,\text{statV}\,, & \text{Gaussian.} \end{cases} \qquad \text{(S-6.32)}$$

(b) Neglecting the resistance of the ionosphere, the current I circulating in the wire, and the corresponding power dissipated by Joule heating P_{diss} are, respectively,

$$I = \frac{\mathcal{E}}{R} = b_{m} \frac{v \ell B_{eq}}{R}\,, \quad \text{and} \quad P_{\text{diss}} = RI^2 = b_{m}^2 \frac{v^2 \ell^2 B_{eq}^2}{R}\,. \qquad \text{(S-6.33)}$$

The power dissipated in the tether by Joule heating must equal minus the work done by the magnetic force on the wire. This can be easily verified, since the magnetic force acting on the wire is

$$\mathbf{F} = b_{m} I \ell\, \hat{\mathbf{r}} \times \mathbf{B}_{eq} = -b_{m}^2 \frac{B_{eq}^2 \ell^2}{R}\, \mathbf{v}\,, \qquad \text{(S-6.34)}$$

and the corresponding work rate is

$$P = \mathbf{F} \cdot \boldsymbol{v} = -b_{\mathrm{m}}^2 \frac{B_{\mathrm{eq}}^2 \ell^2}{R} v^2 = -P_{\mathrm{diss}} .$$ (S-6.35)

If we assume that the tether is a copper wire (conductivity $\sigma \simeq 10^7 \ \Omega^{-1}\mathrm{m}^{-1}$ SI, $\sigma \simeq 9 \times 10^{16} \ \mathrm{s}^{-1}$ Gaussian) of cross section $A=1 \ \mathrm{cm}^2$, the magnitude of the magnetic drag force on the system is

$$F_{\mathrm{drag}} = b_{\mathrm{m}}^2 \frac{B_{\mathrm{eq}}^2 \ell^2}{\ell/(\sigma A)} \, v = \begin{cases} \dfrac{(3.2 \times 10^{-5})^2 \times 1000}{1/(10^7 \times 10^{-4})} \times 8\,000 \simeq 8.2\,\mathrm{N}, & \text{SI} \\[4mm] \dfrac{1}{c^2} \dfrac{(0.32)^2 \times 10^5}{1/(9 \times 10^{16})} \times 8 \times 10^5 \simeq 8.2 \times 10^5 \,\mathrm{dyn}, & \text{Gaussian.} \end{cases}$$
(S-6.36)

This problem gives an elementary description of the principle of the "Tethered Satellite System", investigated in some Space Shuttle missions as a possible generator of electric power for orbiting systems.

S-6.5 Eddy Currents in a Solenoid

(a) We choose a cylindrical coordinate system (r, ϕ, z), with the solenoid axis as z axis. A time-dependent current generates a time-dependent magnetic field, which, in turn, induces a time-dependent contribution to the electric field. The induced electric field is associated to a displacement current density, and, in a conductor, also to a conduction current density $\mathbf{J} = \sigma \mathbf{E}$. Both current densities, in turn, affect the magnetic field. According to our symmetry assumptions, the only non-zero component of the magnetic field is B_z, and the only non-zero component of the electric field is E_ϕ, therefore the only non-zero component of the conduction current density is J_ϕ. Both B_z and E_ϕ depend only on r. In principle, we must solve, for $r < R$, the two following simultaneous equations, obtained from the two Maxwell equations (6.1) and (6.5) writing the curls in cylindrical coordinates (see Table A.1 of the Appendix),

$$\frac{1}{r} \partial_r (r E_\phi) = -b_{\mathrm{m}} \, \partial_t B_z ,$$ (S-6.37)

$$-\partial_r B_z = 4\pi k_{\mathrm{m}} \, \sigma E_\phi + \frac{k_{\mathrm{m}}}{k_{\mathrm{e}}} \, \partial_t E_\phi .$$ (S-6.38)

Solving the simultaneous differential equations (S-6.37) and (S-6.38) is somewhat involved. However, if the angular frequency ω of the driving current is low enough, the *slowly varying current approximation* (SVCA) provides a sufficiently accurate solution of the problem.

In the SVCA, first we neglect the displacement current compared to the conduction current, i.e., we neglect the last term on the right-hand side of (S-6.38). Then, we start by calculating $\mathbf{B}$ as in the static case. Neglecting boundary effects, a DC current

I would generate a uniform magnetic field $\mathbf{B} = \hat{\mathbf{z}} 4\pi k_{\mathrm{m}} \mu_{\mathrm{r}} nI$ inside our solenoid, and $\mathbf{B} \equiv 0$ outside. Thus, inside the solenoid, we would have $\mathbf{B} = \hat{\mathbf{z}} \mu_0 \mu_{\mathrm{r}} nI$ in SI units, and $\mathbf{B} = \hat{\mathbf{z}} 4\pi \mu_{\mathrm{r}} nI/c$ in Gaussian units. If we replace I by $I_0 \cos \omega t$ we obtain

$$\mathbf{B}^{(0)} = \hat{\mathbf{z}} 4\pi k_{\mathrm{m}} \mu_{\mathrm{r}} nI_0 \cos \omega t \,, \tag{S-6.39}$$

which we assume as our zeroth-order approximation for the field inside the solenoid. In the next step of SVCA, we evaluate the first order correction by calculating the electric field $\mathbf{E}^{(1)}$ induced by (S-6.39), and its associated current densities. These current densities, in turn, contribute to the first order correction to the magnetic field. *A posteriori*, our procedure will be justified if the first order correction to the magnetic field, $\mathbf{B}^{(1)}$, is much smaller than $\mathbf{B}^{(0)}$. And so on for the successive correction orders. According to Faraday's law of induction, $\mathbf{B}^{(0)}$ induces an electromotive force

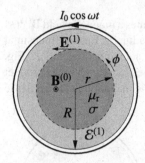

$$\mathcal{E}^{(1)}(r) = -b_{\mathrm{m}} \frac{\mathrm{d}\Phi(\mathbf{B}^{(0)})}{\mathrm{d}t} \tag{S-6.40}$$

Fig. S-6.8

on a circle of radius $r < R$, coaxial to the solenoid (Fig. S-6.8). This corresponds to an azimuthal electric field $\mathbf{E}^{(1)}$ of magnitude

$$\mathbf{E}^{(1)}(r) = k_{\mathrm{m}} b_{\mathrm{m}} 2\pi \mu_{\mathrm{r}} rnI_0 \,\omega \sin \omega t \, \hat{\phi} = \begin{cases} \dfrac{\mu_0 \mu_{\mathrm{r}}}{2} rnI_0 \omega \sin \omega t \hat{\phi} \,, & \text{SI}, \\[2mm] \dfrac{1}{c^2} 2\pi \mu_{\mathrm{r}} rnI_0 \,\omega \sin \omega t \, \hat{\phi} \,, & \text{Gaussian}. \end{cases} \tag{S-6.41}$$

(b) Due to the conductivity σ of the solenoid core, the electric field $\mathbf{E}^{(1)}(r)$ originates an azimuthal current density $\mathbf{J}^{(1)}(r) = \sigma \mathbf{E}^{(1)}(r)$ (eddy currents) in the material. The corresponding Joule dissipation heats up the material. The energy turned into heat per unit volume at each instant t is

$$\mathbf{J}^{(1)}(r) \cdot \mathbf{E}^{(1)}(r) = \sigma \left[E^{(1)}(r) \right]^2 = \sigma \left(k_{\mathrm{m}} b_{\mathrm{m}} 2\pi \mu_{\mathrm{r}} rnI_0 \,\omega \sin \omega t \right)^2 \,, \tag{S-6.42}$$

with a time average

$$\left\langle \mathbf{J}^{(1)}(r) \cdot \mathbf{E}^{(1)}(r) \right\rangle = 2\sigma (k_{\mathrm{m}} b_{\mathrm{m}} \pi \mu_{\mathrm{r}} rnI_0 \,\omega)^2 \,. \tag{S-6.43}$$

The total dissipated power is found by integrating (S-6.43) over the volume of the cylindrical core

$$P_d = \int_{\text{cylinder}} \langle \mathbf{J}^{(1)}(r) \cdot \mathbf{E}^{(1)}(r) \rangle \, d^3x = 2\sigma \, (k_m b_m \, \pi \mu_r n I_0 \, \omega)^2 \int_0^R r^2 \ell \, 2\pi r dr$$

$$= \sigma \pi \ell \left(k_m b_m \, \pi \mu_r n I_0 \, \omega R^2 \right)^2 = \begin{cases} \dfrac{\sigma \mu_0^2 \mu_r^2}{16} \pi n^2 I_0^2 \omega^2 \ell R^4, & \text{SI}, \\[2ex] \dfrac{1}{c^4} \pi^3 \sigma \mu_r^2 n^2 I_0^2 \omega^2 \ell R^4, & \text{Gaussian}. \end{cases}$$

$$\text{(S-6.44)}$$

(c) The induced current density $\mathbf{J}^{(1)}(r) = \sigma \mathbf{E}^{(1)}(r)$ generates a magnetic field $\mathbf{B}^{(1)}(r)$ in the cylindrical volume enclosed by the surface of radius r. Each infinitesimal cylindrical shell between r and $r + dr$ behaves like a solenoid or radius r (Fig. S-6.9), generating a magnetic field whose value is obtained by replacing the product nI by the product $J^{(1)}(r) \, dr$. Thus, the contribution to the magnetic field in r of the infinitesimal shell is

$$d\mathbf{B}_{\text{int}}^{(1)}(r) = \hat{\mathbf{z}} \, 4\pi k_m \mu_r \sigma E^{(1)}(r) \,. \qquad \text{(S-6.45)}$$

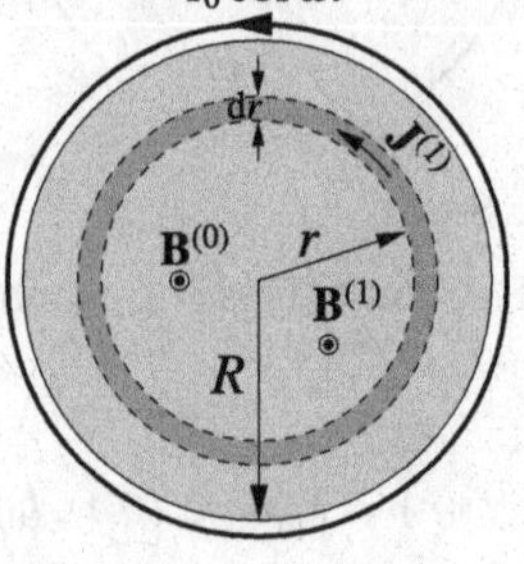

Also all infinitesimal cylindrical shells between r' and $r' + dr'$, with $r < r' < R$, contribute to the field in r, and the resulting first-order correction to the field in r is

$$\mathbf{B}^{(1)}(r) = \int_r^R d\mathbf{B}_{\text{int}}^{(1)}(r) = 4\pi k_m \mu_r \sigma \int_r^R E^{(1)}(r') \, dr' \,.$$

Fig. S-6.9

$$\text{(S-6.46)}$$

If we replace (S-6.41) into (S-6.46) we obtain

$$\mathbf{B}^{(1)}(r) = 8\pi^2 k_m^2 b_m \mu_r^2 \sigma n I_0 \omega \sin \omega t \int_r^R r' \, dr' = \hat{\mathbf{z}} \, 4\pi^2 k_m^2 b_m \, \mu_r^2 \sigma n I_0 \, (R^2 - r^2) \, \omega \sin \omega t$$

$$= \begin{cases} \hat{\mathbf{z}} \dfrac{1}{4} \mu_0^2 \mu_r^2 \sigma n I_0 \, (R^2 - r^2) \, \omega \sin \omega t \,, & \text{SI}, \\[2ex] \hat{\mathbf{z}} \dfrac{4\pi^2}{c^3} \mu_r^2 \sigma n I_0 \, (R^2 - r^2) \, \omega \sin \omega t \,, & \text{Gaussian}. \end{cases} \qquad \text{(S-6.47)}$$

Thus, $\mathbf{B}^{(1)}(r)$ is maximum for $r = 0$, where all infinitesimal cylindrical shells contribute, and zero for $r = R$. Our treatment is justified if $B^{(1)}(0) \ll B^{(0)}$ for all $r < R$ and for all t, i.e., if

$$\frac{\langle B^{(1)}(0) \rangle}{\langle B^{(0)} \rangle} = \pi \, k_m b_m \, \mu_r \sigma \omega \, R^2 \ll 1 \,, \qquad \text{(S-6.48)}$$

where the angle brackets denote the average over time. This gives the condition on ω

$$\omega \ll \frac{1}{\pi k_m b_m \mu_r \sigma R^2} = \begin{cases} \dfrac{4}{\mu_0 \mu_r \sigma R^2}, & \text{SI}, \\[2ex] \dfrac{c^2}{\pi \mu_r \sigma R^2}, & \text{Gaussian}. \end{cases} \tag{S-6.49}$$

Thus, for materials with a high value of the product $\mu_r \sigma$, the frequency must be very low. For instance, iron has a relative magnetic permeability $\mu_r \simeq 5\,000$, and a conductivity $\sigma \simeq 10^7 \,\Omega^{-1} \text{m}^{-1}$ in SI units. Assuming a solenoid with $R = 1\,\text{cm}$, we obtain the following condition on the frequency ν of the driving current

$$\nu = \frac{\omega}{2\pi} \ll \frac{4}{8\pi^2 \times 10^{-7} \times 5 \times 10^3 \times 10^7 \times 10^{-4}} \simeq 0.10\,\text{Hz}, \tag{S-6.50}$$

which is a very low value. Iron is a good material as the core of an electromagnet, due to its high magnetic permeability, but a poor material as the core of a transformer or of an inductor, due to its high conductivity, which gives origin to high eddy-current losses. On the other hand, manganese-zinc ferrite (a ceramic compound containing iron oxides combined with zinc and manganese compounds) also has a relative magnetic permeability $\mu_r \simeq 5000$, but a much lower conductivity, $\sigma \simeq 5\,\Omega^{-1}\text{m}^{-1}$. The condition on the frequency of the driving current is thus

$$\nu \ll \frac{4}{8\pi^2 \times 10^{-7} \times 5 \times 10^3 \times 5 \times 10^{-4}} \simeq 2 \times 10^5\,\text{Hz}, \tag{S-6.51}$$

and ferrite is used in electronics industry to make cores for inductors and transformers, and in various microwave components.

It is also instructive to compare the energy dissipated per cycle, $U_{\text{diss}} = (2\pi/\omega)P_{\text{diss}}$, to the total magnetic energy stored in the solenoid,

$$U_M = \left\langle \frac{b_m \left(B^{(0)}\right)^2}{2k_m \mu_r} \right\rangle \pi R^2 \ell = \begin{cases} \left\langle \dfrac{\left(B^{(0)}\right)^2}{2\mu_0 \mu_r} \right\rangle \pi R^2 \ell, & \text{SI}, \\[3ex] \left\langle \dfrac{\left(B^{(0)}\right)^2}{8\pi \mu_r} \right\rangle \pi R^2 \ell, & \text{Gaussian}. \end{cases} \tag{S-6.52}$$

The ratio is

$$\frac{U_{\text{diss}}}{U_M} \simeq \frac{\pi}{4} k_m b_m \mu_r \sigma \omega R^2. \tag{S-6.53}$$

Thus, the condition (S-6.49) is also equivalent to the requirement that the energy loss per cycle due to Joule heating is small compared to the total stored magnetic energy.

S-6.6   Feynman's "Paradox"

(a) The mutual inductance M between the charged ring and the superconducting ring is, assuming $a \ll R$ (see Problem 6.12),

$$M = 4\pi k_m b_m \frac{\pi a^2}{2R} \, . \tag{S-6.54}$$

Thus, when a current $I(t)$ is circulating in the smaller ring of radius a, the magnetic flux through the charged ring is

$$\Phi_I = M I(t) = 4\pi k_m b_m \frac{\pi a^2}{2R} I(t) \, . \tag{S-6.55}$$

If Φ_I is time-dependent, it gives origin to an induced electric field $\mathbf{E}_I$, whose line-integral around the charged ring is

$$\oint \mathbf{E}_I \cdot d\boldsymbol{\ell} = -b_m \frac{d\Phi_I}{dt} = -4\pi k_m b_m^2 \frac{\pi a^2}{2R} \partial_t I(t) \, . \tag{S-6.56}$$

Due to the symmetry of our problem, field $\mathbf{E}_I$ is azimuthal on the xy plane, and independent of ϕ. Its magnitude on the charged ring is thus

$$E_I = \frac{1}{2\pi R} \oint \mathbf{E}_I \cdot d\boldsymbol{\ell} = -k_m b_m^2 \frac{\pi a^2}{R^2} \partial_t I(t) \, , \tag{S-6.57}$$

and the force exerted on an infinitesimal element $d\ell$ of the charged ring is

$$d\mathbf{f} = \mathbf{E}_I \lambda \, d\ell = -\hat{\phi} k_m b_m^2 \frac{\pi a^2}{R^2} \lambda \, d\ell \, \partial_t I(t) \, , \tag{S-6.58}$$

corresponding to a torque $d\boldsymbol{\tau}$ about the center of the ring

$$d\boldsymbol{\tau} = \mathbf{r} \times d\mathbf{f} = -\hat{\mathbf{z}} k_m b_m^2 \frac{\pi a^2}{R} \lambda \, d\ell \, \partial_t I(t) \, . \tag{S-6.59}$$

The total torque on the charged ring is thus

$$\boldsymbol{\tau} = \int d\boldsymbol{\tau} = -\hat{\mathbf{z}} k_m b_m^2 \frac{\pi a^2}{R} \lambda \, 2\pi R \, \partial_t I(t) = -\hat{\mathbf{z}} k_m b_m^2 \frac{\pi a^2}{R} Q \, \partial_t I(t) \, , \tag{S-6.60}$$

where $Q = 2\pi R \lambda$ is the total charge of the ring. The equation of motion for the charged ring is thus

$$m R^2 \frac{d\omega}{dt} = \tau = -k_m b_m^2 \frac{\pi a^2}{R} Q \, \partial_t I(t) \, , \tag{S-6.61}$$

where mR^2 is the moment of inertia of the ring. The solution for $\omega(t)$ is

$$\omega(t) = -k_m b_m^2 \frac{\pi a^2}{mR^3} Q \int_0^t \partial_t I(t')\, dt' = k_m b_m^2 \frac{\pi a^2}{mR^3} Q \left[I_0 - I(t) \right] , \quad \text{(S-6.62)}$$

and the final angular velocity is

$$\omega_f = k_m b_m^2 \frac{\pi a^2}{mR^3} Q I_0 = \begin{cases} \dfrac{\mu_0 a^2 Q}{4mR^3} I_0 , & \text{SI}, \\[3mm] \dfrac{\pi a^2 Q}{c^3 mR^3} I_0 , & \text{Gaussian}, \end{cases} \quad \text{(S-6.63)}$$

corresponding to a final angular momentum

$$L_f = mR^2 \omega_f = k_m b_m^2 \frac{\pi a^2 Q}{R} I_0 = \begin{cases} \dfrac{\mu_0 a^2 Q}{R} I_0 , & \text{SI}, \\[3mm] \dfrac{\pi a^2 Q}{c^3 R} I_0 , & \text{Gaussian}, \end{cases} \quad \text{(S-6.64)}$$

independent of the mass m of the ring.

(b) The rotating charged ring is equivalent to a circular loop carrying a current $I_{rot} = Q\omega/2\pi$. Thus, after the current in the small ring is switched off, there is still a magnetic field due to the rotation of the charged ring. The final magnetic field at the center of the rings is

$$\mathbf{B}_c = \hat{\mathbf{z}} \frac{k_m}{2} \frac{I_{rot}}{R} = \hat{\mathbf{z}} \frac{k_m}{4\pi} \frac{Q\omega_f}{R}$$

$$= \hat{\mathbf{z}} \frac{k_m^2 b_m^2 a^2 Q^2}{4mR^4} I_0 = \begin{cases} \hat{\mathbf{z}} \dfrac{\mu_0^2 a^2 Q^2}{64\pi^2 mR^4} I_0 , & \text{SI}, \\[3mm] \hat{\mathbf{z}} \dfrac{a^2 Q^2}{4c^4 mR^4} I_0 , & \text{Gaussian}, \end{cases} \quad \text{(S-6.65)}$$

parallel to the initial field $\mathbf{B}_0 = \hat{\mathbf{z}} k_m I_0/(2a)$, in agreement with Lenz's law. We further have

$$\pi a^2 B_c = M I_{rot} , \quad \text{(S-6.66)}$$

where M is the mutual inductance of the rings (S-6.54).

(c) As seen above at point **(b)**, the rotating charged ring generates a magnetic field all over the space. This field modifies the magnetic flux through the rotating ring itself, giving origin to self-induction. Let $\mathcal{L}$ be the "self-inductance" of the rotating ring. The magnetic flux generated by the rotating ring through itself is

$$\Phi_{rot} = \frac{1}{b_m} \mathcal{L} I_{rot} = \frac{1}{b_m} \mathcal{L} \frac{Q\omega}{2\pi} . \quad \text{(S-6.67)}$$

Correspondingly, (S-6.56) for the line integral of the electric field around the charged ring is modified as follows:

$$\oint \mathbf{E}_I \cdot \mathrm{d}\boldsymbol{\ell} = -b_{\mathrm{m}} \left(\frac{\mathrm{d}\Phi_I}{\mathrm{d}t} + \frac{\mathrm{d}\Phi_{\mathrm{rot}}}{\mathrm{d}t} \right) = -\frac{4\pi^2 k_{\mathrm{m}} b_{\mathrm{m}}^2 a^2}{2R} \partial_t I - \mathcal{L} \frac{Q}{2\pi} \frac{\mathrm{d}\omega}{\mathrm{d}t} . \quad \text{(S-6.68)}$$

The torque on the ring becomes

$$\boldsymbol{\tau} = -\hat{\mathbf{z}} \left(\frac{k_{\mathrm{m}} b_{\mathrm{m}}^2 \pi a^2 Q}{R} \partial_t I + \mathcal{L} \frac{Q^2 a^2}{2\pi} \frac{\mathrm{d}\omega}{\mathrm{d}t} \right) , \quad \text{(S-6.69)}$$

and the equation of motion (S-6.61) becomes

$$m R^2 \frac{\mathrm{d}\omega}{\mathrm{d}t} = -\frac{k_{\mathrm{m}} b_{\mathrm{m}}^2 \pi a^2 Q}{R} \partial_t I - \mathcal{L} \frac{Q^2 a^2}{2\pi} \frac{\mathrm{d}\omega}{\mathrm{d}t} ,$$

or

$$\left(m R^2 + \mathcal{L} \frac{Q^2 a^2}{2\pi} \right) \frac{\mathrm{d}\omega}{\mathrm{d}t} = -\frac{k_{\mathrm{m}} b_{\mathrm{m}}^2 \pi a^2 Q}{R} \partial_t I , \quad \text{(S-6.70)}$$

which is equivalent to (S-6.61) if we replace the mass of the charged ring by an effective value

$$m_{\mathrm{eff}} = m + \mathcal{L} \frac{Q^2 a^2}{2\pi R^2} . \quad \text{(S-6.71)}$$

Thus we obtain for the dependence of ω on $I(t)$

$$\omega(t) = k_{\mathrm{m}} b_{\mathrm{m}}^2 \frac{\pi a^2 Q}{m_{\mathrm{eff}} R^3} \left[I_0 - I(t) \right] , \quad \text{(S-6.72)}$$

and for its final value

$$\omega_{\mathrm{f}} = k_{\mathrm{m}} b_{\mathrm{m}}^2 \frac{\pi a^2 Q}{m_{\mathrm{eff}} R^3} I_0 , \quad \text{(S-6.73)}$$

corresponding to a final angular momentum

$$L_{\mathrm{f}} = m R^2 \omega_{\mathrm{f}} = k_{\mathrm{m}} b_{\mathrm{m}}^2 \frac{\pi a^2 Q}{R + \mathcal{L} a^2 Q^2 / (2\pi m R)} I_0$$

$$= \begin{cases} \dfrac{\mu_0}{4\pi} \dfrac{\pi a^2 Q}{R + \mathcal{L} a^2 Q^2 / (2\pi m R)} I_0 , & \text{SI,} \\[3mm] \dfrac{1}{c^3} \dfrac{\pi a^2 Q}{R + \mathcal{L} a^2 Q^2 / (2\pi m R)} I_0 , & \text{Gaussian.} \end{cases} \quad \text{(S-6.74)}$$

The final magnetic flux through the charged ring is

$$\Phi_f = \frac{1}{b_m} \mathcal{L} \frac{Q\omega_f}{2\pi} = k_m b_m \frac{\mathcal{L}a^2 Q^2}{2mR^3 + \mathcal{L}Q^2 a^2 R/\pi} I_0, \qquad \text{(S-6.75)}$$

and the approximations of point **(a)** are valid only if

$$\Phi_f \ll \Phi_0 = 4\pi k_m b_m \frac{\pi a^2}{2R} I_0, \quad \text{or} \quad \frac{\mathcal{L}Q^2}{4\pi^2 mR^2 + 2\pi\mathcal{L}Q^2 a^2} \ll 1. \qquad \text{(S-6.76)}$$

S-6.7 Induced Electric Currents in the Ocean

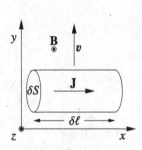

Fig. S-6.10

(a) We choose a Cartesian coordinate system with the y axis parallel to the velocity v of the fluid and the z axis parallel to the magnetic field, as shown in Fig. S-6.10. Due to the motion of the fluid, the charge carriers (mainly the Na^+ and Cl^- ions of the dissolved salt) are subject to a force per unit charge equal to $b_m v \times \mathbf{B}$, parallel to the x axis. This is equivalent to an electric field $\mathbf{E}_{eq} \equiv b_m v \times \mathbf{B}$. The induced current density is thus

$$\mathbf{J} = \sigma \mathbf{E}_{eq} = b_m \sigma v \times \mathbf{B}. \qquad \text{(S-6.77)}$$

(b) Inserting the typical values given in the text into (S-6.77) we obtain

$$J \simeq \begin{cases} 4 \times 1 \times 5 \times 10^{-5} = 2 \times 10^{-4} \, \text{A/m}^2, & \text{SI}, \\ 3.6 \times 10^{10} \times \dfrac{100}{c} \times 0.5 = 60 \, \text{statA/cm}^2, & \text{Gaussian}. \end{cases} \qquad \text{(S-6.78)}$$

(c) We evaluate the force on a fluid element of cylindrical shape, with area of the bases δS and height $|\delta\ell|$, where ℓ is parallel to $\mathbf{J}$ and to the x axis. The current intensity in the cylinder is $I = J\delta S$, and the force acting on it is thus $\delta\mathbf{F} = b_m I \delta\ell \times \mathbf{B} = -b_m BJ \, \delta S\delta\ell \, \hat{\mathbf{y}} = -b_m BJ \, \delta\mathcal{V}\hat{\mathbf{y}}$, where $\delta\mathcal{V}$ is the volume of the cylinder. The mass of the cylinder is $\delta m = \rho \, \delta\mathcal{V}$, with $\rho = 10^3 \, \text{kg/m}^3$ (1 g/cm^3 in Gaussian units), for water. Both v and $\delta\mathbf{F}$ are parallel to the y direction, and the equation of motion can be written in scalar form

$$\delta m \frac{dv}{dt} = \delta F. \qquad \text{(S-6.79)}$$

Replacing the values of δm and δF we obtain

$$\rho \, \delta \mathcal{V} \frac{\mathrm{d}v}{\mathrm{d}t} = -b_{\mathrm{m}} B J \, \delta \mathcal{V} ,$$

$$\rho \frac{\mathrm{d}v}{\mathrm{d}t} = -b_{\mathrm{m}}^2 B^2 \sigma v ,$$ (S-6.80)

where we have divided both sides by $\delta \mathcal{V}$ and replaced J by its expression (S-6.77). The solution is a decreasing exponential $v = v_0 e^{-t/\tau}$ with a time constant

$$\tau = \frac{\rho}{\sigma b_{\mathrm{m}}^2 B^2} \simeq 10^{11} \, \mathrm{s} \simeq 3 \times 10^3 \, \mathrm{yr} .$$ (S-6.81)

S-6.8 A Magnetized Sphere as Unipolar Motor

(a) We recall from Problem 5.10 that the magnetic field inside a uniformly magnetized sphere is uniform and equals

$$\mathbf{B} = \frac{8\pi}{3} \frac{k_{\mathrm{m}}}{b_{\mathrm{m}}} \mathbf{M} = \begin{cases} \dfrac{2\mu_0}{3} \mathbf{M} , & \text{SI}, \\[2mm] \dfrac{8\pi}{3} \mathbf{M} , & \text{Gaussian}. \end{cases}$$ (S-6.82)

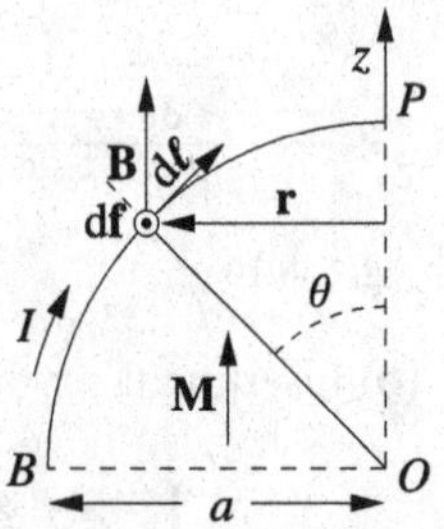

Outside of the sphere we have the same magnetic field that would be generated by a magnetic dipole of moment $\mathbf{m} = \mathbf{M} \, 4\pi a^3/3$, located at the center of the sphere. When an electric current I flows in the circuit, the magnetic force on

Fig. S-6.11

an element $\mathrm{d}\ell$ of the "meridian" wire BP is $\mathrm{d}\mathbf{f} = I\mathrm{d}\ell \times \mathbf{B}$, directed out of paper in Fig. S-6.11. Since the component of $\mathbf{B}$ perpendicular to $\mathrm{d}\ell$ is continuous across the surface of the sphere, there is no ambiguity. The torque $\mathrm{d}\boldsymbol{\tau}$ on the wire element $\mathrm{d}\ell$ is

$$\mathrm{d}\boldsymbol{\tau} = \mathbf{r} \times \mathrm{d}\mathbf{f} = I\mathbf{r} \times (\mathrm{d}\ell \times \mathbf{B}) = \hat{\mathbf{z}} \, I a \sin\theta \, a \, \mathrm{d}\theta \, B \cos\theta$$

$$= \hat{\mathbf{z}} \, I a^2 B \cos\theta \, \sin\theta \, \mathrm{d}\theta ,$$ (S-6.83)

where $\mathbf{r}$ is the distance of $\mathrm{d}\ell$ form the rotation axis of the sphere ($r = a \sin\theta$), and we have used $a \, \mathrm{d}\theta = \mathrm{d}\ell$. The total torque on the meridian wire BP is thus

$$\boldsymbol{\tau} = \int \mathrm{d}\boldsymbol{\tau} = \hat{\mathbf{z}} \, I a^2 B \int_0^{\pi/2} \sin\theta \cos\theta \, \mathrm{d}\theta = \hat{\mathbf{z}} \frac{1}{2} I a^2 B ,$$ (S-6.84)

while the torque on the current-carrying portion AB of the "equatorial" wire is zero, because the magnetic force is radial, as shown in Fig. S-6.12. Thus, (S-6.84) is the total torque on the sphere.

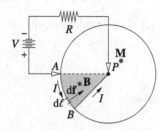

Fig. S-6.12

(b) When the sphere rotates, the total electromotive force $\mathcal{E}_{\text{tot}}$ in the circuit is the sum of the electromotive force of the voltage source and the electromotive force $\mathcal{E}_{\text{rot}}$ due to the rotation of the of the wires

$$\mathcal{E}_{\text{tot}} = V + \mathcal{E}_{\text{rot}} = V - b_{\text{m}}\frac{\mathrm{d}\Phi}{\mathrm{d}t}, \qquad \text{(S-6.85)}$$

where Φ is the flux of the magnetic field through any surface bounded by the closed path $ABPCDEA$ in Fig. S-6.13. Lines PC, CD, DE and EA are coplanar lines, lying on a plane containing also the rotation axis OP of the sphere and the meridian arc PA, while AB is an equatorial arc, and BP a meridian arc, both lying on the surface of the sphere. The flux of **B** through any surface bounded by the closed path $ABPCDEA$ is the same, because $\nabla \cdot \mathbf{B} = 0$. For simplicity, we choose a surface comprising two parts:

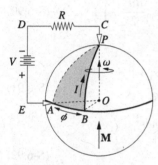

Fig. S-6.13

1. the planar surface $PCDEA$, its perimeter being closed by the arc AP, through which the flux is zero, and
2. the spherical polar triangle PAB shaded in Fig. S-6.13.

The flux through PAB can be easily calculated remembering that the flux of **B** through any closed surface is zero. Consider the closed surface formed by PAB and the three circular sectors OAP, OBP and OAB. The flux through OAP and OBP is zero, thus the flux Φ_{PAB} through PAB and the flux Φ_{OAB} through OAB must be equal (Φ_{OAB} must be taken with the minus sign when evaluating its contribution to the flux through the total closed surface, since the magnetic field *enters* through OAB and exits through PAB), and we have

$$\Phi_{\text{PAB}} = \Phi_{\text{OAB}} = \frac{1}{2}Ba^2\phi, \qquad \text{(S-6.86)}$$

where ϕ is the angle $\widehat{AOB}$. We thus have

$$\mathcal{E}_{\text{tot}} = V - b_{\text{m}}\frac{\mathrm{d}\Phi}{\mathrm{d}t} = V - b_{\text{m}}\frac{Ba^2}{2}\frac{\mathrm{d}\phi}{\mathrm{d}t} = V - b_{\text{m}}\frac{Ba^2}{2}\omega, \qquad \text{(S-6.87)}$$

and the current flowing in the circuit is

$$I = \frac{\mathcal{E}_{\text{tot}}}{R} = \frac{1}{R}\left(V - b_{\text{m}}\frac{Ba^2}{2}\omega\right). \tag{S-6.88}$$

The torque on the sphere is zero when $I = 0$, thus the terminal angular velocity of the sphere is

$$\omega_{\text{t}} = \frac{2V}{b_{\text{m}}Ba^2} = \begin{cases} \dfrac{2V}{Ba^2}, & \text{SI}, \\[2mm] \dfrac{2V}{Ba^2}c, & \text{Gaussian}, \end{cases} \tag{S-6.89}$$

independent of the moment of inertia of the sphere I and of the resistance R of the circuit. The equation of motion for the sphere is

$$I\frac{d\omega}{dt} = \tau = \frac{1}{2}Ba^2 I = \frac{Ba^2}{2R}\left(V - b_{\text{m}}\frac{Ba^2}{2}\omega\right), \tag{S-6.90}$$

which can be rewritten

$$I\frac{d\omega}{dt} + b_{\text{m}}\frac{(Ba^2)^2}{4R}\omega = \frac{Ba^2 V}{2R}. \tag{S-6.91}$$

Assuming that the sphere is at rest at $t = 0$, the solution is

$$\omega(t) = \omega_{\text{t}}\left(1 - e^{-t/\tau}\right), \tag{S-6.92}$$

where ω_{t} is given by (S-6.89), and

$$\tau = \frac{4IR}{b_{\text{m}}(Ba^2)^2}. \tag{S-6.93}$$

S-6.9 Induction Heating

(a) We split the total current density $\mathbf{J}$ appearing in the Maxwell equation for $\nabla \times \mathbf{B}$ into the sum of the free current density $\mathbf{J}_{\text{f}}$, and the magnetization current density $\mathbf{J}_{\text{m}}$

$$\nabla \times \mathbf{B} = 4\pi k_{\text{m}}(\mathbf{J}_{\text{f}} + \mathbf{J}_{\text{m}}) + \frac{k_{\text{m}}}{k_{\text{e}}}\partial_t\mathbf{E}. \tag{S-6.94}$$

In a homogeneous medium the magnetization current density is

$$\mathbf{J}_\mathrm{m} = \frac{1}{b_\mathrm{m}} \nabla \times \mathbf{M} = \frac{\mu_\mathrm{r} - 1}{4\pi k_\mathrm{m} \mu_\mathrm{r}} \nabla \times \mathbf{B} \,, \tag{S-6.95}$$

and (S-6.94) can be rewritten

$$\nabla \times \mathbf{B} = 4\pi k_\mathrm{m} \mathbf{J}_\mathrm{f} + \frac{\mu_\mathrm{r} - 1}{\mu_\mathrm{r}} \nabla \times \mathbf{B} \,, \tag{S-6.96}$$

from which we obtain

$$\nabla \times \mathbf{B} = 4\pi k_\mathrm{m} \mu_\mathrm{r} \mathbf{J}_\mathrm{f} \,. \tag{S-6.97}$$

Now we evaluate the curl of (S-6.97), recalling (A.12) of the Appendix,

$$\nabla \times (\nabla \times \mathbf{B}) = -\nabla^2 \mathbf{B} + \nabla(\nabla \cdot \mathbf{B}) = 4\pi k_\mathrm{m} \mu_\mathrm{r} \nabla \times \mathbf{J}_\mathrm{f} \,, \tag{S-6.98}$$

then, we recall that $\nabla \cdot \mathbf{B} = 0$ and that $\mathbf{J}_\mathrm{f} = \sigma \mathbf{E}$, obtaining

$$-\nabla^2 \mathbf{B} = 4\pi k_\mathrm{m} \mu_\mathrm{r} \sigma \nabla \times \mathbf{E} \,. \tag{S-6.99}$$

According to Maxwell's equation $\nabla \times \mathbf{E} = -b_\mathrm{m} \, \partial_t \mathbf{B}$ we finally have

$$\nabla^2 \mathbf{B} = 4\pi k_\mathrm{m} b_\mathrm{m} \, \mu_\mathrm{r} \, \sigma \, \partial_t \mathbf{B} \,, \quad \text{or} \quad \partial_t \mathbf{B} = \alpha \, \nabla^2 \mathbf{B} \,, \tag{S-6.100}$$

where

$$\alpha = \frac{1}{4\pi k_\mathrm{m} b_\mathrm{m} \mu_\mathrm{r} \sigma} = \begin{cases} \dfrac{1}{\mu_0 \mu_\mathrm{r} \sigma} \,, & \text{SI,} \\[2mm] \dfrac{c^2}{4\pi \mu_\mathrm{r} \sigma} \,, & \text{Gaussian.} \end{cases} \tag{S-6.101}$$

(b) The tangential component of the auxiliary vector $\mathbf{H}$ must be continuous through the $x = 0$ plane, thus, the tangential component of $\mathbf{B}/\mu_\mathrm{r}$ must be continuous. In the vacuum half-space ($x < 0$) we have $\mathbf{B} = \hat{\mathbf{y}} \, B_0 \cos(\omega t)$, correspondingly, the field at $x = 0^+$ (just inside our medium) is

$$\mathbf{B}(0^+, t) = \hat{\mathbf{y}} \, \mu_\mathrm{r} B_0 \cos(\omega t) \,. \tag{S-6.102}$$

In one dimension, (6.6) is rewritten

$$\partial_t B = \alpha \frac{\partial^2 B}{\partial x^2} \,, \tag{S-6.103}$$

and, as an educated guess, we look for a solution of the form $B(x, t) = \mathrm{Re}\left[\tilde{B}(x) \, e^{-i\omega t} \right]$. The differential equation for the time-independent function $\tilde{B}(x)$ is

$$- i\omega \tilde{B} = \alpha \frac{\partial^2 \tilde{B}}{\partial x^2} \;, \qquad (\text{S-6.104})$$

and we look for an exponential solution of the form $\tilde{B}(x) = \tilde{B}(0) \, e^{\gamma x}$, with $\tilde{B}(0)$ and γ two constants to be determined. The boundary condition gives $\tilde{B}(0) = \mu_\mathrm{r} B_0$, and, by substituting into (S-6.104), we have

$$- i\omega \, \mu_\mathrm{r} B_0 \, e^{\gamma x} = \alpha \gamma^2 \mu_\mathrm{r} B_0 \, e^{\gamma x} \;, \quad \text{or} \quad \alpha \gamma^2 = -i\omega \;, \qquad (\text{S-6.105})$$

which leads to

$$\gamma = \pm \sqrt{-i \, \frac{\omega}{\alpha}} = \pm \frac{1-i}{\sqrt{2}} \sqrt{\frac{\omega}{\alpha}} = \pm (1-i) \frac{1}{\ell_s} \qquad (\text{S-6.106})$$

where $(1 - i)/\sqrt{2} = \sqrt{-i}$, and the quantity

$$\ell_s = \sqrt{\frac{2\alpha}{\omega}} = \sqrt{\frac{2}{4\pi k_\mathrm{m} b_\mathrm{m} \mu_\mathrm{r} \sigma \omega}} = \begin{cases} \sqrt{\dfrac{2}{\mu_0 \mu_\mathrm{r} \sigma \omega}} \;, & \text{SI}, \\[3mm] \sqrt{\dfrac{c^2}{2\pi \mu_\mathrm{r} \sigma \omega}} \;, & \text{Gaussian}, \end{cases} \qquad (\text{S-6.107})$$

is a length. We disregard the positive value of γ, which would lead to a magnetic field exponentially increasing with depth into the material, and obtain

$$\mathbf{B} = \hat{\mathbf{y}} \, \mathrm{Re}\left[\mu_\mathrm{r} B_0 \, e^{-(1-i)x/\ell_s - i\omega t} \right] = \hat{\mathbf{y}} \, \mu_\mathrm{r} B_0 \, e^{-x/\ell_s} \cos\left(\frac{x}{\ell_s} - \omega t \right) \;. \qquad (\text{S-6.108})$$

Thus the magnetic field decreases exponentially with depth into the material, with a decay constant ℓ_s. A slab of our material can be considered as semi-infinite if its depth is much larger than ℓ_s.

(c) The electric field $\mathbf{E}(x)$ inside the material can be evaluated from Maxwell's equation $\mathbf{\nabla} \times \mathbf{E} = -b_\mathrm{m} \, \partial_t \mathbf{B}$. Assuming $\mathbf{E}(x, t) = \mathrm{Re}\left[\tilde{\mathbf{E}}(x) \, e^{-i\omega t} \right]$ we have

$$(\mathbf{\nabla} \times \mathbf{E})_y = -\partial_x \mathrm{Re}\left[\tilde{E}_z(x) \, e^{-i\omega t} \right] \;,$$

$$\partial_t B = \mathrm{Re}\left[-i\omega \mu_\mathrm{r} B_0 \, e^{-(1-i)x/\ell_s - i\omega t} \right] \;, \qquad (\text{S-6.109})$$

thus $\mathbf{E}(x, t) = \hat{\mathbf{z}} \operatorname{Re}\left[\tilde{E}_z(x)\,\mathrm{e}^{-\mathrm{i}\omega t}\right]$, with $\partial_x \tilde{E}z = -\mathrm{i}\,b_\mathrm{m}\omega\mu_\mathrm{r} B_0\,\mathrm{e}^{-(1-\mathrm{i})x/\ell_s}$. Integrating with respect to x we obtain

$$\tilde{E}_z = \frac{\mathrm{i}}{1-\mathrm{i}}\,b_\mathrm{m}\omega\mu_\mathrm{r}\ell_s B_0\,\mathrm{e}^{-(1-\mathrm{i})x/\ell_s} = -\frac{1-\mathrm{i}}{2}\,b_\mathrm{m}\omega\mu_\mathrm{r}\ell_s B_0\,\mathrm{e}^{-(1-\mathrm{i})x/\ell_s}\,. \quad (\text{S-6.110})$$

The dissipated power per unit volume, *due to the free currents only*, is thus

$$\langle \mathbf{J}_\mathrm{f} \cdot \mathbf{E}\rangle = \frac{\sigma}{2}\left|\tilde{E}_z\right|^2 = \frac{\sigma}{4}\,b_\mathrm{m}^2\,\mu_\mathrm{r}^2\omega^2\ell_s^2 B_0^2 \mathrm{e}^{-2x/\ell_s} = \frac{\sigma}{4}\,b_\mathrm{m}^2\,\frac{2\,\mu_\mathrm{r}^2\omega^2 B_0^2}{4\pi k_\mathrm{m} b_\mathrm{m}\mu_\mathrm{r}\sigma\omega}\,\mathrm{e}^{-2x/\ell_s}$$

$$= b_\mathrm{m}\frac{\mu_\mathrm{r}\omega B_0^2}{8\pi k_\mathrm{m}}\,\mathrm{e}^{-2x/\ell_s} = \begin{cases} \dfrac{\mu_\mathrm{r}\omega B_0^2}{2\mu_0}\,\mathrm{e}^{-2x/\ell_s}\,, & \text{SI}, \\[2ex] \dfrac{\mu_\mathrm{r}\omega B_0^2}{32\pi^2}\,\mathrm{e}^{-2x/\ell_s}\,, & \text{Gaussian}, \end{cases} \quad (\text{S-6.111})$$

where we have substituted (S-6.107) for ℓ_s in the fraction. The total dissipated power per unit surface of the slab is

$$\int_0^\infty \langle \mathbf{J}_\mathrm{f} \cdot \mathbf{E}\rangle\,\mathrm{d}x = b_\mathrm{m}\frac{\mu_\mathrm{r}\omega B_0^2}{16\pi k_\mathrm{m}}\,\ell_s = \frac{B_0^2}{16\pi}\sqrt{\frac{b_\mathrm{m}\mu_\mathrm{r}\omega}{2k_\mathrm{m}\sigma}}\,. \quad (\text{S-6.112})$$

One might wonder if there is also a contribution of the magnetization volume and surface current densities, $\mathbf{J}_\mathrm{m}$ and $\mathbf{K}_\mathrm{m}$, to the dissipated power. In the presence of the magnetic field (S-6.108), our medium of relative magnetic permeability μ_r acquires a magnetization $\mathbf{M}$

$$\mathbf{M} = \frac{b_\mathrm{m}}{4\pi k_\mathrm{m}}\frac{\mu_\mathrm{r}-1}{\mu_\mathrm{r}}\mathbf{B} = \hat{\mathbf{y}}\frac{b_\mathrm{m}}{4\pi k_\mathrm{m}}(\mu_\mathrm{r}-1)\operatorname{Re}\left[B_0\,\mathrm{e}^{-(1-\mathrm{i})x/\ell_s-\mathrm{i}\omega t}\right]\,, \quad (\text{S-6.113})$$

corresponding to a volume magnetization current density

$$\mathbf{J}_\mathrm{m} = \frac{1}{b_\mathrm{m}}\nabla \times \mathbf{M}\,. \quad (\text{S-6.114})$$

Taking the symmetry of the problem into account, and introducing the complex amplitudes $\tilde{\mathbf{J}}_\mathrm{m}$ and $\tilde{J}_{m z}$ such that $\mathbf{J}_\mathrm{m} = \operatorname{Re}\left(\tilde{\mathbf{J}}_\mathrm{m}\,\mathrm{e}^{-\mathrm{i}\omega t}\right) = \hat{\mathbf{z}}\operatorname{Re}\left(\tilde{J}_{m z}\,\mathrm{e}^{-\mathrm{i}\omega t}\right)$, we have

$$\tilde{J}_{m z} = \frac{\mu_\mathrm{r}-1}{4\pi k_\mathrm{m}\mu_\mathrm{r}}\partial_x\tilde{B} = -\frac{\mu_\mathrm{r}-1}{4\pi k_\mathrm{m}}\frac{1-\mathrm{i}}{\ell_s}B_0\,\mathrm{e}^{-(1-\mathrm{i})x/\ell_s}\,. \quad (\text{S-6.115})$$

The corresponding dissipated power per unit volume is

$$\langle \mathbf{J}_m \cdot \mathbf{E} \rangle = \frac{1}{2} \mathrm{Re} \left(\tilde{J}_{mz} \tilde{E}_z^* \right) = b_m (\mu_r - 1) \frac{\mu_r \omega B_0^2}{8\pi k_m} \, \mathrm{e}^{-2x/\ell_s} = (\mu_r - 1) \langle \mathbf{J}_f \cdot \mathbf{E} \rangle \ ,$$

(S-6.116)

and the total dissipated power per unit surface is

$$\int_0^\infty \langle \mathbf{J}_m \cdot \mathbf{E} \rangle \, \mathrm{d}x = (\mu_r - 1) \int_0^\infty \langle \mathbf{J}_f \cdot \mathbf{E} \rangle \, \mathrm{d}x = b_m (\mu_r - 1) \frac{\mu_r \omega B_0^2}{16\pi k_m} \ell_s \ . \quad \text{(S-6.117)}$$

However, we also have a surface magnetization current density $\mathbf{K}_m$ flowing on the $x = 0$ plane, given by

$$\mathbf{K}_m = \frac{1}{b_m} \mathbf{M}(0^+) \times \hat{\mathbf{n}} = \hat{\mathbf{z}} \frac{\mu_r - 1}{4\pi k_m} \mathrm{Re} \left(B_0 \, \mathrm{e}^{-i\omega t} \right) = \hat{\mathbf{z}} \, K_{mz} \cos(\omega t) \ , \quad \text{(S-6.118)}$$

where $\hat{\mathbf{n}} = -\hat{\mathbf{x}}$ is the outward-pointing unit vector on the $x = 0$ boundary plane. This surface current density corresponds to a "dissipated" power per unit surface

$$\langle \mathbf{K}_m(t) \cdot \mathbf{E}(0, t) \rangle = \frac{1}{2} \mathrm{Re} \left[K_{mz} \tilde{E}_z(0) \right] = -b_m (\mu_r - 1) \frac{\mu_r \omega B_0^2}{16\pi k_m} \ell_s \, , \quad \text{(S-6.119)}$$

which cancels out the contribution (S-6.117). Thus, the total dissipated power in the medium is due to the free current only, and given by (S-6.112). Note that the parallel component of the electric field must be continuous at the boundary between two media, so that $E_z(0, t)$ appearing in (S-6.119) is a well defined quantity.

S-6.10   A Magnetized Cylinder as DC Generator

(a) We can consider the magnetic field as due to the azimuthal magnetization surface current density $\mathbf{K}_m$, flowing on the lateral surface of the cylinder. We have $\mathbf{K}_m = \mathbf{M} \times \hat{\mathbf{n}}/b_m$, where $\hat{\mathbf{n}}$ is the outward unit vector normal to the surface. Thus, the magnetized cylinder is equivalent to a solenoid of the same sizes, with n turns per unit length, current I per turn, and the product $nI = K_m$. Far from the two bases we have an approximately uniform field $\mathbf{B}_0$, independent of the radius and height of the cylinder,

$$\mathbf{B}_0 \simeq 4\pi k_m K_m \hat{\mathbf{z}}$$

$$= 4\pi \frac{k_m}{b_m} \mathbf{M} = \begin{cases} \mu_0 \mathbf{M} \, , & \text{SI,} \\ 4\pi \mathbf{M} \, , & \text{Gaussian.} \end{cases} \quad \text{(S-6.120)}$$

The field at, for instance, the upper base, can be evaluated by considering an "extended" cylinder, obtained by joining an identical, coaxial cylinder, at the base we are considering, as shown in Fig. S-6.14. The total field at the base is now due to both cylinders, and, being far from both bases of the extended cylinder, its value is $\mathbf{B}_0 \simeq 4\pi(k_m/b_m)\,\mathbf{M}$. Both cylinders contribute to this field, and, for symmetry reasons, the z components B_z of both contributions are equal, while the radial components cancel each other. The dashed lines of Fig. S-6.14 represent three $\mathbf{B}$ field lines for each cylinder, one along the axis and two off-axis. Thus, the z component of the field generated by the single cylinder at its base is

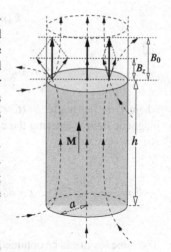

$$B_z = 2\pi\,\frac{k_m}{b_m}\,M = \frac{B_0}{2}\,. \qquad (S\text{-}6.121)$$

Fig. S-6.14

(b) We apply Faraday's law of induction to the flux of the magnetic field through the closed path $AEFGBCDA$, represented by the thick line in Fig. S-6.15. Points A, E, F, G, and B are fixed in the laboratory frame, while points C and D rotate with the magnetized cylinder. We have

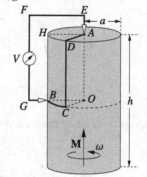

Fig. S-6.15

$$\mathcal{E} = -b_m\,\frac{d\Phi}{dt}\,, \qquad (S\text{-}6.122)$$

where $\mathcal{E}$ is the electromotive force around the closed path, measured by the voltmeter V, and Φ is the flux of the magnetic field through any surface bounded by the closed path. We choose a surface consisting of three parts:

1. the plane surface bounded by the path $AEFGBHA$, fixed in the laboratory frame, through which the flux of $\mathbf{B}$ is zero;
2. the surface bounded by the path $BCDHB$, lying on the lateral surface of the cylinder; and
3. the circular sector AHD on the upper base, where points A and H are fixed, while point D is rotating.

The flux of $\mathbf{B}$ through the two surfaces $BCDH$ and AHD can be calculated analogously to the flux through the polar spherical triangle PAB of Fig. S-6.13, Problem 6.8. We consider the closed surface comprising, in addition to $BCDH$ and AHD, the circular sector OBC and the two rectangles $COAD$ and $BOAH$. The flux must be zero through the total closed surface, and is zero through the two rectangles because $\mathbf{B}$ is parallel to their surfaces. Thus we have

$$\Phi_{AHD} + \Phi_{BCDH} + \Phi_{OBC} = 0, \tag{S-6.123}$$

and

$$\Phi_{AHD} + \Phi_{BCDH} = -\Phi_{OBC} = \frac{1}{2} B_0 \, a^2 \phi, \tag{S-6.124}$$

where ϕ is the angle $\widehat{BOC} = \widehat{HAD}$, and the sign accounts for the fact that the magnetic field is entering the closed surface through OBC. The electromotive force is

$$\mathcal{E} = -b_{\mathrm{m}} \frac{\mathrm{d}\Phi}{\mathrm{d}t} = -b_{\mathrm{m}} \frac{1}{2} B_0 \, a^2 \frac{\mathrm{d}\phi}{\mathrm{d}t} = 2\pi k_{\mathrm{m}} M a^2 \omega = \begin{cases} \dfrac{\mu_0}{2} M a^2 \omega, & \text{SI}, \\[2mm] \dfrac{2\pi}{c} M a^2 \omega, & \text{Gaussian}. \end{cases} \tag{S-6.125}$$

The same result can be obtained by evaluating the electromotive force $\mathcal{E}$ as the integral of $b_{\mathrm{m}}(v \times \mathbf{B}) \cdot \mathrm{d}\ell$ along the path AOB

$$\mathcal{E} = b_{\mathrm{m}} \int_A^O (v \times \mathbf{B}) \cdot \mathrm{d}\ell + b_{\mathrm{m}} \int_O^B (v \times \mathbf{B}) \cdot \mathrm{d}\ell = b_{\mathrm{m}} \int_0^a \omega r B_0 \, \mathrm{d}r$$

$$= b_{\mathrm{m}} \frac{1}{2} B_0 a^2 \omega = 2\pi k_{\mathrm{m}} M a^2 \omega, \tag{S-6.126}$$

since $v = 0$ along the path AO, which lies on the rotation axis of the cylider.

S-6.11   The Faraday Disk and a Self-sustained Dynamo

(a) The magnetic force on the each charge carrier of the rotating disk is $q \, b_{\mathrm{m}} v \times \mathbf{B}$, where q is the charge of the carrier ($-e$ for the electrons), and $v = \omega \times \mathbf{r}$ is the velocity of a charge-carrier at a distance $\mathbf{r}$ from the rotation axis, at rest relative to the disk. At equilibrium, carriers must be at rest relative to the disk, and the magnetic force must be compensated by a static electric field $\mathbf{E}$ such that $\mathbf{E} + b_{\mathrm{m}} v \times \mathbf{B} = 0$. This corresponds to an electric potential drop V between the center and the circumference of the disk

$$V = \varphi(a) - \varphi(0) = -\int_0^a \mathbf{E} \cdot \mathrm{d}\mathbf{r} = b_{\mathrm{m}} \int_0^a \omega r B \, \mathrm{d}r = b_{\mathrm{m}} \, \omega B \frac{a^2}{2}. \tag{S-6.127}$$

The rotating disk is thus a voltage source, known as a *Faraday disk*.
(b) In the presence of the brush contacts at points O and A of Fig. 6.9, the electromotive force $\mathcal{E}$ of he circuit equals the voltage drop V of (S-6.127). The total current I

circulating in the circuit is thus

$$I = \frac{\mathcal{E}}{R} = b_m \frac{\omega B a^2}{2R}. \tag{S-6.128}$$

The power dissipated in the circuit by Joule heating is $P_d = I^2 R = \mathcal{E}^2/R$, and there must an external a torque τ_{ext} providing a mechanical power $P_m = \tau_{ext} \cdot \omega = P_d$ in order to maintain a rotation at constant angular velocity. Thus,

$$\tau_{ext} = \hat{z} b_m^2 \frac{\omega B^2 a^4}{4R}. \tag{S-6.129}$$

Alternatively, the external torque must compensate the torque of the magnetic forces on the disk. **Fig. S-6.16**
Since the current exits the disk through the brush contact A, it is difficult to make assumptions on the symmetry of the current density distribution. However, the problem can be tackled as follows. The torque on an infinitesimal volume element, $r\,d\phi\,dr\,dz$ in cylindrical coordinates (Fig. S-6.16), is $d\tau = b_m\,\mathbf{r} \times (\mathbf{J} \times \mathbf{B})\,r\,d\phi\,dr\,dz$, and the total magnetic torque on the disk is obtained by integrating $d\tau$ over the disk volume

$$\tau_B = b_m \int_0^a dr \int_0^h dz \int_0^{2\pi} r\,d\phi\,\mathbf{r} \times (\mathbf{J} \times \mathbf{B}). \tag{S-6.130}$$

The triple vector product in (S-6.130) can be rewritten

$$\mathbf{r} \times (\mathbf{J} \times \mathbf{B}) = \mathbf{J}(\mathbf{r} \cdot \mathbf{B}) - \mathbf{B}(\mathbf{r} \cdot \mathbf{J}) = -\hat{z}\,B r J_r, \tag{S-6.131}$$

since $(\mathbf{r} \cdot \mathbf{B}) = 0$ because $\mathbf{r}$ and $\mathbf{B}$ are orthogonal to each other, and J_r is the r component of $\mathbf{J}$. We further have

$$\int_0^h dz \int_0^{2\pi} r\,d\phi\,J_r = I, \tag{S-6.132}$$

independently of r, since the double integral is the flux of $\mathbf{J}$ through a lateral cylindrical surface of radius r and height h, as shown in Fig. S-6.17. Thus we have for the torque exerted by the magnetic forces on the disk

$$\tau_B = -\hat{z}\,b_m\,B I \int_0^a r\,dr, \tag{S-6.133}$$

and finally, substituting (S-6.128) for I,

$$\tau_b = -\hat{z}\, b_m\, BI \frac{a^2}{2} = -\hat{z}\, b_m^2 \frac{\omega B^2 a^4}{4R} = -\tau_{\text{ext}}.$$ (S-6.134)

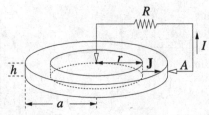

Fig. S-6.17

(c) If the disk acts as the current source for the solenoid we must have

$$B = 4\pi k_m\, nI = 4\pi k_m b_m\, n \frac{\omega B a^2}{2R}$$ (S-6.135)

from which we obtain

$$\omega = \frac{2R}{4\pi k_m b_m\, na^2} = \begin{cases} \dfrac{2R}{\mu_0 na^2}, & \text{SI}, \\[2mm] \dfrac{Rc^2}{2\pi na^2}, & \text{Gaussian}, \end{cases}$$ (S-6.136)

independently of the intensity of the magnetic field B.

S-6.12 Mutual Induction Between Circular Loops

(a) We can assume the magnetic field generated by the current I circulating in loop B to be uniform and equal to $B_0\,\hat{z} = \hat{z}\, 2\pi k_m\, I/b$ all over the surface of loop A, since $a \ll b$. The angle between the axis of loop A and the z axis is $\theta = \omega t$, and the flux of the magnetic field through the surface of loop A is

$$\Phi = B_0\, \pi a^2 \cos \omega t = \frac{2\pi k_m I}{b} \pi a^2 \cos \omega t = \frac{2\pi^2 a^2 k_m I}{b} \cos \omega t.$$ (S-6.137)

Thus, according to Faraday's law of induction, there is an induced electromotive force $\mathcal{E}$ on loop A

$$\mathcal{E} = -\frac{d\Phi}{dt} = \frac{2\pi^2 a^2 k_m I}{b} \omega \sin \omega t,$$ (S-6.138)

and the current circulating in loop A is

$$I_A = \frac{2\pi^2 a^2 k_m I}{Rb} \omega \sin \omega t.$$ (S-6.139)

(b) The power dissipated into Joule heating is

$$P_{\text{diss}} = RI_A^2 = \frac{4\pi^4 a^4 \omega^2 k_m^2 I^2}{Rb^2} \sin^2 \omega t . \tag{S-6.140}$$

(c) The torque acting on loop A is $\boldsymbol{\tau} = \mathbf{m} \times \mathbf{B}_0$, where $\mathbf{m} = \hat{\mathbf{n}} I_A \pi a^2$ is the magnetic moment of loop A, and $\hat{\mathbf{n}}$ is the unit vector perpendicular to its surface, directed so that its tip sees I_A circulating counterclockwise. Thus

$$\tau = \frac{2\pi^2 a^2 k_m I}{Rb} \omega \sin \omega t \, \pi a^2 \frac{2\pi k_m I}{b} \sin \omega t = \frac{4\pi^4 a^4 \omega k_m^2 I^2}{Rb^2} sin^2 \omega t , \tag{S-6.141}$$

and the corresponding mechanical power is

$$P_{\text{mech}} = \boldsymbol{\tau} \cdot \boldsymbol{\omega} = \frac{4\pi^4 a^4 \omega^2 k_m^2 I^2}{Rb^2} sin^2 \omega t = P_{\text{diss}} , \tag{S-6.142}$$

and all the mechanical power needed to keep loop A rotating at constant angular velocity is turned into Joule heating.
(d) The flux through the surface of loop B of the magnetic field generated by the current I circulating in loop A is

$$\Phi_B = M_{AB} I , \tag{S-6.143}$$

where M_{AB} is the coefficient of mutual induction between loop A and loop B. We know that $M_{AB} = M_{BA}$, and from (S-6.137) we have

$$M_{AB} = M_{BA} = \frac{2\pi^2 a^2 k_m}{b} \cos \omega t , \tag{S-6.144}$$

thus

$$\Phi_B = \frac{2\pi^2 a^2 k_m I}{b} \cos \omega t , \tag{S-6.145}$$

and

$$\mathcal{E} = -\frac{d\Phi}{dt} = \frac{2\pi^2 a^2 k_m I}{b} \omega \sin \omega t , \tag{S-6.146}$$

as (S-6.137) and (S-6.138).

S-6.13 Mutual Induction Between a Solenoid and an Internal Loop

(a) Neglecting boundary effects, the magnetic field inside the solenoid is uniform, parallel to the solenoid axis z, and equal to

$$\mathbf{B} = 4\pi k_{\mathrm{m}} n I \,\hat{\mathbf{z}}. \tag{S-6.147}$$

Thus, its flux through the surface $\mathbf{S}$ of the rotating coil is

$$\Phi_a(t) = \mathbf{B} \cdot \mathbf{S}(t) = 4\pi k_{\mathrm{m}} n I \pi a^2 \cos \omega t = 4\pi^2 a^2 k_{\mathrm{m}} n I \cos \omega t = M_{\mathrm{sl}}(t)\, I, \tag{S-6.148}$$

where

$$M_{\mathrm{sl}}(t) = 4\pi^2 a^2 k_{\mathrm{m}} n \cos \omega t \tag{S-6.149}$$

is the coefficient of mutual inductance between solenoid and loop, time dependent because the loop is rotating. The coefficient of mutual inductance is symmetric, $M_{\mathrm{sl}} = M_{\mathrm{ls}}$, i.e., the inductance by the solenoid on the loop equals the inductance by the loop on the solenoid, we shall use this property for the answer to point (c).

(b) The electromotive force acquired by the loop equals the rate of change of the magnetic flux through it,

$$\mathcal{E} = -\frac{d\Phi}{dt} = 4\pi^2 a^2 k_{\mathrm{m}} n I \omega \sin \omega t, \tag{S-6.150}$$

and the current circulating in the loop is

$$I_a = \frac{4\pi^2 a^2 k_{\mathrm{m}} n I \omega}{R} \sin \omega t. \tag{S-6.151}$$

The loop dissipates a power P_{diss} due to Joule heating

$$P_{\mathrm{diss}} = R I_a^2 = \frac{(4\pi^2 a^2 k_{\mathrm{m}} n I \omega)^2}{R} \sin^2 \omega t. \tag{S-6.152}$$

This power must be provided by the work of the torque τ applied to the loop in order to keep it in rotation at constant angular velocity. The time-averaged power is

$$\langle P_{\mathrm{diss}} \rangle = \frac{(4\pi^2 a^2 k_{\mathrm{m}} n I \omega)^2}{2R}, \tag{S-6.153}$$

since $\langle \sin^2 \omega t \rangle = 1/2$.

(c) The magnetic field generated by a magnetic dipole $\mathbf{m}$ is identical to the field generated by a current-carrying loop of radius a and current I_1 such that $b_{\mathrm{m}} \pi a^2 I_1 = m$, at distances $r \gg a$ from the center of the loop. The result of point (a) is valid, in particular, in the case $a \ll b$. In this case we can replace the magnetic dipole by a loop, and use the symmetry property of the mutual-inductance coefficient. The flux Φ_s generated by the dipole through the solenoid is thus

$$\Phi_s = M_{1s}(t)\, I_1 = 4\pi^2 a^2 k_m n I_1 \cos \omega t = 4\pi \frac{k_m}{b_m} n\, m \cos \omega t \,. \qquad (S\text{-}6.154)$$

S-6.14 Skin Effect and Eddy Inductance in an Ohmic Wire

Assuming a very long, straight cylindrical wire, the problem has cylindrical symmetry. We choose a cylindrical coordinate system (r, ϕ, z) with the z axis along the axis of the wire, and expect that the electric field inside the wire can be written as

$$\mathbf{E} = \hat{\mathbf{z}}\, E(r, t) = \hat{\mathbf{z}} \operatorname{Re}\left[\tilde{E}(r)\, e^{i\omega t} \right], \qquad (S\text{-}6.155)$$

where $\tilde{E}(r)$ is the static complex amplitude associated to the electric field. We start from the two Maxwell equations

$$\nabla \times \mathbf{E} = -b_m\, \partial_t \mathbf{B}\,, \qquad \nabla \times \mathbf{B} = 4\pi k_m\, \mathbf{J} + \frac{1}{b_m c^2}\, \partial_t \mathbf{E}\,, \qquad (S\text{-}6.156)$$

where we have assumed $\varepsilon_r = 1$ and $\mu_r = 1$ inside copper. If we substitute $\mathbf{J} = \sigma \mathbf{E}$ into the second of (S-6.156) we obtain

$$\nabla \times \mathbf{B} = 4\pi k_m \sigma \mathbf{E} + \frac{1}{b_m c^2}\, \partial_t \mathbf{E} = 4\pi k_m \sigma \mathbf{E} + \hat{\mathbf{z}}\, \frac{\omega}{b_m c^2} \operatorname{Re}\left[i\tilde{E}(r)\, e^{i\omega t} \right]$$

$$= \begin{cases} \mu_0 \sigma \mathbf{E} + \hat{\mathbf{z}}\, \dfrac{\omega}{c^2} \operatorname{Re}\left[i\tilde{E}(r)\, e^{i\omega t} \right], & \text{SI,} \\[3mm] \dfrac{4\pi\sigma}{c}\, \mathbf{E} + \hat{\mathbf{z}}\, \dfrac{\omega}{c} \operatorname{Re}\left[i\tilde{E}(r) e^{i\omega t} \right], & \text{Gaussian.} \end{cases} \qquad (S\text{-}6.157)$$

In SI units, the conductivity of copper is $\sigma = 5.96 \times 10^7\ \Omega^{-1}\mathrm{m}^{-1}$, and the product $\mu_0 \sigma c^2$ is

$$\mu_0\, \sigma c^2 = 6.77 \times 10^{18}\ \mathrm{s}^{-1}\,. \qquad (S\text{-}6.158)$$

Alternatively, in Gaussian units, the conductivity of copper is $\sigma = 5.39 \times 10^{17}\ \mathrm{s}^{-1}$ and the product $4\pi\sigma$ is $6.77 \times 10^{18}\ \mathrm{s}^{-1}$. Thus the displacement current $\partial_t \mathbf{D}$ is negligible compared to the conduction current $\mathbf{J}$ for frequencies $\nu = \omega/(2\pi) \ll 10^{18}$ Hz, i.e., up to the ultraviolet. In other words, the displacement current can be neglected compared to the conduction current for all practical purposes in good conductors, and we can rewrite the second of (S-6.156) simply as $\nabla \times \mathbf{B} = 4\pi k_m\, \sigma\, \mathbf{E}$. Evaluating the curl of both sides of the first of (S-6.156) we have

$$\nabla \times (\nabla \times \mathbf{E}) = -b_{\mathrm{m}} \partial_t (\nabla \times \mathbf{B}) = -4\pi k_{\mathrm{m}} b_{\mathrm{m}} \sigma \, \partial_t \mathbf{E}, \qquad (\text{S-6.159})$$

which, remembering that

$$\nabla \times (\nabla \times \mathbf{E}) = \nabla(\nabla \cdot \mathbf{E}) - \nabla^2 \mathbf{E}, \qquad (\text{S-6.160})$$

and assuming $\nabla \cdot \mathbf{E} = 0$, turns into a diffusion equation for the electric field $\mathbf{E}$

$$\nabla^2 \mathbf{E} = 4\pi k_{\mathrm{m}} b_{\mathrm{m}} \sigma \, \partial_t \mathbf{E}. \qquad (\text{S-6.161})$$

Introducing our assumption (S-6.155), we have the following equation in cylindrical coordinates for the complex amplitude $\tilde{E}(r)$,

$$\nabla^2 \tilde{E}(r) = \frac{1}{r} \partial_r [r \, \partial_r \tilde{E}(r)] = \mathrm{i}\omega \, 4\pi k_{\mathrm{m}} b_{\mathrm{m}} \, \sigma \, \tilde{E}(r) \qquad (\text{S-6.162})$$

or

$$\frac{1}{r} \partial_r [r \, \partial_r \tilde{E}(r)] = \mathrm{i} \frac{2}{\delta^2} \, \tilde{E}(r), \qquad (\text{S-6.163})$$

where we have introduced the skin depth

$$\delta = \sqrt{\frac{1}{2\pi \, k_{\mathrm{m}} b_{\mathrm{m}} \sigma \omega}} = \begin{cases} \sqrt{\dfrac{2}{\mu_0 \sigma \omega}}, & \text{SI,} \\[3mm] \dfrac{c}{\sqrt{2\pi \sigma \omega}}, & \text{Gaussian.} \end{cases} \qquad (\text{S-6.164})$$

Equation (S-6.163), multiplied by r^2, is Bessel's differential equation with $n = 0$. However, in this context, we prefer to find the approximate solutions for the two limiting cases $\delta \gg r_0$ and $\delta \ll r_0$, where r_0 is the radius of the wire. For the weak skin effect, i.e., for $\delta \gg r_0$, we write the solution of (S-6.163) as a Taylor series

$$\tilde{E}(r) = E_0 \sum_{n=0}^{\infty} a_n \left(\frac{r}{\delta}\right)^n \qquad (\text{S-6.165})$$

which, substituted into the left-hand side of (S-6.163) gives

$$\begin{aligned}
\frac{1}{r} \partial_r [r \, \partial_r \tilde{E}(r)] &= \frac{1}{r} \partial_r \left[r E_0 \sum_{n=0}^{\infty} \frac{a_n n \, r^{n-1}}{\delta^n} \right] = \frac{1}{r} E_0 \, \partial_r \sum_{n=0}^{\infty} \frac{a_n n r^n}{\delta^n} \\
&= \frac{1}{r} E_0 \sum_{n=0}^{\infty} \frac{a_n n^2 r^{n-1}}{\delta^n} = E_0 \sum_{n=0}^{\infty} \frac{a_n n^2 r^{n-2}}{\delta^n},
\end{aligned} \qquad (\text{S-6.166})$$

while the right-hand side is

$$i \frac{2}{\delta^2} \tilde{E}(r) = 2i E_0 \sum_{n=0}^{\infty} \frac{a_n r^n}{\delta^{n+2}} . \tag{S-6.167}$$

Comparing the coefficients of the same powers of r in (S-6.166) and (S-6.167) we obtain the recurrence relation

$$a_{n+2} = \frac{2i}{(n+2)^2} a_n , \tag{S-6.168}$$

which leads to

$$a_{2n} = \frac{i^n}{2^n (n!)^2} \quad \text{and} \quad a_{2n+1} = 0 , \tag{S-6.169}$$

for all $n \geq 0$ and $n \in \mathbb{N}$. We thus have

$$\tilde{E}(r) = E_0 \sum_{n=0}^{\infty} \frac{i^n}{2^n (n!)^2} \left(\frac{r}{\delta}\right)^{2n}$$

$$= E_0 \left[1 + \frac{i}{2} \frac{r^2}{\delta^2} - \frac{1}{16} \frac{r^4}{\delta^4} - \frac{i}{48} \frac{r^6}{\delta^6} + \cdots + \frac{i^n}{2^n (n!)^2} \frac{r^{2n}}{\delta^{2n}} + \cdots \right] . \tag{S-6.170}$$

The complex amplitude I associated to the total current through the wire is

$$I = \int_0^{r_0} J 2\pi r \, dr = 2\pi \sigma \int_0^{r_0} \tilde{E}(r) \, r \, dr$$

$$= 2\pi \sigma E_0 \int_0^{r_0} \left[1 + \frac{i}{2} \frac{r^2}{\delta^2} - \frac{1}{16} \frac{r^4}{\delta^4} - \frac{i}{48} \frac{r^6}{\delta^6} + \cdots + \frac{i^n}{2^n (n!)^2} \frac{r^{2n}}{\delta^{2n}} + \cdots \right] r \, dr$$

$$= 2\pi \sigma E_0 \left[\frac{r_0^2}{2} + \frac{i}{8} \frac{r_0^4}{\delta^2} - \frac{1}{96} \frac{r_0^6}{\delta^4} - \frac{i}{2304} \frac{r_0^8}{\delta^6} + \cdots + \frac{i^n}{2^n (n!)^2 (2n+2)} \frac{r_0^{2n+2}}{\delta^{2n}} \right]$$

$$= \pi r_0^2 \sigma E_0 \left[1 + \frac{i}{4} \frac{r_0^2}{\delta^2} - \frac{1}{48} \frac{r_0^4}{\delta^4} - \frac{i}{1152} \frac{r_0^6}{\delta^6} + \cdots + \frac{i^n}{2^n (n+1)! n!} \frac{r_0^{2n}}{\delta^{2n}} + \cdots \right] . \tag{S-6.171}$$

We can define the impedance per unit length of the wire, $Z_\ell = R_\ell + i\omega L_\ell$ (where R_ℓ is the resistance per unit length, and L_ℓ the self-inductance per unit length), as the ratio of the electric field at the wire surface to the total current through the wire, i.e., as

$$Z_\ell = \frac{1}{\pi r_0^2 \sigma} \underbrace{\left[1 + \frac{i}{2} \left(\frac{r_0}{\delta} \right)^2 - \frac{1}{16} \left(\frac{r_0}{\delta} \right)^4 - \frac{i}{48} \left(\frac{r_0}{\delta} \right)^6 + \cdots \right]}_{A}$$

$$\times \underbrace{\left[1 + \frac{i}{4} \left(\frac{r_0}{\delta} \right)^2 - \frac{1}{48} \left(\frac{r_0}{\delta} \right)^4 - \frac{i}{1152} \left(\frac{r_0}{\delta} \right)^6 + \cdots \right]^{-1}}_{B^{-1}}, \qquad \text{(S-6.172)}$$

where A and B are Taylor expansions in even powers of $r_0/\delta \ll 1$, which we have truncated at the 6th order. The first four expansion coefficients of B^{-1}, i.e., $1, b_1, b_2$, and b_3,

$$B^{-1} = \left[1 + b_1 \left(\frac{r_0}{\delta} \right)^2 + b_2 \left(\frac{r_0}{\delta} \right)^4 + b_3 \left(\frac{r_0}{\delta} \right)^6 + \cdots \right], \qquad \text{(S-6.173)}$$

can be evaluated by requiring that the product BB^{-1} equals 1 with a remainder of the order of $(r_0/\delta)^8$, i.e.,

$$1 = BB^{-1} \simeq \left[1 + \frac{i}{4} \left(\frac{r_0}{\delta} \right)^2 - \frac{1}{48} \left(\frac{r_0}{\delta} \right)^4 - \frac{i}{1152} \left(\frac{r_0}{\delta} \right)^6 + \cdots \right]$$

$$\times \left[1 + b_1 \left(\frac{r_0}{\delta} \right)^2 + b_2 \left(\frac{r_0}{\delta} \right)^4 + b_3 \left(\frac{r_0}{\delta} \right)^6 + \cdots \right] \qquad \text{(S-6.174)}$$

leading to

$$b_1 = -\frac{i}{4}, \quad b_2 = -\frac{1}{24}, \quad \text{and} \quad b_3 = \frac{7i}{1152} \qquad \text{(S-6.175)}$$

Thus we have for Z_ℓ

$$Z_\ell \simeq \frac{1}{\pi r_0^2 \sigma} \left[1 + \frac{i}{2} \left(\frac{r_0}{\delta} \right)^2 - \frac{1}{16} \left(\frac{r_0}{\delta} \right)^4 - \frac{i}{48} \left(\frac{r_0}{\delta} \right)^6 + \cdots \right]$$

$$\times \left[1 - \frac{i}{4} \left(\frac{r_0}{\delta} \right)^2 - \frac{1}{24} \left(\frac{r_0}{\delta} \right)^4 + \frac{7i}{1152} \left(\frac{r_0}{\delta} \right)^6 + \cdots \right]$$

$$\simeq \frac{1}{\pi r_0^2 \sigma} \left[1 + \frac{i}{4} \left(\frac{r_0}{\delta} \right)^2 + \frac{1}{48} \left(\frac{r_0}{\delta} \right)^4 - i \frac{23}{1152} \left(\frac{r_0}{\delta} \right)^6 + \cdots \right]. \qquad \text{(S-6.176)}$$

The zeroth-order term of the expansion,

$$R_\ell^{(0)} = \frac{1}{\pi r_0^2 \sigma}, \qquad \text{(S-6.177)}$$

is simply the direct-current resistance per unit length of the wire. The third term

$$R_\ell^{(1)} = \frac{r_0^2}{48\pi\sigma\delta^4} = \frac{k_m^2 b_m^2 \pi r_0^2 \sigma \omega^2}{12} = \begin{cases} \dfrac{\mu_0^2 \pi r_0^2 \sigma \omega^2}{192}, & \text{SI,} \\[2mm] \dfrac{\pi r_0^2 \sigma \omega^2}{12c^4}, & \text{Gaussian} \end{cases} \qquad \text{(S-6.178)}$$

is the lowest order contribution of the weak skin effect to the resistance increase. The second-order term of the expansion can be interpreted as

$$\frac{i}{4\pi r_0^2 \sigma} \left(\frac{r_0}{\delta}\right)^2 = i\omega L_\ell^{(0)}, \qquad \text{(S-6.179)}$$

leading to

$$L_\ell^{(0)} = \frac{1}{4\pi\sigma\omega r_0^2} \left(\frac{r_0}{\delta}\right)^2 = \frac{1}{2}k_m b_m = \begin{cases} \dfrac{\mu_0}{8\pi}, & \text{SI,} \\[2mm] \dfrac{1}{2c^2}, & \text{Gaussian,} \end{cases} \qquad \text{(S-6.180)}$$

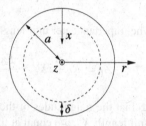

Fig. S-6.18

which is the DC self-inductance per unit length of a straight cylindrical wire, while the sixth-order term is the lowest order contribution of the weak skin-effect to the self-inductance of the cylindrical wire. Thus, at the low-frequency limit, the current depends on the radial coordinate, but no true skin effect is observed. According to (S-6.171), the current is actually stronger on the axis of the wire than at its surface. Things are different at the high-frequency limit.

For a strong skin effect, i.e., for $\delta \ll a$, the electric field is significantly different from zero only close to the wire surface. Thus, it is convenient to introduce the variable $x = a - r$, shown in Fig. S-6.18. We can assume $r \simeq a$ in (S-6.163), and we further have $\partial_r = -\partial_x$, leading to

$$\partial_x^2 \tilde{E} = i\frac{2}{\delta^2} \tilde{E}. \qquad \text{(S-6.181)}$$

Substituting $\tilde{E} = E_0 e^{\alpha x}$, we have

$$\alpha = \pm\sqrt{i\frac{2}{\delta^2}} = \pm\frac{1+i}{\delta}, \qquad \text{(S-6.182)}$$

and the solution corresponding to a field decreasing for increasing x (increasing depth into the wire) is

$$\tilde{E} \simeq E_0 e^{-x/\delta} e^{-ix/\delta} = E_0 e^{-(a-r)/\delta} e^{-i(a-r)/\delta}, \qquad \text{(S-6.183)}$$

where $E_0 e^{i\omega t}$ is the electric field at the wire surface. The complex amplitude corresponding to the total current current through the wire is thus

$$I = \int_0^a J \, 2\pi r \, dr = 2\pi\sigma \int_0^a \tilde{E}(r) r \, dr = 2\pi\sigma E_0 \int_0^a e^{-(a-r)/\delta} \, e^{-i(a-r)/\delta} \, e^{i\omega t} r \, dr$$

$$= 2\pi\sigma E_0 e^{-a(1+i)/\delta + i\omega t} \int_0^a e^{r(1+i)/\delta} r \, dr \, . \tag{S-6.184}$$

Remembering that

$$\int x \, e^{ax} dx = e^{ax} \left(\frac{x}{a} - \frac{1}{a^2} \right) , \tag{S-6.185}$$

and neglecting terms in δ^2, we obtain finally

$$I = \pi a \delta \sigma (1 - i) \, E_0 \, . \tag{S-6.186}$$

The impedance per unit length of the wire, Z_ℓ, can again be defined as

$$Z_\ell = R_\ell + iX_\ell = \frac{E_0}{I} = \frac{1}{\pi a \delta \sigma (1 - i)} = \frac{1}{2\pi a \delta \sigma} + \frac{i}{2\pi a \delta \sigma} , \tag{S-6.187}$$

so that the magnitudes of the resistance per unit length R_ℓ, and of the reactance per unit length X_ℓ, are equal at the high frequency limit:

$$R_\ell = X_\ell = \frac{1}{2\pi a \delta \sigma} \, . \tag{S-6.188}$$

The value of R_ℓ shows that the current actually flows through a thin annulus close to the surface (the "skin" of the wire), of width δ and approximate area $2\pi a \delta$. The reactance per unit length can be considered as due to a self-inductance per unit length L_ℓ, according to $X_\ell = \omega L_\ell$, with

$$L_\ell = \frac{1}{2\pi a \delta \sigma \omega} = \sqrt{\frac{k_m b_m}{2\pi a^2 \sigma \omega}} = \begin{cases} \sqrt{\dfrac{\mu_0}{8\pi^2 a^2 \sigma \omega}} \, , & \text{SI,} \\[4mm] \sqrt{\dfrac{1}{2\pi c^2 a^2 \sigma \omega}} \, , & \text{Gaussian.} \end{cases} \tag{S-6.189}$$

S-6.15 Magnetic Pressure and Pinch Effect for a Surface Current

(a) We use a cylindrical coordinate system (r, ϕ, z), with the cylinder axis as z axis. The field lines of **B** are circles around the z axis because of symmetry. Thus, $B_\phi(r)$ is the only nonzero component of **B**. According to Ampère's law we have

$$B_\phi(r) = \begin{cases} 0, & r < a, \\ 2k_\mathrm{m} \dfrac{I}{r} = 4\pi k_\mathrm{m} K \dfrac{a}{r}, & r > a. \end{cases} \tag{S-6.190}$$

(b) First approach (heuristic). The current $\mathrm{d}I$ flowing in an infinitesimal surface strip parallel to z, of width $a\,\mathrm{d}\phi$, is $\mathrm{d}I = K a\,\mathrm{d}\phi$. The force $\mathrm{d}\mathbf{f}$ exerted by an azimuthal magnetic field $\mathbf{B} \equiv (0, B_\phi, 0)$ on an infinitesimal strip portion of length $\mathrm{d}z$ is

$$\mathrm{d}\mathbf{f} = b_\mathrm{m}\,\mathrm{d}z\,\mathrm{d}I\,\hat{\mathbf{z}} \times \mathbf{B} = -b_\mathrm{m} K a B_\phi\,\mathrm{d}\phi\,\mathrm{d}z\,\hat{\mathbf{r}}, \tag{S-6.191}$$

directed towards the axis, i.e., so to shrink the conducting surface (pinch effect). However, here we must remember that $B_\phi(r)$ is discontinuous at the cylinder surface, being zero inside. Therefore, we replace the value of B_ϕ in (S-6.191) by its "average" value $B_\phi^\mathrm{aver} = [B_\phi(a^+) - B_\phi(a^-)]/2 = 2\pi k_\mathrm{m} K$ (the point is the same as for the calculation of electrostatic pressure on a surface charge layer). Thus, the absolute value of the force acting on an infinitesimal area $\mathrm{d}S = a\,\mathrm{d}\phi\,\mathrm{d}z$ is

$$|\mathrm{d}f| = 2\pi k_\mathrm{m} b_\mathrm{m} K^2\,\mathrm{d}S = \begin{cases} \dfrac{\mu_0}{2} K^2\,\mathrm{d}S, & \text{SI} \\ \dfrac{2\pi}{c^2} K^2\,\mathrm{d}S, & \text{Gaussian}, \end{cases} \tag{S-6.192}$$

and the magnetic pressure on the surface is

$$P = \frac{|\mathrm{d}f|}{\mathrm{d}S} = 2\pi k_\mathrm{m} b_\mathrm{m} K^2 = k_\mathrm{m} b_\mathrm{m} \frac{I^2}{2\pi a^2} = \begin{cases} \dfrac{\mu_0 I^2}{2(2\pi a)^2}, & \text{SI} \\ \dfrac{1}{c^2}\dfrac{I^2}{2\pi a^2}, & \text{Gaussian}. \end{cases} \tag{S-6.193}$$

Second method (rigorous). The magnetic force per infinitesimal volume d^3r, where a current density **J** is flowing in the presence of a magnetic field **B**, is

$$\mathrm{d}^3\mathbf{f} = b_\mathrm{m} \mathbf{J} \times \mathbf{B}\,\mathrm{d}^3r. \tag{S-6.194}$$

Due to the symmetry of our problem, the term $(\mathbf{B} \cdot \boldsymbol{\nabla})\,\mathbf{B}$ appearing in (6.7) is

$$(\mathbf{B} \cdot \nabla) \mathbf{B} = \left(B_\phi \frac{1}{r} \partial_\phi \right) \mathbf{B} = 0 , \qquad \text{(S-6.195)}$$

where we have used the gradient components in cylindrical coordinates of Table A.1, and the fact that the only nonzero component of $\mathbf{B}$, i.e., B_ϕ, is independent of ϕ. The infinitesimal volume element in cylindrical coordinates is $\mathrm{d}^3 r = r \, \mathrm{d}r \, \mathrm{d}\phi \, \mathrm{d}z$, thus

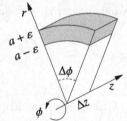

$$\mathrm{d}^3 \mathbf{f} = -\hat{\mathbf{r}} \frac{b_m}{8\pi k_m} \left[\partial_r B_\phi^2(r) \right] r \, \mathrm{d}r \, \mathrm{d}\phi \, \mathrm{d}z . \qquad \text{(S-6.196)}$$

Now we integrate (S-6.196) with respect to $\mathrm{d}r$ between $r = a - \varepsilon$ and $a + \varepsilon$, obtaining the force $\mathrm{d}^2 \mathbf{f}$ acting on the small shaded volume of Fig. S-6.19, delimited by the two cylindrical surfaces $r = a - \varepsilon$ and $r = a + \varepsilon$, **Fig. S-6.19** with infinitesimal azimuthal aperture $\mathrm{d}\phi$, and longitudinal length $\mathrm{d}z$. Integrating by parts we have

$$\int\limits_{a-\varepsilon}^{a+\varepsilon} \left[\partial_r B_\phi^2(r) \right] r \, \mathrm{d}r = \left[r B_\phi^2(r) \right]_{a-\varepsilon}^{a+\varepsilon} - \int\limits_{a-\varepsilon}^{a+\varepsilon} B_\phi^2(r) \, \mathrm{d}r , \qquad \text{(S-6.197)}$$

At the limit $\varepsilon \to 0$, the first term on the right-hand side equals $B_\phi^2(a^+)$, because $B_\phi^2(r) = 0$ for $r < a$. At the same limit $\varepsilon \to 0$, the integral on the right-hand side approaches zero because, according to the mean-value theorem, it equals $2\varepsilon B_\phi^2(\bar{r})$, with $\bar{r}$ some value in the range $(a - \varepsilon, a + \varepsilon)$. We thus have

$$\mathrm{d}^2 \mathbf{f} = -\hat{\mathbf{r}} \frac{b_m}{8\pi k_m} B_\phi^2(a^+) a \, \mathrm{d}\phi \, \mathrm{d}z . \qquad \text{(S-6.198)}$$

where $a \, \mathrm{d}\phi \, \mathrm{d}z$ is the infinitesimal surface element on which $\mathrm{d}^2 \mathbf{f}$ is acting. The pressure is thus

$$P = \frac{b_m}{8\pi k_m} B_\phi^2(a^+) = \frac{b_m}{8\pi k_m} (4\pi k_m K)^2 = 2\pi b_m k_m K^2, \qquad \text{(S-6.199)}$$

in agreement with (S-6.193). Now we prove (6.7):

$$4\pi k_m (\mathbf{J} \times \mathbf{B})_i = [(\nabla \times \mathbf{B}) \times \mathbf{B}]_i = \varepsilon_{ijk} \left(\varepsilon_{jlm} \frac{\partial B_m}{\partial x_l} \right) B_k = \varepsilon_{ijk} \varepsilon_{jlm} \frac{\partial B_m}{\partial x_l} B_k$$

$$= (\delta_{kl}\delta_{im} - \delta_{km}\delta_{il}) B_k \frac{\partial B_m}{\partial x_l} = B_k \frac{\partial B_i}{\partial x_k} - B_k \frac{\partial B_k}{\partial x_i}$$

$$= (\mathbf{B} \times \nabla) B_i - \frac{1}{2} \frac{\partial (B_k B_k)}{\partial x_i} = (\mathbf{B} \times \nabla) B_i - \frac{1}{2} \nabla_i B^2 , \qquad \text{(S-6.200)}$$

where the subscripts i, j, k, l, m range from 1 to 3, and $x_{1,2,3} = x, y, z$, respectively. The symbol ε_{ijk} is the Levi-Civita symbol, defined by $\varepsilon_{ijk} = 1$ if (i, j, k) is a cyclic permutation of $(1, 2, 3)$, $\varepsilon_{i,j,k} = -1$ if (i, j, k) is an anticyclic permutation of $(1, 2, 3)$, and $\varepsilon_{i,j,k} = 0$ if at least two of the subscripts (i, j, k) are equal.

(c) The magnetic energy ΔU_{M} stored in the infinite layer between z and $z + \Delta z$ equals the volume integral

$$\Delta U_{\mathrm{M}} = \int\limits_{\text{layer}} u_{\mathrm{M}} \, \mathrm{d}^3 r = \int\limits_{\text{layer}} \frac{b_{\mathrm{m}}}{8\pi k_{\mathrm{m}}} B^2(r) \, \mathrm{d}^3 r$$

$$= 2\pi \, \Delta z \int\limits_{a}^{\infty} \frac{b_{\mathrm{m}}}{8\pi k_{\mathrm{m}}} B_{\phi}^2(r) \, r \, \mathrm{d}r \ , \tag{S-6.201}$$

which, involving the integral $\int_a^{\infty} r^{-1} \mathrm{d}r$, is infinite. However, if the radius of the cylinder increases by $\mathrm{d}a$, the integrand does not change for $r > a + \mathrm{d}a$, while the integration volume decreases (Fig. S-6.20). Correspondingly, the (infinite) value of the integral decreases by the finite value

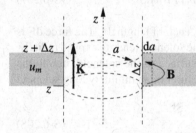

$$\mathrm{d}(\Delta U_{\mathrm{M}}) = -\Delta z \, \frac{b_{\mathrm{m}}}{8\pi k_{\mathrm{m}}} B_{\phi}^2(a^+) \, 2\pi a \, \mathrm{d}a \ . \tag{S-6.202}$$

Fig. S-6.20

Thus, an expansion of the current carrying surface leads to a *decrease* of the magnetic energy. If the system were isolated, the force $\mathrm{d}\mathbf{f}$ acting on the surface element $\mathrm{d}S = a \, \mathrm{d}\phi \, \mathrm{d}z$ would be directed radially outwards, leading to an expansion of the cylinder. However, the system is *not* isolated, because a current source is required to keep the current surface density $\mathbf{K}$ constant. An increase of the radius $\mathrm{d}a$ leads to a decrease of the magnetic flux in the layer equal to $\mathrm{d}(\Delta\Phi) = B_{\phi}(a^+) \, \Delta z \, \mathrm{d}a$ (see figure), which, in turn, implies the appearance of an electromotive force $\Delta\mathcal{E}$. In fact, in order to keep $\mathbf{K}$ constant during the time interval $\mathrm{d}t$ in which the cylinder radius increases by $\mathrm{d}a$, the source must provide to the layer the energy $\mathrm{d}(\Delta U_{\text{source}})$, that compensates the work $\mathrm{d}(\Delta W) = \Delta\mathcal{E} \, I \, \mathrm{d}t$ done by the electromotive force $\Delta\mathcal{E} = -b_{\mathrm{m}} \, \mathrm{d}(\Delta\Phi)/\mathrm{d}t$, so that

$$\mathrm{d}(\Delta U_{\text{source}}) = -b_{\mathrm{m}} I \, \mathrm{d}(\Delta\Phi) = 2\pi b_{\mathrm{m}} a K B_{\phi}(a^+) \, \Delta z \, \mathrm{d}a$$

$$= b_{\mathrm{m}} \, 2\pi a \, \frac{1}{4\pi k_{\mathrm{m}}} B_{\phi}^2(a^+) \, \Delta z \, \mathrm{d}a = -2\mathrm{d}(\Delta U_{\mathrm{m}}) \ . \tag{S-6.203}$$

Thus, the total energy balance for the layer is given by

$$\mathrm{d}(\Delta U_{\text{tot}}) = \mathrm{d}(\Delta U_{\text{source}}) + \mathrm{d}(\Delta U_m) = -\mathrm{d}(\Delta U_m) \ , \tag{S-6.204}$$

and the force per unit surface is

$$P = -\frac{1}{2\pi a \Delta z}\frac{\mathrm{d}(\Delta U_{\text{tot}})}{\mathrm{d}a} = +\frac{1}{2\pi a \Delta z}\frac{\mathrm{d}(\Delta U_m)}{\mathrm{d}a}\,, \tag{S-6.205}$$

in agreement with (S-6.193).

S-6.16 Magnetic Pressure on a Solenoid

(a) The magnetic force $\mathbf{df}$ on an infinitesimal coil arc of length $\mathrm{d}\boldsymbol{\ell}$, carrying a current I, is (Fig. S-6.21)

$$\mathbf{df} = b_{\text{m}}\,I\,\mathrm{d}\boldsymbol{\ell}\times\mathbf{B}\,. \tag{S-6.206}$$

Thus, the force $\mathbf{dF}$ on the surface element $\mathrm{d}\mathbf{S} = \mathrm{d}\boldsymbol{\ell}\times\mathrm{d}\mathbf{z}$ of the solenoid, of width $\mathrm{d}\ell$, is

$$\mathbf{dF} = b_{\text{m}}\,I\,Bn\,\mathrm{d}\boldsymbol{\ell}\times\mathrm{d}\mathbf{z} = b_{\text{m}}\,I\,Bn\,\mathrm{d}\mathbf{S}\,, \tag{S-6.207}$$

since the surface element comprises $n\mathrm{d}z$ coil arcs, each of length $\mathrm{d}\ell$. The force $\mathbf{dF}$ is directed towards the exterior of the solenoid, and the solenoid tends to expand radially.

The magnetic field $\mathbf{B}$ is discontinuous at the surface of the solenoid, due to the presence of the electric current in the coils. At the limit of an infinitely long solenoid we have

$$\mathbf{B} = \mathbf{B}_0 = 4\pi k_{\text{m}}nI\,\hat{\mathbf{z}} = \begin{cases} \mu_0 nI\,\hat{\mathbf{z}}\,, & \text{SI,} \\[2mm] \dfrac{4\pi}{c}\,nI\hat{\mathbf{z}}\,, & \text{Gaussian,} \end{cases} \tag{S-6.208}$$

inside, where $\hat{\mathbf{z}}$ is the unit vector along the solenoid axis, and $\mathbf{B} = 0$ outside. Thus we substitute the average value

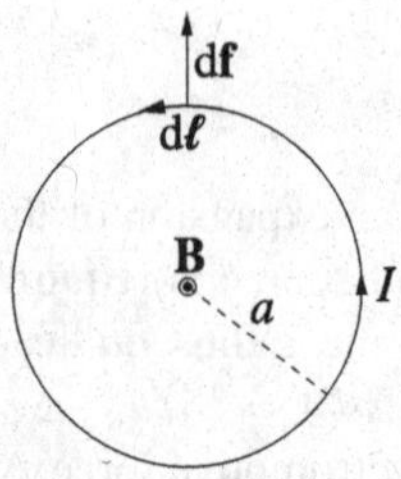

$$\frac{B(a^+) + B(a^-)}{2} = \frac{B_0}{2} = 2\pi k_{\text{m}}\,nI$$

for B in (S-6.207), obtaining

$$\mathbf{dF} = 2\pi b_{\text{m}}k_{\text{m}}\,n^2 I^2 \mathrm{d}\mathbf{S}\,. \tag{S-6.209}$$

Fig. S-6.21

The pressure P on the solenoid surface is obtained by dividing $\mathrm{d}F$ by $\mathrm{d}S$, thus

$$P = \frac{\mathrm{d}F}{\mathrm{d}S} = 2\pi b_{\text{m}}k_{\text{m}}\,n^2 I^2 = \frac{b_{\text{m}}B_0^2}{8\pi k_{\text{m}}} = \begin{cases} \dfrac{\mu_0}{2}n^2 I^2 = \dfrac{B_0^2}{2\mu_0}\,, & \text{SI,} \\[3mm] 2\pi n^2 I^2 = \dfrac{B_0^2}{8\pi}\,, & \text{Gaussian.} \end{cases} \tag{S-6.210}$$

(b) The magnetic energy of the solenoid can be written in terms of the magnetic energy density u_M associated to the magnetic field $\mathbf{B}_0$

$$u_M = \frac{b_m}{8\pi k_m} B_0^2 = \begin{cases} \dfrac{B_0^2}{2\mu_0}, & \text{SI,} \\[2mm] \dfrac{B_0^2}{8\pi}, & \text{Gaussian.} \end{cases} \tag{S-6.211}$$

Neglecting the boundary effects, we obtain the total magnetic energy of the solenoid U_M by multiplying u_M by the solenoid volume

$$U_M = \pi a^2 h \, u_M = \frac{a^2 h b_m B_0^2}{8k_m} = 2\pi^2 a^2 h b_m k_m n^2 I^2 , \tag{S-6.212}$$

thus, if the solenoid radius a increases by $\mathrm{d}a$ the energy U_M increases by

$$\mathrm{d}U_M = 4\pi^2 a h b_m k_m n^2 I^2 \, \mathrm{d}a , \tag{S-6.213}$$

while B_0, given by (S-6.208), and thus u_M, remain constant. This implies an increase in the flux Φ of $\mathbf{B}_0$ through each coil of the solenoid

$$\mathrm{d}\Phi = 2\pi a B_0 \, \mathrm{d}a = 8\pi^2 k_m a \, nI \, \mathrm{d}a , \tag{S-6.214}$$

corresponding to a total electromotive force (the solenoid comprises hn coils)

$$\mathcal{E} = -b_m \frac{\mathrm{d}\Phi}{\mathrm{d}t} = -b_m k_m \, 8\pi^2 a h n^2 I^2 \frac{\mathrm{d}a}{\mathrm{d}t} \tag{S-6.215}$$

that must be compensated by the current source in order to keep I constant. The work $\mathrm{d}W_{\text{source}}$ done by the current source is thus

$$\mathrm{d}W_{\text{source}} = -\mathcal{E}I \, \mathrm{d}t = b_m k_m \, 8\pi^2 a n^2 h I^2 \, \mathrm{d}a . \tag{S-6.216}$$

Thus the total energy of the system solenoid+current source changes by

$$\mathrm{d}U_{\text{tot}} = \mathrm{d}U_M - \mathrm{d}W_{\text{source}} = -4\pi^2 a h b_m k_m n^2 I^2 \, \mathrm{d}a . \tag{S-6.217}$$

The pressure on the solenoid surface is $P = -\mathrm{d}U_{\text{tot}}/\mathrm{d}V$, where $V = \pi a^2 h$ is the volume of the solenoid. Thus

$$P = -\frac{\mathrm{d}U_{\text{tot}}}{\mathrm{d}V} = -\frac{1}{2\pi a h} \frac{\mathrm{d}U_{\text{tot}}}{\mathrm{d}a} = 2\pi b_m k_m \, n^2 I^2 , \tag{S-6.218}$$

in agreement with (S-6.210).

S-6.17   A Homopolar Motor

The motor is schematized in the diagram of Fig. S-6.22,
that displays only one "half" of the circuit because of the
symmetry of the problem. We use cylindrical coordinates
(r, ϕ, z) with the origin O at the center of the cylindrical
magnet, of radius b and length l. The z axis coincides with
the axes of the magnet and of the cell, which here is represented by the voltage source V. The circuit $ACDEF$ is
closed by brush contacts (white arrows in the figure) to
the magnet at points $A \equiv (0, \phi, l/2)$ and $F \equiv (b, \phi, 0)$, so
that the current I can flow through the conducting magnet.
The circuit is free to rotate around the z axis. Let $a > b$
and h be the horizontal and vertical sizes of the circuit,
respectively. We denote by $\mathbf{B} = \mathbf{B}(r, \phi, z)$ the magnetic
field generated by the magnet, independent of ϕ, and with
$B_\phi \equiv 0$. Some field lines of $\mathbf{B}$ are sketched in Fig. S-6.22.

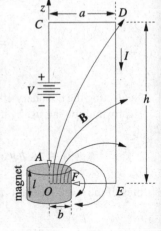

Fig. S-6.22

The magnetic field on the $z = 0$ plane is parallel to the z
axis, directed upwards for $r < b$, and downwards for $r > b$. For simplicity, we approx-
imate $\mathbf{B}(r, \phi, 0) = B_0 \hat{\mathbf{z}}$ for $r < b$, with B_0 independent of r, even if this approximation
is valid only for $l \gg b$.

The voltage source drives a current I through the circuit. When the circuit is at rest we simply have $I = V/R$,
but, when the circuit rotates, we must take into account
the motion of the circuit in the presence of the magnetic
field. Since the magnetic field $\mathbf{B}$ lies on the plane of the
circuit, the force $d\mathbf{f} = I\, d\boldsymbol{\ell} \times \mathbf{B}$ on an infinitesimal segment of the circuit $d\boldsymbol{\ell}$ is perpendicular to the plane of
the circuit (out of paper in the case represented in Fig.
S-6.23). The corresponding infinitesimal torque relative
to the z axis is thus

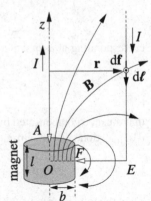

$$d\boldsymbol{\tau} = \mathbf{r} \times d\mathbf{f} = b_\mathrm{m}\, I\, \mathbf{r} \times (d\boldsymbol{\ell} \times \mathbf{B}), \qquad \text{(S-6.219)}$$

where $\mathbf{r}$ is the distance of $d\boldsymbol{\ell}$ from the z axis. The torque

Fig. S-6.23

$d\boldsymbol{\tau}$ is always parallel (or antiparallel) to $\hat{\mathbf{z}}$, independently
of the circuit element $d\boldsymbol{\ell}$ we are considering. For the vector product $d\boldsymbol{\ell} \times \mathbf{B}$ we have

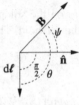

$$d\boldsymbol{\ell} \times \mathbf{B} = -\hat{\boldsymbol{\phi}}\, B\, d\ell \sin\theta = -\hat{\boldsymbol{\phi}}\, B\, d\ell \cos\psi$$
$$= -\hat{\boldsymbol{\phi}}\, \mathbf{B} \cdot \hat{\mathbf{n}}\, d\ell, \qquad \text{(S-6.220)}$$

where θ is the angle between $d\boldsymbol{\ell}$ and $\mathbf{B}$, $\hat{\mathbf{n}}$ is the unit
vector perpendicular to $d\boldsymbol{\ell}$, and $\psi = \theta - \pi/2$ is the angle

Fig. S-6.24

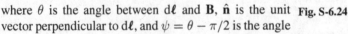

between $\mathbf{B}$ and $\hat{\mathbf{n}}$, as shown in Fig. S-6.24. Since $\hat{\mathbf{r}}$ is perpendicular to $\hat{\phi}$ (unit vectors of the corresponding cylindrical coordinates), we have for the total torque acting on the circuit

$$\boldsymbol{\tau} = b_{\mathrm{m}} I \int_A^F \mathbf{r} \times (\mathrm{d}\boldsymbol{\ell} \times \mathbf{B}) = -\hat{\mathbf{z}}\, b_{\mathrm{m}}\, I \int_A^F \mathbf{B} \cdot \hat{\mathbf{n}}\, r\, \mathrm{d}\ell . \qquad (\text{S-6.221})$$

The last integral of (S-6.221) can calculated, within our approximations, if we first demonstrate that the line integral of $\mathbf{B} \cdot \hat{\mathbf{n}}\, r$ around the closed path $OCDEO$ of Fig. S-6.22 is zero, i.e., that

$$\oint \mathbf{B} \cdot \hat{\mathbf{n}}\, r\, \mathrm{d}\ell = \int_A^F \mathbf{B} \cdot \hat{\mathbf{n}}\, r\, \mathrm{d}\ell + \int_F^O \mathbf{B} \cdot \hat{\mathbf{n}}\, r\, \mathrm{d}\ell + \int_O^A \mathbf{B} \cdot \hat{\mathbf{n}}\, r\, \mathrm{d}\ell = 0 . \qquad (\text{S-6.222})$$

First, we note that the integral along the whole $\overline{OC}$ path is zero, both because r is zero, and because $\mathbf{B}$ is parallel to $\mathrm{d}\boldsymbol{\ell}$, thus perpendicular to $\hat{\mathbf{n}}$. Thus, the integral of (S-6.222) becomes

$$\oint \mathbf{B} \cdot \hat{\mathbf{n}}\, r\, \mathrm{d}\ell = \int_C^D \mathbf{B} \cdot \hat{\mathbf{n}}\, r\, \mathrm{d}r - \int_D^E \mathbf{B} \cdot \hat{\mathbf{n}}\, r\, \mathrm{d}z - \int_E^O \mathbf{B} \cdot \hat{\mathbf{n}}\, r\, \mathrm{d}r$$

$$= \int_C^D \mathbf{B} \cdot \hat{\mathbf{n}}\, r\, \mathrm{d}r + \int_E^D \mathbf{B} \cdot \hat{\mathbf{n}}\, r\, \mathrm{d}z + \int_O^E \mathbf{B} \cdot \hat{\mathbf{n}}\, r\, \mathrm{d}r , \qquad (\text{S-6.223})$$

since $\mathrm{d}\ell = \mathrm{d}r$ along $\overline{CD}$, $\mathrm{d}\ell = -\mathrm{d}z$ along $\overline{DE}$, and $\mathrm{d}\ell = -\mathrm{d}r$ along $\overline{EO}$.

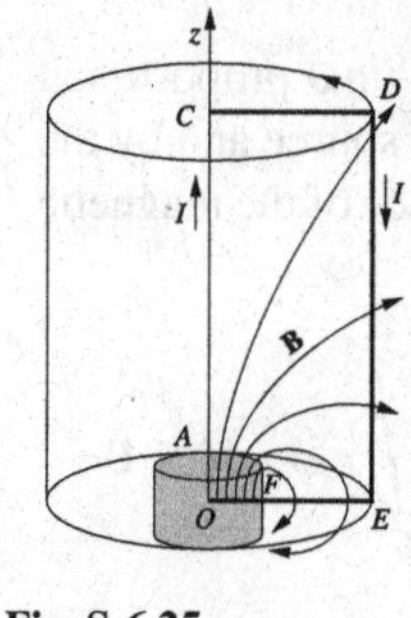

As a next step, we generate a cylinder by rotating the $CDEO$ path around the z axis, as in Fig. S-6.25. The outgoing flux of the magnetic field $\mathbf{B}$ through the total surface of the cylinder is

$$\underbrace{\int \mathbf{B} \cdot \hat{\mathbf{n}}\, \mathrm{d}S}_{\substack{\text{upper} \\ \text{base}}} + \underbrace{\int \mathbf{B} \cdot \hat{\mathbf{n}}\, \mathrm{d}S}_{\substack{\text{lateral} \\ \text{surface}}} + \underbrace{\int \mathbf{B} \cdot \hat{\mathbf{n}}\, \mathrm{d}S}_{\substack{\text{lower} \\ \text{base}}} = 0 ,$$

$$(\text{S-6.224})$$

since $\nabla \cdot \mathbf{B} = 0$. Equation (S-6.224) can be rewritten

Fig. S-6.25

$$0 = \int_0^a \mathbf{B}(r, \phi, h) \cdot \hat{\mathbf{n}} \, 2\pi r \, dr + \int_0^h \mathbf{B}(a, \phi, z) \cdot \hat{\mathbf{n}} \, 2\pi r \, dz$$

$$+ \int_0^a \mathbf{B}(r, \phi, 0) \cdot \hat{\mathbf{n}} \, 2\pi r \, dr = 2\pi \oint \mathbf{B} \cdot \hat{\mathbf{n}} \, r \, d\ell \,, \qquad (S\text{-}6.225)$$

which demonstrates (S-6.222). For the last integral appearing in (S-6.221) we thus have

$$\int_A^F \mathbf{B} \cdot \hat{\mathbf{n}} \, r \, d\ell = - \int_F^O \mathbf{B} \cdot \hat{\mathbf{n}} \, r \, d\ell = \int_0^b B_0 \, r \, dr = \frac{B_0 \, b^2}{2} \,, \qquad (S\text{-}6.226)$$

where we have remembered that the line integrals are zero on the z axis, that $d\ell = -dr$ on the $\overline{FO}$ line, and that, within our approximations, $\mathbf{B} \cdot \hat{\mathbf{n}} = -B_0$, independently of r, on the $\overline{FO}$ line. The torque on the rotating circuit is

$$\boldsymbol{\tau} = -\hat{\mathbf{z}} \, b_{\mathrm{m}} \, I \int_A^F \mathbf{B} \cdot \hat{\mathbf{n}} \, r \, d\ell = -\hat{\mathbf{z}} \, b_{\mathrm{m}} \, I \frac{1}{R} \left(V + b_{\mathrm{m}} \, \omega \frac{B_0 \, b^2}{2} \right) . \qquad (S\text{-}6.227)$$

This is why sliding contacts are needed in a homopolar motor. If the line segment $\overline{FO}$ were rotating with the rest of the circuit, the total torque on the complete circuit around the z axis would be zero, because the torque acting on $\overline{FO}$ would compensate the torque on the rest of the circuit.

If we denote by I the moment of inertia of the rotating circuit, the equation of motion is

$$I \frac{d\omega}{dt} = \tau = -b_{\mathrm{m}} \, I \frac{B_0 \, b^2}{2} - \eta \omega \,, \qquad (S\text{-}6.228)$$

where we have assumed the presence of a frictional torque $\tau_{\mathrm{fr}} = -\eta \omega$ proportional to the angular velocity. The current I is determined by the voltage source and by the electromotive force $\mathcal{E}$, due to the rotation of the circuit in the presence of the magnetic field $\mathbf{B}$,

$$\mathcal{E} = b_{\mathrm{m}} \int_C^F (\boldsymbol{\omega} \times \mathbf{r}) \times \mathbf{B} \cdot d\boldsymbol{\ell} = b_{\mathrm{m}} \int_C^F \omega r \, \hat{\boldsymbol{\phi}} \times \mathbf{B} \cdot d\boldsymbol{\ell} = -b_{\mathrm{m}} \, \omega \int_C^F r \, \hat{\boldsymbol{\phi}} \cdot d\boldsymbol{\ell} \times \mathbf{B}$$

$$= b_{\mathrm{m}} \, \omega \int_C^F r \, \mathbf{B} \cdot \hat{\mathbf{n}} \, d\ell = b_{\mathrm{m}} \, \omega \frac{B_0 \, b^2}{2} \,, \qquad (S\text{-}6.229)$$

where we have used (S-6.220) and (S-6.226) in the last two steps. The current is thus

$$I = \frac{1}{R}\left(V + b_{\mathrm m}\,\omega\,\frac{B_0\,b^2}{2}\right), \tag{S-6.230}$$

and the equation of motion is

$$I\,\frac{\mathrm d\omega}{\mathrm dt} = -b_{\mathrm m}\frac{1}{R}\left(V + b_{\mathrm m}\,\omega\,\frac{B_0\,b^2}{2}\right)\frac{B_0\,b^2}{2} - \eta\omega$$

$$= -b_{\mathrm m}\frac{V\,B_0\,b^2}{2R} - \omega\left(b_{\mathrm m}^2\,\frac{B_0^2\,b^4}{4R} + \eta\right), \tag{S-6.231}$$

with solution

$$\omega = -\frac{2b_{\mathrm m}\,V\,B_0\,b^2}{b_{\mathrm m}^2\,B_0^2\,b^4 + 4R\eta}\left(1 - \mathrm e^{-t/T}\right), \quad \text{where} \quad T = \frac{4RI}{b_{\mathrm m}^2\,B_0^2\,b^4 + 4R\eta}. \tag{S-6.232}$$

If we assume negligible frictional torque, i.e., $\eta \ll b_{\mathrm m}^2 B_0^2 b^4/(4R)$, (S-6.232) reduces to

$$\omega = -\frac{2V}{b_{\mathrm m}\,B_0\,b^2}\left(1 - \mathrm e^{-t/T}\right), \quad \text{where} \quad T = \frac{4RI}{b_{\mathrm m}^2\,B_0^2\,b^4}, \tag{S-6.233}$$

however, inserting "reasonable values" into (S-6.233), such as $V = 1.5\,\mathrm V$, $B_0 = 100\,\mathrm{Gauss} = 10^{-2}\,\mathrm T$ and $b = 0.5\,\mathrm{cm}$ we obtain for the steady state solution

$$\omega_0 = -\frac{2V}{b_{\mathrm m}\,B_0\,b^2} \simeq -1\,200\,\mathrm{rad/s}, \quad \text{i.e.,} \quad \nu_0 \simeq 190\,\mathrm s^{-1}, \tag{S-6.234}$$

which is indeed a very fast rotation! In the absence of friction, the steady state is reached when $V + \mathcal E = 0$, so that $I = 0$ and there is no torque acting on the circuit. The final steady-state kinetic energy of the rotating circuit in these conditions is

$$K_{\mathrm{ss}} = \frac{1}{2}\,I\omega_0^2 = \frac{1}{2}I\,\frac{4V^2}{b_{\mathrm m}^2\,B_0^2\,b^4} = \frac{2V^2 I}{b_{\mathrm m}^2\,B_0^2\,b^4}. \tag{S-6.235}$$

The current flowing in the circuit is

$$I(t) = \frac{1}{R}\left(V + \frac{b_{\mathrm m}\,B_0 b^2\omega}{2}\right) = \frac{V}{R}\,\mathrm e^{-t/T}, \tag{S-6.236}$$

and the total energy provided by the voltage source is

$$U = \int_0^\infty V I\,\mathrm dt = \frac{V^2}{R}\int_0^\infty \mathrm e^{-t/T}\,\mathrm dt = \frac{V^2}{R}\,T = \frac{4V^2 I}{b_{\mathrm m}^2\,B_0^2\,b^4} = 2K_{\mathrm{ss}}, \tag{S-6.237}$$

or twice the final kinetic energy. An amount equal to K_{ss} is dissipated into Joule heat. More realistically, we must take the frictional torque into account. For instance, the steady-state angular velocity is reduced by a factor 10 if we assume $4R\eta = 9 b_m B_0 b^2$. This, assuming $R = 1\,\Omega$, means

$$\eta \simeq 6 \times 10^{-5}\,\text{Nms}\,.\tag{S-6.238}$$

In the presence of friction the steady-state angular velocity is

$$\omega_f = -\frac{2 b_m V B_0 b^2}{b_m^2 B_0^2 b^4 + 4R\eta}\,,\tag{S-6.239}$$

and the power dissipated by friction is

$$P_{fr} = \tau_{fr}\,\omega_f = \eta \omega_f^2 = \eta \left(\frac{2 b_m V B_0 b^2}{b_m^2 B_0^2 b^4 + 4R\eta}\right)^2.\tag{S-6.240}$$

The voltage source drives a current

$$I_f = \frac{V}{R}\left(1 - \frac{b_m^2 B_0^2 b^4}{b_m^2 B_0^2 b^4 + 4R\eta}\right) = \frac{4V\eta}{b_m^2 B_0^2 b^4 + 4R\eta}\,,\tag{S-6.241}$$

and provides a power

$$P_{source} = V I_f = \frac{4V^2\eta}{b_m^2 B_0^2 b^4 + 4R\eta}\,.\tag{S-6.242}$$

The power dissipated into Joule heat is

$$P_J = R I_f^2 = R \left(\frac{4V\eta}{b_m^2 B_0^2 b^4 + 4R\eta}\right)^2,\tag{S-6.243}$$

and we can easily check that

$$P_J + P_{fr} = P_{source}\,.\tag{S-6.244}$$

S-6.18 A Magnetic Cylinder Sliding Inside a Solenoid

(a) Within our assumptions the partially filled solenoid can be considered as a series of two shorter solenoids of inductance L_1 and L_2, respectively. If both solenoids can be assumed to be long with respect to their radii (which seems acceptable as far as both $Z \gg b$ and $(a - Z) \gg b$), the inductances are given by (see Table 5.1)

$$L_1 = \mu_r 4\pi k_m b_m n^2 Z S\,, \quad L_2 = 4\pi k_m b_m n^2 (a - Z) S\,,\tag{S-6.245}$$

since both solenoids have cross section $S = \pi b^2$ and n turns per unit length, but different relative magnetic permittivities (μ_r and 1) and lengths (Z and $a - Z$). The total inductance is

$$L_{\text{tot}} = L_1 + L_2$$

$$= 4\pi k_m b_m n^2 S \left[(\mu_r - 1) Z + a \right] = \begin{cases} \mu_0 \, n^2 S \left[(\mu_r - 1) Z + a \right] , & \text{SI,} \\[2mm] \dfrac{4\pi}{c^2} n^2 S \left[(\mu_r - 1) Z + a \right] , & \text{Gaussian.} \end{cases}$$

$$\text{(S-6.246)}$$

The magnetic energy stored in the solenoid is

$$U = \frac{1}{2} L \, I^2 = 2\pi k_m b_m \, n^2 S \left[(\mu_r - 1) Z + a \right] I_0^2 . \qquad \text{(S-6.247)}$$

With increasing Z the energy increases if $\mu_r > 1$ (paramagnetic material) and decreases if $\mu_r < 1$ (diamagnetic material).

To evaluate the force acting on the cylinder we evaluate the variation of the energy of the system for an infinitesimal virtual variation dZ of the penetration depth. The magnetic corresponding energy variation is

$$dU = 2\pi k_m b_m \, n^2 S (\mu_r - 1) \, I_0^2 \, dZ . \qquad \text{(S-6.248)}$$

To obtain the total energy variation of the system we must add to (S-6.248) the energy lost or absorbed by the current generator in order to keep I constant. A displacement dZ of the cylinder corresponds to a variation of the flux of the magnetic field through the solenoid

$$d\Phi = n \, dZ \, S (B_1 - B_2) = n \, dZ \, S (\mu_r - 1) 4\pi k_m n I_0 = 4\pi k_m n^2 \, S (\mu_r - 1) \, I_0 \, dZ ,$$

$$\text{(S-6.249)}$$

where $n \, dZ$ is the number of turns through which the field changes from B_2 to B_1. If the displacement is performed within a duratio dt, according to Faraday's law the change of flux generates an electromotive force

$$\mathcal{E} = -b_m \frac{d\Phi}{dt} = -b_m \frac{4\pi k_m n^2 \, S (\mu_r - 1) \, I_0 \, dZ}{dt} . \qquad \text{(S-6.250)}$$

This electromotive force must be compensated by the current source in order to keep the current constant. Thus during the time interval dt the source performs a work

$$dW = -\mathcal{E} \, I_0 \, dt = 4\pi k_m b_m n^2 \, S (\mu_r - 1) \, I_0^2 \, dZ = 2 \, dU , \qquad \text{(S-6.251)}$$

at the expense of its internal energy. Thus the energy variation of the whole system is

$$dU_{\text{tot}} = dU - dW = -dU . \qquad \text{(S-6.252)}$$

The force acting on the cylinder can thus be evaluated as

$$F = -\frac{dU_{\text{tot}}}{dZ} = +\frac{dU}{dZ} = 2\pi k_{\mathrm{m}} b_{\mathrm{m}} \, n^2 S(\mu_{\mathrm{r}} - 1) I_0^2 = \begin{cases} \dfrac{\varepsilon_0}{2} n^2 S \, (\mu_{\mathrm{r}} - 1) \, I_0^2 \,, & \text{SI,} \\[2ex] \dfrac{2\pi}{c^2} n^2 S \, (\mu_{\mathrm{r}} - 1) \, I_0^2 \,, & \text{Gaussian.} \end{cases}$$

$$\text{(S-6.253)}$$

Thus $F > 0$ if $\mu_{\mathrm{r}} > 1$, and $F < 0$ if $\mu_{\mathrm{r}} < 1$. A paramagnetic cylinder is pulled into the solenoid, while a diamagnetic cylinder is pushed out.

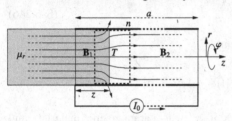

(b) Using expressions (S-6.245) for the contributions of the filled and empty regions of the solenoid to its global inductance implies assuming that the magnetic field **B** is uniform in both regions. However, the two regions have different magnetic fields, and this is in contrast with the boundary conditions at at the cylinder-vacuum interface, where B_z should be continuous.

Fig. S-6.26

Thus there must actually be a transition region T inside our solenoid (dashed box in Fig. S-6.26) where the z component of **B** continuously changes from its value $B_1 = 4\pi k_{\mathrm{m}} \mu_{\mathrm{r}} n I_0$ in the cylinder to its value $B_2 = 4\pi k_{\mathrm{m}} n I_0$ in vacuum. Actually, the transition extends in both directions, but we can reasonably assume that the field at the left of the transition region is practically $\mathbf{B}_1$, and at the right of T is practically $\mathbf{B}_2$. Note that (although not strictly relevant to our problem) the field lines display a refraction angle at the cylinder-vacuum interface because the radial (in cylindical coordinates with the cylinder axis as polar axis) component H_r, not B_r is continuous; also note that $\partial_z B_z$ is discontinous.

When the cylinder penetrates by a further dZ into the solenoid, the transition region shifts globally to the right by an amount dZ. Thus, during this displacement, the volume with field $\mathbf{B}_1$ increases by an amount $S dZ$, surrounded by $n \, dZ$ turns of the coil, while the volume with field $\mathbf{B}_2$ decreases by the same amount. Thus the energy of the system changes by

$$dU = 2\pi k_{\mathrm{m}} b_{\mathrm{m}} \, n^2 S(\mu_{\mathrm{r}} - 1) \, I_0^2 \, dZ \,, \qquad \text{(S-6.254)}$$

identical to (S-6.248). The presence of the transition region simply adds a constant value to the energy (S-6.247), which is lost in differentiation. It is important to note that, analogously to the solution of Problem 3.12, here we are assuming that for a long cylinder a displacement of its left end outside the solenoid (Fig. S-6.26), which "frees" from the magnetic material a volume $S dZ$, yields a negligible variation in the total magnetic energy. Of course, since the strength of the magnetic field decreases with distance from the end of the solenoid, this assumption is consistent with both $Z \gg b$ and $(a - Z) \gg b$ so that the end of the cylinder is sufficiently far away from the solenoid. On a physical ground, the validity of this approximation is more questionable for a paramagnetic material than for a diamagnetic one. In fact, a ferromagnetic material of very high relative magnetic permeability μ_r would channel the field lines inside the cylinder, so that the decrease in the strength of **B** with distance would be slower than in the case of a diamagnetic material which acts to expel the field lines.

In general, the mechanical force on the cylinder will result from the magnetic field B acting on the magnetization currents which flow on its surface and, because of symmetry, are directed in the azimuthal (ϕ) direction and are either positive or negative (i.e., either parallel or anti-parallel to the current flowing in the solenoid loops) for $\mu_r > 1$ and $\mu_r < 1$, respectively. This is the origin of the different sign of the force for paramagnetic and diamagnetic materials. Clearly a force along z cannot be originated by the z components of **B**. Thus the force must be due to the radial components of **B**, which are found at the ends of the cylinder.

S-6.19   Conducting Cylindrical Shell in a Magnetic Field

(a) Under the assumption of slowly-varying currents and fields we can evaluate the induced electric field $\mathbf{E}_1$ from the approximate equation $\nabla \times \mathbf{E} \simeq -b_m \partial_t \mathbf{B}_0$. We also assume that the problem has cylindrical symmetry around the shell axis. This is not necessarily true from the assumption of a "uniform" magnetic field. A magnetic field uniform over the whole space implies rotational symmetry around any axis parallel to the field, leading to inconsistencies for the induced electric field. But we can assume, for instance, that our uniform magnetic field is generated by a large long solenoid coaxial with the conducting shell, so that there is a privileged axis of rotational symmetry. Under these assumptions the electric field is azimuthal, and, considering a circular path C of radius $r < a$ coaxial with the shell, enclosing a surface Σ, we have

$$\oint_C \mathbf{E}_1 \cdot d\boldsymbol{\ell} \simeq -b_m \frac{\partial}{\partial t} \int_\Sigma \mathbf{B}_0 \cdot d\mathbf{S} , \qquad (S\text{-}6.255)$$

leading to

$$2\pi r E_{1\phi} \simeq -b_m \pi r^2 \frac{\partial}{\partial t} (\mathcal{B} \cos \omega t) , \qquad (S\text{-}6.256)$$

and, finally,

$$E_{1\phi} \simeq \frac{1}{2} b_m \omega r \mathcal{B} \sin \omega t . \qquad (S\text{-}6.257)$$

Assuming that both conditions $h \ll a$ and $h \ll \delta_\omega$ hold for the shell thickness h, where δ_ω is the collisional skin depth at frequency ω, the current density inside the shell can be considered uniform, $\mathbf{J}_1 \simeq \sigma \mathbf{E}_1(a)$. Thus the resulting magnetic field $\mathbf{B}_1$ equals the field generated by a solenoid with the product $nI = J_1 h$, and we have

$$\mathbf{B}_1 \simeq \hat{\mathbf{z}} 4\pi k_m J_1 h = \hat{\mathbf{z}} 2\pi k_m b_m \sigma \omega a h \mathcal{B} \sin \omega t = \begin{cases} \hat{\mathbf{z}} \dfrac{\mu_0}{2} \sigma \omega a h \mathcal{B} \sin \omega t \,, & \text{SI,} \\[2mm] \hat{\mathbf{z}} \dfrac{2\pi}{c^2} \sigma \omega a h \mathcal{B} \sin \omega t \,, & \text{Gaussian.} \end{cases}$$

$$\text{(S-6.258)}$$

The condition $B_1 \ll B_0$ implies

$$\frac{\sigma \omega a h}{2\varepsilon_0 c^2} \ll 1 \ \text{(SI)}, \qquad\qquad \frac{2\pi \sigma \omega a h}{c^2} \ll 1 \ \text{(Gaussian)}, \qquad\qquad \text{(S-6.259)}$$

where we have used the identity $\mu_0 = 1/(\varepsilon_0 c^2)$ for the SI units.

A conductivity of $1\ ^{-1}\mathrm{m}^{-1}$ in SI units corresponds to $9 \times 10^9\ \mathrm{s}^{-1}$ in Gaussian units. Thus, in Gaussian units, we have $\sigma \simeq 5.4 \times 10^{17}\ \mathrm{s}^{-1}$. We have $\omega = 2\pi \times 50\ \mathrm{Hz} \simeq 314\ \mathrm{rad/s}$. The condition for $B_1 \ll B_0$ in SI units is

$$ah \ll \frac{2\varepsilon_0 c^2}{\sigma \omega} \simeq 8.5 \times 10^{-5}\ \mathrm{m}^2 \,, \qquad\qquad \text{(S-6.260)}$$

analogously we have in Gaussian units

$$ah \ll \frac{c^2}{2\pi \sigma \omega} \simeq 0.85\ \mathrm{cm}^2 \,. \qquad\qquad \text{(S-6.261)}$$

The displacement current is negligible if $J = \sigma E \gg \partial_t E/(4\pi k_e) \simeq \omega E/(4\pi k_e)$, or

$$\omega \ll 4\pi k_e \sigma = \begin{cases} \dfrac{\sigma}{\varepsilon_0} & \text{(SI)} \\[2mm] 4\pi \sigma & \text{(Gaussian)} \end{cases} \simeq 6.8 \times 10^{18}\ \mathrm{rad/s} \,, \qquad\qquad \text{(S-6.262)}$$

condition well verified in our case ($\omega = 314$).

(b) The field B_1 is discontinuous through the shell surface, since it is zero outside the shell. This originates a discontinuity of the Poynting vector. We have

$$\mathbf{S}(a^-) = \hat{\mathbf{r}} \frac{1}{k_m} E_1 (B_0 + B_1) \,, \qquad \mathbf{S}(a^+) = \hat{\mathbf{r}} \frac{1}{k_m} E_1 B_0 \,. \qquad\qquad \text{(S-6.263)}$$

The difference between the entering and the exiting energy flux per unit length through the shell wall is

$$\Delta \Phi(\mathbf{S}) = 2\pi a \left[S(a^-) - S(a^+) \right] = \frac{2\pi a}{k_m} E_1(a) B_1(a^-)$$

$$= \frac{2\pi a}{k_m} E_1(a)\, (k_m J_1 h) = 2\pi a h E_1(a) J_1 \,, \qquad\qquad \text{(S-6.264)}$$

which equals the power dissipated per unit length of the shell by Joule heating. We have used (S-6.258).

(c) We use complex notation with $\tilde{B}_0 = \mathcal{B}e^{-i\omega t}$. If we denote by $\tilde{B}_t$ the total magnetic field inside the shell, we have $\tilde{B}_t = \tilde{B}_0 + \Delta\tilde{B}$, where $\Delta\tilde{B}$ is the magnetic field generated by the current induced in the shell walls. For the induced electric field $\tilde{E}_\phi$ and the current density $\tilde{J}$ in the shell walls we have

$$\tilde{E}_\phi = -\frac{1}{2}\,b_{\mathrm{m}}r\,\frac{\partial\tilde{B}_t}{\partial t} = \frac{i}{2}\,b_{\mathrm{m}}\omega r\,\tilde{B}_t\,, \quad \text{and} \quad \tilde{J} = \sigma\tilde{E}_\phi(a)\,. \tag{S-6.265}$$

Combining we obtain

$$\tilde{B}_t = \tilde{B}_0 + 4\pi k_{\mathrm{m}}h\,\tilde{J} = \tilde{B}_0 + 4\pi k_{\mathrm{m}}h\sigma\tilde{E}_\phi(a) = \tilde{B}_0 + i\,2\pi k_{\mathrm{m}}b_{\mathrm{m}}ah\omega\sigma\,\tilde{B}_t\,, \tag{S-6.266}$$

from which

$$\tilde{B}_t = \frac{\tilde{B}_0}{1 - i\,2\pi k_{\mathrm{m}}b_{\mathrm{m}}ah\omega\sigma} = \begin{cases} \dfrac{\tilde{B}_0}{1 - i\mu_0 ah\omega\sigma/2}\,, & \text{SI,} \\[2mm] \dfrac{\tilde{B}_0}{1 - i\,2\pi k_{\mathrm{m}}b_{\mathrm{m}}ah\omega\sigma/c^2}\,, & \text{Gaussian.} \end{cases} \tag{S-6.267}$$

At the slowly-varying fields limit we obtain

$$\tilde{B}_t \simeq \tilde{B}_0\,(1 + i\,2\pi k_{\mathrm{m}}b_{\mathrm{m}}ah\omega\sigma)\,, \tag{S-6.268}$$

and, in real notation

$$\Delta B = \mathrm{Re}\left(i\,2\pi k_{\mathrm{m}}b_{\mathrm{m}}ah\omega\sigma\mathcal{B}e^{-i\omega t}\right) = 2\pi k_{\mathrm{m}}b_{\mathrm{m}}ah\omega\sigma\mathcal{B}\sin\omega t\,, \tag{S-6.269}$$

in agreement with (S-6.258).

S-6.20 Electromagnetic Inertia of a Rotating Charged Ring

(a) The rotating ring with rigid charge distribution is equivalent to a circular loop carrying a current $I = \omega Q/2\pi$. Thus the self-induced flux of the magnetic field through the loop is $\Phi_{\mathrm{self}} = cLI$. The time-dependence of $\omega = \omega(t)$ gives origin to an induced electric field $\mathbf{E}_{\mathrm{i}} = E_{\mathrm{i}}(r)\,\hat{\phi}$, azimuthal because of symmetry. According to Faraday's law we have on the ring ($z = 0, r = a$)

$$\oint \mathbf{E}_{\mathrm{i}}\cdot d\boldsymbol{\ell} = 2\pi a E_{\mathrm{i}}(a) = -\frac{1}{c}\frac{d}{dt}\Phi_{\mathrm{self}} = \frac{LQ}{2\pi}\dot{\omega}\,, \tag{S-6.270}$$

which leads to

$$E_{\mathrm{i}}(a) = -\frac{LQ}{(2\pi)^2 a}\dot{\omega} = -\frac{\alpha Q}{(2\pi)^2 c^2}\dot{\omega}\,. \tag{S-6.271}$$

(b) The force exerted by the induced electric field on a ring element of length $d\ell$ and charge $dq = Q\,d\ell/(2\pi a)$ is

$$df = E_i(a)\,dq = E_i(a)\,\frac{Q}{2\pi a}\,d\ell . \qquad \text{(S-6.272)}$$

Thus the torque exerted by the induced electric field on the whole ring is

$$\tau_E = \hat{\mathbf{z}} \oint a\,df = \hat{\mathbf{z}} \oint E_i(a)\,\frac{Q}{2\pi}\,d\ell = E_i(a)\,Qa\,\hat{\mathbf{z}}$$

$$= -\frac{\alpha Q^2 a}{(2\pi)^2 c^2}\,\dot{\omega}\,\hat{\mathbf{z}} = -\frac{LQ^2}{(2\pi)^2}\,\dot{\omega} . \qquad \text{(S-6.273)}$$

The equation of motion for the ring is thus

$$I\dot{\omega} = \tau_{\text{ext}} + \tau_E , \qquad \text{(S-6.274)}$$

which we can rewrite as

$$I'\dot{\omega} \equiv (I + I_{\text{EM}})\,\dot{\omega} = \left(I + \frac{LQ^2}{(2\pi)^2} \right)\dot{\omega} = \tau_{\text{ext}} \qquad \text{(S-6.275)}$$

where I_{EM} can be considered as a contributiobn of electromagnetic nature to the moment of inertia of the ring. This electromagnetic contribution can be written as

$$I_{\text{EM}} = \frac{\alpha}{4\pi^2}\,\frac{Q^2}{ac^2}\,a^2 \equiv m_{\text{EM}}\,a^2 , \qquad \text{(S-6.276)}$$

where m_{EM} is some sort of "electromagnetic mass" of the ring.
(c) The power provided by the induced electric field is

$$P_E = \tau_E \cdot \omega = -\frac{LQ^2}{(2\pi)^2}\,\omega\,\dot{\omega} . \qquad \text{(S-6.277)}$$

The magnetic energy of the ring is

$$U_{\text{mag}} = \frac{1}{2}LI^2 = \frac{L}{2}\left(\frac{\omega Q}{2\pi} \right)^2 , \qquad \text{(S-6.278)}$$

thus we have

$$\frac{dU_{\text{mag}}}{dt} = L\left(\frac{Q}{2\pi} \right)^2 \omega\,\dot{\omega} = -P_E , \qquad \text{(S-6.279)}$$

showing that the work done by the induced electric field is stored as magnetic energy of the system.

S-6.21   A Bar Sliding in a Magnetic Field

(a) Let us denote by $x(t)$ the position of the moving bar and by $v = dx/dt$ its velocity. At a generic time t the flux of the magnetic field through the rectangular circuit comprising the two bars and the portions of wires between them is $\Phi_{B_0}(t) = B_0 h x(t)$, thus the electromotive force around the circuit is

$$\mathcal{E} = -\frac{d\Phi_{B_0}}{dt} = -B_0 h v , \qquad (S\text{-}6.280)$$

and the current flowing in the circuit is

$$I = \frac{\mathcal{E}}{R} = -\frac{B_0 h v}{R} . \qquad (S\text{-}6.281)$$

The moving bar is subject to the magnetic force

$$F = B_0 h I = -\frac{B_0^2 h^2}{R} v , \qquad (S\text{-}6.282)$$

and its equation of motion is

$$m \frac{dv}{dt} = -\frac{B_0^2 h^2}{R} v , \qquad (S\text{-}6.283)$$

with solution for the bar velocity

$$v(t) = v_0 \, e^{-t/\tau} , \quad \text{where} \quad \tau = \frac{mR}{B_0^2 h^2} . \qquad (S\text{-}6.284)$$

The position of the bar at time t is thus

$$x(t) = x_0 + \int_0^t v_0 \, e^{-t'/\tau} dt' = x_0 + v_0 \tau \left(1 - e^{-t/\tau}\right) , \qquad (S\text{-}6.285)$$

and the maximum distance from the origin that the bar will reach is

$$x(\infty) = x_0 + v_0 \tau. \qquad (S\text{-}6.286)$$

(b) Taking the self-inductance of the circuit into account the electromotive force is

$$RI = \mathcal{E} = -\frac{d\Phi_{B_0}}{dt} - L \frac{dI}{dt} = -B_0 h v - L \frac{dI}{dt} , \qquad (S\text{-}6.287)$$

which forms a system of equations with the equation of motion

$$m \frac{dv}{dt} = B_0 h I \ . \tag{S-6.288}$$

From (S-6.288) we obtain for I and dI/dt

$$I = \frac{m}{B_0 h} \frac{dv}{dt} \ , \qquad \frac{dI}{dt} = \frac{m}{B_0 h} \frac{d^2 v}{dt^2} \ , \tag{S-6.289}$$

which, substituted into (S-6.287), lead to the equation

$$\frac{Rm}{B_0 h} \frac{dv}{dt} = -B_0 h v - \frac{Lm}{B_0 h} \frac{d^2 v}{dt^2} \ , \tag{S-6.290}$$

which can be rewritten as

$$\frac{d^2 v}{dt^2} + \frac{R}{L} \frac{dv}{dt} + \frac{B_0^2 h^2}{Lm} v = 0 \ . \tag{S-6.291}$$

This equation is analogous to the equation of a damped harmonic oscillator with undamped frequency $\omega_0 = \sqrt{B_0^2 h^2/(Lm)}$ and viscous damping coefficient $\gamma = R/L$. Equation (S-6.291) is a linear, homogeneous second-order differential equation with constant coefficients, thus its general solution has the form $v(t) = A_1 e^{\alpha_1 t} + A_2 e^{\alpha_2 t}$, where A_1 and A_2 are constants to be determined from the initial conditions. Thus, differently from point **a)**, where one initial condition, in our case the initial velocity v_0, was sufficient to determine the problem, here we need a second condition, for instance the initial acceleration a_0 of the bar, in order to determine the two constants A_1 and A_2. Given v_0 and a_0. we must solve the system of equations

$$A_1 + A_2 = v_0 \ , \tag{S-6.292}$$
$$\alpha_1 A_1 + \alpha_2 A_2 = a_0 \tag{S-6.293}$$

with solutions
$$A_1 = \frac{\alpha_2 v_0 - a_0}{\alpha_2 - \alpha_1} \ , \qquad A_2 = \frac{a_0 - \alpha_1 v_0}{\alpha_2 - \alpha_1} \ . \tag{S-6.294}$$

The exponents $\alpha_{1,2}$ are the roots of the characteristic polinomial

$$\alpha^2 + \frac{R}{L} \alpha + \frac{B_0^2 h^2}{Lm} = 0 \ , \tag{S-6.295}$$

namely
$$\alpha_{1,2} = \frac{-R \mp \sqrt{R^2 - 4L B_0^2 h^2/m}}{2L} \ . \tag{S-6.296}$$

We have three possible cases:

  i. $R > 2B_0h\sqrt{L/m}$, the argument of the square root is positive and we have an *overdamped* decay of v, i.e., v decays without oscillating. At the limit $L \to 0$ we have

$$\lim_{L \to 0} \sqrt{R^2 - 4LB_0^2h^2/m} = R - \frac{2LB_0^2h^2}{Rm}, \qquad \text{(S-6.297)}$$

thus

$$\lim_{L \to 0} \alpha_1 = -\infty, \qquad \lim_{L \to 0} \alpha_2 = -\frac{B_0^2h^2}{Rm}, \qquad \text{(S-6.298)}$$

$$\lim_{L \to 0} A_1 = 0, \qquad \lim_{L \to 0} A_2 = v_0. \qquad \text{(S-6.299)}$$

At this limit $\tau_1 = 1/\alpha_1$ tends to zero, which means immediate decay, and, according to (S-6.299), also the coefficient A_1 tends to zero. As a result, if the initial acceleration $a_0 \neq \alpha_2 v_0$ it rapidly jumps to this value (and for $L = 0$ it is discontinuous). On the other hand the limits for A_2 and α_2 are in agreement with (S-6.284).

  ii. $R = 2B_0h\sqrt{L/m}$, we have $\alpha_1 = \alpha_2 = -R/(2L)$ and v is *critically damped*, this corresponds to the quickest decay.

  iii. $R < 2B_0h\sqrt{L/m}$, the square root is imaginary and v is *underdamped*, it oscillates with a slightly different frequency than in the undamped case, with the oscillation amplitude gradually decreasing to zero.

S-6.22   Magnetic Levitation of a Superconducting Ring

(a) When an electric current I flows in the coil of radius R the magnetic field along its axis is

$$\mathbf{B}(z) = \hat{\mathbf{z}} 2\pi k_m I \frac{R^2}{\left(R^2 + z^2\right)^{3/2}} = \begin{cases} \hat{\mathbf{z}} I \dfrac{\mu_0 R^2}{2\left(R^2 + z^2\right)^{3/2}}, & \text{(SI)}, \\[3mm] \hat{\mathbf{z}} I \dfrac{2\pi R^2}{c\left(R^2 + z^2\right)^{3/2}}, & \text{(Gaussian)}, \end{cases} \qquad \text{(S-6.300)}$$

where $z = 0$ corresponds to the center of the loop. The flux of **B** through the superconducting ring of radius $r \ll R$, coaxial with the loop at a height z above the loop plane, can be approximated as

$$\Phi(z) \simeq \pi r^2 B(z) = I\, 2\pi^2 r^2 k_m \frac{R^2}{\left(R^2 + z^2\right)^{3/2}}, \qquad \text{(S-6.301)}$$

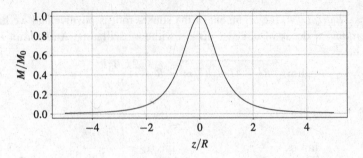

Fig. S-6.27 .

Thus the coefficient of mutual inductance between the coil and the ring is

$$M(z) = \frac{\Phi(z)}{I} = 2\pi^2 r^2 k_m \frac{R^2}{\left(R^2 + z^2\right)^{3/2}} = M_0 \frac{R^3}{\left(R^2 + z^2\right)^{3/2}}$$

$$= M_0 \frac{1}{\left(1 + \zeta^2\right)^{3/2}}, \tag{S-6.302}$$

where we have introduced the dimensionless variable $\zeta = z/R$, while $M_0 = 2\pi^2 r^2$ k_m/R is the mutual inductance at $\zeta = z = 0$, i.e., when coil and ring are coplanar. The plot of $M(\zeta)$ is shown in Fig. S-6.27.

(b) The magnetic flux through the superconducting ring must remain constant, thus, since it was zero for $t < 0$, it must remain zero at any time $t \geq 0$. When the ring is at height z the coil induces a magnetic flux $M(z) I_c/b_m$ through it, which must be canceled by the magnetic field generated by the current $I(z)$ circulating in the superconducting ring. Thus we must have

$$L_r I(z) = -M(z) I_c , \quad \text{leading to} \quad I(z) = -\frac{M(z)}{L_r} I_c . \tag{S-6.303}$$

When the ring is at a height z the magnetic energy of the system coil+ring is

$$U_M(z) = \frac{1}{2} L_c I_c^2 + \frac{1}{2} L_r I^2(z) + M(z) I_c I(z) = \frac{1}{2} L_c I_c^2 + \frac{1}{2} \frac{M^2(z)}{L_r} I_c^2 - \frac{M^2(z)}{L_r} I_0^2$$

$$= \frac{1}{2} L_c I_c^2 - \frac{1}{2} \frac{M^2(z)}{L_r} I_c^2 , \tag{S-6.304}$$

where we have substituted the second of (S-6.303) for $I(z)$. If the height of the ring is increased by an infinitesimal amount dz the magnetic energy of the system increases by an amount

$$dU_M = -\frac{I_c^2}{L_r} M(z) \frac{dM}{dz} dz , \tag{S-6.305}$$

here, and in the following, remember that $dM/dz < 0$ for $z > 0$. In the meantime the current source must perform work in order to keep the current I_c flowing in the coil constant. The contribution of the superconducting ring to the magnetic flux through the coil changes by

$$d\Phi = \frac{1}{b_m} d\left[M(z)I(z)\right] = -\frac{1}{b_m} d\left[\frac{M^2(z)}{L_r} I_c\right] = -\frac{2}{b_m} \frac{I_c}{L_r} M(z) \frac{dM}{dz} dz ,$$

$$(S\text{-}6.306)$$

which, assuming that the change occurs in a time dt, corresponds to an electromotive force

$$\mathcal{E} = -b_m \frac{d\Phi}{dt} = 2 \frac{I_c}{L_r} M(z) \frac{dM}{dz} \frac{dz}{dt} ,$$

$$(S\text{-}6.307)$$

which must be compensated by an opposite electromotive force provided by the current source. During the time dt the current source thus performs a work

$$dW = -\mathcal{E} I_c dt = -2 \frac{I_0^2}{L_r} M(z) \frac{dM}{dz} dz = 2dU_M ,$$

$$(S\text{-}6.308)$$

at the expenses of its internal energy. Thus, when the height of the superconducting ring increases by dz, the total "potential" energy of the system, including the current source, changes by

$$dU_{pot} = -dU_M + mg\, dz ,$$

$$(S\text{-}6.309)$$

where we have taken also the gravitational potential energy of the ring, $mg\, dz$, into account. We can thus define an "effective" potential energy of the system coil+ring+source

$$U_{pot} = \frac{1}{2} \frac{M^2(z)}{L_r} I_c^2 + mgz ,$$

$$(S\text{-}6.310)$$

where we have neglected the term $L_c I_c^2/2$ of (S-6.304) because it is constant. If we measure the energy in units of mgR and the current circulating in the coil I_c in units I_0 defined as

$$I_0 = \frac{\sqrt{L_r R^3 mg}}{\pi^2 r^2 k_m} ,$$

$$(S\text{-}6.311)$$

which is the current for which the effective potential energy equals mgR at $\zeta = 0$, (S-6.310) can be rewritten as

$$U_{pot}(\zeta) = \frac{\alpha^2}{\left(1 + \zeta^2\right)^3} + \zeta ,$$

$$(S\text{-}6.312)$$

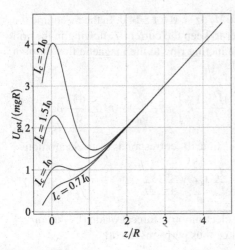

Fig. S-6.28

where $\alpha = I_c/I_0$. This function is plotted in Fig. S-6.28 for four different values of α, namely $\alpha = 0.7,\ 1,\ 1.5,\ 2$. From Fig. S-6.28 we see that at $t = 0^+$, when I_c is already circulating in the coil but the ring is still at rest, the ring cannot be located exactly at $\zeta = 0$ if we want it to start moving. In fact the net force acting on the ring at $\zeta = 0$ is $-mg$, directed downwards. Therefore the ring must be located at a position $\zeta_0 = z_0/R$ immediately higher then the position of the relative maximum of $U_{\mathrm{pot}}(\zeta)$, which tends asymptotically to $\zeta = 0$ for increasing I_c. At $t = 0^+$ we have no kinetic energy, and the total energy of the ring is

$$U\left(\zeta_0\right) = \frac{\alpha^2}{\left(1 + \zeta_0^2\right)^3} + \zeta_0\,, \qquad (\text{S-6.313})$$

while when the ring reaches its maximum height $\zeta_{\max} = z_{\max}/R$, again with no kinetic energy, the total energy is

$$U\left(\zeta_{\max}\right) = \frac{\alpha^2}{\left(1 + \zeta_{\max}^2\right)^3} + \zeta_{\max}\,. \qquad (\text{S-6.314})$$

Thus energy conservation leads to

$$\zeta_{\max} = \frac{\alpha^2}{\left(1 + \zeta_0^2\right)^3} - \frac{\alpha^2}{\left(1 + \zeta_{\max}^2\right)^3} + \zeta_0 \simeq \frac{\alpha^2}{\left(1 + \zeta_0^2\right)^3} + \zeta_0\,, \qquad (\text{S-6.315})$$

where we have neglected $M^2\left(\zeta_{\max}\right)$ with respect to $M^2\left(\zeta_0\right)$.

The maximum and minimum of $U_{\mathrm{pot}}(\zeta)$ occur for ζ values such that

$$0 = \frac{\partial U_{\mathrm{pot}}}{\partial \zeta} = -\frac{6\alpha^2\zeta}{\left(1 + \zeta^2\right)^4} + 1\,, \quad \text{or} \quad \left(1 + \zeta^2\right)^4 - 6\alpha^2\zeta = 0\,. \qquad (\text{S-6.316})$$

We see from Fig. S-6.28 that there must be a limit value $\alpha_{\min}$ under which the maximum and minimum of the potential no longer exist. Physically, this corresponds to the current for which the magnitude of the magnetic force is everywhere smaller than, or equal to, mg, so that the ring cannot raise. Fig. S-6.29 shows the plots of the second of (S-6.316) for three different values of α. We see that (S-6.316) has two real roots, corresponding to the maximum and minimum of U_{pot}, for $\alpha > \alpha_{\text{min}}$, maximum and minimum coalesce for $\alpha = \alpha_{\text{min}}$, and there are no real roots for $\alpha < \alpha_{\text{min}}$. It is possible to evaluate α_{min} by considering that, when maximum and minimum coalesce, we have an inflection point, therefore also the derivative of (S-6.316) must be zero. Thus we must solve the system

$$\begin{cases} (1 + \zeta^2)^4 = 6\alpha_{\text{min}}^2 \zeta \\ 8(1 + \zeta^2)^3 \zeta = 6\alpha_{\text{min}}^2 \end{cases} \quad \text{(S-6.317)}$$

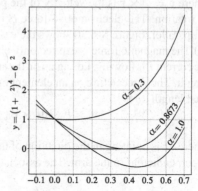

for α_{min} and ζ (the point of inflection). We multiply the second of (S-6.317) by ζ, so that the right-hand sides are equal, and equate the left-hand sides, obtaining

$$(1 + \zeta^2)^4 = 8(1 + \zeta^2)^3 \zeta^2, \quad \text{(S-6.318)}$$

or

Fig. S-6.29

$$1 + \zeta^2 = 8\zeta^2, \quad \text{hence} \quad \zeta = \frac{1}{\sqrt{7}},$$
$$\text{(S-6.319)}$$

for the coordinate of the inflection point. We substitute this value into, for instance, the first of (S-6.317) obtaining

$$\alpha_{\text{min}} = \sqrt{\frac{(1 + \zeta^2)^4}{6\zeta}} = \frac{(8/7)^2}{\sqrt{6/\sqrt{7}}} = 0.867327. \quad \text{(S-6.320)}$$

Thus we must assume $\alpha > \alpha_{\text{min}}$, and the smallest root of (S-6.316), corresponding to the maximum of U_{pot}, is the lower limit for $\zeta_0 = z_0/R$. We see from the plot that, for α of the order of 1 or higher, we can reasonably assume that $\zeta_0 \ll 1$, obtaining

$$\zeta_0 > \frac{1}{6\alpha^2} \quad \text{(S-6.321)}$$

as a reasonable lower limit for ζ_0.

(c) The ring oscillates between $\zeta = \zeta_0$ and $\zeta = \zeta_{\text{max}}$. The profile of the potential well where the motion occurs is given by (S-6.312), plotted in Fig. S-6.28, thus the oscillations are not harmonic.

S-6.23   Electromagnetic Brake with Energy Recovery

(a) Only the arm in contact with the external circuit contributes to the electromotive force $\mathcal{E}$, see Fig. S-6.30. We have

$$\mathcal{E} = b_{\mathrm{m}} \oint (\boldsymbol{v} \times \mathbf{B}_0) \cdot \mathrm{d}\boldsymbol{\ell} = b_{\mathrm{m}} \int_0^a \omega r B_0 \, \mathrm{d}r = b_{\mathrm{m}} \frac{B_0 \, a^2 \omega}{2} \ , \qquad \text{(S-6.322)}$$

where $\boldsymbol{v} = \boldsymbol{\omega} \times \mathbf{r}$ is the velocity of the point of the arm at distance $\mathbf{r}$ from the center of rotation. The electromotive force $\mathcal{E}$ is positive if $\boldsymbol{\omega}$ is parallel to $\mathbf{B}_0$, negative if antiparallel. Figure S-6.30 shows the case of $\boldsymbol{\omega}$ and $\mathbf{B}_0$ parallel.

(b) The magnetic force is exerted only on the rotor arm where the current I is flowing, i.e., again on the arm in contact with the circuit. The force acting on an element of the arm of infinitesimal length $\mathrm{d}\ell$ is

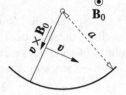

$$\mathrm{d}\mathbf{f} = b_{\mathrm{m}} \, I \mathrm{d}\boldsymbol{\ell} \times \mathbf{B}_0 \ , \qquad \text{(S-6.323)}$$

where $\mathrm{d}\boldsymbol{\ell}$ is pointing in the direction of the current I. Thus, $\mathrm{d}\mathbf{f}$ is perpendicular to the arm. The associated torque $\boldsymbol{\tau}$ is

Fig. S-6.30

$$\boldsymbol{\tau} = b_{\mathrm{m}} \int \mathbf{r} \times (I \mathrm{d}\boldsymbol{\ell} \times \mathbf{B}_0) = -\hat{\mathbf{z}} \, b_{\mathrm{m}} \int_0^a r I B_0 \, \mathrm{d}r = -\hat{\mathbf{z}} \, b_{\mathrm{m}} \frac{B_0 \, a^2 I}{2} \ , \qquad \text{(S-6.324)}$$

where $\hat{\mathbf{z}}$ is the unit vector perpendicular to the plane of the rotor and parallel to $\mathbf{B}_0$. The torque is positive and induces a counterclockwise rotation if I is negative (i.e., if it flows in clockwise direction), and vice versa.

(c) The current flowing in the circuit is

$$I = \frac{V_0 + \mathcal{E}}{R} = \frac{2V_0 + b_{\mathrm{m}} B_0 a^2 \omega}{2R} \ , \qquad \text{(S-6.325)}$$

where V_0 is positive if the voltage source is oriented as in Fig. 6.21, i.e., if it would generate a current flowing counterclockwise.

The equation of motion for the rotor is

$$I \frac{\mathrm{d}\omega}{\mathrm{d}t} = \tau \ , \qquad \text{(S-6.326)}$$

where the torque τ is given by (S-6.324) and I by (S-6.325). We thus have

$$I \frac{\mathrm{d}\omega}{\mathrm{d}t} = -\frac{2 b_{\mathrm{m}} B_0 a^2 V_0 + b_{\mathrm{m}}^2 B_0^2 a^4 \omega}{4R} = -\frac{b_{\mathrm{m}}^2 B_0^2 a^4}{4R} \left(\omega + \frac{2V_0}{b_{\mathrm{m}} B_0 a^2} \right) \ , \qquad \text{(S-6.327)}$$

which leads to

$$\frac{\mathrm{d}\omega}{\mathrm{d}t} + \frac{b_{\mathrm{m}}^2 B_0^2 a^4}{4 I R} \, \omega = -\frac{a^2 V_0 b_{\mathrm{m}} B_0}{2 I R} \ . \qquad \text{(S-6.328)}$$

The kinetic energy of the rotor is $I\omega^2/2$, and differentiating it with respect to time we obtain

$$\frac{d}{dt}\left(\frac{I\omega^2}{2}\right) = I\omega\frac{d\omega}{dt} = -\omega\frac{2b_m B_0 a^2 V_0 + b_m^2 B_0^2 a^4 \omega}{4R}$$

$$= -\frac{b_m B_0 a^2 \omega}{2}\frac{b_m B_0 a^2 \omega + 2V_0}{2R} = -\mathcal{E}I , \tag{S-6.329}$$

where we have used (S-6.327), (S-6.322) and (S-6.325). Using the first of (S-6.325) we have $\mathcal{E} = IR - V_0$ which, inserted into the last of (S-6.329) leads to

$$\frac{d}{dt}\left(\frac{I\omega^2}{2}\right) = -RI^2 + V_0 I . \tag{S-6.330}$$

Since $R > 0$, term $(-RI^2)$ is always negative and corresponds to the power dissipated in the resistance by Joule heating. On the other hand, we are free to choose the sign of V_0 so that the battery either supplies power at the expense of its internal energy, or absorbs power from the kinetic energy of the rotor, thus increasing its internal energy. The first case occurs if $V_0 I > 0$: the battery provides current flowing in the direction needed to stop the rotor at the expense of its internal energy. The second case occurs if $V_0 I < 0$: the battery power has the same sign as the dissipated power, which means that the kinetic energy of the rotor goes partly into Joule heating and partly to increase the internal energy of the battery.

(d) The solution of (S-6.327) is

$$\omega(t) = \omega_0 e^{-t/T} + \omega_\infty \left(1 - e^{-t/T}\right) , \tag{S-6.331}$$

where

$$T = \frac{4IR}{a^4 b_m^2 B_0^2} , \qquad \omega_\infty = -\frac{2V_0}{a^2 b_m B_0} . \tag{S-6.332}$$

If the battery is removed ($V_0 = 0$) and the rotor has an initial angular velocity ω_0, its angular velocity decreases asymptotically ($\omega_\infty = 0$). In the presence of the battery $\omega(t)$ reaches asymptotically the value ω_∞, independently of its initial value ω_0. This corresponds to a stationary state where $\mathcal{E} + V_0 = 0$ and no current flows in the circuit. Note that the condition $|V_0| < \mathcal{E}(\omega_0)$ ensures that $|\omega_\infty| < \omega_0$, so that the final angular velocity is smaller than its initial value.

For $V_0 > 0$, as in Fig. 6.21, the battery voltage has the same sign as $\mathcal{E}$ and contributes to brake the rotor at the expense of its internal energy ($V_0 I > 0$). The angular velocity must change sign in order to reach the asymptotic state in which $\mathcal{E} + V_0 = 0$, thus there is an intermediate time t_0 at which $\omega(t_0) = 0$, with $t_0 = T \ln(1 - \omega_0/\omega_\infty)$. Opening the circuit at t_0 would make the rotor stop.

If $V_0 < 0$, the sign of the battery voltage is opposite to the sign of $\mathcal{E}$, the battery absorbs energy, but it prevents the rotor from reaching zero angular velocity ($\omega_\infty > 0$). Thus, the electromagnetic brake cannot both recover energy and completely stop the rotor. (Note that the case $\mathcal{E}(\omega_0) + V_0 < 0$ would not be meaningful for a brake, because as soon as the circuit is closed the battery would accelerate the wheel!).

(e) The battery power (either supplied or absorbed) is

$$
\begin{aligned}
V_0 I &= \frac{V_0^2 + V_0 \mathcal{E}}{R} = \frac{2V_0^2 + b_\mathrm{m} B_0 a^2 V_0 \omega(t)}{2R} \\
&= \frac{2V_0^2 + b_\mathrm{m} B_0 a^2 V_0 \left[\omega_0 e^{-t/T} + \omega_\infty \left(1 - e^{-t/T} \right) \right]}{2R} \\
&= \frac{2V_0^2 - 2V_0^2 + b_\mathrm{m} B_0 a^2 V_0 \left[\omega_0 e^{-t/T} + (2V_0)/(a^2 b_\mathrm{m} B_0) e^{-t/T} \right]}{2R} \\
&= \frac{b_\mathrm{m} B_0 a^2 V_0 \omega_0 + 2V_0^2}{2R} e^{-t/T} \\
&= \frac{b_\mathrm{m} B_0 a^2 V_0}{2R} (\omega_0 - \omega_\infty) e^{-t/T} .
\end{aligned}
\tag{S-6.333}
$$

The last equality shows again that if $V_0 < 0$ then $V_0 I < 0$ and the battery absorbs energy for a total amount U_abs given by the integral

$$
\begin{aligned}
U_\mathrm{abs} &= \int_0^\infty V_0 I \, \mathrm{d}t = T \frac{b_\mathrm{m} B_0 a^2 V_0 \omega_0 + 2V_0^2}{2R} \\
&= \frac{2I}{a^4 b_\mathrm{m}^2 B_0^2} \left(b_\mathrm{m} B_0 a^2 V_0 \omega_0 + 2V_0^2 \right) .
\end{aligned}
\tag{S-6.334}
$$

The efficiency, i.e., the fraction of absorbed energy is

$$
\mathcal{R} = \frac{U_\mathrm{abs}}{I \omega_0^2/2} = -\left(\hat{V} + 2\hat{V}^2 \right) = -2\hat{V} \left(1 + \hat{V} \right) ,
\tag{S-6.335}
$$

where $\hat{V} = V_0/\mathcal{E}(\omega_0) = 2V_0/ \left(b_\mathrm{m} B_0 a^2 \omega_0 \right)$. Maximum efficiency of $\mathcal{R} = 1/2 = 50\%$ is obtained for $\hat{V} = -1/2$.

S-6.24 A High Frequency Capacitor

(a) In the following we express all fields in complex notation, omitting the common time dependence factor $e^{-i\omega t}$.

To the lowest order, the total charges $\pm Q_0$ on the two plates are obtained from by the continuity equation

$$
\frac{\mathrm{d}Q_0}{\mathrm{d}t} = -I ,
\tag{S-6.336}
$$

which, using $\mathrm{d}Q_0/\mathrm{d}t = -i\omega Q_0$, yields

$$Q_0 = i \frac{I_0}{\omega} . \tag{S-6.337}$$

Since the surface density is assumed to be uniform we have

$$\sigma_0 = \frac{Q_0}{\pi a^2} = i \frac{I_0}{\pi a^2 \omega} , \tag{S-6.338}$$

and the electric field between the plates is

$$E_0 = 4\pi k_e \sigma_0 = i \frac{4 k_e I_0}{a^2 \omega} . \tag{S-6.339}$$

(b) For symmetry reasons the magnetic field is azimuthal, $\mathbf{B} = B(r, t) \hat{\boldsymbol{\phi}}$. The lowest order term is obtained from the equation $c^s (\nabla \times \mathbf{B}_1)_z = \partial_t E_0$ where $s = 1$ and $s = 2$ in c.g.s. and SI units, respectively. Using the expression of the curl in cylindrical coordinates (Table A.1), we have

$$\frac{1}{r} \partial_r (r B_1) = -\frac{i\omega}{c^s} E_0 . \tag{S-6.340}$$

Integrating this equation with respect to r leads to

$$r B_1(r) = -\frac{i\omega}{c^s} E_0 \int_0^r r' dr' = -\frac{i\omega r^2}{2c^s} E_0 , \tag{S-6.341}$$

from which we finally obtain

$$B_1 = \frac{i\omega r}{2a^2 c^s} E_0 . \tag{S-6.342}$$

Of course (S-6.342) can be also quickly obtained using Stokes' theorem to obtain, after a trivial integration along a $\mathbf{B}$-field line (i.e., a circle of radius r), $2\pi r B_1 c^s = -i\omega \pi r^2 E_0$.

(c) The lowest order correction to the electric field is obtained from the equation $c^{2-s} (\nabla \times \mathbf{E}_1)_\phi = -\partial_t \mathbf{B}_1$. Assuming $\mathbf{E}_1 = E_1(r) \hat{\mathbf{z}}$ and using Table A.1 again we obtain $-c^{2-s} \partial_r E_1 = i\omega B_1$. Inserting (S-6.342) and integrating with respect to r we find

$$E_1(r) = E_1(0) - \frac{\omega^2 r^2}{4c^2} E_0 . \tag{S-6.343}$$

This same expression can be obtained using Stokes' theorem applied to a rectangle parallel to the z-axis and extending from $r = 0$ to $r = a$.

The correction E_1 implies a correction $\sigma_1 = E_1/4\pi k_e$ to the surface charge density and, in turn, to the total charge:

$$Q_1 = \int_0^a \sigma_1(r)\, 2\pi r\, dr \ . \tag{S-6.344}$$

The condition $dQ/dt = -I$, with $Q = Q_0 + Q_1$, imposes $Q_1 = 0$ because of (S-6.336). Thus, since $E_1 = 4\pi k_e \sigma_1$, we have

$$0 = \int_0^a \left[E_1(0) - \frac{\omega^2 r^2}{4c^2} E_0 \right] 2\pi r\, dr \ , \tag{S-6.345}$$

which yields

$$E_1(0) = \frac{\omega^2 a^2}{8c^2} E_0 \ . \tag{S-6.346}$$

Thus, up to second order in $(\omega a/c)^2$ we have

$$E(r) \simeq E_0 \left(1 + \frac{1}{2}\frac{\omega^2 a^2}{4c^2} \right) - \frac{\omega^2 r^2}{4c^2} E_0 \ . \tag{S-6.347}$$

We thus see that imposing the total current through the capacitor to be I_0 leads to a "renormalization" of the field on the axis at each order of iteration. On the other hand considering the capacitor in series with an ideal current source may appear less problematic than considering an ideal voltage source since the voltage difference between the two plates cannot be defined because the field is not uniform in the capacitor. That said, in the following we assume to fix the voltage difference on the axis ($r = 0$), i.e. between the two points in which the plates are in contact with the external circuit; this is also the choice in Feynman's book [1]. See also Jackson's book ([2], Sect. 6.9) for a discussion about input currents and voltage in passive linear circuit elements.

(d) Assuming that $E(r = 0) = E_0$ imposes $E_1(0) = 0$, so that the solution for the electric field up to the first order of approximation is

$$E(r) \simeq E_0 \left(1 - \frac{\omega^2 r^2}{4c^2} \right) \ . \tag{S-6.348}$$

This corresponds to the first and second terms of the series (6.8). Now it is sufficient to show that, if we assume that the nth term of the electric field E_n is

$$E_n = E_0 \frac{(-1)^n}{(n!)^2} \left(\frac{\omega r}{2c} \right)^{2n} \ , \tag{S-6.349}$$

then the following $(n + 1)$th term in the SVCA expansion has the same expression but for the replacement $n \to n + 1$.

The $(n + 1)$th term for the **B**-field expansion is obtained from $c^s (\nabla \times \mathbf{B}_{n+1})_z = \partial_t E_n$ which yields

$$\frac{1}{r}\partial_r(r B_{n+1}) = -\frac{i\omega}{c^s} E_n = -E_0 \frac{(-1)^n}{(n!)^2} \frac{i\omega^{2n+1}}{2^{2n} c^{2n+s}} r^{2n} . \tag{S-6.350}$$

Multyplying both sides by r and integrating yields

$$r B_{n+1} = -E_0 \frac{(-1)^n}{(n!)^2} \left(\frac{i\omega^{2n+1}}{2^{2n} c^{2n+s}} \frac{r^{2n+2}}{2n + 2} \right) , \tag{S-6.351}$$

or equivalently

$$B_{n+1} = -E_0 \frac{(-1)^n}{(n!)^2(n+1)} \frac{i\omega^{2n+1} r^{2n+1}}{2^{2n+1} c^{2n+s}} , \tag{S-6.352}$$

Now, the $(n + 1)$th **E**-field terms can be found from $c^{2-s}(\nabla \times \mathbf{E}_{n+1})_\phi = -\partial_t B_{n+1}$, i.e.

$$\partial_r E_{n+1} = -\frac{i\omega B_{n+1}}{c^{2-s}} = E_0 \frac{(-1)^{n+1}}{(n!)^2(n+1)} \frac{\omega^{2n+2} r^{2n+1}}{2^{2n+1} c^{2n+2}} . \tag{S-6.353}$$

Integrating again with the condition $E_{n+1}(0) = 0$ yields

$$E_{n+1} = E_0 \frac{(-1)^{n+1}}{(n!)^2(n+1)} \frac{\omega^{2n+2}}{2^{2n+1} c^{2n+2}} \frac{r^{2n+2}}{2n + 2} \tag{S-6.354}$$

$$= E_0 \frac{(-1)^{n+1}}{(n!)^2(n+1)^2} \frac{\omega^{2n+2} r^{2n+2}}{2^{2n+2} c^{2n+2}} = E_0 \frac{(-1)^{n+1}}{((n+1)!)^2} \left(\frac{\omega r}{2c}\right)^{2(n+1)} , \tag{S-6.355}$$

which corresponds to the $(n + 1)$th term in (6.8). This proves that the series represents the full SVCA solution, which converges to an exact solution of Maxwell's equations described by the Bessel function $J_0(\omega r/2c)$ (see Fig. S-13.16 for a plot of $J_0(x)$, and the discussion by Feynman for the relation to a cylindrical resonant cavity.)

References

1. R. P. Feynman, R. B. Leighton, M. Sands, *The Feynman Lectures on Physics*, Vol. II, Sect. 23–2 (Addison-Wesley Publishing Company, Reading, MA, 2006)
2. J. D. Jackson, *Classical Electrodyamics*, 3rd edn. (Wiley, New York, 1999)

Chapter S-7
Solutions for Chapter 7

S-7.1 Coupled RLC Oscillators (1)

(a) Assuming the two currents I_1 and I_2 to flow clockwise, and applying Kirchoff's mesh rule to the two loops of the circuit, we have

$$L\frac{dI_1}{dt} + \frac{Q_1}{C_1} + \frac{Q_0}{C_0} = 0, \qquad L\frac{dI_2}{dt} + \frac{Q_2}{C_1} - \frac{Q_0}{C_0} = 0, \qquad \text{(S-7.1)}$$

where Q_1 is the charge of the left capacitor, Q_2 the charge of the right capacitor, and Q_0 the charge of capacitor C_0. Charge conservation in the two loops implies

$$\frac{dQ_1}{dt} = I_1, \qquad \frac{dQ_2}{dt} = I_2, \qquad \text{(S-7.2)}$$

while Kirchhoff's junction rule, applied either to junction A or to junction B, leads to

$$\frac{dQ_0}{dt} = I_1 - I_2. \qquad \text{(S-7.3)}$$

Differentiating (S-7.1), substituting (S-7.2) and (S-7.3), and dividing by L, we obtain

$$\frac{d^2 I_1}{dt^2} = -\frac{1}{LC_1}I_1 - \frac{1}{LC_0}(I_1 - I_2), \quad \text{or} \quad \frac{d^2 I_1}{dt^2} = -\omega_1^2 I_1 - \omega_0^2(I_1 - I_2)$$

$$\frac{d^2 I_2}{dt^2} = -\frac{1}{LC_1}I_2 - \frac{1}{LC_0}(I_2 - I_1), \quad \text{or} \quad \frac{d^2 I_2}{dt^2} = -\omega_1^2 I_2 - \omega_0^2(I_2 - I_1), \quad \text{(S-7.4)}$$

where we have introduced the quantities $\omega_0 = 1/\sqrt{LC_0}$ and $\omega_1 = 1/\sqrt{LC_1}$. By substituting $I_1 = A_1 e^{-i\omega t}$ and $I_2 = A_2 e^{-i\omega t}$ from (7.2) into (S-7.4), we obtain

© The Editor(s) (if applicable) and The Author(s), under exclusive license to Springer Nature Switzerland AG 2023
A. Macchi et al., *Problems in Classical Electromagnetism*,
https://doi.org/10.1007/978-3-031-22235-1

$$(\omega_1^2 + \omega_0^2 - \omega^2) A_1 - \omega_0^2 A_2 = 0$$
$$-\omega_0^2 A_1 + (\omega_1^2 + \omega_0^2 - \omega^2) A_2 = 0 . \tag{S-7.5}$$

Non-trivial solutions for this system exist only if the determinant

$$D = D(\omega) = (\omega_1^2 + \omega_0^2 - \omega^2)^2 - \omega_0^4 = (\omega_1^2 - \omega^2)(\omega_1^2 + 2\omega_0^2 - \omega^2) \tag{S-7.6}$$

equals zero. Thus, the frequencies of the normal modes of the circuit are the roots of the equation $D(\omega) = 0$, i.e.,

$$\omega = \omega_1 \equiv \Omega_+ , \qquad \omega = \sqrt{\omega_1^2 + 2\omega_0^2} \equiv \Omega_- . \tag{S-7.7}$$

Substituting these values for ω into (S-7.5) we obtain that $A_1 = A_2$, i.e., $I_1(t) = I_2(t)$, for the mode of frequency Ω_+, and that $A_1 = -A_2$, i.e., $I_1(t) = -I_2(t)$, for the mode of frequency Ω_-.

The normal modes of this simple case, with only two degrees of freedom, can also be evaluated, more simply, by taking the sum and the difference of (S-7.4), obtaining the harmonic oscillator equations

$$\frac{d^2 I_\pm}{dt^2} = -\Omega_\pm^2 I_\pm . \tag{S-7.8}$$

for the variables $I_\pm \equiv I_1 \pm I_2$, then the currents in the two meshes are $I_1 = (I_+ + I_-)/2$ and $I_2 = (I_+ - I_-)/2$, respectively.

When the circuit is in the mode of frequency Ω_+, no current flows through the AB branch (capacitor C_0), where the two currents cancel out because $I_1 = I_2$. Frequency Ω_+ is simply the resonant frequency of a single-loop LC circuit of inductance L and capacitance C_1, i.e., the frequency at which the impedance of the loop is zero

$$Z_{LC}(\omega) = Z_L(\omega) + Z_{C_1}(\omega) = -i\omega L - \frac{1}{i\omega C_1} = 0 . \tag{S-7.9}$$

Since $Z_{LC}(\Omega_+) = 0$, the current flows "freely" through each loop.

For the mode of frequency Ω_-, we have $I_1 = -I_2$, and a current $2I_1$ flows through the AB branch. The effective impedance of the circuit is the series of $Z_{C_0} = (i\omega C_0)^{-1}$ with the parallel of the two impedances Z_{LC},

$$Z = Z_{C_0} + \frac{Z_{LC} Z_{LC}}{Z_{LC} + Z_{LC}} = Z_0 + \frac{Z_{LC}}{2} = -\frac{1}{i\omega C_0} - \frac{1}{2}\left(i\omega L + \frac{1}{i\omega C_1}\right) , \tag{S-7.10}$$

which vanishes if

$$\omega^2 = \frac{1}{L}\left(\frac{2}{C_1} + \frac{1}{C_0}\right) = \Omega_-^2 . \tag{S-7.11}$$

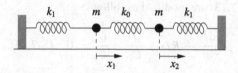

Fig. S-7.1

The circuit is equivalent to the two coupled identical harmonic oscillators of Fig. S-7.1. Each oscillator comprises a mass m, connected to a fixed wall by a spring of Hooke's constant $k_1 = m\omega_1^2$. The two masses are connected to each other by a third spring of Hooke's constant $k_0 = m\omega_0^2$. We assume that all springs have their respective rest lengths when the two masses are at their equilibrium positions. The equations of motions for the two masses are

$$m\frac{d^2x_1}{dt^2} = -k_1x_1 - k_0(x_1 - x_2), \quad \text{or} \quad \frac{d^2x_1}{dt^2} = -\omega_1^2x_1 - \omega_0^2(x_1 - x_2)$$

$$m\frac{d^2x_2}{dt^2} = -k_1x_2 - k_0(x_2 - x_1), \quad \text{or} \quad \frac{d^2x_2}{dt^2} = -\omega_1^2x_2 - \omega_0^2(x_2 - x_1), \quad \text{(S-7.12)}$$

where x_1 and x_2 are the displacements of the two masses from their equilibrium positions. Equations (S-7.12) for x_1 and x_2 are formally equivalent to Eqs. (S-7.4) for I_1 and I_2, and thus have the same solutions. For the mode at frequency Ω_+, the two masses oscillate in phase ($x_1 = x_2$), central spring (k_0) has always its rest length, and does not exert forces on the two masses. Thus, frequency Ω_+ is the characteristic frequency each single harmonic oscillator. For the mode at frequency Ω_-, we have $x_1 = -x_2$ and the two masses oscillate with opposite phases.

(b) The presence of a nonzero resistance R in series with each inductor changes Eq. (S-7.1) into

$$L\frac{dI_1}{dt} + RI_1 + \frac{Q_1}{C_1} + \frac{Q_0}{C_0} = 0, \quad L\frac{dI_2}{dt} + RI_2 + \frac{Q_2}{C_1} - \frac{Q_0}{C_0} = 0, \text{(S-7.13)}$$

By differentiating the equations and proceeding as for (S-7.8) we obtain

$$\frac{d^2I_\pm}{dt^2} = -\Omega_\pm^2 I_\pm - \gamma\frac{dI_\pm}{dt}, \qquad \text{(S-7.14)}$$

with $\gamma = R/L$. These are the equations of two damped oscillators. The amplitudes of the normal modes vary in time as $\exp(-i\Omega_\pm t - \gamma t)$, decaying with a time constant $\tau = \gamma^{-1}$. The damping rate of the normal modes can also be found by looking for solutions in the form $I_{1,2} = A_{1,2}\,e^{-i\omega t}$, but allowing $A_{1,2}$ and ω to have imaginary parts. For the equivalent mechanical system, the same equations are obtained by inserting frictional forces $f_i = -m\gamma dx_i/dt$ in the equations of motion (S-7.12).

(c) Inserting the voltage source, Eqs. (S-7.13) are modified as follows:

$$L\frac{dI_1}{dt} + RI_1 + \frac{Q_1}{C_1} + \frac{Q_0}{C_0} = V_0 e^{-i\omega t}, \quad L\frac{dI_2}{dt} + RI_2 + \frac{Q_2}{C_1} - \frac{Q_0}{C_0} = 0, \quad \text{(S-7.15)}$$

and, by proceeding as for (S-7.8) and (S-7.14), we have

$$\frac{d^2 I_\pm}{dt^2} = -\Omega_\pm^2 I_\pm - \gamma\frac{dI_\pm}{dt} - \frac{i\omega V_0}{L}e^{-i\omega t}, \quad\quad\quad \text{(S-7.16)}$$

which are the equations of two forced oscillators with a driving term $-(i\omega V_0/L)e^{-i\omega t}$. Resonances are observed when $\omega = \Omega_+$ and for $\omega = \Omega_+$, i.e., when the driving frequency equals one of the frequencies of the normal modes.

S-7.2   Coupled *RLC* Oscillators (2)

(a) Proceeding as in Solution S-7.1, we assume I_1 and I_2 to flow clockwise. Applying Kirchoff's mesh rule to both meshes of the circuit we obtain

$$L\frac{dI_1}{dt} + \frac{Q_1}{C} + L_0\left(\frac{dI_1}{dt} - \frac{dI_2}{dt}\right) = 0, \quad L\frac{dI_2}{dt} + \frac{Q_2}{C} - L_0\left(\frac{dI_1}{dt} - \frac{dI_2}{dt}\right) = 0,$$
$$\text{(S-7.17)}$$

again with $I_1 = dQ_1/dt$ and $I_2 = dQ_2/dt$. Differentiating (S-7.17) with respect to t we obtain

$$(L + L_0)\frac{d^2 I_1}{dt^2} + \frac{I_1}{C} - L_0\frac{d^2 I_2}{dt^2} = 0$$

$$(L + L_0)\frac{d^2 I_2}{dt^2} + \frac{I_2}{C} - L_0\frac{d^2 I_1}{dt^2} = 0. \quad\quad\quad \text{(S-7.18)}$$

The sum and difference of the two equations of (S-7.18) give the following equations for the new variables $I_\pm \equiv I_1 \pm I_2$

$$\frac{d^2 I_+}{dt^2} = -\frac{I_+}{LC} \equiv -\Omega_+^2 I_+, \quad \frac{d^2 I_-}{dt^2} = -\frac{I_-}{(L + 2L_0)C} \equiv -\Omega_-^2 I_-, \quad \text{(S-7.19)}$$

which show that $I_\pm$ are the normal oscillation modes of the circuit, and $\Omega_\pm$ the corresponding frequencies.

(b) Inserting $R \neq 0$, Eqs. (S-7.17) turn into

$$L\frac{dI_1}{dt} + RI_1 + \frac{Q_1}{C} + L_0\left(\frac{dI_1}{dt} - \frac{dI_2}{dt}\right) + R(I_1 - I_2) = 0$$

$$L\frac{dI_2}{dt} + RI_2 + \frac{Q_2}{C} - L_0\left(\frac{dI_1}{dt} - \frac{dI_2}{dt}\right) - R(I_1 - I_2) = 0. \qquad \text{(S-7.20)}$$

Performing again the sum and difference of the two equations we obtain

$$\frac{d^2 I_+}{dt^2} = -\gamma_+\frac{dI_+}{dt} - \Omega_+^2 I_+ , \qquad \frac{d^2 I_-}{dt^2} = -\gamma_-\frac{dI_-}{dt} - \Omega_-^2 I_- , \qquad \text{(S-7.21)}$$

with $\gamma_+ = R/L$, and $\gamma_- = 3R/(L + 2L_0)$. These are the equations for two damped oscillators, with different damping rates $\gamma_\pm$.

S-7.3 Coupled RLC Oscillators (3)

(a) Let us denote by Q_1 and Q_2 the charges of the capacitors on the AB and on the DE branches, respectively. According to Kirchhoff's mesh rule we have, for the three meshes of the circuit,

$$L\frac{dI_1}{dt} = -\frac{Q_1}{C} , \qquad L\frac{dI_2}{dt} = \frac{Q_1}{C} - \frac{Q_2}{C} , \qquad L\frac{dI_3}{dt} = \frac{Q_2}{C} , \qquad \text{(S-7.22)}$$

and, according to Kirchoff's junction rule applied to the A and D junctions,

$$\frac{dQ_1}{dt} = I_1 - I_2 , \qquad \frac{dQ_2}{dt} = I_2 - I_3 . \qquad \text{(S-7.23)}$$

Differentiating Eqs. (S-7.22) with respect to t, and substituting dQ_1/dt and dQ_2/dt from (S-7.23), we obtain

$$\frac{d^2 I_1}{dt^2} = \frac{1}{LC}(-I_1 + I_2) ,$$

$$\frac{d^2 I_2}{dt^2} = \frac{1}{LC}(I_1 - 2I_2 + I_3) , \qquad \text{(S-7.24)}$$

$$\frac{d^2 I_3}{dt^2} = \frac{1}{LC}(I_2 - I_3) .$$

Mathematically, the circuit is equivalent to a mechanical system comprising three identical masses m, coupled by two identical springs of Hooke's constant k, as shown in Fig. S-7.2. If we denote by x_1, x_2,

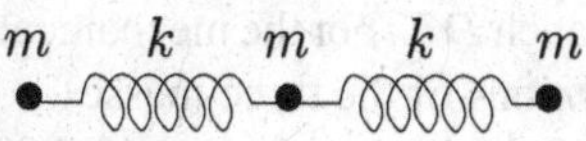

Fig. S-7.2

and x_3 the displacement of each mass from its rest position, the equations of motion for the three masses are

$$\frac{d^2 x_1}{dt^2} = \frac{k}{m} (x_2 - x_1) ,$$

$$\frac{d^2 x_2}{dt^2} = -\frac{k}{m} (x_2 - x_1) + \frac{k}{m} (x_2 - x_3) , \qquad (S\text{-}7.25)$$

$$\frac{d^2 x_3}{dt^2} = -\frac{k}{m} (x_2 - x_3) ,$$

which are identical to (S-7.25), after substituting $I_j \rightarrow x_j$, with $j = 1, 2, 3$, and $1/(LC) \rightarrow k/m$.

(b) The frequencies of the normal modes can be found by looking for solutions of (S-7.25) in the form

$$I_j(t) = A_j \, e^{-i\omega t} . \qquad (S\text{-}7.26)$$

After substituting (S-7.26) and $\omega_0^2 = 1/(LC)$ into (S-7.25), and dividing by the common exponential factor, we obtain the system of linear equations in matrix form

$$\begin{pmatrix} (\omega_0^2 - \omega^2) & -\omega_0^2 & 0 \\ -\omega_0^2 & (2\omega_0^2 - \omega^2) & -\omega_0^2 \\ 0 & -\omega_0^2 & (\omega_0^2 - \omega^2) \end{pmatrix} \begin{pmatrix} A_1 \\ A_2 \\ A_3 \end{pmatrix} = 0 \qquad (S\text{-}7.27)$$

which has non-trivial solutions only if the determinant of the matrix is zero, i.e., if

$$\left(\omega_0^2 - \omega^2\right) \left[\left(2\omega_0^2 - \omega^2\right) \left(\omega_0^2 - \omega^2\right) - \omega_0^4 \right] - \omega_0^4 \left(\omega_0^2 - \omega^2\right) = 0 . \qquad (S\text{-}7.28)$$

Equation (S-7.28) is a cubic equation in ω^2, in the following we shall consider only the corresponding nonnegative values of ω. A first solution is $\omega = \omega_0 = \Omega_1$. If we substitute $\omega = \Omega_1$ into (S-7.27) we obtain $A_1 = -A_3$, and $A_2 = 0$, corresponding to zero current in the central mesh, and I_1 and I_2 oscillating with opposite phases. For the mechanical system of Fig. S-7.2, this solution corresponds to the central mass at rest, while the left and right masses oscillate with opposite phases.

Dividing (S-7.28) by $(\omega_0^2 - \omega^2)$ we obtain the equation

$$- 3\omega_0^2 \omega^2 + \omega^4 = 0 , \qquad (S\text{-}7.29)$$

which has the two solutions $\omega = \sqrt{3}\omega_0 = \Omega_2$ and $\omega = 0 = \Omega_3$. The mode of zero frequency (Ω_3) corresponds to a DC current $I = I_1 = I_2 = I_3$ flowing freely through the inductors, while I_1 and I_2 cancel out in branch AB, and I_2 and I_3 cancel out in branch DE. For the mechanical system, this solution correspond to a pure translational motions of the three masses.

Substituting Ω_2 into (S-7.27) we obtain

$$A_2 = -2A_1, \quad A_3 = A_1, \tag{S-7.30}$$

i.e., I_1 and I_3 have the same amplitude and oscillate in phase, while I_2 oscillates with double amplitude and opposite phase. The two external masses of Fig. S-7.2 oscillate in phase, at constant distance from each other, while the central mass oscillates with opposite phase and double amplitude, so that the center of mass is at rest.

The three quantities

$$\mathcal{J}_0 = I_1 + I_2 + I_3, \quad \mathcal{J}_1 = I_1 - I_3, \quad \mathcal{J}_2 = I_1 - 2I_2 + I_3, \tag{S-7.31}$$

corresponding to the three normal modes of the circuits, oscillate at the frequencies $\Omega_0 = 0$, Ω_1, and Ω_2, respectively.

(c) Taking the finite resistances into account, (S-7.22) become

$$\frac{dI_1}{dt} + RI_1 = -\frac{Q_1}{C}, \quad \frac{dI_2}{dt} + RI_2 = \frac{Q_1}{C} - \frac{Q_2}{C}, \quad \frac{dI_3}{dt} + RI_3 = \frac{Q_2}{C}, \tag{S-7.32}$$

which give for the normal modes

$$\frac{d^2\mathcal{J}_0}{dt^2} + \frac{R}{L}\frac{d\mathcal{J}_0}{dt} = 0,$$

$$\frac{d^2\mathcal{J}_1}{dt^2} + \frac{R}{L}\frac{d\mathcal{J}_1}{dt} + \Omega_1^2 \mathcal{J}_1 = 0, \tag{S-7.33}$$

$$\frac{d^2\mathcal{J}_2}{dt^2} + \frac{R}{L}\frac{d\mathcal{J}_2}{dt} + \Omega_2^2 \mathcal{J}_2 = 0.$$

The solution for $\mathcal{J}_0$ describes a non-oscillating, exponentially decreasing current $\mathcal{J}_0 = C_0 \, e^{-\gamma t}$, with decay rate $\gamma = R/L$. The last two equations describe damped oscillating currents $\mathcal{J}_{1,2} = C_{1,2} \exp(-i\tilde{\Omega}_{1,2}t - \gamma t)$, with

$$\tilde{\Omega}_{1,2} = \sqrt{\omega_{1,2}^2 - \frac{\gamma^2}{4}}, \tag{S-7.34}$$

where we have assumed $\Omega_{1,2} > \gamma/2$.

S-7.4 The LC Ladder Network

(a) Let Q_n be the charge on the nth capacitor. Kirchoff's junction rule at junction D of Fig. 7.4 implies

$$\frac{dQ_n}{dt} = I_{n-1} - I_n, \tag{S-7.35}$$

while Kirchhoff's mesh rule applied to mesh $DEFG$ implies

$$\frac{Q_n}{C} - \frac{Q_{n+1}}{C} = L\frac{dI_n}{dt} . \tag{S-7.36}$$

Now we differentiate (S-7.36) with respect to time, and insert (S-7.35) for the derivatives of Q_n, obtaining

$$\frac{d^2 I_n}{dt^2} = \omega_0^2 \left(I_{n-1} - 2I_n + I_{n+1} \right), \quad \text{where} \quad \omega_0^2 = \frac{1}{LC} . \tag{S-7.37}$$

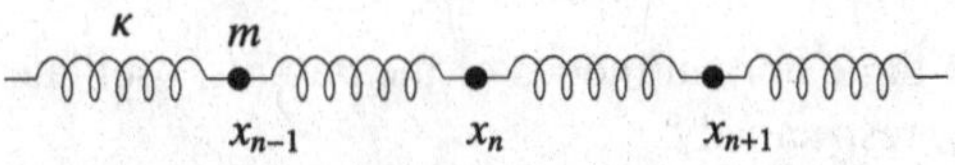

Fig. S-7.3 .

The equivalent mechanical system is a linear sequence of N identical masses m, each pair of consecutive masses being bound to each other by a spring of Hooke's constant κ (we use the Greek letter κ here because we shall need the letter k for the wavevector later on), as shown in Fig. S-7.3. We denote by x_n the displacement of each mass from its equilibrium position, i.e., its position when all springs have their rest length. Thus, the equation of motion of the nth mass is

$$m\frac{d^2 x_n}{dt^2} = -\kappa \left(x_n - x_{n-1} \right) + \kappa \left(x_{n+1} - x_n \right), \tag{S-7.38}$$

which, divided by m, and after introducing $\omega_0^2 = \kappa/m$ becomes

$$\frac{d^2 x_n}{dt^2} = \omega_0^2 \left(x_{n-1} - 2x_n + x_{n+1} \right) , \tag{S-7.39}$$

mathematically equivalent to (S-7.37). This equation can be generalized to the case of a mechanical system where transverse displacements are allowed, in addition to the longitudinal displacements. If the masses can move in three dimensions, and we denote by $\mathbf{r}_n$ the displacement of the nth mass from its equilibrium position, the equation of motion is written

$$\frac{d^2 \mathbf{r}_n}{dt^2} = \omega_0^2 \left(\mathbf{r}_{n-1} - 2\mathbf{r}_n + \mathbf{r}_{n+1} \right), \tag{S-7.40}$$

which is separable into three one-dimensional equations, each identical to (S-7.37).
(b) First, we note that, without loss of generality, we can assume the wavevector k appearing in Eq. (7.3) to be positive ($k > 0$), so that (7.3) represents a wave traveling from left to right. Changing the sign of k simply gives a wave of the same frequency

propagating in the opposite direction, whose dispersion relation is the same as for the forward-propagating wave, because of the inversion symmetry of the problem.

Inserting (7.3) in (S-7.37), and dividing both sides by $Ce^{-i\omega t}$ we obtain

$$- \omega^2\, e^{ikna} = \omega_0^2 \left[e^{ik(n+1)a} - 2e^{ikna} + e^{ik(n-1)a} \right] , \qquad (S\text{-}7.41)$$

where, again, we have substituted $\omega_0^2 = 1/LC$. Dividing both sides by e^{ikna} we obtain

$$\omega^2 = \omega_0^2 \left(2 - e^{ika} - e^{-ika} \right) = 2\omega_0^2 \left(1 - \cos ka \right) = 4\omega_0^2 \sin^2(ka/2) , \qquad (S\text{-}7.42)$$

or, performing the square root,

$$\omega = 2\omega_0 \left| \sin\left(\frac{ka}{2} \right) \right| . \qquad (S\text{-}7.43)$$

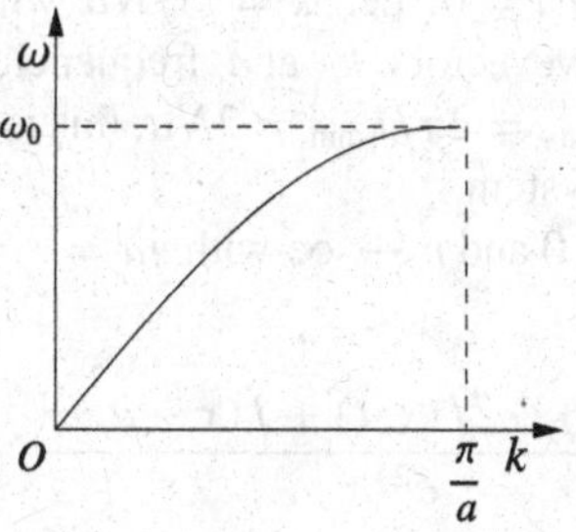

The dispersion relation (S-7.43) is shown in Fig. S-7.4 for $0 < k \le \pi/a$, this range being sufficient to describe all waves propagating in the system. In fact, although (S-7.43) seems to imply that $\omega(k)$ is a periodic function of k, with period $2\pi/a$, the wavevectors k and $k' = k + 2\pi s/a$, with s any integer, actually represent the same wave, since

Fig. S-7.4

$$e^{ik'na} = e^{i(k+2\pi s/a)na} = e^{ikna}e^{2\pi isn} = e^{ikna} , \qquad (S\text{-}7.44)$$

sn being an integer. This is why it is sufficient to consider the range $0 < k \le \pi/a$.

The existence of a maximum wave vector and of a cut-off frequency is related to the discrete periodic nature of the network, which imposes a minimum sampling rate a. The value $k_{max} = \pi/a$ corresponds to $\lambda_{min} = 2\pi/k_{max} = 2a$, and waves with a smaller wavelength cannot exist. In these waves, the current intensity value is repeated every two meshes of the network, as shown in Fig. S-7.5. A wave with a smaller period cannot exist because of the geometry of the network. One can also note that the

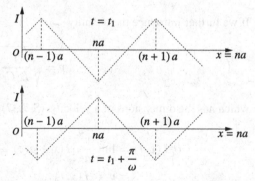

Fig. S-7.5

direction of wave propagation cannot be determined by observing the wave profile a two instants t_1 and $t_1 + \pi/\omega$ (half a period later, upper and lower parts of Fig. S-7.5). This is consistent with the group velocity $v_g(k_{max}) = (\partial_k\omega)(k_{max}) = 0$. The maximum wavevector corresponds to a high cut-off frequency $\omega_{max} = 2\omega_0$. Since higher frequencies cannot be transmitted, the LC network is a low-pass filter.

(c) The general monochromatic solution of frequency ω is a standing wave, i.e., the sum of two waves, one propagating from left to right and the other form right to left

$$I_n(t) = Ae^{ikna-i\omega t} + Be^{-ikna-i\omega t}, \qquad (S\text{-}7.45)$$

where ω and k are related by the dispersion relation (S-7.43). Because of our boundary conditions we must have

$$x_0(t) = 0 \Rightarrow A + B = 0; \qquad x_N(t) = 0 \Rightarrow Ae^{ikNa} + Be^{-ikNa} = 0. \quad (S\text{-}7.46)$$

This gives the condition $e^{ikNa} - e^{-ikNa} = 2i \sin(kNa) = 0$, i.e., $k = \pi l/Na$ with $l = 1, 2, 3, \ldots, N-1, N$. We have N allowed wavevectors k_l and frequencies $\omega_l = \omega(k_l)$. Note that $k_{min} = \pi/Na$ corresponds to $\lambda_{max} = 2\pi/k_{min} = 2Na$, this is a standing wave of wavelength twice the length of the system.

(d) We obtain the limit to a continuous by letting $a \to 0$ and $n \to \infty$ with $na = x = $ constant so that

$$\lim_{a \to 0} \frac{I_{n+1}(t) - 2I_n(t) + I_{n-1}(t)}{a^2} = \lim_{a \to 0} \frac{I(x+a,t) - 2I(x,t) + I(x-a,t)}{a^2}$$

$$= \partial_x^2 I(x,t). \qquad (S\text{-}7.47)$$

At this limit we can define a *capacity per unit length* C_ℓ, and an *inductance per unit length* L_ℓ, of the circuit, such that the capacitance and inductance of a circuit segment of length Δx are, respectively,

$$C = C_\ell \Delta x \quad \text{and} \quad L = L_\ell \Delta x. \qquad (S\text{-}7.48)$$

If we further introduce the quantity

$$v = \sqrt{\frac{1}{L_\ell C_\ell}}, \qquad (S\text{-}7.49)$$

which has the dimensions of a velocity, (S-7.37) is written for the continuous system

$$\partial_t^2 I(x,t) = \lim_{a \to 0} \frac{v^2}{a^2} [I(x+a,t) - 2I(x,t) + I(x-a,t)]$$

$$= v^2 \partial_x^2 I(x,t). \qquad (S\text{-}7.50)$$

This is the equation for a wave propagating with velocity v, independent of the wave frequency ω. At the limit of a continuous system there is no dispersion. This is the case of ideal transmission lines, like parallel wires and coaxial cables with no resistance. See Problem 7.6 for the case of a realistic transmission line with resistive losses where, however, dispersion can be eliminated.

S-7.5   The CL Ladder Network

(a) We have the same electric potential on the lower horizontal branch of each mesh, and we assume it to be zero. The voltage drop across the nth capacitor is

$$V_{n-1} - V_n = \frac{Q_n}{C}. \tag{S-7.51}$$

The current in the nth inductor is $I_n - I_{n+1}$, corresponding to a voltage drop across the inductor $L \, (dI_n/dt - dI_{n+1}/dt)$. Thus we have

$$V_{n-1} = L \left(\frac{dI_{n-1}}{dt} - \frac{dI_n}{dt} \right), \quad V_n = L \left(\frac{dI_n}{dt} - \frac{dI_{n+1}}{dt} \right), \tag{S-7.52}$$

which, inserted into (S-7.51), give

$$L \left(\frac{dI_{n-1}}{dt} - 2\frac{dI_n}{dt} + \frac{dI_{n+1}}{dt} \right) = \frac{Q_n}{C}. \tag{S-7.53}$$

Differentiating (S-7.53) with respect to time, and using $\frac{dQ_n}{dt} = I_n$, we obtain

$$L \left(\frac{d^2 I_{n-1}}{dt^2} - 2\frac{d^2 I_n}{dt^2} + \frac{d^2 I_{n+1}}{dt^2} \right) = \frac{I_n}{C}, \tag{S-7.54}$$

which is Eq. (7.4).

(b) By substituting $I_n = A e^{ikna - i\omega t}$ and $I_{n\pm 1} = A e^{ik(n\pm 1)a - i\omega t}$ into (S-7.54), defining $\omega_0^2 = (LC)^{-1}$, and dividing both sides by $LA \, e^{ikna - i\omega t}$, we obtain

$$-\omega^2 \left(e^{ika} - 2 + e^{-ika} \right) = \omega_0^2. \tag{S-7.55}$$

The left-hand side can be rewritten

$$-\omega^2 \left(e^{ika} - 2 + e^{-ika} \right) = -\omega^2 \left[2\cos(ka) - 2 \right] = -2\omega^2 \left[\cos(ka) - 1 \right]$$

$$= -2\omega^2 \left[\cos(ka) - \cos^2\left(\frac{ka}{2}\right) - \sin^2\left(\frac{ka}{2}\right) \right]$$

$$= -2\omega^2 \left[\cos^2\left(\frac{ka}{2}\right) - \sin^2\left(\frac{ka}{2}\right) - \cos^2\left(\frac{ka}{2}\right) - \sin^2\left(\frac{ka}{2}\right) \right]$$

$$= 4\omega^2 \sin^2\left(\frac{ka}{2}\right). \tag{S-7.56}$$

Substituting into (S-7.55) we have

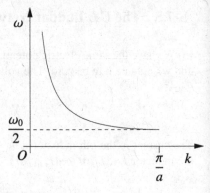

$$\omega^2 = \frac{\omega_0^2}{4\sin^2(ka/2)} , \qquad \text{(S-7.57)}$$

or

$$\omega = \frac{\omega_0}{2|\sin(ka/2)|} . \qquad \text{(S-7.58)}$$

Fig. S-7.6

Figure S-7.6 shows the plot of the dispersion relation. Compare this behavior with the dispersion relation shown in Fig. S-7.4 for an LC network, where capacitors and inductors are swapped with respect to the present case (Problem 7.4). In the LC network $2\omega_0$ is an upper cut-off frequency. Here, in the CL network, we have a lower cut-off frequency $\omega_0/2$, and the CL ladder network acts as a low-pass filter.

S-7.6 A Non-dispersive Transmission Line

(a) The voltage drop from x to $x + dx$ is

$$V(x, t) - V(x + dx, t) = \partial_t I(x, t) L + I(x, t) R , \qquad \text{(S-7.59)}$$

which yields, after replacing R by $R_\ell dx$ and L by $L_\ell dx$,

$$\partial_x V = -L_\ell \partial_t I - R_\ell I . \qquad \text{(S-7.60)}$$

The charge associated to the capacitance per unit length is $Q = Q(x, t) = C V(x, t)$ and charge conservation yields

$$\partial_t Q(x, t) = I(x - dx, t) - I(x, t) - I_L(x, t) , \qquad \text{(S-7.61)}$$

with the leakage current given by

$$I_L = I_L(x, t) = V(x, t)/R_p = V(x, t)G_\ell dx . \qquad \text{(S-7.62)}$$

We thus obtain, by eliminating Q and replacing C by $C_\ell dx$,

$$C_\ell \partial_t V = -\partial_x I - G_\ell V . \qquad \text{(S-7.63)}$$

Now we eliminate V by calculating

$$\partial_x^2 I = -C_\ell \partial_t \partial_x V - G_\ell \partial_x V$$
$$= +L_\ell C_\ell \partial_t^2 I + C_\ell R_\ell \partial_t I + G_\ell L_\ell \partial_t I + G_\ell R_\ell I \,, \tag{S-7.64}$$

which yields Eq. (7.6).

(b) By substituting (7.7) in (7.6) we obtain

$$-k^2 + \frac{\omega^2}{v_0^2} = -i\omega(R_\ell C_\ell + L_\ell G_\ell) + R_\ell G_\ell \,, \tag{S-7.65}$$

where $v_0^2 = (L_\ell C_\ell)^{-1}$. Thus, the wavevector k is a complex number. Writing $k = k_r + ik_i$ we obtain

$$k_r^2 - k_i^2 = \frac{\omega^2}{v_0^2} - R_\ell G_\ell \,, \tag{S-7.66}$$

$$2k_r k_i = \omega(R_\ell C_\ell + L_\ell G_\ell) \,. \tag{S-7.67}$$

The wave is thus evanescent, i.e.

$$I(x, t) = I_0 e^{-k_i x} e^{ik_r x - i\omega t} \,, \tag{S-7.68}$$

where the acceptable values for k_i are positive. Since in general $k_r = k_r(\omega)$ if $R_\ell \neq 0$ or $G_\ell \neq 0$, resistive effects make the line to be dispersive, so that a wavepacket is distorted along its propagation.

(c) If we assume that $k_i^2 = R_\ell G_\ell$ in Eq. (S-7.66), then $k_r = \omega/v_0$, which means that the propagation is non-dispersive: the phase velocity $v_p = \omega/k_r = v_0$ is independent on frequency. In addition, since k_i does not depend on ω the evancescence length (k_i^{-1}) is also frequency-independent. By substituing $k_i = (R_\ell G_\ell)^{1/2}$ and $k_r = \omega(L_\ell C_\ell)^{1/2}$ in Eq. (S-7.67) we obtain the condition

$$2(R_\ell G_\ell)^{1/2}(L_\ell C_\ell)^{1/2} = R_\ell C_\ell + L_\ell G_\ell \,. \tag{S-7.69}$$

Taking the square of both sides and rearranging the terms yields $(R_\ell C_\ell - L_\ell G_\ell)^2 = 0$ which brings the simple, equivalent condition

$$R_\ell C_\ell = L_\ell G_\ell \,. \tag{S-7.70}$$

This is the condition for a non-dispersive or *distortionless* transmission line due to O. Heaviside.

If the input current at one side of the line, say $x = 0$, is given by

$$I(0, t) = I_0(t) = \int \tilde{I}_0(\omega) e^{-i\omega t} d\omega \,, \tag{S-7.71}$$

where $\tilde{I}_0(\omega)$ is the Fourier transform, then the current along the line will be given by

$$I(x,t) = \int \tilde{I}_0(\omega) e^{ik_r x - i\omega t} e^{-k_i x} d\omega = e^{-k_i x} \int \tilde{I}_0(\omega) e^{-i\omega(t - x/v_0)} d\omega$$

$$= e^{-k_i x} I_0(t - x/v_0) , \tag{S-7.72}$$

since k_i is independent on ω. This is equivalent to state that the general solution of Eq. (7.6) with the condition (S-7.70) has the form (7.8) with $v = v_0$ and $\kappa = k_i$.

The same conclusion may be obtained by direct substitution of (7.8) into Eq. (7.6). The partial derivatives are given by

$$\partial_t I = -v e^{-\kappa x} f'(x - vt) ,$$
$$\partial_t^2 I = v^2 e^{-\kappa x} f''(x - vt) ,$$
$$\partial_x I = -\kappa e^{-\kappa x} f(x - vt) + e^{-\kappa x} f'(x - vt) ,$$
$$\partial_x^2 I = \kappa^2 e^{-\kappa x} f(x - vt) - 2\kappa e^{-\kappa x} f'(x - vt) + e^{-\kappa x} f''(x - vt) , \tag{S-7.73}$$

where $f'(x) = df(x)/dx$ and $f''(x) = d^2 f(x)/dx^2$. Thus Eq.(7.6) becomes

$$(\kappa^2 - R_\ell G_\ell) f + (v(R_\ell C_\ell + L_\ell G_\ell) - 2\kappa) f' + (1 - L_\ell C_\ell v^2) f'' = 0 \tag{S-7.74}$$

For this equation to be true for arbitrary f, the coefficients of f, f' and f'' must be all zero. Thus

$$\kappa^2 = R_\ell G_\ell , \qquad 2\kappa = v(R_\ell C_\ell + L_\ell G_\ell) , \qquad v^2 = (L_\ell C_\ell)^{-1} , \tag{S-7.75}$$

which bring again the conditions on the line parameters found above.

S-7.7 An "Alternate" LC Ladder Network

(a) Let Q_n be the charge of the capacitor at the right of mesh n. Applying Kirchoff's mesh rule to the even and odd meshes of the ladder network we have, respectively,

$$-\frac{Q_{2n-1}}{C} + L_2 \frac{dI_{2n}}{dt} + \frac{Q_{2n}}{C} = 0 , \qquad -\frac{Q_{2n}}{C} + L_1 \frac{dI_{2n+1}}{dt} + \frac{Q_{2n+1}}{C} = 0 , \tag{S-7.76}$$

while Kirchhoff's junction rule gives

$$\frac{dQ_{2n-1}}{dt} = I_{2n-1} - I_{2n} , \qquad \frac{dQ_{2n}}{dt} = I_{2n} - I_{2n+1} . \tag{S-7.77}$$

Differentiating (S-7.76) with respect to time, and inserting (S-7.77), we obtain

$$L_2 \frac{d^2 I_{2n}}{dt^2} = \frac{1}{C} \left(\frac{dQ_{2n-1}}{dt} - \frac{d^2 Q_{2n}}{dt^2} \right) = \frac{1}{C} (I_{2n-1} - 2I_{2n} + I_{2n+1})$$

$$L_1 \frac{d^2 I_{2n+1}}{dt^2} = \frac{1}{C} \left(\frac{dQ_{2n}}{dt} - \frac{d^2 Q_{2n+1}}{dt^2} \right) = \frac{1}{C} (I_{2n} - 2I_{2n+1} + I_{2n+2}), \quad \text{(S-7.78)}$$

identical to (7.9).

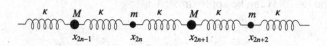

Fig. S-7.7 .

A mechanical equivalent to our network is the one-dimensional sequence of masses and springs shown in Fig. S-7.7, where the masses have, alternately, the values M and m, while all springs are identical, with Hooke's constant κ. If we denote by x_{2n+1} the positions of the odd masses M, and by and x_{2n} the positions of even masses m, the equations of motion for the system are

$$m \frac{d^2 x_{2n}}{dt^2} = -\kappa(x_{2n} - x_{2n+1}) + \kappa(x_{2n-1} - x_{2n}) = \kappa(x_{2n-1} - 2x_{2n} + x_{2n+1})$$

$$M \frac{d^2 x_{2n+1}}{dt^2} = -\kappa(x_{2n+1} - x_{2n+2}) + \kappa(x_{2n} - x_{2n+1}) = \kappa(x_{2n} - 2x_{2n+1} + x_{2n+2}),$$

$$\text{(S-7.79)}$$

which, after the substitutions $m \to L_2, M \to L_1, x \to I$, and $\kappa \to 1/C$, are identical to (S-7.78).

(b) Substituting (7.10) into (S-7.78), and dividing both sides by $e^{-i\omega t}$, we obtain

$$-\omega^2 L_2 I_e e^{i(2n)ka} = \frac{1}{C} \left(I_o e^{i(2n-1)ka} - 2I_e e^{i(2n)ka} + I_o e^{i(2n+1)ka} \right)$$

$$-\omega^2 L_2 I_o e^{i(2n+1)ka} = \frac{1}{C} \left(I_e e^{i(2n)ka} - 2I_o e^{i(2n+1)ka} + I_e e^{i(2n+2)ka} \right). \quad \text{(S-7.80)}$$

Now we define the two angular frequencies $\omega_o = 1/\sqrt{L_1 C}$ and $\omega_e = 1/\sqrt{L_2 C}$, and divide (S-7.80) by $e^{i(2n)ka}$, obtaining

$$(2\omega_e^2 - \omega^2) I_e - 2\omega_e^2 \cos(ka) I_o = 0$$

$$2\omega_o^2 \cos(ka) I_e - (2\omega_o^2 - \omega^2) I_o = 0. \quad \text{(S-7.81)}$$

This system of linear equations has non-trivial solutions if and only if its determinant is zero, i.e., if

$$(2\omega_e^2 - \omega^2)(2\omega_o^2 - \omega^2) - 4\omega_o^2 \omega_e^2 \cos^2(ka) = 0, \quad \text{(S-7.82)}$$

the solution of the above quadratic equation in ω^2 is

$$\omega^2 = \omega_e^2 + \omega_o^2 \pm \sqrt{(\omega_e^2 + \omega_o^2)^2 - 4\omega_o^2 \omega_e^2 \sin^2(ka)}. \qquad \text{(S-7.83)}$$

Both solutions are physically acceptable: the system allows for *two* types of propagating waves, described by two different dispersion relations.

At the limit $L_2 \ll L_1$ (or $m \ll M$, for the equivalent mechanical system) we have $\omega_o^2 \ll \omega_e^2$, and (S-7.83) can be approximated as

$$\omega^2 \simeq \omega_e^2 + \omega_o^2 \pm \omega_e^2 \sqrt{1 + 2\frac{\omega_o^2}{\omega_e^2} - 4\frac{\omega_o^2}{\omega_e^2}\sin^2(ka)} \qquad \text{(S-7.84)}$$

where we have disregarded the fourth-order term ω_o^4/ω_e^4 inside the square root. If we further use the approximation $\sqrt{1+x} \simeq 1 + x/2$, valid for $x \ll 1$, (S-7.84) becomes

$$\omega^2 \simeq \omega_e^2 + \omega_o^2 \pm \omega_e^2 \left\{ 1 + \frac{\omega_o^2}{\omega_e^2}\left[1 - 2\sin^2(ka)\right] \right\}, \qquad \text{(S-7.85)}$$

corresponding to the two dispersion relations

$$\omega \simeq \begin{cases} \sqrt{2(\omega_e^2 + \omega_o^2) - 2\omega_o^2 \sin^2(ka)} \\ \sqrt{2}\omega_o \sin(ka) \end{cases} \qquad \text{(S-7.86)}$$

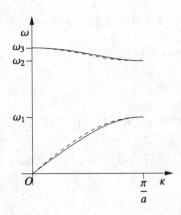

Fig. S-7.8

The lower branch can propagate for frequencies between 0 and $\omega_1 = \sqrt{2}\omega_o$, while the upper branch lies between $\omega_2 = \omega_e\sqrt{2(1 - \omega_o^2/\omega_e^2)}$ and $\omega_3 = \sqrt{2}\omega_e$. Thus, there is a gap of "forbidden" frequencies between ω_1 and ω_2. Figure S-7.8 shows the exact solution (continuous lines), and the approximate solution (dashed lines), still in good agreement, for $\omega_o^2/\omega_e^2 = 0.25$.

Of course, the two branches are present also in the case of the alternating mechanical oscillators, and provide a model for an effect known in solid state physics. The vibrations of a lattice formed by identical ions have a single branch (Problem 7.4), with a dispersion relation similar to the lower branch, which is named "acoustic branch". In an ionic crystal, formed by two ion species alternating on the sites of the lattice, we observe also the upper branch, named "optical branch".

S-7.8 Resonances in an LC Ladder Network

(a) According to Problem 7.4, the current flowing in the nth mesh is

$$\frac{d^2 I_n}{dt^2} = \omega_0^2 (I_{n+1} - 2I_n + I_{n-1}).\tag{S-7.87}$$

We are looking for a propagating wave solution, and define the phase

$$\phi \equiv ka,\tag{S-7.88}$$

where a is the length of a single mesh, to be substituted into (S-7.45), writing $I_n(t)$ as

$$I_n(t) = A e^{in\phi - i\omega t}.\tag{S-7.89}$$

Substituting (S-7.89) into (S-7.87), and dividing by $e^{-i\omega t}$, we get

$$-\omega^2 e^{in\phi} = \omega_0^2 \left[e^{i(n+1)\phi} - 2e^{in\phi} + e^{i(n-1)\phi} \right],\tag{S-7.90}$$

from which we obtain the dispersion relation

$$\omega^2 = \omega_0^2 (2 - e^{i\phi} - e^{-i\phi}) = 2\omega_0^2 (1 - \cos\phi) = 4\omega_0^2 \sin^2(\phi/2),\tag{S-7.91}$$

whose inverse is

$$\sin\left(\frac{\phi}{2}\right) = \frac{\omega}{2\omega_0}, \quad \text{or} \quad \phi = 2\arcsin\left(\frac{\omega}{2\omega_0}\right),\tag{S-7.92}$$

that shows that ϕ is a real number if $\omega < 2\omega_0$.

Due to the presence of the current source, (S-7.89) holds if the current in the 0th mesh is

$$I_0(t) = I_s e^{-i\omega t},\tag{S-7.93}$$

thus we must have $A = I_s$, and the final expression for $I_n(t)$ is

$$I_n(t) = I_s e^{in\phi - i\omega t},\tag{S-7.94}$$

where ϕ is given by (S-7.92)

(b) If $\omega > 2\omega_0$ the current wave cannot propagate in the ladder. We look for a solution of the form suggested by the hint. Substituting (7.12) into (S-7.87) we obtain

$$-\omega^2 \alpha^{-n} = \omega_0^2 \left[\alpha^{-(n+1)} - 2\alpha^{-n} + \alpha^{-(n-1)} \right],\tag{S-7.95}$$

which, multiplied by α^n / ω_0^2, turns into

$$\alpha^2 + \left(\frac{\omega^2}{\omega_0^2} - 2\right)\alpha + 1 = 0 . \tag{S-7.96}$$

The solutions are

$$\alpha = 1 - \frac{\omega^2}{2\omega_0^2} \pm \sqrt{\left(1 - \frac{\omega^2}{2\omega_0^2}\right)^2 - 1} . \tag{S-7.97}$$

We must have $|\alpha| < 1$ for an infinite ladder, otherwise the current would grow indefinitely in successive meshes. Thus, we keep the solution with the plus sign, because $\omega > 2\omega_0$ implies that all solutions of (S-7.97) are negative, obtaining

$$I_n(t) = I_s(-1)^n|\alpha|^n \mathrm{e}^{-\mathrm{i}\omega t}, \qquad |\alpha| = \frac{\omega^2}{2\omega_0^2} - 1 - \sqrt{\left(\frac{\omega^2}{2\omega_0^2} - 1\right)^2 - 1} , \quad \text{(S-7.98)}$$

that we can rewrite as

$$I_n(t) = I_s\,\mathrm{e}^{-\gamma n - \mathrm{i}\omega t} , \quad \text{where} \quad \gamma = \mathrm{i}\pi + \ln|\alpha| . \tag{S-7.99}$$

(c) We consider the case of the propagating wave ($\omega < 2\omega_0$) first. If the ladder comprises N meshes numbered as in Fig. 7.10, the boundary condition at the right end is $I_N(t) \equiv 0$ (mesh number N does not exist!). The most general solution is the sum of two counterpropagating waves

$$I_n(t) = A\mathrm{e}^{\mathrm{i}n\phi - \mathrm{i}\omega t} + B\mathrm{e}^{-\mathrm{i}n\phi - \mathrm{i}\omega t} . \tag{S-7.100}$$

Imposing the conditions $I_0 = I_s$ and $I_N = 0$, we obtain

$$A + B = I_s, \qquad A\mathrm{e}^{\mathrm{i}N\phi} + B\mathrm{e}^{-\mathrm{i}N\phi} = 0, \tag{S-7.101}$$

with solutions

$$A = +\frac{\mathrm{i}}{2}\,I_s\,\frac{\mathrm{e}^{-\mathrm{i}N\phi}}{\sin(N\phi)}, \qquad B = -\frac{\mathrm{i}}{2}\,I_s\,\frac{\mathrm{e}^{+\mathrm{i}N\phi}}{\sin(N\phi)}, \tag{S-7.102}$$

where $\phi = \phi(\omega)$ depends on ω according to (S-7.92). We observe resonances when $\sin(N\phi) = 0$, i.e., for $\phi = m\pi/N$ with m an integer. Remembering (S-7.88)

$$N = m\,\frac{\pi}{\phi} = m\,\frac{\pi}{ka} = m\,\frac{\pi}{a}\,\frac{\lambda}{2\pi} = m\,\frac{1}{a}\,\frac{\lambda}{2} , \tag{S-7.103}$$

and, multiplying both sides by a

$$L = Na = m\,\frac{\lambda}{2} , \tag{S-7.104}$$

where L is the total length of the ladder network. This corresponds to the case when the frequency of the current source equals the frequency of one of the standing waves allowed in the network, i.e., when the length of the ladder network is an integer multiple of a half wavelength.

If $\omega > 2\omega_0$, the general solution is given by

$$I_n(t) = A\alpha_+^n e^{-i\omega t} + B\alpha_-^n e^{-i\omega t}, \tag{S-7.105}$$

where $\alpha_\pm = \alpha_\pm(\omega)$ are the two solutions of (S-7.97), now also the case $|\alpha| > 1$ is allowed, because $|\alpha|^n$ cannot diverge if n is limited. The boundary conditions are

$$A + B = I_s, \qquad A\alpha_+^N + B\alpha_-^{+N} = 0, \tag{S-7.106}$$

with solutions

$$A = +I_s \frac{\alpha_-^N}{\alpha_-^N - \alpha_+^N}, \qquad B = -I_s \frac{\alpha_+^N}{\alpha_-^N - \alpha_+^N}. \tag{S-7.107}$$

The A and B coefficients diverge if $\alpha_- = \alpha_+ = 1$, i.e., if $\omega = 2\omega_0$. Thus, for $\omega > 2\omega_0$ there are no resonances, but the response of the system diverges as the frequency approaches the cut-off value, i.e. as $\omega \to 2\omega_0$.

S-7.9 Cyclotron Resonances (1)

(a) The rotating electric field can be written as

$$\mathbf{E} = \mathbf{E}(t) = E_0 \left(\hat{\mathbf{x}} \cos \omega t \pm \hat{\mathbf{y}} \sin \omega t \right), \tag{S-7.108}$$

where the positive (negative) sign indicates counterclockwise (clockwise) rotation. From the equation of motion

$$m \frac{d\mathbf{v}}{dt} = q \left(\mathbf{E} + \frac{\mathbf{v}}{c} \times \mathbf{B} \right), \tag{S-7.109}$$

we see that $dv_z/dt = 0$, thus, if we assume that $v_z(0) = 0$, the motion occurs in the (x, y) plane. The equations of motion along the x and y axes are

$$\frac{dv_x}{dt} = +q v_y \frac{B_0}{mc} + \frac{q E_0}{m} \cos \omega t$$
$$\frac{dv_y}{dt} = -q v_x \frac{B_0}{mc} \pm \frac{q E_0}{m} \sin \omega t. \tag{S-7.110}$$

In principle, we can differentiate both equations with respect to time, and then substitute the expressions for $dv_{x,y}/dt$, thus obtaining two uncoupled second-order differential equations for a driven harmonic oscillator.

But we prefer a a more "elegant" approach, introducing the complex variable $\zeta = v_x + iv_y$. The velocity is thus represented by a complex vector in the $(\text{Re}\zeta, \text{Im}\zeta)$ plane. Adding the second of (S-7.110), multiplied by i, to the first, we obtain

$$\frac{d\zeta}{dt} = -i\omega_c\,\zeta + \frac{qE_0}{m}\,e^{\pm i\omega t}, \tag{S-7.111}$$

where $\omega_c = qB_0/mc$ is the cyclotron (or Larmor) frequency. The solution of the associated homogeneous equation is

$$\zeta(t) = A\,e^{-i\omega_c t}, \tag{S-7.112}$$

where A is an arbitrary complex constant. Equation (S-7.112) describes the motion in the absence of the electric field, when the velocity rotates clockwise with frequency ω_c in the ζ plane. We then search for a particular integral of the inhomogeneous equation in the form

$$\zeta = \zeta_0\,e^{\pm i\omega t},$$

and find, by direct substitution,

$$\zeta_0 = -i\,\frac{qE_0}{m\,(\omega_c \pm \omega)}. \tag{S-7.113}$$

Thus the general solution of (S-7.111) is

$$\zeta(t) = A\,e^{-i\omega_c t} - i\,\frac{qE_0}{m\,(\omega_c \pm \omega)}\,e^{\pm i\omega t}. \tag{S-7.114}$$

Assuming $\omega_c > 0$, we observe a resonance at $\omega = \omega_c$ only if the field rotates clockwise. In this case the electric field accelerates the particle along the direction of its "natural" motion.

(b) At resonance ($\omega = \omega_c$), we search for a non-periodic solution of the type

$$\zeta(t) = \zeta_R(t)\,e^{-i\omega t}, \tag{S-7.115}$$

which, substituted into (S-7.111), gives

$$\left(\frac{d\zeta_R}{dt}\right) e^{-i\omega t} - i\omega\,\zeta_R\,e^{-i\omega t} = -i\omega_c\,\zeta_R\,e^{-i\omega t} + \frac{qE_0}{m}\,e^{-i\omega t}, \tag{S-7.116}$$

and since $\omega_c = \omega$

$$\left(\frac{d\zeta_R}{dt}\right) = \frac{qE_0}{m}. \tag{S-7.117}$$

The solution of (S-7.117) is

$$\zeta_R(t) = \zeta(0) + \frac{qE_0}{m} t, \qquad (\text{S-7.118})$$

which gives

$$\zeta(t) = \left[\zeta(0) + \frac{qE_0}{m} t\right] e^{-i\omega t}. \qquad (\text{S-7.119})$$

The trajectory is a spiral, with the radial velocity increasing linearly with time.
(c) Introducing a viscous force $\mathbf{f}_v = -m\gamma v$, we obtain the following equation for ζ

$$\frac{d\zeta}{dt} = -i\omega_c \zeta - \gamma \zeta + \frac{qE_0}{m} e^{-i\omega t}. \qquad (\text{S-7.120})$$

The solution has the form of (S-7.114) with ω_c replaced by $(\omega_c - i\gamma)$, i.e.,

$$\zeta = -i \frac{qE_0}{m(\omega_c - \omega - i\gamma)} e^{-i\omega t} + A e^{-i\omega_c t - \gamma t}, \qquad (\text{S-7.121})$$

where the second term undergoes an exponential decay, and any memory of the initial conditions is lost after a transient phase, while the periodic part of the solution does not diverge at resonance, due to the presence of $i\gamma$ in the denominator. Thus, the steady-state solution at resonance is

$$\zeta_R = \frac{qE_0}{m\gamma} e^{-i\omega t}. \qquad (\text{S-7.122})$$

The average dissipated power is the time average of the instantaneous dissipated power over a period

$$P = \langle \mathbf{f} \cdot v \rangle = \langle q\mathbf{E} \cdot v \rangle. \qquad (\text{S-7.123})$$

The components of the particle velocity in the steady state are

$$v_x = \text{Re}(\zeta) = \frac{qE_0\gamma}{m\left[(\omega_c - \omega)^2 + \gamma^2\right]} \cos\omega t - \frac{qE_0(\omega_c - \omega)}{m\left[(\omega_c - \omega)^2 + \gamma^2\right]} \sin\omega t,$$

$$v_y = \text{Im}(\zeta) = -\frac{qE_0\gamma}{m\left[(\omega_c - \omega)^2 + \gamma^2\right]} \sin\omega t - \frac{qE_0(\omega_c - \omega)}{m\left[(\omega_c - \omega)^2 + \gamma^2\right]} \cos\omega t.$$

$$(\text{S-7.124})$$

Thus, inserting (S-7.124) and the relations

$$E_x = E_0 \cos\omega t, \quad E_y = -E_0 \sin\omega t, \quad \langle \cos^2\omega t \rangle = \langle \sin^2\omega t \rangle = \frac{1}{2},$$

$$\langle \cos\omega t \sin\omega t \rangle = 0,$$

into (S-7.123), we obtain for the average power dissipated

$$P = \frac{q^2 E_0^2 \gamma}{m \left[(\omega_c - \omega)^2 + \gamma^2 \right]} . \tag{S-7.125}$$

At resonance

$$P = \frac{q^2 E_0^2}{m\gamma} . \tag{S-7.126}$$

S-7.10 Cyclotron Resonances (2)

(a) The equations of motion are

$$\frac{dv_x}{dt} = +\omega_c v_y + \frac{q E_0}{m} \cos \omega t , \qquad \frac{dv_y}{dt} = -\omega_c v_x , \tag{S-7.127}$$

where $\omega_c = q B_0 / m$. By differentiating (S-7.127) with respect to time, and substituting the values for $\dot{v}_x$ and $\dot{v}_y$ from (S-7.127) itself, we obtain the two equations

$$\frac{d^2 v_x}{dt^2} = +\omega_c \frac{dv_y}{dt} - \frac{q E_0 \omega}{m} \sin \omega t = -\omega_c^2 v_x - \frac{q E_0 \omega}{m} \sin \omega t ,$$

$$\frac{d^2 v_y}{dt^2} = -\omega_c \frac{dv_x}{dt} = -\omega_c^2 v_y - \frac{q E_0 \omega_c}{m} \cos \omega t , \tag{S-7.128}$$

each of which describes the velocity of a driven harmonic oscillator. The steady state solutions are

$$v_x = \frac{q E_0 \omega}{m (\omega^2 - \omega_c^2)} \sin \omega t , \qquad v_y = \frac{q E_0 \omega_c}{m (\omega^2 - \omega_c^2)} \cos \omega t . \tag{S-7.129}$$

We observe a resonance if $\omega = |\omega_c|$, independently on the signs of q and B_0. With respect to Problem 7.9, where a rotating electric field was assumed, here a resonance is always found because the linearly oscillating electric field can be decomposed into two counter-rotating fields of the same amplitude, of which one will excite the resonance.

(b) In the presence of a frictional force $\mathbf{f} = -m\gamma \mathbf{v}$ the equations of motion become

$$\frac{dv_x}{dt} = +\omega_c v_y - \gamma v_x + \frac{q E_0}{m} \cos \omega t , \qquad \frac{dv_y}{dt} = -\omega_c v_x - \gamma v_y , \tag{S-7.130}$$

and cannot be uncoupled by the procedure of point (a). Analogously to Problem 7.9, we introduce the complex quantity $\zeta = v_x + i v_y$, obtaining the single equation

$$\frac{d\zeta}{dt} = -i\omega_c\zeta - \gamma\zeta + \frac{qE_0}{2m}\left(e^{i\omega t} + e^{-i\omega t}\right), \tag{S-7.131}$$

where we have used Euler's formula for the cosine. Differently from Problem 7.9, now we search for a steady-state solution of the form

$$\zeta = Ae^{-i\omega t} + Be^{i\omega t}, \tag{S-7.132}$$

where A and B are two complex constants to be determined. By direct substitution into (S-7.131) we have

$$-i\omega Ae^{-i\omega t} + i\omega Be^{i\omega t} = -(i\omega_c + \gamma)Ae^{-i\omega t} - (i\omega_c + \gamma)Be^{i\omega t} + \frac{qE_0}{2m}\left(e^{i\omega t} + e^{-i\omega t}\right),$$

which is separable into two equations relative, respectively, to the terms rotating clockwise and counterclockwise in the complex plane

$$-i\omega A = -(i\omega_c + \gamma)A + \frac{qE_0}{2m}, \quad i\omega B = -(i\omega_c + \gamma)B + \frac{qE_0}{2m}. \tag{S-7.133}$$

The solutions for A and B are

$$A = \frac{qE_0}{2m[i(\omega_c - \omega) + \gamma]} = \frac{qE_0\gamma}{2m[(\omega_c - \omega)^2 + \gamma^2]} - i\frac{qE_0(\omega_c - \omega)}{2m[(\omega_c - \omega)^2 + \gamma^2]}$$

$$B = \frac{qE_0}{2m[i(\omega_c + \omega) + \gamma]} = \frac{qE_0\gamma}{2m[(\omega_c + \omega)^2 + \gamma^2]} - i\frac{qE_0(\omega_c + \omega)}{2m[(\omega_c + \omega)^2 + \gamma^2]}, \tag{S-7.134}$$

from which we obtain the stationary-state velocity components of the particle

$$v_x = [\text{Re}(A) + \text{Re}(B)]\cos\omega t + [\text{Im}(A) - \text{Im}(B)]\sin\omega t$$
$$v_y = [\text{Im}(A) + \text{Im}(B)]\cos\omega t - [\text{Re}(A) - \text{Re}(B)]\sin\omega t. \tag{S-7.135}$$

The average absorbed power is

$$P = \langle q\mathbf{v}\cdot\mathbf{E}\rangle = \langle qv_xE_x\rangle = q[\text{Re}(A) + \text{Re}(B)]E_0\frac{1}{2}$$

$$= \frac{q^2E_0^2\gamma}{4m[(\omega_c - \omega)^2 + \gamma^2]} + \frac{q^2E_0^2\gamma}{4m[(\omega_c + \omega)^2 + \gamma^2]}, \tag{S-7.136}$$

since $E_x = E_0\cos\omega t$, $\langle\cos^2\omega t\rangle = 1/2$, and $\langle\cos\omega t\sin\omega t\rangle = 0$. Thus, again we observe a resonance at $\omega = |\omega_c|$, i.e whatever the signs of q and B_0. Assuming $\gamma \ll \omega_c$ the power absorbed at resonance is

$$P_{\text{max}} \simeq \frac{q^2E_0^2}{4m\gamma}. \tag{S-7.137}$$

S-7.11   A Quasi-Gaussian Wave Packet

We need to evaluate the inverse transform

$$f(x) = A \int\limits_{-\infty}^{+\infty} e^{-L^2(k-k_0)^2} e^{i\phi(k)} e^{ikx} dk$$

$$\simeq \int\limits_{-\infty}^{+\infty} \exp\Big[-L^2(k-k_0)^2 + i\phi_0 + i\phi_0'(k-k_0)+$$

$$+\frac{i}{2}\phi_0''(k-k_0)^2 + i(k-k_0)x + ik_0x\Big] dk, \qquad (S\text{-}7.138)$$

where, for brevity, we wrote x instead of $(x - vt)$, and ϕ_0, ϕ_0', ... instead of $\phi(k_0)$, $\phi'(k_0)$, ... By using (7.1) we obtain

$$f(x) \simeq A e^{ik_0x+i\phi_0} \int\limits_{-\infty}^{+\infty} \exp\Big[-L^2(k-k_0)^2\Big(1 - i\frac{\phi_0''}{2L^2}\Big) + i(k-k_0)(x+\phi_0')\Big] dk$$

$$= C \exp\Big[-\frac{(x+\phi_0')^2}{4L^2(1-i\phi_0''/2L^2)}\Big], \qquad (S\text{-}7.139)$$

where C is a constant, whose value is not relevant for our purposes. By substituting

$$\frac{1}{1-i\phi_0''/(2L^2)} = \frac{1+i\phi_0''/(2L^2)}{1+\phi_0''^2/(4L^4)} \qquad (S\text{-}7.140)$$

we obtain the wave packet profile as

$$f(x-vt) = C \exp\Big\{-\frac{(x-vt+\phi_0')^2[1+i\phi_0''/(2L^2)]}{L^2[1+\phi_0''^2/(4L^4)]}\Big\}. \qquad (S\text{-}7.141)$$

We thus see that the packet is wider than the purely Gaussian case, since $L^2[1+\phi_0''^2/(4L^4)] > L^2$. In addition, the center of the packet is shifted from $(x - vt)$ to $(x - vt + \phi_0')$, and there is an aperiodic (anharmonic) modulation due to the factor $(i\phi_0''/2L^2)$ in the numerator of the exponent.

S-7.12 A Wave Packet Traveling along a Weakly Dispersive Line

(a) There is no dispersion if $b = 0$. In these conditions the signal propagates at velocity v keeping its shape:

$$f(x - vt) = Ae^{-i\omega_0(t-x/v)}e^{-(t-x/v)^2/\tau^2} . \tag{S-7.142}$$

(b) The phase velocity and the group velocity are, by definition,

$$v_\phi = \frac{\omega}{k} = v(1 + bk) , \qquad v_g = \frac{\partial \omega}{\partial k} = v(1 + 2bk) . \tag{S-7.143}$$

We can write v_ϕ and v_g as functions of ω by first inverting (7.17), obtaining for $k = k(\omega)$

$$k = \sqrt{\frac{1}{(2b)^2} + \frac{\omega}{bv}} - \frac{1}{2b} . \tag{S-7.144}$$

Then we expand the square root to the second order in ω/v, obtaining

$$k \simeq \frac{\omega}{v} - \frac{\omega^2 b}{v^2}. \tag{S-7.145}$$

The same result can also be obtained by an iterative procedure, by inserting the first order value for k, i.e., $k = \omega/v$, into the bracket at the right hand side of (7.17). Thus, the phase and group velocities to the first order are, using (S-7.143),

$$v_{\phi0} \simeq v + b\omega_0 , \qquad v_{g0} \simeq v + 2b\omega_0 . \tag{S-7.146}$$

(c) The peak of the signal propagates at the group velocity, thus $t_x = x/v_{g0}$. The spectral width of the wave packet may be estimated as $\Delta\omega \simeq 1/\tau$, which corresponds to a spread in the propagation velocity of its Fourier components

$$\Delta v \simeq v \left(\frac{2b}{v} \Delta\omega \right) \simeq \frac{2b}{\tau} . \tag{S-7.147}$$

Thus the spread of the wave packet in time and space can be estimated as

$$\Delta t \simeq \frac{\partial t_x}{\partial v_g} \Delta v = t_x \frac{\Delta v}{v_g} , \qquad \Delta x \simeq v_g \Delta t = \frac{2bx}{v_g \tau} . \tag{S-7.148}$$

(d) We approximate

$$k(\omega) \simeq k_0 + k_0'(\omega - \omega_0) + \frac{1}{2}k_0''(\omega - \omega_0)^2 , \qquad (S\text{-}7.149)$$

where

$$k_0' = \left.\frac{\partial k}{\partial \omega}\right|_{\omega_0} \simeq \frac{1}{v} - 2\frac{\omega_0 b}{v^2} \simeq \frac{1}{v_g} , \qquad k_0'' = \left.\frac{\partial^2 k}{\partial \omega^2}\right|_{\omega_0} \simeq -2\frac{b}{v^2} . \qquad (S\text{-}7.150)$$

The spectrum of the wave packet (i.e., its Fourier transform) is

$$\tilde{f}(\omega) = \sqrt{\pi}\tau A e^{-[\omega-\omega_0]^2 \tau^2/4} . \qquad (S\text{-}7.151)$$

Since we are only interested in the behavior of the function, we evaluate the following integral forgetting proportionality constants,

$$
\begin{aligned}
f(x,t) &\sim \int \exp\left[ik(\omega)x - i\omega t - \frac{(\omega - \omega_0)^2 \tau^2}{4} \right] d\omega \\
&\sim \int \exp\left[ik_0 x + ik_0' x(\omega - \omega_0) + i\frac{k_0'' x}{2}(\omega - \omega_0)^2 - \frac{(\omega - \omega_0)^2 \tau^2}{4} \right] d\omega \\
&\sim \exp(ik_0 x - i\omega_0 t) \int \exp\left[-i(t - k_0' x)\,\omega' + \left(-\frac{\tau^2}{4} + i\frac{k_0'' x}{2} \right)\omega'^2 \right] d\omega' \\
&\sim \exp\left[ik_0 x - i\omega_0 t - \frac{(t - k_0' x)^2}{\tau^2 - 2ik_0'' x} \right] .
\end{aligned}
\qquad (S\text{-}7.152)
$$

The factor which describes the envelope of the wave packet (recalling that $k_0' = 1/v_{g0}$) is

$$
\begin{aligned}
\exp\left[-\frac{(t - x/v_g)^2}{\tau^2 - 2ik_0'' x} \right] &= \exp\left[-(t - x/v_g)^2 \frac{\tau^2 + 2ik_0'' x}{\tau^4 + (2k_0'' x)^2} \right] \\
&= \exp\left\{ -(t - x/v_g)^2 \frac{1 + 2ik_0'' x/\tau^2}{\tau^2[1 + (2k_0'' x/\tau)^2]} \right\} \quad (S\text{-}7.153)
\end{aligned}
$$

The temporal width of the wave packet increases during the propagation as

$$\Delta t(x) = \tau\sqrt{1 + \left(\frac{2k_0'' x}{\tau} \right)^2} . \qquad (S\text{-}7.154)$$

Chapter S-8
Solutions for Chapter 8

S-8.1 Poynting Vector(s) in an Ohmic Wire

For symmetry reasons, the magnetic field is azimuthal and depends only on the radial coordinate r. Applying Ampère's law to a circular path of radius $r < a$ around the wire axis yields.

$$2\pi r B = \frac{4\pi}{c}(\pi r^2 J) , \qquad (S-8.1)$$

which leads to

$$\mathbf{B} = \frac{2\pi}{c} r \sigma E \hat{\phi} , \qquad (S-8.2)$$

where $\hat{\phi}$ is the azimuthal unit vector. Thus, the Poynting vector at a distance r from the axis is

$$\mathbf{S} = \frac{c}{4\pi}\mathbf{E} \times \mathbf{B} = -\mathbf{r}\frac{\sigma}{2}E^2 . \qquad (S-8.3)$$

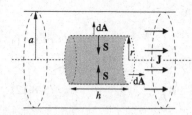

Fig. S-8.1

The energy flux $\Phi_S \equiv \Phi(\mathbf{S})$ through the surface of a cylinder of radius $r < a$ and length h and coaxial to the wire is thus (Fig. S-8.1)

$$\Phi_S = \oint \mathbf{S} \cdot d\mathbf{A} = -2\pi r h S(r)$$
$$= -\pi r^2 h \sigma E^2 , \qquad (S-8.4)$$

where $d\mathbf{A}$ is the vector surface element of the cylinder. The energy flows inwards, and is entirely dissipated into Joule heating inside the cylinder volume, as we can check by calculating

A. Macchi et al., *Problems in Classical Electromagnetism*,
https://doi.org/10.1007/978-3-031-22235-1

$$W = \int \mathbf{J} \cdot \mathbf{E} \, dV = \pi r^2 h J E = \pi r^2 h \sigma E^2 \,, \tag{S-8.5}$$

where the integral is extended to the volume the cylinder. The equality $W = -\Phi_S$ satisfies Poynting's theorem since there is no variation in time of the EM energy.

Note that, in the approximation of an infinitely long wire, the electric field is uniform also for $r > a$ (in the case of a finite wire of length $2h \gg a$, this is a good approximation in the central region for $r \ll h$, see Problem 4.9), while the magnetic field $B = 2\pi J a^2 / rc$. Within this approximation, $\mathbf{S} = -(a^2 \sigma E^2 / 2r)\hat{\mathbf{r}}$ for $r > a$, so that the energy flux is independent of r and it is still equal to minus the total dissipated power:

$$\Phi_S = -2\pi r h S(r) = -\pi a^2 h \sigma E^2 \quad (r > a). \tag{S-8.6}$$

(b) We must show that $\mathbf{\nabla} \cdot (\mathbf{S} - \mathbf{S'}) = 0$, i.e., that $\mathbf{S} - \mathbf{S'} = \mathbf{\nabla} \times \mathbf{f}$, where $\mathbf{f}$ is a vector function of the coordinates. Let us substitute $\mathbf{E} = -\mathbf{\nabla}\varphi$ into (8.7)

$$\mathbf{S} = \frac{c}{4\pi}\mathbf{E} \times \mathbf{B} = -\frac{c}{4\pi}\mathbf{\nabla}\varphi \times \mathbf{B}. \tag{S-8.7}$$

Now from the vector identity

$$\mathbf{\nabla} \times (\varphi\mathbf{B}) = \mathbf{\nabla}\varphi \times \mathbf{B} + \varphi\mathbf{\nabla} \times \mathbf{B} = \mathbf{\nabla}\varphi \times \mathbf{B} + \varphi\left(\frac{4\pi}{c}\mathbf{J}\right) \tag{S-8.8}$$

we obtain

$$\mathbf{\nabla}\varphi \times \mathbf{B} = \mathbf{\nabla} \times (\varphi\mathbf{B}) - \varphi\left(\frac{4\pi}{c}\mathbf{J}\right), \tag{S-8.9}$$

which can be substituted into (S-8.7), leading to

$$\mathbf{S} = \varphi\mathbf{J} - \frac{c}{4\pi}\mathbf{\nabla} \times (\varphi\mathbf{B}). \tag{S-8.10}$$

Thus, we are free to redefine the Poynting vector as

$$\mathbf{S'} = \varphi\mathbf{J}, \tag{S-8.11}$$

since

$$\mathbf{S} - \mathbf{S'} = \mathbf{\nabla} \times \left(-\frac{c}{4\pi}\varphi\mathbf{B}\right). \tag{S-8.12}$$

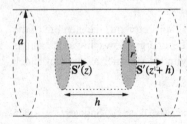

We can show that $\mathbf{S'}$ is equivalent to $\mathbf{S}$ by computing its flux through the same cylindrical surface as above. Since $\mathbf{S'}$ is parallel to the wire axis, only the two base surfaces contribute to the flux $\Phi_{S'} \equiv \Phi(\mathbf{S'})$.

$$\Phi_{S'} = \pi a^2[-S(z) + S(z+h)]$$
$$= \pi a^2 J[\varphi(z+h) - \varphi(z)] . \tag{S-8.13}$$

Since $\varphi = -Ez$, we finally obtain

$$\Phi_{S'} = \pi a^2[-S(z) + S(z+h)] = -\pi a^2 h J E , \tag{S-8.14}$$

which gives again minus the total dissipated power.

S-8.2 Poynting Vector(s) in a Capacitor

(a) The magnetic field has azimuthal symmetry, i.e., $\mathbf{B} = B(r)\,\hat{\phi}$, and can be evaluated from the equation $c\nabla \times \mathbf{B} = \partial_t \mathbf{E}$, which, with our assumption $E = E_0 t/\tau$, leads to

$$B(r) = \frac{r}{2c}\,\partial_t E = \frac{r}{2c\tau}\,E_0 . \tag{S-8.15}$$

(b) The corresponding Poynting vector $\mathbf{S}$ is

$$\mathbf{S} = \frac{c}{4\pi}E\,\hat{\mathbf{z}} \times (B\,\hat{\phi}) = -\frac{r}{8\pi}\,(E\partial_t E)\left(\hat{\mathbf{z}} \times \hat{\phi}\right) = -\frac{1}{2}\,\partial_t\left(\frac{E^2}{8\pi}\right)\mathbf{r} . \tag{S-8.16}$$

We evaluate the flux of $\mathbf{S}$ through the smallest closed cylindrical surface enclosing our capacitor, shown in Fig. S-8.2. Since $\mathbf{S}$ is radial, only the lateral surface of the cylinder contributes to the flux, and we have

$$\Phi(\mathbf{S}) = -2\pi a h S(a) = -\pi a^2 h \partial_t\left(\frac{E^2}{8\pi}\right) . \tag{S-8.17}$$

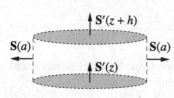

Fig. S-8.2

Quantity $(E^2 + B^2)/8\pi$ is the energy density associated to the EM field, and, since in our case B does not depend on time, is also the total EM energy density inside the capacitor. Thus, $\Phi(\mathbf{S})$ equals minus the time derivative of the energy stored in the capacitor. For a general dependence of $E_z(t)$ on time, $\mathbf{B}$ is also time-dependent, and the flux of $\mathbf{S}$ equals the time derivative of the electrostatic energy to the first order, within the slowly varying current approximation.

(c) The electric potential is $\varphi = -Ez$. By substituting $\mathbf{E} = -\nabla\varphi$ into (8.7) we obtain

$$\mathbf{S} = \frac{c}{4\pi}(-\nabla\varphi) \times \mathbf{B} = -\frac{c}{4\pi}\left[\nabla \times (\varphi\mathbf{B}) - \varphi\nabla \times \mathbf{B}\right] . \tag{S-8.18}$$

Thus, the vector

$$\mathbf{S}' \equiv \frac{c}{4\pi} \varphi \nabla \times \mathbf{B} = \frac{1}{4\pi} \varphi \, \partial_t \mathbf{E} = \mathbf{S} + \frac{c}{4\pi} \nabla \times (\varphi \mathbf{B}) \qquad (S\text{-}8.19)$$

equals $\mathbf{S}$ plus the curl of a vector function, and is thus another suitable Poynting vector. Since $\mathbf{S}'$ is perpendicular to the capacitor plates, its flux through our closed cylindrical surface is (see Fig. S-8.2)

$$\Phi(\mathbf{S}') = \pi a^2 \left[S'(z+h) - S'(z) \right] = -\pi a^2 h \left(\frac{E \partial_t E}{4\pi} \right) = -\pi a^2 h \partial_t \left(\frac{E^2}{8\pi} \right) , \qquad (S\text{-}8.20)$$

in agreement with (S-8.17).

S-8.3   Poynting's Theorem in a Solenoid

(a) We take a cylindrical coordinate system with the z axis along the solenoid axis. Inside an infinite solenoid the magnetic field is uniform and equals $\mathbf{B} = B \hat{\mathbf{z}} = (4\pi/c) \, n I \, \hat{\mathbf{z}}$. According to Faraday's law of induction, the rate of change of $\mathbf{B} = \mathbf{B}(t)$, due to the time dependence of $I = I(t)$, generates an electric field $\mathbf{E}$ associated to the induced electromotive force. For symmetry reasons, the field lines of $\mathbf{E}$ are circles coaxial to the solenoid, i.e., we have $\mathbf{E} = E(r) \, \hat{\boldsymbol{\phi}}$. Applying Faraday's law to a circle of radius $r < a$, coaxial to the solenoid, we have

$$2\pi r E(r) = -\pi r^2 \frac{1}{c} \, \partial_t B = -\pi r^2 \frac{4\pi n I_0}{c^2 \tau} , \qquad (S\text{-}8.21)$$

from which $E(r) = -2\pi n I_0 r / (c^2 \tau)$.
(b) The Poynting vector inside the solenoid $(r < a)$ is

$$\mathbf{S} = \frac{c}{4\pi} \mathbf{E} \times \mathbf{B} = -\frac{2\pi \, (n I_0)^2 \, rt}{(c\tau)^2} \, (\hat{\boldsymbol{\phi}} \times \hat{\mathbf{z}}) = -\frac{2\pi \, (n I_0)^2 t}{(c\tau)^2} \, \mathbf{r} . \qquad (S\text{-}8.22)$$

Thus, the flux of $\mathbf{S} = \mathbf{S}(r)$ through the surface of a closed cylinder of radius r and height h is nonzero only through the lateral surface, and we have

$$\Phi(\mathbf{S}) = 2\pi r h \, \mathbf{S} \cdot \hat{\mathbf{r}} = -\left(\frac{2\pi n I_0 r}{c\tau} \right)^2 ht . \qquad (S\text{-}8.23)$$

The magnetic energy enclosed by the cylinder surface is

$$U_\mathrm{M} = u_\mathrm{M} V = \frac{B^2}{8\pi} \pi r^2 h = 2\pi^2 r^2 h \left(\frac{n I_0 t}{c\tau} \right)^2 , \qquad (S\text{-}8.24)$$

where V is the volume of the cylinder, thus

$$\frac{dU_M}{dt} = 4\pi^2 r^2 ht \left(\frac{nI_0}{c\tau}\right)^2 = -\Phi(\mathbf{S}) , \qquad (\text{S-8.25})$$

according to Poynting's theorem, since the electric field is constant in time, and $\mathbf{J} \cdot \mathbf{E} = 0$ for $r < a$, i.e. inside the solenoid.

(c) Outside the solenoid ($r > a$ we have $\mathbf{B} = 0$. Correspondingly, also $\mathbf{S} = 0$ and $\Phi(\mathbf{S}) = 0$. The rate of change of the magnetic energy is given by (S-8.25) with $r = a$, and must equal the volume integral of $\mathbf{J} \cdot \mathbf{E}$, which is the work done by the induced field on the current flowing in the coils (notice that this is different from the electric field driving the current and causing Joule heating inthe coils, see Problem 13.18). In our representation, the current is distributed on the surface $r = a$, thus $\mathbf{J} d^3 r$ is replaced by $nI dS = nIa \, d\phi dz$ in the integral, and E is evaluated at $r = a$. We thus obtain

$$\int_V \mathbf{J} \cdot \mathbf{E} \, d^3 r = \int_S nI \, E(a) \, dS = -2\pi ah \left(nI_0 \frac{t}{\tau}\right) \left(\frac{2\pi nI_0 a}{c^2 \tau}\right)$$

$$= -4\pi a^2 ht \left(\frac{nI_0}{c\tau}\right)^2 = -\left.\frac{dU_M}{dt}\right|_{r=R} . \qquad (\text{S-8.26})$$

S-8.4 Poynting Vector in a Capacitor with Moving Plates

(a) We use a cylindrical coordinate system (r, ϕ, z), with the z axis along the symmetry axis of the capacitor, and the origin on the fixed plate. Thus, in the limit of our approximations, the electric field is uniform and parallel to $\hat{\mathbf{z}}$ inside the capacitor, whose capacitance is

$$C = \frac{\pi a^2}{4\pi h(t)} = \frac{a^2}{4(h_0 + vt)} . \qquad (\text{S-8.27})$$

In the case of the isolated plates the charge is constant and equal to Q_0, while the voltage between the plates V and the electric field E between the plates are, respectively,

$$V = \frac{Q_0}{C} = Q_0 \frac{4(h_0 + vt)}{a^2} , \qquad E = \frac{V}{h} = \frac{4Q_0}{a^2} . \qquad (\text{S-8.28})$$

In the case of constant voltage between the plates, $V = V_0$, the charge Q of the capacitor and the electric field E are, respectively.

$$Q = CV_0 = V_0 \frac{a^2}{4(h_0 + vt)} , \qquad E = \frac{V_0}{h_0 + vt} . \qquad (\text{S-8.29})$$

In the case of constant charge, the electrostatic force between the plates F_{es} is also constant and equals

$$F_{es} = -Q\frac{E}{2} = -\frac{2Q_0^2}{a^2},$$ (S-8.30)

while in the case of constant voltage we have

$$F_{es} = -Q\frac{E}{2} = -V_0^2\frac{a^2}{8(h_0 + vt)^2},$$ (S-8.31)

in both cases the minus signs means that the force is attractive. In both cases the applied external force $\mathbf{F}_{mech}$ must cancel the electrostatic force, i.e., we must have $F_{mech} = -F_{es}$, for the plates to move at constant velocity.

(b) The electrostatic energy can be written as

$$U = \frac{1}{2}\frac{Q^2}{C} = \frac{1}{2}CV^2,$$ (S-8.32)

so that at constant charge we have

$$U = Q_0^2\frac{2(h_0 + vt)}{a^2}, \qquad \frac{dU}{dt} = \frac{2vQ_0^2}{a^2} > 0,$$ (S-8.33)

while at constant voltage we have

$$U = V_0^2\frac{a^2}{8(h_0 + vt)}, \qquad \frac{dU}{dt} = -\frac{a^2vV_0^2}{8(h_0 + vt)^2} < 0.$$ (S-8.34)

(c) At constant charge, the electric field $\mathbf{E} = E_0\hat{\mathbf{z}}$ is also constant, with $E_0 = Q_0/(\pi a^2)$, therefore the displacement current density $\mathbf{J}_D = \partial_t\mathbf{E}/c$ is zero. Also the conduction current density $\mathbf{J}_C$ is zero between the plates (actually, there is a conduction current localized on the moving plate, we shall come back to this point below), so that also the magnetic field $\mathbf{B}$ is zero between the plates.

At constant voltage, the electric field is $\mathbf{E} = \hat{\mathbf{z}}V_0/h(t) = \hat{\mathbf{z}}V_0/(h_0 + vt)$, implying the presence of a displacement current along $\hat{\mathbf{z}}$. The magnetic field can be calculated by taking the path integral of $\mathbf{B}$ over a circumference of radius $r < a$ coaxial with, and located between, the plates, which equals the flux of the displacement current through the enclosed circle. Due to the cylindrical symmetry of the system, the only nonzero component of $\mathbf{B}$ is azimuthal, $\mathbf{B} = B(r, t)\hat{\boldsymbol{\phi}}$, and calculating its path integral over the circle of radius r corresponding to a field line we have for $B = B(r, t)$

$$2\pi r B = -\frac{\pi r^2}{c}\partial_t E = -\frac{\pi r^2}{c}\frac{V_0 v}{(h_0 + vt)^2},$$ (S-8.35)

so that

$$B = -\frac{r}{2c}\frac{V_0\,v}{(h_0+vt)^2}. \tag{S-8.36}$$

(d) At constant charge we have $\mathbf{B} = 0$, and the Poynting vector $\mathbf{S} = (c/4\pi)\mathbf{E} \times \mathbf{B}$ is also zero. In this case, (S-8.30) and (S-8.33) tell us that the rate of work done against the electric force W_{mech}

$$W_{\text{mech}} = \mathbf{F}_{\text{mech}} \cdot \boldsymbol{v} = -\mathbf{F}_{\text{es}} \cdot \boldsymbol{v} = \frac{2Q_0^2 v}{a^2} \tag{S-8.37}$$

equals the rate of change the electrostatic energy dU/dt. This rate of work must also equal minus the integral of $\mathbf{J} \cdot \mathbf{E}$ over the whole space, according to Poynting's theorem. We verify this at the end of this answer.

At constant voltage, the Poynting vector is radial, $\mathbf{S} = S\hat{\mathbf{r}}$, and, according to (S-8.29) and (S-8.36), we have

$$S = -\frac{c}{4\pi}E_z B_\phi = \frac{V_0^2\,v\,r}{8\pi\,(h_0+vt)^3}. \tag{S-8.38}$$

Evaluating the flux of $\mathbf{S}$ through the minimum closed surface enclosing the capacitor, of lateral surface $2\pi a\,(h_0 + vt)$, we obtain

$$\Phi_S = 2\pi a(h_0+vt)\frac{V_0^2\,v\,a}{8\pi\,(h_0+vt)^3} = \frac{a^2 V_0^2\,v}{4\,(h_0+vt)^2}. \tag{S-8.39}$$

Through (S-8.31) and (S-8.34) we can verify that

$$-\Phi_S = \frac{dU}{dt} + Fv. \tag{S-8.40}$$

Note also that, in this case, Φ_S equals the power *absorbed* by the voltage source. In fact, the current flowing through the circuit is

$$I = \frac{dQ}{dt} = -\frac{a^2 v}{4\,(h_0+vt)^2}\,V_0, \tag{S-8.41}$$

where we have inserted the first of (S-8.29), corresponding to a power absorption by the source

$$W = -V_0 I = \frac{a^2 v}{4\,(h_0+vt)^2}\,V_0^2 = \Phi_s. \tag{S-8.42}$$

We avoided so far to discuss the role of the conduction current circulating in the plates [the following discussion will require some familiarity with the distributions $\delta(x)$ and $\Theta(x)$, where $\Theta(x)$ is the Heaviside step function, defined by $\Theta(x) = 1$ for $x > 0$ and $\Theta(x) = 0$ for $x < 0$; notice that $d\Theta(x)/dx = \delta(x)$]. Let us consider the constant charge case. Since the upper plate has a charge Q_0 distributed on the surface $z = -h_0 + vt$ and moves with velocity v, there is actually a current density

$$\mathbf{J}_C = \frac{Q_0}{\pi a^2}\, v\, \delta(z - h_0 - vt) \,. \tag{S-8.43}$$

On the other hand, the electric field between the plates may be written as

$$\mathbf{E} = -\frac{4Q_0}{a^2} [\Theta(z) - \Theta(z - h_0 - vt)]\hat{\mathbf{z}} \,, \tag{S-8.44}$$

where $\Theta(z)$ is the Heaviside step function, defined by $\Theta(z) = 1$ for $z > 0$ and $\Theta(z) = 0$ for $z < 0$. This expression takes into account the fact that at each time t the field exists only in the $0 < z < vt$ region, so it is actually a time-dependent field. Since $d\Theta(z)/dz = \delta(z)$, the displacement current is

$$\mathbf{J}_D = \frac{1}{c}\partial_t \mathbf{E} = -\frac{4Q_0 v}{a^2 c}\, \delta(z - h_0 - vt)\,\hat{\mathbf{z}} = -\frac{4\pi}{c}\, \mathbf{J}_C \,, \tag{S-8.45}$$

so that the source term for the magnetic field $(4\pi/c)\,\mathbf{J}_C + \mathbf{J}_D$ is zero. It also follows that

$$\mathbf{J}_C \cdot \mathbf{E} = -\frac{1}{4\pi}(\partial_t \mathbf{E}) \cdot \mathbf{E} = -\frac{1}{8\pi}\partial_t E^2 \,, \tag{S-8.46}$$

which ensures energy conservation, since the work done on the current equals the rate of change of the electrostatic energy. In detail, we have

$$\partial_t E^2 = \left(4\pi\sigma_{up}\right)^2 \partial_t \Theta(z - h_0 - vt) = -v \left(\frac{4Q_0}{a^2}\right)^2 \delta(z - h_0 - vt) \tag{S-8.47}$$

thus

$$\int \mathbf{J}_C \cdot \mathbf{E}\, d^3 r = \frac{v (4\pi\sigma)^2}{8\pi} \int \delta(z - h_0 - vt)\, d^3 r$$

$$= 2\pi^2 a^2 \sigma^2 v = \frac{2v Q_0^2}{a^2} = \frac{dU}{dt} \,. \tag{S-8.48}$$

S-8.5 Radiation Pressure on a Perfect Mirror

(a) We consider the case of perpendicular incidence
first, and choose a Cartesian reference frame with
the x axis perpendicular to the mirror surface. The
incident plane wave packet has duration τ (with
$\tau \gg 2\pi/\omega$, the laser period), corresponding to a
length $c\tau$, and propagates along $\hat{\mathbf{x}}$ (Fig. S-8.3). We
want to calculate how much momentum is trans-
ferred to an area A of the mirror surface during the
reflection of the whole wave packet. The momentum transferred per unit time and area is the pressure
exerted by the radiation.

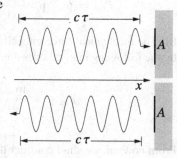

Fig. S-8.3

The momentum density of an EM field is $\mathbf{S}/c^2$,
where $\mathbf{S} = c\mathbf{E} \times \mathbf{B}/4\pi$ is the Poynting vector. Thus the total momentum delivered by
the incident wave packet on the area A is

$$\mathbf{p}_i = \left\langle \frac{\mathbf{S}_i}{c^2} \right\rangle c\tau A = \hat{\mathbf{x}}\,\frac{I}{c}\,\tau A \qquad (S\text{-}8.49)$$

where the angle brackets denote the average over one cycle, $\mathbf{S}_i$ is the Poynting vector
of the incident packet, and $I = |\langle\mathbf{S}_i\rangle|$ is the intensity of the incident pulse (the average
flux of energy per unit time and area), according to Poynting's theorem of energy
conservation.

The reflected wave packet carries a total momentum, over the area A,

$$\mathbf{p}_r = \left\langle \frac{\mathbf{S}_r}{c^2} \right\rangle c\tau A = -\hat{\mathbf{x}}\,\frac{I}{c}\,\tau A \qquad (S\text{-}8.50)$$

where $\mathbf{S}_r = -\mathbf{S}_i$ is the Poynting vector of the reflected packet. The momentum transferred to the mirror over the surface area A during the time interval τ is thus

$$\Delta\mathbf{p} = \mathbf{p}_i - \mathbf{p}_r = |\Delta\mathbf{p}|\,\hat{\mathbf{x}} \qquad (S\text{-}8.51)$$

and the corresponding pressure is

$$P_{\text{rad}} = \frac{|\Delta\mathbf{p}|}{\tau A} = 2\,\frac{I}{c}. \qquad (S\text{-}8.52)$$

Using a similar heuristic argument, it is quite straightforward to find the radiation
pressure for oblique incidence at an angle θ from the normal to the mirror surface. In
fact, in this case the momentum transferred to the mirror along the normal is

$$\Delta\mathbf{p} = \mathbf{p}_i - \mathbf{p}_r = 2\hat{\mathbf{x}}\,\frac{I}{c^2}\,c\tau A \cos\theta = 2\hat{\mathbf{x}}\,\frac{I}{c}\,\tau A \cos\theta\,, \qquad (S\text{-}8.53)$$

and the area of incidence is now $A/\cos\theta$. Thus

$$P_{\text{rad}} = |\mathbf{p}_i - \mathbf{p}_r| \frac{\cos\theta}{\tau A} = 2\frac{I}{c}\tau A\cos\theta \frac{\cos\theta}{\tau A} = 2\frac{I}{c}\cos^2\theta . \quad \text{(S-8.54)}$$

(b) The mechanical force on a closed system of charges, currents and fields is given by the following integral over the volume of the system

$$\mathbf{F}_{\text{mech}} = \frac{d\mathbf{p}_{\text{mech}}}{dt} = \int_V \left(\varrho\mathbf{E} + \frac{1}{c}\mathbf{J}\times\mathbf{B}\right) d^3r. \quad \text{(S-8.55)}$$

From now on, we shall consider the case of perpendicular incidence only, and leave the case of oblique incidence as a further exercise for the reader. In the present case, $\varrho = 0$ everywhere and only the magnetic term contributes. Thus, in plane geometry the time-averaged force on a planar surface of area A is

$$\langle F_{\text{mech}}\rangle = \int_0^{+\infty} \left\langle\frac{1}{c}\mathbf{J}\times\mathbf{B}\right\rangle A\,dx \quad \text{(S-8.56)}$$

and is directed along $\hat{\mathbf{x}}$ for symmetry reasons.

The current in a perfect mirror is localized on the surface, where the magnetic field is discontinuous. Here we assume that the wave fields $\mathbf{B}$ and $\mathbf{E}$ are parallel to $\hat{\mathbf{z}}$ and $\hat{\mathbf{y}}$, respectively. Let $\mathbf{E}_i(x, t) = \hat{\mathbf{y}}E_i\cos(kx - \omega t)$ be the incident electric field. The total field $\mathbf{E}(x, t)$ is the sum of $\mathbf{E}_i$ and the field $\mathbf{E}_r(x, t) = -\hat{\mathbf{y}}E_i\cos(-kx - \omega t)$ of the reflected wave, so that $\mathbf{E}(0, t) = 0$. The total fields for $x < 0$ thus have the form of standing waves

$$E_y(x, t) = 2E_i\sin(kx)\sin\omega t , \quad \text{(S-8.57)}$$
$$B_z(x, t) = 2E_i\cos(kx)\cos\omega t . \quad \text{(S-8.58)}$$

The discontinuity of B_z leads to a surface current $J_y = K_y\,\delta(x)$ where

$$K_y = -\frac{c}{4\pi}\left[B_z(0^+, t) - B_z(0^-, t)\right] = \frac{c}{4\pi}B_z(0^-, t) = \frac{c}{2\pi}E_i\cos\omega t \quad \text{(S-8.59)}$$

where we have used Stokes' theorem and $B_z(0^+, t) = 0$. The force per unit surface, i.e., the pressure, is given by the surface current multiplied by the mean value of the field across the current layer (the argument is identical to the one used for calculating the electrostatic pressure on a surface charge layer in electrostatics):

$$P_{\text{rad}} = \left\langle K_y \frac{1}{2c} \left[B_z(0^+, t) + B_z(0^-, t) \right] \right\rangle = \frac{c}{8\pi} \left\langle B_z^2(0^-, t) \right\rangle$$

$$= \frac{c}{8\pi} \left(\frac{1}{2} 4 |E_i|^2 \right) = 2\frac{I}{c}, \tag{S-8.60}$$

since $B_z(0^-, t) = 2E_i \cos \omega t$, and $I = (c/4\pi) \left(|E_i|^2/2 \right)$. This is equivalent to evaluate the integral in (S-8.56) as

$$\int_{0^-}^{+\infty} J_y B_z \, dx = -\int_{0^-}^{+\infty} \left(\frac{c}{4\pi} \partial_x B_z - \frac{1}{4\pi} \partial_t E_y \right) B_z \, dx = -\frac{c}{4\pi} \int_{0^-}^{+\infty} \frac{1}{2} \partial_x B_z^2 \, dx$$

$$= \frac{c}{8\pi} B_z^2(0^-, t), \tag{S-8.61}$$

where we used the fact that $E_y = 0$ and $\partial_t E_y = 0$ for $x \geq 0^-$.

(c) The momentum conservation theorem (8.8) states that, for a closed system of charges, currents and EM fields bounded by a closed surface S, the following balance equation holds:

$$\frac{d}{dt} (\mathbf{p}_{\text{mech}} + \mathbf{p}_{\text{EM}})_i = \oint_S \sum_j T_{ij} \, \hat{n}_j \, d^2r, \tag{S-8.62}$$

where $i, j = x, y, z$, T_{ij} is the Maxwell stress tensor, and $\hat{n}_j$ is the j component of the outward-pointing unit vector locally normal to S. Thus, the integral on the right-hand side is the outward the flux of the vector $T \cdot \hat{n}$ through S. In (S-8.62), $\mathbf{p}_{\text{mech}}$ is the mechanical momentum of the system, while the momentum associated to the EM field is

$$\mathbf{p}_{\text{EM}} = \int_V \mathbf{g} \, d^3r, \tag{S-8.63}$$

where $\mathbf{g} = \mathbf{S}/c^2$ is the momentum density (8.9), and the integral is evaluated over the volume bounded by S.

In our case, we take the front surface A of the mirror and close it by adding a surface extending deep in the mirror, where the fields are zero. Thus, the amount of EM momentum which flows into the mirror (and "transformed" into mechanical momentum) is given by the integral

$$\int_A \sum_j T_{ij} \, \hat{n}_j d^2r = A \sum_j T_{ij}(0^-, t) \, \hat{n}_j. \tag{S-8.64}$$

The radiation pressure on the mirror is the time-averaged momentum flow per unit area,

$$P_{\text{rad}} = \left\langle \sum_j T_{1j}(0^-, t)\,\hat{n}_j \right\rangle = -\langle T_{11}(0^-, t)\rangle, \qquad (S\text{-}8.65)$$

since, in our case, $\hat{n} = (-1, 0, 0)$. Thus we actually need to evaluate $T_{11}(0, t)$ only:

$$T_{11}(0, t) = -\frac{1}{8\pi} B_z^2(0^-, t) . \qquad (S\text{-}8.66)$$

The radiation pressure is thus

$$P_{\text{rad}} = -\langle T_{11}(0, t)\rangle = \frac{1}{8\pi} \langle B_z^2(0^-, t)\rangle = \frac{1}{4\pi}|E_i|^2 = 2\frac{I}{c} . \qquad (S\text{-}8.67)$$

S-8.6 A Light Beam

(a) The divergence of the electric field in vacuum is zero. With our geometry, this means that, since we have assumed $E_y = 0$, we have

$$0 = \nabla \cdot \mathbf{E} = \partial_x E_x + \partial_z E_z . \qquad (S\text{-}8.68)$$

From (S-8.68) and (8.14) we obtain

$$\partial_z E_z = -\partial_x E_x = 2E_0 \frac{x}{k_0 r_0^2} e^{-r^2/r_0^2} \cos(k_0 z - \omega t), \qquad (S\text{-}8.69)$$

which gives, by integrating with respect to z,

$$E_z = E_0 \frac{2x}{k_0 r_0^2} e^{-r^2/r_0^2} \sin(k_0 z - \omega t) . \qquad (S\text{-}8.70)$$

By proceeding analogously for $\mathbf{B}$ we find

$$B_z = B_0 \frac{2y}{k_0 r_0^2} e^{-r^2/r_0^2} \sin(k_0 z - \omega t) . \qquad (S\text{-}8.71)$$

The z-components have maxima near $r = r_0$ and their amplitude is (as an order of magnitude)

$$\max(E_z, B_z) \simeq \frac{1}{k_0 r_0}(E_0, B_0) . \qquad (S\text{-}8.72)$$

Thus the larger the pulse width r_0 with respect to the wavelength $\lambda_0 = 2\pi/k_0$, the lower the amplitude of the longitudinal components.

(b) Let us verify if the longitudinal fields are consistent with Maxwell's equations. First, we check if $\partial_t E_z = c\,(\nabla \times \mathbf{B})_z = c\,\partial_x B_y$:

$$\partial_x B_y = -B_0 \frac{2x}{r_0^2} e^{-r^2/r_0^2} \cos(k_0 z - \omega t)\,, \tag{S-8.73}$$

$$\partial_t E_z = -E_0 \omega \frac{2x}{k_0 r_0^2} e^{-r^2/r_0^2} \cos(k_0 z - \omega t)\,, \tag{S-8.74}$$

thus $(\omega/k_0)E_0 = B_0 c$ must hold. Analogously we check if $\partial_t B_z = -c(\nabla \times \mathbf{E})_z = \partial_y E_x$:

$$\partial_y E_x = -E_0 \frac{2y}{r_0^2} e^{-r^2/r_0^2} \cos(k_0 z - \omega t)\,, \tag{S-8.75}$$

$$\partial_t B_z = -B_0 \omega \frac{2y}{k_0 r_0^2} e^{-r^2/r_0^2} \cos(k_0 z - \omega t)\,, \tag{S-8.76}$$

thus $E_0 = (\omega/k_0 c)B_0$ must hold as well. Combining the two conditions we obtain

$$\omega = k_0 c\,, \qquad E_0 = B_0/c\,, \tag{S-8.77}$$

which are the same relations as for a plane monochromatic wave.
(c) The Poynting vector is

$$\mathbf{S} = \frac{c}{4\pi} \mathbf{E} \times \mathbf{B} = \frac{c}{4\pi} \left(-\hat{\mathbf{x}} E_z B_y - \hat{\mathbf{y}} E_x B_z + \hat{\mathbf{z}} E_x B_y \right)\,, \tag{S-8.78}$$

and the averages of its components over one cycle are

$$\langle S_x \rangle = -\frac{c}{4\pi} \frac{2x}{k r_0^2} E_0^2 e^{-2r^2/r_0^2} \langle \sin(kz - \omega t)\cos(kz - \omega t) \rangle = 0\,, \tag{S-8.79}$$

$$\langle S_y \rangle = -\frac{c}{4\pi} \frac{2y}{k r_0^2} E_0^2 e^{-2r^2/r_0^2} \langle \cos(kz - \omega t)\sin(kz - \omega t) \rangle = 0\,, \tag{S-8.80}$$

$$\langle S_z \rangle = \frac{c}{4\pi} E_0^2 e^{-2r^2/r_0^2} \langle \cos^2(kz - \omega t) \rangle = \frac{c}{8\pi} E_0^2 e^{-2r^2/r_0^2}\,. \tag{S-8.81}$$

Thus, we can define the local intensity and the total power of the beam as

$$I(r) = \langle S_z \rangle\,, \qquad P = \int_0^\infty I(r)\, 2\pi r\, dr\,. \tag{S-8.82}$$

(d) Since

$$\nabla^2 \begin{pmatrix} E_x \\ B_y \end{pmatrix} = (\partial_x^2 + \partial_y^2) \begin{pmatrix} E_x \\ B_y \end{pmatrix}$$

$$= \left[\frac{4}{r_0^2} \left(\frac{x^2 + y^2}{r_0^2} - 1 \right) - k_0^2 \right] \begin{pmatrix} E_x \\ B_y \end{pmatrix} ,$$

$$= -\frac{\omega^2}{c^2} \left[1 - \frac{4}{(k_0 r_0)^2} \left(\frac{r^2}{r_0^2} - 1 \right) \right] \begin{pmatrix} E_x \\ B_y \end{pmatrix} \qquad \text{(S-8.83)}$$

it is evident that $(\nabla^2 + \omega^2/c^2)(E_x, B_y) \neq 0$ because of an "extra" term proportional to $\sim (k_0 r_0)^{-2}$. Thus the expression (8.14) may be accurate only if $r_0 \gg 1/k = \lambda/2\pi$, i.e., if the beam is much wider than the (main) wavelength.

One may thus wonder why the longitudinal components E_z and B_z obtained from the "wrong" E_x and B_y in (8.14) seemed to satisfy Maxwell's equations. Actually this is true only up to first order in $(k_0 r_0)^{-1}$. In fact, by noting that

$$\partial_t B_x = -c(\nabla \times \mathbf{E})_x = c\partial_y E_z \neq 0 , \qquad \text{(S-8.84)}$$

$$\partial_t E_y = c(\nabla \times \mathbf{B})_y = -c\partial_x B_z \neq 0 , \qquad \text{(S-8.85)}$$

one should consider additional components E_y and B_x with a maximum amplitude

$$\max(E_y, B_x) \sim \frac{1}{(k_0 r_0)^2}(E_0, B_0) . \qquad \text{(S-8.86)}$$

This is consistent with (8.14) and the associated E_z, B_z being accurate up to the order $\sim (k_0 r_0)^{-2}$.

It may have been apparent *a priori* that (8.14) is not a solution of Maxwell's equations since it is known that a beam with finite width actually undergoes diffraction. The width of a Gaussian beam doubles after a typical distance, called Rayleigh length, $r_R = k r_0^2$. This corresponds to an aperture angle

$$\theta_d \simeq \frac{r_0}{z_R} = \frac{1}{k r_0} \simeq \frac{\lambda}{r_0} . \qquad \text{(S-8.87)}$$

It might be interesting to note that this result may be inferred by the values for the longitudinal field components obtained at point **(a)**. In fact, the beam may be obtained as a linear superposition of plane waves of the same frequency but different wavevectors. For a monochromatic plane wave, the electric and magnetic field are perpendicular to the wavevector **k**. Thus, if as an estimate $E_z/E_y \sim 2/(k_0 r_0)$ one infers that the beam must be a superposition of plane waves with a typical ratio $k_x/k_z \sim 2/(k_0 r_0)$ which determines the typical angular spread of the wavevector spectrum, hence the spreading angle of the beam.

S-8.7   Intensity and Angular Momentum of a Light Beam

(a) First, we define the shorthand symbols $C = \cos(kz - \omega t)$, $S = \sin(kz - \omega t)$, and $E_0' = \partial_r E_0(r)$, that we shall use throughout the solution of the problem. We have for the intensity of the beam

$$I(r) \equiv S_z = \frac{c}{4\pi}(E_x B_y - E_y B_x) = \frac{c}{4\pi} E_0^2(r)\,[CC - (-SS)]$$

$$= \frac{c}{4\pi} E_0^2(r)\,[C^2 + S^2] = \frac{c}{4\pi} E_0^2(r)\,. \tag{S-8.88}$$

(b) The divergence of the fields in vacuum must be zero. For the electric field we have

$$0 = \nabla \cdot \mathbf{E} = \partial_x E_x + \partial_y E_y + \partial_z E_z \,, \tag{S-8.89}$$

thus

$$\partial_z E_z = -\partial_x E_x - \partial_y E_y = -\frac{x}{r} E_0'(r)\, C + \frac{y}{r} E_0'(r)\, S,$$

and, integrating with respect to z,

$$E_z = -\frac{1}{kr} E_0'(r)\,[x\,S + y\,C] \,. \tag{S-8.90}$$

Analogously, we can evaluate B_z:

$$\partial_z B_z = -\partial_x B_x - \partial_y B_y = -\frac{x}{rc} E_0'(r)\, S - \frac{y}{rc} E_0'(r)\, C \,,$$

$$B_z = +\frac{1}{krc} E_0'(r)\,[x\,C - y\,S] \,. \tag{S-8.91}$$

(c) The x and y components of the Poynting vector are

$$S_x = \frac{c}{4\pi}(E_y B_z - E_z B_y)$$

$$= \frac{c}{4\pi}\left\{ (-E_0 S)\left[\frac{E_0'}{krc}(x\,C - y\,S) \right] - \left[-\frac{E_0'}{kr}(x\,S + y\,C) \right]\frac{1}{c}E_0 C \right\}$$

$$= \frac{c}{4\pi}\frac{E_0 E_0'}{kr}\left(-x\,SC + y\,S^2 + x\,SC + y\,C^2\right) = \frac{c}{4\pi} E_0 E_0' \frac{y}{kr}\,. \tag{S-8.92}$$

$$S_y = \frac{c}{4\pi}(E_z B_x - E_x B_z)$$

$$= \frac{c}{4\pi}\left\{ \left[-\frac{E_0'}{kr}(x\,S + y\,C) \right]\frac{1}{c}E_0 S - (E_0 C)\left[\frac{E_0'}{krc}(x\,C - y\,S) \right] \right\}$$

$$= \frac{c}{4\pi}\frac{E_0 E_0'}{kr}\left(-x\,S^2 - y\,CS - x\,C^2 + y\,CS\right) = -\frac{c}{4\pi} E_0 E_0' \frac{x}{kr}\,. \tag{S-8.93}$$

Since we have

$$\frac{c}{4\pi} E_0(r) E_0'(r) = \frac{c}{8\pi} \partial_r E_0^2(r) = \frac{1}{2} \partial_r I(r) , \tag{S-8.94}$$

the Poynting vector can be written

$$\mathbf{S} = \left(\frac{y}{2kr} \partial_r I(r), \ -\frac{x}{2kr} \partial_r I(r), \ I(r) \right) . \tag{S-8.95}$$

Assuming a Gaussian beam, we have $E_0(r) \propto e^{-r^2/r_0^2}$, and $S_{x,y} \propto S_z/(kr_0) \propto \theta_d S_z$, with θ_d the diffraction angle of (S-8.87).

(d) We have

$$\ell_z = \frac{1}{c^2}(x \, S_y - y \, S_x) = \frac{-x^2 - y^2}{2krc^2} \partial_r I(r) = -\frac{r}{2kc^2} \partial_r I(r)$$

$$= -\frac{r}{2c\omega} \partial_r I(r). \tag{S-8.96}$$

We eventually obtain the total angular momentum by integrating the above expression by parts,

$$L_z = \int\limits_0^\infty \ell_z(r) \, 2\pi r \, dr = -\int\limits_0^\infty \frac{r}{2c\omega} \partial_r I(r) \, 2\pi r \, dr$$

$$= \frac{1}{c\omega} \int\limits_0^\infty I(r) \, 2\pi r \, dr = \frac{W}{c\omega} . \tag{S-8.97}$$

S-8.8   Feynman's Paradox Solved

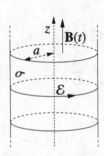

Fig. S-8.4

(a) We use a cylindrical coordinate system (r, ϕ, z) with the cylinder axis as z axis. The induced electric field $\mathbf{E}_{ind}$ has azimuthal symmetry, i.e. $\mathbf{E}_{ind} = E_\phi(r, t)\hat{\phi}$, and can be obtained from Faraday's law by equating its line integral over the circumference of radius r to the temporal derivative of the magnetic field flux through the circle (Fig. S-8.4):

$$E_\phi = -\frac{r}{2c} \partial_t B_{ext}(t) . \tag{S-8.98}$$

(We assumed the slowly varying current approximation, whose validity is ensured by the $t_f \gg a/c$ condition.)

On an infinitesimal surface element of the cylindrical surface $dS = a d\phi dz$ the induced electric field exerts a force

$$\mathbf{df} = \hat{\boldsymbol{\phi}} df = \hat{\boldsymbol{\phi}} \sigma E_\phi(r = a) \, dS = -\hat{\boldsymbol{\phi}} \sigma \frac{a}{2c} \, \partial_t B_{\text{ext}}(t) \, dS \, , \qquad \text{(S-8.99)}$$

where $\sigma = Q/(2\pi a)$ is the surface charge density. The corresponding mechanical torque is $d\boldsymbol{\tau} = \hat{\mathbf{z}} a \, df$. By integrating over the whole surface of the cylinder we obtain for the total torque

$$\boldsymbol{\tau} = -\frac{\pi a^3 h \sigma}{c} \, \partial_t \mathbf{B}_{\text{ext}}(t) \, . \qquad \text{(S-8.100)}$$

The equation of motion for the rotation of the cylinder is

$$I \frac{d\boldsymbol{\omega}}{dt} = \boldsymbol{\tau} = -\frac{\pi a^3 h \sigma}{c} \, \partial_t \mathbf{B}_{\text{ext}}(t) \, , \qquad \text{(S-8.101)}$$

which has the solution (the total time derivative being trivially equivalent to the partial derivative when applied to $\mathbf{B}_{\text{ext}}(t)$)

$$\boldsymbol{\omega}(t) = -\frac{\pi a^3 h}{Ic} \sigma \left[\mathbf{B}_{\text{ext}}(t) - \mathbf{B}_{\text{ext}}(0) \right] = -\frac{a^2 Q}{2Ic} \left[\mathbf{B}_{\text{ext}}(t) - \mathbf{B}_{\text{ext}}(0) \right] \, . \text{ (S-8.102)}$$

The angular momentum $L_c(t) = I\boldsymbol{\omega}(t)$. The final values are dependent only on the initial value of $\mathbf{B}_{\text{ext}}$ and not on its temporal profile, i.e.

$$\boldsymbol{\omega}(t_f) = -\frac{a^2}{2Ic} Q \mathbf{B}_0 \, . \qquad \text{(S-8.103)}$$

(b) The rotation of the charged cylinder leads to a surface current $\mathbf{K}$ at $r = a$,

$$\mathbf{K} = \sigma \boldsymbol{v} = \sigma a \omega \, \hat{\boldsymbol{\phi}} \, . \qquad \text{(S-8.104)}$$

Such current generates an uniform magnetic field $\mathbf{B}_{\text{ind}}$ inside the long cylinder (equivalent to a solenoid where $nI = K$),

$$\mathbf{B}_{\text{ind}} = \frac{4\pi}{c} K \hat{\mathbf{z}} = \frac{4\pi}{c} \sigma a \boldsymbol{\omega} \, . \qquad \text{(S-8.105)}$$

We now proceed as in point a) but adding the induced field $\mathbf{B}_{\text{ind}}$ to the external field $\mathbf{B}_{\text{ext}}$:

$$I\frac{d\omega}{dt} = \tau = -\frac{\pi a^3 h\sigma}{c}\partial_t\left(\mathbf{B}_{ext}(t) + \mathbf{B}_{ind}(t)\right)$$

$$= -\frac{a^2 Q}{2c}\partial_t\left(\mathbf{B}_{ext}(t) + \frac{4\pi}{c}\sigma a\omega\right)$$

$$= -\frac{a^2 Q}{2c}\partial_t\mathbf{B}_{ext}(t) - \frac{a^2 Q^2}{hc^2}\frac{d\omega}{dt}\,,\qquad\text{(S-8.106)}$$

which can be rewritten as

$$I'\frac{d\omega}{dt} = \tau = -\frac{\pi a^3 h\sigma}{c}\partial_t\mathbf{B}_{ext}(t)\,,\qquad\text{(S-8.107)}$$

$$I' = I + \frac{a^2 Q^2}{hc^2}\,.\qquad\text{(S-8.108)}$$

Equation (S-8.107) is identical to Eq. (S-8.101) but for the replacement $I \to I'$, which means that the effects of the rotation-induced magnetic field $\mathbf{B}_{ind}$ are equivalent to an additional inertia of the cylinder. The final velocity becomes

$$\omega'(t_f) = -\frac{a^2}{2I'c}Q\,\mathbf{B}_0\,.\qquad\text{(S-8.109)}$$

Notice that the total magnetic field does not vanish inside the cylinder at $t = t_f$, being equal to the induced field

$$\mathbf{B}_{tot}(t_f) = \mathbf{B}_{ind}(t_f) = \frac{4\pi}{c}\sigma a\omega'(t_f)\,.\qquad\text{(S-8.110)}$$

(c) For a magnetic field $\mathbf{B} = B_z\hat{\mathbf{z}}$ and a configuration with cylindrical symmetry the density of EM angular momentum in Eq. (8.18) becomes

$$\boldsymbol{\ell} \equiv \mathbf{r} \times \mathbf{g} = -\frac{1}{4\pi}r E_r B_z\hat{\mathbf{z}}\,.\qquad\text{(S-8.111)}$$

The contribution of the induced electric field E_ϕ vanishes in the vector product. However, the angular momentum is not zero because of the radial electrostatic field inside the cylinder, which is easily found from Gauss's theorem:

$$E_r(r) = \begin{cases} \dfrac{2\lambda}{r} = -\dfrac{2Q}{hr} & (r < a) \\[2mm] 0 & (r > a) \end{cases}\,.\qquad\text{(S-8.112)}$$

Thus, $\boldsymbol{\ell} \neq 0$ inside the cylinder ($r < a$). The total EM angular momentum is thus given by

$$\mathbf{L}_{\mathrm{EM}} = \frac{1}{4\pi} B_z \int\limits_0^a r \, \frac{2\lambda}{r} 2\pi r h \, \mathrm{d}r = \frac{1}{2} B_z \lambda h a^2 \hat{\mathbf{z}} = \frac{Qa^2}{2} \mathbf{B} . \qquad \text{(S-8.113)}$$

Notice that $\mathbf{B}$ represents the total field inside the cylinder and that the equation for $\mathbf{L}_{\mathrm{EM}}$ is valid at any time. Now, Eq. (S-8.102) can be rewritten (using the total field) as

$$I\omega(t) + \frac{a^2 Q}{2c} \mathbf{B}_{\mathrm{ext}}(t) = \frac{a^2 Q}{2c} \mathbf{B}_{\mathrm{ext}}(0) , \qquad \text{(S-8.114)}$$

which is equivalent to

$$\mathbf{L}_c(t) + \mathbf{L}_{\mathrm{EM}}(t) = \mathbf{L}_{\mathrm{EM}}(0) , \qquad \text{(S-8.115)}$$

thus showing the conservation of the total angular momentum of the system, since $\mathbf{L}_c(0) = 0$. The "paradox" thus consists in ignoring that a static EM field can contain a finite angular momentum. Similar considerations hold for Problem 6.6 where, however, the EM angular momentum is more difficult to calculate.[4]

S-8.9   Magnetic Monopoles

(a) We build a magnetic dipole $\mathbf{m}$ by locating two magnetic charges (magnetic monopoles) $+q_m$ and $-q_m$ at a distance $\mathbf{h}$ from each other, so that $\mathbf{m} = q_m \mathbf{h}$. The magnetic field at distances $r \gg h$ can be evaluated from (8.19), using the same approximations as for the field of an electric dipole, obtaining

$$\mathbf{B}_{\mathrm{dip}} = \alpha \, \frac{(\mathbf{m} \cdot \hat{\mathbf{r}}) \, \hat{\mathbf{r}} - \mathbf{m}}{r^3} . \qquad \text{(S-8.116)}$$

On the other hand, the field of a usual magnetic dipole $\mathbf{m} = I\,\mathbf{S}$, consisting of a small circular loop of surface $\mathbf{S}$ carrying a current I, with the head of $\mathbf{S}$ pointing so that it "sees" I circulating counterclockwise, is

$$\mathbf{B}_{\mathrm{dip}} = k_{\mathrm{m}} \frac{(\mathbf{m} \cdot \hat{\mathbf{r}}) \, \hat{\mathbf{r}} - \mathbf{m}}{r^3} . \qquad \text{(S-8.117)}$$

Comparing the formulas, we obtain $\alpha = k_{\mathrm{m}}$, i.e., $\alpha = \mu_0 / 4\pi = 1/4\pi\varepsilon_\theta c^2$ in SI units, and $\alpha = 1/c$ in Gaussian units.

The magnetic force an electric charge q_e, moving with velocity $\boldsymbol{v}$ in the presence of a magnetic field $\mathbf{B}$, is $\mathbf{f}_L = q_e \, b_{\mathrm{m}} \, \boldsymbol{v} \times \mathbf{B}$. The force exerted by a magnetic field $\mathbf{B}$

---

[4] The present version of Feynman's "paradox" is taken from J. Belcher and K. McDonald (http://cosmology.princeton.edu/~mcdonald/examples/feynman_cylinder.pdf) who further discuss subtle aspects of this problem.

on a magnetic monopole of charge q_m is $\mathbf{f}_m = q_m \mathbf{B}$. Thus the physical dimensions of the magnetic charge q_m are

$$[q_m] = [q_e \, b_{\mathrm{m}} \, v] = \begin{cases} [q_e \, v], & \text{SI} \\ [q_e], & \text{Gaussian} \end{cases} \tag{S-8.118}$$

i.e., the same physical dimensions as an electric charge in Gaussian units, and the dimensions of an electric charge times a velocity in SI units.

(b) In analogy with the equation $\nabla \cdot \mathbf{E} = 4\pi k_e \, \rho_e$, where ρ_e is the volume density of electric charge, Maxwell's equation $\nabla \cdot \mathbf{B} = 0$ is modified as

$$\nabla \cdot \mathbf{B} = 4\pi k_{\mathrm{m}} \, \rho_m = \begin{cases} \mu_0 \, \rho_m & \text{SI} \\ \dfrac{4\pi}{c} \, \rho_m & \text{Gaussian,} \end{cases} \tag{S-8.119}$$

where ρ_m is the volume density of magnetic charge. Equation (S-8.119) can be proved by first observing that, in the presence of magnetic charges, Gauss's law for the magnetic field is

$$\oint \mathbf{B} \cdot d\mathbf{S} = 4\pi k_{\mathrm{m}} \, Q_m = 4\pi k_{\mathrm{m}} \int \rho_m \, d^3 x \,, \tag{S-8.120}$$

where the flux of B is evaluated through any closed surface, and Q_m is the net magnetic charge inside the surface, then applying the divergence theorem.

The conservation of magnetic charge is expressed by the continuity equation

$$\nabla \cdot \mathbf{J}_m = -\partial_t \rho_m \,. \tag{S-8.121}$$

Maxwell's equation for $\nabla \times \mathbf{E}$ (describing Faraday's law of induction) must be completed in order to take the magnetic current density into account, by writing

$$\nabla \times \mathbf{E} = \eta \, \mathbf{J}_m - b_{\mathrm{m}} \, \partial_t \mathbf{B} \,. \tag{S-8.122}$$

The constant η can be determined, for instance, by applying the divergence operator to both sides of the equation, remembering the divergence of the curl of any vector field is always zero,

$$0 = \nabla \cdot (\nabla \times \mathbf{E}) = \eta \nabla \cdot \mathbf{J}_m - b_{\mathrm{m}} \, \partial_t \nabla \cdot \mathbf{B} = \eta \nabla \cdot \mathbf{J}_m - 4\pi k_{\mathrm{m}} \, \partial_t \rho_m \,, \tag{S-8.123}$$

from which $\eta = -4\pi k_{\mathrm{m}}$ follows. We thus obtain

$$\nabla \times \mathbf{E} = -4\pi k_{\mathrm{m}} \mathbf{J}_m - b_{\mathrm{m}} \partial_t \mathbf{B} = \begin{cases} -\mu_0 \mathbf{J}_m - \partial_t \mathbf{B} \,, & \text{SI} \\ -\dfrac{4\pi}{c} \, \mathbf{J}_m - \dfrac{1}{c} \mathbf{B} \,, & \text{Gaussian.} \end{cases} \tag{S-8.124}$$

(c) We choose a cylindrical reference frame (r, ϕ, z) with the z axis coinciding with the axis of the beam. Because of the cylindrical symmetry of our magnetic charge distribution, the only non-zero component of the magnetic field is B_r. Applying Gauss's law to a cylindrical surface coaxial with the beam we obtain

$$B_r = \begin{cases} 2\pi k_m n q_m r, & r \leqslant a \\ \dfrac{2\pi k_m n q_m a^2}{r}, & r \geqslant a. \end{cases} \tag{S-8.125}$$

The electric field $\mathbf{E}$ is solenoidal and can be obtained by applying Kelvin-Stokes theorem to a circular path of radius r coaxial with the beam

$$\oint \mathbf{E} \cdot d\boldsymbol{\ell} = 2\pi r E_\phi = \int \boldsymbol{\nabla} \times \mathbf{E} \cdot d\mathbf{S} = \begin{cases} -\pi r^2 4\pi k_m n q_m v, & r \leqslant a \\ -\pi r^2 4\pi k_m n q_m \dfrac{a^2}{r} v, & r \geqslant a, \end{cases} \tag{S-8.126}$$

leading finally to

$$E_\phi = \begin{cases} 2\pi k_m n q_m v r & \text{if } r \leq a \\ \dfrac{2\pi k_m n q_m v a^2}{r} & \text{if } r \geq a. \end{cases} \tag{S-8.127}$$

Thus, for instance for $r \leqslant a$, we have

$$E_\phi = \begin{cases} \dfrac{\mu_0 n q_m v r}{2}, & \text{SI} \\ \dfrac{2\pi n q_m v r}{c}, & \text{Gaussian.} \end{cases} \tag{S-8.128}$$

S-8.10 An Interstellar Light Sail

(a) The energy of an electromagnetic field filling the volume V is

$$U = \int_V \left(\frac{E^2}{8\pi} + \frac{B^2}{8\pi} \right) d^3 r, \tag{S-8.129}$$

and its momentum is

$$\mathbf{p} = \int_V \left(\frac{1}{4\pi c} \mathbf{E} \times \mathbf{B} \right) d^3 r. \tag{S-8.130}$$

Approximating the wave packet to a plane wave we have

$$B = E , \qquad \mathbf{E} \times \mathbf{B} = \hat{\mathbf{x}} E B , \tag{S-8.131}$$

where $\hat{\mathbf{x}}$ is the unit vector in the propagation direction of the wave. Thus

$$U = \int_V \frac{E^2}{4\pi} d^3r , \qquad \mathbf{p} = \int_V \frac{E^2}{4\pi c} d^3r = \hat{\mathbf{x}} \frac{U}{c} . \tag{S-8.132}$$

(b) We use energy and momentum conservation, denoting by U_i and $\mathbf{p}_i$ the energy and momentum of the wave packet before reflection from the mirror, respectively, and by U_f and $\mathbf{p}_f$ the same quantities after reflection. We have

$$U_i = \frac{1}{2} M v^2 + U_f , \tag{S-8.133}$$

$$\mathbf{p}_i = M \boldsymbol{v} + \mathbf{p}_f . \tag{S-8.134}$$

Since we are dealing with a one-dimensional motion, we can use the magnitudes of momenta and velocities, remembering that $\mathbf{p}_f$ is antiparallel to $\mathbf{p}_i$, while $\boldsymbol{v}$ is parallel. Thus (S-8.134) is rewritten

$$p_i = M v - p_f . \tag{S-8.135}$$

Inserting the second of (S-8.132) we obtain

$$U_f = M v c - U_i , \tag{S-8.136}$$

which can be substituted into (S-8.133) leading to

$$v^2 + 2cv - \frac{4U_i}{M} = 0 , \tag{S-8.137}$$

with solutions

$$v = -c \pm \sqrt{c^2 + \frac{4U_i}{M}} , \tag{S-8.138}$$

of which only the solution with the + sign is acceptable. Thus we have, at the first order in U_i/Mc^2,

$$v = c\sqrt{1 + \frac{4U_i}{Mc^2}} - c \simeq c\left(1 + \frac{2U_i}{Mc^2}\right) - c = \frac{2U_i}{Mc} . \tag{S-8.139}$$

It is also possible to treat the relativistic case, using the conservation of four-momentum

$$P_i + P_i^{\text{mir}} = P_f + P_f^{\text{mir}} , \tag{S-8.140}$$

where P_i and P_f are the four-momenta of the wave packet before and after reflection, respectively, and P_i^{mir} and P_f^{mir} are the initial and final four-momenta of the mirror. We have

$$\frac{U_i}{c} + Mc = \frac{U_f}{c} + \gamma Mc , \qquad \text{(S-8.141)}$$

$$p_i = \gamma M v - p_f , \qquad \text{(S-8.142)}$$

where $\gamma = 1/\sqrt{1 - v^2/c^2}$. According to the second of (S-8.132) we have $p_i = U_i/c$ and $p_f = U_f/c$, which, substituted into (S-8.142) lead to

$$\frac{U_f}{c} = \gamma M v - \frac{U_i}{c} . \qquad \text{(S-8.143)}$$

Substituting U_f/c into (S-8.141) we obtain

$$\gamma(c + v) - \frac{Mc^2 + 2U_i}{Mc} = 0 , \qquad \text{(S-8.144)}$$

dividing by γ we have

$$c + v = \frac{Mc^2 + 2U_i}{Mc} \sqrt{1 - \frac{v^2}{c^2}} , \qquad \text{(S-8.145)}$$

and squaring left and right hand sides we obtain the quadratic equation for v

$$\left(2 + \frac{4U_i}{Mc^2} + \frac{4U_i^2}{M^2c^4}\right) v^2 + 2cv - \left(\frac{4U_i}{M} + \frac{4U_i^2}{M^2c^2}\right) = 0 . \qquad \text{(S-8.146)}$$

We are interested in the positive solution

$$v = c \frac{1 + U_i/(Mc^2)}{1 + U_i/(Mc^2) + Mc^2/(2U_i)} \simeq \frac{2U_i}{Mc} \qquad \text{(S-8.147)}$$

at the limit $U_i \ll Mc^2$, in agreement with (S-8.139).

(c) The energy required to accelerate the sail up to $v = 0.3\,c$ can be evaluated from (S-8.139). For $v = 0.3$ c we have that

$$v = \frac{2U_i}{Mc} \quad \text{leads to} \quad U_i = \frac{Mcv}{2} = \frac{0.3Mc^2}{2} , \qquad \text{(S-8.148)}$$

inserting $M = 10$ g we obtain

$$U_i \simeq \frac{0.3}{2} \times 10^{-2} \text{ kg} \times \left(3 \times 10^8\right)^2 \text{ m}^2/\text{s}^2 = 1.35 \times 10^{14} \text{ J} . \qquad \text{(S-8.149)}$$

If W is the power of the laser beam, the time t required to provide the energy U_i is

$$t = \frac{U_i}{W} = \frac{1.35 \times 10^{14} \text{ J}}{10^{19} \text{ W}} = 1.35 \times 10^4 \text{ s} . \tag{S-8.150}$$

In the non-relativistic approximation during the time t the sail travels a distance

$$\ell_a = \frac{1}{2}at^2 = \frac{1}{2}vt \simeq \frac{1}{2}0.3 \times 10^8 \frac{\text{m}}{\text{s}} \times 1.35 \times 10^4 \text{ s} \simeq 6 \times 10^{11} \text{ m}. \tag{S-8.151}$$

The Rayleigh length ℓ_R for a (quasi-) monochromatic beam of wavelength λ and waist surface S is

$$\ell_R \simeq \frac{S}{\lambda} . \tag{S-8.152}$$

If we require that the laser beam is not larger than the sail at t we need a wavelength such that $\ell_a = \ell_R$ for a beam waist of surface $S = L^2/2$. Thus we must have

$$\lambda = \frac{L^2/2}{\ell_a} = \frac{8 \text{ m}^2}{6 \times 10^{11} \text{ m}} = 1.3 \times 10^{-11} \text{ m} . \tag{S-8.153}$$

At the moment no laser operating at such a small wavelength has been developed, this is one of the many difficulties of the project.

S-8.11   Radiation Pressure at Oblique Incidence

From the energy-momentum conservation theorem of Maxwell's equation we know that the wavepacket delivers a flow of energy per unit surface $I \equiv |\mathbf{S}|$ where $\mathbf{S} = (c/4\pi)\,\mathbf{E} \times \mathbf{B}$ is the Poynting vector, while the density of electromagnetic momentum is $\mathbf{g} = \mathbf{S}/c^2$. Thus, the incident wavepacket of intensity I, cross-sectional area A and duration τ contains a total momentum $\mathbf{p}_i = (I/c^2)(Ac\tau)\,\hat{\mathbf{n}}_i$ where $\hat{\mathbf{n}}_i = (\cos\theta_i, \sin\theta_i)$ is the direction of incidence. Under the assumption of a quasi-monochromatic field (so that $\hat{\mathbf{n}}_i = \mathbf{k}_i/k$, with $k = \omega/c$) the reflected wavepacket has the same shape as the incident one, and since its intensity is RI, its total momentum is $\mathbf{p}_r = R(I/c^2)(Ac\tau)\,\hat{\mathbf{n}}_r$ with $\hat{\mathbf{n}}_r = (-\cos\theta_i, \sin\theta_i)$ according to the law of reflection. Thus, the amount of momentum $\Delta\mathbf{p}$ delivered to the medium is

$$\Delta\mathbf{p} = \mathbf{p}_i - \mathbf{p}_r = \frac{I}{c}A\tau \left((1+R)\cos\theta_i, \; (1-R)\sin\theta_i \right) . \tag{S-8.154}$$

The total force exerted per unit surface is obtained dividing $\Delta\mathbf{p}$ by the wavepacket duration τ and the area over which the wavepacket impinges, which is equal to $A/\cos\theta_i$,

$$\mathbf{P} = (P_\perp, P_\parallel) = \frac{\Delta\mathbf{p}}{c\tau(A/\cos\theta_i)} . \tag{S-8.155}$$

so that

$$P_\perp = (1 + R)\frac{I}{c}\cos^2\theta_i ,\tag{S-8.156}$$

$$P_\parallel = (1 - R)\frac{I}{c}\sin\theta_i \cos\theta_i .\tag{S-8.157}$$

This result is independent of A and τ and thus it is appropriate to describe the limit of a plane, monochromatic wave. The flow of parallel momentum $P_\parallel$ is non-vanishing only in the presence of absorption ($R < 1$) and for oblique incidence ($\theta_i \neq 0$).

Both (S-8.156) and (S-8.157) can also be derived by evaluating the electromagnetic momentum flux through the surface, with the help of Fresnel formulas for the reflection coefficient and Maxwell stress tensor to compute the flux: see Reference [1].

S-8.12   A Square Law Detector

(a) The radiation pressure P of a circularly polarized plane wave with electric field components of amplitude $E_y = E_z = E$ is independent of time and is obtained from (S-8.156) of Solution S-8.11 by substituting $R = 1$ and $\cos\theta_i = 1$:

$$P = \frac{1}{2\pi}E^2 .\tag{S-8.158}$$

The equation of motion of the mirror can be written as

$$M\frac{d^2\xi}{dt^2} + \kappa\xi = \frac{1}{2\pi}D^2E^2 ,\tag{S-8.159}$$

where ξ is the mirror displacement and κ the spring constant. It leads to

$$\xi = \frac{1}{2\pi}\frac{E^2}{\kappa} = \frac{1}{2\pi}\frac{E^2}{\omega_0^2 M} .\tag{S-8.160}$$

where $\omega_0^2 = \kappa/M$.

(b) We can choose the additional plane wave to be polarized in the y direction. Then, neglecting an irrelevant initial phase difference, the radiation pressure is now given by

$$\begin{aligned}
P &= \frac{1}{2\pi}\left[\left(E\cos\omega t + E_1\cos\omega_1 t\right)^2 + E^2\sin^2\omega t\right] \\
&= \frac{1}{2\pi}\left[E^2 + 2EE_1\cos\omega t\cos\omega_1 t + E_1^2\cos^2\omega_1 t\right] ,
\end{aligned}\tag{S-8.161}$$

i.e.,

$$P = \frac{1}{2\pi} \left\{ E^2 + \frac{E_1^2}{2} + EE_1 \left[\cos(\omega - \omega_1)t + \cos(\omega + \omega_1)t \right] + \frac{E_1^2}{2} \cos 2\omega_1 t \right\} ,$$

$$\text{(S-8.162)}$$

The equation of motion of the mirror is linear and thus we can consider the different frequency terms in (S-8.162) separately. The terms proportional to EE_1 give the following oscillatory displacements

$$\xi_-(t) = \frac{EE_1}{2\pi M(\omega_0^2 - (\omega - \omega_1)^2)} \cos(\omega - \omega_1)t \simeq \frac{EE_1}{2\pi M\omega_0^2} \cos(\omega - \omega_1)t,$$

$$\text{(S-8.163)}$$

$$\xi_+(t) = \frac{EE_1}{2\pi M(\omega_0^2 - (\omega + \omega_1)^2)} \cos(\omega + \omega_1)t ,$$

$$\text{(S-8.164)}$$

while the term proportional to E_1^2 gives

$$\xi_2(t) = \frac{E_1^2}{4\pi M \left(\omega_0^2 - 4\omega_1^2\right)} \cos 2\omega_1 t .$$

$$\text{(S-8.165)}$$

For the chosen frequency inequalities $\xi_- \gg \xi_+$.

(c) The displacements $\xi_\pm(t)$ depends linearly on the amplitude of the small field E_1 while ξ_2 has a quadratic dependence. In addition the displacement $\xi_-(t)$ corresponds to a low frequency oscillation while ξ_2 oscillates at a frequency larger than ω_0, thus the denominator in the expression of $\xi_-(t)$ is much smaller than the denominator of $\xi_2(t)$.

In conclusion the presence of the larger plane wave converts the frequency of the detector response to the smaller wave while it increases its amplitude.

S-8.13 Poynting Vector for a Rotating Charged Spherical Shell

(a) We denote by $Q = 4\pi a^2 \sigma$ the total electric charge of the rotating spherical shell. The electric and magnetic fields are, in Gaussian units,

$$\mathbf{E}(\mathbf{r}) = \begin{cases} 0 , \\ \dfrac{Q}{r^2} \hat{\mathbf{r}} , \end{cases} \qquad \mathbf{B}(\mathbf{r}) = \begin{cases} \dfrac{8\pi}{3c} \sigma a \omega , & r < a , \\ \dfrac{3(\mathbf{m} \cdot \mathbf{r})\mathbf{r}}{r^5} - \dfrac{\mathbf{m}}{r^3} , & r > a , \end{cases} \qquad \text{(S-8.166)}$$

where we have used the results of Solution S-5.12 and

$$\mathbf{m} = \frac{4\pi}{3c} a^4 \sigma \omega = \frac{Qa^2}{3c} \omega .$$

$$\text{(S-8.167)}$$

The Poynting vector $\mathbf{S} = c\,\mathbf{E} \times \mathbf{B}/(4\pi)$ is thus zero inside the shell, since $\mathbf{E} = 0$ for $r < a$, while we have for $r > a$

$$\mathbf{S} = -\frac{c}{4\pi}\left(\frac{Q}{r^2}\hat{\mathbf{r}}\right) \times \left(\frac{Qa^2}{3cr^3}\,\boldsymbol{\omega}\right) = -\frac{Q^2a^2}{12\pi r^5}\,\hat{\mathbf{r}} \times \boldsymbol{\omega} = \frac{Q^2a^2}{12\pi r^5}\sin\theta\,\omega\,\hat{\boldsymbol{\phi}}\,,$$

$$\text{(S-8.168)}$$

for the last expression we have used a (r, θ, ϕ) spherical coordinate system with the origin at the center of the spherical shell and the polar axis parallel to $\boldsymbol{\omega}$, as shown in Fig. S-8.13 We have (see Table A.1 of the Appendix)

$$\boldsymbol{\nabla} \cdot \mathbf{S} = \frac{1}{r\sin\theta}\frac{\partial}{\partial\phi}\left(\frac{Q^2a^2}{12\pi r^5}\sin\theta\,\omega\right) = 0\,, \qquad \text{(S-8.169)}$$

and, in particular, there is no energy flux along r.

(b) The momentum density of the electromagnetic field is

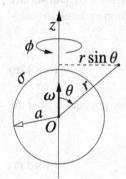

Fig. S-8.5

$$\mathbf{g} = \frac{\mathbf{S}}{c^2} = \frac{Q^2a^2}{12\pi c^2 r^5}\sin\theta\,\omega\,\hat{\boldsymbol{\phi}}\,. \qquad \text{(S-8.170)}$$

The distance of the point (r, θ, ϕ) from the polar axis is $r\sin\theta$, see Fig. S-8.5, thus the angular momentum density $\mathcal{L}$ with respect to the polar axis is

$$\mathcal{L} = \hat{\mathbf{z}}\,g\,r\sin\theta = \frac{Q^2a^2}{12\pi c^2 r^4}\sin^2\theta\,\omega\,, \qquad \text{(S-8.171)}$$

where $\hat{\mathbf{z}}$ is the unit vector along the polar axis. The total angular momentum $\mathbf{L}$ of the electromagnetic field is obtained by integrating (S-8.171) over the whole volume outside the spherical shell

$$\mathbf{L} = \int_{\text{ext}}\mathcal{L}\,\mathrm{d}^3r = \frac{Qa^2\omega}{12\pi c^2}\int_a^\infty r^2\,\mathrm{d}r\,\frac{1}{r^4}\int_0^\pi \mathrm{d}\theta\int_0^{2\pi}\sin\theta\,\mathrm{d}\phi\,\sin^2\theta$$

$$= \frac{Q^2a^2\omega}{12\pi c^2}\frac{1}{a}2\pi\frac{4}{3} = \frac{2Q^2a\,\omega}{9c^2}\,, \qquad \text{(S-8.172)}$$

where we have used

$$\int\sin^3\theta\,\mathrm{d}\theta = \frac{1}{3}\cos^3\theta - \cos\theta\,, \qquad \int_0^\pi\sin^3\theta\,\mathrm{d}\theta = \frac{4}{3}\,. \qquad \text{(S-8.173)}$$

Equation (S-8.172) corresponds to an electromagnetic contribution to the moment of inertia of the spherical shell

$$\mathcal{I}^{\mathrm{em}} = \frac{2Q^2 a}{9c^2} \,.$$

(S-8.174)

(c) Let us consider an infinitesimal surface element $a^2 \sin\theta \, d\theta \, d\phi$ of the spherical shell. It carries a charge $dQ = \sigma a^2 \sin\theta \, d\theta \, d\phi$, corresponding to an electromagnetic contribution to the canonical momentum

$$d\mathbf{p}^{\mathrm{em}} = \sigma a^2 \sin\theta \, d\theta \, d\phi \, \frac{\mathbf{A}(a, \theta, \phi)}{c} = \frac{1}{c} \sigma a^2 \sin\theta \, d\theta \, d\phi \, \frac{4\pi}{3} \sigma a \, \boldsymbol{\omega} \times \mathbf{a}$$

$$= \hat{\boldsymbol{\phi}} \, \frac{4\pi}{3c} \sigma^2 a^4 \omega \sin^2\theta \, d\theta \, d\phi$$

(S-8.175)

where we have used the vector potential $\mathbf{A}$ for $r = a$ given by (S-5.105). Our surface element distance from the polar axis is $a \sin\theta$, thus its contribution to the electromagnetic part of the canonical angular momentum is

$$d\mathbf{L}^{\mathrm{em}} = \hat{\mathbf{z}} \, \frac{4\pi}{3c} \sigma^2 a^5 \omega \sin^3\theta \, d\theta \, d\phi \,,$$

(S-8.176)

and the total electromagnetic angular momentum is obtained by integrating over the shell surface

$$\mathbf{L}^{\mathrm{em}} = \hat{\mathbf{z}} \, \frac{4\pi}{3c} \sigma^2 a^5 \omega \int_0^{2\pi} d\phi \int_0^{\pi} d\theta \, \sin^3\theta = \hat{\mathbf{z}} \, \frac{2Q^2 a \, \omega}{9c^2} \,,$$

(S-8.177)

in agreement with (S-8.172). We have used $4\pi a^2 \sigma = Q$. A general proof of the equivalence between (S-8.172) and (S-8.177) is given under general stationary conditions using vector calculus identities in [2].

References

1. L.D. Landau, E.M. Lifshitz, *The Classical Theory of Fields*, 2nd edn. Chap. 47 (Elsevier, Oxford, 1975)
2. https://www.physics.princeton.edu/~kirkmcd/examples/spin.pdf

Chapter S-9
Solutions for Chapter 9

S-9.1 The Fields of a Current-Carrying Wire

(a) In the S reference frame the wire generates an azimuthal magnetic field $\mathbf{B} = B_\phi(r)\hat{\phi}$. In cylindrical coordinates we have $B_\phi = B_\phi(r) = (2I/rc)$. The Lorentz force on the charge q is

$$\mathbf{F} = q\,\frac{v}{c} \times \mathbf{B} = \hat{\mathbf{r}}\,F_r = -\hat{\mathbf{r}}\,q\,B_\phi(r)\,\frac{v}{c} = -\hat{\mathbf{r}}\,\frac{2qIv}{rc^2}\,. \tag{S-9.1}$$

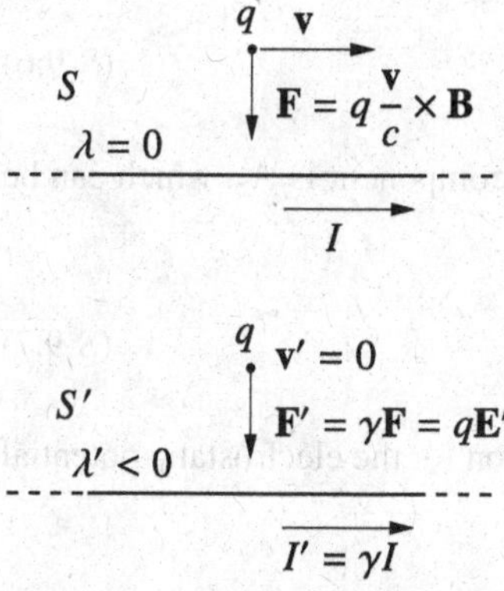

Fig. S-9.1

The S' frame moves with velocity v with respect to S. Applying the Lorentz transformations, in S' the force on q is $\mathbf{F}' = \hat{\mathbf{r}}\,F_r' = \hat{\mathbf{r}}\,\gamma F_r$ (where $\gamma = \sqrt{1-\beta^2}$, and $\beta = v/c$ with $v = |\mathbf{v}|$), see Fig S-9.1. Since q is at rest in S', the force $\mathbf{F}'$ is due to the electric field $\mathbf{E}'$ only, with $\mathbf{E}' = \hat{\mathbf{r}}\,E_r' = \hat{\mathbf{r}}\,F_r'/q$. This corresponds to the transformation $\mathbf{E}'_\perp = \hat{\mathbf{r}}\,E_r' = -\hat{\mathbf{r}}\,\gamma\beta B_\phi$ or, in vector form,

$$\mathbf{E}'_\perp = \gamma\,(\boldsymbol{\beta} \times \mathbf{B})\,, \tag{S-9.2}$$

where the subscript "$\perp$" refers to the direction perpendicular to $\mathbf{v}$. At the limit $|\mathbf{v}| \ll c$ (for which $\mathbf{F} = \mathbf{F}'$) we get $\mathbf{E}'_\perp \simeq \boldsymbol{\beta} \times \mathbf{B}$, which is correct up to first order in $\beta = v/c$, and may be called the "Galilei" transformation of the field. The electric field[5]

---

[5] In general, the complete transformation is $\mathbf{E}'_\perp(\mathbf{r}', t) = \gamma\boldsymbol{\beta} \times \mathbf{B}[\mathbf{r}(\mathbf{r}', t'), t(\mathbf{r}', t')]$, where $\mathbf{r} = \mathbf{r}(\mathbf{r}', t')$ and $t = t(\mathbf{r}', t')$, according to the Lorentz transformations of the coordinates. Since in cylindrical coordinates B_ϕ depends on r only, and for the coordinates in the plane transverse to the boost velocity $\mathbf{r}'_\perp = \mathbf{r}_\perp$, in the present case we have the trivial transformation $r' = r$.

© The Editor(s) (if applicable) and The Author(s), under exclusive license to Springer Nature Switzerland AG 2023
A. Macchi et al., *Problems in Classical Electromagnetism*,
https://doi.org/10.1007/978-3-031-22235-1

$$E'_r = -\gamma\beta B_\phi(r') = -2\gamma\beta I/(r'c) \tag{S-9.3}$$

is generated by a uniform linear charge density $\lambda' = -\beta\gamma I/c$ on the wire, as can be easily verified by applying Gauss's law. Thus *the wire is negatively charged in S'*.[6]

Since the force is purely magnetic in S and purely electric in S', at this point we cannot say much about the magnetic field in S'.

(b) We know that $J = (\rho c, \mathbf{J})$ is a four-vector. The cross-section W of the wire is invariant for a Lorentz boost along the wire axis, thus the linear charge density $\lambda = W\rho$ and the electric current $\mathbf{I} = W\mathbf{J}$ transform like ρ and $\mathbf{J}$. Therefore the linear charge density of the wire in S' is

$$\lambda' = \gamma\left(\lambda - \beta\frac{I}{c}\right) = -\gamma\beta\frac{I}{c}, \tag{S-9.4}$$

which, according to Gauss's law, generates the radial electric field $E'_r = 2\lambda'/r$, in agreement with our result of point **a)**. We also obtain the current intensity in S',

$$I' = \gamma(I - \beta c\lambda) = \gamma I, \tag{S-9.5}$$

which generates the magnetic field $B'_\phi = 2I'/(r'c) = \gamma B_\phi$.

The same results can be obtained through the transformation of the four-potential $(\phi, \mathbf{A})$. In S, we have obviously $\phi = 0$, since there is no net charge, while the vector potential $\mathbf{A}$ satisfies the equation

$$\nabla^2\mathbf{A} = -\frac{4\pi}{c}\mathbf{J} \tag{S-9.6}$$

Thus, $\mathbf{A}$ is parallel to the wire and its only non-zero component is A_z, which can be evaluated from the equation

$$\nabla^2 A_z = -\frac{4\pi}{c}I\,\delta(r). \tag{S-9.7}$$

This is mathematically identical to the Poisson equation for the electrostatic potential of a uniformly charged wire, thus the solution is

$$A_z = -\frac{2I}{c}\ln\left(\frac{r}{a}\right), \tag{S-9.8}$$

---

[6] It might seem that the law of charge conservation is violated in the transformation from S to S'. Actually, this a consequence of the somewhat "pathological" nature of currents which are not closed in a loop, as in the case of an infinite wire. In fact, strictly speaking, the infinite current-carrying wire is not a steady system, since charges of opposite sign are accumulating at the two "ends" of the wire, i.e., at $z = \pm\infty$. If we introduce "return" currents to close the loop in S, e.g., if we assume the wire to be the inner conductor of a coaxial cable, or if we add a second wire carrying the current $-I$ at some distance, we find that the return currents would appear as opposite charge densities in S', as required by charge conservation.

where a is an arbitrary constant. It is straightforward to verify that $B_\phi = -\partial_r A_z$.

The scalar potential in S' is

$$\phi' = \gamma(\phi - \beta A_z) = -\gamma\beta A_z = -\frac{2\gamma\beta I}{c} \ln\left(\frac{r}{a}\right), = -2\lambda' \ln\left(\frac{r}{a}\right), \quad \text{(S-9.9)}$$

where $\lambda' = -\beta\gamma I/c$. The electric field is evaluated from $\mathbf{E}' = -\nabla\phi'$, obtaining the same result of point (a). For the vector potential in S', trivially $A_z' = \gamma(A_z - \beta\phi) = \gamma A_z$ from which we get $B_\phi' = \gamma B_\phi$ again.

These results are in agreement with the explicit formulas for the transformation of the EM field (9.3), which, in our case, lead to $\mathbf{E}' = \gamma\boldsymbol{\beta} \times \mathbf{B}$ and $\mathbf{B}' = \gamma\mathbf{B}$.

(c) Let us first consider the linear charge densities of both ions ($\lambda_i = Zen_i W$) and electrons ($\lambda_e = -en_e W$) in S, where n_i and n_e are the ion and electron volume densities, respectively. Since there is no net charge on the wire in S, we have $\lambda_i = -\lambda_e$.

Let us evaluate the charge densities λ_i' and λ_e' in S' from relativistic kinematics. In S, a wire segment of length ΔL carries an ion charge $\Delta Q = \lambda_i \Delta L$. In S', the segment has the same charge as in S (the charge is a Lorentz invariant), but the length undergoes a Lorentz contraction, $\Delta L' = \Delta L/\gamma$. Thus we have a higher charge density $\lambda_i' = \Delta Q/\Delta L' = \gamma\lambda_i$. This is a quite general result: in a frame where a fluid moves at velocity $\boldsymbol{v}$, the fluid has a higher density (by a factor γ) than in its rest frame.

On the other hand, the electrons are *not* at rest in S: they move along the wire with a velocity $\boldsymbol{v}_e < 0$ such that $I = -en_e v_e W = \lambda_e v_e = -\lambda_i v_e$. Thus, their density is already *higher* by a factor $\gamma_e = 1/\sqrt{1 - v_e^2/c^2}$ *than the density* λ_{e0} *in the rest frame of the electrons*: we have $\lambda_{e0} = \lambda_e/\gamma_e$. In S', the electrons drift with a velocity v_e'

$$v_e' = \frac{v_e - v}{1 - v_e v/c^2}, \quad \text{(S-9.10)}$$

according to Lorentz transformations. Thus, the electron density in S' is

$$\lambda_e' = \gamma_e' \lambda_{e0} = \frac{\gamma_e'}{\gamma_e} \lambda_e, \quad \text{(S-9.11)}$$

where $\gamma_e' = 1/\sqrt{1 - v_e'^2/c^2}$. The expression for γ' can be put in a more convenient form by some algebra:

$$\gamma'_e = \cfrac{1}{\sqrt{1 - \cfrac{(v_e - v)^2}{c^2\left(1 - \frac{v_e v}{c^2}\right)^2}}} = \sqrt{\cfrac{\left(1 - \frac{v_e v}{c^2}\right)^2}{\left(1 - \frac{v_e v}{c^2}\right)^2 - \cfrac{(v_e - v)^2}{c^2}}}$$

$$= \left(1 - \frac{v_e v}{c^2}\right) \cfrac{1}{\sqrt{1 - 2\frac{v_e v}{c^2} + \frac{v_e^2 v^2}{c^4} - \frac{v_e^2}{c^2} + 2\frac{v_e v}{c^2} - \frac{v^2}{c^2}}}$$

$$= \left(1 - \frac{v_e v}{c^2}\right) \cfrac{1}{\sqrt{\left(1 - \frac{v_e^2}{c^2}\right)\left(1 - \frac{v^2}{c^2}\right)}}$$

$$= \left(1 - \frac{v_e v}{c^2}\right) \gamma_e \gamma . \tag{S-9.12}$$

We thus obtain for the *total* charge density in S'

$$\lambda' = \lambda'_i + \lambda'_e = \lambda_i \left(\gamma - \frac{\gamma'_e}{\gamma_e}\right) = \lambda_i \gamma \left(1 - 1 + \frac{v_e v}{c^2}\right) = \lambda_i \gamma \frac{v_e v}{c^2}$$

$$= -\gamma v \frac{I}{c^2} , \tag{S-9.13}$$

as previously found on the basis of Lorentz transformations for the forces, charge and current densities, and EM fields.

It might be interesting to remark that there is an issue of charge conservation already in the S frame. The wire is electrically neutral, thus its ion and electron charge densities are exactly equal and opposite when it is disconnected from any voltage or current source, and in the absence of external fields. Now assume that we drive a steady current I through the wire, keeping the conduction electrons in motion with a velocity v_e along the wire axis. If the wire is still electrically neutral, as we assumed, the absolute values of the charge densities of ions and electrons must still be equal and opposite. However, while the charge density of the ions, at rest, has not changed, the charge density of the moving electrons undergoes a "relativistic increase" by a factor γ_e. If the total charge density does not change (the wire must still be neutral), some electrons must have left the wire.[7] We can explain where the missing electrons have gone only by recalling that the wire is not "open", but must be part of a closed current loop, with specific boundary conditions and how the circuit is closed.

---

[7] Of course, the effect is negligibly small for ordinary conduction in metals, for which the typical electron velocities v_e are of the order of $10^{-10} c$. On the other hand, this issue if very important for relativistic hydrodynamics, i.e., for contexts where fluids move at velocities close to c.

S-9.2 The Fields of a Plane Capacitor

(a) We choose a Cartesian coordinate system with the y axis perpendicular to the plates, so that the lower plate is at $y = 0$ and the upper plate at $y = h$, and the x axis parallel to $\boldsymbol{v}$, so that $\boldsymbol{v} = \beta c\, \hat{\mathbf{x}}$. The only non-vanishing component of the EM field in S is $E_y = 4\pi\sigma$. By applying a Lorentz transformation we find for the fields in S'

$$E_y' = \gamma E_y = 4\pi\gamma\sigma \;, \qquad B_z' = -\beta\gamma E_y = -4\pi\beta\gamma\sigma \;. \tag{S-9.14}$$

(b) In S' the electric field E_y' is generated by the surface charge densities $\pm\sigma' = \pm E_y'/4\pi = \pm\gamma\sigma$ on the capacitor plates. Similarly, the magnetic field B_z' is generated by the two surface current densities $\pm\mathbf{K}' = \pm K_x'\,\hat{\mathbf{x}}$ with $K_x' = cB_z'/(4\pi) = -\beta\gamma\sigma c$, flowing on the two capacitor plates.

These results are in agreement with the Lorentz transformation of the four-vector

$$K_\mu = (c\sigma, \mathbf{K}) \;. \tag{S-9.15}$$

We can check that K_μ is actually a four-vector, by imagining two volume four-current densities $J_\mu = (c\rho, \pm\mathbf{J})$ distributed over the two thin layers, $|y| < \delta/2$ and $|h - y| < \delta/2$, around the capacitor plates, such that $\sigma = \rho\delta$ and $\mathbf{K} = \mathbf{J}\delta$. Since δ is invariant for transformations with velocity parallel to $\mathbf{J}$, it follows that also $K_\mu \equiv J_\mu\delta$ transforms as a four-vector:

$$\sigma' = \gamma(\sigma - \beta K_x/c) = \gamma\sigma \;, \qquad K_x' = \gamma(K_x - \beta c\sigma) = -\beta\gamma\sigma c \;. \tag{S-9.16}$$

(c) In S there is a perpendicular force per unit surface $p = \sigma E_y/2 = 2\pi\sigma^2$ on the internal surfaces of the plates, such that the plates attract each other. In S', the force per unit surface is the sum of two terms of electrostatic and magnetic nature, respectively,

$$p' = \frac{1}{2}\sigma' E_y' + \frac{1}{2}K_x' B_z' = 2\pi\sigma^2\gamma^2 - 2\pi\sigma^2\beta^2\gamma^2 = 2\pi\sigma^2\gamma^2(1 - \beta^2) = 2\pi\sigma^2$$
$$= p \;. \tag{S-9.17}$$

The invariance of p is also proven from the equivalent expression

$$p' = \frac{1}{8\pi}E_y'^2 - \frac{1}{8\pi}B_z'^2 = \frac{1}{8\pi}(\mathbf{E}'^2 - \mathbf{B}'^2) \;, \tag{S-9.18}$$

which is a Lorentz invariant.

In S, the total force is $F = pA$. In S', due to the Lorentz contraction of lengths, $A' = (L/\gamma)\, L = A/\gamma$, so that $F' = p'A' = pA/\gamma = F/\gamma$.

S-9.3 The Fields of a Solenoid

(a) We choose a Cartesian reference frame with the solenoid axis as z axis, and the x axis such that $v = v_x \hat{x}$. In addition, we shall also use a cylindrical reference frame sharing the z axis with the Cartesian frame, and with the azimuthal coordinate ϕ such that the $\phi = 0$ plane coincides with the xz plane. In S, the magnetic field inside the solenoid is longitudinal and uniform, $\mathbf{B} = B \hat{z}$, with $B = 4\pi n I / c$, and the force on q is $\mathbf{F} = q\, \boldsymbol{v} \times \mathbf{B}/c = -q\beta B\, \hat{y}$.

In the S' frame the charge q is at rest, thus the force on it must be due to an electric field only. According to the Lorentz transformations of the fields we have

$$
\begin{aligned}
E'_x &= E_x = 0\,, & B'_x &= B_x = 0\,, \\
E'_y &= \gamma(E_y - \beta B_z) = -\gamma\beta B\,, & B'_y &= \gamma(B_y + \beta E_z) = 0\,, \\
E'_z &= \gamma(E_z + \beta B_y) = 0\,, & B'_z &= \gamma(B_z - \beta E_y) = \gamma B\,, & \text{(S-9.19)}
\end{aligned}
$$

and the force on q is thus $\mathbf{F}' = q E'_y \hat{y} = -q\gamma\beta\, B_z \hat{y} = \gamma \mathbf{F}$.

(b) Since we are assuming $\beta \ll 1$, we have $\gamma = 1/\sqrt{1 - \beta^2} = 1 + \beta^2/2 + \cdots \simeq 1$ up to the first order in β, and we can neglect the relativistic contraction of lengths. Thus the cross-section of the solenoid remains circular in S' to within our approximations. The electric field outside the solenoid is zero (we discuss this point further below), thus the electric field component perpendicular to the solenoid winding surface is discontinuous, implying the presence of surface charge density σ'. We have from Gauss's theorem

$$
\sigma' = \frac{E'_\perp}{4\pi} = \frac{E'_y}{4\pi}\sin\phi = -\beta \frac{B}{4\pi}\sin\phi = -\beta n \frac{I}{c}\sin\phi\,, \qquad \text{(S-9.20)}
$$

where the subscript $\perp$ means perpendicular to the solenoid winding surface.

This result is in agreement with the transformation laws for the four-vector $K_\mu = (c\sigma, \mathbf{K})$, where $\mathbf{K}$ is the surface current density on the walls of the solenoid (see Problem 9.2). In S we have $\mathbf{K} = nI\hat{\phi} = nI(-\hat{x}\sin\phi + \hat{y}\cos\phi)$, and in S'

$$
\sigma' = \gamma\left(\sigma - \beta \frac{K_x}{c}\right) \simeq -\beta \frac{K_x}{c} = \beta n \frac{I}{c}\sin\phi\,. \qquad \text{(S-9.21)}
$$

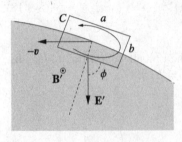

A surface charge density varying as $\sin\phi$ on the lateral surface of an infinite cylinder generates a uniform electrostatic field inside the cylinder, as seen in the solution of Problem 3.11. But there we also saw that surface charge density generates a "two-dimensional dipole" field *outside* the cylinder. This might seem in contradiction with the fact that, since the external EM field is zero in the S frame,

Fig. S-9.2

it must be zero in S' as well. But there are *not only static fields* in S', because the transverse motion of the solenoid generates a time-dependent magnetic field, which, in turn, is related to a non-conservative electric field and to boundary conditions which are different from the static case. We start by noting that an electric field that is uniform (and nonzero) inside the solenoid, and zero outside, is not conservative. Let us choose a rectangular path C of sides a and b crossing the solenoid winding as in Fig. S-9.2. The path C is at rest in S', while the solenoid moves toward the left with velocity $-v$. At $t = 0$ the upper side of length a is tangent to the winding at its central point, and a is sufficiently small for the enclosed winding arc to be well approximated by a straight line segment. We also have $b \ll a$. The field $\mathbf{E}'$ is not conservative because the line integral of $\mathbf{E}'$ along C does not vanish:

$$\oint_C \mathbf{E}' \cdot d\boldsymbol{\ell} = E'_\parallel a = E'_y a \cos\phi . \tag{S-9.22}$$

This is consistent with the fact that the flux of $\mathbf{B}'$ through the rectangle enclosed by the path C is time-dependent. The winding arc enclosed by the rectangle moves towards the lower side of length a with velocity $v \cos\phi$, and the flux of $\mathbf{B}'$ through the rectangle is

$$\Phi_C(\mathbf{B}') = \int_C \mathbf{B}' \cdot d\mathbf{S} = B'a[b - (v \cos\phi)t] , \tag{S-9.23}$$

corresponding to a line integral

$$-\frac{1}{c}\frac{d\Phi_C(\mathbf{B}')}{dt} = \frac{v}{c}B'a\cos\phi = E'y a\cos\phi , \tag{S-9.24}$$

in agreement with (S-9.22).

S-9.4 The Four-Potential of a Plane Wave

(a) The fields of the plane wave may be written in complex notation as

$$\mathbf{E} = \hat{\mathbf{y}} E_0 e^{ikx-i\omega t} , \qquad \mathbf{B} = \hat{\mathbf{z}} B_0 e^{ikx-i\omega t} , \tag{S-9.25}$$

with $E_0 = B_0$. A vector potential of the form $\mathbf{A} = \hat{\mathbf{y}} A_0 e^{ikx-i\omega t}$ generates an electric field along $\hat{\mathbf{y}}$ and a magnetic field along $\hat{\mathbf{z}}$ given by

$$\mathbf{E} = -\frac{1}{c}\partial_t\mathbf{A} - \nabla\varphi , \qquad \mathbf{B} = \nabla \times \mathbf{A} . \tag{S-9.26}$$

In the absence of electric charges we have $\varphi \equiv 0$, and we obtain from (S-9.26)

$$A_0 = -\frac{ic}{\omega} E_0, \qquad A_0 = -\frac{i}{k} B_0, \qquad (S\text{-}9.27)$$

which are equivalent since $\omega = kc$. The vector potential $\mathbf{A} = \hat{\mathbf{y}} A_0 \, e^{ikx - i\omega t}$ obviously satisfies the wave equation in vacuum, and also respects the Lorenz gauge condition. (b) The Lorentz transformations from S to S' give $\omega' = \gamma \omega$ (transverse Doppler effect), and $k'_x = k_x = k$, $k'_y = -\omega' v/c^2 = -\gamma \beta k_x$. The nonzero components of the fields in S' are $E'_y = E_y$, $E'_x = \gamma \beta B_z$, and $B'_z = \gamma B_z$. We may thus write

$$\mathbf{E}' = (\hat{\mathbf{x}} \gamma \beta + \hat{\mathbf{y}}) E_0 \, e^{i(k'_x x' + k'_y y' - \omega' t')}, \qquad \mathbf{B}' = \hat{\mathbf{z}} \gamma B_0 e^{i(k'_x x' + k'_y y' - \omega' t')}, \qquad (S\text{-}9.28)$$

The polarization is linear and directed along the unit vector $\varepsilon = \beta \hat{\mathbf{x}} + \hat{\mathbf{y}}/\gamma$.
(c) Assuming $\varphi' = 0$, we have in S'

$$\mathbf{E}' = -\frac{1}{c} \partial'_t \mathbf{A}', \qquad \mathbf{B}' = \nabla' \times \mathbf{A}', \qquad (S\text{-}9.29)$$

which are both satisfied if we choose

$$\mathbf{A}' = -\frac{c}{\omega'} \mathbf{E}' = \left(\hat{\mathbf{x}} \beta + \hat{\mathbf{y}} \frac{1}{\gamma} \right) A_0 \, e^{i(k'_x x' + k'_y y' - \omega' t')}, \qquad (S\text{-}9.30)$$

being $\omega' = \gamma \omega$.
(d) The Lorentz transformation from s to S' for the four-potential $A_\mu = (\varphi, \mathbf{A}) = (0, 0, A_y, 0)$ gives

$$\bar{A}'_\mu = (-\gamma \beta A_y, 0, \gamma A_y, 0) \equiv (\bar{\varphi}', 0, \bar{A}'_y, 0). \qquad (S\text{-}9.31)$$

The fields derived from this four-potential are

$$\bar{E}'_x = -\partial'_x \bar{\varphi}' = -ik'_x \bar{\varphi}' = ikc\gamma\beta A_y = \gamma\beta(i\omega A_y) = \gamma\beta E_y = E'_x, \qquad (S\text{-}9.32)$$

$$\bar{E}'_y = -\frac{1}{c}\partial'_t \bar{A}'_y - \partial'_y \bar{\varphi}' = i\frac{\omega'}{c}\bar{A}'_y - ik'_y \bar{\varphi}' = i\left(\gamma \frac{\omega}{c}\right)\gamma A_y - i(-\gamma\beta k)(-\gamma\beta c)A_y$$

$$= i\frac{\omega}{c}\gamma^2 \left(1 - \beta^2\right) A_y = E_y = E'_y, \qquad (S\text{-}9.33)$$

$$\bar{B}'_z = \partial'_x \bar{A}'_y = ik'_x \bar{A}'_y = ik\gamma A_y = \gamma B_z = B'_z, \qquad (S\text{-}9.34)$$

in agreement with the results of point (c).
(e) The expressions $A'_\mu = (0, \mathbf{A}')$ and $\bar{A}'_\mu = (\bar{\varphi}', \bar{\mathbf{A}}')$ are two possible choices for the four-potential. Thus they must differ at most by a gauge transformation, i.e., there must be a scalar function $f = f(x', t')$ such that

$$\mathbf{A}' = \bar{\mathbf{A}}' + \nabla' f, \qquad \varphi' = \bar{\varphi}' - \frac{1}{c}\partial'_t f. \qquad (S\text{-}9.35)$$

Since $\varphi' = 0$ we find $\partial_t' f/c = \bar{\varphi}'$, i.e.,

$$f = \frac{c}{\omega'}\bar{\varphi}' = -\frac{ic}{\omega'}\gamma\beta A_y \,, \tag{S-9.36}$$

Now, since

$$\mathbf{V}'f = (\hat{\mathbf{x}}\,ik_x' + \hat{\mathbf{y}}\,ik_y')f = \left(\hat{\mathbf{x}}\beta - \hat{\mathbf{y}}\gamma\beta^2\right)A_y \,, \tag{S-9.37}$$

we also have that

$$\bar{\mathbf{A}}' + \mathbf{V}'f = \left[\hat{\mathbf{x}}\beta + \hat{\mathbf{y}}\left(\gamma - \gamma\beta^2\right)\right]A_y = \left(\hat{\mathbf{x}}\beta + \hat{\mathbf{y}}\frac{1}{\gamma}\right)A_y = \mathbf{A}'. \tag{S-9.38}$$

S-9.5   The Force on a Magnetic Monopole

(a) In the reference frame S', where the magnetic monopole is at rest ($\mathbf{v}' = 0$), the magnetic field is

$$\mathbf{B}' = -\gamma\frac{\mathbf{v}}{c}\times\mathbf{E} \,, \tag{S-9.39}$$

thus the force on the monopole is $\mathbf{F}' = q_m\mathbf{B}'$. On the other hand we must have $\mathbf{F}' = \gamma\mathbf{F}$, since $\mathbf{F}$ is perpendicular to $\mathbf{v}$, so that in the laboratory frame S we have

$$\mathbf{F} = \frac{q_m}{\gamma}\mathbf{B}' = -q_m\frac{\mathbf{v}}{c}\times\mathbf{E} \,, \tag{S-9.40}$$

which proves (9.8).

(b) The equation of motion for a magnetic monopole in the presence of a uniform electric field $\mathbf{E} = \hat{\mathbf{z}}E$ alone is identical to the equation of motion at for a an electric charge in the presence of a uniform magnetic field $\mathbf{B} = \hat{\mathbf{z}}B$, after replacing $-q_m\mathbf{E}$ by $q\mathbf{B}$. The solution is a helicoidal motion, with a constant drift velocity parallel to $\mathbf{E}$, and a constant angular velocity $\boldsymbol{\omega}_m = \hat{\mathbf{z}}q_mE/mc$. (Notice that, for a magnetic monopole, the angular velocity vector is parallel to $\mathbf{E}$, while it is antiparallel to $\mathbf{B}$ in the case of an electric charge.)

In the case of crossed electric and magnetic fields, the condition $E > B$ ensures that there is a reference frame S' where the magnetic field vanishes. In fact, taking a Lorentz boost with $\boldsymbol{\beta} = (\mathbf{E}\times\mathbf{B})/E^2$ we have

$$\mathbf{B}' = \gamma(\mathbf{B} - \boldsymbol{\beta}\times\mathbf{E}) = \gamma\left(\mathbf{B} + \frac{E^2\mathbf{B} - (\mathbf{E}\cdot\mathbf{B})\mathbf{E}}{E^2}\right) = 0 \,, \tag{S-9.41}$$

since $\mathbf{E}\cdot\mathbf{B} = 0$. Thus, in the boosted frame there is only the electric field

$$\mathbf{E}' = \gamma(\mathbf{E} + \boldsymbol{\beta} \times \mathbf{B}) = \gamma \left(\mathbf{E} - \frac{B^2}{E^2} \mathbf{E} \right) = \frac{\mathbf{E}}{\gamma} , \qquad \text{(S-9.42)}$$

since $\gamma = 1/\sqrt{1 - \beta^2} = 1/\sqrt{1 - B^2/E^2}$. Thus the motion in S' is a circular orbit with angular frequency $\omega' = (q_m E/\gamma c)$. By transforming back to the laboratory frame S we add a drift velocity $-c\boldsymbol{\beta}$, and the trajectory in S is a cycloid.

S-9.6 Reflection from a Moving Mirror

(a) As an ansatz, we write the total electromagnetic field as the sum of the fields of the incident wave and the fields of a reflected wave of the same frequency and polarization, but opposite direction

$$\mathbf{E}(x, t) = \hat{\mathbf{y}}\, E_y(x, t), \qquad \mathbf{B} = \hat{\mathbf{z}}\, B_z(x, t) ,$$
$$E_y(x, t) = \mathrm{Re}\left(E_i\, \mathrm{e}^{ikx - i\omega t} + E_r\, \mathrm{e}^{-ikx - i\omega t} \right) , \qquad \text{(S-9.43)}$$
$$B_z(x, t) = \mathrm{Re}\left(E_i\, \mathrm{e}^{ikx - i\omega t} - E_r\, \mathrm{e}^{-ikx - i\omega t} \right) . \qquad \text{(S-9.44)}$$

The amplitude of the reflected wave E_r must be determined by the boundary condition at the mirror surface $x = 0$. We may already know that the electric field component parallel to the bounding surface between two media is continuous across the surface, i.e., that $E_\parallel(0^-) = E_\parallel(0^+)$. However, here we prefer to derive this result in detail, because this will help the discussion of the reflection at the surface of a *moving* mirror, which we shall consider in the following. Evaluating the line integral of $\mathbf{E}$ over a closed rectangular loop across the boundary, as in Fig. S-9.3, yields

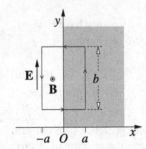

Fig. S-9.3

$$\oint \mathbf{E} \cdot d\boldsymbol{\ell} = [E_y(a, t) - E_y(-a, t)]b$$

$$= -\frac{1}{c} \frac{d\Phi(\mathbf{B})}{dt} = -\frac{b}{c} \int_{-a}^{0} \partial_t B_z \, dx$$

$$= \frac{i\omega b}{c} \int_{-a}^{0} B_z \, dx = \frac{i\omega}{c} \bar{B}_z\, ab , \qquad \text{(S-9.45)}$$

where $\bar{B}_z$ is the mean value of B_z in the $(-a, a)$ interval. If B_z is finite, the "rightmost RHS" of (S-9.45) vanishes at the limit $a \to 0$, and $E_y(0^+, t) = E_y(0^-, t)$.

For a perfect mirror we must have $E_y(0^+, t) = 0$, and the boundary condition implies that also $E_y(0^-, t) = 0$. Thus we obtain

$$E_y(0, t) = (E_i + E_r) e^{-i\omega t} = 0, \qquad E_r = -E_i. \tag{S-9.46}$$

The total electric field for $x \leqslant 0$ is thus a standing wave

$$E_y = E_i \left(e^{ikx - i\omega t} - e^{-ikx - i\omega t} \right) = 2i E_y \sin(kx) e^{-i\omega t}, \tag{S-9.47}$$

with nodes where $\sin kx = 0$ and maximum amplitude $2E_i$. Recalling that $\omega / k = c$, the magnetic field of the wave is

$$B_z = 2E_i \cos(kx) e^{-i\omega t}. \tag{S-9.48}$$

Thus, B_z is *discontinuous* at the $x = 0$ surface. This implies the presence of a surface current density $\mathbf{K} = \hat{\mathbf{y}} K_y(t)$ at $x = 0$, corresponding to a volume current density $\mathbf{J} = \mathbf{K} \delta(x) = \hat{\mathbf{y}} K_y(t) \delta(x)$. By evaluating the line integral of $\mathbf{B}$ over a closed path crossing the mirror surface we find the boundary condition

$$B_z(0^+, t) - B_z(0^-, t) = \frac{4\pi}{c} K_y(t), \tag{S-9.49}$$

and the surface current density on the surface of a perfect mirror is

$$K_y(t) = -\frac{c}{4\pi} B_z(0^-, t) = -\frac{c E_i}{2\pi} e^{-i\omega t}. \tag{S-9.50}$$

(b) Let $\beta = v/c$ (in what follows v, and, consequently, β, may have both positive or negative values, depending on whether the wave and the mirror velocity are parallel or antiparallel, respectively). We know that $(\omega/c, \mathbf{k})$ is a four-vector, and that $\mathbf{k}$ is parallel to v. Thus the frequency of the incident wave in S' is

$$\omega_i' = \gamma(\omega - \mathbf{k} \cdot \mathbf{v}) = \gamma \omega (1 - \beta), \tag{S-9.51}$$

where $k = \omega/c$ has been used. The magnitude of the incident wave vector in S' is $k_i' = \omega_i'/c$. If $v > 0$ $(v < 0)$ we have $\omega_i' < \omega$ $(\omega_i' > \omega)$.

The Lorentz transformations give the following amplitudes for the fields in S'

$$E_{iy}' = \gamma \left(E_{iy} - \beta B_{iz} \right) = \gamma(1 - \beta) E_i, \tag{S-9.52}$$

$$B_{iz}' = \gamma \left(B_{iz} - \beta E_{iy} \right) = \gamma(1 - \beta) E_i, \tag{S-9.53}$$

since $B_{iz} = E_{iy}$. In the S' frame the reflected wave has frequency $\omega_r' = \omega_i'$, and field amplitudes $E_{ry}' = -E_{iy}'$, $B_{rz}' = B_{iz}'$.

(c) The frequency ω_r of the reflected wave in the laboratory frame S can be evaluated by applying the inverse transformation from S' to S

$$\omega_r = \gamma(\omega_r' + \mathbf{k}_r \cdot \boldsymbol{v}) = \gamma(\omega_r' - k_r v) = \gamma\omega_r'(1-\beta) = \omega\gamma^2(1-\beta)^2$$

$$= \omega\frac{1-\beta}{1+\beta}. \qquad (S\text{-}9.54)$$

The electric and magnetic field amplitudes of the reflected wave in S' are $E_r' = -E_i' = -\gamma(1-\beta)E_i$ and $B_r' = B_i' = \gamma(1-\beta)E_i$. We thus have in S

$$E_{r\,y} = \gamma\left(E_{r\,y}' + \beta B_{r\,z}'\right) = -\gamma(1-\beta)\,E_i' = -\gamma^2(1-\beta)^2 E_i = -\frac{1-\beta}{1+\beta}E_i\,,$$
$$(S\text{-}9.55)$$

$$B_{r\,z} = \gamma\left(B_{r\,z}' + \beta E_{r\,y}'\right) = \gamma(1-\beta)\,E_i' = \gamma^2(1-\beta)^2 E_i = \frac{1-\beta}{1+\beta}E_i\,. \quad (S\text{-}9.56)$$

If $\beta < 0$ we have $|E_r| > |E_i|$: in S the reflected wave has a higher amplitude than the incident wave.

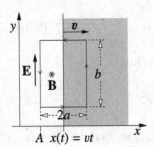

Fig. S-9.4

(d) The complete expressions for the fields in S are

$$E_y(x,t) = E_i\,\mathrm{e}^{ikx-i\omega t} - \frac{1-\beta}{1+\beta}E_i\,\mathrm{e}^{-ik_r x - i\omega_r t}\,,$$
$$(S\text{-}9.57)$$

$$B_z(x,t) = E_i\,\mathrm{e}^{ikx-i\omega t} + \frac{1-\beta}{1+\beta}E_i\,\mathrm{e}^{-ik_r x - i\omega_r t}\,,$$
$$(S\text{-}9.58)$$

thus, also E_y has a finite value at the mirror surface $x(t) = vt$, and is therefore *discontinuous*:

$$E_y[x(t),t] = \frac{2\beta}{1+\beta}E_i\,\mathrm{e}^{-i(1-\beta)\omega t}\,, \qquad (S\text{-}9.59)$$

$$B_z[x(t),t] = \frac{2}{1+\beta}E_i\,\mathrm{e}^{-i(1-\beta)\omega t}\,. \qquad (S\text{-}9.60)$$

This can be seen by considering again the line integral of the electric field $\mathbf{E}$ along a closed rectangular path of sides $2a$ and b, at rest in S. We assume that the left vertical side of the path is on the $x = A$ line, that at time t the mirror surface cuts the two horizontal sides, as in Fig. S-9.4, and that $a \ll \lambda$, where λ is the wavelength in S. The flux of the magnetic field through the rectangular path at time t is thus

$$\Phi(t) \simeq B_z[x(t),t][x(t) - A]\,b = B_z[x(t),t)]\,(vt - A)\,b, \qquad (S\text{-}9.61)$$

so that

$$-\frac{1}{c}\frac{d\Phi(t)}{dt} \simeq -\frac{1}{c}\left[\partial_t B_z[x(t), t](vt - A)b + B_z[x(t), t]vb\right]. \tag{S-9.62}$$

At the limit $a \to 0$, $A \to vt$, the first term of the right-hand side vanishes, and we are left with

$$-\frac{1}{c}\frac{d\Phi(t)}{dt} \simeq -\frac{1}{c}B_z[x(t), t]vb = -B_z[x(t), t]\beta b. \tag{S-9.63}$$

On the other hand, the line integral of $\mathbf{E}$ along the closed rectangular path of Fig. S-9.4 is

$$\oint \mathbf{E} \cdot d\boldsymbol{\ell} = -E_y[x(t), t]b = -\frac{2\beta}{1+\beta}bE_i\,\mathrm{e}^{-i(1-\beta)\omega t} = -B_z[x(t), t]\beta b. \tag{S-9.64}$$

S-9.7 Oblique Incidence on a Moving Mirror

(a) We choose a Cartesian reference frame S where v is parallel to the x axis, the mirror surface lies on the yz plane and the wave vector $\mathbf{k}_i$ of the incident wave lies in the xy plane. The Lorentz transformations to the frame S' give

$$k'_{ix} = \gamma\left(k_{ix} - \omega\frac{v}{c^2}\right) = \gamma\frac{\omega}{c}(\cos\theta_i - \beta), \tag{S-9.65}$$

$$k'_{iy} = k_{iy}, \tag{S-9.66}$$

$$\omega'_i = \gamma(\omega_i - k_x v) = \gamma\omega_i(1 - \beta\cos\theta_i), \tag{S-9.67}$$

$$\tan\theta'_i = \frac{k'_{iy}}{k'_{ix}} = \frac{k_{ix}\tan\theta_i}{\gamma(\omega_i/c)(\cos\theta_i - \beta)} = \frac{k_{ix}\sin\theta_i}{\gamma k_{ix}(\cos\theta_i - \beta)} = \frac{\sin\theta_i}{\gamma(\cos\theta_i - \beta)}, \tag{S-9.68}$$

where, as usual, $\beta = v/c$ and $\gamma = 1/\sqrt{1 - \beta^2}$. In S' the reflection angle θ'_r equals the incidence angle θ'_i, thus

$$k'_{rx} = -k'_{ix}, \quad k'_{ry} = k'_{iy}, \quad \omega'_r = \omega'_i. \tag{S-9.69}$$

(b) By performing the Lorentz transformations back to the laboratory frame S we obtain

$$k_{\mathrm{ry}} = k_{\mathrm{iy}} , \tag{S-9.70}$$

$$k_{\mathrm{rx}} = \gamma \left(k'_{\mathrm{rx}} + \omega' \frac{v}{c^2} \right) = -\gamma^2 \left[k_{\mathrm{ix}} \left(1 + \beta^2 \right) - 2\omega_{\mathrm{i}} \frac{\beta}{c} \right]$$

$$= -2\gamma^2 \frac{\omega}{c} \left[\left(1 + \beta^2 \right) \cos \theta_i - 2\beta \right] , \tag{S-9.71}$$

$$\omega_{\mathrm{r}} = \gamma(\omega'_{\mathrm{r}} + k'_{\mathrm{rx}} v) = \gamma^2 \left[\omega \left(1 + \beta^2 \right) - 2k_{\mathrm{ix}} v \right] = \gamma^2 \frac{\omega}{c} \left(1 + \beta^2 - 2\beta \cos \theta_i \right) , \tag{S-9.72}$$

from which

$$\tan \theta_r \equiv -\frac{k_{\mathrm{ry}}}{k_{\mathrm{rx}}} = \frac{\sin \theta_i}{\gamma^2 \left[2\beta - \left(1 + \beta^2 \right) \cos \theta_i \right]} , \tag{S-9.73}$$

For $\cos \theta_i = v/c = \beta$ the denominator of the "rightmost right-hand side" of (S-9.68) is zero, and the incidence angle θ'_i in S' is a right angle. This means that, in S', the incident wave propagates parallel to the mirror surface, without hitting the mirror, and no reflection occurs. For incidence angles such that $\cos \theta_i > \beta$, all the above formulas are meaningless, since they would imply $k'_{\mathrm{ix}} < 0$, i.e., that the wave is incident on the other side of the mirror.

S-9.8   Pulse Modification by a Moving Mirror

(a) The number of oscillations in the wave packet is a relativistic invariant, and the Lorentz transformations are linear in the EM fields. Thus, in the reference frame S', where the mirror is at rest, the incident wave packet is still square and comprises the same number of oscillations. On the other hand, as already seen in Problem 9.6, the frequency ω'_i and the amplitude E'_i are

$$\omega'_i = \gamma(1 - \beta) \omega_i , \quad E'_i = \gamma(1 - \beta) E_i , \tag{S-9.74}$$

where $\beta = v/c$. In S', the reflected packet has the same shape, duration, and frequency of the incident packet, but opposite amplitude and direction.

$$E'_r = -E'_i , \quad \omega'_r = \omega'_i , \quad \tau'_r = \tau'_i = N \frac{2\pi}{\omega'_i} = N \frac{2\pi}{\gamma(1 - \beta)\omega_i} = \frac{\tau_i}{\gamma(1 - \beta)} . \tag{S-9.75}$$

Back-transforming to S (see also Problem 9.6) we have

$$E_r = -\frac{1 - \beta}{1 + \beta} E_i , \quad \omega_r = \gamma(1 - \beta) \omega'_r = \gamma^2(1 - \beta^2) \omega_i = \frac{1 - \beta}{1 + \beta} \omega_i \tag{S-9.76}$$

The duration of the reflected wave packet is thus

$$\tau_r = N \frac{2\pi}{\omega_r} = N \frac{2\pi}{\omega} \frac{1+\beta}{1-\beta} = \frac{1+\beta}{1-\beta} \tau_i . \tag{S-9.77}$$

If $\beta > 0$, i.e., if the mirror velocity is parallel to the packet propagation direction, the reflected packet has a longer duration than the incident packet, while the reflected packet is shorter if the mirror velocity is antiparallel. **(b)** The energy per unit surface of each packet is given by its intensity I times its duration τ. The intensity is proportional to the square of the electric field amplitude, thus the relation between the reflected and incident intensities is

$$I_r = \left(\frac{1-\beta}{1+\beta}\right)^2 I_i , \tag{S-9.78}$$

and the relation between the energies per unit surface of the whole reflected and incident packets is

$$U_r = I_r \tau_r = \left(\frac{1-\beta}{1+\beta}\right)^2 I_i \frac{1+\beta}{1-\beta} \tau_i = \frac{1-\beta}{1+\beta} I_i \tau_i = \frac{1-\beta}{1+\beta} U_i . \tag{S-9.79}$$

We see that $U_r \neq U_i$, hence some work per unit surface is needed in order to keep the mirror moving at constant velocity, namely

$$W = U_r - U_i = -\frac{2\beta}{1+\beta} U_i . \tag{S-9.80}$$

Thus a mirror with $\beta < 0$, i.e., moving in the direction opposite to the incident wave packet, transfers some energy to the packet.
(c) As a first step, we determine the distribution of the current density $\mathbf{J}$. Since all the fields are null inside the mirror, i.e., for $x > x(t) = vt$, the current must be localized on the mirror surface, $\mathbf{J}(x,t) = \mathbf{K}(t)\,\delta(x - vt)$. We can evaluate the surface current density $\mathbf{K}(t)$ on the mirror surface by considering the fields close to the surface. By calculating the line integral of $\mathbf{B}$ over a closed rectangular path, fixed in S, of sides b, parallel to $\mathbf{B}$ and to the mirror surface, and $2a$, perpendicular to, and crossing the mirror surface, as in Fig. S-9.5, we obtain

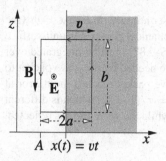

Fig. S-9.5

$$\oint_{\text{path}} \mathbf{B} \cdot d\boldsymbol{\ell} = \frac{4\pi}{c} \Phi(\mathbf{J}) + \frac{1}{c} \frac{d\Phi(\mathbf{E})}{dt} , \tag{S-9.81}$$

where $\Phi(\mathbf{J})$ and $\Phi(\mathbf{E})$ are the fluxes through the surface delimited by the path Jv and $\mathbf{E}$, respectively. At the limit $a \to 0$ and $A \to vt$, we have

$$\oint_{\text{path}} \mathbf{B} \cdot d\boldsymbol{\ell} \simeq B(vt)\,b \,, \qquad \Phi(\mathbf{J}) = K(t)\,b \,, \tag{S-9.82}$$

$$\frac{d\Phi(\mathbf{E})}{dt} \simeq \partial_t E(vt)\,(vt - A)\,b + E(vt)\,bv \simeq E(vt)\,bv \,. \tag{S-9.83}$$

From the knowledge of E and B at the mirror surface (Problem 9.6) we obtain

$$\begin{aligned} K(t) &= \frac{c}{4\pi} \left[B(vt) - \frac{v}{c} E(vt) \right] = \frac{c}{4\pi} \left(1 - \beta^2 \right) \frac{2E_i}{1 + \beta} e^{-i(1-\beta)\omega t} \\ &= \frac{cE_i}{2\pi} (1 - \beta) e^{-i(1-\beta)\omega t} \,. \end{aligned} \tag{S-9.84}$$

Thus, $\mathbf{K}$ and $\mathbf{E}$ are in phase. In order to evaluate the total mechanical work per unit surface on the mirror, we first switch back to the real quantities

$$K(t) = \frac{cE_i}{2\pi}(1 - \beta)\cos[(1 - \beta)\omega t] \,, \tag{S-9.85}$$

$$E(vt) = \frac{2\beta}{1 + \beta} E_i \cos[(1 - \beta)\omega t] \,, \tag{S-9.86}$$

and evaluate the integral over the mirror depth

$$\int_{vt}^{\infty} \mathbf{J} \cdot \mathbf{E}\,dx = \frac{1}{2}\,K(t)E(vt) = \frac{cE_i^2}{2\pi} \frac{\beta(1 - \beta)}{1 + \beta} \cos^2[(1 - \beta)\omega t] \,. \tag{S-9.87}$$

We have inserted the factor $1/2$ to account for the discontinuity of E at $x = vt$ (see also Problem 2.12). Equation (S-9.87) gives the mechanical power per unit surface exerted on the mirror. To find the mechanical work, (S-9.87) must be integrated over the time interval for which $K(t) \neq 0$, i.e., for the time needed by the wave packet to undergo a complete reflection. If the front of the wave packet reaches the mirror at $t = 0$, the end of the packet will leave the mirror at $t = \tau/(1 - \beta)$, which is different from the pulse duration τ because the mirror moves while the wave train is reflected. We thus need the integral

$$\int_0^{\tau/(1-\beta)} \cos^2[(1 - \beta)\omega t]\,dt = \frac{1}{\omega(1 - \beta)} \int_0^{\omega\tau} \cos^2 x\,dx = \frac{\pi N}{(1 - \beta)\,\omega} \,, \tag{S-9.88}$$

since $\omega\tau = 2\pi N$, and the integral of $\cos^2 x$ over one period equals π. We thus obtain

$$W = \int \frac{1}{2} K(t) E(vt) \, dt = \frac{cE_i^2}{2\pi} \frac{\beta(1-\beta)}{1+\beta} \frac{\pi N}{(1-\beta)\omega} = \frac{cE_i^2}{4\pi} \frac{\beta}{1+\beta} \tau \quad \text{(S-9.89)}$$

$$= 2I_i \tau \frac{\beta}{1+\beta} = \frac{2\beta}{1+\beta} U_i , \quad \text{(S-9.90)}$$

in agreement with (S-9.80).

The work W divided by the reflection time gives the mechanical power per unit surface

$$\mathcal{P} = W \frac{1-\beta}{\tau} = \frac{2\beta(1-\beta)}{1+\beta} I_i = \frac{2(1-\beta)}{1+\beta} I_i \frac{v}{c} , \quad \text{(S-9.91)}$$

which must be equal to the the pressure exerted on the moving mirror times its velocity v. We thus obtain that the radiation pressure on a moving mirror is

$$P_{\text{rad}} = \frac{\mathcal{P}}{v} = \frac{2I_i}{c} \frac{1-\beta}{1+\beta} , \quad \text{(S-9.92)}$$

a result which can also be obtained in different ways (see Problems 13.7 and 13.8).

S-9.9   Boundary Conditions on a Moving Mirror

(a) We can assume the wave to be linearly polarized along y, without loss of generality. We choose the origin of the frame S', where the mirror is at rest, so that the mirror surface is on the $x' = 0$ plane. In S' the total fields at the mirror surface are

$$\mathbf{E}'_s(t') = \hat{\mathbf{y}}' E'_s(t') \equiv 0 ,$$
$$\mathbf{B}'_s(t') = \hat{\mathbf{z}}' B'_s(t') e^{-i\omega'_i t'} = -\hat{\mathbf{z}}' 2E'_i e^{-i\omega'_i t'}, \quad \text{(S-9.93)}$$

respectively, where

$$E'_i = \gamma(1-\beta) E_i , \quad \omega'_i = \gamma(1-\beta)\omega_i , \quad \text{(S-9.94)}$$

are the amplitude and frequency of the incident wave in S', as seen in Problem 9.6. Notice that $\mathbf{E}'$ is continuous at $x' = 0$, while $\mathbf{B}'$ is not. By transforming the field amplitudes at the mirror surface back to S we obtain

$$E_s = \gamma(E'_s + \beta B'_s) = \gamma\beta B'_s = -2\gamma^2\beta(1-\beta)E_i \ ,$$
$$B_s = \gamma(B'_s - \beta E'_s) = \gamma B'_s = -2\gamma^2(1-\beta)E_i \ , \qquad \text{(S-9.95)}$$

where $\beta = v/c$ and $\gamma = 1/\sqrt{1-\beta^2}$. Thus, in general, in S we have both $E_s \neq 0$ and $B_s \neq 0$, while the fields are zero inside the mirror.

(b) The EM fields are related to the vector potential by

$$\mathbf{E} = -\frac{1}{c}\partial_t\mathbf{A} \ , \qquad \mathbf{B} = \nabla \times \mathbf{A}. \qquad \text{(S-9.96)}$$

Thus, the only nonzero component of the vector potential is A_y, and we have

$$E_s = -\frac{1}{c}\partial_t A_y \ , \qquad B_s = \partial_x A_y \ . \qquad \text{(S-9.97)}$$

The total derivative of $\mathbf{A}$ appearing in (9.9) can be rewritten

$$\left.\frac{d A}{d t}\right|_{x=x(t)} = [\partial_t A_y + v\,\partial_x A_y]_{x=x(t)} = cE_s - vB_s = c\,(E_s - \beta B_s) = 0 \ ,$$
$$\text{(S-9.98)}$$

according to (S-9.95). Thus the Eqs. (S-9.93) and (S-9.95) imply $d\mathbf{A}/dt = 0$ on the mirror surface in S.

(c) The total vector potential in S is the sum of the vector potentials of the incident and the reflected waves,

$$\mathbf{A}(x,t) = \hat{\mathbf{y}}\left[A_i\,e^{ik_i x - i\omega_i t} + A_r e^{-ik_r x - i\omega_r t}\right] = \hat{\mathbf{y}}\left[A_i\,e^{ik_i(x-ct)} + A_r\,e^{-ik_r(x+ct)}\right] \ , \quad \text{(S-9.99)}$$

where $A_i = icE_i/\omega_i$, $k_i = \omega_i/c$, and $k_r = \omega_r/c$. The boundary condition gives

$$0 = A_y(vt,t) = A_i\,e^{-ik_i(c-v)t} + A_r\,e^{-ik_r(c+v)t} \ . \qquad \text{(S-9.100)}$$

This equation is satisfied if

$$A_r = -A_i \ , \qquad \frac{k_r}{k_i} = \frac{\omega_r}{\omega_i} = \frac{c-v}{c+v} = \frac{1-\beta}{1+\beta} \ . \qquad \text{(S-9.101)}$$

For the total electric field we find

$$E_y = -\frac{1}{c}\partial_t A_y = i\frac{\omega_i}{c}A_i\,e^{ik_i x - i\omega_i t} - i\frac{\omega_r}{c}A_r\,e^{-ik_r x - i\omega_r t}$$
$$= E_i\,e^{ik_i x - i\omega_i t} - \frac{1-\beta}{1+\beta}E_i\,e^{-ik_r x - i\omega_r t} \ . \qquad \text{(S-9.102)}$$

S-9.10   Lorentz Transformations on a Coaxial Cable

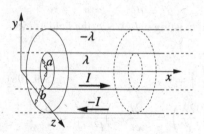

Fig. S-9.6

(a) We choose the symmetry axis of the coaxial cable as x axis of the laboratory reference frame S, as shown in Fig. S-9.6. We integrate the four-current of the inner conductor $J^\alpha = (c\varrho, \mathbf{J})$ over the yz plane, obtaining the quantity $G^\alpha = (c\lambda, \mathbf{I})$. This quantity transforms like a four-vector under boosts along the x axis, because surfaces parallel to the yz plane are invariant under such transformations.

Now we consider a reference frame S' that moves at a constant velocity $\mathbf{V}$ relative to S, with $\mathbf{V}$ parallel to the x axis. In S' we have, according to the Lorentz transformations,

$$\lambda' = \gamma\left(\lambda - \frac{VI}{c^2}\right), \quad I' = \gamma(I - V\lambda), \quad\quad (S\text{-}9.103)$$

where, as usual, $\gamma = 1/\sqrt{1 - V^2/c^2}$. Thus the conditions $\lambda' = 0$ and $I' = 0$ can be achieved at the velocities

$$V_\lambda = \frac{\lambda c^2}{I}, \quad \text{and} \quad V_I = \frac{I}{\lambda}, \quad\quad (S\text{-}9.104)$$

respectively. These velocities identify two different reference frames, that we denote by S' and S'', which cannot exist simultaneously. Equations (S-9.104) and the condition $V < c$ imply that the reference frame S', where $\lambda' = 0$, can exist only if $I/\lambda > c$, while the reference frame S'', where $I'' = 0$ (for the sake of clarity, from here on we denote quantities measured in S'' by double primes), can exist only if $I/\lambda < c$. In the two frames we have

$$\gamma' = \frac{I}{\sqrt{I^2 - \lambda^2 c^2}}, \quad \lambda' = 0, \quad I' = \gamma'\left(\frac{I^2 - \lambda^2 c^2}{I}\right) = \sqrt{I^2 - \lambda^2 c^2},$$
$$(S\text{-}9.105)$$

$$\gamma'' = \frac{\lambda c}{\sqrt{\lambda^2 c^2 - I^2}}, \quad I'' = 0, \quad \lambda'' = \gamma''\left(\frac{\lambda^2 c^2 - I^2}{\lambda c^2}\right) = \frac{\sqrt{\lambda^2 c^2 - I^2}}{c}.$$
$$(S\text{-}9.106)$$

The same results can be obtained more easily by noting that $I^2 - \lambda^2 c^2$ is a Lorentz invariant.

(b) Due to the symmetry of the problem, it is more convenient to use a cylindrical coordinate system (r, ϕ, x), with x as polar axis. In the laboratory reference frame S we have

$$\mathbf{E} = \frac{2\lambda}{r}\,\hat{\mathbf{r}}\,, \qquad \mathbf{B} = \frac{2I}{cr}\,\hat{\phi}\,. \tag{S-9.107}$$

In S' we have $r' = r$, $\theta' = \theta$, in S'' we have $r'' = r$, $\theta'' = \theta$.

In S', moving with velocity $V_\lambda = \lambda c^2/I$ and existing if $I/\lambda > c$, the field as evaluated from the Lorentz transformations are

$$E'_r = \gamma'(E_r - \frac{V_\lambda}{c}\,B_\phi) = \gamma'\left(\frac{2\lambda}{r} - \frac{\lambda}{I}\frac{2I}{r}\right) = 0\,, \tag{S-9.108}$$

$$B'_\phi = \gamma'\left(B_\phi - \frac{V_\lambda}{c}\,E_r\right) = \gamma'\left(\frac{2I}{cr} - \frac{\lambda c}{I}\frac{2\lambda}{r}\right) = \gamma'\frac{2I}{cr}\frac{I^2 - \lambda^2 c^2}{I^2} = \frac{B_\phi}{\gamma'}\,. \tag{S-9.109}$$

alternatively we obtain, evaluating the fields from the sources,

$$E'_r = \frac{2\lambda'}{r} = 0 \tag{S-9.110}$$

$$B'_\phi = \frac{2I'}{cr} = \frac{2\sqrt{I^2 - \lambda^2 c^2}}{cr} = \frac{2I}{\gamma' cr} = \frac{B_\phi}{\gamma'}\,, \tag{S-9.111}$$

In S'', moving with velocity $V_I = I/\lambda$ and existing if $I/\lambda < c$, the Lorentz transformations lead to

$$E''_r = \gamma''\left(E_r - \frac{V_I}{c}\,B_\phi\right) = \gamma''\left(\frac{2\lambda}{r} - \frac{I}{\lambda c}\frac{2I}{cr}\right) = \gamma'\frac{2\lambda}{r}\frac{\lambda^2 c^2 - I^2}{\lambda^2 c^2} = \frac{E_r}{\gamma''}\,, \tag{S-9.112}$$

$$B''_\phi = \gamma''\left(B_\phi - \frac{V_I}{c}\,E_r\right) = \gamma''\left(\frac{2I}{cr} - \frac{I}{c\lambda}\frac{2\lambda}{r}\right) = 0\,, \tag{S-9.113}$$

alternatively we obtain, evaluating the fields from the sources,

$$E''_r = \frac{2\lambda''}{r} = \frac{2\sqrt{\lambda^2 c^2 - I^2}}{rc} = \frac{2}{rc}\frac{\lambda c}{\gamma''} = \frac{E_r}{\gamma''}\,, \tag{S-9.114}$$

$$B''_\phi = \frac{2I''}{cr} = 0\,. \tag{S-9.115}$$

The same results can be obtained by noting that $E^2 - B^2$ is a Lorentz invariant. Thus, for instance, in the frame S'', where I'', and therefore $\mathbf{B}''$, are zero, we must have

$$E''^2 = E^2 - B^2\,, \tag{S-9.116}$$

from which we obtain

$$E'' = \sqrt{E^2 - B^2} = \sqrt{\frac{4\lambda^2}{r^2} - \frac{4I^2}{c^2r^2}} = \frac{2}{r}\sqrt{\lambda^2 - \frac{I^2}{c^2}} = \frac{2}{cr}\sqrt{\lambda^2c^2 - I^2}$$

$$= \frac{2}{cr}\frac{\lambda c}{\gamma''} = \frac{E}{\gamma''} , \qquad\qquad\qquad (S\text{-}9.117)$$

where we have used

$$\sqrt{\lambda^2c^2 - I^2} = \frac{\lambda c}{\gamma''} = \frac{cr}{2}\sqrt{E^2 - B^2} . \qquad (S\text{-}9.118)$$

(c) In the laboratory reference frame S the surface electrostatic force on the inner conductor is

$$P_{es}(a) = \frac{1}{2}\sigma E(a^+) = \frac{1}{2}\frac{\lambda}{2\pi a}\frac{2\lambda}{a} = \frac{\lambda^2}{2\pi a^2} = \frac{1}{8\pi}E^2(a^+) , \qquad (S\text{-}9.119)$$

and points outwards. We have used $\sigma = \lambda/(2\pi a)$. The surface magnetic force is

$$P_{mag}(a) = \frac{1}{2c}\mathbf{K} \times \mathbf{B}(a^+) \cdot \mathbf{r} = -\frac{1}{2c}\frac{I}{2\pi a}\frac{2I}{ca} = -\frac{I^2}{2\pi c^2a^2} = \frac{1}{8\pi}B^2(a^+) ,$$
$$(S\text{-}9.120)$$

where we have used $\mathbf{K} = \hat{\mathbf{x}}\,I/(2\pi a)$, $\mathbf{K}$ being the surface current density. This force points inwards, and the total force per unit surface is thus

$$P(a) = P_{es}(a) + P_{mag}(a) = \frac{\lambda^2}{2\pi a^2} - \frac{I^2}{2\pi c^2a^2} = \frac{1}{2\pi c^2a^2}\left(\lambda^2c^2 - I^2\right) ,$$
$$(S\text{-}9.121)$$

which is a Lorentz invariant. An analogous procedure for the external conductor leads to

$$P(b) = -\frac{1}{2\pi c^2b^2}\left(\lambda^2c^2 - I^2\right) . \qquad (S\text{-}9.122)$$

Thus the internal and external conductors repel each other because of the magnetic force, and attract each other because of the electric force. If $\lambda c > I$ attraction is dominant, the internal conductor tends to expand, and the external conductor tends to shrink. If $\lambda c < I$ repulsion is dominant

(d) In the laboratory reference frame S the total density of electric and magnetic density in the $a < r < b$ region is

$$u = \frac{E^2}{8\pi} + \frac{B^2}{8\pi} = \frac{\lambda^2}{2\pi r^2} + \frac{I^2}{2\pi c^2r^2} = \frac{1}{2\pi r^2}\left(\lambda^2 + \frac{I^2}{c^2}\right) . \qquad (S\text{-}9.123)$$

The density of electromagnetic momentum in the same region is

$$\mathbf{g} = \frac{1}{4\pi c}\mathbf{E} \times \mathbf{B} = \frac{\lambda I}{\pi c^2 r^2}\hat{\mathbf{x}} . \qquad (S\text{-}9.124)$$

In S', where λ' and $\mathbf{E}'$ are zero, we have

$$u' = \frac{1}{8\pi} B'^2 \,. \tag{S-9.125}$$

In S'', where I'' and $\mathbf{B}''$ are zero, we have

$$u'' = \frac{1}{8\pi} E''^2 \,, \tag{S-9.126}$$

while both $\mathbf{g}'$ and $\mathbf{g}''$ are zero.

S-9.11 Lorentz Transformations on a Rectangular Metal Pipe

(a) Applying the Lorentz transformations to the four-vector $(\rho c, \mathbf{J})$ we obtain

$$J'_x = \gamma J_x = \pm\gamma J \,, \quad J'_y = J_y = \pm J \,, \quad \rho' = -\gamma\frac{V}{c^2}J_x = \mp\gamma\frac{V}{c^2}J \,, \tag{S-9.127}$$

for the current density the plus sign applies to the left and upper walls, the minus sign applies to the right and lower walls of Fig. 9.8. For the charge density the plus sign applies to the lower, the minus sign to the upper wall. Thus the current density is unchanged on the left and right walls, while it increases by a factor γ on the upper and lower walls, where also a charge density appears.

Applying the transformations for the electromagnetic field we obtain

$$B'_z = \gamma B_z = \gamma B_0 \,, \quad E'_y = \gamma\frac{V}{c} B_z = \gamma\frac{V}{c} B_0 \,, \tag{S-9.128}$$

The fields are zero outside the pipe. In S' the cross section of the pipe has sides $a' = a/\gamma$ and $b' = b$. The vertical walls have thickness $h' = h/\gamma$ and the horizontal walls have unchanged thickness h.

(b) Calculating the energy and electromagnetic momentum densities in S and in S' we obtain

$$u = \frac{B_z^2}{8\pi} \,, \qquad u' = \frac{B_z'^2}{8\pi} + \frac{E_y'^2}{8\pi} = \gamma^2\frac{B_0^2}{8\pi}\left(1 + \frac{V^2}{c^2}\right) \,, \tag{S-9.129}$$

$$\mathbf{g} = \frac{c}{4\pi} \mathbf{E} \times \mathbf{B} = 0 \,, \qquad \mathbf{g}' = \frac{c}{4\pi} \mathbf{E}' \times \mathbf{B}' = \gamma^2\frac{c}{4\pi}\frac{V}{c} B_0^2 \hat{\mathbf{x}}' \,. \tag{S-9.130}$$

(c) The discontinuity condition for the magnetic field at the upper wall is obtained by applying Ampére's circuital law to the rectangular loop L_1 of Fig. S-9.7. The long side of L_1 has length ℓ and is parallel to the z' axis, its short side, which tends to zero at our limit, is parallel to the y' axis. We have

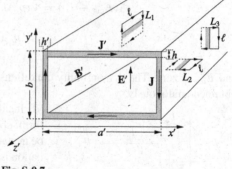

$$\left(B'_z - B_{\text{out}}\right)\ell = \frac{4\pi}{c}\,J'\ell h \quad \text{(S-9.131)}$$

$$B'_z - B_{\text{out}} = \frac{4\pi}{c}\,\gamma J h = B'_z\,,$$

Fig. S-9.7

Therefore $B_{\text{out}} = 0$. The discontinuity condition for **B** at the right wall is obtained by applying Ampére's circuital law to the rectangular loop L_2 of Fig. S-9.7. The long side of L_2 has length ℓ and is parallel to the z' axis, the short side is parallel to the x' axis, and tends to zero at our limit. We have

$$
\begin{aligned}
\left(B_{\text{out}} - B'_z\right)\ell &= -\frac{4\pi}{c}\,J\ell h' + \frac{1}{c}\frac{\partial \Phi(E')}{\partial t} = -\frac{\ell B_0}{\gamma} + \frac{1}{c}\frac{\partial}{\partial t}\left(\ell X' E'\right) \\
&= -\frac{\ell B_0}{\gamma} - \frac{1}{c}\ell V\left(\gamma\frac{V}{c}B_0\right) = -\frac{\ell B_0}{\gamma} - \gamma\ell\frac{V^2}{c^2}B_0 = \frac{\ell B_0}{\gamma} - \gamma\ell\left(1 - \frac{1}{\gamma^2}\right)B_0 \\
&= -\frac{\ell B_0}{\gamma} - \gamma\ell B_0 + \frac{\ell B_0}{\gamma} = -\ell B'_z\,, \quad\quad\quad\quad\quad\quad\quad\text{(S-9.132)}
\end{aligned}
$$

and B_{out} is again zero. Here $\Phi(E')$ is the flux of the electric field through L_2. The flux is not constant because L_2 is considered as at rest in S', while the pipe is moving with velocity $-V\,\hat{x}'$. The discontinuity conditions for **B**' at the left and lower wall are obtained analogously.

For the condition on **E**' at the right wall of the rectangular pipe we evaluate its line integral around Loop L_3 of Fig. S-9.7. We have;

$$\oint \mathbf{E}' \cdot d\mathbf{s} = \ell\left(E' - E_{\text{out}}\right) = -\frac{1}{c}\frac{d\Phi(B')}{dt} = \frac{1}{c}\,V\ell B' = \ell\,\frac{V}{c}\,\gamma B_0 = \ell E'\,, \tag{S-9.133}$$

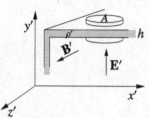

Fig. S-9.8

and E_{out} is zero. Here $\Phi(B')$ is the flux of the magnetic fiels through the loop L_3. Again, the flux changes in time because L_3 is at rest in S' while the pipe is moving.

For the condition on **E**' at the upper wall we apply Gauss's theorem to a shallow cylindrical closed surface of base surface A crossing the upper wall, as shown in Fig. S-9.8. We have

$$A(E_{\text{out}} - E') = 4\pi\rho' Ah = -4\pi Ah\gamma \frac{V}{c^2} J$$

$$= -\gamma B_0 A \frac{V}{c} = -E'A \,, \tag{S-9.134}$$

and E_{out} is zero. The discontinuity conditions on $\mathbf{E}'$ at the right and lower walls are obtained analogously.

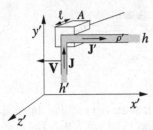

(d) Inside each wall the current density is uniform, so that $\nabla' \cdot \mathbf{J}' = \partial\rho'/\partial t' = 0$, as it must be in stationary conditions. At the edges the situation is slightly more complex. As an example consider the cuboid-shaped closed surface A of Fig. S-9.9, which wraps a portion of the upper left edge of length ℓ. The entering and exiting charge fluxes due to the current densities are

Fig. S-9.9

$$\Phi_{Q,\text{in}} = J\ell\frac{h}{\gamma} \,, \quad \text{and} \quad \Phi_{Q,\text{out}} = J'\ell h = \gamma J\ell h \,, \tag{S-9.135}$$

respectively. Thus we have

$$\Phi_{Q,\text{out}} - \Phi_{Q,\text{in}} = J\ell h \frac{\gamma^2 - 1}{\gamma} = J\ell h \frac{\beta^2}{\gamma(1-\beta^2)} = \gamma J\ell h \frac{V^2}{c^2} \,, \tag{S-9.136}$$

where we have used $\beta = V/c$. Thus the charge Q_A inside the closed surface A is not constant, and we have

$$\frac{\partial Q_A}{\partial t'} = -\gamma J\ell h \frac{V^2}{c^2} \,. \tag{S-9.137}$$

This is apparent inconsistency is due to the fact that the closed surface A is at rest in S', while the pipe is moving with velocity $-V\,\hat{\mathbf{x}}'$. According to (S-9.127) the upper wall of the pipe has a negative charge density $\rho' = -\gamma JV/c^2$ which is entering A through its right face. This leads to a charge chamge rate

$$\frac{\partial Q_A}{\partial t'} = -V\ell h\rho' = -V\ell h\gamma J \frac{V}{c^2} = -\gamma J\ell h \frac{V^2}{c^2} \,, \tag{S-9.138}$$

in agreement with (S-9.137).

S-9.12 Force on an Ohmic Wire

(a) Here we use a cylindrical coordinate system (r, ϕ, z) with the wire axis as z axis. The electric field $\mathbf{E}_0$ is continuous through the wire surface, thus the current density inside the wire is $\mathbf{J} = \sigma \mathbf{E}_0$ and the current flowing in the wire is $I = \pi a^2 J$. The power per unit length dissipated in the wire by Joule heating is

$$P_{\mathrm{d}} = \pi a^2 \, \mathbf{J} \cdot \mathbf{E} = \pi a^2 \sigma E_0^2 \ . \tag{S-9.139}$$

The magnetic field $\mathbf{B}_J$ has only the $\hat{\phi}$ component $B_{J\phi}$ because of symmetry, with

$$B_{J\phi}(r) = \frac{2I}{rc} = \frac{2\pi a^2 J}{rc} = \frac{2\pi a^2 \sigma}{rc} E_0 \ . \tag{S-9.140}$$

The Poynting vector at the wire surface is

$$\mathbf{S}(a) = \frac{c}{4\pi} \, \mathbf{E} \times \mathbf{B}_J(a) = -\hat{\mathbf{r}} \, \frac{a\sigma}{2} \, E_0^2 \ . \tag{S-9.141}$$

The flux per unit length of the Poynting vector through the wire surface is

$$\Phi(\mathbf{S}) = -S(a) \, 2\pi a = -\pi a^2 \sigma E_0^2 = -P_{\mathrm{d}} \ . \tag{S-9.142}$$

(b) In the laboratory frame we use a Cartesian coordinate system (x, y, z) with the axis of the wire as z axis, and S' moving along the x axis, thus $\mathbf{V} = \hat{\mathbf{x}} V$. We have, at the first order in $\beta = V/c$, $\mathbf{E}' \simeq \mathbf{E}_0'$ and $\mathbf{B}' \simeq -(\mathbf{V}/c) \times \mathbf{E}_0 + \mathbf{B}_J \equiv \mathbf{B}_0' + \mathbf{B}_J$. Note that here $\mathbf{B}_J$ is written as a function of the transformed coordinates ($x' \simeq x - Vt$, $y' = y$, $z' = z$), so that $r = \sqrt{(x' + Vt')^2 + y'^2}$. Further, we have $\mathbf{J}' = \mathbf{J}$ and $\rho' = 0$ because $\mathbf{J}$ is perpendicular to $\mathbf{V}$.

(c) The force per unit length acting on the wire is

$$\mathbf{F}' \doteq \mathbf{J}' \times \mathbf{B}_0' \, \pi a^2 \simeq -I E_0 \mathbf{V} \ , \quad \text{where} \quad I = \pi a^2 J \ , \tag{S-9.143}$$

we have neglected the term in $\mathbf{J}' \times \mathbf{B}_J$ which gives origin to a "shrinking" force on the wire with zero vector average value. Equation (S-9.143) is in agreement with the transformation of the four-force per unit volume exerted by the external field on the wire. In the laboratory frame S we have

$$f_\mu = \left(\frac{\mathbf{J} \cdot \mathbf{E}_0}{c}, 0, 0, 0 \right) \ . \tag{S-9.144}$$

In the moving frame S' we have, at the first order in β

$$f_\mu' \simeq \left(\frac{\mathbf{J} \cdot \mathbf{E}_0}{c}, -\mathbf{V} \, (\mathbf{J} \cdot \mathbf{E}_0), 0, 0 \right) \ . \tag{S-9.145}$$

428 S-9 Solutions for Chapter 9

We obtain F' by multiplying f_1' by the cross section πa^2 of the wire, which is invariant at the limit $\beta \ll 1$. In fact the cross section of the wire lies in the xy plane, $\mathbf{V}$ is parallel to the x axis, thus we have $dy' = dy$ while, in principle, dx' is affected by length contraction, which, however, is zero at the first order in β. Note that the norm of the four-vector f_μ is invariant only at the first order in β.

The thermal rest energy of the wire increases with time because of Joule heating, and this increases the inertia of the wire. This is why a force is needed in S', where the wire is in uniform motion.

Chapter S-10
Solutions for Chapter 10

S-10.1 Cyclotron Radiation

(a) The electric dipole moment $\mathbf{p} = -e\mathbf{r}$ rotates in the xy plane with frequency ω_L, which is also the frequency of the emitted radiation. The dipole approximation is valid if the dimensions of the radiating source are much smaller than the emitted wavelength λ. Here this corresponds to the condition $2r_L = 2v/\omega_L \ll \lambda = 2\pi c/\omega_L$, always true for non-relativistic velocities.

The rotating dipole can be written as $\mathbf{p} = p_0\,(\hat{\mathbf{x}}\cos\omega_L t + \hat{\mathbf{y}}\sin\omega_L t)$. For the electric field of the dipole radiation observed in a direction of unit vector $\hat{\mathbf{n}}$, we have $\mathbf{E} \propto -(\mathbf{p}\times\hat{\mathbf{n}})\times\hat{\mathbf{n}}$. If $\hat{\mathbf{n}} = \hat{\mathbf{z}}$, then $\mathbf{E} \propto \hat{\mathbf{x}}\cos\omega_L t + \hat{\mathbf{y}}\sin\omega_L t$ (circular polarization); if $\hat{\mathbf{n}} = \hat{\mathbf{x}}$ or $\hat{\mathbf{n}} = \hat{\mathbf{y}}$, we vave $\mathbf{E} \propto -\hat{\mathbf{y}}\sin\omega_L t$ and $\mathbf{E} \propto -\hat{\mathbf{x}}\cos\omega_L t$, respectively (linear polarization).

Since $\ddot{\mathbf{r}} = \mathbf{v}\times\boldsymbol{\omega}_L$ (where $\boldsymbol{\omega}_L = \hat{\mathbf{z}}\omega_L$), the radiated power can be written as

$$P_{\text{rad}} = \frac{2}{3}\frac{|e\ddot{\mathbf{r}}|^2}{c^3} = \frac{2}{3}\frac{e^2 v^2 \omega_L^2}{c^3}\,. \tag{S-10.1}$$

(b) We assume that the energy loss due to radiation is small enough to cause a variation of the orbit radius $\Delta r_c \ll r_c$ during a single period, so that, during a single period, the motion is still approximately circular. Thus the magnitude of the electron velocity $v = v(t)$ can be written as $v \simeq \omega_L r$, where $r = r(t)$ is the radius of the orbit at time t. The electron energy is

$$U = \frac{m_e v^2}{2} = \frac{m_e \omega_L^2 r^2}{2}\,, \tag{S-10.2}$$

and the equation for the energy loss, $dU/dt = -P_{\text{rad}}$, becomes

$$\frac{d}{dt}\left(\frac{m_e \omega_L^2 r^2}{2}\right) = -\frac{2}{3c^3}\left(e^2\omega_L^4 r^2\right) = -\frac{2r_e m_e \omega_L^4}{3c} r^2\,, \tag{S-10.3}$$

A. Macchi et al., *Problems in Classical Electromagnetism*,
https://doi.org/10.1007/978-3-031-22235-1

where $r_e = e^2/(m_e c^2)$ is the classical electron radius. Substituting the relation $d(r^2)/dt = 2r\, dr/dt$ into (S-10.3) we obtain

$$\frac{dr}{dt} = -\frac{2r_e\omega_L^2}{3c} r \equiv -\frac{r}{\tau}\,, \quad \text{with} \quad \tau = \frac{3c}{2r_e\omega_L^2} = \frac{3m_e c^3}{2e^2\omega_L^2}\,, \quad \text{(S-10.4)}$$

whose solution is

$$r(t) = r(0)\, e^{-t/\tau}\,, \quad \text{(S-10.5)}$$

and the trajectory of the electron is a spiral with a decay time τ. Inserting the expressions for r_e and ω_L we have

$$\tau = \frac{3}{2}\frac{m_e^3 c^5}{e^4 B_0^2} = \frac{5.2 \times 10^5}{B_0^2}\, \text{s} \quad \text{(S-10.6)}$$

where the magnetic field B_0 is in G. The condition $\tau \gg \omega_L^{-1}$ implies

$$\frac{3}{2}\frac{m_e^3 c^5}{e^4 B_0^2} \gg \frac{m_e c}{e B_0}\,, \quad \text{or} \quad B_0 \ll \frac{3}{2}\frac{m_e^2 c^4}{e^3} = 9.2 \times 10^{13}\, \text{G}\,, \quad \text{(S-10.7)}$$

a condition well verified in all experimental conditions: such high fields can be found only on neutron stars! (see Problem 10.5)

(c) We insert a frictional force $\mathbf{f}_{\text{fr}} = -m_e \eta\, \mathbf{v}$ into the equation of motion, obtaining

$$m_e \frac{d\mathbf{v}}{dt} = -\frac{e}{c}\mathbf{v} \times \mathbf{B}_0 - m_e \eta\, \mathbf{v}\,. \quad \text{(S-10.8)}$$

This corresponds to the following two coupled equations for the the x and y components the electron velocity

$$\dot{v}_x = -\omega_L v_y - \eta v_x\,, \qquad \dot{v}_y = \omega_L v_x - \eta v_y\,. \quad \text{(S-10.9)}$$

An elegant method to solve these equations is to combine the x and y coordinates of the electron into a single complex variable $R = x + iy$, and the velocity components into the complex variable $V = v_x + iv_y$. The two Eqs. (S-10.9) are thus combined into the single complex equation

$$\dot{V} = (i\omega_L - \eta)V\,, \quad \text{with solution} \quad V = V(0)\, e^{i\omega_L t - \eta t} = v_0\, e^{i\omega_L t - \eta t}\,. \quad \text{(S-10.10)}$$

For the electron position we have

$$R = \int V\, dt + C = \frac{v_0}{i\omega_L - \eta}\, e^{i\omega_L t - \eta t} + C = -\frac{(\eta + i\omega_L)\, v_0}{\omega_L^2 + \eta^2}\, e^{i\omega_L t - \eta t} + C\,, \quad \text{(S-10.11)}$$

where C is a complex constant depending on our choice of the origin of the coordinates. We choose $C = 0$, and rewrite (S-10.11) as

$$R = -\frac{v_0}{\sqrt{\omega_L^2 + \eta^2}} (\cos\phi + i\sin\phi) \, e^{i\omega_L t - \eta t} = -\frac{v_0}{\sqrt{\omega_L^2 + \eta^2}} \, e^{i(\omega_L t + \phi) - \eta t} \, ,$$

(S-10.12)

where

$$\cos\phi = \frac{\eta}{\sqrt{\omega_L^2 + \eta^2}} \, , \quad \sin\phi = \frac{\omega_L}{\sqrt{\omega_L^2 + \eta^2}} \, , \quad \phi = \arctan\left(\frac{\omega_L}{\eta}\right) \, . \quad \text{(S-10.13)}$$

Going back to the real quantities we have

$$v_x = \mathrm{Re}(V) = v_0 \, e^{-\eta t} \cos\omega_L t \, , \tag{S-10.14}$$

$$v_y = \mathrm{Im}(V) = v_0 \, e^{-\eta t} \sin\omega_L t \, , \tag{S-10.15}$$

$$x = \mathrm{Re}(R) = -\frac{v_0}{\sqrt{\omega_L^2 + \eta^2}} \, e^{-\eta t} \cos(\omega_L t + \phi) \, , \tag{S-10.16}$$

$$y = \mathrm{Im}(R) = -\frac{v_0}{\sqrt{\omega_L^2 + \eta^2}} \, e^{-\eta t} \sin(\omega_L t + \phi) \, . \tag{S-10.17}$$

Thus, the velocity rotates with frequency ω_L, while its magnitude decays exponentially, $|v(t)| = v_0 \, e^{-\eta t}$. For the radius of the trajectory we have

$$r(t) = |R(t)| = \frac{v_0}{\sqrt{\omega_L^2 + \eta^2}} \, e^{-\eta t} \, . \tag{S-10.18}$$

Thus, choosing $\eta = 1/\tau$, the motion with frictional force is identical to the motion with radiative power loss, and

$$\mathbf{f}_{\mathrm{fr}} \cdot \boldsymbol{v} = -m_e \eta v^2 = -\frac{m_e v^2}{\tau} = -m_e v^2 \frac{2e^2 \omega_L^2}{3m_e c^3} = -\frac{2e^2 v^2 \omega_L^2}{3c^3} = -P_{\mathrm{rad}} \, . \quad \text{(S-10.19)}$$

A drawback of this approach is that the frictional coefficient inserted here is not universal but is dependent on the force on the electron (in this case, via the dependence on ω_L). See Problem 10.12 for a more general approach to radiation friction.

S-10.2 Atomic Collapse

(a) An electron describing a circular orbit of radius a_0 (Bohr radius) around a proton corresponds to a counterrotating electric dipole $\mathbf{p}(t)$ of magnitude $p_0 = ea_0$. The angular velocity of the orbit ω can be evaluated by considering that the centripetal acceleration is due to the Coulomb force,

$$\omega^2 a_0 = \frac{1}{m_e} \frac{e^2}{a_0^2} , \tag{S-10.20}$$

from which we obtain

$$\omega = \sqrt{\frac{e^2}{m_e a_0^3}} = 4.1 \times 10^{16} \text{ rad/s} . \tag{S-10.21}$$

Actually, the strongest emission from the hydrogen atom occurs at a frequency smaller by about one order of magnitude.

Since $\mathbf{p}$ is perpendicular to $\boldsymbol{\omega}$, we have $\ddot{\mathbf{p}} = (\mathbf{p} \times \boldsymbol{\omega}) \times \boldsymbol{\omega}$ and $|\ddot{\mathbf{p}}|^2 = \left(\omega^2 p_0\right)^2$ (the same result can be obtained by considering the rotating dipole as the superposition of two perpendicularly oscillating dipoles). Thus the radiated power is

$$P_{\text{rad}} = \frac{2}{3c^3} |\ddot{\mathbf{p}}|^2 = \frac{2}{3} \frac{\omega^4 e^2 a_0^2}{c^3} = \frac{2}{3} \frac{e^2 r_e^2 c}{a_0^4} , \tag{S-10.22}$$

where r_e is the classical electron radius.

(b) We assume that, due to the emission of radiation, the electron loses its energy according to $dU/dt = -P_{\text{rad}}$, where $U = K + V$ is the total electron energy, K and V being the kinetic and potential energy, respectively. If the energy lost per period is small with respect to the total energy, we may assume that the electron the orbit is almost circular during a period, with the radius *slowly* decreasing with time, $r = r(t)$ with $\dot{r}/r \ll \omega$.

Since the velocity is $v = r\omega$, the total energy can be written as a function of a:

$$U = K + V = \frac{m_e v^2}{2} - \frac{e^2}{r} = -\frac{e^2}{2r} . \tag{S-10.23}$$

Therefore

$$\frac{dU}{dt} \simeq -\frac{e^2}{2} \frac{d}{d} \left(\frac{1}{r}\right) = \frac{e^2}{2r^2} \frac{dr}{dt} . \tag{S-10.24}$$

Since

$$P_{\text{rad}} = \frac{2}{3} \frac{e^2 r_e^2 c}{r^4} \tag{S-10.25}$$

the equation $dU/dt = -P_{\text{rad}}$ can be written as

$$\frac{e^2}{2r^2} \frac{dr}{dt} = -\frac{2}{3} \frac{e^2 r_e^2 c}{r^4} \quad \Rightarrow \quad r^2 \frac{dr}{dt} = -\frac{4}{3} r_e^2 c \quad \Rightarrow \quad \frac{1}{3} \frac{dr^3}{dt} = \frac{4}{3} r_e^2 c \tag{S-10.26}$$

The solution, assuming $r(0) = a_0$, is

$$r^3 = a_0^3 - 4r_e^2 c t,$$ (S-10.27)

giving for the time need by the electron to fall on the nucleus

$$\tau = \frac{a_0^3}{4r_e^2 c} \simeq 1.6 \times 10^{-11} \text{ s}.$$ (S-10.28)

This is a well-known result, showing that a classical "Keplerian" atom is not stable. It is however interesting to notice that the value of τ is of the same order of magnitude of the lifetime of the first excited state, i.e., of the time by which the excited state decays to the ground state emitting radiation.

S-10.3 Radiative Damping of the Elastically Bound Electron

(a) The solution of (10.5) with the given initial conditions and $\eta = 0$ is

$$\mathbf{r} = \mathbf{s}_0 \cos \omega_0 t.$$ (S-10.29)

The corresponding average radiated power in the dipole approximation is

$$P_{\text{rad}} = \frac{2}{3c^3} \langle -e|\ddot{\mathbf{r}}|^2 \rangle = \frac{2e^2}{3c^3} \omega_0^4 s_0^2 \langle \cos^2 \omega_0 t \rangle = \frac{e^2}{3c^3} \omega_0^4 s_0^2.$$ (S-10.30)

The radiated power is emitted at the expense of the energy of the oscillating electron. Thus, the total mechanical energy of the electron must decrease in time, and the harmonic-oscillator solution of (S-10.29) cannot be exact. Assuming that the energy of the oscillator decays very slowly, i.e., with a decay constant $\tau \gg \omega_0^{-1}$, we can approximate (S-10.29) as

$$\mathbf{r} \simeq \mathbf{s}(t) \cos \omega_0 t.$$ (S-10.31)

where $s(t)$ is a decreasing function of time to be determined. Consequently, we must replace s_0 by $s(t)$ also in Eq. (S-10.30) for the actual average radiated power.

(b) At time t, the total energy of the oscillating electron is $U(t) = m_e \omega_0^2 s^2(t)$. The time decay constant τ is defined as

$$\tau = \frac{U(t)}{P_{\text{rad}}(t)} = \frac{3m_e c^3}{2e^2 \omega_0^2} = \frac{3c}{2r_e \omega_0^2},$$ (S-10.32)

and is thus independent of t. Since the classical electron radius is $r_e \simeq 2.82 \times 10^{-15}$ m, the condition $\tau > 2\pi/\omega_0$ leads to

$$\omega_0 < \frac{3}{4\pi} \frac{c}{r_e} \simeq 3 \times 10^{22} \text{ rad/s} . \tag{S-10.33}$$

For a comparison, estimating ω_0 as the frequency of the 1S←2P Lyman-alpha emission line of the hydrogen atom, we have $\omega_0 \simeq 3 \times 10^{16}$ rad/s.

(c) We look for a solution of the form $\mathbf{r} = \text{Re}(\mathbf{s}_0 \, e^{-i\omega t})$, with *complex* ω. Substituting this into (10.5), the characteristic equation becomes

$$\omega^2 + i\eta\omega + \omega_0^2 = 0 , \tag{S-10.34}$$

whose solution is

$$\omega = -i\frac{\eta}{2} \pm \sqrt{\omega_0^2 - \frac{\eta^2}{4}} \simeq -i\frac{\eta}{2} \pm \omega_0 , \tag{S-10.35}$$

where we have neglected the terms of the order $(\eta/\omega_0)^2$ and higher. Thus, the approximated solution for the electron position is

$$\mathbf{r} \simeq \mathbf{s}_0 \, e^{-\eta t/2} \cos\omega_0 t . \tag{S-10.36}$$

Actually, this approximation gives an initial velocity $\dot{\mathbf{r}}(0) = -\eta \, \mathbf{s}_0/2$ instead of zero. However, this discrepancy can be neglected if $\eta \ll \omega_0$. The maximum speed reached by the electron is $v_{\max} \simeq \omega_0 s_0$, and $\eta s_0/2 \ll \omega_0 s_0$.

The time-dependent total energy of the electron and average radiated power are

$$U(t) \simeq \frac{m_e}{2} \omega_0^2 s_0^2 \, e^{-\eta t} , \quad \text{and} \quad P_{\text{rad}}(t) \simeq \frac{e^2}{3c^3} \omega_0^4 s_0^2 \, e^{-\eta t} . \tag{S-10.37}$$

The condition $dU/dt = -P_{\text{rad}}$ leads to

$$\eta = \frac{2r_e\omega_0^2}{3c} = \frac{1}{\tau} . \tag{S-10.38}$$

S-10.4 Radiation Emitted by Orbiting Charges

(a) Let us denote by $\mathbf{r}_1$ and $\mathbf{r}_2$ the location vectors of the two charges with respect to the center of their common circular orbit. In polar coordinates we have

$$\mathbf{r}_1 \equiv [R, \phi_1(t)] , \quad \text{and} \quad \mathbf{r}_2 \equiv [R, \phi_2(t)] . \tag{S-10.39}$$

Defining $\Delta\phi = \phi_2 - \phi_1$ and choosing an appropriate origin of time, the equations of motion in polar coordinates are

$$\mathbf{r}_1 \equiv \left(R, \omega t - \frac{\Delta\phi}{2} \right) , \quad \text{and} \quad \mathbf{r}_2 \equiv \left(R, \omega t + \frac{\Delta\phi}{2} \right) . \tag{S-10.40}$$

In Cartesian coordinates we have

$$\mathbf{r}_1 \equiv [x_1(t), y_1(t)] , \quad \text{and} \quad \mathbf{r}_2 \equiv [x_2(t), y_2(t)] , \tag{S-10.41}$$

with, respectively,

$$x_1(t) = R \cos\left(\omega t - \frac{\Delta\phi}{2}\right) , \qquad y_1(t) = R \sin\left(\omega t - \frac{\Delta\phi}{2}\right) , \tag{S-10.42}$$

$$x_2(t) = R \cos\left(\omega t + \frac{\Delta\phi}{2}\right) , \qquad y_2(t) = R \sin\left(\omega t + \frac{\Delta\phi}{2}\right) . \tag{S-10.43}$$

The dipole moment of the system is $\mathbf{p} = q(\mathbf{r}_1 + \mathbf{r}_2)$, with Cartesian components

$$p_x = q R \left[\cos\left(\omega t - \frac{\Delta\phi}{2}\right) + \cos\left(\omega t + \frac{\Delta\phi}{2}\right)\right] = 2q R \cos\left(\frac{\Delta\phi}{2}\right) \cos\omega t , \tag{S-10.44}$$

$$p_y = q R \left[\sin\left(\omega t - \frac{\Delta\phi}{2}\right) + \sin\left(\omega t + \frac{\Delta\phi}{2}\right)\right] = 2q R \cos\left(\frac{\Delta\phi}{2}\right) \sin\omega t , \tag{S-10.45}$$

i.e., $\mathbf{p}$ has constant magnitude $p = 2q R \cos(\Delta\phi/2)$, and rotates in the $z = 0$ plane with angular frequency ω.

(b) In the dipole approximation, the electric field of the radiation emitted along a direction of unit vector the $\hat{\mathbf{n}}$ is parallel to the vector

$$(\mathbf{p} \times \hat{\mathbf{n}}) \times \hat{\mathbf{n}} = \mathbf{p}_\perp . \tag{S-10.46}$$

Since for a dipole rotating in the $z = 0$ plane

$$(\mathbf{p} \times \hat{\mathbf{x}}) \times \hat{\mathbf{x}} \text{ is parallel to } \hat{\mathbf{y}} , \quad \text{and} \quad (\mathbf{p} \times \hat{\mathbf{y}}) \times \hat{\mathbf{y}} \text{ is parallel to } \hat{\mathbf{x}} , \tag{S-10.47}$$

the polarization of the radiation observed in the $\hat{\mathbf{x}}$ ($\hat{\mathbf{y}}$) direction is linear and along $\hat{\mathbf{y}}$ ($\hat{\mathbf{x}}$). For radiation observed the $\hat{\mathbf{z}}$ direction

$$(\mathbf{p} \times \hat{\mathbf{z}}) \times \hat{\mathbf{z}} \text{ is parallel to } \mathbf{p} , \tag{S-10.48}$$

and the observed polarization is circular.

The total radiated power is

$$P_{\text{rad}} = \frac{2}{3c^3} |\ddot{\mathbf{p}}|^2 = \frac{4q^2 R^2 \omega^4}{3c^3} \cos^2\left(\frac{\Delta\phi}{2}\right) , \tag{S-10.49}$$

which obviously vanishes when $\mathbf{p} = 0$, i.e, for $\Delta\phi = \pi$ (charges on opposite ends of a rotating diameter), and has a maximum for $\Delta\phi = 0$ (superposed charges).

(c) In this case charges are superposed to each other every half turn. We choose the coordinates and the time origin so that the charges are superposed at $t = 0$ we have $\mathbf{r}_1 = \mathbf{r}_2 = (R, 0)$. Thus the trajectories can be written as

$$r_1 = r_2 = R, \qquad \phi_1(t) = \omega t, \qquad \phi_2(t) = -\omega t, \qquad \text{(S-10.50)}$$

in polar coordinates, and as

$$x_1(t) = R \cos \omega t, \qquad\qquad y_1(t) = R \sin \omega t,$$
$$x_2(t) = R \cos \omega t, \qquad\qquad y_2(t) = -R \sin \omega t, \qquad \text{(S-10.51)}$$

in Cartesian coordinates. The total dipole moment is thus $\mathbf{p} = (2qR \cos \omega t)\,\hat{\mathbf{x}}$. No radiation is emitted along x, while the radiation emitted along all other directions is linearly polarized. The total average radiated power is

$$P_{\text{rad}} = \frac{2}{3c^3} |\ddot{\mathbf{p}}|^2 = \frac{4q^2 R^2 \omega^4}{3c^3}. \qquad \text{(S-10.52)}$$

(d) With an appropriate choice of the time origin the equations of motion of the three charges can be written, in polar coordinates, as

$$r_1 = r_2 = r_3 = R, \qquad\qquad \phi_1(t) = \omega t,$$
$$\phi_2(t) = \omega t + \Delta\phi_2, \qquad\qquad \phi_3(t) = \omega t + \Delta\phi_3, \qquad \text{(S-10.53)}$$

and, in Cartesian coordinates,

$$x_i = R \cos \phi_i(t), \qquad y_i = R \sin \phi_i(t), \qquad (i = 1, 2, 3). \qquad \text{(S-10.54)}$$

The electric dipole moment vanishes if the three charges are on the vertices of a rotating equilateral triangle ($\Delta\phi_2 = -\Delta\phi_3 = 2\pi/3$), and has its maximum value when the three charges are overlapped ($\Delta\phi_2 = \Delta\phi_3 = 0$).

(e) The magnetic dipole moment for a point charge q, traveling at angular velocity ω on a circular orbit of radius R, is defined by

$$\mathbf{m} = \frac{1}{2c} \int \mathbf{r} \times \mathbf{J}\, d^3x = \frac{qR^2\omega}{2c}, \qquad \text{(S-10.55)}$$

and is constant (notice that $\mathbf{m}$ is proportional to the angular momentum of the orbiting charge). Thus the magnetic dipole does not contribute to radiation, because the radiation fields are proportional to $\ddot{\mathbf{m}}$.

This problem explains why a circular coil carrying a constant current does not radiate, although we may consider the current as produced by charges moving on circular orbits, and thus subject to acceleration.

S-10.5 Spin-Down Rate and Magnetic Field of a Pulsar

(a) Due to the nonzero angle α between the magnetic moment and the rotation axis of the pulsar, the component of $\mathbf{m}$ perpendicular to ω rotates with frequency ω. Thus the Pulsar thus emits magnetic dipole radiation of frequency ω. The total power is

$$P = \frac{2}{3c^3}|\ddot{\mathbf{m}}_\perp|^2 = \frac{2}{3}\frac{m_\perp^2 \omega^4}{c^3} , \qquad (S\text{-}10.56)$$

where $m_\perp = m \sin \alpha$.

(b) The mechanical energy is $U = I\omega^2/2$ where $I = 2MR^2/5 \simeq 1.1 \times 10^{43}$ g cm^2 is the moment of inertia of the pulsar, assuming a uniform mass distribution over the volume of a sphere of radius R. Assuming that the energy loss is due to radiation emission only, we can write

$$\frac{dU}{dt} = \frac{d}{dt}\left(\frac{I\omega^2}{2}\right) = I\omega\dot{\omega} = -P , \qquad (S\text{-}10.57)$$

and, substituting (S-10.56), we have

$$I\omega\dot{\omega} = -\frac{2}{3}\frac{m_\perp^2 \omega^4}{c^3} \quad \Rightarrow \quad \frac{\dot{\omega}}{\omega^3} = -\frac{2m_\perp^2}{3Ic^3} . \qquad (S\text{-}10.58)$$

By integrating over time from 0 to t we obtain

$$\frac{1}{2\omega^2(t)} - \frac{1}{2\omega^2(0)} = \frac{2m_\perp^2}{3Ic^3} t , \qquad (S\text{-}10.59)$$

and thus

$$\omega(t) = \frac{\omega(0)}{\sqrt{1+\dfrac{t}{\tau}}} , \quad \text{where} \quad \tau = \frac{3Ic^3}{4m_\perp^2\omega^2(0)} . \qquad (S\text{-}10.60)$$

(c) We can rewrite $\dot{\omega}/\omega^3$ as $\dot{\omega}/\omega^3 = -T\dot{T}/4\pi^2$, where $T = 2\pi/\omega$ is the rotation period of the pulsar, and $\dot{T} = -2\pi\dot{\omega}/\omega^2$. Thus we can obtain the magnetic dipole moment $m = m_\perp$ of the pulsar as a function of the experimentally measured parameters from (S-10.58):

$$m = \sqrt{\frac{3Ic^3}{8\pi^2}T\dot{T}} \simeq 3.3 \times 10^{36}\sqrt{T\dot{T}} \text{ erg/G} , \qquad (S\text{-}10.61)$$

where T is in seconds. The magnetic field immediately outside of the pulsar surface is the field of a magnetic dipole located at the pulsar center:

$$\mathbf{B} = \frac{3(\hat{\mathbf{r}} \cdot \mathbf{m}) \hat{\mathbf{r}} - \mathbf{m}}{r^3} , \tag{S-10.62}$$

and thus $B_{\text{max}} = 2m/R^3$. Thus we obtain the practical formula

$$B_{\text{max}} \simeq 6.6 \times 10^{21} \sqrt{T\dot{T}} \text{ G} , \tag{S-10.63}$$

Inserting the experimental values for T and $\dot{T}$ we obtain

$$B_{\text{max}} \approx (9.6 \pm 0.25) \times 10^{16} \text{ G} . \tag{S-10.64}$$

S-10.6   A Bent Dipole Antenna

(a) If we divide the antenna into a series of infinitesimal resistors, each of length dz and resistance $dR = (R/a) \, dz$, we can write the dissipated power as

$$P_{\text{diss}} = \int \langle I^2 \rangle \, dR = \int\limits_{-a}^{+a} \frac{I_0^2}{2} \left(1 - \frac{|z|}{a} \right)^2 \frac{R}{a} \, dz = \frac{I_0^2 R}{3} . \tag{S-10.65}$$

(b) The linear charge density on the antenna q_ℓ can be obtained from the continuity equation $\partial_t q_\ell = -\partial_z I$, obtaining

$$q_\ell = \pm \frac{i I_0}{a\omega} e^{-i\omega t}, \tag{S-10.66}$$

where the signs $+$ and $-$ apply to $z > 0$ and $z < 0$, respectively. The linear charge density is uniform (independent of z) on each half of the antenna. For symmetry reasons, the only non-vanishing component of the electric dipole $\mathbf{p}$ is along z and it is given by

$$p_z = \int\limits_{-a}^{+a} z \, q_\ell \, dz = 2 \int\limits_{0}^{+a} \frac{i I_0}{a\omega} e^{-i\omega t} z \, dz = \frac{i I_0 a}{\omega} e^{-i\omega t} . \tag{S-10.67}$$

(c) The average radiated power, in the dipole approximation, is

$$P_{\text{rad}} = \frac{1}{3c^3} \langle |\ddot{p}_z|^2 \rangle = \frac{I_0^2 a^2 \omega^2}{6c^3} = \frac{2\pi^2 a^2 I_0^2}{3c\lambda^2} , \qquad (S\text{-}10.68)$$

where $\lambda = 2\pi c/\omega$ is the radiation wavelength. Thus

$$\frac{P_{\text{rad}}}{P_{\text{diss}}} = \frac{2\pi^2 a^2}{c\lambda^2 R} , \qquad (S\text{-}10.69)$$

where we recall that R has the dimensions of the inverse of a velocity in Gaussian units.

(d) The angular distribution of the radiated power is proportional to $\sin^2 \theta$ where θ is the angle between the observation direction and $\mathbf{p}$. Thus the emitted radiation intensity is zero along the z axis and maximum for observation in the xy plane.

(e) The bent antenna has a linear charge density $\pm(iI_0/a\omega)e^{-i\omega t}$ on its horizontal and vertical arms, respectively. Thus the electric dipole moment has two components

$$p_x = \int\limits_0^a \frac{iI_0}{a\omega} e^{-i\omega t} x \, dx = \frac{iI_0 a}{2\omega} e^{-i\omega t} , \qquad (S\text{-}10.70)$$

$$p_z = -\int\limits_{-a}^0 \frac{iI_0}{a\omega} e^{-i\omega t} z \, dz = \frac{iI_0 a}{2\omega} e^{-i\omega t} . \qquad (S\text{-}10.71)$$

Since the components are perpendicular to each other, the cycle-averaged radiated power can be calculated as the sum of the powers from each dipole:

$$P_{\text{rad}} = \frac{1}{3c^3} \langle |\ddot{p}_x|^2 + |\ddot{p}_z|^2 \rangle = \frac{1}{12c^3} (I_0 \omega a)^2 , \qquad (S\text{-}10.72)$$

which is one half of the value for the linear antenna, while the dissipated power P_{diss} does not change.

The electric dipole of the bent antenna lies along the diagonal direction, which thus corresponds to the direction of zero emitted intensity. The intensity is maximum in the plane perpendicular to the dipole.

S-10.7 A Receiving Circular Antenna

(a) We choose a Cartesian reference frame such that the wave is propagating in the z direction, its electric field $\mathbf{E}$ is along the x axis, and its magnetic field $\mathbf{B}$ is along the y axis. The current I flowing in the antenna is $I = \mathcal{E}_{\text{circ}}/R$, where $\mathcal{E}_{\text{circ}} = -(1/c)\, d\Phi(\mathbf{B})/dt$ is the electromotive force, and $\Phi(\mathbf{B})$ is the flux of $\mathbf{B}$ through the circle delimited by the antenna. Since we have assumed $\lambda \gg a$, $\mathbf{B}$ is practically uniform over the whole surface of the circle, and $\Phi(\mathbf{B}) \simeq \pi a^2 \mathbf{B} \cdot \hat{\mathbf{n}}$, where $\hat{\mathbf{n}}$ is unit vector perpendicular to the circle surface. Thus the circular antenna must lie on the xz plane in order to maximize $\Phi(\mathbf{B})$. With a proper choice of the time origin the magnetic field on the circle surface can be written as $\mathbf{B} \simeq \hat{\mathbf{y}} B_0 \cos \omega t$, and

$$\mathcal{E}_{\text{circ}} = \pi a^2 \frac{\omega}{c} E_0 \sin(\omega t) , \tag{S-10.73}$$

since $B_0 = E_0$ in Gaussian units.

(b) The electromotive force on a linear antenna parallel to the x axis is practically $\mathcal{E}_{\text{lin}} = \ell E_0 \cos(\omega t + \phi)$, where ℓ is rhe length of the antenna and ϕ is a phase angle. The ratio of the average electromotive force of the circular antenna to the average electromotive force of the linear antenna is thus

$$\frac{\langle \mathcal{E}_{\text{circ}} \rangle}{\langle \mathcal{E}_{\text{lin}} \rangle} \simeq \frac{\langle \mathcal{E}_{\text{circ}} \rangle}{E_0 \ell} = \frac{\pi a^2 \omega}{\ell c} = 2\pi^2 \frac{a^2}{\ell \lambda} . \tag{S-10.74}$$

In the range $10^2 \, \text{cm} < \lambda < 10^3 \, \text{cm}$, and with our assumptions $\ell \simeq 50 \, \text{cm}$ and $a \simeq 25 \, \text{cm}$, this ratio varies between 2.5 and 0.25. The circular antenna is more convenient for shorter wavelengths.

(c) The radiation emission from the circular antenna is dominated by the magnetic dipole term. The dipole moment of the antenna is

$$\mathbf{m} = \frac{1}{c} I \pi a^2 \hat{\mathbf{n}} , \tag{S-10.75}$$

where I is the current circulating in the antenna due to the electromotive force induced by the incident wave. The corresponding time-averaged radiated power is

$$P_{\text{rad}} = \frac{2}{3c^3} \left(|\ddot{\mathbf{m}}|^2 \right) = \frac{2}{3c^5} (\pi a^2)^2 \omega^4 \left\langle I^2 \right\rangle$$

$$= \frac{(\pi a^2)^4 \omega^6}{3c^7 R^2} E_0^2 . \tag{S-10.76}$$

In Gaussian units, the intensity of the incoming wave is $I = c E_0^2 / 4\pi$, and (S-10.76) can be rewritten

$$P_{\text{rad}} = \frac{4\pi (\pi a^2)^4 \omega^6}{3c^8 R^2} I = \frac{2(2\pi)^7 (\pi a^2)^4}{3c^2 R^2 \lambda^6} I . \tag{S-10.77}$$

The factor multiplying I,

$$\sigma_{\text{scatt}} = \frac{2(2\pi)^7 (\pi a^2)^4}{3c^2 R^2 \lambda^6} , \tag{S-10.78}$$

has the dimensions of a surface (R has the dimensions of an inverse velocity in Gaussian units), and is the radiative scattering cross section for our circular antenna, in the magnetic dipole approximation.

The time-averaged power dissipated by Joule heating is $P_{\text{diss}} = R\langle I^2 \rangle$, so that

$$\frac{P_{\text{rad}}}{P_{\text{diss}}} = \frac{2}{3c^5 R}(\pi a^2)^2 \omega^4 = \frac{2(2\pi)^4 (\pi a^2)^2}{3Rc\lambda^4} . \qquad \text{(S-10.79)}$$

S-10.8   Polarization of Scattered Radiation

(a) We choose a Cartesian reference frame with the origin located on the scattering particle, and the z axis parallel to the wave vector of $\mathbf{k}$ the incident wave. In order to have complete rotational symmetry around the z axis it is convenient to assume that the incident wave is circularly polarized. The electric field of the incoming wave can thus be written

$$\mathbf{E_i} = E_0(\hat{\mathbf{x}} \pm i\hat{\mathbf{y}})\,e^{ikz - i\omega t} . \qquad \text{(S-10.80)}$$

Thus, the dipole moment of the scatterer is $\mathbf{p} = \alpha \mathbf{E_i} = \alpha E_0(\hat{\mathbf{x}} \pm i\hat{\mathbf{y}})\,e^{ikz - i\omega t}$.

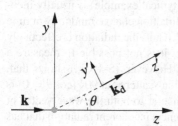

Because of the rotational symmetry of the problem around the z axis, it is sufficient to consider the scattered radiation with the wave vector $\mathbf{k_d}$ lying in the yz plane and forming an angle θ with the z axis, as shown in Fig. S-10.1. Disregarding a proportionality factor depending on α and θ, the electric field $\mathbf{E_d}$ of the scattered radiation can be written

Fig. S-10.1

$$\mathbf{E_d} \propto -(\mathbf{p} \times \hat{\mathbf{n}}) \times \hat{\mathbf{n}}$$
$$\propto -[(\hat{\mathbf{x}} \pm i\hat{\mathbf{y}}) \times \hat{\mathbf{n}}] \times \hat{\mathbf{n}} , \qquad \text{(S-10.81)}$$

where $\hat{\mathbf{n}} = (0, \sin\theta, \cos\theta)$ is the unit vector parallel to $\mathbf{k_d}$. Now, recalling that

$$(\hat{\mathbf{x}} \pm i\hat{\mathbf{y}}) \times \hat{\mathbf{n}} = (\pm i\cos\theta, -\cos\theta, \sin\theta), \qquad \text{(S-10.82)}$$
$$[(\hat{\mathbf{x}} \pm i\hat{\mathbf{y}}) \times \hat{\mathbf{n}}] \times \hat{\mathbf{n}} = (-1, \mp i\cos^2\theta, \pm i\sin\theta\cos\theta), \qquad \text{(S-10.83)}$$

we find that

$$\mathbf{E_d} \propto \left(1, \pm i\cos^2\theta, \mp i\sin\theta\cos\theta\right) . \qquad \text{(S-10.84)}$$

Since an observer would measure the polarization of the scattered radiation with respect to the direction $\hat{\mathbf{n}}$, we calculate the components of the field in the rotated coordinate system (x', y', z'), rotated by an angle θ around the x axis, so that $x' = x$ and z' is along $\hat{\mathbf{n}}$:

$$E'_{dx} = E_{dx} \propto 1, \tag{S-10.85}$$

$$E'_{dy} = E_{dy} \cos\theta - E_{dz} \sin\theta \propto \pm I_i \cos^3\theta \pm i \sin^2\theta \cos\theta$$

$$= \pm i \cos\theta \left(\cos^2\theta + \sin^2\theta \right) = \pm i \cos\theta, \tag{S-10.86}$$

$$E'_{dz} = E_{dy} \sin\theta + E_{dz} \cos\theta \propto \pm i \sin\theta \cos^2\theta \mp i \sin\theta \cos^2\theta = 0. \tag{S-10.87}$$

The last equality is a check that the radiation field is transverse. We thus obtain

$$\mathbf{E_d} \propto \hat{\mathbf{x}}' \pm i \cos\theta \, \hat{\mathbf{y}}', \tag{S-10.88}$$

which gives the dependence of the polarization on the scattering angle θ. In addition, the angular distribution or the radiated power is given by

$$\frac{dP_{\text{scatt}}}{d\Omega} \propto |\mathbf{E_d}|^2 \propto 1 + \cos^2\theta. \tag{S-10.89}$$

(b) The radiation from most sources (sunlight is a typical example) is usually incoherent. This means that its phase and electric field direction change randomly at time intervals not much longer than the oscillation period. Thus, the radiation is effectively unpolarized at direct observation, in the sense that it is not possible to measure a definite polarization because of its fast variations. However, (S-10.88) shows that, independently of the source polarization, the radiation scattered at 90° ($\cos\theta = 0$) is always linearly polarized (in the direction perpendicular to both the wave vector of the incoming light and the observation direction). Hence, incoherent radiation that has undergone scattering (as the blue light from the sky) tends to be polarized, even if the radiation from the primary source (in this case the sun) is unpolarized. A measurement of the polarization might help, then, to localize the position of the Sun on a cloudy day.

S-10.9 Polarization Effects on Thomson Scattering

(a) Equation (10.7) leads to the following two equations for the velocity components of the electron, v_x and v_y,

$$m_e \dot{v}_x = -eE_0 \cos\theta \cos(kz - \omega t), \quad m_e \dot{v}_y = -eE_0 \sin\theta \sin(kz - \omega t), \tag{S-10.90}$$

where m_e is the electron mass. We search for a steady-state solution of the form

$$v_x = V_{0x} \sin(kz - \omega t) , \quad v_y = V_{0y} \cos(kz - \omega t) , \qquad \text{(S-10.91)}$$

with V_{0x} and V_{0y} two real constants to be determined. Substituting into (S-10.90) we obtain

$$V_{0x} = \frac{eE_0 \cos\theta}{m_e \omega} , \quad V_{0y} = -\frac{eE_0 \sin\theta}{m_e \omega} . \qquad \text{(S-10.92)}$$

The second derivative of the electric dipole moment of the electron with respect to time is

$$\ddot{\mathbf{p}} = -e\dot{\mathbf{v}} = -\frac{e^2 E_0}{m_e} \left[(\hat{\mathbf{x}} \cos\theta \cos(kz - \omega t) + \hat{\mathbf{y}} \sin\theta \sin(kz - \omega t) \right] , \qquad \text{(S-10.93)}$$

and the electron radiates at frequency ω. The polarization for scattered radiation propagating in a generic direction of unit vector $\hat{\mathbf{n}}$ direction is parallel to the projection of the dipole moment onto the plane perpendicular to $\hat{\mathbf{n}}$, i.e., to $\mathbf{p}_\perp = (\ddot{\mathbf{p}} \times \hat{\mathbf{n}}) \times \hat{\mathbf{n}}$. Thus, we observe linear polarization parallel to $\hat{\mathbf{y}}$ for the radiation emitted along $\hat{\mathbf{x}}$, and linear polarization parallel to $\hat{\mathbf{x}}$ for the radiation emitted along $\hat{\mathbf{y}}$, and elliptical polarization for the radiation emitted along $\hat{\mathbf{z}}$.

If $0 < \theta < \pi/4$, so that $\sin\theta < \cos\theta$, we choose the observation-direction unit vector $\hat{\mathbf{n}} = (\sin\psi, 0, \cos\psi)$, lying in the xz plane, and forming an angle ψ with the z axis, as shown in Fig. S-10.2, where $\mathbf{k}_i$ is the wave vector of the incident wave. Now

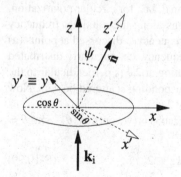

we choose a Cartesian reference frame x', y', z', with $y' \equiv y$ and z' along $\hat{\mathbf{n}}$, so that the scattered radiation of interest is propagating along z'. If we perform an orthogonal projection onto the $x'y'$ plane of an ellipse lying on the xy plane, of half axes $\cos\theta$ parallel to x, and $\sin\theta$ parallel to y, we obtain an ellipse of half-axes $\cos\theta \cos\psi$ along x', and $\sin\theta$ along y'. Thus we observe a circular polarization if $\cos\theta \cos\psi = \sin\theta$, i.e., if $\cos\psi = \tan\theta$. Analogously, if $\pi/4 < \theta < \pi/2$, so that $\sin\theta > \cos\theta$, we choose the observation-direction unit vector $\hat{\mathbf{n}} = (0, \sin\psi, \cos\psi)$, lying in the yz plane, and we observe circular polar-

Fig. S-10.2

ization if $\sin\theta \cos\psi = \cos\theta$, i.e., if $\cos\psi = \cot\theta$.

(b) The average total scattered power is

$$P = \frac{2}{3c^3} \left\langle |\ddot{\mathbf{p}}|^2 \right\rangle = \frac{2e^4}{3m_e^2 c^3} \left\langle |\mathbf{E}|^2 \right\rangle , \qquad \text{(S-10.94)}$$

where

$$\left\langle |\mathbf{E}|^2 \right\rangle = \left\langle E_x^2 + E_y^2 \right\rangle = \frac{1}{2} E_0^2 (\cos^2\theta + \sin^2\theta) = \frac{1}{2} E_0^2 . \qquad \text{(S-10.95)}$$

Thus, the total scattered power is independent of θ and can be written as

$$P = \frac{e^4 E_0^2}{3m_e^2 c^3} = \frac{cE_0^2}{3} r_e^2 = \frac{4\pi}{3} r_e^2 I , \qquad \text{(S-10.96)}$$

where

$$r_e = \frac{e^2}{m_e c^2} , \quad \text{and} \quad I = \frac{cE_0^2}{4\pi} ,$$

are the classical electron radius and the intensity of the incident wave, respectively.
(c) The magnetic field of the wave is

$$\mathbf{B} = E_0 \left[-\hat{\mathbf{x}} \sin\theta \sin(kz - \omega t) + \hat{\mathbf{y}} \cos\theta \cos(kz - \omega t) \right] . \qquad \text{(S-10.97)}$$

The only non-vanishing component of $\mathbf{v} \times \mathbf{B}$ is in the $\hat{\mathbf{z}}$ direction, and the magnetic force on the electron can be written as

$$\begin{aligned} F_z &= -\frac{e}{c} (\mathbf{v} \times \mathbf{B})_z = -\frac{e}{c} (v_x B_y - v_y B_x) \\ &= -\frac{e^2 E_0^2}{2cm_e\omega} (\cos^2\theta - \sin^2\theta) \sin(2kz - 2\omega t) , \end{aligned} \qquad \text{(S-10.98)}$$

this quantity vanishes for $\theta = \pi/4$, when $\cos\theta = \sin\theta$, i.e., for circular polarization.
(d) The magnetic force F_z drives dipole oscillations along the z axis at frequency 2ω. Thus, in addition to the scattered radiation of frequency ω discussed at points **(a)** and **(b)**, we observe also scattered radiation of frequency 2ω, angularly distributed as $\sin^2\psi$ around the z axis. Since the dipole oscillating at 2ω is perpendicular to the dipole oscillating at ω, we can simply add the corresponding scattered powers. Now we want to evaluate the power emitted at frequency 2ω.

The equation of motion for the electron along the z axis is (we put $\cos^2\theta - \sin^2\theta = \cos 2\theta$)

$$m_e \dot{v}_z = F_z = -\frac{e^2 E_0^2}{2cm_e\omega} \cos 2\theta \sin(2kz - 2\omega t) . \qquad \text{(S-10.99)}$$

Once more, we search for a steady-state solution of the form

$$v_z = V_{0z} \cos(2kz - 2\omega t) , \qquad \text{(S-10.100)}$$

with V_{0z} a constant. Substituting into (S-10.99) we obtain

$$V_{0z} = -\frac{e^2 E_0^2}{4cm_e^2\omega^2} \cos(2\theta) \qquad \text{(S-10.101)}$$

and

$$\dot{v}_z = -\frac{e^2 E_0^2}{2cm_e^2\omega} \cos(2\theta) \sin(2kz - 2\omega t) . \qquad \text{(S-10.102)}$$

The total average power emitted by the dipole oscillating at 2ω is

$$P_{2\omega} = \frac{2}{3c^3} \left\langle |\ddot{\mathbf{p}}_{2\omega}|^2 \right\rangle = \frac{2}{3c^3} \left\langle |e\dot{v}_z|^2 \right\rangle = \frac{e^6 E_0^4}{12c^5 m_e^4 \omega^2} \cos^2(2\theta)$$

$$= \frac{4\pi}{3} \frac{e^2 E_0^2}{4c^2 m_e^2 \omega^2} \cos^2(2\theta) r_e^2 I$$

$$= \frac{4\pi}{3} \frac{V_{0z}}{c} \cos^2(2\theta) r_e^2 I \ . \tag{S-10.103}$$

S-10.10 Scattering and Interference

(a) With a proper choice of the time origin, the electric field of the incident plane wave at $x = \pm d/2$ can be written as

$$\mathbf{E}_i \left(\pm \frac{d}{2}, t \right) = E_0 \, e^{\pm ikd/2 - i\omega t} \, \hat{\mathbf{z}} \ , \tag{S-10.104}$$

and the phase difference between the two scatterers is

$$\phi_+ - \phi_- = kd \ . \tag{S-10.105}$$

We denote by $r_\pm$ the optical paths between the observation point P and the scatterers located at $(\pm d/2, 0, 0)$, as shown in Fig. 10.5. The difference between the two optical paths is

$$\Delta r = (r_+ - r_-) \simeq -d \sin \theta \ , \tag{S-10.106}$$

where θ is the angle between the y axis and the line joining the origin to P, as shown in Fig. 10.5. The approximation is valid for $L \gg d$. The phase difference between the two scattered waves in P is obtained by combining (S-10.105) and (S-10.106),

$$\Delta\phi = kd(1 - \sin\theta) \ . \tag{S-10.107}$$

(b) If we neglect the difference between the magnitudes of the scattered fields $\mathbf{E}_+$ and $\mathbf{E}_-$ in P, $\mathbf{E}_\pm$ being the field of the wave scattered at $(\pm d/2, 0, 0)$, the total scattered intensity I_s in P is proportional to

$$I_s \propto |\mathbf{E}_+ + \mathbf{E}_-|^2 \propto |\mathbf{E}_+|^2 \left| e^{ikd(1 - \sin\theta)/2} + e^{-ikd(1 - \sin\theta)/2} \right|^2$$

$$\propto \frac{1}{r^2} \cos^2 \left[\frac{kd}{2} (1 - \sin\theta) \right] \ . \tag{S-10.108}$$

Since $r \cos\theta = L$, we can also write

$$I_s \propto \frac{\cos^2 \theta}{L^2} \cos^2 \left[\frac{kd}{2}(1 - \sin \theta) \right] . \tag{S-10.109}$$

We denote by $u = (kd/2)(1 - \sin \theta)$ the argument of the second $\cos^2$ appearing in (S-10.109). For $-\pi/2 \leqslant \theta \leqslant \pi/2$ the variable u varies continuously and monotonically from kd to 0. If $kd \ll 1$ (i.e., if $d \ll \lambda/2\pi$), then $\cos^2 u \simeq 1$ e $I_s(\theta) \sim \cos^2 \theta$, as if a single scatterer was present. If $kd < \pi/2$ the function $\cos^2 u$ has no zeros, meaning that interference fringes are not observed if the distance between the scatterers is less than $\lambda/4$. If

$$\frac{\pi}{2} < \frac{kd}{2} < (n+1) \frac{\pi}{2} ,$$

with n an integer number and $n \geqslant 1$, the function $\cos^2 u$ has n zeros, and one observes n scattered-intensity minima and $n + 1$ maxima as θ varies from $-\pi/2$ to $+\pi/2$. The intensity of the maxima is modulated by the function $\cos^2 \theta$.

S-10.11   Optical Beats Generating a "Lighthouse Effect"

(a) On the $z = 0$ plane the electric fields $\mathbf{E}_\pm$ emitted by the two dipoles are parallel to $\hat{\mathbf{z}}$ (perpendicular to the plane), and their amplitudes are independent of ϕ. Since for each dipole $\mathbf{E}_\pm \propto -\omega_\pm^2 \mathbf{p}_0$, the field amplitudes are $E_+ \simeq E_-$, equal to each other up to the first order in $\delta\omega/\omega_0$. The difference between the optical paths from the two dipoles to P is $\delta r \simeq d \sin \phi = (\pi c/\omega_0) \sin \phi$, which yields a phase difference of $\pi \sin \phi$. The total field may be thus written as

$$\begin{aligned} E &= E_0 \cos[(\omega_0 + \delta\omega/2)t + \pi \sin \phi/2] + E_0 \cos[(\omega_0 - \delta\omega/2)t - \pi \sin \phi/2] \\ &= 2E_0 \cos(\omega_0 t) \cos(\delta\omega t + \pi \sin \phi) . \end{aligned} \tag{S-10.110}$$

(b) The EM energy flux in the radiation zone is given by Poynting's vector $\mathbf{S}$, which is proportional to the square modulus of the electric field. Thus

$$S \propto 4 \cos^2(\omega_0 t) \cos^2(\delta\omega t + \pi \sin \phi). \tag{S-10.111}$$

Using the "fast", or "instantaneous", detector, only the factor $\cos^2(\omega_0 t)$ is averaged, and the measured signal is proportional to

$$\langle S \rangle \propto 2 \cos^2(\delta\omega t + \pi \sin \phi) , \quad \text{since} \quad \langle \cos^2(\omega_0 t) \rangle = \frac{1}{2} . \tag{S-10.112}$$

At time t, the direction of maximum flux intensity is determined by the condition

$$\delta\omega t + \pi \sin \phi = \begin{cases} 0 \\ \pi \end{cases} , \tag{S-10.113}$$

which means that the direction of maximum flux ϕ_{max} *rotates* in the $z = 0$ plane, similarly to a lighthouse beam, according to

$$\phi_{max}(t) = \arcsin\left(-\frac{\delta\omega}{\pi}t\right). \qquad (\text{S-10.114})$$

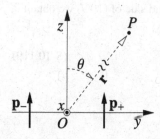

Fig. S-10.3

If the EM flux is measured with the "slow" detector, i.e., averaging over times longer than $2\pi/\delta\omega$, both $\cos^2$ terms of (S-10.111) are averaged to $1/2$, and the total flux is the sum of the two independent fluxes from the two dipoles.

(c) Now the observation point P is on the $x = 0$ plane, at a distance **r** from the origin, as in Fig. S-10.3. The angle between the z axis and **r** is θ. Within our approximations, the intensities of the two electric fields $\mathbf{E}_+$ and $\mathbf{E}_-$ in P are equal and proportional to $\sin\theta$. Thus the two separate intensities are dependent on θ, while they are independent of ϕ on the $z = 0$ plane. The amplitude of the Poynting vector is proportional to

$$S \propto 2\sin^2\theta\cos^2(\omega_0 t)\cos^2(\delta\omega t + \pi\sin\theta). \qquad (\text{S-10.115})$$

Thus the "fast" detector still measures a rotation of the direction of maximum emission, but the intensity is modulated by a $\sin^2\theta$ factor.

S-10.12 Radiation Friction Force

(a) We insert (10.11) for $\mathbf{F}_{rad}$ into (10.9), obtaining

$$\int_t^{t+T} \mathbf{F}_{rad}(t)\cdot\mathbf{v}(t)\,dt = m_e\tau\int_t^{t+T}\frac{d^2\mathbf{v}(t)}{dt^2}\cdot\mathbf{v}(t)dt$$

$$= m_e\tau\left[\frac{d\mathbf{v}(t)}{dt}\cdot\mathbf{v}(t)\right]_t^{t+T} - m_e\tau\int_t^{t+T}\left|\frac{d\mathbf{v}(t)}{dt}\right|^2 dt, \qquad (\text{S-10.116})$$

where we have used integration by parts in the second line. The first term vanishes since the motion is periodic[8]:

$$\left[\frac{d\mathbf{v}(t)}{dt}\cdot\mathbf{v}(t)\right]_t^{t+T} = \frac{1}{2}\left[\frac{d}{dt}v^2(t+T) - \frac{d}{dt}v^2(t)\right] = 0. \qquad (\text{S-10.117})$$

---

[8] Actually it is not strictly necessary for the motion to be periodic, it is sufficient that $dv^2(t)/dt$ vanishes at the initial and final instants of the time interval considered.

We thus obtain

$$\int\limits_{t}^{t+T} \mathbf{F}_{rad}(t) \cdot \mathbf{v}(t)\, dt = -m_e \tau \int_t^{t+T} \left|\frac{d\mathbf{v}}{dt}\right|^2 dt \ . \tag{S-10.118}$$

Substituting Larmor's formula (10.10) into the right-hand side of (10.9) we obtain

$$-\int\limits_0^t P_{rad}(t')\, dt' = -\int\limits_0^t \frac{2e^2}{3c^2} \left|\frac{d\mathbf{v}}{dt'}\right|^2 dt' \ , \tag{S-10.119}$$

and (10.9) is verified if we choose

$$\tau = \frac{2e^2}{3m_e c^2} \ . \tag{S-10.120}$$

Apart from the 2/3 factor, τ is the time needed by light to travel a distance equal to the classical electron radius $r_e = 2.82 \times 10^{-13}$ cm, and we have $\tau \sim 10^{-23}$ s.

(b) After substituting (10.12) into (10.8), we search for a steady-state solution of the form $\mathbf{v}(t) = \mathbf{v}_0\, e^{-i\omega t}$, and find

$$\mathbf{v}_0 = -\frac{i\, e\mathbf{E}_0}{m_e \omega (1 + i\omega\tau)} \ . \tag{S-10.121}$$

Analogously, the steady-state solution of (10.13) is

$$\mathbf{v}_0 = -\frac{i\, e\mathbf{E}_0}{m_e \omega \left(1 + i\dfrac{\eta}{\omega}\right)} \ . \tag{S-10.122}$$

The two solutions are identical if we choose $\eta = \omega^2 \tau$. The same result can be obtained by a direct comparison of F_{rad} to the frictional force $-m_e \eta \mathbf{v}$.

Equation (10.8) represents the first attempt to derive an expression for the "radiation friction" or "radiation reaction" force which is deeply related to the back-action of the electron on itself, since the electron interacts with the electric field it generates (self-force9. However (10.8) is considered unsatisfactory for for two reasons: (i) it increases the order of the equation of motion, and, consequently, one needs a further initial condition for the acceleration; and (ii) it has unphysical "runaway" solutions in the absence of an external field, such as $\mathbf{a}(t) = \mathbf{a}_0\, e^{t/\tau}$ with $\mathbf{a} = d\mathbf{v}/dt$. This problem has a long and still open history. Additional discussion may be found in textbooks and in the literature, also in very recent works related to highly relativistic electrons in ultraintense laser fields (for which the radiation friction effect becomes important).

S-10.13   Radiation of an Electron Falling on a Proton

(a) The initial conditions imply that the motion occurs on the straight half-line joining the proton to the initial position of the electron. The equation of conservation of mechanical energy is

$$\frac{m_e v^2}{2} - \frac{e^2}{r} = 0,$$ (S-10.123)

where m_e is the electron mass and e the elementary charge. From (S-10.123) we obtain

$$v = -\sqrt{\frac{2e^2}{m_e r}} = -c\sqrt{\frac{2r_e}{r}},$$ (S-10.124)

where $r_e = e^2/(m_e c^2)$ is the classical electron radius.

Thus the condition for non-relativistic motion is no longer valid for r close to r_e, so that r_e is a rough approximation of r_1. The condition $v < c/\sqrt{10}$ leads to $r_1 = 20 r_e$.

(b) The dipole moment is $p = -er$, its second derivative with respect to time is thus proportional to the electron acceleration,

$$\ddot{p} = -e\ddot{r} = -e\left(-\frac{e^2}{m_e r^2}\right) = \frac{e^3}{m_e r^2},$$ (S-10.125)

and the radiated power as a function of r is

$$P = \frac{2}{3c^3}|\ddot{p}|^2 = \frac{2e^6}{3m_e^2 r^4 c^3} = \frac{2e^2 r_e^2 c}{3r^4}.$$ (S-10.126)

(c) The energy lost by radiation $U_r(r)$ when the electron is at distance r from the proton is obtaimed by integrating (S-10.126) over the path

$$U_r(r) = \int_\infty^{t(r)} P \, dt = \int_\infty^r P(r') \frac{dr'}{|v(r')|} = \frac{2e^2 r_e^2 c}{3} \int_\infty^r \frac{1}{r'^4} \left(\frac{1}{c}\sqrt{\frac{r'}{2r_e}}\right) dr'$$

$$= \frac{2e^2 r_e^{3/2}}{3\sqrt{2}} \int_\infty^r r'^{-7/2} \, dr' = \frac{2\sqrt{2}\, e^2 r_e^{3/2}}{15 \, r^{5/2}}.$$ (S-10.127)

The condition $U_r(r) \ll K(r) = e^2/r$ is thus verified for

$$r \gg r_e \left(\frac{2\sqrt{2}}{15}\right)^{2/3} = 0.329 \, r_e.$$ (S-10.128)

If we choose $U_r(r_2) = K(r_2)/10$ we have $r_2 \simeq 1.04\, r_e$. Since $r_2 < r_1$, an accurate estimate of the radiative energy loss requires a relativistic treatment of the electron motion.

S-10.14   Scattering by a Perfectly Conducting Sphere (1)

(a) Since the radius of the sphere, a, is much smaller than the radiation wavelength, λ, we can consider the electric field of the incident wave as uniform over the whole volume of the sphere. As shown in Problems 1.1 and 2.1, the "electron sea" is displaced by an amount $\boldsymbol{\delta}$ with respect to the ion lattice in order to keep the total electric field equal to zero inside the sphere. According to (S-2.2) we have

$$\boldsymbol{\delta} = -\hat{\mathbf{y}}\, \frac{3}{4\pi \varrho_0}\, E_0 \cos(\omega t)\,, \tag{S-10.129}$$

where ϱ_0 is the volume charge density of the ion lattice. This corresponds to a volume polarization $\mathbf{P}$

$$\mathbf{P} = -\rho \boldsymbol{\delta} = \hat{\mathbf{y}}\, \frac{3}{4\pi}\, E_0 \cos(\omega t)\,, \tag{S-10.130}$$

and to a total dipole moment of the conducting sphere

$$\mathbf{p} = \frac{4\pi}{3}\, a^3\, \mathbf{P} = \mathbf{E}_0\, a^3 = \hat{\mathbf{y}}\, a^3 E_0 \cos(\omega t)\,. \tag{S-10.131}$$

The scattered, time-averaged power is thus

$$W_{\text{scatt}}^{(\text{el})} = \frac{1}{3c^3}\, |\ddot{\mathbf{p}}|^2 = \frac{\omega^4 a^6}{3c^3}\, E_0^2\,. \tag{S-10.132}$$

The intensity of the incident wave is $I = (c/8\pi)E_0^2$, so we obtain for the scattering cross section

$$\sigma_{\text{scatt}}^{(\text{el})} = \frac{W_{\text{scatt}}^{(\text{el})}}{I} = \frac{8\pi}{3}\, \frac{\omega^4 a^6}{c^4} = 128\pi^4 (\pi a^2) \left(\frac{a}{\lambda}\right)^4\,. \tag{S-10.133}$$

(b) Due to the condition $a \ll \lambda$, also the magnetic field of the wave can be considered as uniform over the sphere volume

$$\mathbf{B}(t) = \hat{\mathbf{z}}\, B_0 \cos(\omega t) = \hat{\mathbf{z}}\, E_0 \cos(\omega t)\,. \tag{S-10.134}$$

Analogously to what seen above for the electric polarization, the sphere must acquire also a uniform magnetization $\mathbf{M}$ in order to cancel the magnetic field of the wave at its interior. According to (S-5.74) of Problem 5.10, we must have

$$M(t) = -\hat{z} \frac{3}{8\pi} B_0 \cos(\omega t) = -\hat{z} \frac{3}{8\pi} E_0 \cos(\omega t) , \quad \text{(S-10.135)}$$

corresponding to a magnetic dipole moment of the sphere

$$m = \frac{4\pi a^3}{3} M = -\hat{z} \frac{a^3}{2} E_0 \cos(\omega t) , \quad \text{(S-10.136)}$$

Thus the power scattered by the magnetic dipole is one fourth of the electric dipole contribution:

$$W_{\text{scatt}}^{(\text{magn})} = \frac{1}{3c^3} |\ddot{m}|^2 = \frac{\omega^4 a^6}{12c^3} E_0^2 . \quad \text{(S-10.137)}$$

The total cross section is thus 5/4 times the value due to the electric dipole only:

$$\sigma_{\text{scatt}}^{(\text{el,magn})} = 160\pi^4 (\pi a^2) \left(\frac{a}{\lambda}\right)^4 . \quad \text{(S-10.138)}$$

A discussion on how the magnetic dipole term contributes to the angular distribution of the scattered radiation can be found in Reference [2].

S-10.15 Scattering by a Perfectly Conducting Sphere (2)

According to (S-10.131) and (S-10.136) of Solution S-10.14 the electric and magnetic dipole moments induced by the impinging wave on the conducting sphere are, respectively,

$$p_s = a^3 E_i , \quad \text{and} \quad m_s = -\frac{a^3 B_i}{2} = -\frac{1}{2} p_s \times \hat{n} . \quad \text{(S-10.139)}$$

The ratio of their magnitudes is thus $p_s/m_s = 2$.

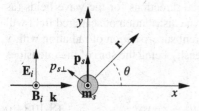

Fig. S-10.4

(a) The electric field of the scattered radiation in the far-field region is

$$E_{\text{rad}} = \frac{1}{rc^2} (\ddot{p}_s \times \hat{r} + \ddot{m}_s) \times \hat{r} , \quad \text{(S-10.140)}$$

and for a monochromatic wave of frequency ω we have

$$E_{\text{rad}} = -\frac{\omega^2}{rc^2} (p_s \times \hat{r} + m_s) \times \hat{r} . \quad \text{(S-10.141)}$$

For the radiation scattered in the xy plane we have, in Cartesian coordinates, $\hat{\mathbf{r}} \equiv$ $(\cos\theta, \sin\theta, 0)$. If the wave is linearly polarized along y, as in Fig. S-10.4, we have

$$\mathbf{p}_s \times \hat{\mathbf{r}} + \mathbf{m}_s = (p_s \cos\theta + m_s)\,\hat{\mathbf{z}} = p_s\left(\cos\theta - \frac{1}{2}\right)\hat{\mathbf{z}} \qquad (S\text{-}10.142)$$

and $\left|\hat{\mathbf{z}} \times \hat{\mathbf{r}}\right| = 1$. In the figure $p_{s\perp}$ is the component of the induced electric dipole perpendicular to the direction of propagation of the scattered radiation. Thus

$$|\mathbf{E}_{\text{rad}}|^2 \propto \left(\cos\theta - \frac{1}{2}\right)^2. \qquad (S\text{-}10.143)$$

We have zero intensity for $\cos\theta = 1/2$, i.e., $\theta = \pi/3$, and maximum intensity for $\theta = \pi$, i.e., for backward scattering.

In the case of z polarization, as in Fig. S-10.5, we have

$$\left(\mathbf{p}_s \times \hat{\mathbf{r}}\right) \times \hat{\mathbf{r}} = -\mathbf{p}_s\,,$$
$$\mathbf{m}_s \times \hat{\mathbf{r}} = \hat{\mathbf{z}}\, m_s \sin(\theta + \pi/2)$$
$$= \hat{\mathbf{z}}\,\frac{1}{2}\, p_s \cos\theta\,.$$
$$(S\text{-}10.144)$$

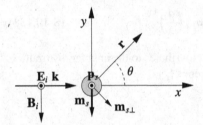

Fig. S-10.5

In the figure $m_{s\perp}$ is the component of the induced magnetic dipole perpendicular to the direction of propagation of the scattered radiation. Thus

$$|\mathbf{E}_{\text{rad}}|^2 \propto \left(1 - \frac{1}{2}\cos\theta\right)^2. \qquad (S\text{-}10.145)$$

We still have maximum intensity for backscattering, while we have minimum intensity for forward scattering ($\theta = 0$), which, however, does not correspond to zero intensity. **(b)** For unpolarized light there are no privileged directions for the wave fields (as far as the measurement is time-averaged), thus the distribution of scattered light will have cylindrical symmetry. We have an incoherent superposition of radiation with y and z polarizations, resulting in in the total intensity being the sum of the intensities (S-10.143) and (S-10.145)

$$|\mathbf{E}_{\text{rad}}|^2 \propto \frac{5}{4}\left(1 + \cos^2\theta\right) - 2\cos\theta\,. \qquad (S\text{-}10.146)$$

Chapter S-11
Solutions for Chapter 11

S-11.1 Wave Propagation in a Conductor at High and Low Frequencies

(a) We determine the conductivity of the metal by searching for a steady-state solution in complex form, $v = \tilde{v} e^{-i\omega t}$, of (11.10) in the presence of an oscillating electric field $\mathbf{E}(\mathbf{r}, t) = \tilde{\mathbf{E}} e^{-i\omega t}$. We find

$$\tilde{v} = -\frac{ie}{m_e(\omega + i\eta)} \tilde{\mathbf{E}} \, , \qquad (S\text{-}11.1)$$

corresponding to a current density

$$\tilde{\mathbf{J}} = -en_e \tilde{v} = \frac{ie^2 n_e}{m_e \omega(\omega + i\eta)} \tilde{\mathbf{E}} = \frac{i\omega_p^2}{4\pi(\omega + i\eta)} \tilde{\mathbf{E}} \equiv \sigma(\omega) \tilde{\mathbf{E}} \, , \qquad (S\text{-}11.2)$$

where ω_p is the plasma frequency of the metal. At the limits of high frequencies $\omega \gg \eta$, and of low frequencies $\omega \ll \eta$, we have

$$\sigma(\omega) \simeq \begin{cases} \dfrac{i\omega_p^2}{4\pi\omega} \, , & \text{for } \omega \gg \eta \, , \\[3mm] \dfrac{\omega_p^2}{4\pi\eta} \, , & \text{for } \omega \ll \eta \, . \end{cases} \qquad (S\text{-}11.3)$$

The DC conductivity is thus $\sigma_{DC} = \sigma(0) = \omega_p^2/4\pi\eta$. In a metal, typically we have $\omega_p \sim 10^{16}$ s^{-1}, since $n_e \sim 10^{23}$ cm^{-3} and $\eta \sim 10^{13}$ s^{-1}. It is thus a very good approximation to assume σ to be purely imaginary for optical frequencies, i.e., for $\omega \sim 10^{15}$ s^{-1}, and to be purely real and equal to σ_{DC} (i.e. independent of frequency) for microwaves and longer wavelengths.

(b) Assuming plane geometry and monochromatic waves, in the absence of sources at $x = +\infty$, the electric field of the wave for $x > 0$ can be written as (in complex notation)

© The Editor(s) (if applicable) and The Author(s), under exclusive license to Springer
Nature Switzerland AG 2023
A. Macchi et al., *Problems in Classical Electromagnetism*,
https://doi.org/10.1007/978-3-031-22235-1

$$E(x,t) = E_t\, e^{ikx-i\omega t}\,, \tag{S-11.4}$$

where the wave vector k is determined by the general dispersion relation (11.4) in a medium where the refractive index $n = n(\omega)$ or, equivalently, the permittivity $\varepsilon = \varepsilon(\omega) = n^2$ are known. For an incident wave of amplitude E_i, the electric field at the surface is given by the Fresnel formula

$$E_t = \frac{2}{1+n}\, E_i\,. \tag{S-11.5}$$

The permittivity $\varepsilon(\omega)$ is related to the complex conductivity of the medium by (11.3). Inserting (S-11.3) for $\sigma(\omega)$, if $\omega \gg \eta$ we have $\varepsilon \simeq 1 - \omega_p^2/\omega^2$, and k^2 is real, so that the wave is propagating. For $k^2 < 0$, i.e., for $\omega_p > \omega$, we have $ikx = -|k|x = -x/\ell_p$, with $\ell_p = c/\sqrt{\omega_p^2 - \omega^2}$, and the wave is evanescent:

$$E(x,t) = E_t\, e^{-x/\ell_p - i\omega t} \tag{S-11.6}$$

(the solution $\propto e^{x/\ell_p}$ has been disregarded as unphysical because it is divergent for $x \to \infty$). For a metal, the condition $\omega < \omega_p$ implies that the metal is reflecting for frequencies in the optical range, while it becomes transparent for ultraviolet frequencies.

If $\omega \ll \eta$, we have that also $\sigma_{DC} \ll \eta$, so that $\varepsilon \simeq 4\pi i\sigma_{DC}/\omega$ is an imaginary number. In this case, since $k = \pm(1+i)/\ell_c$ with $\ell_c = \sqrt{\omega\sigma_{DC}/2}c$, the evanescent solution is

$$E(x,t) = E_t\, e^{-x/\ell_c - ix/\ell_c - i\omega t}\,. \tag{S-11.7}$$

(c) The net flux of energy through the surface is given by the time average of the x-component of the Poynting vector $\mathbf{S} = (c/4\pi)\mathbf{E} \times \mathbf{B}$ at $x = 0$. We obtain the magnetic field of the wave from the relation $\partial_t \mathbf{B} = -c\nabla \times \mathbf{E}$. Thus the complex field amplitudes for $x \geqslant 0$ can be written as

$$\tilde{E}_y = E_t\, e^{ik_0 nx}\,, \qquad \tilde{B}_z = nE_t\, e^{ik_0 nx}\,, \tag{S-11.8}$$

where $k_0 = \omega/c$. Thus we need to evaluate

$$\langle S_x(0)\rangle = \frac{1}{2}\frac{c}{4\pi}\mathrm{Re}\left[\tilde{E}_y(0)\tilde{B}_z^*(0)\right] = \frac{c}{8\pi}|E_t|^2\mathrm{Re}(n)\,. \tag{S-11.9}$$

At the limit $\omega \gg \eta$, n is purely imaginary, as found above, and $\langle S_x(0)\rangle = 0$, and there is no energy dissipated into the metal (it can be easily shown that the reflection coefficient obtained from the Fresnel formulas has unity modulus, i.e., all the incident energy is reflected). At the limit $\omega \gg \eta$ we obtain

$$\langle S_x(0)\rangle = \frac{c}{8\pi}|E_t|^2\sqrt{\frac{2\pi\sigma_{DC}}{\omega}} \simeq \frac{c}{16\pi}|E_i|^2\sqrt{\frac{\omega}{2\pi\sigma_{DC}}}\,, \tag{S-11.10}$$

where in the latter expression $|1 + n|^2 \simeq |n|^2 = 2(2\pi\sigma_{DC}/\omega)$ has been assumed.

The energy dissipated per unit volume is

$$\langle \mathbf{J} \cdot \mathbf{E} \rangle = \frac{1}{2}\text{Re}\left(\sigma\tilde{E}_y\tilde{E}_y{}^*\right) = \frac{|E_t|^2}{2}\text{Re}\left(\sigma e^{ik_0 nx}e^{-ik_0 n^*x}\right)$$

$$= \frac{|E_t|^2}{2}\text{Re}(\sigma)\exp[-2k_0\text{Im}(n)x] . \qquad \text{(S-11.11)}$$

If σ is imaginary then there is no dissipation, consistently with what found above. In the $\omega \ll \eta$ regime, the total energy dissipated per unit surface is given by the integral

$$\int_0^\infty \langle \mathbf{J} \cdot \mathbf{E} \rangle \, dx = \frac{|E_t|^2}{2}\frac{\sigma_{DC}}{2k_0\text{Im}(n)} = \frac{|E_t|^2}{2}\frac{\sigma_{DC}}{2\omega_c\sqrt{2\pi\sigma_{DC}/\omega}}$$

$$= \frac{c}{8\pi}|E_t|^2\sqrt{\frac{2\pi\sigma_{DC}}{\omega}} , \qquad \text{(S-11.12)}$$

which is equal to the EM energy flux of (S-11.10).

S-11.2 Energy Densities in a Free Electron Gas

(a) We use the complex representation for all fields, $A(x, t) = \text{Re}\left(\tilde{A}e^{ikx-i\omega t}\right)$, where A is the considered field. For the electric field of the wave we have $\tilde{\mathbf{E}} = \mathbf{E}_0$, where E_0 can be considered as a real quantity. The equation of motion for an electron, neglecting the nonlinear magnetic term, is

$$m_e\frac{d^2\mathbf{r}}{dt^2} = m_e\frac{d\mathbf{v}}{dt} = -e\mathbf{E} , \qquad \text{(S-11.13)}$$

which has the steady-state solution for the electron velocity and position

$$\tilde{\mathbf{v}} = -\frac{ie}{m_e\omega}\mathbf{E}_0 , \qquad \tilde{\mathbf{r}} = \frac{e}{m_e\omega^2}\mathbf{E}_0 . \qquad \text{(S-11.14)}$$

The polarization density is

$$\tilde{\mathbf{P}} = -en_e\tilde{\mathbf{r}} = -\frac{n_e e^2}{m_e\omega^2}\mathbf{E}_0 = -\frac{1}{4\pi}\frac{\omega_p^2}{\omega^2}\mathbf{E}_0 \equiv \chi(\omega)\mathbf{E}_0 , \qquad \text{(S-11.15)}$$

corresponding to a dielectric permittivity of the medium

$$\varepsilon(\omega) = 1 + 4\pi\chi(\omega) = 1 - \frac{\omega_p^2}{\omega^2} . \qquad \text{(S-11.16)}$$

Using (11.4), the dispersion relation is obtained as

$$\omega^2 = \frac{k^2 c^2}{\varepsilon(\omega)} = \omega_p^2 + k^2 c^2 \, . \tag{S-11.17}$$

The phase and group velocities are

$$v_\varphi = \frac{\omega}{k} = \frac{c}{\sqrt{1 - \dfrac{\omega_p^2}{\omega^2}}} \, , \qquad v_g = \frac{\partial \omega}{\partial k} = c \sqrt{1 - \frac{\omega_p^2}{\omega^2}} \, , \tag{S-11.18}$$

so that both v_φ and v_g are real if $\omega > \omega_p$, and $v_\varphi v_g = c^2$. Finally, using the equation $c\nabla \times \mathbf{E} = -\partial_t \mathbf{B}$. i.e., $ikc\tilde{\mathbf{E}} = i\omega\tilde{\mathbf{B}}$, we obtain $E_0 = (v_\varphi/c)B_0$.

(b) From the definition of the EM energy density

$$u_{\text{EM}} = \left\langle \frac{1}{8\pi}(\mathbf{E}^2 + \mathbf{B}^2) \right\rangle = \frac{1}{16\pi}(E_0^2 + B_0^2) = \frac{1}{16\pi}E_0^2 \left(1 + \frac{c^2}{v_\varphi^2} \right)$$

$$= \frac{1}{16\pi}E_0^2 \left(2 - \frac{\omega_p^2}{\omega^2} \right) \, . \tag{S-11.19}$$

(c) From the definition of the kinetic energy density

$$u_{\text{K}} = \left\langle n_e \frac{m_e}{2} v^2 \right\rangle = n_e \frac{m_e}{2} \frac{1}{2} \left| \frac{e E_0}{m_e \omega} \right|^2 = \frac{1}{4} \frac{n_e e^2}{m_e \omega^2} E_0^2$$

$$= \frac{1}{16\pi}E_0^2 \frac{\omega_p^2}{\omega^2} \, . \tag{S-11.20}$$

Thus

$$u = u_{\text{EM}} + u_{\text{K}} = \frac{1}{8\pi}E_0^2 \, , \tag{S-11.21}$$

independently of n_e.

(d) In our case (11.11) can be rewritten

$$v_g E_t^2 = c(E_0^2 - E_r^2) \, . \tag{S-11.22}$$

Using Fresnel formulas as functions of the phase velocity $v_\varphi = c/n$, with $n = \sqrt{\varepsilon_r}$, we obtain

$$E_r = \frac{v_\varphi - c}{v_\varphi + c} E_0 \, , \qquad E_t = \frac{2v_\varphi}{v_\varphi + c} E_0 \, , \tag{S-11.23}$$

leading to

$$4v_g v_\varphi^2 = 4c^2 v_\varphi , \qquad \text{(S-11.24)}$$

which is equivalent to $v_g v_\varphi = c^2$.

S-11.3 Longitudinal Waves

(a) We obtain from Maxwell's equations, assuming $\mathbf{B} = 0$,

$$0 = \nabla \times \mathbf{B} = \frac{1}{c}(4\pi \mathbf{J} + \partial_t \mathbf{E}) = \frac{1}{c}(4\pi \, \partial_t \mathbf{P} + \partial_t \mathbf{E}) = \frac{1}{c}\,\partial_t(4\pi \mathbf{P} + \mathbf{E}) .$$
$$\text{(S-11.25)}$$

where $\mathbf{P}$ is the polarization density of the medium and $\mathbf{J} = \partial_t \mathbf{P}$ the associated polarization current. Assuming all fields to have an harmonic dependence $\sim e^{-i\omega t}$, we have $\mathbf{P} = \chi(\omega)\mathbf{E}$ with $\chi = [\varepsilon_r(\omega) - 1]/(4\pi)$. Now, using (11.12), we can write

$$0 = -i\omega(4\pi \mathbf{P} + \mathbf{E}) = -i\omega \left\{ [\varepsilon_r(\omega) - 1]\mathbf{E} + \mathbf{E} \right\} = -i\omega \, \varepsilon_r(\omega) \, \mathbf{E} , \quad \text{(S-11.26)}$$

implying $\varepsilon_r(\omega) = 0$.

(b) The *total* charge and current densities in the medium can be obtained from $\mathbf{E}$ using the equations

$$\rho = \frac{1}{4\pi}\nabla \cdot \mathbf{E} , \qquad \mathbf{J} = -\frac{1}{4\pi}\,\partial_t \mathbf{E} , \qquad \text{(S-11.27)}$$

which also imply the continuity equation $4\pi \, \partial_t \rho = -\nabla \cdot \mathbf{J}$. For $\mathbf{E}$ given by (11.12) we obtain

$$\rho = \frac{ik}{4\pi}\,E_0 \, e^{ikx - i\omega t} , \qquad \mathbf{J} = \hat{\mathbf{x}}\,\frac{i\omega}{4\pi}\,E_0 \, e^{ikx - i\omega t} . \qquad \text{(S-11.28)}$$

(c) Assuming electrons moving with negligible friction, the equation of motion for the single electron is

$$m_e \frac{d^2\mathbf{r}}{dt^2} = -m_e\omega_0^2 \, \mathbf{r} - e\mathbf{E} , \qquad \text{(S-11.29)}$$

where m_e is the electron mass, and $\mathbf{r}$ is the distance of the electron from its equilibrium position. For a monochromatic field $\mathbf{E} = \mathbf{E}_0 \, e^{-i\omega t}$ the stationary solution is

$$\mathbf{r} = \frac{e\mathbf{E}}{m_e(\omega^2 - \omega_0^2)} . \qquad \text{(S-11.30)}$$

The polarization density of the medium is

$$\mathbf{P} = -en_e \, \mathbf{r} = -\frac{n_e e^2}{m_e(\omega^2 - \omega_0^2)}\,\mathbf{E} \equiv \chi(\omega)\mathbf{E} , \qquad \text{(S-11.31)}$$

where n_e is the number of electrons per unit volume, and

$$\chi(\omega) = -\frac{\bar{n}_e\, e^2}{m_e(\omega^2 - \omega_0^2)} = -\left(\frac{1}{4\pi}\right)\frac{\omega_p^2}{\omega^2 - \omega_0^2}\,, \qquad \text{(S-11.32)}$$

is the dielectric susceptibility of the medium, and $\omega_p = \sqrt{4\pi n_e e^2/m_e}$ is its plasma frequency. The dielectric permittivity is thus

$$\varepsilon_r(\omega) = 1 + 4\pi\chi(\omega) = 1 - \frac{\omega_p^2}{\omega^2 - \omega_0^2}\,, \qquad \text{(S-11.33)}$$

and the longitudinal-wave condition $\varepsilon_r(\omega) = 0$ leads to

$$\omega = \sqrt{\omega_p^2 + \omega_0^2}\,. \qquad \text{(S-11.34)}$$

It is important to notice that the wavevector $\mathbf{k}$ is *not* determined by this equation; it may have any value, and the phase velocity may thus be arbitrary (lower or greater than c). Longitudinal waves in condensed matter physics are also called *polaritons*. In a free electron medium where $\omega_0 = 0$ (a simple metal, a ionized gas or a plasma), we have $\omega = \omega_p$; in this case the waves are called plasma waves or *plasmons*.

S-11.4   Transmission and Reflection by a Thin Conducting Foil

(a) Since the problem of determining the transmission and reflection coefficient is linear, and the medium is isotropic, the choice of polarization is arbitrary. For definiteness, we assume linear polarization, with the electric field $\mathbf{E}$ of the incoming wave parallel to the y axis, and the magnetic $\mathbf{B}$ parallel to the z axis.

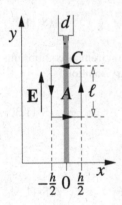

We apply Stokes's theorem to a closed rectangular path C, delimiting a surface area A, twice: once for $\mathbf{E}$ and once for $\mathbf{B}$. In both cases the base of the path extends from $x = -h/2$ to $x = +h/2$, while the height, of length ℓ, is parallel to the y axis for the electric field $\mathbf{E}$, as shown in Fig. S-11.1, and to the z axis for the magnetic field $\mathbf{B}$. For the electric field we have

$$\oint_C \mathbf{E}\cdot d\mathbf{l} = \left[E\left(+\frac{h}{2}\right) - E\left(-\frac{h}{2}\right)\right]\ell \qquad \text{(S-11.35)}$$

$$= +\mathrm{i}\frac{\omega}{c}\int_A \mathbf{B}\cdot d\mathbf{A} = \mathrm{i}\frac{\omega}{c}\bar{B}\,\ell h\,, \qquad \text{(S-11.36)}$$

Fig. S-11.1

where $\bar{B}$ is the amplitude of **B** at some point of the surface A, according to the mean value theorem. Since **B** is limited, at the limit $h \to 0$ we have $\bar{B}h \to 0$, and the first of (11.13) is proved. For the magnetic field we have

$$\oint_C \mathbf{B} \cdot d\mathbf{l} = \left[B\left(+\frac{h}{2}\right) - B\left(-\frac{h}{2}\right) \right] \ell = \int_A \left(\frac{4\pi}{c} \mathbf{J} - i\frac{\omega}{c} \mathbf{E} \right) \cdot d\mathbf{A}$$

$$= \int_A \frac{4\pi}{c} K \, \delta(x) \, dx \, dz - i \int_A \frac{\omega}{c} \mathbf{E} \cdot d\mathbf{A}$$

$$= \frac{4\pi}{c} \ell \bar{K} + -i\frac{\omega}{c} \bar{E} \, \ell h \,, \tag{S-11.37}$$

where, in the second line, we have replaced **J** by $\mathbf{K} \, \delta(x)$, and, in the third line, $\bar{K}$ is a value assumed by K somewhere on the segment of length ℓ. Since, again, $\bar{E}$ is limited, the product $\bar{E} \, \ell h \to 0$ as $h \to 0$, and the second of (11.13) is proved.

(b) The most general expression for the field is the sum of the incident and the reflected wave for $x < 0$, and the transmitted wave only for $x > 0$:

$$E(x, t) = \begin{cases} E_i \, e^{ikx - i\omega t} + E_r \, e^{-ikx - i\omega t} \,, & x < 0 \,, \\ E_t \, e^{ikx - i\omega t} \,, & x > 0 \,. \end{cases} \tag{S-11.38}$$

The amplitudes E_r and E_t must be determined as functions of E_i and other parameters, by imposing (11.13) as boundary conditions. Noticing that $K = \sigma d E(0) = \sigma d E_t$ and that $c \partial_x E(x, t) = -\partial_t B(x, t)$, we have

$$E_t - E_i - E_r = 0, \qquad E_t - E_i + E_r = -4\pi \frac{\sigma d}{c} E_t \,, \tag{S-11.39}$$

so that, writing $2\pi\sigma d/c = \eta$ as a shorthand, we have

$$E_r = -\frac{\eta}{1 + \eta} E_i \,, \qquad E_t = \frac{1}{1 + \eta} E_i \,. \tag{S-11.40}$$

(c) At the limit $\nu \gg \omega$ the conductivity is given by $\sigma = n_e e^2 / m_e \nu$ and is a real number (Ohmic conductor). The mechanical power P is the cycle average of $\mathbf{J} \cdot \mathbf{E}$ integrated over the volume of the foil, thus we obtain (per unit surface)

$$P = \frac{1}{2} |E(0)^2| \, d = \frac{1}{2} \frac{\sigma d}{(1 + \eta)^2} E_i^2 = \frac{c}{4\pi} \frac{\eta}{(1 + \eta)^2} E_i^2 \,, \tag{S-11.41}$$

(notice that E_i can be taken as a real quantity).

At the limit $\nu \ll \omega$ the conductivity is $\sigma = i n_e e^2 / m_e \omega = i\omega_p^2 / 4\pi\omega$ and is thus imaginary, corresponding to a real permittivity $\varepsilon = 1 - \omega_p^2 / \omega^2$. Accordingly, **J** and **E** have opposite phase, and $\langle \mathbf{J} \cdot \mathbf{E} \rangle = 0$, as can be directly verified.

(d) The energy flux through the foil is given by the difference between the values of the Poynting flux at the two surfaces (here we switch back to real fields for simplicity),

$$S(0^+) - S(0^-) = \frac{c}{4\pi} \left[E(0^+)B(0^+) - E(0^-)B(0^-) \right] . \qquad \text{(S-11.42)}$$

Inserting the boundary conditions we may write

$$E(0^+)B(0^+) - E(0^-)B(0^-) = E(0) \left[B(0^+) - B(0^-) \right] = -E(0)\frac{4\pi}{c} Jd , \quad \text{(S-11.43)}$$

so that

$$S(0^+) - S(0^-) = -J E(0)d = -K E(0) , \qquad \text{(S-11.44)}$$

i.e., the energy flux through the foil equals the mechanical power dissipated in the foil (all quantities have been defined per unit surface).

Alternatively, we may compute the energy flux directly and compare it to the mechanical power. For the cycle-averaged Poynting vector at the two surfaces we have

$$\langle S(0^+) \rangle = \frac{c}{4\pi} \langle E^2(0^+) \rangle = \frac{2\pi}{c} |E_t|^2 = \frac{2\pi}{c} \frac{1}{|1+\eta|^2} E_i^2 , \qquad \text{(S-11.45)}$$

$$\langle S(0^-) \rangle = \varepsilon_0 c^2 \langle E(0^-)B(0^-) \rangle = \frac{2\pi}{c} \text{Re} \left[(E_i + E_r)(E_i^* - E_r^*) \right]$$

$$= \frac{2\pi}{c} \frac{1}{|1+\eta|^2} \text{Re} \left(2\eta^* + 1 \right) E_i^2 . \qquad \text{(S-11.46)}$$

If $\nu \ll \omega$, then η is purely imaginary and $S(0^-) = S(0^+)$: there is no net energy flux inside the foil, consistently with the vanishing of the mechanical power.

If $\nu \gg \omega$, then η is real and the net flux of energy is

$$\langle S(0^+) \rangle - \langle S(0^-) \rangle = \frac{2\pi}{c} \frac{1 - (2\eta + 1)}{(1+\eta)^2} E_i^2 = -\frac{c}{4\pi} \frac{\eta}{(1+\eta)^2} E_i^2 , \text{(S-11.47)}$$

which is equal to minus the absorbed power (S-11.41).

S-11.5   Anti-reflection Coating

(a) In the absence of sources at $x = +\infty$, the general solution can be written as (omitting the common time dependence $e^{-i\omega t}$)

$$E = \begin{cases} E_i\, e^{ikx} + E_r\, e^{-ikx} & (x < 0), \\ E_+\, e^{in_1 kx} + E_-\, e^{-in_1 kx} & (0 < x < d), \\ E_t\, e^{in_2 kx} & (x > d), \end{cases} \qquad \text{(S-11.48)}$$

where $k = \omega/c$, E_i is the amplitude of the incident wave, E_r the amplitude of the wave reflected at $x = 0$, E_+ and E_- the amplitudes of the waves propagating along $+\hat{\mathbf{x}}$ and $-\hat{\mathbf{x}}$, respectively, in the $0 < x < d$ layer, and E_t the amplitude of the wave propagating along $+\hat{\mathbf{x}}$ in the $x > d$ half-space. The subscripts of the electric fields E in (S-11.48) are in agreement with the subscripts of the wave vectors $\mathbf{k}$ in Fig. 11.3.
(b) The matching conditions require the electric field and its derivative with respect to x (which is proportional to the magnetic field) to be continuous at the planes $x = 0$ and $x = d$. We thus obtain

$$E_i + E_r = E_+ + E_- , \qquad \text{(S-11.49)}$$
$$E_i - E_r = n_1(E_+ - E_-) , \qquad \text{(S-11.50)}$$
$$E_+\, e^{+in_1 kd} + E_-\, e^{-in_1 kd} = E_t\, e^{in_2 kd} , \qquad \text{(S-11.51)}$$
$$n_1\left(E_+\, e^{+in_1 kd} - E_-\, e^{-in_1 kd}\right) = n_2 E_t\, e^{in_2 kd} . \qquad \text{(S-11.52)}$$

(c) Since we require that there is no reflected wave in vacuum, E_r must be zero. Posing $E_r = 0$ in (S-11.49)–(S-11.52), the latter can be regarded as an homogeneous linear system in E_i, E_+, E_- and E_t. Such system has non-trivial solutions only if its determinant is zero, i.e. if

$$e^{2in_1 kd} = \frac{n_1 + n_2}{n_1 - n_2}\frac{n_1 - 1}{n_1 + 1}. \qquad \text{(S-11.53)}$$

In the case of a layer of thickness d with vacuum at both sides, $n_2 = 1$ and the right-hand side of (S-11.53) equals unity, thus $e^{2in_1 kd} = 1$. This implies $2n_1 kd = 2m\pi$, with m any integer. Thus, there is no reflected wave when the layer thickness is $d = m\lambda/2n_1$ (since $k = 2\pi/\lambda$), i.e. when the "optical depth" nd equals an half-integer number of wavelengths.
(d) In the general case, the left-hand side of (S-11.53) is a complex number of modulus 1, while the right-hand side is always real number if n_1 and n_2 are real as we assumed. Thus, we have solutions only if $e^{2in_1 kd} = \pm 1$. The case $e^{2in_1 kd} = +1$ is the case of $n_2 = 1$, considered above at the end of point **c)**. In the second case $e^{2in_1 kd} = -1$ we have the condition

$$2n_1 kd = (2m + 1)\pi , \qquad \frac{n_1 + n_2}{n_1 - n_2}\frac{n_1 - 1}{n_1 + 1} = -1 , \qquad \text{(S-11.54)}$$

the second equation implying $n_2 = \sqrt{n_1}$. The thickness of the layer must be

$$d = (2m + 1)\frac{\lambda}{4n_1} , \qquad \text{(S-11.55)}$$

with m, again, any integer. The smallest possible thickness is $d = \lambda/(4n_1)$, corresponding to $m = 0$. This shows that, with a suitable choice of materials and of layer thickness, we can produce an "anti-reflection" coating on an optical element (such as a window or lens) from which we do not want any reflection to occur.

S-11.6 Birefringence and Waveplates

(a) The incident wave can be considered as the superposition of two waves having, respectively, P and S polarization, i.e., one having the electric field lying in the xy plane, and the other parallel to z. The difference between the refractive indices for P and S polarization, n_p and n_s, gives origin to two different refraction angles, $\theta_{t,p}$ and $\theta_{t,s}$, according to Snell's law. With our assumptions, the refraction angles are

$$\sin\theta_{t,p} = \frac{\sin\theta_i}{n_p} = \sin(\theta_t - \alpha) \,, \qquad \sin\theta_{t,s} = \frac{\sin\theta_i}{n_s} = \sin(\theta_t + \alpha) \,, \quad \text{(S-11.56)}$$

at the limit $\alpha \ll 1$ we can approximate $\sin\alpha \simeq \alpha$ and $\cos\alpha \simeq 1$, obtaining

$$\sin(\theta_t \pm \alpha) = \sin\theta_t \cos\alpha \pm \cos\theta_t \sin\alpha \simeq \sin\theta_t \pm \alpha\cos\theta_t \,. \qquad \text{(S-11.57)}$$

The refractive indices are $n_p = \bar{n} + \delta n$, and $n_s = \bar{n} - \delta n$, respectively, with $\delta n \ll \bar{n}$. For P polarization we have, up to the first order in $\delta n/\bar{n}$,

$$\sin\theta_t - \alpha\cos\theta_t = \frac{\sin\theta_i}{\bar{n} + \delta n} \simeq \frac{\sin\theta_i}{\bar{n}}\left(1 - \frac{\delta n}{\bar{n}}\right) \,, \qquad \text{(S-11.58)}$$

and, analogously, for S polarization we have

$$\sin\theta_{t,s} \simeq \frac{\sin\theta_i}{\bar{n}}\left(1 + \frac{\delta n}{\bar{n}}\right) \,. \qquad \text{(S-11.59)}$$

The above results lead to

$$\bar{n} = \frac{\sin\theta_i}{\sin\theta_t} \,, \quad \text{and} \quad \alpha = 2\frac{\sin\theta_i}{\cos\theta_t}\frac{\delta n}{\bar{n}^2} = 2\,\delta n\frac{\sin^2\theta_t}{\cos\theta_t \sin\theta_i} \,. \qquad \text{(S-11.60)}$$

(b) In order to have exiting circularly polarized light, the exiting P- and S-polarized components must be phase-shifted by $\delta\phi = \pi/2$. This can obtained making use of the difference between the two optical path lengths, $n_p d$ and $n_s d$. The condition for circularly polarized light is thus

$$\delta\phi = k\,2\delta n\,d = \frac{4\pi\,\delta n\,d}{\lambda} \doteq \frac{\pi}{2} \,, \qquad \text{(S-11.61)}$$

i.e. $d = \lambda/(8\delta n)$. This is called a *quarter-wave plate*. If $\delta\phi = \pi$ instead, i.e., if $d = \lambda/(4\delta n)$, there is a relative change of sign between the two components, which leads to a polarization rotation of $\pi/2$; this is an *half-wave plate*.

S-11.7   Magnetic Birefringence and Faraday Effect

(a) Neglecting the effect of the magnetic field of the wave, much smaller than the external field $\mathbf{B}_0$, the equation of motion for the electrons is

$$m_e \frac{d^2\mathbf{r}}{dt^2} = -e\mathbf{E} - e\frac{v}{c} \times \mathbf{B}_0 - m_e\omega_0^2\mathbf{r} . \qquad (S-11.62)$$

The electric field of the circularly polarized EM wave can be written, in complex notation, as

$$\mathbf{E}_\pm = E(\hat{\mathbf{x}} \pm i\hat{\mathbf{y}})\, e^{ikz - i\omega t} , \qquad (S-11.63)$$

where the plus and minus signs correspond to left-handed (clockwise) and right-handed (counter-clockwise) circular polarizations, respectively. We look for solutions of (S-11.62) of the form

$$\mathbf{r}_\pm = r_\pm(\hat{\mathbf{x}} \pm i\hat{\mathbf{y}})\, e^{ikz - i\omega t} , \qquad \mathbf{v}_\pm = v_\pm(\hat{\mathbf{x}} \pm i\hat{\mathbf{y}})\, e^{ikz - i\omega t} , \qquad (S-11.64)$$

with $v_\pm = -i\omega r_\pm$. The vector product $\mathbf{v}_\pm \times \mathbf{B}_0$ is

$$\mathbf{v}_\pm \times \mathbf{B}_0 = v_\pm B_0\,(\hat{\mathbf{x}} \pm i\hat{\mathbf{y}}) \times \hat{\mathbf{z}} = v_\pm B_0(-\hat{\mathbf{y}} \pm i\hat{\mathbf{x}}) = \pm i v_\pm B_0\,(\hat{\mathbf{x}} \pm i\hat{\mathbf{y}}) , \quad (S-11.65)$$

thus (S-11.62) leads to the equation for $\mathbf{r}_\pm$

$$(\omega_0^2 - \omega^2)\, r_\pm = -\frac{e}{m_e}E \mp ie\frac{v_\pm}{m_e c}B_0 = -\frac{e}{m_e}E \mp \omega\omega_c r_\pm , \qquad (S-11.66)$$

where $\omega_c = eB_0/m_e c$ is the cyclotron frequency. The solution for $r_\pm$ is

$$r_\pm = \frac{eE}{m_e\left(\omega^2 - \omega_0^2 \mp \omega\omega_c\right)} . \qquad (S-11.67)$$

Thus, we have a different polarization of the medium $\mathbf{P}_\pm$, and a corresponding different dielectric susceptibility $\chi_\pm$, for each each circular-polarization state of the EM wave,

$$\mathbf{P}_\pm = -en_e\mathbf{r}_\pm \equiv \chi_\pm\mathbf{E}_\pm . \qquad (S-11.68)$$

In turn, this gives two different dielectric constants $\varepsilon_\pm = 1 + 4\pi\chi_\pm$

$$\varepsilon_\pm = 1 - \frac{\omega_p^2}{\omega^2 - \omega_0^2 \mp \omega\omega_c}, \tag{S-11.69}$$

where $\omega_p = \sqrt{4\pi e^2 n_e/m_e}$ is the plasma frequency of the medium. The propagation of the wave requires $\varepsilon_\pm > 0$, i.e., $\omega > \omega_{co\pm}$, where the two cutoff frequencies $\omega_{co\pm}$ depend on the polarization of the wave

$$\omega_{co\pm} = \sqrt{\omega_0^2 + \omega_p^2 + \frac{\omega_c^2}{4}} \pm \frac{\omega_c}{2}. \tag{S-11.70}$$

The magnetized medium is thus birefringent. For waves of frequency in the range $\omega_{co-} < \omega < \omega_{co+}$, only one state of circular polarization can propagate in the medium, while we have an evanescent wave for the opposite polarization. The two resonant frequencies $\omega_{res\pm}$, defined by $\chi(\omega_{res\pm}) \to \infty$, also depend on polarization:

$$\omega_{res\pm} = \sqrt{\omega_0^2 + \frac{\omega_c^2}{4}} \pm \frac{\omega_c}{2}. \tag{S-11.71}$$

Notice that in the case $\omega_0 = 0$, i.e., for a magnetized free-electron medium, there is a single resonance at $\omega = \omega_c$, for *only one* circular polarization (see Problem 7.9).

The knowledge of the permittivity (or, equivalently, of the refraction index) for the two independent states of circular polarization is sufficient to study the propagation of a transverse wave of arbitrary polarization, since the latter can be always expressed as a linear superposition of circularly polarized states. Notice that if we had searched for linearly polarized solutions, we would have found a *mixing* of polarization vectors directed along $\hat{\mathbf{x}}$ and $\hat{\mathbf{y}}$, i.e. the permittivity would have been a matrix instead of a number. It can be shown that such matrix can be diagonalized, with circularly polarized states as eigenvectors and (S-11.69) as eigenvalues.

(b) The linearly polarized wave can be considered as a superposition of the two states of circular polarization, so that at $z = 0$ the electric field of the wave can be written

$$\mathbf{E}(z = 0, t) = \hat{\mathbf{x}} E e^{-i\omega t} = \frac{E}{2} \left[(\hat{\mathbf{x}} + i\hat{\mathbf{y}}) + (\hat{\mathbf{x}} - i\hat{\mathbf{y}}) \right] e^{-i\omega t}. \tag{S-11.72}$$

The two circularly polarized components travel at different phase velocities $v_\pm = c/n_\pm$, where $n_\pm = \sqrt{\varepsilon_\pm}$ is the refractive index associated to each polarization state. At $z = \ell$, the electric field of the wave is

$$\mathbf{E}(z = \ell, t) = \frac{E}{2} \left[(\hat{\mathbf{x}} + i\hat{\mathbf{y}}) e^{ik_+\ell} + (\hat{\mathbf{x}} - i\hat{\mathbf{y}}) e^{-ik_-\ell} \right] e^{-i\omega t}, \tag{S-11.73}$$

where $k_\pm = \omega/v_\pm = (\omega/c)n_\pm$. To first order in ω_c/ω, we can write $n_\pm \simeq n_0 \pm \delta n$, where $n_0 = n(\omega_c = 0)$ and

$$\delta n = \frac{\omega\omega_c\omega_p^2}{2n_0(\omega^2 - \omega_0^2)^2}. \tag{S-11.74}$$

Thus, the wave vectors for the two polarizations can be written $k_\pm \simeq k_0 \pm \delta k$, where $k_0 = (\omega/c)\,n_0$ and $\delta k = (\omega/c)\,\delta n$. The electric field at $z = \ell$ can be rewritten as

$$\mathbf{E}(z = \ell, t) = \frac{E}{2}\left[(\hat{\mathbf{x}} + i\hat{\mathbf{y}})\,e^{i\,\delta k\,\ell} + (\hat{\mathbf{x}} - i\hat{\mathbf{y}})\,e^{-i\,\delta k\,\ell}\right] e^{i k_0 \ell - i\omega t}$$

$$\propto \hat{\mathbf{x}}\cos(\delta k\,\ell) - \hat{\mathbf{y}}\sin(\delta k\,\ell). \tag{S-11.75}$$

The polarization has thus rotated by an angle $\phi = \delta k\,\ell$, proportional to the intensity of the magnetic field.

S-11.8   Whistler Waves

The dielectric permittivity of a magnetized free electron gas for circularly polarized transverse waves, propagating along the magnetic field, is (see Problem 11.7)

$$\varepsilon = \varepsilon_\pm(\omega) = 1 - \frac{\omega_p^2}{\omega(\omega \mp \omega_c)}, \tag{S-11.76}$$

where $\omega_p = \sqrt{4\pi e^2 n_e/m_e}$ is the plasma frequency of the medium, $\omega_c = eB_0/m_e c$ is the cyclotron (Larmor) frequency, and the plus and minus signs refer to left-handed (counterclockwise) and right-handed (clockwise) circular polarizations, respectively. Since, in general, the dispersion relation is $\omega^2 = k^2 c^2/\varepsilon(\omega)$, (11.16) implies that $\varepsilon = c^2/\alpha\omega$. For $\omega \ll \omega_c$ and $\omega \ll \omega_p^2/\omega_c$, (S-11.76) reduces to

$$\varepsilon_\pm \simeq \pm\frac{\omega_p^2}{\omega\omega_c}. \tag{S-11.77}$$

Wave propagation requires $\varepsilon > 0$. Thus, only left-handed polarized waves can propagate in the presence of a dispersion relation given by (11.16), with $\alpha = c^2\omega_c/\omega_p^2$.

Assuming the values of n_e and B_0 given in the text, we estimate $\omega_p \simeq 5.6 \times 10^6\,\mathrm{s}^{-1}$ and $\omega_c \simeq 8.8 \times 10^6\,\mathrm{s}^{-1}$. A typical frequency for which (S-11.77) holds is $\omega \sim 10^5\,\mathrm{s}^{-1}$.
(b) First, we notice that, in general, (11.16) implies $v_g = \partial_k\omega = 2\alpha k = 2\omega/k = 2v_\varphi$. Thus, the phase velocity depends on frequency as

$$v_\varphi = \frac{\omega}{k} = \sqrt{\alpha\omega} = \sqrt{\frac{\omega_c\omega}{\omega_p^2}}\,c \ll c. \tag{S-11.78}$$

For $\omega = 10^5$, and the above values of ω_p and ω_c, we obtain $v_\varphi \simeq 0.03c$.
(c) With a spectral range from ω_1 to $2\omega_1$, the frequency components travel with velocities differing by a factor up to 2, so that the wave packet generated by the lightning will spread out and increase its length during its propagation. The higher

frequencies travel faster, and are thus received earlier by the observer, than the slower frequencies. This is the origin of name "whistlers".

In order to estimate the spread of the packet after a distance $L = 10^9$ cm, we assume that the center of the wave packet travels with a group velocity $v_g \simeq 0.06\,c$, reaching a distance L after a time $\tau = L/v_g = 0.56$ s. The "extreme" frequencies ω_1 and ω_2 will have group velocities $v_1 \simeq 0.04\,c$ and $v_2 \simeq 0.08\,c$, respectively, and the pulse duration may be roughly estimated as the difference $\Delta\tau = \tau_1 - \tau_2 = L/v_1 - L/v_2 \simeq (0.83 - 0.42)$ s $= 0.41$ s, provided that the duration at the emission is much shorter than $\Delta\tau$. This rough estimate neglects the deformation of the wave packet due to the strong dispersion.

S-11.9 Wave Propagation in a "Pair" Plasma

Actually, it is convenient to calculate the dispersion relation in the presence of an external magnetic field $\mathbf{B}_0$ first, then, the answer to point (a) is simply obtained as a special case with $\mathbf{B}_0 = 0$. We assume $\mathbf{B}_0 = B_0\,\hat{\mathbf{z}}$ and a wave linearly polarized along $\hat{\mathbf{x}}$ in a Cartesian reference frame xyz. The differential equations for the velocities of positrons, $\mathbf{v}_+$, and electrons, $\mathbf{v}_-$, are respectively

$$\frac{\mathrm{d}v_{x\pm}}{\mathrm{d}t} = \pm\frac{e}{m_{\mathrm{e}}c}(E_x + v_{y\pm}B_0)\,, \qquad \frac{\mathrm{d}v_{y\pm}}{\mathrm{d}t} = \mp\frac{e}{m_{\mathrm{e}}c}(v_{x\pm}B_0)\,, \quad \text{(S-11.79)}$$

where we have assumed $v_{z\pm} = 0$. Differentiating the first of (S-11.79) once more with respect to t, and substituting the second of (S-11.79) for $\mathrm{d}v_{y\pm}/\mathrm{d}t$, we obtain

$$\frac{\mathrm{d}^2 v_{x\pm}}{\mathrm{d}t^2} = \mp\mathrm{i}\omega\frac{e}{m_{\mathrm{e}}}E_x \pm \frac{eB_0}{m_{\mathrm{e}}c}\frac{\mathrm{d}v_{y\pm}}{\mathrm{d}t} = \mp\mathrm{i}\omega\frac{e}{m_{\mathrm{e}}}E_x + \omega_{\mathrm{c}}^2 v_{x\pm}\,, \quad \text{(S-11.80)}$$

where $\omega_{\mathrm{c}} = \sqrt{eB_0/m_{\mathrm{e}}c}$ is the cyclotron frequency. Substituting $E_x = E_0\,\mathrm{e}^{-\mathrm{i}\omega t}$ we obtain

$$v_{x\pm} = \mp\mathrm{i}\omega\frac{e}{m_{\mathrm{e}}(\omega_{\mathrm{c}}^2 - \omega^2)}E_0\,. \qquad \text{(S-11.81)}$$

Analogously, for $\mathbf{v}_{y\pm}$ we have

$$v_{y\pm} = \pm\frac{eB_0}{m_{\mathrm{e}}}v_{x\mp} = -\mathrm{i}\omega\frac{e}{m_{\mathrm{e}}}E_0\,, \qquad \text{(S-11.82)}$$

which has the same value for both electrons and positrons. The components of the current density are thus

$$J_x = n_0\, e\, (v_{x+} - v_{x-}) = -\frac{2i\omega n_0 e^2}{m_e(\omega_c^2 - \omega^2)}\, E_0 \,,$$

$$J_y = n_0\, e\, (v_{y+} - v_{y-}) = 0 \,. \tag{S-11.83}$$

The dielectric permittivity of the pair plasma, $\varepsilon(\omega)$, is obtained from the usual definitions $\mathbf{J} = \sigma \mathbf{E} = -i\omega \chi \mathbf{E}$ and is

$$\varepsilon(\omega) = 1 - \frac{2\omega_p^2}{\omega^2 - \omega_c^2} \,. \tag{S-11.84}$$

The same result can be obtained for circular polarization, both for left-handed and right-handed waves, confirming that there is no magnetically induced birefringence in a pair plasma. This is different from the case of a medium containing free electrons only, considered in Problem 11.7.

For case **(a)**, where $\mathbf{B}_0 = 0$, we set $\omega_c = 0$, and obtain a cut-off frequency at $\omega = 2\omega_p$.

For case **(b)**, there is a resonance at $\omega = \omega_c$, while wave propagation is forbidden for frequencies in the range $\omega_c < \omega < \sqrt{\omega_c^2 + 2\omega_p^2}$.

S-11.10　Surface Waves

(a) In a dielectric medium described by $\varepsilon = \varepsilon(\omega)$, a monochromatic EM field of frequency ω satisfies the Helmoltz equation. Thus we have for the magnetic field

$$\left(\nabla^2 + \varepsilon \frac{\omega^2}{c^2} \right) B_z = 0 \,. \tag{S-11.85}$$

Substituting (11.17) for B_z into the Helmholtz equation, we obtain

$$q^2 - k^2 + \frac{\omega^2}{c^2}\varepsilon = 0 \,. \tag{S-11.86}$$

(b) From the equation $c\nabla \times \mathbf{B} = 4\pi \mathbf{J} + \partial_t \mathbf{E}$ and the definition of ε we obtain (for monochromatic waves in complex notation) $c\nabla \times \mathbf{B} = -i\omega\varepsilon\, \mathbf{E}$. By substituting (11.17) for $\mathbf{B}$ we obtain

$$-i\omega\varepsilon\, \mathbf{E} = (\hat{\mathbf{x}}\partial_y - \hat{\mathbf{y}}\partial_x)B_z c = (ik\hat{\mathbf{x}} - q\hat{\mathbf{y}})B_z c \,, \tag{S-11.87}$$

which gives for the electric field

$$\mathbf{E} = -(k\hat{\mathbf{x}} + iq\hat{\mathbf{y}})\frac{c}{\varepsilon\omega}\, B_z \,. \tag{S-11.88}$$

(c) From the definition of $\mathbf{S} = c\mathbf{E} \times \mathbf{B}/(4\pi)$ we find that $\mathbf{S}$ has components both along x and along y, given by

$$S_x = \frac{c}{4\pi} E_y B_z = \frac{q c B_0^2}{4\pi\varepsilon\omega} e^{2qx} \cos(ky - \omega t) \sin(ky - \omega t), \quad \text{(S-11.89)}$$

$$S_y = -\frac{c}{4\pi} E_x B_z = \frac{k c B_0^2}{4\pi\varepsilon\omega} e^{2qx} \cos^2(ky - \omega t). \quad \text{(S-11.90)}$$

However, averaging over one oscillation period we obtain $\langle S_x \rangle = 0$, thus the net energy flux is in the y-direction only, since $\langle S_y \rangle \neq 0$.

(d) The tangential component of the magnetic field at the interface between two media must be continuous. Thus, from $B_z(0^-) = B_z(0^+)$ we get $B_1 = B_2$.

(e) Also the tangential component of the electric must be continuous at the interface, thus $E_y(0^-) = E_y(0^+)$. Using the results of points (b) and (d) we obtain

$$\frac{q_1}{\varepsilon_1} = -\frac{q_2}{\varepsilon_2}. \quad \text{(S-11.91)}$$

Since both $q_1 > 0$ and $q_2 > 0$, ε_1 and ε_2 must have opposite signs.

(f) Using the relationship $(q_1/\varepsilon_1)^2 = (q_2/\varepsilon_2)^2$ and the result of point (a) we obtain

$$\varepsilon_2^2 \left(k^2 - \frac{\omega^2}{c^2} \varepsilon_1 \right) = \varepsilon_1^2 \left(k^2 - \frac{\omega^2}{c^2} \varepsilon_2 \right), \quad \text{(S-11.92)}$$

from which it follows that

$$\omega^2 = k^2 c^2 \frac{\varepsilon_2^2 - \varepsilon_1^2}{\varepsilon_2^2 \varepsilon_1 - \varepsilon_1^2 \varepsilon_2} = k^2 c^2 \frac{\varepsilon_2 + \varepsilon_1}{\varepsilon_2 \varepsilon_1}. \quad \text{(S-11.93)}$$

Since wave can propagate only if $k^2 > 0$, and $\varepsilon_1 \varepsilon_2 < 0$, we get the additional condition $\varepsilon_1 + \varepsilon_2 < 0$.

(g) Since $\varepsilon_2 < -\varepsilon_1 = -1$ must hold, we may choose a metal, or a free electron gas, or an ideal plasma ..., for which $\varepsilon_2 = 1 - \omega_p^2/\omega^2$, and a frequency such that $\omega < \omega_p/\sqrt{2}$.

The above described EM modes are *surface waves* (also named surface *plasmons*). These waves propagate along the surface of a conductor and are evanescent along the perpendicular direction, so that the EM energy is confined in a narrow layer, thinner than the wavelength in vacuum. Surface waves are a building block of *plasmonics*, a discipline oriented to develop optical and electronic devices on a nanometric scale.[9]

---

[9] See e.g. W. L. Barnes et al, "*Surface plasmon subwavelength optics*", Nature **424**, 824 (2003); E. Ozbay, "*Plasmonics: merging photonics and electronics at nanoscale dimensions*", Science **311**, 189 (2006).

S-11.11   Mie Resonance and a "Plasmonic Metamaterial"

(a) The incident field can be written, in complex notation, as

$$\mathbf{E}_i = \mathbf{E}_i(x, t) = \mathbf{E}_0\, e^{ikx - i\omega t} \,. \tag{S-11.94}$$

Since $a \ll \lambda$, the electric field can be considered as uniform over the volume of the sphere, thus $\mathbf{E}_i \simeq \mathbf{E}_0\, e^{-i\omega t}$, assuming the center of the sphere to be located at $x = 0$. Now we introduce a spherical coordinate system (r, θ, ϕ) with the origin at the center of the sphere, and the zenith direction parallel to $\mathbf{E}_i$. At the surface of the sphere, $r = a$, we have the usual boundary conditions at the interface between two media

$$E_\perp(a^+, \theta) - E_\perp(a^-, \theta) = 4\pi\sigma(\theta)\,, \qquad E_\parallel(a^+, \theta) - E_\parallel(a^-, \theta) = 0\,, \tag{S-11.95}$$

where $\sigma(\theta)$ is the surface charge density on the sphere, independent of ϕ within our approximations. The problem is thus analogous to the case of a dielectric sphere in a static uniform external field, treated in Problem 3.4. We can extend the results for the internal field and polarization to the case of an oscillating field as follows

$$\mathbf{E}_{\text{int}} = \frac{3\mathbf{E}_0}{\varepsilon_r(\omega) + 2}\,, \qquad \mathbf{P} = \chi\mathbf{E}_{\text{int}} = \frac{3(\varepsilon_r(\omega) - 1)}{4\pi(\varepsilon_r(\omega) + 2)}\mathbf{E}_0\,. \tag{S-11.96}$$

The difference with the electrostatic case is that now ε_r depends on frequency, and is not necessarily positive and greater than one, so that the internal field $\mathbf{E}_{\text{int}}$ can be greater than the external applied field $\mathbf{E}_0$. A resonance appears when the real part of the denominator vanishes. Setting $\eta = 0$ for simplicity, the resonance condition is

$$\varepsilon_r(\omega) + 2 = 3 - \frac{\omega_p^2}{\omega^2 - \omega_0^2} = 0\,, \tag{S-11.97}$$

which yields

$$\omega^2 = \omega_0^2 + \frac{\omega_p^2}{3}\,. \tag{S-11.98}$$

The physical meaning of the resonance is particularly clear for $\omega_0 = 0$, e.g., for a metallic (nano)sphere in a high-frequency (optical) field. In this case the resonance frequency is

$$\omega = \frac{\omega_p}{\sqrt{3}}\,, \tag{S-11.99}$$

that is the natural frequency of the collective "Mie oscillations" of the electron sphere treated in Problem 1.5, also known as the lowest-order surface plasmon of the sphere. The resonance thus corresponds to the excitation of this oscillation mode.

(b) The *macroscopic* polarization is given by the dipole moment of each nanosphere, $\mathbf{p}_{sphere} = \mathbf{P}V$, with $V = (4\pi/3)a^3$ the volume of the sphere, times the number of nanospheres per unit volume, n_s:

$$\mathbf{P}_{macro} = n_s \mathbf{p}_{sphere} = -\frac{3n_s V \omega_p^2}{3\omega^2 - \omega_p^2} \mathbf{E}_0. \tag{S-11.100}$$

This is equivalent to a *macroscopic* dielectric function

$$\varepsilon_r(\omega) = 1 - \frac{3n_s V \omega_p^2}{3\omega^2 - \omega_p^2}. \tag{S-11.101}$$

Wave propagation requires ε_r to be positive, i.e.,

$$\omega < \frac{\omega_p}{\sqrt{3}}, \qquad \omega > \frac{\omega_p}{\sqrt{3}}\sqrt{1 + 3n_s V}. \tag{S-11.102}$$

This is a simple example of an artificial "metamaterial", where the plasmonic properties of the nanostructures composing the material determine the optical response.

S-11.12 Wave Incident at Brewster's Angle

(a) The magnetic field $\mathbf{B}$ must be continuous through the vacuum-medium interface, thus $B_t = B_i = B_0 = E_0$, where the subscripts i and t stand for *incident* and *transmitted*, respectively. On the other hand, the amplitude of the transmitted electric field is $E_t = B_t/n = E_0/n$.

(b) The Poynting vector in vacuum, i.e. for $x < 0$, is, averaged over one period, $|\bar{S}| = cE_i B_i/(4\pi) = cE_0^2/(4\pi)$ (we use bars to denote time-averaged quantities). In the medium, i.e. for $x > 0$, we have $|\bar{S}| = cE_t B_t/(4\pi) = cE_0^2/(4\pi n)$. The discontinuity of the energy flux through the interface is due to the difference between the x components of the Poynting vector

$$\Delta \Phi_E = \bar{S}_x(0^+) - \bar{S}_x(0^-) = \bar{S}(0^+)\cos\theta_t - \bar{S}(0^-)\cos\theta_i$$

$$= \frac{1}{n}\bar{S}(0^-)\cos(\theta_B - \pi/2) - \bar{S}(0^-)\cos\theta_B = \bar{S}(0^-)\left(-\frac{\sin\theta_B}{n} - \cos\theta_B\right), \tag{S-11.103}$$

where $\theta_i = \theta_B$ and $\theta_i + \theta_t = \pi/2$. Angle θ_i is the incidence angle, equal to Brewster's angle θ_B, while θ_t is the transmission angle, see Fig. 11.8. According to Snell's law we have

$$\frac{\sin \theta_B}{n} = \sin \theta_t = \cos \theta_i = -\cos \theta_B , \qquad \text{(S-11.104)}$$

which leads to $\Delta \Phi_E = 0$. On the other hand, the y component of S is discontinuous.
(c) The surface charge density σ can be obtained from Gauss's law

$$\sigma = \frac{E_x(0^+) - E_x(0^-)}{4\pi} = \frac{E_0 \sin \theta_i - E_t \sin \theta_t}{4\pi} = \frac{E_0 \sin \theta_i}{4\pi}\left(1 - \frac{1}{n^2}\right) , \qquad \text{(S-11.105)}$$

where

$$E_x(0^-) = -E_0 \sin \theta_i \quad \text{and} \quad E_x(0^+) = -E_t \sin \theta_t = \frac{E_0}{n}\frac{\sin \theta_i}{n} = -E_0 \frac{\sin \theta_i}{n^2} . \qquad \text{(S-11.106)}$$

(d) The difference between the electromagnetic momentum fluxes through the interface can be evaluated from Maxwell's stress tensor

$$\Delta \Phi_P = \bar{T}_{xx}(0^+) - \bar{T}_{xx}(0^-) = -\frac{1}{8\pi}\left[E_x^2(0^+) - E_x^2(0^-)\right]$$

$$= \frac{1}{8\pi} E_0^2 \sin^2 \theta_i \left(1 - \frac{\sin^2 \theta_i}{n^4}\right) . \qquad \text{(S-11.107)}$$

since E_y and B_z are continuous at the interface. Thus the pressure is of "electrostatic" nature

$$\Delta \Phi_P = \frac{1}{2}\sigma \left(E_x(0^+) + E_x(0^-)\right) . \qquad \text{(S-11.108)}$$

(e) Because of the law of reversibility of the optical paths and of the symmetry of our problem (see Fig. S-11.2) there is no reflection also at $x = d$. Thus the fields of the wave exiting at $x = d$ equal the fields of the wave entering at $x = 0$, except for a phase shift φ. Taking into account the continuity of $\mathbf{B} = B_z \hat{\mathbf{z}}$, we have

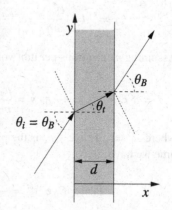

$$B_z = \frac{E_0}{c} e^{ik \sin \theta_B y - i\omega t} \times \begin{cases} e^{ik \cos \theta_B x} & (x < 0) \\ e^{ink \cos \theta_t x} & (0 < x < d) \\ e^{ik \cos \theta_B x + i\varphi} & (x > d) \end{cases}$$

$$\text{(S-11.109)}$$

where $k = \omega/c$ is the wave vector in vacuum. The phase shift is due to the condition $B_z(x = d^-) = B_z(x = d^+)$, thus we have

Fig. S-11.2

$$\varphi = kd(n \cos \theta_t - \cos \theta_B) . \qquad \text{(S-11.110)}$$

In this case the global force on the dielectric medium is zero.

S-11.13   Electromagnetic Wave in a Conducting Medium

(a) According to (11.3) we have

$$\varepsilon_r = 1 + i\,\frac{4\pi\sigma}{\omega}\,,\tag{S-11.111}$$

which can easily be obtained from the general relation $\varepsilon_r = 1 + 4\pi\chi$ combined with the two relations $\mathbf{J} = \sigma\mathbf{E}$ and $\mathbf{J} = \partial_t\mathbf{P}$ applied to an oscillating monochromatic electric field $\mathbf{E} = \hat{\mathbf{y}}\,E_0\,e^{-i\omega t}$, where, for simplicity and without loss of generality, we have assumed a linear polarization parallel to the y axis. For the refractive index we have $n = \sqrt{\varepsilon_r}$, and, at the limit $\sigma \gg \omega$,

$$n = \sqrt{1 + i\,\frac{4\pi\sigma}{\omega}} \simeq \sqrt{i\,\frac{4\pi\sigma}{\omega}} = \frac{1+i}{\sqrt{2}}\sqrt{\frac{4\pi\sigma}{\omega}} = (1+i)\sqrt{\frac{2\pi\sigma}{\omega}}\,,\tag{S-11.112}$$

thus we have $n_0 = \sqrt{2\pi\sigma/\omega}$.

Alternatively, still assuming a linear polarization parallel to the y axis, we can use the classical free-electron model, evaluating the stationary solution of the equation of motion

$$\ddot{y} = -\frac{e}{m_e}\,E_0\,e^{-i\omega t} - \eta\dot{y}\,,\quad\text{which is}\quad y = \frac{e}{m_e(\omega^2 + i\omega\eta)}\,E_0\,e^{-i\omega t}\,.\tag{S-11.113}$$

At the limit $\eta \gg \omega$ we have

$$y = -i\,\frac{e}{m_e\omega\eta}\,E_0\,e^{-i\omega t}\,,\quad\text{and}\quad \dot{y} = \frac{e}{m_e\eta}\,E_0\,e^{-i\omega t}\,.\tag{S-11.114}$$

Assuming n_e electrons per unit volume, we have for the current density

$$J = \sigma\,E_0\,e^{-i\omega t} = -e\,n_e\,\dot{y} = \frac{n_e e^2}{m_e\eta}\,E_0\,e^{-i\omega t} = \frac{\omega_p^2}{4\pi\eta}\,E_0\,e^{-i\omega t}\,,\quad\text{and}\quad \sigma = \frac{\omega_p^2}{4\pi\eta}\,,\tag{S-11.115}$$

where $\omega_p = \sqrt{4\pi n_e e^2/m_e}$ is the plasma frequency. For the polarization per unit volume we have

$$P = \chi\,E_0\,e^{-i\omega t} = -n_e e y = i\,\frac{n_e e^2}{m_e\omega\eta}\,E_0\,e^{-i\omega t}\,,\tag{S-11.116}$$

from which we obtain

$$\chi = i\,\frac{n_e e^2}{m_e\omega\eta} = i\,\frac{\omega_p^2}{4\pi\omega\eta} = i\,\frac{\sigma}{\omega}\,,\tag{S-11.117}$$

in agreement with (S-11.111) and $\varepsilon_r = 1 + 4\pi\chi$.

(b) For simplicity we assume again a linear polariztion parallel to the y axis. The electromagnetic fields for $x \geq 0$ are, in complex notation,

$$E_y = E_t \, e^{iknx - i\omega t} \,, \qquad B_z = nE_t \, e^{iknx - i\omega t} \,, \qquad \text{(S-11.118)}$$

where $k = \omega/c$. The energy flux is

$$\langle S_x(0) \rangle = \left\langle \frac{c}{4\pi} \mathbf{E} \times \mathbf{B} \right\rangle = \frac{1}{2} \mathrm{Re} \left(\frac{c}{4\pi} E_y(0) B_z^*(0) \right) = \frac{c}{8\pi} |E_t|^2 \, \mathrm{Re(n)}$$

$$= \frac{c}{8\pi} |E_t|^2 \sqrt{\frac{2\pi\sigma}{\omega}} = \frac{c}{4} \sqrt{\frac{\sigma}{2\pi\omega}} |E_t|^2 \,. \qquad \text{(S-11.119)}$$

The power dissipated by Joule heating per unit volume is

$$W_v = \langle J_y E_y^* \rangle = \langle \sigma E_y E_y^* \rangle = \frac{1}{2} \mathrm{Re} \left(\sigma \left| E_y \right|^2 \right) = \frac{\sigma}{2} |E_t|^2 \, e^{-2k \, \mathrm{Im(n)} \, x} \,. \qquad \text{(S-11.120)}$$

The power dissipated per unit surface is obtained by integrating W_v in dx from zero to infinity

$$W_s = \int_0^\infty W_v \, dx = \frac{\sigma}{2} |E_t|^2 \int_0^\infty e^{-2k \, \mathrm{Im(n)} \, x} dx = \frac{\sigma}{4k \, \mathrm{Im(n)}} |E_t|^2$$

$$= \frac{c}{4} \sqrt{\frac{\sigma}{2\pi\omega}} |E_t|^2 = \langle S_x(0) \rangle \,. \qquad \text{(S-11.121)}$$

S-11.14   Wave Reflection at the Ionosphere

(a) If $\omega < \omega_p$, then $\varepsilon_r < 0$ and the refractive index $n = \sqrt{\varepsilon_r}$ is a purely imaginary number. In this case there is total reflection (dissipation is neglected) at any angle of incidence. This is a consequence of the dispersion relation $\omega^2 = k^2 c^2 / \varepsilon_r$ where we can write the square of the wavevector $\mathbf{k}$ as $k^2 = k_x^2 + k_z^2$ with x and z the horizontal and vertical directions, respectively. Since k_x is real (because the parallel component of $\mathbf{k}$ must be conserved at the interface between the air and the ionized gas) the condition $\varepsilon_r < 0$ implies $k_z^2 < 0$, i.e., k_z is purely imaginary and the wave is evanescent in the gas layer.

If $\omega > \omega_p$, total reflection may still occur at large enough values of the incidence angle θ. In fact, being $k_x = (\omega/c) \sin \theta$, the dispersion relation can be rewritten as

$$k_z^2 = \frac{\omega^2}{c^2} \varepsilon_r - k_x^2 = \frac{\omega^2}{c^2} - \frac{\omega_p^2}{c^2} - \frac{\omega^2}{c^2} \sin^2 \theta = \frac{\omega^2 \cos^2 \theta - \omega_p^2}{c^2} \,, \qquad \text{(S-11.122)}$$

so that, at incidences such that $\cos\theta < \omega_p/\omega$, we have $k_z^2 < 0$. This is equivalent to consider Snell's law (11.5), according to which the limit angle for total reflection is $\sin\theta_{\text{lim}} = \mathsf{n}$.

In our problem the signal must be incident at an angle $\theta = \arctan[L/(2h)]$ on the ionosphere in order to reach the receiver R. The larger the value of L the larger the angle θ, thus total reflection of high frequencies is favored by large values of L. This effect might compensate the reduction of the signal with increasing distance.

(b) The electric field emitted by the dipole antenna lies in the plane of incidence, thus we must take the Fresnel coefficient $R_P(\theta)$ for the ratio of the electric field of the reflected wave to the electric field of the incident wave for P-polarization, given by the first of (11.8) (remember that in Fresnel's formulas the refractive index is, in general, a complex number, so that the formula can be used for any value of ω/ω_p). The ratio of the reflected power to the incident power is $R_P(\theta) = |E_r/E_i|^2$. Thus we have for the intensity reaching the receiver located at a distance L

$$I(L) = \frac{3}{4\pi} P_a \frac{\sin^2\theta}{r^2} R_p(\theta) , \qquad\qquad (\text{S-11.123})$$

where P_a is the total emitted power, and $r = 2\sqrt{L^2/4 + h^2} = L/\sin\theta$ is the distance traveled by the radio signal (because of specular reflection, for the receiver the signal is identical to that received from a virtual source located at height $2h$ above the emitter).

(c) An electron density $n_e = 10^{12}$ m^{-3} corresponds to a plasma frequency $\omega_p = 6.3 \times 10^9$ Hz, thus $2\pi c/\omega_p \simeq 0.3$ m. Consistently, radio waves with wavelenghts above this value are typically used for ionosphere-assisted transmission.

The effect of the magnetic field is relevant at frequencies close to the cyclotron frequency ω_c, for which resonance occurs. We have

$$\omega_c = \frac{eB}{m_e c} \simeq 1.8 \times 10^6 \text{ Hz} , \qquad\qquad (\text{S-11.124})$$

corresponding to wavelengths of the order of $2\pi c/\omega_c \simeq 10^3$ m. Thus the related effects, like an anisotropy of the refractive index and cyclotron resonance (Problem 11.7) are observed only for waves "long" compared to "standard" radio waves.

S-11.15   Waves in a Dipole Chain

Since $a \ll 2\pi c/\omega$ we can use (S-3.23) of Problem 3.4 for the induced dipole of the nth sphere

$$\mathbf{p}_n' = a^3 \frac{\varepsilon_r - 1}{\varepsilon_r + 2} \mathbf{E}_n \equiv \alpha\mathbf{E}_n , \qquad\qquad (\text{S-11.125})$$

where $\mathbf{E}_n$ is the external field at $x = nd$, i.e., the field generated by the spheres different from the nth sphere. Since the distance between the particles is much less than $2\pi c/\omega$, the field generated by a particle on the two neighboring ones will be given by the static expression of the dipole field. By considering only the left and right nearest neighbors, we can write the field at $x = nd$ as

$$\mathbf{E}_n = \frac{3(\mathbf{p}_{n-1} \cdot \hat{\mathbf{x}})\, \hat{\mathbf{x}} - \mathbf{p}_{n-1}}{d^3} + \frac{3(\mathbf{p}_{n+1} \cdot \hat{\mathbf{x}})\, \hat{\mathbf{x}} - \mathbf{p}_{n+1}}{d^3}, \tag{S-11.126}$$

so that

$$\mathbf{p}_n = \alpha \mathbf{E}_n = \frac{\alpha}{d^3}[3(\mathbf{p}_{n-1} \cdot \hat{\mathbf{x}})\, \hat{\mathbf{x}} - \mathbf{p}_{n-1} + 3(\mathbf{p}_{n+1} \cdot \hat{\mathbf{x}})\hat{\mathbf{x}} - \mathbf{p}_{n+1}]$$

$$= \begin{cases} \dfrac{2\alpha}{d^3}(\mathbf{p}_{n-1} + \mathbf{p}_{n+1}) & (\mathbf{p}_n \parallel \hat{\mathbf{x}}) \\[2mm] -\dfrac{\alpha}{d^3}(\mathbf{p}_{n-1} + \mathbf{p}_{n+1}) & (\mathbf{p}_n \perp \hat{\mathbf{x}}) \end{cases} \tag{S-11.127}$$

By inserting the expression (11.21) for $\mathbf{p}_n$ and dividing both sides by $\overline{\mathbf{p}}\, e^{iknd - i\omega t}$ (the division is allowed because all dipoles are parallel to one another, independently of $\overline{\mathbf{p}}$ being parallel or perpendicular to $\hat{\mathbf{x}}$), we obtain the relation

$$1 = \frac{2\alpha}{d^3} \begin{cases} 2\cos kd & (\mathbf{p}_n \parallel \hat{\mathbf{x}}) \\ -\cos kd & (\mathbf{p}_n \perp \hat{\mathbf{x}}) \end{cases} \tag{S-11.128}$$

Inserting $\varepsilon_{\mathrm{r}} = 1 - \omega_{\mathrm{p}}^2/\omega^2$ into the second of (S-11.125) we obtain

$$\alpha = a^3 \frac{\omega_{\mathrm{p}}^2}{3\omega^2 - \omega_{\mathrm{p}}^2}, \tag{S-11.129}$$

which, inserted into (S-11.128) leads to

$$\omega^2 = \frac{\omega_{\mathrm{p}}^2}{3}\left(1 - \frac{2a^3}{d^3}\begin{cases} 2\cos kd & (\mathbf{p}_n \parallel \hat{\mathbf{x}}) \\ -\cos kd & (\mathbf{p}_n \perp \hat{\mathbf{x}}) \end{cases}\right) \tag{S-11.130}$$

As typical for a discrete system (Problem 7.4) with inversion symmetry, the dispersion relation needs to be analyzed for $0 < k < \pi/d$ only. We see that the slope of the $\omega = \omega(k)$ relation is the opposite between longitudinal waves ($\mathbf{p}_n \parallel \hat{\mathbf{x}}$) and transverse waves ($\mathbf{p}_n \perp \hat{\mathbf{x}}$). For the latter, $\partial\omega/\partial k < 0$, which means that the group and phase velocities are in opposite directions (like the "CL" newtork of Problem 7.5).

Chapter S-12
Solutions for Chapter 12

S-12.1 The Coaxial Cable

(a) Since the capacitance has been defined assuming *static* conditions and boundary effects are negligible for an "infinite" wire, we evaluate the capacitance per unit length of the cable, C, as for a cylindrical capacitor assuming the charge density to be constant in time and uniformly distributed. For symmetry reasons the electrostatic field between the two conductors is radial and independent of z and ϕ, and it is obtained easily from Gauss's law as

$$\mathbf{E} = \frac{2\lambda}{r}\,\hat{\mathbf{r}}, \qquad a < r < b. \tag{S-12.1}$$

Thus, the potential drop between the two conductors is

$$V = -\int_a^b E_r(r)\,dr = -2\lambda \ln\left(\frac{b}{a}\right), \tag{S-12.2}$$

so that we obtain

$$C = \frac{\lambda}{|V|} = \frac{1}{2\ln(b/a)}. \tag{S-12.3}$$

Similarly, a static current I uniformly distributed on the inner conductor generates a magnetic field

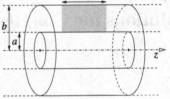

$$\mathbf{B} = B_\phi(r)\,\hat{\phi} = \frac{2I}{cr}\,\hat{\phi}\,. \qquad \text{(S-12.4)}$$

The inductance per unit length of the cable can be obtained by evaluating the flux of $\mathbf{B}$ **Fig. S-12.1** through a rectangle of width Δz, lying on a plane containing the z axis, and extending from $r = a$ to $r = b$, as highlighted in Fig. S-12.1. The flux is

$$\Phi(\mathbf{B}) = \int_a^b B_\phi(r)\,\Delta z\,dr = \frac{2I}{c}\ln\left(\frac{b}{a}\right)\Delta z\,, \qquad \text{(S-12.5)}$$

corresponding to an inductance per unit length $\mathcal{L}$

$$\mathcal{L} = \frac{\Phi(\mathbf{B})}{I c\,\Delta z} = \frac{2}{c^2}\ln\left(\frac{b}{a}\right)\,. \qquad \text{(S-12.6)}$$

The same result can be obtained by calculating the magnetic energy in a cable section of length Δz, and inductance $\Delta z\mathcal{L}$,

$$\frac{1}{2}\Delta z\mathcal{L} I^2 \equiv \Delta z\int_a^b \frac{B^2}{8\pi}\,2\pi r\,dr = \frac{I^2}{c^2}\ln\left(\frac{b}{a}\right)\Delta z\,. \qquad \text{(S-12.7)}$$

(b) The coaxial cable is a continuous system with finite capacitance and inductance per unit length, thus we know from Problem 7.4 that a current signal propagates along the wire according to the wave equation (S-7.50), with velocity

$$v = \frac{1}{\sqrt{\mathcal{L}C}} = c\,. \qquad \text{(S-12.8)}$$

The general solution for the propagating current signal is thus $I(z,t) = I(z-vt)$, and propagation occurs with no dispersion. The associated charge signal $\lambda(z,t)$ is related to $I(z,t)$ by the continuity equation,

$$\partial_t\lambda(z,t) = -\partial_z I(z,t) = -I'(z-ct)\,, \qquad \text{(S-12.9)}$$

where I' denotes the derivative of I with respect to its argument. Since $\partial_t\lambda(z-ct) = -c\lambda'(z-ct)$, we obtain

$$\lambda(z,t) = \lambda(z - ct) = \frac{1}{c} I(z - ct) . \qquad \text{(S-12.10)}$$

(c) A transverse electric field $\mathbf{E}$ must be radial for symmetry reasons, $\mathbf{E} = E_r(r, z, t)\,\hat{\mathbf{r}}$. Applying Gauss's law to a cylindrical surface of radius $a < r < b$, infinitesimal height Δz, and coaxial to the cable, we find $E_r = 2\lambda(z, t)/r$. Again for symmetry reasons, a transverse magnetic field must be azimuthal, $\mathbf{B} = B_\phi(r, z, t)\,\hat{\boldsymbol{\phi}}$. Applying Stokes' law to a circle of radius $a < r < b$, coaxial to the cable, we obtain $B_\phi = 2I(z, t)/rc$. The displacement current does not contribute to the flux through the circle, since $\mathbf{E}$ is radial. Thus, the fields of have the same dependence on λ and I as the static fields, the only difference being that here both $\lambda = \lambda(z, t)$ and $I = I(z, t)$ depend on z and t. Notice that it is such peculiar character of the TEM configuration which allows to use the capacitance and inductance calculated for static fields to obtain the propagation velocity of electromagnetic signals along the cable, a result also true for any transmission line in TEM mode.

We can check that the above fields constitute a solution to Maxwell's equations by verifying that

$$\nabla \times \mathbf{E} = \partial_z E_r\,\hat{\boldsymbol{\phi}} = \frac{2}{r}\partial_z \lambda(z - ct)\,\hat{\boldsymbol{\phi}} = \frac{2}{r}\lambda'(z - ct)\,\hat{\boldsymbol{\phi}}$$

$$= \frac{2}{rc} I'(z - ct)\,\hat{\boldsymbol{\phi}} = -\frac{2}{rc^2}\,\partial_t I(z - ct)\,\hat{\boldsymbol{\phi}}$$

$$= -\frac{1}{c}\,\partial_t \mathbf{B} . \qquad \text{(S-12.11)}$$

(d) The source at $z = 0$ must do a work $W(t)$ in order to drive the current between the inner and outer conductors,

$$W(t) = V(0, t)\,I(0, t) = -2c\lambda^2(0, t)\ln\left(\frac{b}{a}\right) . \qquad \text{(S-12.12)}$$

The local flux of energy at any point (r, ϕ, z), with $a < r < b$ and $z > 0$, is

$$S(r, z, t) = \frac{c}{4\pi}\,\mathbf{E} \times \mathbf{B} = \hat{\mathbf{z}}\,\frac{c}{4\pi}\,\frac{2\lambda(z - ct)}{r}\,\frac{2I(z - ct)}{rc}$$

$$= \hat{\mathbf{z}}\,\frac{c}{\pi r^2}\,\lambda^2(z - ct) , \qquad \text{(S-12.13)}$$

corresponding to a total flow of energy at z

$$\Phi(z, t) = \int_a^b S_z\,2\pi r\,dr = 2c\lambda^2(z - ct)\ln\left(\frac{b}{a}\right) = -W(z - ct) . \text{(S-12.14)}$$

This shows that the energy flow is sustained by the source.

(e) The expressions for the fields, and for the capacitance and inductance per unit length, are, in the presence of generic values of ε and μ,

$$E_r = 2\frac{\lambda}{\varepsilon r} , \qquad\qquad B_\phi = \frac{2\mu I}{rc} , \qquad\qquad \text{(S-12.15)}$$

$$C = \frac{\varepsilon}{2\ln(b/a)} , \qquad\qquad \mathcal{L} = \frac{2}{c^2}\ln\left(\frac{b}{a}\right) , \qquad\qquad \text{(S-12.16)}$$

corresponding to a wave velocity $v = c/\sqrt{\varepsilon\mu} < c$. In general, however, both ε and μ can depend on frequency, and the cable becomes a dispersive transmission line with phase velocity $v_\phi(\omega) = c/\sqrt{\varepsilon(\omega)\,\mu(\omega)}$.

S-12.2 Electric Power Transmission Line

(a) The continuity equation is $\partial_t \lambda = -\partial_z I$. Writing λ in the form $\lambda = \lambda_0\,e^{ikz-i\omega t}$, we obtain

$$-i\omega\lambda_0 = -ikI_0 , \quad \text{or} \quad \lambda_0 = \frac{k}{\omega}I_0 = \frac{I_0}{v_\varphi} , \qquad\qquad \text{(S-12.17)}$$

where v_φ is the phase velocity of the signal.
(b) The electric field $\mathbf{E}$ can be calculated by applying Gauss's law to a cylindrical surface coaxial to the wire, obtaining

$$E_r(r, z, t) = \frac{2\lambda(z, t)}{r} . \qquad\qquad \text{(S-12.18)}$$

The magnetic field $\mathbf{B}$ can be obtained from the equation $c\nabla \times \mathbf{B} = 4\pi\mathbf{J} + \partial_t\mathbf{E}$. If we choose a circle of radius r coaxial to the wire and apply Stokes' theorem we have

$$\oint \mathbf{B} \cdot d\boldsymbol{\ell} = \frac{1}{c}\int (4\pi\mathbf{J} + \partial_t\mathbf{E}) \cdot d\mathbf{S} . \qquad\qquad \text{(S-12.19)}$$

The $\partial_t\mathbf{E}$ term is radial and thus does not contribute to the flux at the right-hand side, so that

$$2\pi r B_\phi = 4\pi\frac{I}{c} , \quad \text{and} \quad B_\phi(r, z, t) = \frac{2I(z, t)}{rc} . \qquad\qquad \text{(S-12.20)}$$

The equations for $E_r(r, \phi, z)$ and $B_\phi(r, \phi, z)$ have the same form as in the static case of a wire with constant and uniform charge density and current, respectively. We also have $|E_r|/|B_\phi| = c/v_\varphi$. These are a typical properties of the TEM (transverse electromagnetic) mode for the transmission lines. Maxwell's equation $c\nabla \times \mathbf{E} = -\partial_t\mathbf{B}$ gives $c\partial_z E_r = -\partial_t B_\varphi$ leads to

$$ik\lambda_0 = i\omega\frac{I_0}{c^2} \quad \Rightarrow \quad k\frac{I_0}{v_\varphi} = \omega\frac{I_0}{c^2} \quad \Rightarrow \quad \frac{\omega}{v_\varphi^2} = \frac{\omega}{c^2} \quad \Rightarrow \quad v_\varphi = c \,, \quad \text{(S-12.21)}$$

where we have used (S-12.17) and $k = \omega/v_\varphi$.

In SI units we have

$$E_r = \frac{\lambda}{2\pi\varepsilon_0} \,; \qquad B_\phi = \frac{\mu_0 I}{2\pi r} \,, \qquad \frac{|E_r|}{|B_\phi|} = \frac{c^2}{v_\varphi}. \qquad \text{(S-12.22)}$$

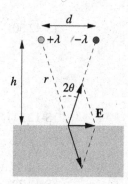

Fig. S-12.2

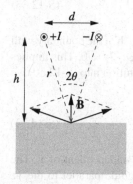

Fig. S-12.3

(c) Consider a line on the midplane, at a distance h from the plane containing the two wires, as in Figs. S-12.2 and S-12.3. The distance of the line from each wire is $r = \sqrt{h^2 + d^2/4}$. The electric and magnetic fields generated by the two wires sum up to

$$\mathbf{E} = 2\hat{\mathbf{x}}\frac{2I_0}{rc}\sin\theta\,e^{-i\omega t} \,, \qquad \text{(S-12.23)}$$

$$\mathbf{B} = 2\hat{\mathbf{y}}\frac{2I_0}{rc}\sin\theta\,e^{-i\omega t} \,, \qquad \text{(S-12.24)}$$

with $\hat{\mathbf{x}}$ and $\hat{\mathbf{y}}$ the unit vectors parallel and perpendicular to the plane containing the wires, respectively.

Since $\sin\theta = d/(2r)$, we obtain

$$|E_x| = |B_y| = \frac{2I_0 d}{r^2 c} \,. \qquad \text{(S-12.25)}$$

The corresponding expressions in SI units are

$$|E_x| = |B_y|c = \frac{I_0 d}{2\pi c\varepsilon_0 r^2} = \frac{\mu_0 I_0 dc}{2\pi r^2}\text{(S-12.26)}$$

Thus

$$|B_y| = \frac{4\pi \times 10^{-7} \times 10^3 \times 5}{2\pi \times (30^2 + 5^2/2^2)}\,\text{T} \simeq 10^{-6}\,\text{T} \,,$$
$$\text{(S-12.27)}$$

and

$$|E_x| \simeq 3 \times 10^2\,\text{V/m} \,. \qquad \text{(S-12.28)}$$

For a comparison, the average magnetic field at the Earth surface is $\sim 5 \times 10^{-5}$ T, while the electric field is $\sim 1.5 \times 10^2$ V/m. Possible screening effects by the Earth's surface have been neglected.

S-12.3 TEM and TM Modes in an "Open" Waveguide

(a) Inserting (12.3) into the wave equation for **B**

$$\left(\nabla^2 - \frac{1}{c^2}\partial_t^2\right) B_z = 0 , \qquad (S\text{-}12.29)$$

and recalling that $\partial_z B_z = 0$, we obtain the following relation between k_x, k_y and ω

$$k_x^2 + k_y^2 - \frac{\omega^2}{c^2} = 0 . \qquad (S\text{-}12.30)$$

(b) The electric field of the wave can be obtained from

$$\partial_t \mathbf{E} = c\nabla \times \mathbf{B} = c\,(\hat{\mathbf{x}}\,\partial_y - \hat{\mathbf{y}}\,\partial_x)\,B_z,$$
$$-i\omega \mathbf{E} = cB_0\left[-\hat{\mathbf{x}}\,k_y\sin(k_y y) - \hat{\mathbf{y}}\,ik_x\cos(k_y y)\right]\mathrm{e}^{ik_x x - i\omega t} , \qquad (S\text{-}12.31)$$

which leads to

$$E_x = -i\,\frac{k_y c}{\omega}\,B_0\sin(k_y y)\,\mathrm{e}^{ik_x x - i\omega t} , \qquad (S\text{-}12.32)$$

$$E_y = \frac{k_x c}{\omega}\,B_0\cos(k_y y)\,\mathrm{e}^{ik_x x - i\omega t} . \qquad (S\text{-}12.33)$$

(c) The parallel component $E_\parallel$ of the electric field **E** must vanish at the boundary with a perfectly conducting surface, thus we must have $E_x(y = \pm a/2) = 0$. This implies that $\sin(k_y a/2) = 0$, and $k_y a = 2m\pi$, with $m \in \mathbb{N}$. By substitution into (S-12.30) we obtain

$$\omega^2 = k_x^2 c^2 + \left(\frac{\pi c}{a}\right)^2 (2m)^2 . \qquad (S\text{-}12.34)$$

The $m = 0$ mode corresponds to $E_x = 0$ and to E_y and B_z independent of y. The fields are thus uniform over any cross-section of the waveguide parallel to the yz plane, and we have $\omega = k_x c$. This is the TEM mode typical of transmission lines. The $m = 1$ mode has frequency

$$\omega = \sqrt{k_x^2 c^2 + \left(\frac{2\pi c}{a}\right)^2} > \frac{2\pi c}{a} \equiv \omega_{\text{co}} , \qquad (S\text{-}12.35)$$

where $\omega_{\text{co}} \equiv 2\pi c/a$ is the cut-off frequency.

(d) The energy flux is given by Poynting's vector, parallel to the $z = 0$ plane,

$$\mathbf{S} = \frac{c}{4\pi}\mathbf{E} \times \mathbf{B} = \frac{c}{4\pi}(E_y B_z\,\hat{\mathbf{x}} - E_x B_z\,\hat{\mathbf{y}}) . \qquad (S\text{-}12.36)$$

By averaging over one full cycle we find $\langle S_y \rangle = 0$, i.e., there is no net energy flux along y. Averaging S_x over one cycle we obtain

$$\langle S_x \rangle = \frac{c^2}{8\pi} \frac{k_x}{\omega} B_0^2 \cos^2(k_y y) \,. \tag{S-12.37}$$

The group velocity of the wave is

$$v_g = \partial_k \omega = \frac{k_x c^2}{\sqrt{k_x^2 c^2 + \omega_{co}^2}} = \frac{k_x c^2}{\omega} \,, \tag{S-12.38}$$

thus we can also write

$$\langle S_x \rangle = v_g \frac{B_0^2}{8\pi} \cos^2(k_y y) \,. \tag{S-12.39}$$

S-12.4 Square and Triangular Waveguides

(a) The electric field must satisfy the wave equation in vacuum

$$\left(\nabla^2 - \frac{1}{c^2} \partial_t^2 \right) \mathbf{E} = 0 \,, \tag{S-12.40}$$

and, substituting (12.4) for $\mathbf{E}$, we obtain the time-independent Helmoltz's equation for the only nonzero component of the electric field, $\tilde{E}_x$,

$$\left(\partial_x^2 + \partial_y^2 - k_z^2 + \frac{\omega^2}{c^2} \right) \tilde{E}_x = 0 \,. \tag{S-12.41}$$

In vacuum we must also have $\nabla \cdot \mathbf{E} = 0$, this condition is automatically satisfied if we assume that $\tilde{E}_x$ is independent of x, $\tilde{E}_x = \tilde{E}_x(y)$, and (S-12.41) reduces to

$$\left(\partial_y^2 - k_z^2 + \frac{\omega^2}{c^2} \right) \tilde{E}_x(y) = 0 \,. \tag{S-12.42}$$

According to the boundary conditions, the parallel component of $\mathbf{E}$ must be zero at the perfectly reflecting walls of the waveguide $y = 0$ and $y = a$. This condition is satisfied if we assume

$$\tilde{E}_x(y) = E_{0x} \sin(k_y y) \,, \quad \text{with} \quad k_y = n \frac{\pi}{a} \,, \quad n = 1, 2, 3, \ldots \,, \tag{S-12.43}$$

where E_{0x} is an arbitrary, constant amplitude. The electric field of our $\hat{\mathbf{x}}$ polarized wave can thus be written

$$\mathbf{E} = \hat{\mathbf{x}}\,\tilde{E}_x(y)\,e^{ik_z z - i\omega t} = \hat{\mathbf{x}}\,E_{0x}\sin\left(\frac{n\,\pi}{a}y\right)e^{ik_z z - i\omega t} . \tag{S-12.44}$$

Substituting (S-12.43) for $\tilde{E}_x$ into (S-12.42) leads to

$$\left(-k_y^2 - k_z^2 + \frac{\omega^2}{c^2}\right)E_{0x} = 0 , \tag{S-12.45}$$

which, disregarding the trivial case $E_{0x} = 0$, is true only if

$$k_y^2 + k_z^2 - \frac{\omega^2}{c^2} = 0 , \quad \text{or} \quad k_z = \sqrt{\frac{\omega^2}{c^2} - k_y^2} = \sqrt{\frac{\omega^2}{c^2} - n^2\frac{\pi^2}{a^2}} . \tag{S-12.46}$$

The wave can propagate only if k_z is real, thus we must have

$$\omega > n\,\frac{\pi c}{a} . \tag{S-12.47}$$

The cutoff frequency ω_a is the lowest value of ω at which wave propagation occurs. Since we must have $n \geqslant 1$, we have $\omega_a = \pi c/a$. If we choose a frequency such that $\pi c/a < \omega < 2\pi c/a$, only the $n = 1$ mode can propagate in the guide.

The cross-section of the waveguide being square, the conditions for a $\hat{\mathbf{y}}$ polarized TE wave are obtained by interchanging the roles of x and y in all the above formulae, and the electric field is

$$\mathbf{E} = \hat{\mathbf{y}}\,E_{0y}\sin(k_x x)\,e^{ik_z z - i\omega t} = \hat{\mathbf{y}}\,E_{0y}\sin\left(\frac{m\,\pi}{a}x\right)e^{ik_z z - i\omega t} , \tag{S-12.48}$$

with, again, E_{0y} an arbitrary amplitude, $m = 1, 2, 3, \ldots$ and the same dispersion relation as between ω and k_z as above. Modes with $m = n$ are degenerate, sharing the same wavevector k_z.

In general, a monochromatic TE wave propagating in the guide will be a superposition of the two polarizations. The electric field will be

$$\mathbf{E} = \left[\hat{\mathbf{x}}\,E_{0x}\sin\left(\frac{n\pi}{a}y\right) + \hat{\mathbf{y}}\,E_{0y}\sin\left(\frac{n\pi}{a}x\right)\right]e^{ik_z z - i\omega t} . \tag{S-12.49}$$

(b) In the case of the triangular waveguide, the parallel component of the electric field $\mathbf{E}$ must be zero on the three $x = 0$, $y = 0$, and $y = x$ planes. A field of the form (S-12.49) already satisfies the boundary conditions at the $x = 0$ and $y = 0$ planes. The additional condition at the $y = x$ plane is $\mathbf{E}(x, x) \cdot \hat{\mathbf{n}} = 0$, where $\hat{\mathbf{n}} = (-1, 1, 0)/\sqrt{2}$ is the unit vector perpendicular to the $y = x$ plane. Thus

$$E_{0x}\sin\left(\frac{n\pi}{a}x\right) - E_{0y}\sin\left(\frac{n\pi}{a}x\right) = 0 , \tag{S-12.50}$$

which is satisfied if $E_{0y} = E_{0x} \equiv E_0$, so that we eventually obtain

$$\mathbf{E} = E_0 \left[\hat{\mathbf{x}} \sin \left(\frac{n\pi}{a} y \right) + \hat{\mathbf{y}} \sin \left(\frac{n\pi}{a} x \right) \right] e^{ik_z z - i\omega t} . \qquad \text{(S-12.51)}$$

S-12.5   Waveguide Modes as an Interference Effect

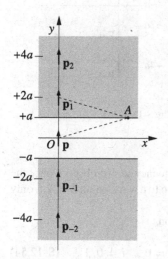

Fig. S-12.4

(a) The electrostatic potential ϕ must be zero on the two conducting planes at $y = \pm a$, and the electric fields at $y = a^-$ and $y = -a^+$ must be perpendicular to their surfaces (parallel to $\hat{\mathbf{y}}$). The real dipole $\mathbf{p}$ is located at the origin of our coordinate system, thus, we need an image dipole equal to $\mathbf{p}$, located at $(0, 2a, 0)$ and represented by $\mathbf{p}_1$ in Fig. S-12.4, in order to fulfill these conditions at the generic point A of the $y = +a$ plane. Analogously, the real dipole $\mathbf{p}$ requires a further image dipole $\mathbf{p}$ located at $(0, -2a, 0)$, represented by $\mathbf{p}_{-1}$ in Fig. S-12.4, in order to fulfill the conditions at the $y = -a$ conducting plane. But now the three dipoles $\mathbf{p}$, $\mathbf{p}_1$, and $\mathbf{p}_{-1}$ together do not generate a potential equal to zero on either plane. We can readjust the potential at $y = +a$ by adding a new image dipole equal to $\mathbf{p}$, symmetrical to $\mathbf{p}_{-1}$, at $(0, 4a, 0)$, represented by $\mathbf{p}_2$. But this requires adding a further image dipole $\mathbf{p}_{-2}$, and so on. Thus, the exact solution requires two infinite sets of equal image dipoles, $\mathbf{p}_n$ and $\mathbf{p}_{-n}$, with $n = 1, 2, 3, \ldots$, located respectively at $(0, 2na, 0)$ and $(0, -2na, 0)$. The resulting electrostatic potential between the plates is finite because, for high n values, the contribution of $\pm n$th dipole is proportional to $(2na)^{-2}$.

(b) In order to fulfill the boundary conditions, all the image dipoles must oscillate in phase with the real dipole. Consider the radiation emitted by each dipole in the $\hat{\mathbf{n}} \equiv (\sin \theta, \cos \theta, 0)$ direction following we consider wavevectors lying in the $z = 0$ plane, but our considerations apply to wavevectors lying in any plane containing the y axis, due to the rotational symmetry of the problem. The optical path difference between the waves emitted by two neighboring dipoles (real or images) is $\Delta \ell = 2a \cos \theta$, as shown in Fig. S-12.5 for the case of the real dipole $\mathbf{p}$ and the image $\mathbf{p}_1$. This corresponds to a phase difference $\Delta \varphi = k \Delta \ell$, and the condition for constructive interference is

$$k \Delta \ell = \frac{2\omega a}{c} \cos \theta = 2\pi m ,$$

$$\theta = \arccos \left(m \frac{\pi c}{\omega a} \right) , \qquad \text{(S-12.52)}$$

with $m = 0, 1, 2, \dots$. Due to the mirror symme-
try of the system for reflections through the $y = 0$ plane (actually, *antisymmetry*, since all dipoles
are inverted by the reflection), if an angle θ satis-
fies (S-12.52) for constructive interference, so does
$\pi - \theta$. In other words, at large distance from the
oscillating dipole, each interference order $m > 0$
corresponds to the superposition of two waves with
wavevectors $\mathbf{k}_\pm \equiv (\sin\theta, \pm\cos\theta, 0)\,\omega/c$, respectively.

The $m = 0$ condition corresponds to $\theta = \pi/2$,
and the waves travels along the x axis. For $m > 0$,
we can write

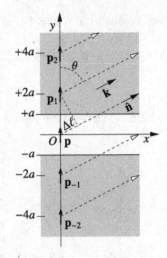

$$k_x = \frac{\omega}{c}\sin\theta = \frac{\omega}{c}\sqrt{1 - \cos^2\theta} = \frac{\omega}{c}\sqrt{1 - \left(m\frac{\pi c}{\omega a}\right)^2}$$

$$= \sqrt{\frac{\omega^2}{c^2} - \left(m\frac{\pi}{a}\right)^2}\,, \qquad\qquad \text{(S-12.53)}$$

Fig. S-12.5

and k_x is real only if $\omega > m\pi c/a$. Thus, given a frequency ω, we observe only the
modes with $m < \omega a/(\pi c)$. If $\omega < \pi c/a$, corresponding to a wavelength $\lambda > 2a$, only
the mode $m = 0$ can propagate.

(c) Both magnetic fields must satisfy the wave equation,

$$(c^2\nabla^2 - \partial_t^2)\,\mathbf{B}_i = (c^2\partial_x^2 + c^2\partial_y^2 - \partial_t^2)\,\mathbf{B}_i = 0\,, \quad i = 0, 1\,, \quad \text{(S-12.54)}$$

from which we obtain, denoting by $\mathbf{k}_0$ and $\mathbf{k}_1$ the respective wavevectors,

$$k_{0x}^2\,c^2 = \omega^2\,, \qquad k_{1x}^2\,c^2 + k_{1y}^2\,c^2 = \omega^2\,. \qquad \text{(S-12.55)}$$

(d) Assuming electric fields of the form $\mathbf{E} = \tilde{\mathbf{E}}\,e^{-i\omega t}$, where $\tilde{\mathbf{E}}$ depends on the space
coordinates only, Maxwell's equation in vacuum, $\partial_t\mathbf{E} = c\nabla \times \mathbf{B}$, gives

$$-i\omega\mathbf{E} = c(\hat{\mathbf{x}}\,\partial_y B_z - \hat{\mathbf{y}}\,\partial_x B_z)\,. \qquad \text{(S-12.56)}$$

For the wave of type "0" we obtain

$$\mathbf{E}_0 = \hat{\mathbf{y}}\,\frac{k_{0x}c}{\omega}\,B_0\,e^{ik_{0x}x - i\omega t} = \hat{\mathbf{y}}\,B_0\,e^{ik_{0x}x - i\omega t}\,. \qquad \text{(S-12.57)}$$

For the wave of type "1" we obtain

$$\mathbf{E}_1 = \frac{ic}{\omega}B_1\left[\hat{\mathbf{y}}\,k_{1x}\cos(k_{1y}y) - \hat{\mathbf{x}}\,ik_{1y}\sin(k_{1y}y)\right]e^{ik_{1x}x - i\omega t}\,. \qquad \text{(S-12.58)}$$

(e) The "type-0" wave has the three vectors $\mathbf{E}_0$, $\mathbf{B}$ and $\mathbf{k}$ perpendicular to one another, analogously to a plane wave in the free space (TEM mode). Further, $\mathbf{E}_0$ is perpendicular to the two conducting surfaces, automatically satisfying the boundary conditions. Thus, the frequency ω and the wavevector $\mathbf{k} = \hat{\mathbf{x}} k_{0x}$, with $k_{0x} = \omega/c$, are subject to no constraint.

On the other hand, the electric field of the "type-1" wave has a component E_x parallel to the two conducting surfaces, in addition to the transverse E_y component (the mode is TM rather than TEM). The boundary conditions at $y = \pm a$ require that $E_x(y = \pm a) = 0$. Thus we must have $\sin(\pm k_y a) = 0$, or $k_y = m\pi/a$, with $m = 1, 2, 3, \ldots$, leading to

$$k_x = \sqrt{\frac{\omega^2}{c^2} - \left(m\frac{\pi}{a}\right)^2}. \tag{S-12.59}$$

The m-th mode can propagate only if the corresponding k_x is real, and has a lower cut-off frequency $\omega_{co}(m) = 2\pi mc/a$.

A comparison to point (b) shows that the type-0 wave (TEM mode) corresponds to the $m = 0$ interference order, while the type-1 waves (TM modes) correspond to the > 0 interference orders. Actually, more precisely, we need not single dipoles, but "dipole layers", spread parallel to the z axis, in order to generate waves with fields independent of z. If the real dipole of points (a) and (b) is parallel, rather than perpendicular, to the conducting planes, the different boundary conditions would lead to TE, rather than TM modes [1].

S-12.6   Propagation in an Optical Fiber

(a) The electric field (12.6) corresponds to the sum of two plane waves of the same frequency and different wavevectors, $\mathbf{k}_1$ and $\mathbf{k}_2$, propagating in the medium. For both waves the dispersion relation is $\omega = kc/\mathsf{n}$, where $\mathsf{n} = \mathsf{n}(\omega)$ is the refractive index of the medium. Both waves impinge on the medium-vacuum interface at the angle θ, and the condition for total reflection is, according to Snell's law,

$$\sin\theta > \frac{1}{\mathsf{n}}. \tag{S-12.60}$$

(b) The internal reflections at the $y = \pm a/2$ planes turn the wave of type "1" into a wave of type "2", and vice versa. Thus the field amplitudes of the two waves at the interface are related by the amplitude reflection coefficient r

$$E_2(x, y = +a/2, t) = \mathsf{r}E_1(x, y = +a/2, t),$$
$$E_1(x, y = -a/2, t) = \mathsf{r}E_2(x, y = -a/2, t). \tag{S-12.61}$$

For total reflection there is no transmission of energy through the $y = \pm a/2$ planes, thus the amplitudes of the incident and the reflected fields must be equal but for a change of phase. For S-polarization ($\mathbf{E}$ parallel to the interface, as in our case) r is written, according to the Fresnel equations,

$$ r = \frac{n \cos \theta - i\sqrt{n^2 \sin^2 \theta - 1}}{n \cos \theta + i\sqrt{n^2 \sin^2 \theta - 1}} , \tag{S-12.62}$$

and, if $n \sin \theta > 1$, the square roots are real and $|r| = 1$. Thus we can write

$$ r = e^{i\delta} = \cos \delta + i \sin \delta , \qquad \tan \frac{\delta}{2} = -\frac{\sqrt{n^2 \sin^2 \theta - 1}}{n \cos \theta} . \tag{S-12.63}$$

Substituting $r = e^{i\delta}$ into (S-12.61) we obtain the following conditions at the $y = \pm a/2$ planes

$$ E_2 e^{-ik_y a/2} = E_1 e^{+ik_y a/2} e^{i\delta} , \qquad E_1 e^{-ik_y a/2} = E_2 e^{+ik_y a/2} e^{i\delta} . \tag{S-12.64}$$

By calculating the determinant of the homogeneous system for E_1 and E_2 we obtain the condition

$$ 1 = e^{2i(k_y a + \delta)} , \tag{S-12.65}$$

true if

$$ 2k_y a + 2\delta = 2m\pi , \qquad m = 0, 1, 2, \ldots . \tag{S-12.66}$$

The implicit relation determining the allowed frequencies is

$$ k_x^2 = \frac{\omega^2}{c^2} n^2 - k_y^2 > 0 . \tag{S-12.67}$$

If $n \sin \theta \gg 1$ then $\delta \simeq -2\theta$, and if $\theta \to \pi/2$ then

$$ k_y \to (m+1)\frac{\pi}{a} . \tag{S-12.68}$$

(c) All the above results are valid also for P-polarization, where the electric field of the wave lies in the xy plane. Only (S-12.62) must be replaced by

$$ r_{\parallel} = e^{i\delta_{\parallel}} = \frac{-n^2 \cos \theta + i\sqrt{\sin^2 \theta - n^2}}{n^2 \cos \theta + i\sqrt{\sin^2 \theta - n^2}} , \tag{S-12.69}$$

corresponding to a different dependence of r and δ on θ.

S-12.7 Wave Propagation in a Filled Waveguide

(a) The electric field $\mathbf{E}$ of a monochromatic EM wave of frequency ω propagating in a medium of refractive index $n = n(\omega)$ satisfies Helmholtz's equation

$$\left(\nabla^2 + n^2(\omega)\frac{\omega^2}{c^2}\right)\mathbf{E} = 0 . \qquad (S\text{-}12.70)$$

We are considering a TE mode with $\mathbf{E} = \hat{\mathbf{z}}\,E_z(y)\,e^{ikx-i\omega t}$, thus we have

$$\left(\partial_y^2 - k^2 + n^2(\omega)\frac{\omega^2}{c^2}\right)E_z(y) = 0 , \qquad (S\text{-}12.71)$$

whose general solution has the form s $E_z(y) = A\cos(qy) + B\sin(qy)$, with A and B two arbitrary constants. The electric field being parallel to the conducting walls at $y = \pm a/2$, the boundary conditions are $E_z(y = \pm a/2) = 0$, from which we obtain

$$E_z(y) = E_0 \begin{cases} \cos(q_n y) , & n = 1, 3, 5 \ldots \\ \sin(q_n y) , & n = 2, 4, 6 \ldots \end{cases} , \qquad q_n = n\frac{\pi}{a} , \qquad (S\text{-}12.72)$$

and (S-12.71) turns into

$$q_n^2 + k^2 - n^2(\omega)\frac{\omega^2}{c^2} = 0 . \qquad (S\text{-}12.73)$$

The wave can propagate only if k is real, i.e., if $\omega > q_n c \equiv \omega_n$.

In the case of a plasma

$$q_n^2 + k^2 - \frac{\omega^2 - \omega_p^2}{c^2} = 0 , \qquad (S\text{-}12.74)$$

and the cut-off frequencies are

$$\omega_n' = \sqrt{q_n^2 c^2 + \omega_p^2} . \qquad (S\text{-}12.75)$$

(b) The incident wave must be in the $n = 1$ mode, and its electric field is

$$\mathbf{E_i} = \hat{\mathbf{z}}\,E_0 \cos(q_1 y)\,e^{ik_1 x - i\omega t} , \qquad (S\text{-}12.76)$$

where $k_1 = \sqrt{\omega^2/c^2 - q_1^2}$. The total electric field is the sum of the incident field $\mathbf{E_i}$ and the reflected field $\mathbf{E_r}$ for $x < 0$, while only the transmitted field $\mathbf{E_t}$ is present in the $x > 0$ region. The boundary condition at $x = 0$ is $(E_{iz} + E_{rz})|_{x=0} = E_{tz}|_{x=0}$, thus all the waves must have the same dependence on t and y. The total field must thus be

$$E_z = \begin{cases} \left(E_0\, e^{ik_1 x} + E_r\, e^{-ik_1 x}\right) \cos(q_1 y)\, e^{-i\omega t}\,, & x < 0\,, \\ E_t \cos(q_1 y)\, e^{ik_t x - i\omega t}\,, & x > 0\,, \end{cases} \tag{S-12.77}$$

where $k_t = \sqrt{n(\omega)\,\omega^2/c^2 - q_1^2}$. The boundary condition on the electric field yields

$$E_0 + E_r = E_t\,. \tag{S-12.78}$$

In addition, the magnetic field must be also continuous at $x = 0$. From $\partial_t \mathbf{B} = -c\boldsymbol{\nabla} \times \mathbf{E}$ we obtain

$$B_x = \frac{i\pi c}{\omega a} \times \begin{cases} \left(E_0\, e^{ik_1 x} + E_r\, e^{-ik_1 x}\right) \sin(q_1 y)\, e^{-i\omega t}\,, & x < 0\,, \\ E_t \sin(q_1 y)\, e^{ik_t x - i\omega t}\,, & x > 0\,, \end{cases} \tag{S-12.79}$$

$$B_y = -\frac{c}{\omega} \times \begin{cases} k\left(E_0\, e^{ik_1 x} - E_r\, e^{-ik_1 x}\right) \cos(q_1 y)\, e^{-i\omega t}\,, & x < 0\,, \\ k_t E_t \cos(q_1 y)\, e^{ik_t x - i\omega t}\,, & x > 0\,, \end{cases} \tag{S-12.80}$$

We notice that the continuity of B_x is ensured by the condition $E_0 + E_r = E_t$, while the continuity of B_y yields

$$k_1(E_0 - E_r) = k_t E_t\,. \tag{S-12.81}$$

Eventually, we obtain

$$E_r = \frac{k_1 - k_t}{k_1 + k_t} E_0\,, \qquad E_t = \frac{2k_1}{k_1 + k_t} E_0\,, \tag{S-12.82}$$

which are identical to Fresnel's formulas for S-polarization. In fact, the field of the incoming wave (S-12.76) can be written as

$$\mathbf{E}_i = \hat{\mathbf{z}}\, E_0 \cos(q_1 y)\, e^{ik_1 x - i\omega t} = \hat{\mathbf{z}}\, \frac{E_0}{2} \left(e^{iq_1 y} + e^{-iq_1 y}\right) e^{ik_1 x - i\omega t}$$

$$= \hat{\mathbf{z}}\, \frac{E_0}{2} e^{ik_1 x + iq_1 y - i\omega t} + \hat{\mathbf{z}}\, \frac{E_0}{2} e^{ik_1 x - iq_1 y - i\omega t}\,, \tag{S-12.83}$$

which is the superposition of two z-polarized plane waves of equal amplitude, and wavevectors of equal magnitude, but opposite y component, $\mathbf{k} = \hat{\mathbf{x}}\, k_1 \pm \hat{\mathbf{y}}\, q_1$. Thus both plane waves impinge on the vacuum-medium interface at the same incidence angle $|\theta| = \arctan(q_1/k_1)$.

S-12.8   Schumann Resonances

(a) Substituting the electric field (12.9) into the periodic boundary conditions (12.8) we obtain

$$e^{ik_x L} = 1, \qquad e^{ik_y L} = 1, \tag{S-12.84}$$

solved by

$$k_x = m\frac{2\pi}{L}, \qquad k_y = n\frac{2\pi}{L}, \qquad m, n = 0, 1, 2, \dots \tag{S-12.85}$$

where m and n are not allowed to be zero simultaneously, and $L = 2\pi R_\oplus$. Since the wave equations gives us $\omega^2 = k^2 c^2$, we have

$$\omega^2 = \left(\frac{c}{R_\oplus}\right)^2 (m^2 + n^2). \tag{S-12.86}$$

The lowest frequency corresponds to $m = 1, n = 0$ or $m = 0, n = 1$, and its value is

$$\nu_{min} = \frac{\omega_{min}}{2\pi} = \frac{\omega_{10}}{2\pi} = \frac{c}{2\pi R_\oplus} \simeq 7.5\,\text{s}^{-1}, \tag{S-12.87}$$

corresponding to a wavelength $\lambda_{max} = 2\pi R_\oplus \simeq 40\,000$ km, the length of a great circle of the Earth. The experimentally observed value is $\nu_{min} \simeq 8\,\text{s}^{-1}$.

(b) An ohmic conductor can be considered as perfectly reflecting at a frequency ω if its conductivity $\sigma(\omega)$, assumed to be real, fulfills the condition $\sigma(\omega) \gg \omega/4\pi k_e$, where $k_e = 1$ in Gaussian units, and $k_e = 1/(4\pi\varepsilon_0)$ in SI units. Heuristically, the condition corresponds to the conduction current $\mathbf{J}$ being much larger than Maxwell's displacement current. Since $\varepsilon_0 = 8.854 \times 10^{-12}$ SI units, and $\sigma/\omega \approx 0.6\,\text{s}\,\Omega^{-1}\text{m}^{-1}$, sea water can be considered as a perfect conductor in the frequency range of the Schumann resonances. In Gaussian units, the low-frequency conductivity of sea water is $\sigma \simeq 4 \times 10^{10}\,\text{s}^{-1}$.

A discussion of Schumann resonances based on a "realistic" spherical geometry can be found in Reference [2], Section 8.9 and Problem 8.7. Nevertheless, our simplified approach reveals the essential point that the characteristic length L of the system, which determines the maximum wavelength for a standing wave ($\lambda \approx L$), is the Earth's circumference, rather than the height of the ionosphere above the the Earth's surface.

S-12.9   A One-Dimensional Cavity Fed by an Antenna (1)

(a) In the absence of the antenna our cavity has normal modes with wavelengths, angular frequencies and wavevectors

$$\lambda_m = \frac{2a}{m} , \quad \omega_m = m\frac{\pi c}{a} , \quad k_m = m\frac{\pi}{a} , \tag{S-12.88}$$

with $m \geq 1$ an integer, so that the cavity length a is an integer number of half wavelengths. These modes are steady-state solutions for the electromagnetic field in the cavity, i.e., permanent oscillations if perfectly conducting walls are assumed. The polarization of these modes can be any superposition of y and z. The amplitudes of the electric field E and the magnetic field B for the mode with $m = 1$ are shown in Fig. S-12.6. In the presence of the antenna with driving current in the $\hat{y}$ direction, the electric and magnetic fields have only the $E_y = E_y(x,t)$ and $B_z = B_z(x,t)$ components, respectively. Further, we must have $E_y(\pm a/2, t) \equiv 0$ because the cavity walls are perfectly conducting. The presence of the surface current $K(t)$ on the $x = 0$ plane implies that B_z has a jump discontinuity at $x = 0$, see, for instance, Problems (8.5) and (11.4),

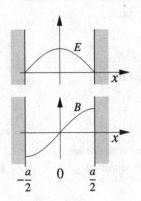

Fig. S-12.6

$$B_z(0^+, t) - B_z(0^-, t) = \frac{4\pi}{c}K(t) , \tag{S-12.89}$$

as shown in Fig. S-12.7, while E_y is continuous at $x = 0$. The general steady solutions for the fields have the forms $E_y(x,t) = E_y(x)\,\mathrm{e}^{-i\omega t}$ and $B_z(x,t) = B_z(x)\,\mathrm{e}^{-i\omega t}$, from now on we shall omit the time dependence factor $\mathrm{e}^{-i\omega t}$, common to all fields. The solutions in the $0 < x < a/2$ region can be written as

$$E_y(x) = E_+\mathrm{e}^{ikx} + E_-\mathrm{e}^{-ikx} , \quad (x > 0) \quad \text{(S-12.90)}$$
$$B_z(x) = E_+\mathrm{e}^{ikx} - E_-\mathrm{e}^{-ikx} , \quad (x > 0) \quad \text{(S-12.91)}$$

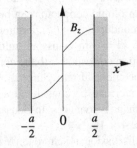

Fig. S-12.7

where $k = \omega/c$, and E_+ and E_- are constants to be determined. Due to the boundary conditions at the walls and at the antenna we can assume that E_y is an even function of x, while B_z is an odd function, so that in the $-a/2 < x < 0$ region we have

$$E_y(x) = E_+\mathrm{e}^{-ikx} + E_-\mathrm{e}^{ikx} , \quad B_z(x) = -E_+\mathrm{e}^{-ikx} + E_-\mathrm{e}^{ikx} . \quad (x < 0) \tag{S-12.92}$$

The boundary conditions at the walls and at the antenna are

$$E_+\mathrm{e}^{ika/2} + E_-\mathrm{e}^{-ika/2} = 0 , \tag{S-12.93}$$

$$E_+ - E_- = \frac{2\pi}{c}K_0 , \tag{S-12.94}$$

leading to the following values for E_+ and E_-

$$E_\pm = \pm \frac{\pi}{c} K_0 \frac{e^{\mp ika/2}}{\cos(ka/2)} . \tag{S-12.95}$$

The singularities for $ka/2 = (n + 1/2)\pi$, with $n \geq 0$ an integer, correspond to the normal cavity modes of wavelengths $\lambda_n = 2\pi/k_n = a/(n+1)$. These are the normal modes of (S-12.88) with m odd, with the difference that here $B_z(x, t)$ is still an odd function of x, but discontinuous at $x = 0$ according to (S-12.89) and as shown in Fig. S-12.7. The singularities appear because these modes do not require the antenna in order to oscillate, therefore, in the absence of losses, the power supplied by the antenna is continuously added to the oscillation energy of the cavity The equivalents of the free-cavity modes with m even cannot be excited by the antenna because their electric fields are odd functions of x, thus with $E_y(0, t) \equiv 0$ at the location of the antenna. In the presence of the antenna the electric field is symmetrized, see Fig. S-12.8, but it remains zero at the location of the antenna.

(b) In order to evaluate the instantaneous power P_s fed by the antenna per unit surface we first evaluate the real part of the electric field at $x = 0$ (note that, without loss of generality, we can assume K_0 to be real)

$$E_y(0, t) = \text{Re}\left[\frac{\pi}{c}\frac{K_0}{\cos(ka/2)}e^{-i\omega t}\left(e^{-ika/2} - e^{ika/2}\right)\right] = \frac{\pi}{c}K_0\frac{\sin(ka/2)}{\cos(ka/2)}\text{Re}[-2ie^{-i\omega t}]$$

$$= -\frac{2\pi}{c}K_0 \tan\left(\frac{ka}{2}\right)\sin\omega t , \tag{S-12.96}$$

and then evaluate the power per unit surface as

$$P_s(t) = K(t)\, E_y(0, t) = (K_0 \cos\omega t)\left[-\frac{2\pi}{c}K_0 \tan\left(\frac{ka}{2}\right)\sin\omega t\right]$$

$$= -\frac{\pi}{c}K_0^2 \tan\left(\frac{ka}{2}\right)\sin 2\omega t . \tag{S-12.97}$$

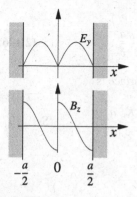

$-\dfrac{a}{2} \quad 0 \quad \dfrac{a}{2}$

Fig. S-12.8

The generator exchanges energy with the field at a 2ω rate to sustain the non-normal modes. The time average of the fed power is $\langle P_s(t)\rangle = 0$ because there is no dissipation in the system. Note that $P_s(t) \equiv 0$ for modes such that $\tan(ka/2) = 0$, corresponding to $\lambda = a/j$ with $j \geq 1$ an integer. These modes have $E_y(0, t) \equiv 0$, thus a node at the location of the antenna and cannot absorb energy. The amplitudes of the electric and magnetic fields of the first of these modes, corresponding to $ka/2 = \pi$, is shown in Fig. S-12.8. Note that $E_y(x)$ is a continuous and symmetric function of x, as required, but its derivative $\partial E_y/\partial x$ is discontinuous at $x = 0$.

S-12.10   A One-Dimensional Cavity Fed by an Antenna (2)

(a) Due to the symmetry of the problem it is sufficient to consider the fields for $x \geq 0$. For $x > a/2$, i.e., inside the right wall, we have, see, for instance, Problems (6.9), (6.14) and (11.1),

$$E_y(x) = E_s e^{-(1-i)(x-a/2)/\delta} , \qquad B_z(x) = \frac{i+1}{k\delta} E_s e^{-(1-i)(x-a/2)/\delta} \qquad (x > a/2) ,$$
$$\text{(S-12.98)}$$

where E_s is the electric field at the surface, to be determined from the continuity conditions for E_y and B_z at $x = a/2$. The boundary condition at the antenna is still (S-12.89), thus we have

$$E_+ e^{ika/2} + E_- e^{-ika/2} = E_s , \qquad\qquad\qquad \text{(S-12.99)}$$

$$E_+ e^{ika/2} - E_- e^{-ika/2} = \frac{i+1}{k\delta} E_s , \qquad\qquad \text{(S-12.100)}$$

$$E_+ - E_- = \frac{2\pi}{c} K_0 . \qquad\qquad\qquad\qquad \text{(S-12.101)}$$

We first apply these equations to the case $ka = \pi(2n+1)$, corresponding to the normal modes of the cavity. Since $e^{\pm i\pi(n+1/2)} = \pm(-1)^n i$, the system reduces to

$$(-1)^n i E_+ - (-1)^n i E_- = E_s , \qquad\qquad \text{(S-12.102)}$$

$$(-1)^n i E_+ + (-1)^n i E_- = \frac{i+1}{k\delta} E_s , \qquad\qquad \text{(S-12.103)}$$

$$E_+ - E_- = \frac{2\pi}{c} K_0 , \qquad\qquad\qquad\qquad \text{(S-12.104)}$$

with solutions

$$E_\pm = \pm \frac{\pi}{c} K_0 \left(1 \pm \frac{i+1}{k\delta} \right) . \qquad\qquad \text{(S-12.105)}$$

The electric field at $x = 0$ is

$$E(0) = \frac{2\pi}{c} K_0 \frac{i+1}{k\delta} = \frac{2(i+1)}{(2n+1)} \frac{K_0}{c} \frac{a}{\delta} , \qquad \text{(S-12.106)}$$

and the average power supplied by the antenna per unit surface is

$$\langle P_s(t) \rangle = \frac{1}{2} \text{Re}(K_0 E_0^*) = -\frac{\pi}{c} \frac{K_0^2}{k\delta} = -\frac{1}{2n+1} \frac{K_0^2}{c} \frac{a}{\delta} . \qquad \text{(S-12.107)}$$

(b) In the case of an arbitrary value of $\omega = kc$, we have to solve the linear system of three equations in the three unknowns E_+, E_- and E_s given by (S-12.99)–(S-12.101). The determinant is

$$\Delta = -\frac{2(1+i)}{k\delta}\cos(ka/2) + 2i\sin(ka/2) , \tag{S-12.108}$$

and applying Cramer's rule we eventually obtain

$$E_\pm = \pm\frac{2\pi}{c\Delta}K_0\left(\frac{1+i}{k\delta}\mp 1\right)e^{\mp ika/2} , \qquad E_s = -\frac{4\pi K_0}{c\Delta} . \tag{S-12.109}$$

Thus we have

$$E_y(0) = -\frac{2\pi i}{c}K_0\frac{\sin(ka/2) + k\delta\frac{1+i}{2}\cos(ka/2)}{\cos(ka/2) - k\delta\frac{1+i}{2}\sin(ka/2)} . \tag{S-12.110}$$

The average power supplied by the antenna per unit surface is thus

$$\langle P_s(t)\rangle = \frac{1}{2}\text{Re}(K_0 E_0^*) = \frac{\pi}{c}K_0^2\,\text{Im}\left[\frac{\sin(ka/2) + k\delta\frac{1+i}{2}\cos(ka/2)}{\cos(ka/2) - k\delta\frac{1+i}{2}\sin(ka/2)}\right] \tag{S-12.111}$$

$$= -\frac{\pi}{2c}K_0^2\frac{k\delta}{[\cos(ka/2) + (k\delta/2)\sin(ka/2)]^2 + (k\delta/2)^2\sin^2(ka/2)} ,$$

and is non-vanishing and finite for all modes, including the normal modes, because of dissipation inside the walls. For resonant modes with $\cos(ka/2) = 0$ and $|\sin(ka/2)| = 1$, we obtain $\langle P_s(t)\rangle = (\pi/c)\,(K_0^2/k\delta)$, which diverges for $\delta \to 0$ since the resonance is not damped by dissipation at this limit. The behavior of (S-12.111) for the first three resonances in the case of a cavity with $a = 4$ cm is shown for $\delta = 6 \times 10^{-3}$ cm, a typical skin depth for Al in the microwave region, in Fig. S-12.9. The three resonances occur at $\lambda_1 = 2a = 8$ cm, $\lambda_3 = 2a/3 = 2.6667$ cm and $\lambda_5 = 2a/5 = 1.6$ cm, $k_m = 2\pi/\lambda_m$. Thus $k_1\delta = 4.71 \times 10^{-3}$, $k_3\delta = 1.41 \times 10^{-2}$ and $k_5 = 2.36 \times 10^{-2}$. Note that, in reality, δ is frequency dependent. Figure S-12.10 shows the effect of a larger skin depth, namely $\delta = 1$ mm, meaning a very poor conductor. Broadenings and shifts of the resonances are clearly visible, particularly at higher modes. Shifts and broadenings of resonances are very common phenomena in physics, for instance they occur for the harmonic oscillator in the presence of losses. For modes with $\sin(ka/2) = 0$, we obtain $\langle P_s(t)\rangle = (\pi K_0^2/2c)\,k\delta$ which is much smaller than near resonance for good conductors since $k\delta \ll 1$.

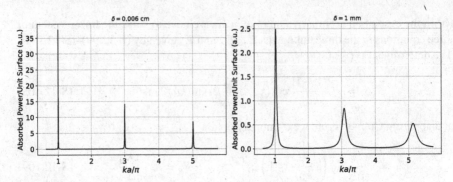

Fig. S-12.9   **Fig. S-12.10**

References

1. R.P. Feynman, R.B. Leighton, M. Sands, *The Feynman Lectures on Physics*, Vol. II, Sect. 24–8 (Addison-Wesley Publishing Company, Reading, MA, 2006)
2. J.D. Jackson, *Classical Electrodinamics*, Sects. 9.2 and 9.4, 3rd edn. (Wiley, New York, 1998)

Chapter S-13
Solutions for Chapter 13

S-13.1 Electrically and Magnetically Polarized Cylinders

(a) Long cylinders. In the "magnetic case", the parallel component of the auxiliary field, $\mathbf{H} = \mathbf{B}/[(\mu_0)\,\mu_r]$ (here, and the following, the parentheses mean that μ_0 appears in SI units only, not in Gaussian units) is continuous at the lateral surface of the cylinder. Thus the magnetic field inside the cylinder, $\mathbf{B}_i$, is

$$\mathbf{B}_i = \mu_r \mathbf{B}_0 \,. \tag{S-13.1}$$

The interface condition for the electric field is that the parallel component of $\mathbf{E}$ must be continuous at the lateral surface, thus we have for the internal field

$$\mathbf{E}_i = \mathbf{E}_0 \,. \tag{S-13.2}$$

These results are consistent with the analogy between the equations for $\mathbf{E}$ in electrostatics and $\mathbf{H}$ in magnetostatics and in the absence of free currents, i.e. $\nabla \times \mathbf{E} = 0$ and $\nabla \times \mathbf{H} = 0$.

(b) Flat cylinders. In the "magnetic case", the perpendicular component of $\mathbf{B}$ is continuous at the bases, thus we have

$$\mathbf{B}_i = \mathbf{B}_0 \,. \tag{S-13.3}$$

In the "electric case", the perpendicular component of the auxiliary vector $\mathbf{D}$ must be continuous at the interface, thus internal field is

$$\mathbf{E}_i = \frac{1}{\varepsilon_r} \mathbf{E}_0 \,. \tag{S-13.4}$$

These results are consistent with the analogy between the equations for $\mathbf{B}$ and for $\mathbf{D}$ in electrostatics and in the absence of free charges, i.e. $\nabla \cdot \mathbf{B} = 0$ and $\nabla \cdot \mathbf{D} = 0$.

© The Editor(s) (if applicable) and The Author(s), under exclusive license to Springer Nature Switzerland AG 2023
A. Macchi et al., *Problems in Classical Electromagnetism*,
https://doi.org/10.1007/978-3-031-22235-1

(c) Let us assume (S-13.2) as zero-order solution for the case of the "long" dielectric cylinder. According to (3.1) the cylinder acquires a uniform electric polarization

$$\mathbf{P} = \frac{\varepsilon_r - 1}{4\pi k_e} \mathbf{E}_i = \frac{\varepsilon_r - 1}{4\pi k_e} \mathbf{E}_0 \, , \qquad (\text{S-13.5})$$

corresponding to two bound surface charge densities $\sigma_b = \mathbf{P} \cdot \hat{\mathbf{n}} = \pm P$ at the cylinder bases. When evaluating the field at the cylinder center, due to the condition $a \ll h$ the total bound charges on the two bases can be approximated by two point charges $\pm Q$, with

$$Q = \pi a^2 P = \frac{a^2(\varepsilon_r - 1)}{4k_e} E_0 = \begin{cases} \pi a^2 \varepsilon_0 \, (\varepsilon_r - 1) \, E_0 \, , & \text{SI}, \\ \dfrac{a^2(\varepsilon_r - 1)}{4} \varepsilon_r \, E_0 \, , & \text{Gaussian}, \end{cases} \qquad (\text{S-13.6})$$

located at distances $\pm h/2$. Thus, at the cylinder center we have an additional field

$$E_b \simeq -2k_e \frac{Q}{(h/2)^2} = -2 \, (\varepsilon_r - 1) \, E_0 \left(\frac{a}{h}\right)^2 \, , \qquad (\text{S-13.7})$$

corresponding to a second-order correction. The electric field up to the second order in (a/h) is thus

$$\mathbf{E}_i^{(2)} = \mathbf{E}_i + \mathbf{E}_b = \mathbf{E}_0 \left[1 - 2(\varepsilon_r - 1) \left(\frac{a}{h}\right)^2 \right] \, . \qquad (\text{S-13.8})$$

In the corresponding "magnetic case", the formal analogy between $\mathbf{H}$ and $\mathbf{E}$ leads to a second-order correction to the auxiliary field $\mathbf{H}_i$ at the cylinder center

$$\mathbf{H}_b = -2(\mu_r - 1) \, \mathbf{H}_0 \left(\frac{a}{h}\right)^2 \, , \qquad (\text{S-13.9})$$

where $H_0 = B_0/(\mu_0)$. Because of the formal analogy between $\mathbf{H}$ and $\mathbf{E}$, the correction to $\mathbf{H}$ at the center of the cylinder can be interpreted as due to the presence of *fictitious equivalent magnetic charges* $Q_m = \pm \pi a^2 M$ on the two cylinder bases. The fictitious magnetic charge densities $\sigma_m = \pm M$ at the two bases are associated to the magnetization $\mathbf{M} = \chi_m \mathbf{H}_0$, where χ_m is given by (5.22) in terms of μ_r. Each magnetic charge gives origin to an auxiliary field

$$\mathbf{H} = \begin{cases} \dfrac{1}{4\pi} \dfrac{Q_m}{r^2} \hat{\mathbf{r}} \, , & \text{SI}, \\ \dfrac{1}{c} \dfrac{Q_m}{r^2} \hat{\mathbf{r}} \, , & \text{Gaussian.} \end{cases} \qquad (\text{S-13.10})$$

Recalling that, in SI units, $\mathbf{B} = \mu_0(\mathbf{H} + \mathbf{M})$, we obtain for the magnetic field at the cylinder center

$$\mathbf{B} = \mu_0(\mathbf{H}_0 + \mathbf{M}) + \mu_0\mathbf{H}_b = \mu_r H_0 + \mu_0 H_b \equiv \mathbf{B}_i + \mathbf{B}_b \,, \qquad (\text{S-13.11})$$

and the second-order correction is

$$\mathbf{B}_b = -2(\mu_r - 1)\,\mathbf{B}_0 \left(\frac{a}{h}\right)^2 \,. \qquad (\text{S-13.12})$$

In Gaussian units we have $\mathbf{B} = \mathbf{H} + 4\pi\mathbf{M}$, the second order correction remaining the same as in (S-13.12).

Notice that it would have been *wrong* to write

$$\mathbf{B}_b = (\mu_0)\mu_r\mathbf{H}_b \quad (\text{wrong!}) \,, \qquad (\text{S-13.13})$$

as it would have been wrong to write

$$E_b \simeq -2\frac{k_e}{\varepsilon_r}\frac{Q}{(h/2)^2} = -2\frac{(\varepsilon_r - 1)}{\varepsilon_r}\,E_0\left(\frac{a}{h}\right)^2 \quad (\text{wrong!}) \,, \qquad (\text{S-13.14})$$

instead of (S-13.7), because we are considering the fields generated by polarization charges, and inserting μ_r or ε_r would mean taking the effects of the medium polarization into account twice.

Alternatively, we can recall that the zero-order approximation of the cylinder magnetization is

$$\mathbf{M} = \chi_m\mathbf{H}_i = \chi_m\frac{\mathbf{B}_i}{(\mu_0)\,\mu_r} = \chi_m\frac{\mathbf{B}_0}{(\mu_0)} \,, \quad (\text{S-13.15})$$

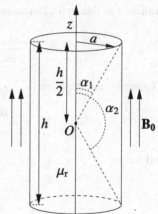

again, μ_0 appearing in SI units only. The magnetization is associated to a surface magnetization current density $\mathbf{K}_m = \mathbf{M} \times \hat{\mathbf{n}}/b_m$ on the lateral surface of the cylinder

$$\mathbf{K}_m = \frac{\chi_m}{b_m}\frac{B_0}{(\mu_0)}\,\hat{\boldsymbol{\phi}} \,, \qquad (\text{S-13.16})$$

where $\hat{\boldsymbol{\phi}}$ is the azimuthal unit vector of the cylindrical coordinates with the cylinder axis as longitudinal axis.

Thus, the cylinder is equivalent to a finite solenoid of **Fig. S-13.1** height h and radius a, with the product nI equal to a K_m. The magnetic field of a finite solenoid on its axis is

$$B_M = 2\pi k_m n I \left(\cos\alpha_1 - \cos\alpha_2\right) = 2\pi k_m K_m \left(\cos\alpha_1 - \cos\alpha_2\right)$$

$$= 2\pi k_m \frac{\chi_m}{b_m} \frac{B_0}{(\mu_0)} \left(\cos\alpha_1 - \cos\alpha_2\right) = (\mu_r - 1) \frac{B_0}{2} \left(\cos\alpha_1 - \cos\alpha_2\right),$$

$$\text{(S-13.17)}$$

where the angles α_1 and α_2 are shown in Fig. S-13.1. At the solenoid center we have

$$\cos\alpha_1 = -\cos\alpha_2 = \frac{h/2}{\sqrt{a^2 + (h/2)^2}} \simeq 1 - \frac{1}{2}\left(\frac{2a}{h}\right)^2 = 1 - 2\left(\frac{a}{h}\right)^2 , \quad \text{(S-13.18)}$$

thus

$$B_M \simeq (\mu_r - 1) B_0 \left[1 - 2\left(\frac{a}{h}\right)^2\right] . \quad \text{(S-13.19)}$$

The total field at the cylinder center equals the external field B_0 plus the field due to the cylinder magnetization

$$B(0) = B_0 + B_M = \mu_r B_0 - 2\mu_r B_0 \left(\frac{a}{h}\right)^2 + 2 B_0 \left(\frac{a}{h}\right)^2$$

$$= \mu_r B_0 - 2(\mu_r - 1) B_0 \left(\frac{a}{h}\right)^2 , \quad \text{(S-13.20)}$$

in agreement with (S-13.19).

The correction to the field at the center of the magnetic "flat" cylinder can be evaluated as due to a circular loop of radius a carrying an electric current $I_s = K_m h$:

$$B_b = \frac{2\pi k_m I_s}{a} = 2\pi k_m K_m \frac{h}{a} = 2\pi \frac{k_m}{b_m} M \frac{h}{a} . \quad \text{(S-13.21)}$$

At zeroth order we have

$$\mathbf{H}_i \simeq \frac{\mathbf{B}_0}{(\mu_0)\,\mu_r} , \quad \text{thus} \quad \mathbf{M} \simeq \chi_m \frac{\mathbf{B}_0}{(\mu_0)\,\mu_r} , \quad \text{(S-13.22)}$$

and we get

$$B_b = \frac{\mu_r - 1}{2\mu_r} \frac{a}{h} B_0 . \quad \text{(S-13.23)}$$

The auxiliary field $\mathbf{H}$ is given by (5.19), thus we have, up to the second order

$$H_i + H_b = \frac{B_0 + B_b}{(\mu_0)\,\mu_r} = \begin{cases} \dfrac{B_0 + B_b}{\mu_0} - M = H_i + \dfrac{B_b}{\mu_0} & \text{SI,} \\[2ex] B_0 + B_b - 4\pi M = H_i + B_b & \text{Gaussian.} \end{cases} \quad \text{(S-13.24)}$$

Thus we have

$$H_b = \frac{B_b}{(\mu_0)} = \frac{\mu_r - 1}{2\mu_r} \frac{a}{h} H_0 . \qquad \text{(S-13.25)}$$

Due to the formal analogy between **H** and **E** we have for the flat dielectric cylinder

$$E_b = \frac{\varepsilon_r - 1}{2\varepsilon_r} \frac{a}{h} E_0 . \qquad \text{(S-13.26)}$$

S-13.2 Oscillations of a Triatomic Molecule

(a) The equations of motion for the two lateral masses are

$$m\ddot{x}_1 = -k(x_1 - x_c + \ell) , \qquad m\ddot{x}_2 = -k(x_2 - x_c - \ell) ; \qquad \text{(S-13.27)}$$

from (13.1) we obtain for the position of the central mass

$$x_c = -\frac{m}{M}(x_1 + x_2) , \qquad \text{(S-13.28)}$$

which, substituted into (S-13.27) after dividing by m, leads to a system of two equations of motion involving x_1 and x_2 only

$$\ddot{x}_1 = -k\left(\frac{1}{m} + \frac{1}{M}\right)x_1 - \frac{k}{M}x_2 - \frac{k}{m}\ell , \qquad \text{(S-13.29)}$$

$$\ddot{x}_2 = -k\left(\frac{1}{m} + \frac{1}{M}\right)x_2 - \frac{k}{M}x_1 + \frac{k}{m}\ell . \qquad \text{(S-13.30)}$$

Adding and subtracting these equations we obtain

$$\ddot{x}_1 + \ddot{x}_2 = -k\left(\frac{1}{m} + \frac{2}{M}\right)(x_1 + x_2) = -k\frac{M_{\text{tot}}}{mM}(x_1 + x_2) \qquad \text{(S-13.31)}$$

$$\ddot{x}_1 - \ddot{x}_2 = -\frac{k}{m}(x_1 - x_2 + 2\ell) , \qquad \text{(S-13.32)}$$

where $M_{\text{tot}} = M + 2m$ is the total mass of the molecule. Thus, introducing the new variables

$$x_+ = x_1 + x_2 \quad \text{and} \quad x_- = x_1 - x_2 + 2\ell , \qquad \text{(S-13.33)}$$

we obtain the following equations for the normal longitudinal modes of the molecule

$$\ddot{x}_\pm = -\omega_\pm^2 x_\pm , \quad \text{where} \quad \omega_+ = \sqrt{\frac{kM_{\text{tot}}}{mM}} \quad \text{and} \quad \omega_- = \sqrt{\frac{k}{m}} . \qquad \text{(S-13.34)}$$

Frequency ω_+ corresponds to an antisymmetric motion of the masses: while the lateral masses move, for instance, to the right by the same amount, the central mass moves to the left, and vice versa, so that $x_{cm} = 0$. Frequency ω_- corresponds to a symmetric motion: the lateral masses perform opposite oscillations, while the central mass does not move.

(b) The electric dipole moment of the molecule is parallel to the molecular axis and its magnitude is

$$p = -qx_1 + 2qx_c - qx_2 = -q\left(1 + \frac{2m}{M}\right)(x_1 + x_2) = -q\frac{M_{tot}}{M}x_+ \ . \quad \text{(S-13.35)}$$

Thus, the dipole oscillates in the antisymmetric mode at frequency ω_+. The dipole moment is zero when the molecule oscillates in the symmetric mode, and radiation at frequency ω_- is due only to quadrupole emission, which is weaker than dipole emission.

(c) The initial conditions for x_+ are

$$x_+(0) = x_1(0) + x_2(0) = d_1 + d_2, \qquad \dot{x}_+(0) = 0 \ , \quad \text{(S-13.36)}$$

thus for $t > 0$

$$x_+(t) = (d_1 + d_2)\cos\omega_+t \ . \quad \text{(S-13.37)}$$

The symmetric mode is also excited, but does not contribute to the dipole radiation. The instantaneous radiated power is

$$P = \frac{2}{3c^3}|\ddot{p}|^2 = \frac{2q^2}{3c^3}\left(\frac{M_{tot}}{M}\right)^2 \omega_+^4(d_1 + d_2)^2\cos^2\omega_+t \ . \quad \text{(S-13.38)}$$

S-13.3 Impedance of an Infinite Ladder Network

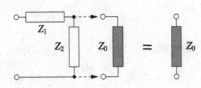

(a) Our infinite network is a sequence of identical sections. As we did for Problem 4.10, we note that adding a further L-section to the left of of Fig. 13.3 does not change the impedance of the ladder network. Thus we must have (see Fig. S-13.2)

Fig. S-13.2

$$Z_0 = Z_1 + \frac{Z_2 Z_0}{Z_2 + Z_0} \ , \quad \text{(S-13.39)}$$

from which $Z_0^2 - Z_1 Z_0 - Z_1 Z_2 = 0$ follows. The solution is

$$Z_0 = \frac{Z_1}{2} + \sqrt{\frac{Z_1^2}{4} + Z_1 Z_2} \,, \qquad \text{(S-13.40)}$$

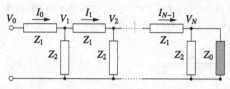

Fig. S-13.3

The other solution of the quadratic equation has been discarded because in the case of real, positive impedances (the purely resistive case of Problem 4.10) it would give an unphysical negative value. Thus, a finite ladder of N sections, terminated by an impedance Z_0 as shown in Fig. S-13.3, is equivalent to the infinite ladder.

(b) In Fig. 13.3, current I_n flows through the Z_1 impedance of the $(n+1)$-th section, thus, the voltage drop across the impedance, $V_n - V_{n+1}$, must equal $I_n Z_1$. On the other hand, I_n is input into the semi-infinite ladder network starting at node n, thus we must have $I_n = V_n / Z_0$. The two conditions give

$$V_n - V_{n+1} = \frac{V_n}{Z_0} Z_1 \,, \qquad \text{(S-13.41)}$$

so that we obtain for the ratio of the voltages at adjacent nodes

$$\alpha \equiv \frac{V_{n+1}}{V_n} = 1 - \frac{Z_1}{Z_0} \,. \qquad \text{(S-13.42)}$$

If $V_0(t) = V_0 \, \mathrm{e}^{-\mathrm{i}\omega t}$ is the input voltage, we have $V_n = \alpha^n V_0 \, \mathrm{e}^{-\mathrm{i}\omega t}$ at the n-th node. For a purely resistive network we have

$$Z_0 \equiv R_0 = \frac{R_1}{2} + \sqrt{\frac{R_1^2}{4} + R_1 R_2} \,, \qquad \text{(S-13.43)}$$

which is a real number, and $\alpha = 1 - R_1/R_0 < 1$. At each successive node the signal is damped by a factor α.

(c) For the LC network we have

$$Z_0 = -\frac{\mathrm{i}\omega L}{2} + \sqrt{-\frac{\omega^2 L^2}{4} + \frac{\mathrm{i}\omega L}{\mathrm{i}\omega C}} = -\frac{\mathrm{i}\omega L}{2} + \sqrt{\frac{L}{C} - \frac{\omega^2 L^2}{4}}$$

$$= \sqrt{\frac{4L^2}{4LC} - \frac{\omega^2 L^2}{4}} - \frac{\mathrm{i}\omega L}{2} = \frac{L}{2}\sqrt{\frac{4}{LC} - \omega^2} - \frac{\mathrm{i}\omega L}{2}$$

$$= \frac{L}{2}\left(\sqrt{\omega_{\mathrm{co}}^2 - \omega^2} - \mathrm{i}\omega\right) \,, \qquad \text{(S-13.44)}$$

where $\omega_{\mathrm{co}} \equiv 2/\sqrt{LC}$. Thus

$$\alpha = 1 - \frac{Z_1}{Z_0} = 1 + \frac{2i\omega L}{L\left(\sqrt{\omega_{co}^2 - \omega^2} - i\omega\right)} = \frac{\sqrt{\omega_{co}^2 - \omega^2} + i\omega}{\sqrt{\omega_{co}^2 - \omega^2} - i\omega} . \qquad \text{(S-13.45)}$$

If $\omega < \omega_{co}$, the square roots are real and α is the ratio of a complex number to its own complex conjugate, therefore $|\alpha| = 1$, and we can write $\alpha = e^{i\phi}$ with

$$\tan\left(\frac{\phi}{2}\right) = \frac{\omega}{\sqrt{\omega_{co}^2 - \omega^2}} . \qquad \text{(S-13.46)}$$

Thus the voltage at node n is $V_n = V_0\, e^{in\phi - i\omega t}$, and the signal propagates along the network without damping. The above equation also gives the dispersion relation

$$\omega = \omega_{co}\left|\sin\left(\frac{\phi}{2}\right)\right| . \qquad \text{(S-13.47)}$$

This is analogous to the dispersion relation (S-7.43) found in Problem 7.4, when we substitute ϕ for ka.

If $\omega > \omega_{co}$, Z_0 is a purely imaginary number,

$$Z_0 = \pm i\sqrt{\omega^2 - \omega_{co}} , \qquad \text{(S-13.48)}$$

and α is real

$$\alpha = \frac{\pm\sqrt{\omega^2 - \omega_{co}^2} + \omega}{\pm\sqrt{\omega^2 - \omega_{co}^2} - \omega} . \qquad \text{(S-13.49)}$$

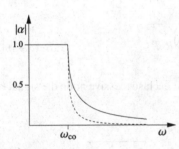

Inserting the negative root into (S-13.49) leads to $|\alpha| < 1$, and the signal is damped. The positive root would lead to an unphysical $|\alpha| > 1$, implying an amplification of the signal along the network, without an external energy source.

Thus the LC network behaves as a low-pass filter, since signals at frequencies $\omega > \omega_{co}$ are attenuated by a factor $|\alpha|^N$ after N nodes. The dependence of the network transmission on frequency approaches an ideal low-pass filter, for which transmission is zero for $\omega > \omega_{co}$, at high numbers of circuit sections N. Figure S-13.4 shows $|\alpha|$ (solid line) and $|\alpha|^2$ (dashed line) as a functions of the signal frequency ω.

Fig. S-13.4

(d) For the CL network (Problem 7.5) we proceed analogously to point c) for the LC network, and obtain

$$Z_0 = \frac{i}{2\omega C} + \sqrt{-\frac{1}{4\omega^2 C^2} + \frac{\omega L}{\omega C}} = \frac{1}{2C}\left[\sqrt{\frac{1}{\omega_{co}^2} - \frac{1}{\omega^2}} + \frac{i}{\omega}\right], \quad (S\text{-}13.50)$$

and

$$\alpha = \frac{\sqrt{\omega_{co}^{-2} - \omega^{-2}} - i/\omega}{\sqrt{\omega_{co}^{-2} - \omega^{-2}} + i/\omega}. \quad (S\text{-}13.51)$$

We have undamped propagation for $|\alpha| = 1$, i.e., when $\omega > \omega_{co}$. For $\omega < \omega_{co}$ the signals are damped, and the network acts as a high-pass filter.

S-13.4   Discharge of a Cylindrical Capacitor

(a) We use cylindrical coordinates (r, ϕ, z). For symmetry reasons, assuming $h \gg b$, the electric field between the capacitor plates is radial, and easily evaluated from Gauss's law as

$$E_r = E_r(r) = \frac{2Q_0}{hr}, \quad \text{(Gaussian units).} \quad (S\text{-}13.52)$$

The potential difference V across the plates is

$$V = \left|\int_a^b \mathbf{E} \cdot d\mathbf{s}\right| = \frac{2Q_0}{h}\int_a^b \frac{dr}{r} = \frac{2Q_0}{h}\ln(b/a), \quad (S\text{-}13.53)$$

and the capacity of our cylindrical capacitor is

$$C = \frac{Q_0}{V} = \frac{h}{2\ln(b/a)} \quad (S\text{-}13.54)$$

The initial electrostatic energy is $U_{es}(0) = Q_0^2/2C$.

After the plates are connected through the resistor at $t = 0$, the system is an RC circuit, and the capacitor charge at time t is

$$Q(t) = Q_0 e^{-t/\tau}, \quad \text{where} \quad \tau = RC = \frac{Rh}{2\ln(b/a)} \quad (S\text{-}13.55)$$

Assuming that the charge densities remain uniform over the plates during the discharge, the absolute value of the charge of each plate between its bottom, $z = 0$, and any height $z < h$ (see Fig. S-13.5 for the case of the inner plate of Fig. 13.4) is

$$\Delta Q(z, t) = Q(t) \frac{z}{h} . \qquad \text{(S-13.56)}$$

The decay of the charge implies a current flowing over each plate, along the $\hat{\mathbf{z}}$ direction. Let $I_a(z, t)$ and $I_b(z, t)$ be the currents in the inner and outer plate, respectively, which can obtained from the continuity equation: for the inner plate

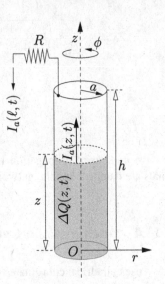

Fig. S-13.5

$$I_a(z, t) = -\frac{\mathrm{d}[\Delta Q(z, t)]}{\mathrm{d}t} = \frac{Q(t)}{\tau} \frac{z}{h} = \frac{Q_0}{\tau} \frac{z}{h} \mathrm{e}^{-t/\tau} . \qquad \text{(S-13.57)}$$

Since, in the assumption of uniform charge densities, the charge on the outer plate is $-\Delta Q(z, t)$, then $I_b(z, t) = -I_a(z, t)$.

We can evaluate $\mathbf{B}$ in the $a < r < b$ region from Maxwell's equation

$$\nabla \times \mathbf{B} = \frac{4\pi}{c} \mathbf{J} + \frac{1}{c} \partial_t \mathbf{E} . \qquad \text{(S-13.58)}$$

The only nonzero component of $\mathbf{J}$ is along z and the only nonzero component of $\mathbf{E}$ is along r, given by

$$\mathbf{E} = \hat{\mathbf{r}} \frac{2Q(t)}{hr} , \qquad \text{(S-13.59)}$$

while $\mathbf{B}$ must be independent of ϕ because of the symmetry of our problem. Thus, according to the curl components in cylindrical coordinates of Table A.1 of the Appendix we have

$$(\nabla \times \mathbf{B})_r = -\partial_z B_\phi = \frac{1}{c} \partial_t E_r , \quad (\nabla \times \mathbf{B})_z = \frac{1}{r} \partial_r (r B_\phi) = \frac{4\pi}{c} J_z , \qquad \text{(S-13.60)}$$

and we see that the only nonzero component of $\mathbf{B}$ is B_ϕ, which can be evaluated from either of (S-13.60). We choose the second of (S-13.60), and apply Stokes' theorem to a circle C of radius $a < r < b$, coaxial to the capacitor and located at height $0 < z < h$,

$$\oint_C \mathbf{B}(r, z, t) \cdot \mathrm{d}\ell = 2\pi r B_\phi(r, z, t) = \frac{4\pi}{c} I_a(z, t) , \qquad \text{(S-13.61)}$$

$$B_\phi(r, z, t) = \frac{2}{c} \frac{I_a(z, t)}{r} = \frac{2}{ch\tau} \frac{z}{r} Q_0 \mathrm{e}^{-t/\tau} . \qquad \text{(S-13.62)}$$

(b) The Poynting vector is

$$\mathbf{S} = \frac{c}{4\pi} \mathbf{E} \times \mathbf{B} = \hat{\mathbf{z}} \frac{Q_0^2}{\pi h^2 \tau} \frac{z}{r^2} e^{-2t/\tau} , \qquad a < r < b , \qquad \text{(S-13.63)}$$

and $\mathbf{S} = 0$ if $r < a$ or $r > b$. The flux of $\mathbf{S}$ through a plane perpendicular to z at height $0 < z < h$ is thus

$$\Phi_S(z, t) = \frac{Q_0^2 z}{\pi h^2 \tau} e^{-2t/\tau} \int_a^b \frac{1}{r^2} 2\pi r \, dr = \frac{2Q_0^2 z \ln(b/a)}{h^2 \tau} e^{-2t/\tau} . \quad \text{(S-13.64)}$$

The electrostatic energy associated to the volume between the bottom of the capacitor $(z = 0)$ and height z at time t is

$$\Delta U_{es}(z, t) = \frac{z}{h} \frac{Q^2(t)}{2C} = \frac{z}{h} \frac{Q_0^2 e^{-2t/\tau}}{2C} = \frac{z}{h} \frac{Q_0^2 \ln(b/a)}{h} e^{-2t/\tau} , \quad \text{(S-13.65)}$$

because the electric field does not depend on z. Thus we have

$$\frac{d \left[\Delta U_{es}(z, t) \right]}{dt} = -\frac{2\Delta U_{es}(z, t)}{\tau} = -\Phi_S(z, t) . \qquad \text{(S-13.66)}$$

(c) The assumptions of slowly varying currents and of uniform charge density are closely related. In fact, the capacitor can be viewed as a portion of a coaxial cable along which charge and current signals are propagating in TEM mode, at velocity c. In these conditions, the charge density can be assumed as uniform if the propagation of the signals is "instantaneous" with respect to the duration of the discharge, i.e., if the propagation time $h/c \ll \tau$. This is equivalent to assuming that the wavelengths corresponding to the frequency spectrum of the signal are much larger than h, so that the field can be considered as uniform along z.

We can reach the same conclusion by checking that the electric field $\mathbf{E}_1$, generated by the magnetic induction, is much smaller than the electrostatic field $\mathbf{E}_0$. From Maxwell's equation

$$\nabla \times \mathbf{E}_1 \simeq -\frac{1}{c} \partial_t \mathbf{B} , \qquad \text{(S-13.67)}$$

where the only nonzero component of $\mathbf{B}$ is B_ϕ, we obtain

$$\partial_z E_{1r} = -\frac{1}{c} \frac{2}{ch\tau^2} \frac{z}{r} Q_0 e^{-t/\tau}$$

$$E_{1r} = \frac{Q_0}{c^2 h \tau^2} \frac{z^2}{r} e^{-t/\tau} = \frac{1}{2} \left(\frac{z}{c\tau} \right)^2 E_{0r} . \qquad \text{(S-13.68)}$$

where E_{0r} is from the second of (S-13.59). Thus $E_{1r} \ll E_{0r}$ if $h \ll c\tau$.

S-13.5   Fields Generated by Spatially Periodic Surface Sources

(a) In this case fields and potential are electrostatic. The potential $\varphi = \varphi(x, y)$ is a solution of the 2D Laplace's equation

$$(\partial_x^2 + \partial_y^2)\varphi = 0 \quad \text{for} \quad y \neq 0, \tag{S-13.69}$$

and, due to the symmetry of the source, must be an even function of y. We attempt to find a solution by the method of separation of variables, i.e. we look for a solution $\varphi = X(x)Y(y)$, where X depends only on x and Y only on y. Equation (S-13.69) becomes

$$X''(x)Y(y) + X(x)Y''(y) = 0, \tag{S-13.70}$$

where the double primes denote the second derivatives. Dividing by $X(x)Y(y)$ we obtain

$$\frac{Y''(y)}{Y(y)} = -\frac{X''(x)}{X(x)}, \tag{S-13.71}$$

which must hold for every x, y, implying that both sides of the equation must equal some constant value, which, for convenience, we denote by α^2,

$$\frac{Y''(y)}{Y(y)} = \alpha^2, \quad \frac{X''(x)}{X(x)} = -\alpha^2, \tag{S-13.72}$$

whose solutions are

$$Y(y) = A_y\, e^{+\alpha y} + B_y\, e^{-\alpha y}, \quad \text{and} \quad X(x) = A_x\, e^{+i\alpha x} + B_x\, e^{-i\alpha x}, \tag{S-13.73}$$

where A_x, A_y, B_x, and B_y are constants to be determined. Discarding the solutions that diverge for $|y| \to \infty$, and fitting the x dependence to the dependence of σ, which implies $\alpha = k$, we obtain

$$\varphi = \varphi_0\, e^{-k|y|} \cos(kx), \tag{S-13.74}$$

where φ_0 is a constant to be determined. The nonzero components of the electric field are

$$E_x = -\partial_x\varphi = k\varphi_0\, e^{-k|y|} \sin(kx),$$
$$E_y = -\partial_y\varphi = \operatorname{sgn}(y)\, k\varphi_0\, e^{-k|y|} \cos(kx). \tag{S-13.75}$$

The component E_x is continuous at the $y = 0$ plane, as expected, since $\nabla \times \mathbf{E} = 0$. We can obtain the relation between E_y at $y = 0$ and the surface charge density by using Gauss's law,

$$E_y(x, y = 0^+) - E_y(x, y = 0^-) = 4\pi\sigma(x) , \tag{S-13.76}$$

from which we obtain the value of φ_0, namely $\varphi_0 = 2\pi\sigma_0/k$, and, finally

$$\varphi = \frac{2\pi\sigma_0}{k} e^{-k|y|} \cos(kx) . \tag{S-13.77}$$

(b) Here we have magnetostatic fields. Due to the analogy between the Poisson equations for the vector potential $\nabla^2 \mathbf{A} = -4\pi\mathbf{J}/c$, and for the scalar potential $\nabla^2\varphi = -4\pi\rho$, we can use (S-13.77) for obtaining the vector potential $\mathbf{A}$ as

$$\mathbf{A} = \hat{\mathbf{z}} A_0 e^{-k|y|} \cos(kx) , \quad \text{where} \quad A_0 = \frac{2\pi K_0}{kc} . \tag{S-13.78}$$

The nonzero component of the magnetic field are

$$B_x = \partial_y A_z = -\operatorname{sgn}(y) kA_0 e^{-k|y|} \cos(kx) = -\operatorname{sgn}(y) \frac{2\pi K_0}{c} e^{-k|y|} \cos(kx) ,$$

$$B_y = -\partial_x A_z = kA_0 e^{-k|y|} \sin(kx) = \frac{2\pi K_0}{c} e^{-k|y|} \sin(kx) . \tag{S-13.79}$$

Thus, B_y is continuous at $y = 0$, as expected from $\nabla \cdot \mathbf{B} = 0$. Further we have

$$B_x(x, y = 0^+) - B_x(x, y = 0^-) = -\frac{4\pi}{c} K_0 \cos(kx) , \tag{S-13.80}$$

in agreement with Ampère's law.

(c) Since $\sigma = 0$, also the scalar potential is zero, $\varphi = 0$. The inhomogeneous wave equation for the vector potential $\mathbf{A}$ is, in the Lorentz gauge condition,

$$\nabla^2 \mathbf{A} - \frac{1}{c^2} \frac{\partial^2 \mathbf{A}}{\partial t^2} = -\frac{4\pi}{c} \mathbf{J} = -\hat{\mathbf{z}} \frac{4\pi}{c} \delta(y) K_0 e^{-i\omega t} \cos(kx) . \tag{S-13.81}$$

As an educated guess, we search for a solution of the form

$$\mathbf{A} = \hat{\mathbf{z}} A_0 e^{-q|y|-i\omega t} \cos(kx) , \tag{S-13.82}$$

which, for $y \neq 0$ leads to

$$\left(-k^2 + q^2 + \frac{\omega^2}{c^2}\right) \mathbf{A} = 0 , \quad \text{or} \quad q^2 = k^2 - \left(\frac{\omega}{c}\right)^2 . \tag{S-13.83}$$

Thus, if $\omega < kc$, q is real and $\mathbf{A}$ decays exponentially with $|y|$. If $\omega > kc$, q is imaginary and the waves propagates, $\mathbf{A}$ being proportional to $e^{i|q||y|-i\omega t}$. If we integrate (S-13.81) in dy from $-h$ to $+h$ we obtain

$$\lim_{h \to 0} \int_{-h}^{+h} \left(\frac{\partial^2 \mathbf{A}}{\partial x^2} + \frac{\partial^2 \mathbf{A}}{\partial y^2} - \frac{1}{c^2} \frac{\partial^2 \mathbf{A}}{\partial t^2} \right) dy = -\hat{\mathbf{z}} \frac{4\pi}{c} K_0 e^{-i\omega t} \cos(kx) . \quad \text{(S-13.84)}$$

Now, both $\partial^2 \mathbf{A}/\partial x^2$ and $\partial^2 \mathbf{A}/\partial t^2$ are continuous at $y = 0$ and don't contribute to the integral at the limit $h \to 0$. Thus, the left-hand side of (S-13.84) is

$$\lim_{h \to 0} \int_{-h}^{+h} \frac{\partial^2 \mathbf{A}}{\partial y^2} \, dy = \lim_{h \to 0} \left[\partial_y \mathbf{A} \right]_{-h}^{+h} = -\hat{\mathbf{z}} \, A_0 q \cos(kx) \lim_{h \to 0} \left[\text{sgn}(y) \, e^{-q|y| - i\omega t} \right]_{-h}^{+h}$$

$$= -\hat{\mathbf{z}} \, 2A_0 e^{-i\omega t} q \cos(kx) , \quad \text{(S-13.85)}$$

which must equal the right-hand side of (S-13.84), leading to

$$A_0 = \frac{2\pi}{qc} K_0 \quad \text{(S-13.86)}$$

which, at the static limit $\omega \to 0$, $q \to k$, equals (S-13.78).

The nonzero components of the magnetic field are

$$B_x = \partial_y A_z = -\text{sgn}(y) \frac{2\pi}{c} K_0 e^{-q|y| - i\omega t} \cos(kx) ,$$

$$B_y = -\partial_x A_z = \frac{2\pi k}{qc} K_0 e^{-q|y| - i\omega t} \sin \omega t , \quad \text{(S-13.87)}$$

which, at the static limit $\omega \to 0$, $q \to k$, equal (S-13.79). The electric field is obtained from $\mathbf{E} = -\partial_t \mathbf{A}/c = i\omega \mathbf{A}/c$, and its only nonzero component is

$$E_z = -\frac{2\pi i \omega K_0}{qc^2} e^{-q|y| - i\omega t} \cos(kx) . \quad \text{(S-13.88)}$$

(d) In this context, given a function $f = f(x, t)$, we denote its time average by angle brackets, and its space average by a bar, as follows

$$\langle f \rangle = \frac{\omega}{2\pi} \int_{-\pi/\omega}^{+\pi/\omega} f \, dt , \qquad \overline{f} = \frac{k}{2\pi} \int_{-\pi/k}^{+\pi/k} f \, dx . \quad \text{(S-13.89)}$$

Thus we write the average power dissipated per unit time and unit surface on the $y = 0$ plane as

$$\overline{\langle K_z E_z \rangle} = \frac{1}{2} \text{Re} \left[K_0 \left(\frac{2\pi i \omega}{qc} K_0 \right)^* \right] \overline{\cos^2(kx)} = \frac{\pi \omega}{4c} |K_0|^2 \text{Re} \left(\frac{-i}{q} \right) . \text{(S-13.90)}$$

If q is real we have $\langle K_z E_z \rangle = 0$, consistently with the fields being evanescent for $|y| \to \infty$. There is no energy flow out of the $y = 0$ plane, and the work done by the currents is zero on average. On the other hand, if q is imaginary, we have

$$\langle K_z E_z \rangle = -\frac{\pi \omega |K_0|^2}{2|q|c} , \qquad (S\text{-}13.91)$$

which equals minus the flux of electromagnetic energy out of the $y = 0$ plane. In fact, the averaged Poynting vector is

$$\langle S_y \rangle = \frac{c}{4\pi} \langle E_z B_x \rangle = \frac{1}{2} \frac{c}{4\pi} \mathrm{Re} \left[\frac{2\pi i \omega K_0}{2qc} \left(\mathrm{sgn}(y) \, q^* \frac{2\pi K_0^*}{q^* c} \right) \right] \overline{\cos^2(kx)}$$

$$= \mathrm{sgn}(y) \frac{\pi \omega |K_0|^2}{4|q|c} , \qquad (S\text{-}13.92)$$

where we have used $\mathrm{Re}(i/q) = 1/|q|$ (for imaginary q). The flux of energy out of the $y = 0$ plane is thus $2 \left| \langle S_y \rangle \right| = -\langle K_z E_z \rangle$.

S-13.6 Energy and Momentum Flow Close to a Perfect Mirror

(a) The total electric field in front of the mirror is the sum of the fields of the incident ($\mathbf{E}_i$) and of the reflected ($\mathbf{E}_r$) waves, which have equal amplitude and frequency, but opposite polarizations and wavevectors,

$$\begin{aligned}
\mathbf{E} = \mathbf{E}_i + \mathbf{E}_r &= \hat{\mathbf{y}} \, E_\epsilon \left[\cos(kx - \omega t) - \cos(-kx - \omega t) \right] \\
&\quad - \hat{\mathbf{z}} \epsilon E_\epsilon \left[\sin(kx - \omega t) - \sin(-kx - \omega t) \right] \\
&= \hat{\mathbf{y}} \, E_\epsilon [\cos(kx)\cos(\omega t) + \sin(kx)\sin(\omega t) - \cos(kx)\cos(\omega t) + \sin(kx \sin(\omega t)] \\
&\quad - \hat{\mathbf{z}} \epsilon E_\epsilon [\sin(kx)\cos(\omega t) - \cos(kx)\sin(\omega t) + \sin(kx)\cos(\omega t) + \cos(kx)\sin(\omega t)] \\
&= \hat{\mathbf{y}} \, 2E_\epsilon \sin(kx)\sin(\omega t) - \hat{\mathbf{z}} \, 2\epsilon E_\epsilon \sin(kx)\cos(\omega t) , \qquad (S\text{-}13.93)
\end{aligned}$$

where $E_\epsilon \equiv E_0/\sqrt{1 + \epsilon^2}$. We can obtain the magnetic field from Maxwell's equation

$$\begin{aligned}
\partial_t \mathbf{B} = -c \nabla \times \mathbf{E} &= \hat{\mathbf{y}} \, c \, \partial_x E_z - \hat{\mathbf{z}} \, c \, \partial_x E_y \\
&= -\hat{\mathbf{y}} \, 2\epsilon E_\epsilon \, ck \cos(kx)\cos(\omega t) - \hat{\mathbf{z}} \, 2E_\epsilon \, ck \cos(kx)\sin(\omega t) , \qquad (S\text{-}13.94)
\end{aligned}$$

which yields, after integration in dt,

$$\begin{aligned}
\mathbf{B} &= -\hat{\mathbf{y}} \, 2\epsilon E_\epsilon \frac{ck}{\omega} \cos(kx)\sin(\omega t) + \hat{\mathbf{z}} \, 2E_\epsilon \frac{ck}{\omega} \cos(kx)\cos(\omega t) \\
&= -\hat{\mathbf{y}} \, 2\epsilon E_\epsilon \cos(kx)\sin(\omega t) + \hat{\mathbf{z}} \, 2E_\epsilon \cos(kx)\cos(\omega t) , \qquad (S\text{-}13.95)
\end{aligned}$$

where we have used $k = \omega/c$. The Poynting vector is

$$
\begin{aligned}
\mathbf{S} &= \frac{c}{4\pi} \mathbf{E} \times \mathbf{B} = \hat{\mathbf{x}} \frac{c}{4\pi} \left(E_y B_z - E_z B_y \right) \\
&= \hat{\mathbf{x}} \frac{cE_\epsilon^2}{\pi} \left[\sin(kx)\cos(kx)\sin(\omega t)\cos(\omega t) - \epsilon^2 \sin(kx)\cos(kx)\cos(\omega t)\sin(\omega t) \right] \\
&= \hat{\mathbf{x}} \frac{cE_\epsilon^2}{\pi} \sin(kx)\cos(kx)\sin(\omega t)\cos(\omega t) \left(1 - \epsilon^2\right) \\
&= \hat{\mathbf{x}} \frac{c}{4\pi} E_\epsilon^2 \left(1 - \epsilon^2\right) \sin(2kx)\sin(2\omega t) \,.
\end{aligned}
\tag{S-13.96}
$$

Thus $\mathbf{S} = 0$ if $\epsilon = 1$, corresponding to circular polarization. In such a case, $\mathbf{E}$ is parallel to $\mathbf{B}$. In general, also when $\mathbf{S} \neq 0$, we have $\langle \mathbf{S} \rangle = \mathbf{0}$, and there is no net energy flow.
(b) From the definition of T_{ij} we find

$$
\begin{aligned}
F_x = T_{xx} &= \frac{1}{8\pi} \mathbf{B}^2(0^-) = \frac{1}{4\pi} E_\epsilon^2(\cos^2 \omega t + \epsilon^2 \sin^2 \omega t) \\
&= \frac{2I}{c} \left[1 + \frac{1 - \epsilon^2}{1 + \epsilon^2} \cos 2\omega t \right] \,.
\end{aligned}
\tag{S-13.97}
$$

The oscillating (at 2ω) component vanishes for circular polarization. The average of F_x is the radiation pressure on the mirror (Problem 9.8), which does not depend on polarization.

S-13.7 Laser Cooling of a Mirror

(a) A plane wave of intensity I exerts a radiation pressure $2I/c$ on a perfectly reflecting surface. Thus the total force on the mirror, directed along the x axis of Fig. 13.5, is

$$
F = \frac{2A}{c}(I_1 - I_2) \,.
\tag{S-13.98}
$$

If $I_1 > I_2$ we have $F > 0$.
(b) The amplitudes of the electric fields of the two waves, in the mirror rest frame S', are

$$
E_1' = \gamma(E_1 - \beta B_1) = \gamma(1 - \beta) E_1 = \sqrt{\frac{1 - \beta}{1 + \beta}} E_1 \,,
\tag{S-13.99}
$$

$$
E_2' = \gamma(E_2 + \beta B_2) = \gamma(1 + \beta) E_2 = \sqrt{\frac{1 + \beta}{1 - \beta}} E_2 \,,
\tag{S-13.100}
$$

where $\beta = v/c, \gamma = 1/\sqrt{1 - \beta^2}$, and $E_1 = B_1, E_2 = B_2$ in Gaussian units. The intensity of a plane wave is $I = (c/4\pi) |\mathbf{E} \times \mathbf{B}| = c E^2/4\pi$, thus we have

$$I_1' = \frac{1 - \beta}{1 + \beta} I_1, \qquad I_2' = \frac{1 + \beta}{1 - \beta} I_2. \qquad (\text{S-13.101})$$

Since we have assumed $I_1 = I_2$, the total force is

$$F' = \frac{2A}{c}(I_1' - I_2') = \frac{2A}{c} \frac{(1 - \beta)^2 - (1 + \beta)^2}{1 - \beta^2} = -8A \beta\gamma^2 \frac{I}{c}. \qquad (\text{S-13.102})$$

(c) From the answer to point (b) we have $F' < 0$, the direction of the force is opposite to the direction of v. At the limit $v \ll c$, the force in the laboratory frame is equal to the force in the mirror frame, and we have

$$\mathbf{F} \simeq \mathbf{F}' \simeq -8A \frac{I}{c^2} v, \qquad (\text{S-13.103})$$

which is a viscous force. Under the action of this force, the mirror velocity will decrease exponentially in time

$$v(t) = v(0) e^{-t/\tau}, \quad \text{where} \quad \tau = \frac{Mc^2}{8AI}. \qquad (\text{S-13.104})$$

This effect has some analogies with the "laser-cooling" techniques, used in order to cool atoms down to temperatures of the order of 10^{-6} K. These include, for instance, Doppler cooling and Sisyphus cooling. The cooling of a macroscopic mirror by radiation pressure has also been studied [1] for possible applications in experiments of optical interferometry of ultra-high precision, e.g., for the detection of gravitational waves.

S-13.8   Radiation Pressure on a Thin Foil

(a) It is instructive to solve this problem by three different methods. For definiteness we assume a linearly polarized incident wave, with electric field $\mathbf{E}_i = \hat{\mathbf{y}} E_i e^{ik_i x - i\omega t}$, where $k_i = \omega/c$; generalization to arbitrary polarization is straightforward.

First method (heuristic): we assume the incident plane wave to be a square pulse of arbitrary but finite duration τ, and thus length $c\tau$. The momentum of the wave packet impinging on the surface A of the foil is, neglecting boundary effects,

$$\mathbf{p}_i = \frac{\langle \mathbf{S}_i \rangle}{c^2} c\tau A = \hat{\mathbf{x}} \frac{\langle |\mathbf{E}_i|^2 \rangle}{4\pi c} \tau A = \hat{\mathbf{x}} \frac{E_i^2}{8\pi c} \tau A = \hat{\mathbf{x}} \frac{I}{c} \tau A, \qquad (\text{S-13.105})$$

where $I = \langle |\mathbf{S}_i| \rangle = c \left(|\mathbf{E}_i|^2 \right)/(4\pi) = cE_i^2/(8\pi)$ is the intensity of the incident wave. The reflected and transmitted wave packets have momenta

$$\mathbf{p}_r = \frac{\langle \mathbf{S}_r \rangle}{c^2} c\tau A = -\hat{\mathbf{x}}\, R\, \frac{E_i^2}{8\pi c}\tau A = -\hat{\mathbf{x}}\, R\, \frac{I}{c}\tau A \ , \qquad (\text{S-13.106})$$

$$\mathbf{p}_t = \frac{\langle \mathbf{S}_t \rangle}{c^2} c\tau A = +\hat{\mathbf{x}}\, T\, \frac{E_i^2}{8\pi c}\tau A = +\hat{\mathbf{x}}\, T\, \frac{I}{c}\tau A \ , \qquad (\text{S-13.107})$$

respectively, where $R = |r|^2$, $T = |t|^2$, and $R + T = 1$ because of energy conservation. The maount of momentum transfered from the incident wave packet to the foil is

$$\Delta \mathbf{p} = \mathbf{p}_i - (\mathbf{p}_r + \mathbf{p}_t) \ , \qquad (\text{S-13.108})$$

resulting in a pressure pushing the foil toward positive x values (because $\Delta \mathbf{p} > 0$)

$$P_{\text{rad}} = \frac{|\Delta \mathbf{p}|}{\tau A} = \left[1 - (-R + T) \right] \frac{I}{c} = 2R\, \frac{I}{c} \ . \qquad (\text{S-13.109})$$

Second method: we calculate the average force on the foil, parallel to $\hat{\mathbf{x}}$, directly as

$$\langle \mathbf{F} \rangle = \int_0^d \langle \mathbf{J} \times \mathbf{B} \rangle \, A \, \mathrm{d}x \ , \qquad (\text{S-13.110})$$

where we have assumed the left surface of the foil located at $x = 0$, and the right surface located at $x = d$. For a very small thickness d we can write

$$A \int_0^d \langle \mathbf{J} \times \mathbf{B} \rangle \, \mathrm{d}x = \frac{1}{2} Ad \, \langle J(t) \left[B(0^+) + B(0^-) \right] \rangle$$

$$= -\frac{Ac}{8\pi} \langle \left[B^2(0^+) - B^2(0^-) \right] \rangle \ , \qquad (\text{S-13.111})$$

where we have substituted $J(t) = -\left[B(0^+) - B(0^-) \right] c/(4\pi d)$. Since we have $|B(0^+)| = |E_t| = |t\, E_i|$, and $|B(0^-)| = |E_i - E_r| = |(1 - r)\, E_i|$, we can write the radiation pressure on the foil as

$$P_{\text{rad}} = \frac{\langle |\mathbf{F}| \rangle}{A} = -\frac{E_i^2}{16\pi c} \left(|t|^2 - |1 - r|^2 \right) = -\frac{I}{2c} \left(|t|^2 - |1 - r|^2 \right) \ . \quad (\text{S-13.112})$$

Introducing the shorthand $\alpha = (\omega_p^2 d)/(2\omega c)$ in (13.5), so that $\eta = i\alpha$, we have

$$t = \frac{1}{1 + i\alpha}, \qquad\qquad T = |t|^2 = \frac{1}{1 + \alpha^2}, \qquad \text{(S-13.113)}$$

$$r = -\frac{i\alpha}{1 + i\alpha}, \qquad\qquad R = |r|^2 = \frac{\alpha^2}{1 + \alpha^2}, \qquad \text{(S-13.114)}$$

$$|1 - r|^2 = \frac{1 + 5\alpha^2 + 4\alpha^4}{(1 + \alpha^2)^2}, \qquad |t|^2 - |1 - r|^2 = -\frac{4\alpha^2}{1 + \alpha^2}, \qquad \text{(S-13.115)}$$

and finally

$$P_{\text{rad}} = \frac{I}{2c^2} \frac{4\alpha^2}{1 + \alpha^2} = 2R \frac{I}{c}. \qquad \text{(S-13.116)}$$

Third method: we calculate the flow of EM momentum directly using Maxwell's stress tensor T_{ij}. The theorem of EM momentum conservation states that

$$\frac{d p_i}{dt} = \oint_S T_{ij} \, n_j \, dS \qquad \text{(S-13.117)}$$

(summation over the repeated index is implied), where $\hat{n}$ is the unit vector normal to the surface S which envelops the thin foil, and $\mathbf{p}$ is the total momentum (EM and mechanical) of the foil. Since in a steady state the EM contribution is constant, the RHS of (S-13.117) equals the variation of mechanical momentum, i.e., the force.

Taking into account that the electric field has only the component E_y and the magnetic field only the component B_z, and that $\hat{n} = \mp\hat{x}$ on the left ($x = 0^-$) and right ($x = 0^+$) surfaces, respectively, the only relevant component of T_{ij} is T_{xx}, and

$$\frac{d\mathbf{p}}{dt} = \left[T_{xx}(0^+) - T_{xx}(0^-) \right] A. \qquad \text{(S-13.118)}$$

For $T_{xx}(0^+)$ and $T_{xx}(0^-)$ we have

$$T_{xx}(0^+) = -\frac{1}{8\pi} \langle E^2(0^+) + B^2(0^+) \rangle = -\frac{1}{4\pi} \langle |E_t|^2(0^+) \rangle$$

$$= -T \frac{E_i^2}{8\pi},$$

$$T_{xx}(0^-) = -\frac{1}{8\pi} \langle E^2(0^-) + B^2(0^-) \rangle = -\frac{\langle E_i^2 \rangle}{8\pi} \left[|(1 + r)|^2 + |(1 - r)|^2 \right]$$

$$= -\frac{E_i^2}{16\pi} \left(1 + |r|^2 + rr^* + 1 + |r|^2 - rr^* \right)$$

$$= -(1 + R) \frac{E_i^2}{8\pi}. \qquad \text{(S-13.119)}$$

Thus

$$\frac{d\mathbf{p}}{dt} = \frac{E_i^2}{8\pi}(-T + 1 + R)A = 2R\frac{I}{c}A,$$ (S-13.120)

which yields (13.6) again.

(b) From the Lorentz transformation of the fields we obtain the intensity of the incident wave in the S' frame, where the foil is at rest,

$$I' = \frac{1-\beta}{1+\beta}I,$$ (S-13.121)

and the force on the foil in S' is $F' = 2AI'/c$. Since for a force parallel to v we have $F = F'$, in the laboratory frame S we can write

$$F = F' = 2\frac{1-\beta}{1+\beta}\frac{I}{c}A.$$ (S-13.122)

(c) The radiation pressure must be multiplied by a factor $R = R(\omega')$ in the frame S', where the frequency is $\omega' = \sqrt{(1-\beta)/(1+\beta)}\,\omega$. Thus

$$F = 2\frac{1-\beta}{1+\beta}R(\omega')\frac{I}{c}A, \qquad \omega' = \sqrt{\frac{1-\beta}{1+\beta}}\,\omega.$$ (S-13.123)

S-13.9   Thomson Scattering in the Presence of a Magnetic Field

(a) We write the fields in the complex notation. Within our assumptions, the equation of motion for the electron is

$$m_e\frac{d\mathbf{v}}{dt} = -e\left(\mathbf{E} + \frac{\mathbf{v}}{c} \times \mathbf{B}_0\right),$$ (S-13.124)

where $-e$ and m_e are the charge and mass of the electron, respectively. The solution has already been evaluated in Problem 7.10, and is

$$v_x = \frac{\omega_c}{\omega_c^2 - \omega^2}\frac{e}{m_e}E_i\,e^{-i\omega t}, \qquad v_y = \frac{i\omega}{\omega_c^2 - \omega^2}\frac{e}{m_e}E_i\,e^{-i\omega t},$$ (S-13.125)

and $v_z = 0$.

(b) The cycle-averaged radiated power is

$$\langle P \rangle = \frac{e^4}{3m_e^2 c^3}|E_i|^2\frac{\omega^2}{\left(\omega_c^2 - \omega^2\right)^2}\left(\omega_c^2 + \omega^2\right),$$ (S-13.126)

which is maximum at the cyclotron resonance, $\omega = \omega_c$. At the low-frequency limit $\omega/\omega_c \ll 1$ we have $\langle P \rangle \propto \omega^2/\omega_c^2$, while at the high-frequency limit $\omega/\omega_c \gg 1$ the power is independent of frequency ("white" spectrum).

The orbit of the electron is elliptical, consequently the angular distribution and polarization of the scattered radiation are analogous to what found for an electron in the presence of an elliptically polarized wave, in the absence of external magnetic fields, as discussed in Problem 10.9. According to (S-13.125) we have $v_x/v_y = -i\omega_c/\omega$. At the limit $\omega \ll \omega_c$ we have $\langle |v_x| \rangle \gg \langle |v_y| \rangle$, the major axis of the elliptical orbit of the electron is thus parallel to $\hat{\mathbf{x}}$, and the strongest radiation intensity is observed on the yz plane. At the opposite limit, $\omega \gg \omega_c$, we have $\langle |v_x| \rangle \ll \langle |v_y| \rangle$, the major axis of the orbit is parallel to $\hat{\mathbf{y}}$, and the strongest radiation intensity is observed on the xz plane.

S-13.10   Undulator Radiation

(a) According to Maxwell's equation $\nabla \cdot \mathbf{B} = 0$, we must have

$$\partial_x B_x = -\partial_y B_y = -(\partial_y b) \cos(kx) , \qquad (S\text{-}13.127)$$

which, after integration in dx, leads to

$$B_x = -(\partial_y b) \frac{\sin(kx)}{k} , \qquad (S\text{-}13.128)$$

where we have set to zero the integration constant. In static conditions, and in the absence of electric currents, we have $\nabla \times \mathbf{B} = 0$, thus we must also have

$$0 = \partial_x B_y - \partial_y B_x = -kb(y) \sin(kx) + (\partial_y^2 b) \frac{\sin(kx)}{k} , \qquad (S\text{-}13.129)$$

which, divided by $\sin(kx)$, reduces to

$$\partial_y^2 b(y) = k^2 b(y) . \qquad (S\text{-}13.130)$$

The even solution (S-13.130) is

$$b(y) = B_0 \cosh(ky) , \qquad (S\text{-}13.131)$$

where B_0 is a constant to be determined. Thus the two nonzero components of $\mathbf{B}$ are

$$B_x = -B_0 \sin(kx) \sinh(ky) , \qquad B_y = B_0 \cos(kx) \cosh(ky) , \quad (S\text{-}13.132)$$

and on the z axis, where $x = 0$ and $y = 0$, we have

$$\mathbf{B}(0, 0, z) = \hat{\mathbf{y}} B_0 \,. \qquad (S\text{-}13.133)$$

(b) The Lorentz transformations from the laboratory frame S to S' give for the fields in S'

$$B'_x = B_x[x(x', t'), y'] = -B_0 \sinh(ky') \sin[k\gamma(x' + vt')] \,, \qquad (S\text{-}13.134)$$
$$B'_y = \gamma B_y[x(x', t'), y'] = \gamma B_0 \cosh(ky') \cos[k\gamma(x' + vt')] \,, \qquad (S\text{-}13.135)$$
$$E'_z = \gamma v B_y[x(x', t'), y'] = \gamma v B_0 \cosh(ky') \cos[k\gamma(x' + vt')] \qquad (S\text{-}13.136)$$

where $\gamma = 1/\sqrt{1 - v^2/c^2}$. Since the boost is parallel to the x axis, we have $y' = y$.

Disregarding the magnetic force in S', the electron oscillates along $\hat{\mathbf{z}}'$ under the action of the electric field $E' = E'_z(0, 0, t') = \gamma v B_0 \cos(\omega' t')$, where $\omega' = k\gamma v$. Thus, in S', we observe a Thomson scattering, and the electron emits electric-dipole radiation of frequency ω'.

(d) Transforming back to S, the frequencies of the radiation emitted in the forward $(+)$ and backward $(-)$ directions are

$$\omega_\pm = \gamma(1 \pm \beta)\,\omega' = \gamma(1 \pm \beta)k\gamma v = kc\gamma^2\beta(1 \pm \beta) \,, \qquad (S\text{-}13.137)$$

where $\beta = v/c$.

In S', the electron does not emit radiation along its direction of oscillation, i.e., along $\hat{\mathbf{z}}'$. This corresponds to two "forbidden" wavevectors $\mathbf{k}' \equiv (0, 0, \omega'/c)$ and $\mathbf{k}' \equiv (0, 0, -\omega'/c)$. By a back transformation to S we obtain

$$k_x = \gamma\left(k'_x \pm \frac{\omega'}{c}\beta\right) = \pm\gamma\beta\frac{\omega'}{c} \,, \quad k_y = 0 \,, \quad k_z = k'_z = \frac{\omega'}{c} \,, \qquad (S\text{-}13.138)$$

thus, in S, we have no radiation emission at the angles $\pm\theta$ in the xz plane such that

$$\tan\theta = \frac{k_z}{k_x} = \frac{1}{\gamma\beta} \,. \qquad (S\text{-}13.139)$$

The "undulator radiation", emitted by high-energy electrons injected along a periodically modulated magnetic field, is at the basis of free-electron lasers emitting coherent radiation in the X-ray frequency range.

S-13.11   Electromagnetic Torque on a Conducting Sphere

(a) We can write the electric field of the wave as

$$E(z, t) = E_0[\hat{x} \cos(kz - \omega t) - \hat{y} \sin(kz - \omega t)]$$
$$= \text{Re}[E_0(\hat{x} + i\hat{y}) e^{i(kz - \omega t)}], \tag{S-13.140}$$

where $k = \omega/c = 2\pi/\lambda$. Since $a \ll \lambda$, we can consider the electric and magnetic fields of the wave as uniform over the volume of the sphere, and neglect the magnetic induction effects. Thus, the sphere can be considered as located in a uniform rotating electric field

$$\mathbf{E}_0(t) = \text{Re}\left(\tilde{\mathbf{E}}_0 e^{-i\omega t}\right), \quad \text{where} \quad \tilde{\mathbf{E}}_0 = E_0(\hat{x} + i\hat{y}). \tag{S-13.141}$$

In the presence of oscillating fields, the complex electric permittivity of a medium of real conductivity σ is defined as

$$\tilde{\varepsilon}(\omega) = 1 + \frac{4\pi i\sigma}{\omega}. \tag{S-13.142}$$

Thus, our problem is analogous to Problem 3.4, where we considered a dielectric sphere in a uniform external electric field. The internal electric field and the dipole moment of the sphere are

$$\tilde{\mathbf{E}}_{\text{int}} = \frac{3}{\tilde{\varepsilon} + 2} \tilde{\mathbf{E}}_0 = \frac{3\tilde{\mathbf{E}}_0}{3 + 4\pi i\sigma/\omega} = -\frac{3i\omega t_d}{1 - 3i\omega t_d} \tilde{\mathbf{E}}_0, \tag{S-13.143}$$

$$\tilde{\mathbf{p}} = \mathbf{P}V = \chi \tilde{\mathbf{E}}_{\text{int}} V = \frac{3V}{4\pi} \frac{\tilde{\varepsilon} - 1}{\tilde{\varepsilon} + 2} \tilde{\mathbf{E}}_0 = \frac{3V}{4\pi} \frac{4\pi i\sigma/\omega}{3 + 4\pi i\sigma/\omega} \tilde{\mathbf{E}}_0,$$

$$= \frac{3V}{4\pi} \frac{4\pi i\sigma}{3\omega + 4\pi i\sigma} \tilde{\mathbf{E}}_0 = \frac{3V}{4\pi} \frac{i/t_d}{3\omega + i/t_d} \tilde{\mathbf{E}}_0 = \frac{3V}{4\pi} \frac{i}{3\omega t_d + i} \tilde{\mathbf{E}}_0$$

$$= \frac{3V}{4\pi} \frac{1 + 3i\omega t_d}{(3\omega t_d)^2 + 1} \tilde{\mathbf{E}}_0 \tag{S-13.144}$$

where $V = 4\pi a^3/3$ is the volume of the sphere, and $t_d = 1/(4\pi\sigma)$. By writing the complex numerator in terms of its modulus and argument we have

$$1 + 3i\omega t_d = \sqrt{1 + (3\omega t_d)^2} e^{i\phi}, \quad \text{where} \quad \phi = \arctan(3\omega t_d), \tag{S-13.145}$$

and, substituting into (S-13.144) we obtain

$$\tilde{\mathbf{p}} = \frac{3V}{4\pi} \frac{\tilde{\mathbf{E}}_0}{\sqrt{1 + (3\omega t_d)^2}} e^{i\phi}, \tag{S-13.146}$$

and, for the real quantity,

$$
\mathbf{p} = \mathrm{Re}\left[\frac{3V}{4\pi}\frac{E_0}{\sqrt{1+(3\omega t_{\mathrm{d}})^2}}\,(\hat{\mathbf{x}}+\mathrm{i}\hat{\mathbf{y}})\,\mathrm{e}^{-\mathrm{i}(\omega t-\phi)}\right]
$$

$$
= \frac{3V}{4\pi}\frac{E_0}{\sqrt{1+(3\omega t_{\mathrm{d}})^2}}[\hat{\mathbf{x}}\cos(\omega t-\phi)+\hat{\mathbf{y}}\sin(\omega t-\phi)]\,.\quad\text{(S-13.147)}
$$

Thus the dipole moment of the sphere rotates with a phase delay ϕ relative to the electric field of the wave.

(b) The torque acting on an electric dipole $\mathbf{p}$ in the presence of an electric field $\mathbf{E}$ is $\boldsymbol{\tau} = \mathbf{p}\times\mathbf{E}$. In our case, the angle between $\mathbf{p}$ and $\mathbf{E}_0$ is constant in time and equal to ϕ, thus the torque is

$$
\boldsymbol{\tau} = \hat{\mathbf{z}}\,|\mathbf{p}||\mathbf{E}_0|\sin\phi = \hat{\mathbf{z}}\,\frac{3V}{4\pi}\frac{E_0^2}{\sqrt{1+(3\omega t_{\mathrm{d}})^2}}\sin\phi\,.\qquad\text{(S-13.148)}
$$

The same result can be obtained by evaluating

$$
\boldsymbol{\tau} = \frac{1}{2}\,\mathrm{Re}\left(\tilde{\mathbf{p}}\times\tilde{\mathbf{E}}^*\right) = \mathrm{Re}\left[\frac{3V}{8\pi}\frac{E_0^2\,\mathrm{e}^{\mathrm{i}\phi}}{\sqrt{1+(3\omega t_{\mathrm{d}})^2}}\,(\hat{\mathbf{x}}+\mathrm{i}\hat{\mathbf{y}})\times(\hat{\mathbf{x}}-\mathrm{i}\hat{\mathbf{y}})\right]
$$

$$
= \mathrm{Re}\left[\frac{3V}{8\pi}\frac{E_0^2\,(\cos\phi+\mathrm{i}\sin\phi)}{\sqrt{1+(3\omega t_{\mathrm{d}})^2}}\,(-2\mathrm{i}\,\hat{\mathbf{z}})\right] = \hat{\mathbf{z}}\,\frac{3V}{4\pi}\frac{E_0^2}{\sqrt{1+(3\omega t_{\mathrm{d}})^2}}\sin\phi\,.
$$
$$
\text{(S-13.149)}
$$

S-13.12 Surface Waves in a Thin Foil

(a) As an educated guess, we search for solutions for the unknown quantities E_x, and B_z of the form

$$
E_x(x,y,t) = \tilde{E}_x(x)\,\mathrm{e}^{\mathrm{i}ky-\mathrm{i}\omega t}\,,\qquad B_z(x,y,t) = \tilde{B}_z(x)\,\mathrm{e}^{\mathrm{i}ky-\mathrm{i}\omega t}\,,\qquad\text{(S-13.150)}
$$

where $\tilde{E}_x(x)$, and $\tilde{B}_z(x)$ are complex functions to be determined. According to (13.10), E_y is symmetric (even) for reflection across the $x=0$ plane. Since in vacuum we have $\nabla\cdot\mathbf{E} = \partial_x E_x + \partial_y E_y = 0$, we obtain

$$
\partial_x\tilde{E}_x = -\mathrm{i}k E_0\,\mathrm{e}^{-q|x|}\,,\qquad\text{(S-13.151)}
$$

which, after integration in $\mathrm{d}x$, leads to

$$\tilde{E}_x(x) = \mathrm{sgn}(x)\,\frac{ik}{q}\,E_0\,\mathrm{e}^{-q|x|} = \begin{cases} -\dfrac{ik}{q}\,E_0\,\mathrm{e}^{qx}\,, & x<0,\\[2mm] \dfrac{ik}{q}\,E_0\,\mathrm{e}^{-qx}\,, & x>0. \end{cases} \tag{S-13.152}$$

Thus, the E_x component is antisymmetric (odd) for reflection across the $x = 0$ plane. Since our fields are independent of z, Maxwell's equation $\nabla \times \mathbf{E} = -\partial_t \mathbf{B}/c$ reduces to

$$-\frac{1}{c}\,\partial_t B_z = \partial_x E_y - \partial_y E_x = \mathrm{sgn}(x)\left(\frac{k^2}{q} - q\right) E_0\,\mathrm{e}^{-q|x|}\,\mathrm{e}^{\mathrm{i}(ky-\omega t)}\,, \tag{S-13.153}$$

which, after integration in dt and division by $-\mathrm{e}^{\mathrm{i}(ky-\omega t)}/c$, leads to

$$\tilde{B}_z = \mathrm{sgn}(x)\,\frac{\mathrm{i}c}{q\omega}\left(q^2 - k^2\right) E_0\,\mathrm{e}^{-q|x|}\,, \tag{S-13.154}$$

thus $\tilde{B}_z$, like $\tilde{E}_x$, is an odd function of x.

We can obtain the surface charge density $\sigma(y, t)$ and the surface current density $\mathbf{K}(y, t)$ on the foil from the boundary conditions at the $x = 0$ plane. Figure S-13.6 shows the surface current $\mathbf{K}$ and the magnetic field close to the foil.

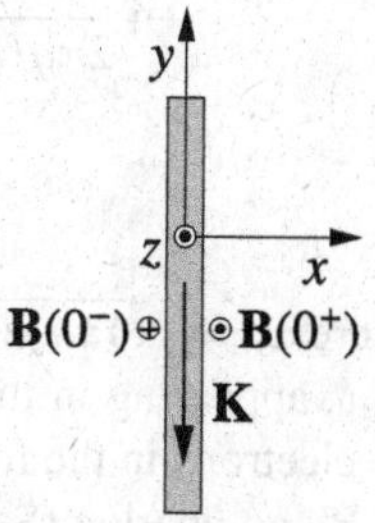

Fig. S-13.6

$$\begin{aligned}\sigma(y, t) &= \frac{1}{4\pi}\left[E_x\left(x = 0^+, y, t\right) - E_x\left(x = 0^-, y, t\right)\right]\\ &= \mathrm{i}\,\frac{2k}{q}\,E_0\,\mathrm{e}^{\mathrm{i}(ky-\omega t)}\,, \end{aligned} \tag{S-13.155}$$

$$\begin{aligned}K_y(y, t) &= -\frac{c}{4\pi}\left[B_z\left(x = 0^+, y, t\right) - B_z\left(x = 0^-, y, t\right)\right]\\ &-\mathrm{i}\,\frac{c^2}{2\pi q\omega}\left(q^2 - k^2\right) E_0\,\mathrm{e}^{\mathrm{i}(ky-\omega t)}\,, \end{aligned} \tag{S-13.156}$$

while the z component of $\mathbf{K}$ is zero because its presence would imply a nonzero y component of $\mathbf{B}$.

(b) The time-averaged Poynting vector can be written as

$$\langle \mathbf{S} \rangle = \frac{c}{4\pi}\langle \mathbf{E} \times \mathbf{B} \rangle = \frac{c}{8\pi}\left[\hat{\mathbf{x}}\,\mathrm{Re}(\tilde{E}_y \tilde{B}_z^*) - \hat{\mathbf{y}}\,\mathrm{Re}(\tilde{E}_x \tilde{B}_z^*)\right], \tag{S-13.157}$$

where

$$\tilde{E}_y \tilde{B}_z^* = -\,\mathrm{sgn}(x)\frac{\mathrm{i}c}{q\omega}\left(q^2 - k^2\right)|E_0|^2\,\mathrm{e}^{-2q|x|}, \tag{S-13.158}$$

$$\tilde{E}_x \tilde{B}_z^* = \frac{kc}{q^2\omega}\left(q^2 - k^2\right)|E_0|^2\,\mathrm{e}^{-2q|x|}\,. \tag{S-13.159}$$

We thus obtain $\langle S_x \rangle = 0$ because $\tilde{E}_y \tilde{B}_z^*$, is purely imaginary, and the energy flow is in the $\hat{\mathbf{y}}$ direction only:

$$\langle \mathbf{S} \rangle = -\hat{\mathbf{y}} \frac{kc}{8\pi q^2 \omega} \left(q^2 - k^2\right) |E_0|^2 e^{-2q|x|} . \tag{S-13.160}$$

(c) Form Helmholtz's equation, we obtain

$$q^2 - k^2 + \frac{\omega^2}{c^2} = 0 . \tag{S-13.161}$$

(d) From (S-13.156) we can write, within our approximations,

$$\mathbf{J} = \frac{\mathbf{K}}{\ell} = -\hat{\mathbf{y}} \, \mathrm{i} \, \frac{c^2}{2\pi q \ell \omega} \left(q^2 - k^2\right) E_0 \, e^{i(ky - \omega t)} \tag{S-13.162}$$

and, combining with (13.11), we obtain

$$-\mathrm{i} \frac{c^2}{2\pi q \ell \omega} \left(q^2 - k^2\right) E_0 \, e^{i(ky - \omega t)} = 4\pi \mathrm{i} \frac{\omega_\mathrm{p}^2}{\omega} E_0 \, e^{i(ky - \omega t)} ,$$

$$q^2 - k^2 = -8\pi^2 \frac{\omega_\mathrm{p}^2}{c^2} q\ell . \tag{S-13.163}$$

where $\omega_\mathrm{p} = \sqrt{4\pi n_e e^2 / M_e}$ is the plasma frequency of the foil material. The product $2n_e \ell$, appearing in the expression $\omega_\mathrm{p}^2 \ell = 4\pi n_e \ell e^2$, is the surface number density of the electrons in the foil, which is the relevant parameter in this problem.

(e) By comparing (S-13.161) and (S-13.163) we obtain

$$\omega^2 = 8\pi^2 \omega_\mathrm{p}^2 q\ell = \Omega q c , \quad \text{where} \quad \Omega = \frac{8\pi^2 \omega_\mathrm{p}^2 \ell}{c} . \tag{S-13.164}$$

Solving (S-13.161) for q yields

$$cq^2 + \Omega q - k^2 c = 0 \quad \Rightarrow \quad q = \frac{\sqrt{\Omega^2 + (2kc)^2} - \Omega}{2c} , \tag{S-13.165}$$

where the root sign has been chosen so to have $q > 0$, as required by the boundary conditions, and in agreement with (13.10). Eventually, we obtain the dispersion relation:

$$\omega^2 = c^2 k^2 - c^2 q^2 = \frac{1}{2} \left[\Omega \sqrt{\Omega^2 + (2kc)^2} - \Omega^2 - (2kc)^2 \right] . \tag{S-13.166}$$

S-13.13   The Fizeau Effect

(a) In the rest frame of the medium, S', we have $\omega'/k' = c/n$. The Lorentz transformations from the laboratory frame S to S' lead to

$$\omega' = \gamma(\omega - uk) \simeq (\omega - uk), \quad k' = \gamma\left(k - \frac{u\omega}{c^2}\right) \simeq \left(k - \frac{u\omega}{c^2}\right), \quad \text{(S-13.167)}$$

since $\gamma \simeq 1$ up to the first order in $\beta = u/c$. Dividing the two equations side by side we obtain

$$\frac{c}{n} = \frac{\omega'}{k'} \simeq \frac{\omega - uk}{k - u\omega/c^2} = \frac{v_\varphi - u}{1 - v_\varphi u/c^2}, \quad \text{(S-13.168)}$$

where, in the last step, we have divided numerator and denominator by k, and substituted the phase velocity in the laboratory frame, $v_\varphi = \omega/k$. Multiplying the first and last term by $1 - v_\varphi u/c^2$ we obtain

$$\frac{c}{n} - u\frac{v_\varphi}{cn} = v_\varphi - u \;\Rightarrow\; v_\varphi\left(1 + \frac{u}{cn}\right) = \frac{c}{n} + u \;\Rightarrow\; v_\varphi = \frac{c(c + nu)}{cn + u}$$

$$\Rightarrow v_\varphi = c\,\frac{1 + n\beta}{n + \beta} \simeq c\left(\frac{1}{n} + \frac{n^2 - 1}{n^2}\,\beta\right), \quad \text{(S-13.169)}$$

where, in the last step, we have approximated the fraction by its first-degree Taylor polynomial in β. The phase velocity in the laboratory frame S is thus

$$v_\varphi = \frac{c}{n} + u\left(1 - \frac{1}{n^2}\right). \quad \text{(S-13.170)}$$

The experiment was performed in 1851, with light propagating in flowing water parallel to the water velocity. Fizeau expected to measure a phase velocity equal to the phase velocity of light in water, c/n, plus the flow velocity of water, u, i.e., $v_\varphi = (c/n) + u$, while the experimental result was in agreement with (S-13.170). This found a satisfactory explanation only 54 years later, in 1905, when Einstein published his theory of special relativity.

(b) Equation (S-13.170) takes into account the first-order correction to v_φ in $\beta = u/c$ for a non-dispersive medium. If the medium is dispersive according to a known law $n = n(\omega)$, we must also take into account that the frequency ω' observed in the rest-frame of the medium is different from the radiation frequency ω in the laboratory frame. We want to calculate the first-order correction to (S-13.170) in $\Delta\omega = \omega' - \omega$. We need to correct only the fist term of the right-hand side of (S-13.170), since the second term is already first-order, and a correction to it would be second-order. The first-order Doppler effect gives us

$$\Delta\omega = \omega' - \omega \simeq -\omega\,\frac{n(\omega)\,u}{c}, \quad \text{(S-13.171)}$$

since the light velocity in the medium is $c/n(\omega)$, an the medium is traveling away from the light source. Thus we have

$$
\begin{aligned}
\frac{c}{n(\omega')} &\simeq \frac{c}{n(\omega)} + \Delta\omega\, \partial_\omega \left(\frac{c}{n(\omega)} \right) \\
&= \frac{c}{n(\omega)} + \left(-\omega\, \frac{n(\omega)\, u}{c} \right) \left[-\frac{c}{n^2(\omega)}\, \partial_\omega n(\omega) \right] \\
&= \frac{c}{n(\omega)} + \omega\, \frac{u}{n(\omega)}\, \partial_\omega n(\omega)\,,
\end{aligned}
\tag{S-13.172}
$$

and the first-order expression for the phase velocity in the case of a dispersive medium is

$$
v_\varphi(\omega) = \frac{c}{n(\omega)} + u \left[1 - \frac{1}{n^2(\omega)} + \frac{\omega}{n(\omega)}\, \partial_\omega n(\omega) \right] + O(u^2)\,.
\tag{S-13.173}
$$

(c) The refractive index of the free electron medium is $n(\omega) = (1 - \omega_p^2/\omega^2)^{1/2}$, where ω_p is the plasma frequency. Thus we have inside the square brackets of (S-13.173)

$$
1 - \frac{1}{n^2(\omega)} = -\frac{1}{1 - \omega_p^2/\omega^2} = \frac{\omega_p^2}{\omega^2 - \omega_p^2}\,,
\tag{S-13.174}
$$

and

$$
\frac{\omega}{n(\omega)}\, \partial_\omega n(\omega) = -\frac{\omega}{(1 - \omega_p^2/\omega^2)^{1/2}}\, \frac{1}{(1 - \omega_p^2/\omega^2)^{3/2}}\, \frac{\omega_p^2}{\omega^3} = -\frac{\omega_p^2}{\omega^2 - \omega_p^2}\,,
\tag{S-13.175}
$$

so that the two first-order corrections to $v_\varphi(\omega)$ cancel out, and the phase velocity is independent of the flow velocity of the medium up to the second order in β.

S-13.14   Lorentz Transformations for Longitudinal Waves

(a) The Lorentz transformations for the wave frequency and wavevector are, in the case of a boost along $\hat{\mathbf{x}}$,

$$
\omega'_L = \gamma\,(\omega_L - V k_L)\,, \qquad k'_L = \gamma \left(k_L - \frac{V \omega_L}{c^2} \right)\,,
\tag{S-13.176}
$$

where $\gamma = 1/\sqrt{1 - V^2/c^2}$. In the special case where the boost velocity equals the phase velocity, $V = v_\varphi = \omega_L/k_L$, we have $\omega'_L = 0$, and the fields are independent of time (*static*) in S'. Further, recalling that $k_L = \omega_L/v_\varphi$, we have

$$k'_{\rm L} = \frac{1}{\sqrt{1 - v_\varphi^2/c^2}} \left(\frac{\omega_{\rm L}}{v_\varphi} - \frac{v_\varphi\,\omega_{\rm L}}{c^2} \right) = \frac{1}{\sqrt{1 - v_\varphi^2/c^2}} \frac{\omega_{\rm L}}{v_\varphi} \left(1 - \frac{v_\varphi^2}{c^2} \right)$$

$$= \frac{\omega_{\rm L}}{v_\varphi}\sqrt{1 - v_\varphi^2/c^2} = \frac{k_{\rm L}}{\gamma}\,. \tag{S-13.177}$$

If S' moves with velocity $\hat{\mathbf{x}}\,V = \hat{\mathbf{x}}\,v_\varphi$ relative to S, the fields in S' are obtained from (9.3) and are

$$\mathbf{E}' = \mathbf{E}'(x') = \hat{\mathbf{x}}\,E_0\,e^{ik'_{\rm L}x'}\,, \qquad \mathbf{B}' = 0\,, \tag{S-13.178}$$

i.e. $\mathbf{E}'$ is constant in time. The charge and current densities in S' can be obtained either by Lorentz transformations or directly from the equations

$$\rho' = \frac{1}{4\pi}\,\nabla'\cdot\mathbf{E}' = \frac{1}{4\pi}\,\partial_{x'}E'_x \quad \text{and} \quad 4\pi\mathbf{J}' + \partial_t'\mathbf{E}' = 0\,, \tag{S-13.179}$$

which lead to

$$\rho' = \frac{ik'_{\rm L}}{4\pi}\,E_0\,e^{ik'_{\rm L}x'}\,, \qquad \mathbf{J}' = 0\,. \tag{S-13.180}$$

(b) The Lorentz transformations for the case $V = c^2/v_\varphi = c^2 k_{\rm L}/\omega_{\rm L}$, and $v_\varphi > c$, lead to the following values for $k'_{\rm L}$ and $\omega'_{\rm L}$

$$k'_{\rm L} = \gamma\left(k_{\rm L} - \frac{V\omega_{\rm L}}{c^2} \right) = 0\,,$$

$$\omega'_{\rm L} = \gamma\left(\omega_{\rm L} - \frac{k_{\rm L}^2 c^2}{\omega} \right) = \gamma\omega_{\rm L}\left(1 - \frac{V^2}{c^2} \right) = \frac{\omega_{\rm L}}{\gamma}\,, \tag{S-13.181}$$

which imply that the fields propagate in space with infinite phase velocity, oscillating with uniform phase at frequency $\omega'_{\rm L}$. The fields are

$$\mathbf{E}' = \mathbf{E}'(t') = \hat{\mathbf{x}}\,E_0\,e^{-i\omega' t'}\,, \qquad \mathbf{B}' = 0\,, \tag{S-13.182}$$

i.e. $\mathbf{E}'$ is uniform in space. We also obtain $\rho' = 0$, and $\mathbf{J}' = \mathbf{J}/\gamma$.
(c) The Lorentz transformations of the wavevector and the frequency for a boost along the y axis are

$$k'_{{\rm L}x} = k_{{\rm L}x} = k_{\rm L}\,,$$

$$k'_{{\rm L}y} = \gamma\left(k_{{\rm L}y} - \frac{V\omega_{\rm L}}{c^2} \right) = -\gamma\frac{V\omega_{\rm L}}{c^2}\,,$$

$$\omega'_{\rm L} = \gamma(\omega_{\rm L} - Vk_{{\rm L}y}) = \gamma\omega_{\rm L}\,. \tag{S-13.183}$$

All fields and currents depend on space and time through a factor $e^{i(k'_{Lx}x'+k'_{Ly}y'-\omega'_L t')}$, thus, the propagation direction forms an angle

$$\theta' = \arctan(k'_{Ly}/k'_{Lx}) = -\arctan(\gamma V v_\varphi/c^2) \qquad (S\text{-}13.184)$$

with the x' axis. The wave has field amplitudes

$$E'_x = \gamma\left(E_x + \frac{V}{c}B_z\right) = \gamma E_0 , \qquad (S\text{-}13.185)$$

$$B'_z = \gamma\left(B_z - \frac{V}{c^2}E_x\right) = -\gamma\frac{V}{c^2}E_0 , \qquad (S\text{-}13.186)$$

all other field components being zero. Thus, in a frame moving transversally to the propagation direction, the wave is no longer purely longitudinal and electrostatic.

S-13.15 Lorentz Transformations for a Transmission Cable

(a) The continuity equation for a linear charge density is written $\partial_t \lambda = -\partial_z I$. Inserting the expressions for λ and I of (13.14) we obtain

$$-i\omega\lambda_0 = -ikI_0 , \quad \Rightarrow \quad I_0 = \frac{\omega}{k}\lambda_0 = v_\varphi\lambda_0 . \qquad (S\text{-}13.187)$$

(b) The dispersion relation is

$$\omega = v_\varphi k = \frac{c}{n}k = \frac{c}{\sqrt{\varepsilon}}k , \qquad (S\text{-}13.188)$$

with

$$v_\varphi = \frac{c}{\sqrt{\varepsilon}} \quad \text{and} \quad k = \frac{\omega}{v_\varphi} = \frac{\omega\sqrt{\varepsilon}}{c} . \qquad (S\text{-}13.189)$$

The electric field can be evaluated by applying Gauss's law to a cylindrical surface coaxial to the wire, of radius r and height h. Since the field is transverse, and we have cylindrical symmetry around the wire, the only nonzero component of $\mathbf{E}$ is E_r

$$\mathbf{E}(r, z, t) = \hat{\mathbf{r}}\frac{2\lambda}{\varepsilon r} = \hat{\mathbf{r}}\,E_r(r)\,e^{ikz-i\omega t} , \quad \text{where} \quad E_r(r) = \frac{2\lambda_0}{\varepsilon r} . \qquad (S\text{-}13.190)$$

The magnetic field can be evaluated by applying Stokes' theorem to a circle of radius r, coaxial to the wire. Because of symmetry, the only nonzero component of $\mathbf{B}$ is the azimuthal component B_ϕ

$$\mathbf{B}(r, z, t) = \hat{\phi} \frac{2I}{rc} = \hat{\phi} B_\phi(r) e^{ikz - i\omega t}, \quad \text{where} \quad B_\phi(r) = \frac{2I_0}{rc} = \frac{2\omega\lambda_0}{krc},$$
$$\text{(S-13.191)}$$

that can be rewritten as

$$B_\phi(r) = \frac{\varepsilon\omega}{kc} E_r(r) = \frac{\varepsilon v_\varphi}{c} E_r(r) = \frac{c}{v_\varphi} E_r(r). \qquad \text{(S-13.192)}$$

(c) The wave frequency ω' in the frame S', moving at the phase velocity $\hat{\mathbf{z}} v_\varphi$ relative to the laboratory frame S, is

$$\omega' = \gamma(\omega - v_\varphi k) = 0, \qquad \text{(S-13.193)}$$

where we have used the second of (S-13.189). Thus the fields are static in S'. For our Lorentz boost we have

$$\beta = \hat{\mathbf{z}} \frac{v_\varphi}{c} = \frac{\hat{\mathbf{z}}}{\sqrt{\varepsilon}}, \quad \gamma = \frac{1}{\sqrt{1 - 1/\varepsilon}} = \sqrt{\frac{\varepsilon}{\varepsilon - 1}}, \qquad \text{(S-13.194)}$$

and the wave vector k' in S' can be written

$$k' = \gamma\left(k - \frac{v_\varphi \omega}{c^2}\right) = \sqrt{\frac{\varepsilon}{\varepsilon - 1}} \left(\frac{\omega\sqrt{\varepsilon}}{c} - \frac{\omega}{c\sqrt{\varepsilon}}\right) = \frac{\omega}{c}\sqrt{\varepsilon - 1}. \qquad \text{(S-13.195)}$$

The (z', t')-dependence (actually, only z'-dependence) of our physical quantities in S' will thus be through a factor $e^{ik'z'}$. The amplitude of linear charge density in S' is

$$\lambda_0' = \gamma\left(\lambda_0 - \frac{v_\varphi}{c^2} I_0\right) = \gamma\left(\lambda_0 - \frac{v_\varphi^2}{c^2}\lambda_0\right) = \gamma\left(\frac{1}{\gamma}\right)^2 \lambda_0 = \frac{\lambda_0}{\gamma}. \qquad \text{(S-13.196)}$$

The amplitude of the current in S' is

$$I_0' = \gamma\left(I_0 - v_\varphi \lambda_0\right) = 0. \qquad \text{(S-13.197)}$$

The field amplitudes transform according to (9.3), thus we have

$$E_r' = \gamma(E_r - \beta B_\phi) = \gamma\left(E_r - \frac{v_\varphi}{c}\frac{c}{v_\varphi} E_r\right) = 0, \qquad \text{(S-13.198)}$$

$$B_\phi' = \gamma\left(B_\phi - \frac{v_\varphi^2}{c^2} E_r\right) = \gamma\left(B_\phi - \frac{v_\varphi}{c}\frac{v_\varphi}{c} B_\phi\right) = \frac{B_\phi}{\gamma}. \qquad \text{(S-13.199)}$$

It might seem surprising that, in S', we have $\lambda' \neq 0$ and $E_r' = 0$, while $I' = 0$ and $B_\phi' \neq 0$. The reason is that we must take into account also the polarization charge of the medium in contact with the wire, $\lambda_p(z, t)$, the presence of a polarization current,

$J_p(r, z, t)$, and their Lorentz transformations. In the laboratory frame S we must have $\lambda(z, t) + \lambda_p(z, t) = \lambda(z, t)/\varepsilon$, thus

$$\lambda_p(z, t) = -\frac{\varepsilon - 1}{\varepsilon} \lambda_0 \, e^{ikz - i\omega t} = -\frac{\lambda_0}{\gamma^2} e^{ikz - i\omega t} = \lambda_0^{(p)} \, e^{ikz - i\omega t} \,, \qquad \text{(S-13.200)}$$

where $\lambda_0^{(p)} = -\lambda_0/\gamma^2$. The electric field (S-13.190) generates a polarization of the medium

$$\mathbf{P}(z, r, t) = \hat{\mathbf{r}} \frac{\varepsilon - 1}{4\pi} E_r(r) \, e^{ikz - i\omega t} = \hat{\mathbf{r}} \, P_r(r) \, e^{ikz - i\omega t} \,, \qquad \text{(S-13.201)}$$

where

$$P_r(r) = \frac{\varepsilon - 1}{4\pi} \frac{2\lambda_0}{\varepsilon r} = \frac{\varepsilon - 1}{\varepsilon} \frac{\lambda_0}{2\pi r} = \frac{1}{\gamma^2} \frac{\lambda_0}{2\pi r} \,. \qquad \text{(S-13.202)}$$

A time-dependent polarization is associated to a polarization current density

$$\mathbf{J}_p = \partial_t \mathbf{P} = -\hat{\mathbf{r}} \, i\omega P_r(r) \, e^{ikz - i\omega t} = \hat{\mathbf{r}} \, J_r(r) \, e^{ikz - i\omega t} \qquad \text{(S-13.203)}$$

where

$$J_r(r) = -i\omega P_r(r) = -i \frac{\omega}{\gamma^2} \frac{\lambda_0}{2\pi r} \,. \qquad \text{(S-13.204)}$$

Thus, $\mathbf{J}_p$ is radial in S. According to the first of (9.1), we have a polarization four-current

$$J_\mu^{(p)}(r, z, t) = \left[c \, \varrho_0^{(p)}(r), \, \hat{\mathbf{r}} \, J_r(r) \right] e^{ikz - i\omega t} \,, \qquad \text{(S-13.205)}$$

where, for instance

$$\varrho_0^{(p)}(r) = \begin{cases} \dfrac{\lambda_0^{(p)}}{\pi r_0^2} \,, & \text{if } r < r_0 \\[2mm] 0 & \text{if } r > r_0 \end{cases} \,, \quad \text{so that} \quad \lambda_0^{(p)} = \int\limits_0^\infty \varrho_0^{(p)}(r) \, 2\pi r \, dr \,, \quad \text{(S-13.206)}$$

and we are interested in the limit $r_0 \to 0$. We can thus write

$$J_\mu^{(p)}(r, z, t) = G_\mu(r) \, e^{ikz - i\omega t} \,, \quad \text{where} \quad G_\mu = \left[c \, \varrho_0^{(p)}, \, \hat{\mathbf{r}} \, J_r(r) \right] \,. \qquad \text{(S-13.207)}$$

The four-vector G_μ transforms according to (9.2), thus we have in S'

$$G_0' = \gamma (G_0 - \boldsymbol{\beta} \cdot \mathbf{G}) = \gamma G_0 \,, \qquad \text{(S-13.208)}$$

since the spacelike component of G_μ, being radial, is perpendicular to $\boldsymbol{\beta}$. The amplitude of the linear polarization charge density in S' is

$$\lambda_0^{(p)\prime} = \gamma \int_0^\infty \frac{G_0}{c} 2\pi r \, dr = \gamma \lambda_0^{(p)} = -\gamma \frac{\lambda_0}{\gamma^2} = -\frac{\lambda_0}{\gamma} , \qquad \text{(S-13.209)}$$

which cancels (S-13.196), therefore we have $\mathbf{E}' = 0$. The radial component of $\mathbf{J}_p$ does not contribute to the magnetic field, thus we are interested in

$$G'_\parallel = \gamma \left(G_\parallel - \beta G_0 \right) = -\gamma \beta G_0 = -\gamma \beta \, \varrho_0^{(p)}(r) \, c , \qquad \text{(S-13.210)}$$

which corresponds to a polarization current in S' of amplitude

$$I_0^{(p)\prime} = \int_\infty G'_\parallel 2\pi r' \, dr' = -\gamma v_\varphi \lambda_0^{(p)} = \gamma v_\varphi \frac{\lambda_0}{\gamma^2} = \frac{I_0}{\gamma} \qquad \text{(S-13.211)}$$

in agreement with (S-13.199).

S-13.16   A Waveguide with a Moving End

(a) The electric field of the TE_{10} must be parallel to the two conducting planes, thus it must vanish on them, and be of the form $\mathbf{E}(x, y, t) = \hat{\mathbf{z}} \, E_0 \cos(\pi y/a) f(x, t)$. The dispersion relation is

$$\omega^2 = \omega_{co}^2 + k^2 c^2 , \quad \text{where} \quad \omega_{co} = \frac{\pi c}{a} \qquad \text{(S-13.212)}$$

is the cutoff frequency of the waveguide. In our terminated waveguide, the global electric field is the superposition of the fields of the wave incident on the terminating wall at $x = 0$, and of the reflected wave. Incident and reflected wave have equal amplitudes, thus

$$\mathbf{E}(x, y, t) = \hat{\mathbf{z}} \, E_0 \cos \left(\frac{\pi y}{a} \right) \sin(kx) \, e^{-i\omega t} , \qquad \text{(S-13.213)}$$

where the phase has been chosen so that $\mathbf{E}(0, y, t) = 0$. The magnetic field can be obtained from the relation $\partial_t \mathbf{B} = -c \nabla \times \mathbf{E}$, and has the components

$$B_x = -\frac{ic}{\omega} \partial_y E_z = \frac{i\pi c}{\omega a} E_0 \sin \left(\frac{\pi y}{a} \right) \sin(kx) \, e^{-i\omega t} , \qquad \text{(S-13.214)}$$

$$B_y = \frac{ic}{\omega} \partial_x E_z = -\frac{kc}{\omega} E_0 \cos \left(\frac{\pi y}{a} \right) \cos(kx) \, e^{-i\omega t} . \qquad \text{(S-13.215)}$$

Note that $B_x(0, y, t) = 0$, as required.

(b) In the frame S' where the waveguide termination is at rest ($v' = 0$), the incident wave has frequency and wavevector

$$\omega_i' = \gamma(\omega - \beta kc) , \qquad k_i' = \gamma(k - \beta\omega/c) , \qquad \text{(S-13.216)}$$

where $\beta = v/c$. Since we assumed $\beta < kc/\omega$, we have $k_i' > 0$ (note that $\omega_i' > 0$ anyway because $k < \omega/c$). In S' the reflected wave has frequency and wavevector

$$\omega_r' = \omega_i' , \qquad k_r' = -k_i' . \qquad \text{(S-13.217)}$$

By transforming back into the laboratory frame S we obtain

$$\omega_r = \gamma(\omega_r' + \beta k_r' c) = \gamma^2 \left[(1 + \beta^2)\omega - 2\beta kc\right] , \qquad \text{(S-13.218)}$$

$$k_r = \gamma\left(-k + \beta\frac{\omega}{c}\right) = \gamma^2 \left[-(1 + \beta^2)k + 2\beta\frac{\omega}{c}\right] , \qquad \text{(S-13.219)}$$

As a check, at the limit $a \to \infty$ we have $\omega_{co} \to 0$ and $k \to \omega/c$, and we obtain (S-9.54) of Problem 9.6 for the frequency reflected by a moving mirror. With some algebraic manipulations we obtain

$$\omega_r^2 - c^2 k_r^2 = \gamma^4 \left\{\left[(1 + \beta^2)\omega - 2\beta kc\right]^2 - c^2 \left[-(1 + \beta^2)k + 2\beta\frac{\omega}{c}\right]^2\right\}$$

$$= \gamma^4 \left[(1 + \beta^2)\omega - 2\beta kc + (1 + \beta^2)kc - 2\beta\omega\right] \times$$

$$\times \left[(1 + \beta^2)\omega - 2\beta kc - (1 + \beta^2)kc + 2\beta\omega\right]$$

$$= \gamma^2 \left[(1 - \beta)^2\omega + (1 - \beta)^2 kc\right]\left[(1 + \beta)^2\omega + (1 + \beta)^2 kc\right]$$

$$= (\omega + kc)(\omega - kc) = \omega^2 - k^2 c^2 . \qquad \text{(S-13.220)}$$

(c) If $v > kc^2/\omega$, in S' we have $k_i' < 0$, the incident wave propagates parallel to $-\hat{\mathbf{x}}'$, and cannot reach the waveguide termination. In these conditions there is no reflected wave. The condition is equivalent to $v > v_g$, the group velocity in the waveguide.

S-13.17 A "Relativistically" Strong Electromagnetic Wave

(a) The equations of motion for p_x, p_y, and p_z in the presence of the electromagnetic fields of the wave are

$$\frac{dp_x}{dt} = -eE_x + \frac{e}{c} v_z B_y ,$$
(S-13.221)

$$\frac{dp_y}{dt} = -eE_y - \frac{e}{c} v_z B_x ,$$
(S-13.222)

$$\frac{dp_z}{dt} = -\frac{e}{c} v_x B_y + \frac{e}{c} v_y B_z .$$
(S-13.223)

In general the magnetic contribution is not negligible, since v_z is not necessarily much smaller than c. However, if we assume $v_z = 0$, the magnetic force vanishes. In these conditions the solutions of (S-13.221)–(S-13.222) are

$$\frac{dp_x}{dt} = -eE_0 \cos \omega t , \qquad\qquad \frac{dp_y}{dt} = -eE_0 \sin \omega t ,$$
(S-13.224)

$$p_x = -\frac{eE_0}{\omega} \sin \omega t , \qquad\qquad p_y = +\frac{eE_0}{\omega} \cos \omega t .$$
(S-13.225)

Inserting these solutions into (S-13.223) we have

$$\frac{dp_z}{dt} = \frac{e}{m_e \gamma}(-p_x B_y + p_y B_x) =$$

$$= \frac{e}{m_e \gamma} \frac{eE_0^2}{\omega c}(-\sin \omega t \cos \omega t + \cos \omega t \sin \omega t) = 0 ,$$
(S-13.226)

so that a p_z is constant in time. Either assuming $v_z = 0$ as initial condition or by a proper change of reference frame, $v_z = 0$ is a self-consistent assumption.

(b) Since $\mathbf{p}^2 = p_x^2 + p_y^2 = (eE_0/\omega)^2$ does not depend on time, the Lorentz factor $\gamma = \sqrt{1 + p^2/(m_e c)^2} = \sqrt{1 + [eE_0/(m_e \omega c)]^2}$ is a constant. This implies

$$\frac{d\mathbf{p}}{dt} = m_e \frac{d(\gamma v)}{dt} = \gamma m_e \frac{dv}{dt} .$$
(S-13.227)

The equations of motion have the same form as in the non-relativistic case if we make the replacement $m_e \rightarrow \gamma m_e$. The relativistic behavior can be obtained by attributing an "effective mass" γm_e, dependent on the wave intensity, to the electron.

(c) Accordingly, the refractive index for the relativistic case can be simply obtained by replacing $m_e \rightarrow m_e \gamma$ into the non-relativistic expression, so that $\omega_p = \sqrt{(4\pi n_e e^2/m_e)} \rightarrow \omega_p/\sqrt{\gamma}$. We thus obtain

$$n^2(\omega) = 1 - \frac{4\pi n_e e^2}{m_e \gamma \omega^2} = 1 - \frac{\omega_p^2}{\gamma \omega^2} .$$
(S-13.228)

(d) The dispersion relation corresponding to $n^2(\omega)$ in (S-13.228) is

$$\omega^2 = k^2 c^2 + \frac{\omega_p^2}{\gamma} .$$
(S-13.229)

The cutoff frequency is $\omega_{co} = \omega_p/\sqrt{\gamma}$ and depends on the wave amplitude. Since $\gamma > 1$, a plasma can be opaque to a low-intensity wave for which $\omega_p > \omega$, but transparent to a high-intensity wave of the same frequency if $\gamma > \omega_p/\omega$.

It should be stressed, however, that the concept of a refractive index dependent on the wave intensity deserves some care. What we have discussed above is just a special case of "relativistically induced transparency", applying to a plane, monochromatic, infinite wave. In the case of a real light beam, of finite duration and extension, different parts of the beam can have different amplitudes, and thus can have different phase velocities, resulting in a complicated nonlinear dispersion.[10] However, (S-13.229) can be of help to a *qualitative* discussion of some important nonlinear effects observed for a relativistically strong wave. An important example is the propagation of a strong beam of finite width, for which the effective refractive index is higher at the boundaries (where the intensity is lower and γ is smaller) than on the beam axis. This can compensate diffraction, analogously to what occurs in an optical fiber (see Problem 12.6), and can cause *self-focusing*.

S-13.18 Electric Current in a Solenoid

(a) This problem originated from the question: "can the electric field in a solenoid have circular field lines, as seems to be required for driving the current in each turn of the coil?" The answer is obviously *no* in static conditions, since $\nabla \times \mathbf{E}$ must be zero. But a *uniform* field is sufficient to drive the current, since the coil of a real solenoid does not consist of single circular circular loops perpendicular to the axis (each loop would require its own current source, in this case!) The winding of a real solenoid is actually a helix, of small, but nonzero pitch. The current is driven by the component of $\mathbf{E}$ *parallel* to the wire, equal to (assuming $\mathbf{E} = E\hat{\mathbf{z}}$)

$$E_\parallel = E \sin \theta \simeq E \frac{a}{\pi b} . \qquad (S\text{-}13.230)$$

The perpendicular component $E_\perp$ is compensated by the electrostatic fields generated by the surface charge distribution of the wire, analogously to Problem 3.11. Thus, the current density and intensity in the wire are

$$J = \sigma E_\parallel \simeq \sigma E \frac{a}{\pi b} , \qquad I = J\pi a^2 \simeq \frac{\sigma a^3}{b} E . \qquad (S\text{-}13.231)$$

Neglecting boundary effects, the current generates a uniform field $\mathbf{B}^{(\text{int})} = \hat{\mathbf{z}} B_z$, with

---

[10] In some cases, nonlinearity effects can compensate dispersion for particular wavepacket shapes, these special solutions can propagate without changing their envelope shape, and are known as *solitons*.

$$B_z = \frac{4\pi n I}{c} = \frac{4\pi I}{2ac} \simeq \frac{2\pi a^2 \sigma}{bc} E , \qquad \text{(S-13.232)}$$

inside the solenoid, since $n = 1/(2a)$ is the number of turns per unit length. The field outside the solenoid, $\mathbf{B}^{(\text{ext})}$, is generated by the total current I flowing parallel to $\hat{\mathbf{z}}$. Thus in the external central region $0 < r \ll h$, $|z| \ll h$, the field is azimuthal, $\mathbf{B}^{(\text{ext})} = \hat{\boldsymbol{\phi}} B_\phi$, with

$$B_\phi \simeq \frac{2I}{cr} = \frac{2\pi J a^2}{cr} \simeq \frac{2\sigma a^3}{bcr} E , \qquad b < r \ll h , \quad |z| \ll h , \text{(S-13.233)}$$

where the z origin is located at the center of the solenoid.

(b) In the external central region $0 < r \ll h$, $|z| \ll h$, the fields E_z and B_ϕ are associated to a a Poynting vector

$$\mathbf{S} = \frac{c}{4\pi} \mathbf{E} \times \mathbf{B} = -\hat{\mathbf{r}} \frac{c}{4\pi} E_z B_\phi = -\hat{\mathbf{r}} \frac{\sigma a^3}{2\pi br} E^2 , \qquad \text{(S-13.234)}$$

with an entering flux through the lateral surface of a coaxial cylinder of length ℓ

$$\Phi_{\text{in}} = 2\pi r \ell \frac{\sigma a^3}{2\pi br} E^2 = \frac{\sigma a^3 \ell}{b} E^2 . \qquad \text{(S-13.235)}$$

The power dissipated by Joule heating in a solenoid portion of length ℓ is obtained by multiplying the power dissipated in a single turn

$$W_{\text{turn}} = I^2 R = \left(\frac{\sigma a^3}{b} E \right)^2 \frac{2\pi b}{\pi a^2 \sigma} = \frac{2\sigma a^4}{b} E^2 , \qquad \text{(S-13.236)}$$

by the number of turns $\ell/(2a)$

$$W(\ell) = \frac{2\sigma a^4}{b} E^2 \frac{\ell}{2a} = \frac{\sigma a^3 \ell}{b} E^2 , \qquad \text{(S-13.237)}$$

in agreement with Poynting's theorem.

S-13.19   An Optomechanical Cavity

(a) In the following we omit the vector notation for the electric fields, since the results are independent of the polarization. The general expression for the electric field of a monochromatic plane wave propagating along x is, in complex notation,

$$E(x, t) = E(x)e^{-i\omega t} = \left(E_1 e^{+ikx} + E_2 e^{-ikx} \right) e^{-i\omega t} , \qquad \text{(S-13.238)}$$

where $k = \omega/c$. The boundary conditions at the two perfectly conducting walls are $E(\pm d/2) = 0$, thus we must have

$$E_1 e^{+ikd/2} + E_2 e^{-ikd/2} = 0 , \qquad E_1 e^{-ikd/2} + E_2 e^{+ikd/2} = 0 . \quad \text{(S-13.239)}$$

This system of two equations has nontrivial solutions for E_1 and E_2 only if the determinant is zero,

$$e^{ikd} - e^{-ikd} = 2i \sin(kd) = 0, \quad \text{(S-13.240)}$$

from which we obtain

$$kd = n\pi \quad (n = 1, 2, 3, \ldots) \qquad \omega = kc = n\frac{\pi c}{d} , \quad \text{(S-13.241)}$$

$$E_2 = -E_1 e^{in\pi} = (-1)^{n+1} E_1 . \quad \text{(S-13.242)}$$

Thus, the electric field of the n-th mode is

$$E_n(x) = \frac{E_0}{2} \left[e^{in\pi x/d} + (-1)^{n+1} e^{-in\pi x/d} \right] . \quad \text{(S-13.243)}$$

The magnetic field can be obtained from $\partial_t \mathbf{B} = -\nabla \times \mathbf{E}$:

$$B_n(x) = \frac{E_0}{2} \left[e^{in\pi x/d} - (-1)^{n+1} e^{-in\pi x/d} \right] . \quad \text{(S-13.244)}$$

(b) The field is thus the superposition of two plane monochromatic waves of equal frequency ω and amplitude $E_0/2$, propagating in opposite directions. The radiation pressure on each reflecting wall is thus the pressure exerted by a normally incident wave of intensity $I = c|E_0/2|^2/8\pi$, evaluated in Problem 8.5,

$$P = \frac{2I}{c} = \frac{|E_0|^2}{16\pi} . \quad \text{(S-13.245)}$$

(c) The energy per unit surface inside the cavity is independent of time and can be evaluated as

$$U = \int_{-d/2}^{+d/2} \frac{1}{8\pi} \left(|\mathbf{E}^2| + |\mathbf{B}^2| \right) dx . \quad \text{(S-13.246)}$$

We have

$$\left|\mathbf{E}^2\right| = \frac{|E_0|^2}{2}\left[1 + (-1)^{n+1}\cos\left(\frac{2n\pi x}{d}\right)\right],\qquad\text{(S-13.247)}$$

$$\left|\mathbf{B}^2\right| = \frac{|E_0|^2}{2}\left[1 - (-1)^{n+1}\cos\left(\frac{2n\pi x}{d}\right)\right].\qquad\text{(S-13.248)}$$

Integrating over x, the oscillating terms of both expressions average to zero, and we finally have

$$U = \frac{|E_0|^2 d}{16\pi}\pi = Pd\ .\qquad\text{(S-13.249)}$$

(d) At mechanical equilibrium, the force due to the radiation pressure on the walls must balance the recoil force of the springs. Assuming that each wall is displaced by δ from its equilibrium position in the absence of fields, we have

$$PS = K\delta = M\Omega^2\delta\ ,\qquad\text{(S-13.250)}$$

where $\Omega = \sqrt{K/M}$ is the free oscillation frequency of the walls. Thus

$$\frac{M\Omega^2\delta}{S} = P = \frac{|E_0|^2}{16\pi}\ ,\qquad\text{(S-13.251)}$$

from which we obtain $\delta = \alpha\left|E_0^2\right|$ where

$$\alpha = \frac{S}{16\pi M\Omega^2} = \frac{S}{16\pi K}\ .\qquad\text{(S-13.252)}$$

The length of the cavity is now $d + 2\delta$, and the resonance condition is

$$d + 2\delta = n\frac{\lambda_n}{2} = n\frac{\pi c}{\omega_n}\ ,\quad (n = 1, 2, \dots)\ ,\qquad\text{(S-13.253)}$$

where the mode frequencies are

$$\omega_n = \frac{2\pi c}{\lambda_n} = \frac{n\pi c}{d + 2\alpha|E_0|^2}\ .\qquad\text{(S-13.254)}$$

This is a simple classical example of a resonant cavity where the frequency and amplitude of the wave depend on each other (and on the cavity length), the link being due to radiation pressure effects; this is called an *optomechanical* cavity [4].

S-13.20   Radiation Pressure on an Absorbing Medium

We assume the incident wave to be linearly polarized parallel to $\hat{y}$ for definiteness (the generalization to a different polarization is straightforward). The electric field of the wave is thus $\mathbf{E}(x, t) = \hat{y}\, E_y(x, t)$, with

$$E_y(x, t) = \begin{cases} \mathrm{Re}\left(E_i\, e^{ikx-i\omega t} + E_r\, e^{-ikx-i\omega t}\right), & (x < 0), \\ \mathrm{Re}\left(E_t e^{iknx-i\omega t}\right), & (x > 0), \end{cases} \quad \text{(S-13.255)}$$

where $E_i = \sqrt{8\pi I_i/c}$, and

$$E_r = \frac{1-n}{1+n}\, E_i, \qquad E_t = \frac{2}{1+n}\, E_i \qquad \text{(S-13.256)}$$

(Fresnel formulas at normal incidence). The magnetic field of the wave can be obtained from $\partial_t \mathbf{B} = -\nabla \times \mathbf{E}$, we have $\mathbf{B}(x, t) = \hat{z}\, B_z(x, t)$, with

$$B_z(x, t) = \begin{cases} \mathrm{Re}\left(E_i\, e^{ikx-i\omega t} - E_r\, e^{-ikx-i\omega t}\right), & (x < 0), \\ \mathrm{Re}\left(nE_t\, e^{iknx-i\omega t}\right), & (x > 0). \end{cases} \quad \text{(S-13.257)}$$

The field for $x > 0$ is exponentially decaying, since

$$E_t\, e^{ik(n_1+in_2)x-i\omega t} = E_t\, e^{ikn_1-i\omega t}\, e^{-kn_2 x}, \qquad \text{(S-13.258)}$$

the decay length being $(kn_2)^{-1} = \lambda/(2\pi n_2) \gg \lambda/n_1$, where $\lambda = 2\pi c/\omega$ is the wavelength in vacuum.

The cycle-averaged value of the Poynting vector at the $x = 0$ plane gives the flux of electromagnetic energy entering the medium. Since the field decays with increasing x, there is no net flux of energy for $x \to \infty$, and all the energy entering the medium is eventually absorbed. Using (S-13.255) and (S-13.257) we find

$$\langle S_x(0^+) \rangle = \left\langle \frac{c}{4\pi} E_y(0^+, t) B_z(0^+, t) \right\rangle = \frac{1}{2} \frac{c}{4\pi} \mathrm{Re}(E_t n^* E_t^*) = \frac{c}{8\pi} |E_t|^2 \mathrm{Re}(n^*)$$

$$= \frac{c}{8\pi} n_1 |E_t|^2 = \frac{c}{8\pi} n_1 \frac{4}{|1+n|^2} |E_i|^2 = \frac{4n_1}{|1+n|^2} I_i \equiv A I_i. \quad \text{(S-13.259)}$$

The reflection coefficient $R = |E_r/E_i|^2 = |1-n|^2/|1+n|^2$. Thus

$$1 - R = 1 - \left|\frac{1-n}{1+n}\right|^2 = \frac{2(n+n^*)}{|1+n|^2} = \frac{4\mathrm{Re}(n)}{|1+n|^2} = \frac{4n_1}{|1+n|^2} = A \quad \text{(S-13.260)}$$

(b) The pressure on the medium is the flow of electromagnetic momentum through the $x = 0$ surface. Such flow is given, in the present conditions, by $P_{rad} = -\langle T_{xx}(x = 0) \rangle$ where T_{ij} is Maxwell stress tensor (see Problem 8.5). Since

$$T_{xx}(0, t) = T_{11}(0, t) = -\frac{1}{8\pi} \left(E^2(0, t) + B^2(0, t) \right) \qquad (\text{S-13.261})$$

we obtain

$$\langle T_{11}(0, t)\rangle = -\frac{1}{16\pi} |E_t|^2 \left(1 + |n|^2\right) = -\frac{1}{4\pi} |E_i|^2 \frac{1 + |n|^2}{|1 + n|^2} = \frac{1}{4\pi} |E_i|^2 \frac{|1 + n|^2 - 2\mathrm{Re}(n)}{|1 + n|^2}$$

$$= -\frac{1}{8\pi} |E_i|^2 \left(2 - \frac{4n_1}{|1 + n|^2}\right) = -\frac{I_i}{c}(1 + R) \equiv -P_{\mathrm{rad}}. \qquad (\text{S-13.262})$$

The same result can also be obtained by calculating the total average force per unit surface exerted on the medium by the electromagnetic field

$$P_{\mathrm{EM}} = \int\limits_{0}^{+\infty} \langle (\mathbf{J} \times \mathbf{B})_x \rangle \, \mathrm{d}x , \qquad (\text{S-13.263})$$

since the electric term gives no contribution. The current density $\mathbf{J}$ inside the medium can be obtained from the equation $\mathbf{J} = (c\nabla \times \mathbf{B} - \partial_t \mathbf{E})/4\pi$, obtaining

$$J_y = \mathrm{Re}\left(-\frac{ikcn}{4\pi} \frac{n}{c} E_t \, e^{-iknx - i\omega t} + \frac{i\omega}{4\pi} E_t \, e^{-iknx - i\omega t}\right)$$

$$= \mathrm{Re}\left(\frac{i\omega}{4\pi} \left(1 - n^2\right) E_t \, e^{-iknx - i\omega t}\right). \qquad (\text{S-13.264})$$

A further way to obtain this result is recalling the relation between conductivity and dielectric permittivity for complex fields, i.e.,

$$\sigma(\omega) = -\frac{i\omega}{4\pi} \left[\varepsilon_r(\omega) - 1\right] = -\frac{i\omega}{4\pi} \left[n^2(\omega) - 1\right] . \qquad (\text{S-13.265})$$

We thus have

$$\langle J_y B_z \rangle = \frac{1}{2} \frac{\omega}{4\pi} \mathrm{Re}\left\{\left[i\left(1 - n^2\right) E_t \, e^{(-ikn_1 - n_2)x}\right]\left[n^* E_t^* e^{(+ikn_1 - n_2)x}\right]\right\}$$

$$= \frac{1}{8\pi} |E_t|^2 \mathrm{Re}\left[i\left(1 - n^2\right) n^*\right] e^{-2n_2 x} . \qquad (\text{S-13.266})$$

Now

$$\mathrm{Re}\left[i\left(1 - n^2\right) n^*\right] = \mathrm{Re}\left[\left(1 - n_1^2 + n_2^2 - 2in_1 n_2\right)(in_1 + n_2)\right] = n_2\left(1 + n_1^2 + n_2^2\right)$$

$$= n_2\left(1 + |n|^2\right) , \qquad (\text{S-13.267})$$

thus, by substituting in (S-13.263) and comparing to (S-13.262) we obtain

$$P_{\text{EM}} = \int_0^\infty \langle J_y B_z \rangle \, dx = \frac{1}{2} \frac{1}{4\pi} |E_t|^2 n_2 \left(1 + |n|^2\right) \int_0^\infty e^{-2n_2 x} dx$$

$$= \frac{1}{2\pi} |E_i|^2 \frac{1 + |n|^2}{1 - |n|^2} n_2 \frac{1}{2n_2} = P_{\text{rad}} \,. \tag{S-13.268}$$

S-13.21   Radiation and Scattering from a Linear Molecule

(a) At the initial time $t = 0$, we assume the center of mass of the molecule to be at rest at the origin of our Cartesian coordinate system. The center of mass will remain at rest, since the net force acting on the molecule is zero. However, the field $\mathbf{E}_0$ exerts a torque $\boldsymbol{\tau}_0 = \mathbf{p}_0 \times \mathbf{E}_0$, and the molecule rotates around the z axis. The equation of motion is $I\ddot{\theta} = \tau_0$, or

$$I\ddot{\theta} = -p_0 E_0 \sin \theta \,, \tag{S-13.269}$$

where $\theta = \theta(t)$ is the angle between $\mathbf{p}_0$ and the x axis. The potential energy of the molecule is

$$V(\theta) = -\mathbf{p}_0 \cdot \mathbf{E}_0 + C = -p_p E_0 \cos \theta + C \,, \tag{S-13.270}$$

where C is an arbitrary constant. The molecule has two equilibrium positions, at $\theta = 0$ (stable), and $\theta = \pi$ (unstable), respectively. For small oscillations around the stable equilibrium position we can approximate $\sin \theta \simeq \theta$, and (S-13.269) turns into the equation for the harmonic oscillator

$$\ddot{\theta} \simeq -\frac{p_0 E_0}{I} \theta \equiv -\omega_0^2 \theta \,, \quad \text{where} \quad \omega_0^2 = \frac{E_0 p_0}{I} \,. \tag{S-13.271}$$

Thus, if the molecule starts at rest at a small initial angle $\theta(0) = \theta_0$, we have $\theta(t) \simeq \theta_0 \cos \omega_0 t$. The potential energy of the molecule can be approximated as

$$V(\theta) \simeq -p_0 E_0 \left(1 - \frac{\theta^2}{2}\right) + C = \frac{1}{2} p_0 E_0 \theta^2 = \frac{1}{2} I \omega_0^2 \theta^2 \,, \tag{S-13.272}$$

where we have chosen $C = p_0 E_0$, in order to have $V(0) = 0$. The kinetic energy of the molecule is

$$K(\dot{\theta}) = \frac{1}{2} I \dot{\theta}^2 \,. \tag{S-13.273}$$

(b) In our coordinate system the instantaneous dipole moment has components

$$p_x = p_0 \cos \theta \simeq p_0 \,, \qquad p_y = p_0 \sin \theta \simeq p_0 \theta_0 \cos(\omega_0 t) \,, \tag{S-13.274}$$

so that, for small oscillations, the radiation emitted by the molecule is equivalent to the radiation of an electric dipole parallel to $\hat{\mathbf{y}}$, and of frequency ω_0. The radiation is linearly polarized, and the angular distribution of the emitted power is $\sim \cos^2 \alpha$, where α is the observation angle relative to $\mathbf{E}_0$. Thus, the radiated power per unit solid angle is maximum in the xz plane and vanishes in the $\hat{\mathbf{y}}$ direction. The time-averaged total emitted power is

$$P_{\text{rad}} = \frac{1}{3c^3} |\ddot{\mathbf{p}}|^2 = \frac{1}{3c^3} \omega_0^4 p_0^2 \theta_0^2 . \tag{S-13.275}$$

We assume that the decay time is much longer than the oscillation period, so that we can write

$$\theta(t) \simeq \theta_s(t) \cos \omega_0 t , \tag{S-13.276}$$

with $\theta_s(0) = \theta_0$, and $\theta_s(t)$ decaying in time so slowly that it is practically constant over a single oscillation. In these conditions the total energy of the molecule during a single oscillation period can be written

$$U(t) = K(\dot{\theta}) + V(\theta) \simeq \frac{1}{2} I \omega_0^2 \theta_s^2(t) . \tag{S-13.277}$$

The rate of energy loss due to the emitted radiation is

$$\frac{dU}{dt} = \omega_0^2 I \theta_s \frac{d\theta_s}{dt} = -P_{\text{rad}}(\theta_s) , \tag{S-13.278}$$

from which we obtain

$$\frac{d\theta_s}{dt} = -\frac{1}{3c^3} \frac{\omega_0^2 p_0^2}{I} \theta_s . \tag{S-13.279}$$

Thus the oscillation amplitude decays exponentially in time

$$\theta_s(t) = \theta_0 \, e^{-t/\tau} , \quad \text{with} \quad \tau = \frac{3I c^3}{\omega_0^2 p_0^2} . \tag{S-13.280}$$

(c) Since $kd \ll 1$, the electric field of the wave can be considered as uniform over the molecule, and we can write $\mathbf{E}_1(0, t) \simeq \mathbf{E}_1 e^{-i\omega t}$ in complex notation. The torque exerted by the wave is $\tau_1 = \mathbf{p}_0 \times \mathbf{E}_1$. The complete equation of motion for the molecule is thus

$$I\ddot{\theta} = -p_0 E_0 \sin \theta - p_0 E_1 \cos \theta \, e^{-i\omega t} , \tag{S-13.281}$$

which, at the limit of small oscillations ($\sin \theta \simeq \theta$, $\cos \theta \simeq 1$) becomes

$$\ddot{\theta} = -\omega_0^2 \theta - \omega_1^2 e^{-i\omega t} , \quad \text{with} \quad \omega_1^2 = \frac{E_1 p_0}{I} = \omega_0^2 \frac{E_1}{E_0} . \quad (S\text{-}13.282)$$

The general solution of (S-13.282) is the sum of the homogeneous solution considered at point **(a)**, which describes free oscillations, and of a particular solution of the complete equation. A particular solution can be found in the form $\theta(t) = \theta_f e^{-i\omega t}$, which, substituted into (S-13.282), gives

$$\theta_f = \frac{\omega_1^2}{\omega^2 - \omega_0^2} . \quad (S\text{-}13.283)$$

For simplicity, we neglected the possible presence of friction in (S-13.271). However, in principle a friction term such as $-\dot{\theta}/\tau$ should appear because of the energy loss by radiation. In the presence of the plane wave the friction term is relevant only close to the $\omega = \omega_0$ resonance, because $\tau^{-1} \ll \omega_0$.

(d) After a transition time of the order of τ possible initial oscillations at ω_0 are damped, and the the molecule reaches a steady state where it oscillates at frequency ω. Assuming, as in **(b)**, small-amplitude oscillations, we have an oscillating dipole component $p_y \simeq p_0 \theta_f e^{-i\omega t}$. The scattered power is

$$P_{\text{scatt}} = \frac{1}{3c^3} |\ddot{p}_y|^2 = \frac{p_0^2}{3c^3} \frac{\omega^4 \omega_1^4}{(\omega^2 - \omega_0^2)^2} = \frac{p_0^4 E_1^2}{3I^2 c^3} \frac{\omega^4}{(\omega^2 - \omega_0^2)^2} . \quad (S\text{-}13.284)$$

The intensity of the wave is $I = (c/4\pi) E_1^2$, thus the scattering cross section is

$$\sigma_{\text{scatt}} = \frac{P_{\text{scatt}}}{I} = \frac{4\pi p_0^4}{3I^2 c^4} \frac{\omega^4}{(\omega^2 - \omega_0^2)^2} . \quad (S\text{-}13.285)$$

An order-of-magnitude estimate for a simple molecule such as H_2 can be performed by noticing that $p_0 \sim ed$ and $I \sim md$, with $m \sim m_p$ the mass of the nuclei, so that $(p_0^4/I^2 c^4) \sim (e^2/m_p c^2)^2$.

S-13.22 Radiation Drag Force

(a) The electric field of the wave in complex notation is

$$\mathbf{E} = \hat{\mathbf{y}} \operatorname{Re} \left(E_0 e^{ikx - i\omega t} \right) . \quad (S\text{-}13.286)$$

Neglecting the magnetic field, the particle oscillates in the $\hat{\mathbf{y}}$ direction without changing its x and z coordinates. Thus, assuming the particle to be initially located at the origin of our Cartesian system, and looking for a solution of the form $\mathbf{v} = \operatorname{Re}(\mathbf{v}_0 e^{-i\omega t})$, we obtain by substitution into (13.21):

$$v_0 = \hat{y} \frac{iq}{m(\omega + i\nu)} E_0 . \tag{S-13.287}$$

(b) The power developed by the electromagnetic force is $q\mathbf{E} \cdot \mathbf{v}$. Thus

$$P_{\text{abs}} = \langle q\mathbf{E} \cdot \mathbf{v} \rangle = \frac{q}{2} \text{Re} \left(E_0 v_0^* \right) = \frac{q^2}{2m} \frac{\nu}{\omega^2 + \nu^2} |E_0|^2. \tag{S-13.288}$$

(c) The electric dipole moment of the particle is $\mathbf{p} = q\mathbf{r}$. Using Larmor's formula for the radiated power we obtain

$$P_{\text{rad}} = \frac{2}{3c^3} \langle \ddot{\mathbf{p}}^2 \rangle = \frac{2q^2}{3c^3} \langle \ddot{\mathbf{r}}^2 \rangle = \frac{q^4}{3m^2c^3} \frac{\omega^2}{\omega^2 + \nu^2} |E_0|^2. \tag{S-13.289}$$

Assuming $P_{\text{rad}} = P_{\text{abs}}$, we obtain

$$\nu = \frac{2q^2\omega^2}{3mc^3}. \tag{S-13.290}$$

(d) We must evaluate

$$F_x = \left\langle \frac{q}{c} v_y B_z \right\rangle , \tag{S-13.291}$$

where for v_y we use the result of (a), while the amplitude of the magnetic field is $B_0 = E_0$. Thus we have

$$F_x = \frac{q}{2c} \text{Re} \left(v_0 E_0^* \right) = \frac{P_{\text{abs}}}{c}. \tag{S-13.292}$$

Thus, the ratio between the energy and the momentum absorbed by the particle from the electromagnetic field equals c.

(e) The radiation from a cluster smaller than one wavelength is coherent and thus scales as N^2, so does the total force. The cluster mass scales as N, thus the acceleration scales as $N^2/N = N$. In other terms, a cluster of many particles may be accelerated much more efficiently than a single particle: the higher the number of particles (within the limits of our approximations), the stronger the acceleration. This is the basis of a concept of "coherent" acceleration using electromagnetic waves, formulated by Veksler [3].

S-13.23   A Coaxial Cable with a Faraday Generator

(a) Let us consider a cylinder of radius r and height h, with $a < r < b$, internal to and coaxial with the conducting toroid, as shown in Fig. S-13.7. The flux of the current density $\mathbf{J}$ through the lateral surface of the cylinder is

$$I = \int_{\text{lat. surf.}} \mathbf{J} \cdot d\mathbf{S} = 2\pi r h J_r , \qquad (S\text{-}13.293)$$

where we have used the symmetry of our problem. From (S-13.293) we obtain

$$\mathbf{J}(r) = \frac{I}{2\pi h r} \hat{\mathbf{r}} . \qquad (S\text{-}13.294)$$

We have assumed that the current I is parallel to the z axis in the outer shield of the coaxial cable, and antiparallel in the inner conductor. Obviously, changing the sign of I would change the sign of $\mathbf{J}$. Here and in what follows the values of B and ω are positive if they are parallel to the $\hat{\mathbf{z}}$ unit vector, I is positive if the current is parallel to $\hat{\mathbf{z}}$ in the outer shield, J and E are positive if parallel to $\hat{\mathbf{r}}$.

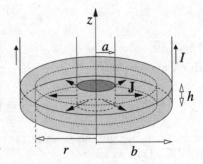

Fig. S-13.7

(b) The potential difference between the outer and inner lateral surfaces of the toroid is

$$V = - \int_a^b E(r)\, dr = - \int_a^b \eta J_r(r)\, dr$$

$$= -\frac{\eta I}{2\pi h} \int_a^b \frac{dr}{r} = -\frac{\eta I}{2\pi h} \ln\left(\frac{b}{a}\right) , \qquad (S\text{-}13.295)$$

and the corresponding resistance is

$$R = \left|\frac{V}{I}\right| = \frac{\eta}{2\pi h} \ln\left(\frac{b}{a}\right) . \qquad (S\text{-}13.296)$$

(c) The surface charge densities on the lateral surfaces of the toroid can be evaluated from Gauss's law, remembering thet the electric field is zero for both $r < a$ and $r > b$,

$$\sigma_a = \frac{1}{4\pi k_e} \left[E(a^+) - E(a^-) \right] = \frac{\eta I}{8\pi^2 k_e ha} = \begin{cases} \dfrac{\varepsilon_0 \eta I}{2\pi ha}, & \text{(SI)}, \\[2mm] \dfrac{\eta I}{8\pi^2 ha}, & \text{(Gaussian)}, \end{cases}$$

$$\text{(S-13.297)}$$

$$\sigma_b = \frac{1}{4\pi k_e} \left[E(b^+) - E(b^-) \right] = -\frac{\eta I}{8\pi^2 k_e hb} = \begin{cases} -\dfrac{\varepsilon_0 \eta I}{2\pi hb}, & \text{(SI)}, \\[2mm] -\dfrac{\eta I}{8\pi^2 hb}, & \text{(Gaussian)}. \end{cases}$$

$$\text{(S-13.298)}$$

Note that the inner and outer lateral surfaces carry the same total charge, with opposite sign: $Q_a = -Q_b = \eta I/(4\pi k_e)$. For the volume charge density ρ inside the toroid we have

$$\rho = \frac{1}{4\pi k_e} \nabla \cdot \mathbf{E} = \frac{1}{4\pi k_e} \nabla \cdot (\eta \mathbf{J}) = \frac{1}{4\pi k_e} \eta \nabla \cdot \mathbf{J} = 0 . \qquad \text{(S-13.299)}$$

Alternatively, Eqs. (S-13.297)–(S-13.298) can be considered as a consequence of (S-13.299), since we can write the electric field as

$$\mathbf{E} = \hat{\mathbf{r}} \frac{\eta I}{2\pi hr} H(r - a) H(b - r) , \quad \text{where} \quad H(x) = \begin{cases} 0 & \text{if } x \le 0, \\ 1 & \text{if } x > 0, \end{cases}$$

$$\text{(S-13.300)}$$

is the Heaviside step function. The derivative of $H(x)$ is the Dirac delta function $\delta(x)$, so that, for instance, for r close to a we have

$$\rho(r) = \frac{1}{4\pi k_e} \nabla \cdot \mathbf{E} = \frac{\eta I}{8\pi^2 k_e ha} \delta(r - a) , \qquad \text{(S-13.301)}$$

which corresponds to the surface charge density of (S-13.297). We have used Table A.1 for the divergence in cylindrical coordinates.

(d) The rotation about the z axis preserves the cylindrical symmetry of the problem. In stationary conditions we still have $\nabla \cdot \mathbf{J} = 0$, and (S-13.293) and (S-13.294) still hold, independently of the presence of the magnetic field $\mathbf{B}$. The simultaneous presence of $\mathbf{B}$ and rotation adds a new contribution to the Lorentz force acting on the single charge carrier q due to the local azimuthal velocity $\mathbf{V}(r) = \boldsymbol{\omega} \times \mathbf{r}$, namely

$$\mathbf{f}_B = b_m q \mathbf{V}(r) \times \mathbf{B} = \pm b_m q \omega B \mathbf{r} = \begin{cases} \pm q \omega B \mathbf{r}, & \text{(SI)}, \\[2mm] \pm \dfrac{q \omega B \mathbf{r}}{c}, & \text{(Gaussian)}, \end{cases}$$

$$\text{(S-13.302)}$$

where the $+$ sign holds if $\boldsymbol{\omega}$ and $\mathbf{B}$ are parallel, the $-$ sign if they are antiparallel. On the other hand, the only nonzero component of the electric field is E_r, so that $\mathbf{E} = E_r \hat{\mathbf{r}}$, because of the symmetry of the problem and of the condition $\nabla \times \mathbf{E} = 0$,

B being constant in time. Remembering that the resistivity η is the ratio of the total Lorentz force per unit charge to the current density, we have

$$J_r = \frac{E_r \pm b_\mathrm{m}\omega Br}{\eta}, \tag{S-13.303}$$

which leads to

$$E_r = \eta J_r \mp b_\mathrm{m}\omega Br = \eta \frac{I}{2\pi hr} \mp b_\mathrm{m}\omega Br. \tag{S-13.304}$$

If ω and **B** are parallel, the magnetic contribution to the Lorentz force is parallel to **E**, and a smaller electric field is sufficient to maintain the electric current. If ω and **B** are antiparallel, a larger electric field is needed to compensate the magnetic contribution. The volume charge density is, remembering that in stationary conditions $\nabla \cdot \mathbf{J} = 0$,

$$\rho(r) = \frac{1}{4\pi k_\mathrm{e}} \nabla \cdot \mathbf{E} = \frac{1}{4\pi k_\mathrm{e}} \nabla \cdot (\mp b_\mathrm{m}\omega B\mathbf{r}) = \mp \frac{1}{4\pi k_\mathrm{e}} \frac{1}{r} \frac{\partial}{\partial r}\left(b_\mathrm{m}\omega Br^2\right)$$

$$= \mp \frac{b_\mathrm{m}\omega B}{2\pi k_\mathrm{e}} = \begin{cases} \mp 2\varepsilon_0\omega B, & \text{(SI)}, \\[2mm] \mp \dfrac{\omega B}{2\pi c}, & \text{(Gaussian)}, \end{cases} \tag{S-13.305}$$

where, again, we have used Table A.1. The upper sign holds if the vectors ω and **B** are parallel, the lower sign if they are antiparallel. The surface charge densities on the lateral surfaces of the toroid are

$$\sigma_a = \frac{1}{4\pi k_\mathrm{e}}\left[E(a^+) - E(a^-)\right] = \frac{\eta I}{8\pi^2 k_\mathrm{e} ha} \mp \frac{b_\mathrm{m}\omega Ba}{4\pi k_\mathrm{e}} = \begin{cases} \dfrac{\varepsilon_0\eta I}{2\pi ha} \mp \varepsilon_0\omega Ba, & \text{(SI)}, \\[2mm] \dfrac{\eta I}{8\pi^2 ha} \mp \dfrac{\omega Ba}{4\pi c}, & \text{(Gaussian)}, \end{cases}$$
$$\tag{S-13.306}$$

$$\sigma_b = \frac{1}{4\pi k_\mathrm{e}}\left[E(b^+) - E(b^-)\right] = -\frac{\eta I}{8\pi^2 k_\mathrm{e} hb} \pm \frac{b_\mathrm{m}\omega Bn}{4\pi k_\mathrm{e}} = \begin{cases} -\dfrac{\varepsilon_0\eta I}{2\pi hb} \pm \varepsilon_0\omega Bb, & \text{(SI)}, \\[2mm] -\dfrac{\eta I}{8\pi^2 hb} \pm \dfrac{\omega Bb}{4\pi c}, & \text{(Gaussian)}. \end{cases}$$
$$\tag{S-13.307}$$

The presence of a nonzero volume charge density ρ in the rotating toroid implies the presence of an azimuthal component of the current density $J_\phi(r) = \rho(r)\mathbf{V}(r)$. In principle this current contributes to the total magnetic field, however this contribution is negligible for realistic rotation velocities. Further, J_ϕ does not contribute to the torque exerted by the magnetic forces on the toroid, see next point.

(e) The torque τ exerted by the magnetic field on the toroid involves only the azimuthal component of the force per unit volume, i.e.,

$$f_{v\phi} = b_{\rm m}(\mathbf{J} \times \mathbf{B})_\phi = -b_{\rm m}J_r(r)B_z = -\frac{b_{\rm m}IB}{2\pi hr} , \qquad \text{(S-13.308)}$$

independently of the azimuthal component of $\mathbf{J}$. The total torque $\boldsymbol{\tau} = \tau\,\hat{\mathbf{z}}$ is obtained by integrating the moment of the force per unit volume $\mathbf{f}_v$ with respect to the symmetry axis of our system over the toroid volume

$$\boldsymbol{\tau} = \int \mathbf{r} \times \mathbf{f}_v\, r\, d\phi\, dr\, dz = \hat{\mathbf{z}} \int_a^b r f_{v\phi}\, 2\pi hr\, dr$$

$$= -\hat{\mathbf{z}}\,\frac{b_{\rm m}IB}{2\pi h} \int_a^b 2\pi hr\, dr = -\hat{\mathbf{z}}\,\frac{b_{\rm m}IB}{2} \left(b^2 - a^2\right) . \qquad \text{(S-13.309)}$$

The mechanical power needed to keep the toroid in rotation at constant angular velocity is

$$P_{\rm m} = -\boldsymbol{\tau} \cdot \boldsymbol{\omega} = \frac{b_{\rm m}IB\omega}{2} \left(b^2 - a^2\right) . \qquad \text{(S-13.310)}$$

(f) The power dissipated by Joule heating in the rotating toroid is

$$P_{\rm d} = \int \eta J_r^2 r\, d\phi\, dr\, dz = \int_a^b \eta J_r^2 2\pi hr\, dr = \frac{\eta I^2}{2\pi h} \ln\left(\frac{b}{a}\right) = RI^2 , \qquad \text{(S-13.311)}$$

as we could expect. Requiring $P_{\rm d} = P_{\rm m}$ we obtain

$$I = \frac{b_{\rm m}B\omega}{2R} \left(b^2 - a^2\right) = \begin{cases} \dfrac{B\omega}{2R} \left(b^2 - a^2\right) , & \text{(SI)}, \\[2mm] \dfrac{B\omega}{2Rc} \left(b^2 - a^2\right) , & \text{(Gaussian)}, \end{cases} \qquad \text{(S-13.312)}$$

meaning that the electromotive force in the toroid is $\mathcal{E} = b_{\rm m}B\omega\left(b^2 - a^2\right)/2$. Note that

$$\mathcal{E} = b_{\rm m} \int_a^b \mathbf{V} \times \mathbf{B} \cdot d\mathbf{r} = b_{\rm m} \int_a^b (\mathbf{r} \times \boldsymbol{\omega}) \times \mathbf{B} \cdot d\mathbf{r} = b_{\rm m}\frac{B\omega\left(b^2 - a^2\right)}{2} . \qquad \text{(S-13.313)}$$

S-13.24 Reflection and Transmission from a Moving Transparent Medium

(a) In the following we use signed wave vectors, $k > 0$ denoting a wave propagating parallel to the x axis and $k < 0$ a wave propagating antiparallel to the x axis.

Introducing the quantities $\beta = V/c$ and $\gamma = 1/\sqrt{1 - \beta^2}$, and applying the Lorentz transformations to the incident-wave four-vector we obtain for S'

$$\omega_i' = \gamma(1 - \beta)\,\omega_i \,, \qquad k_i' = \gamma(1 - \beta)\,k_i = \omega_i'/c \,, \qquad \text{(S-13.314)}$$

where $k_i = \omega/c$ is the wave vector in S.

The reflected wave in S' has frequency $\omega_r' = \omega_i'$ and wave vector $k_r' = -k_i'$. Transforming back to the laboratory reference frame S we obtain

$$\omega_r = \gamma(1 - \beta)\,\omega_r' = \frac{1 - \beta}{1 + \beta}\,\omega_i \,, \qquad k_r' = \gamma(1 - \beta)\,k_r' = -\frac{1 - \beta}{1 + \beta}\,k_i \,. \quad \text{(S-13.315)}$$

The phase velocity of the reflected wave is

$$v_r = \frac{\omega_r}{k_r} = -\frac{\omega_i}{k_i} = -c, \qquad \text{(S-13.316)}$$

where the minus sign simply mean that the wave is propagating in the $-\hat{\mathbf{x}}$ direction.

In S' the transmitted wave has frequency $\omega_t' = \omega_i'$ and wave vector $k_t' = nk_i' = n\omega_i'/c$. Transforming back to S we obtain

$$\omega_t = \gamma(\omega_t' + \beta c k_t') = \gamma(1 + \beta n)\,\omega_i' = \gamma^2(1 + \beta n)(1 - \beta)\omega_i \,, \qquad \text{(S-13.317)}$$

$$k_t = \gamma\left(k_t' + \frac{\beta}{c}\omega_t'\right) = \gamma(n + \beta)\,\frac{\omega_i'}{c} = \gamma^2(n + \beta)(1 - \beta)\,k_i \,. \qquad \text{(S-13.318)}$$

The phase velocity is

$$v_t = \frac{\omega_t}{k_t} = \frac{1 + \beta n}{n + \beta}\,\frac{\omega_i}{k_i} = \frac{1 + \beta n}{n + \beta}\,c \,. \qquad \text{(S-13.319)}$$

(b) First we apply the Lorentz transformations in order to obtain the fields of the incident wave in S'

$$E_i' = \gamma(1 - \beta)E_i \,, \qquad B_i' = E' \,. \qquad \text{(S-13.320)}$$

The electric fields E_r' and E_t' of the reflected and transmitted wave. respectively, can be obtained from the Fresnel equations (11.6) and (11.7) by replacing n_1 by 1, n_2 by n and $\cos\theta_i = \cos\theta_t = 1$, obtaining

$$E_r' = \frac{1 - n}{1 + n}\,E_i' \,, \qquad E_t' = \frac{2}{1 + n}\,E_i' \,. \qquad \text{(S-13.321)}$$

The magnetic field of the reflected wave is $B_r' = -E_r'$. Transforming back to S we obtain

$$E_r = \gamma(E'_r + \beta B'_r) = \gamma(1 - \beta)E'_r = \frac{1-\beta}{1+\beta}\frac{1-n}{1+n}E_i \,,$$

$$B_r = -E_r \,. \tag{S-13.322}$$

In S' the magnetic field of the transmitted wave is $B'_t = nE'_t$, where E'_t is given by the second of (S-13.321). Transforming back to S we obtain

$$E_t = \gamma(E'_t + \beta B'_t) = \frac{2\gamma}{1+n}E'_i(1+\beta n) = \frac{2\gamma^2(1-\beta)(1+\beta n)}{1+n}E_i \,,$$

$$\tag{S-13.323}$$

$$B_t = \gamma(B'_t + \beta E'_t) = \frac{2\gamma}{1+n}E'_i(n+\beta) = \frac{2\gamma^2(1-\beta)(n+\beta)}{1+n}E_i \,, \tag{S-13.324}$$

so that

$$\frac{E_t}{B_t} = \frac{1+\beta n}{n+\beta} = \frac{v_t}{c} \,, \tag{S-13.325}$$

where v_t is the phase velocity of the transmitted wave in S.

(c) The refracrive index has a definite value $n(\omega'_i)$ in S', thus we must replace n by $n(\omega'_i)$ in all the above formulae.

S-13.25   The Electromotive Force in Two Different Frames

(a) The flux of the magnetic field through the square coil can be approximated as $\Phi = a^2 B(X)$, where X is the x coordinate of the center of the loop. This approximation is valid because of the condition $a \ll L$. Further, we are assuming that the self-inductance of the coil is negligible. Thus, according to Farady's law, the electromotive force on the coil is

$$\mathcal{E} = -\frac{1}{c}\frac{d\Phi}{dt} \simeq -\frac{1}{c}\frac{d}{dt}\left[a^2\frac{B_0}{L}X(t)\right] = -\frac{B_0 a^2}{cL}V \,. \tag{S-13.326}$$

Alternatively, $\mathcal{E}$ can be expressed as the line integral of $\mathbf{E} + (\mathbf{V}/c) \times \mathbf{B}$ around the perimeter of the coil

$$\mathcal{E} = \oint\left(\mathbf{E} + \frac{\mathbf{V}}{c} \times \mathbf{B}\right) \cdot d\ell = \frac{1}{c}\oint \mathbf{V} \times \mathbf{B} \cdot d\ell$$

$$= \frac{V}{c}\left[-B_z\left(X + \frac{a}{2}\right) + B_z\left(X - \frac{a}{2}\right)\right]a = -\frac{B_0 a^2}{cL}V \,, \tag{S-13.327}$$

since $\mathbf{E}$ is zero in the laboratory frame, and, again, $X(t)$ is the x coordinate of the center of the loop. Note that $\mathbf{V} \times \mathbf{B}$ is directed along $-\hat{\mathbf{y}}$ in the coil sides parallel

to the y axis, $B_z(X + a/2) > B_z(X - a/2)$, thus the resulting electromotive force is
negative.
(b) The electric current flowing in the coil is

$$I = \frac{\mathcal{E}}{R} = -\frac{B_0 a^2}{cRL} V ,$$
(S-13.328)

the minus sign meaning that it is flowing clockwise. The Lorentz force on the loop is

$$\mathbf{F}_{\mathrm{L}} = \frac{1}{c} \oint I (d\boldsymbol{\ell} \times \mathbf{B}) = \frac{I}{c} \left[-\frac{B_0}{L} \left(X + \frac{a}{2} \right) + \frac{B_0}{L} \left(X - \frac{a}{2} \right) \right] a \hat{\mathbf{x}}$$

$$= -\frac{I}{c} \frac{B_0 a^2}{L} \hat{\mathbf{x}} = -\frac{B_0^2 a^4}{c^2 L^2 R} V \hat{\mathbf{x}} = -\frac{\mathcal{E}^2}{RV} \hat{\mathbf{x}} ,$$
(S-13.329)

the minus sign indicates that the Lorentz force is directed along $-\hat{\mathbf{x}}$, thus opposite
to $\mathbf{V}$. In order to maintain the velocity $\mathbf{V}$ constant we must apply an external force
$\mathbf{F} = -\mathbf{F}_{\mathrm{L}}$. The work done by $\mathbf{F}$ per unit time is

$$P = \mathbf{F} \cdot \mathbf{V} = \frac{\mathcal{E}^2}{R} ,$$
(S-13.330)

equal to the power dissipated in the coil by Joule heating.
(c) Under the assumption $V \ll c$ the Lorentz transformations reduce to the Galilean
transformations, with $\gamma \simeq 1$, $t' \simeq t$ and $x' \simeq x - Vt$. Thus in the frame S' we have
an electric field $\mathbf{E}' = (\mathbf{V}/c) \times \mathbf{B}$ parallel to the y axis, while $B'(x') \simeq B(x' + Vt)$,
thus

$$E_y'(x') = -\frac{V}{c} B_z'(x') = -V \frac{B_0}{cL} (x' + Vt) .$$
(S-13.331)

According to Faraday's law we have for the electromotive force $\mathcal{E}'$ in S'

$$\mathcal{E}' = -\frac{1}{c} \frac{d\Phi'}{dt} = -\frac{1}{c} \frac{d}{dt} \left[a^2 B_z(X', t) \right] = -a^2 \frac{\partial B_z'(X', t)}{\partial t}$$

$$= -a^2 \frac{B_0}{cL} V = \mathcal{E} .$$
(S-13.332)

Evaluating the electromotive force as the line integral of $\mathbf{E}' + (\mathbf{V}'/c) \times \mathbf{B}'$ we obtain

$$\mathcal{E}' = \oint \left(\mathbf{E}' + \frac{\mathbf{V}'}{c} \times \mathbf{B}' \right) \cdot d\boldsymbol{\ell}' = \oint \mathbf{E}' \cdot d\boldsymbol{\ell}' = \left[E_y' \left(X' + \frac{a}{2} \right) - E_y' \left(X' - \frac{a}{2} \right) \right] a$$

$$= -\frac{V}{c} \left[B_z \left(X' + \frac{a}{2} \right) - B_z \left(X' - \frac{a}{2} \right) \right] a = -\frac{B_0 a^2}{cL} V = \mathcal{E} ,$$
(S-13.333)

since $\mathbf{V}' = 0$ in the rest frame of the coil. We have $\mathbf{F}_{\mathrm{L}}' = \mathbf{F}_{\mathrm{L}}$ and $\mathbf{F}' = \mathbf{F}$, and apparently
$\mathbf{F}'$ does no work because the coil is at rest in S'. However, the force $\mathbf{F}$ on the coil is

exerted by something or someone. For the sake of simplicity, let us assume that the experimenter is pushing the coil while walking on the laboratory floor. According to Newton's third law a force $-\mathbf{F}$ is exerted on the floor, which is at rest in the laboratory frame S. Also the magnetic field sources are resting on the floor. In S' the coil is at rest, but the floor, and the field sources, are moving with velocity $-\mathbf{V}$. Thus there is a work done per unit time

$$P' = (-\mathbf{F}) \cdot (-\mathbf{V}) = FV = P \, , \qquad \text{(S-13.334)}$$

equal to the power dissipated by Joule heating. In other words, in S' moving the sources of the magnetic field requires work because of the presence of the coil, which acts as an electromagnetic brake.

Note: all the above formulae are in Gaussian units. For SI units just remove c everywhere.

S-13.26   Electron Orbits in a Magnetic Field

(a) The equation of motion for the electron is

$$m_e \frac{d\boldsymbol{v}}{dt} = -\frac{e\,\boldsymbol{v}}{c} \times \mathbf{B} \, , \qquad \text{(S-13.335)}$$

where $-e$ is the electron charge, m_e its mass and $\boldsymbol{v}$ its velocity. The Cartesian components of (S-13.335) are

$$\frac{dv_x}{dt} = -\frac{e\,v_y}{m_e c} B_z(x) \, , \qquad \text{(S-13.336)}$$

$$\frac{dv_y}{dt} = \frac{e\,v_x}{m_e c} B_z(x) \, , \qquad \text{(S-13.337)}$$

$$\frac{dv_z}{dt} = 0 \, . \qquad \text{(S-13.338)}$$

The motion along z is decoupled from the motion in the xy plane and, for the sake of simplicity, we can assume $v_z = 0$. Now we introduce the vector potential $\mathbf{A} = \hat{\mathbf{y}}\, A_y(x)$ such that $B_z(x) = (\boldsymbol{\nabla} \times \mathbf{A})_z = dA_y(x)/dx$. Then we have from (S-13.337)

$$\frac{dv_y}{dt} = \frac{e}{m_e c} v_x \frac{dA_y(x)}{dx} = \frac{e}{m_e c} \frac{dA_y(x)}{dt} \, , \qquad \text{(S-13.339)}$$

where $d/dt = \partial/\partial t = \boldsymbol{v} \cdot \boldsymbol{\nabla}$ gives the time variation along the electron trajectory. Thus we have

$$\frac{d}{dt}\left[v_y - \frac{eA_y(x)}{m_e c} \right] = 0 \, . \qquad \text{(S-13.340)}$$

The constant of motion $\mathcal{P}_y = m_e v_y - eA_y(x)/c$ is the y component of the canonical momentum [5] and arises from the invariance of the magnetic configuration along y.
(b) Inserting $v_y = \mathcal{P}_y/m_e + eA_y(x)/(m_e c)$ into (S-13.336) we obtain

$$m_e \frac{dv_x}{dt} = -\left[\frac{\mathcal{P}_y}{m_e} + \frac{eA_y(x)}{m_e c}\right]\frac{eB_z(x)}{c}$$

$$= -\frac{d}{dx}\left[\frac{\mathcal{P}_y}{m_e}\frac{eA_y(x)}{c} + \frac{1}{2m_e}\left(\frac{eA_y(x)}{c}\right)^2\right], \qquad \text{(S-13.341)}$$

where the term in square brackets, which we recast in the form $V_{\cdot}(x) = [C\,\hat{A} + \hat{A}^2/2]$, with $C = \mathcal{P}_y/m_e$ and $\hat{A} = eA_y(x)/(m_e c)$, plays the role of an effective potential. Then (S-13.341) reads

$$m_e \frac{dv_x}{dt} = -\frac{d}{dx}V(x). \qquad \text{(S-13.342)}$$

(c) For $B_z(x) = B_0\,(x/L)$ we have $A_y = B_0 x^2/(2L)$, and

$$\hat{A} = \frac{eA_y(x)}{m_e c} = \frac{eB_0 x^2}{2cm_e L} = \frac{\Omega_{ce} x^2}{2L}, \qquad \text{(S-13.343)}$$

with $\Omega_{ce} = eB_0/(m_e c)$ the reference electron cyclotron frequency. We may recast (S-13.342) in a dimensionless form by normalizing lengths over L and times over Ω_{ce}^{-1}. Then (S-13.342) becomes

$$m_e \frac{d^2 X(\tau)}{d\tau^2} = -\frac{d}{dX}\overline{V}(X) = -\frac{d}{dX}\left[\overline{\mathcal{P}}\frac{X^2}{2} + \frac{X^4}{4}\right], \qquad \text{(S-13.344)}$$

where $X(\tau) = x/L$ gives the normalized electron position at normalized time $\tau = t\,\Omega_{ce}$ and

$$\overline{\mathcal{P}} = \frac{\mathcal{P}_y}{m_e \Omega_{ce} L} = \frac{v_y}{\Omega_{ce} L} - \frac{eA_y(x)}{m_e \Omega_{ce} L} = \frac{v_y}{\Omega_{ce} L} - \frac{x^2}{2L^2} = \frac{v_y}{\Omega_{ce} L} - \frac{X^2}{2} \qquad \text{(S-13.345)}$$

is the only dimensionless parameter of the problem.

The shape of the potential $\overline{V}(X)$ depends on the sign of $\overline{\mathcal{P}}$ which, in its turn, depends on the initial conditions $v_y(t = 0)$ and $x(t = 0)$. In particular, keeping $v_y(t = 0)$ fixed, $\overline{\mathcal{P}}$ becomes large and negative for large $|x(t = 0)|$, i.e., for initial positions far from the magnetic null line at $x = 0$.

If $\overline{\mathcal{P}} \geq 0$ the potential $\overline{V}(X)$ has a single minimum at $X = 0$, see Fig. S-13.8,

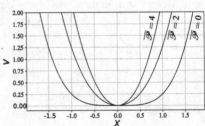

Fig. S-13.8

and the electron makes unharmonic oscillations centered at $X = 0$. These oscillations reduce to harmonic oscillations with dimensional frequency $\omega = \Omega_{ce}\sqrt{\overline{\mathcal{P}}}$ for small amplitudes $X_0^2 \ll \overline{\mathcal{P}}$, i.e., for initial conditions such that $[x(t = 0)]^2 \ll L\, v_y(t = 0)/\Omega_{ce}$, where $v_y(t = 0)/\Omega_{ce}$ has the form of a Larmor radius. From (S-13.345) we see that for these orbits the electron velocity along y oscillates with frequency $2\Omega_{ce}\sqrt{\overline{\mathcal{P}}}$ in addition to a constant term, called a drift velocity:

$$\frac{v_y}{\Omega_{ce}L} = \overline{\mathcal{P}} + \frac{X^2}{2} = \overline{\mathcal{P}} + \frac{X_0^2}{2}\cos^2\left(\Omega_{ce}\sqrt{\overline{\mathcal{P}}}\right) = \overline{\mathcal{P}} + \frac{X_0^2}{4} + \frac{X_0^2}{4}\cos\left(2\Omega_{ce}\sqrt{\overline{\mathcal{P}}}\,t\right)$$

$$= \overline{\mathcal{P}} + \frac{X_0^2}{4} + \frac{X_0^2}{4}\cos\left(2\sqrt{\overline{\mathcal{P}}}\,\tau\right), \tag{S-13.346}$$

thus

$$Y(\tau) = \frac{y(\tau)}{L} = Y_0 + \int_0^\tau \left[w + \frac{X_0^2}{4}\cos\left(2\sqrt{\overline{\mathcal{P}}}\vartheta\right)\right]d\vartheta$$

$$= w\tau + \frac{X_0^2}{8\sqrt{\overline{\mathcal{P}}}}\sin\left(2\sqrt{\overline{\mathcal{P}}}\,\tau\right), \tag{S-13.347}$$

where $Y_0 = y(t = 0)/L$ and $w = \overline{\mathcal{P}} + X_0^2/4$ is the drift velocity.

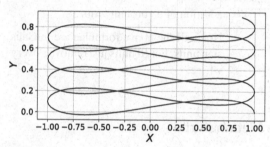

Fig. S-13.9

The resulting electron orbit is called "figure 8 orbit"[11] and is shown in Fig. S-13.9. The same type of orbits occurs in the case of unharmonic oscillations along X but the corresponding "figure-8" trajectories become distorted.

If $\overline{\mathcal{P}} < 0$ the potential $\overline{V}(X)$ has the so called "Mexican hat shape" with a single local maximum at $X = 0$, and two symmetric minima located at $X = \pm\sqrt{2|\overline{\mathcal{P}}|}$, as shown in Fig. S-13.10. In this case two classes of orbits are possible depending on the value of the particle normalized effective energy $\mathcal{W} = [dX(\tau)/d\tau]^2/2 + \overline{V}(X)$. If $\mathcal{W} < 0$ the electrons remain trapped inside the wells at $X = \pm\sqrt{(2|\overline{\mathcal{P}}|}$ and cannot cross the null line $x = 0$, see Fig. S-13.11. These orbits are essentially cyclotron orbits in a non vanishing magnetic field with a drift along y, as follows (S-13.337), that, as for the "figure-8" orbits, arises from the inhomogeneity of the magnetic field. The frequency of the harmonic oscil-

---

[11] Analogously shaped orbits are found in the case of a relativistic electron moving in a linearly polarized monochromatic plane wave discussed in Problem 13.17. Similar types of orbits called "Speiser orbits" play an important role in space and laboratory plasmas [6].

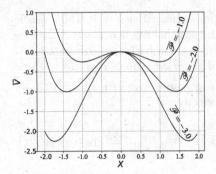

Fig. S-13.10

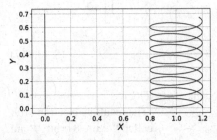

Fig. S-13.11

lations of the orbits trapped near the bottom of the wells corresponds, in dimensional units, to the value of the cyclotron-frequency computed with the value of the magnetic field at the point around which the electron is oscillating.

On the contrary if $\mathcal{W} > 0$ the orbits cross the null line $x = 0$ and are similar to the orbits found for $\overline{\mathcal{P}} > 0$.

The special case when $\overline{\mathcal{P}} < 0$ and $\mathcal{W} = 0$ corresponds to orbits that approach the null line of the magnetic field in a logarithmically diverging time.

You may wonder why the effective energy can be negative. It is a question of definition, the electron in reality has only kinetic energy which is by definition positive. A direct transformation from the real energy would lead to redefine $V(x)$ as $V(x) = \left[C^2 + 2C \, \hat{A} + \hat{A}^2 \right]/2 = \left[C + \hat{A} \right]^2 /2$ which simply amounts to the addition of a constant term.

Note: all the above formulae are in Gaussian units. For SI units just remove c everywhere.

S-13.27 A Parallel-Wire Metamaterial

(a) The total field inside the cylinder $\mathbf{E}_{tot}$ equals the external field $\mathbf{E}_0$ plus the field due to the cylinder polarization $\mathbf{E}_{pol}$, which we assune to be uniform. Thus we have

$$\mathbf{P} = \frac{\varepsilon_r - 1}{4\pi} \mathbf{E}_{tot} = \frac{\varepsilon_r - 1}{4\pi} \left(\mathbf{E}_0 + \mathbf{E}_{pol} \right) = \frac{\varepsilon_r - 1}{4\pi} \left(\mathbf{E}_0 - 2\pi\mathbf{P} \right) , \qquad \text{(S-13.348)}$$

where we have used (S-3.103) of Solution S-3.11. Solving for $\mathbf{P}$ we get

$$\mathbf{P} = \frac{\varepsilon_r - 1}{2\pi(\varepsilon_r + 1)} \mathbf{E}_0 . \qquad \text{(S-13.349)}$$

(b) The surface charge density σ of the conducting cylinder equals the surface polarization charge density of a uniformly transversally polarized cylinder which, in the presence of the same external field $\mathbf{E}_0$, cancels the electric field at its interior. Using (S-3.103) we have

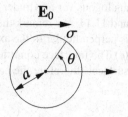

$$P = \frac{E_0}{2\pi}, \qquad (S\text{-}13.350)$$

Fig. S-13.12

which is the limit of (S-13.349) as $\varepsilon_r \to \infty$. Thus we have

$$\sigma = \mathbf{P} \cdot \hat{\mathbf{r}} = \frac{E_0}{2\pi}\cos\theta, \qquad (S\text{-}13.351)$$

where θ is the angle shown in Fig. S-13.12.

(c) Since $a \ll \lambda$ the polarization of each cylinder is given by (S-13.349), with $\mathbf{E}_0$ replaced by the electric field of the wave, which, locally, we can write as $\mathbf{E}_0\,e^{i\omega t}$. Since the cylinders fill a fraction f of the volume given by

$$f = \frac{\pi a^2}{b^2} \qquad (S\text{-}13.352)$$

the macroscopic polarization density of the metamaterial is

$$\mathbf{P}_{\mathrm{met}} = \frac{\pi a^2}{b^2}\frac{\varepsilon_r - 1}{2\pi(\varepsilon_r + 1)}\,\mathbf{E}_0\,e^{i\omega t} = \frac{\varepsilon_r^{\mathrm{met}\perp} - 1}{4\pi}\,\mathbf{E}_0\,e^{i\omega t}, \qquad (S\text{-}13.353)$$

where $\varepsilon_r^{\mathrm{met}\perp}$ is the dielectric permittivity of the metamaterial for polarization perpendicular to the cylinder axes. Solving for $\varepsilon_r^{\mathrm{met}\perp}$ we obtain

$$\varepsilon_r^{\mathrm{met}\perp} = 1 + 2\pi\left(\frac{a}{b}\right)^2\frac{\varepsilon_r - 1}{\varepsilon_r + 1} \qquad (S\text{-}13.354)$$

(d) The dielectric permittivity of the metamaterial is

$$\varepsilon_r^{\mathrm{met}\perp}(\omega) = 1 - 2\pi\left(\frac{a}{b}\right)^2\frac{\omega_p^2}{2\omega^2 - \omega_p^2} \qquad (S\text{-}13.355)$$

The wave can propagate in the metamaterial if $\varepsilon_r^{\mathrm{met}\perp} > 0$, thus

$$\omega^2 > \frac{\omega_p^2}{2} + \pi\left(\frac{a}{b}\right)^2\omega_p^2 \simeq \frac{\omega_p^2}{2}, \qquad (S\text{-}13.356)$$

since $\pi\,(a/b)^2 \ll 1$. We have resonance for $\omega = \omega_p/\sqrt{2}$. This resonance arises from the collective oscillation of electrons perpendicularly to the cylinder axis, and it is the analogous for a cylinder of the Mie oscillation for a sphere, as seen in Problems 1.5 and 11.11. We remark that this oscillation can be excited only when the external field is perpendicular to the axis of the cylinder

(e) The interface conditions require the electric field to be continuous at the cylinder surfaces, therefore the polarization inside each cylinder is

$$\mathbf{P} = \frac{\varepsilon_r - 1}{4\pi}\mathbf{E}_0 .$$
(S-13.357)

As above, the spatially-averaged polarization inside the metamaterial is obtained by multiplying the polarization inside a cylinder by f.

In the case of a metal with $\varepsilon_r = 1 - \omega_p^2/\omega^2$ we have

$$\varepsilon_r^{\text{met}\parallel} = 1 - \left(\frac{a}{b}\right)^2 \frac{\omega_p^2}{4\omega^2} = 1 - \frac{\omega_{p\,\text{eff}}^2}{\omega^2} , \quad \text{where} \quad \omega_{p\,\text{eff}} = \frac{a}{2b}\omega_p ,$$
(S-13.358)

thus, for polarization parallel to the wires, the effective plasma frequency (and cutoff frequency) $\omega_{p\,\text{eff}}$ of the metamaterial is smaller than the plasma frequency of the metal by a factor $a/(2b)$. Our metamaterial is obviously anisotropic.

S-13.28   Motion in an Inhomogeneous Magnetic field

(a) In a static magnetic field the energy of the particle, and thus the magnitude of its velocity, are constant. As long as the particle remains in the $x > 0$ region it will perform a circular motion with frequency and radius

$$\omega_1 = \frac{qB_1}{m\gamma_0 c} , \qquad r_1 = \frac{v_0}{\omega_1} ,$$
(S-13.359)

where $\gamma_0 = \sqrt{1 - v_0^2/c^2}$ (for non-relativistic motion we set $\gamma_0 = 1$).

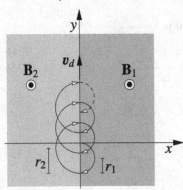

At $t = t_1 = \pi/\omega_1$, the particle enters the $x < 0$ region with velocity $-v_0$ and again performs a circular motion with frequency and radius

$$\omega_2 = \frac{qB_2}{m\gamma_0 c} , \qquad r_2 = \frac{v_0}{\omega_2} .$$
(S-13.360)

At $t = t_2 = t_1 + \pi/\omega_2$, the particle reenters the $x > 0$ region, again with velocity v_0. However, between $t = 0$ and $t = t_2$ it has drifted along $\hat{\mathbf{y}}$ by $\delta y = 2r_2 - 2r_1$. This corresponds to an average drift velocity along $\hat{\mathbf{y}}$

Fig. S-13.13

$$v_d = \frac{2(r_2 - r_1)}{\pi/\omega_1 + \pi/\omega_2} \hat{\mathbf{y}} = \frac{2v_0}{\pi} \frac{\omega_1 - \omega_2}{\omega_2 + \omega_1} \hat{\mathbf{y}}$$

$$= \frac{2v_0}{\pi} \frac{B_1 - B_2}{B_2 + B_1} \hat{\mathbf{y}} . \tag{S-13.361}$$

If for example $q > 0$, $v_0 > 0$, and $B_1 > B_2 > 0$ then $r_2 > r_1$ and $v_d > 0$; this is the case shown in Fig. S-13.13.

(b) In the S' frame there is an electric field

$$\mathbf{E}' \simeq \frac{v_d}{c} \times \mathbf{B} = \begin{cases} \dfrac{v_d}{c} B_2 \hat{\mathbf{x}} & (x < 0) \\ \dfrac{v_d}{c} B_1 \hat{\mathbf{x}} & (x > 0) \end{cases} \tag{S-13.362}$$

while at the limit $v_d \ll c$ the magnetic field $\mathbf{B}' \simeq \mathbf{B}$. Thus, on each side of the $x = 0$ plane the particle motion will be the superposition of a circular cyclotron orbit and a drift in the $\mathbf{E}' \times \mathbf{B}'$ direction, i.e., along $\hat{\mathbf{y}}' = \hat{\mathbf{y}}$. We expect the trajectory to be closed.

At $t' = t = 0$, the particle has initial velocity $\mathbf{v}'(0) = (v_0, -v_d)$ and is located at $(x' = x = 0, \ y' = y = 0)$. The equations of motion for the velocity $\mathbf{v}'$ are

$$\frac{dv_x'}{dt} = \frac{q}{m}\left(E_1' + \frac{v_y'}{c} B_1\right) = \frac{1}{c}\frac{q}{m} B_1 \left(v_y' + v_d\right) , \qquad \frac{dv_y'}{dt} = -\frac{q}{m}\frac{v_x'}{c} B_1 ,$$

$$\tag{S-13.363}$$

with solution

$$v_x' = v_0 \cos\omega_1 t' , \qquad v_y' = -v_0 \sin\omega_1 t' - v_d \quad (x' > 0) \tag{S-13.364}$$

The corresponding trajectory is

$$x' = \frac{v_0}{\omega_1} \sin\omega_1 t' , \qquad y' = \frac{v_0}{\omega_1}(\cos\omega_1 t' - 1) - v_d t' \quad (x' > 0). \tag{S-13.365}$$

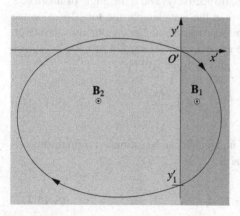

At $t' = t_1 = \pi/\omega_1$, the particle enters the $x' < 0$ region with velocity $\mathbf{v}'(t_1) = (-v_0, -v_d)$ at the position $(0, y_1')$ with $y_1' = -(2v_0 + \pi v_d)/\omega_1$. The solution for the velocity for $t' > t_1$ is analogous to the solution for $0 < t' < t_1$:

$$v_x' = -v_0 \cos\omega_2(t' - t_1) ,$$
$$v_y' = +v_0 \sin\omega_2(t' - t_1) - v_d \quad (x' < 0).$$
$$\tag{S-13.366}$$

Fig. S-13.14

For the position we obtain

$$x' = -\frac{v_0}{\omega_2} \sin \omega_2 (t' - t_1) \,,$$

$$y' = -\frac{v_0}{\omega_2} \left[\cos \omega_2 (t' - t_1) - 1\right]$$

$$- v_d (t' - t_1) + y_1' \,, \quad (x' < 0). \tag{S-13.367}$$

Thus, at $t' = t_2 = t_1 + \pi/\omega_2$, the particle is at $(x' = 0, y' = y_2')$ where $y_2' = y'(t_2)$ is

$$y_2' = \frac{2v_0}{\omega_2} - \frac{\pi v_d}{\omega_2} + y_1' = \frac{2v_0}{\omega_2} - \frac{\pi v_d}{\omega_2} - \frac{2v_0}{\omega_1} - \frac{\pi v_d}{\omega_1} = 2v_0 \left(\frac{1}{\omega_2} - \frac{1}{\omega_1}\right) - \pi v_d \left(\frac{1}{\omega_2} - \frac{1}{\omega_1}\right)$$

$$= 2v_0 \left(\frac{1}{\omega_2} - \frac{1}{\omega_1}\right) + 2v_0 \frac{\omega_1 - \omega_2}{\omega_2 + \omega_1} \frac{\omega_2 + \omega_1}{\omega_1 \omega_2} = 2v_0 \left(\frac{1}{\omega_2} - \frac{1}{\omega_1}\right) + 2v_0 \left(\frac{1}{\omega_1} - \frac{1}{\omega_2}\right) = 0 \,. \tag{S-13.368}$$

We have thus verified that $v'(0) = v'(t_2)$ and $(x'(0), y'(0)) = (x'(t_2), y'(t_2))$, as it must be for a closed orbit with period t_2. Figure S-13.14 shows the particle orbit as observed in S' for the case $B_1 = 4B_2$.

S-13.29   Solar Sail

(a) The pressure of electromagnetic radiation of intensity I on a planar surface is

$$p = (1 + R) \frac{I}{c} \cos^2 \theta \,, \tag{S-13.369}$$

independently of the wavelength, with R the reflection coefficient ($R = 1$ or 0 for perfectly reflecting or absorbing surfaces, respectively) and θ the angle of incidence. Since the grain is assumed to be larger than the wavelength, the total force on the grain can be estimated by a simple integration over the angle; for our aims, however, it will be sufficient to estimate the force as $\simeq \pi a_g^2 I/c$. If a grain of mass m_g is at a distance r from the center of the Sun, the gravity force will be

$$f_g = -G \frac{m_g M_\odot}{r^2} = -G \left(\rho_g \frac{4\pi}{3} a_g^3\right) \frac{M_\odot}{r^2} \,. \tag{S-13.370}$$

The sunlight intensity also scales with r^2 and thus it can be estimated from the knowledge of its value at the distance of the Earth,

$$I = I_E \frac{r_E^2}{r^2} \,, \tag{S-13.371}$$

yielding a total radiative force on the grain

$$f_r = \pi a_g^2 \frac{I_E}{c} \frac{r_E^2}{r^2} . \tag{S-13.372}$$

The dust grain will be swept away if $|f_r| > |f_g|$, which gives (independently of the distance from the Sun r)

$$\frac{I_E}{c} r_E^2 > G \rho_g \frac{4}{3} a_g M_\odot , \tag{S-13.373}$$

or, equivalently,

$$a_g < \frac{3 I_E r_E^2}{4 \rho_g c G M_\odot} \simeq 6.7 \times 10^{-7} \text{ m} , \tag{S-13.374}$$

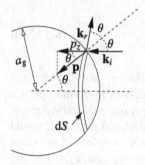

Fig. S-13.15

where $\rho_g \simeq 10^3 \text{ kg m}^{-3}$ was used. Actually the value of a_g falls at the upper limit of the spectral range considered, thus the assumption of a grain larger than the wavelength is not a very good approximation.

For the sake of completeness let us calculate in detail the total force exerted by the sunlight pressure on the grain, assumed spherical. The rays of sunlight incident at an angle θ exert a force per unit surface $\mathbf{p} = -p\,\hat{\mathbf{r}}_g$, with $\hat{\mathbf{r}}_g$ the unit vector along the direction from the center of the grain to the point of incidence. This force must first be integrated over the area $dS = 2\pi a_g^2 \sin\theta\, d\theta$ shown in Fig. S-13.15. The component parallel to the sunlight direction is $p_z = p \cos\theta = p_0 \cos^3\theta$ where $p_0 = (1 + R) I/c$, while the perpendicular components cancel out. Thus the total force is

$$f_r = \int p_x dS = 2\pi a_g^2 p_0 \int_0^{\pi/2} \cos^3\theta \sin\theta\, d\theta = \pi a_g^2 (1 + R) \frac{I}{2c} , \tag{S-13.375}$$

we see that our above estimate $f_r = \pi a_g^2 I/c$ is exact for $R = 1$, which means a perfectly reflecting grain.

(b) For a thin sail the mass is $m_s = \rho_s a_s^2 d_s$, and replaces m_g in (S-13.370). Thus we have the condition

$$\frac{2 I_E}{c} r_E^2 a_s^2 > G \rho_s a_s^2 d_s M_\odot , \tag{S-13.376}$$

which is independent of the value of a_s and yields

$$d_s < \frac{2 I_E r_E^2}{\rho_s c G M_\odot} \simeq 7.8 \times 10^{-7} \text{ m} , \tag{S-13.377}$$

Note that values much smaller than the relevant wavelengths could invalidate the assumption of a perfectly reflecting sail, the reflectivity being dependent on thickness.

(c) Based on the above estimates we may assume the radiation boost force to be twice the gravity force, yielding a total force on the sail

$$f_r = a_s^2 \frac{I_E}{c} \frac{r_E^2}{r^2} = -\frac{dU_s}{dr} . \tag{S-13.378}$$

with $U_s = U_S(r) = a_s^2 I_E r_E^2 / c / r$. The final velocity v_f can thus be obtained as

$$\frac{m_s}{2} v_f^2 = U(r_E) = a_s^2 \frac{I_E}{c} r_E , \tag{S-13.379}$$

which yields

$$\frac{v_f}{c} = \sqrt{\frac{2 I_E}{\sigma_s c^3} r_E} , \tag{S-13.380}$$

where we have substituted $m_s / a_s^2 = \rho_s d_s = \sigma_s$. This yields $v_f \simeq 4.8 \times 10^{-3} c \simeq 1.4 \times 10^6$ m s^{-1}. For a comparison, this is nearly five orders of magnitude greater than the maximum velocity reached by the Voyager probe. A minimal sail weight of 10 g, due to essential miniaturized ("on-chip") equipment placed on board, would require a sail surface of about 10^5 m^2. The sail might reach v_f values of the order of 10^7 m s^{-1}, sufficient to reach the Alpha Centauri system in a century, if initially placed reasonably near to the Sun [7]. However it is worth noting that, close to the Sun, the solar wind (consisting of energetic particles like electrons, protons, alpha particles, but also atoms like C, N, O, Ne, Mg, ...) exerts a pressure higher than the radiation pressure. At a distance of 1 AU from the Sun ($r = r_E$) the pressure of the solar wind is in the range 1–6×10^{-9} Pa. For instance, the effect of the solar wind is dominant in creating the tail of a comet.

S-13.30   Canonical Momentum

(a) If we differentiate P_y with respect to time applying the chain rule we obtain

$$\frac{dP_y}{dt} = m \frac{dv_y}{dt} + \frac{q}{c} \frac{dx}{dt} \frac{\partial A_y}{\partial x} = m \frac{dv_y}{dt} + \frac{q}{c} v_x \frac{\partial A_y}{\partial x} = m \frac{dv_y}{dt} + \frac{q}{c} v_x B_z , \tag{S-13.381}$$

since, with our choice of the vector potential $\mathbf{A}$, we have $B_z = \partial A_y / \partial x$. On the other hand our particle is subject only to the magnetic force, thus its equation of motion is

$$m \frac{d\mathbf{v}}{dt} = \frac{q}{c} \mathbf{v} \times \mathbf{B} , \quad \text{with projection on the } y \text{ axis} \quad m \frac{dv_y}{dt} = -\frac{q}{c} v_x B_z , \tag{S-13.382}$$

which, substituted into (S-13.381), leads to $dP_y/dt = 0$.

(b) In the case of a uniform magnetic field $\mathbf{B}_0 = \hat{\mathbf{z}}\,B_0$ the particle describes a circular orbit with angular frequency and radius

$$\omega = \frac{q\,B_0}{mc}\,, \qquad r = \frac{cm v_0}{q\,B_0}\,, \tag{S-13.383}$$

respectively, where v_0 is the particle speed. Assuming $q > 0$ and $B_0 > 0$ the rotation occurs clockwise. Thus the particle coordinates as functions of time are

$$x(t) = x_0 + r\cos\omega t = x_0 + \frac{cm v_0}{q\,B_0}\cos\omega t\,,$$

$$y(t) = y_0 - r\sin\omega t = y_0 - \frac{cm v_0}{q\,B_0}\sin\omega t\,, \tag{S-13.384}$$

where x_0 and y_0 are the coordinates of the orbit center (center of gyration), while the velocity components are

$$v_x(t) = -v_0\sin\omega t\,, \qquad v_y(t) = -v_0\cos\omega t\,. \tag{S-13.385}$$

Writing the vector potential as $\mathbf{A}^{(1)} = \hat{\mathbf{y}}\,B_0 x$ we have for P_y

$$P_y\hat{\mathbf{y}} = \left[m v_y + \frac{q}{c}A_y^{(1)}(x)\right]\hat{\mathbf{y}} = \left[-m v_0\cos\omega t + \frac{q}{c}B_0\left(x_0 + \frac{cm v_0}{q\,B_0}\cos\omega t\right)\right]\hat{\mathbf{y}}$$

$$= \left(\frac{q}{c}B_0 x_0\right)\hat{\mathbf{y}} = \left(\frac{q}{c}B_0\,\hat{\mathbf{z}}\right)\times(x_0\,\hat{\mathbf{x}})\,, \tag{S-13.386}$$

proportional to the x coordinate of the orbit center.

(c) The uniform magnetic field $\hat{\mathbf{z}}\,B_0$ can be obtained also from the vector potential $\mathbf{A}^{(2)}(y) = -\hat{\mathbf{x}}\,B_0 y$. With this choice we have for the x component of the canonical momentum

$$P_x\hat{\mathbf{x}} = \left[m v_x + \frac{q}{c}A_x^{(2)}(y)\right]\hat{\mathbf{x}} = \left[-m v_0\sin\omega t - \frac{q}{c}B_0\left(y_0 - \frac{cm v_0}{q\,B_0}\sin\omega t\right)\right]\hat{\mathbf{x}}$$

$$= \left(-\frac{q}{c}B_0 y_0\right)\hat{\mathbf{x}} = \left(\frac{q}{c}B_0\,\hat{\mathbf{z}}\right)\times(y_0\,\hat{\mathbf{y}})\,. \tag{S-13.387}$$

proportional to the y coordinate of the center of gyration. Note that the difference $\mathbf{A}^{(1)} - \mathbf{A}^{(2)} = B_0\left(\hat{\mathbf{y}}\,x + \hat{\mathbf{x}}\,y\right) = \nabla(B_0 xy)$ is the gradient of a scalar function of the coordinates, as required by the gauge invariance of the vector potential.

S-13.31 Classical Zeeman Effect

(a) In the absence of the magnetic field the only centripetal force is the Coulomb attraction between electron and nucleus, thus we have for our circular orbit

$$m_e \omega_0^2 r = \frac{e^2}{r^2}, \quad \text{with} \quad \omega_0 = \frac{e}{\sqrt{m_e r^3}}, \tag{S-13.388}$$

where e is the elementary charge and m_e the electron mass. Rigorously, we should use the reduced mass of the proton-electron system but, given that the proton mass is 1836 times larger than the electron mass, we can neglect this correction.

In the presence of a magnetic field $\mathbf{B}_0$ parallel to the angular momentum also the magnetic force is centripetal, and we have

$$m_e \omega^2 r = \frac{e^2}{r^2} + \frac{e \omega r B_0}{c}. \tag{S-13.389}$$

Subtracting (S-13.388) from (S-13.389) we obtain

$$m_e \left(\omega^2 - \omega_0^2 \right) = \frac{e \omega B_0}{m_e c}, \quad \text{or} \quad (\omega - \omega_0)(\omega + \omega_0) = \frac{e \omega B_0}{m_e c}, \tag{S-13.390}$$

where we have divided both sides by $m_e r$. Assuming $\delta \omega = (\omega - \omega_0) \ll \omega_0$ we can approximate $(\omega + \omega_0) \simeq 2\omega$, obtaining

$$\delta \omega = \frac{e}{2 m_e c} B_0, \tag{S-13.391}$$

the factor $e/(2 m_e c)$ is called the *gyromagnetic ratio*, it is written $e/(2 m_e)$ in SI units.

In the case of $\mathbf{B}_0$ antiparallel to the orbital angular momentum we simply have

$$\delta \omega = -\frac{e}{2 m_e c} B_0. \tag{S-13.392}$$

For a hydrogen atom we can assume $r = 5 \times 10^{-9}$ cm ($a_0 = 5.291772 \times 10^{-9}$ is the Bohr radius), the elementary charge is $e = 4.803 \times 10^{-10}$ esu and the electron mass is $m_e = 9.11 \times 10^{-28}$ g. This leads to $\omega_0 = 4.5 \times 10^{16}$ rad/s. The gyromagnetic ratio is 8.787×10^6 rad/(s G), thus a even a small ratio $\delta \omega / \omega_0 = 10^{-3}$ requires a field $B_0 = 5 \times 10^6$ G = 500 T, much higher than the highest fields available in most laboratories.

In 1896 Pieter Zeeman observed that the frequency of a spectral line was split into three components when a magnetic field was applied to the emitting atoms (*normal Zeeman effect*). One component was unperturbed, one was shifted by $\delta \omega$ and one by $-\delta \omega$. This effect was originally interpreted as being due to atoms whose orbital angular momenta were perpendicular to (unperturbed component), parallel or antiparallel to ($\pm \delta \omega$) the magnetic field. In the following year Thomas Preston observed spectral lines split into more than three components by a magnetic field (*anomalous Zeeman effect*). Actually the normal Zeeman effect turns out to be a special case of the anomalous effect, and quantum mechanics is required for a rigorous treatment.

(b) Let us assume axial symmetry around the orbit axis. The time variation of the magnetic field corresponds to the presence of an induced azimuthal electric field

$\mathbf{E} = \hat{\phi} E_\phi$. The line integral of $\mathbf{E}$ over the electron orbit is, according to Faraday's law of induction,

$$\oint \mathbf{E} \cdot d\ell = 2\pi r E_\phi = -\frac{1}{c}\frac{d\Phi}{dt} = -\pi r^2 \frac{1}{c}\frac{dB}{dt} , \qquad \text{(S-13.393)}$$

where Φ is the flux of the magnetic field $\mathbf{B}$ through the area enclosed by the orbit. Due to our assumed axial symmetry the electric field on the orbit is thus

$$E_\phi = \frac{1}{2\pi r}\left(-\pi r^2 \frac{1}{c}\frac{dB}{dt}\right) = -\frac{r}{2c}\frac{dB}{dt} , \qquad \text{(S-13.394)}$$

directed clockwise in Fig. 13.17. This field exerts a torque on the electron

$$\boldsymbol{\tau} = \mathbf{r} \times (-e)\,\mathbf{E} = \hat{\mathbf{z}}\,r(-e)\,E_\phi = \hat{\mathbf{z}}\,\frac{r^2 e}{2c}\frac{dB}{dt} , \qquad \text{(S-13.395)}$$

and the torque changes the orbital angular momentum of the electron $\mathbf{L} = \hat{\mathbf{z}}\,L_z$ according to the law

$$\frac{dL_z}{dt} = r^2 m_{\mathrm{e}}\frac{d\omega}{dt} = \frac{r^2 e}{2c}\frac{dB}{dt} , \quad \text{from which} \quad \frac{d\omega}{dt} = \frac{e}{2m_{\mathrm{e}}c}\frac{dB}{dt} . \qquad \text{(S-13.396)}$$

The change of the electron angular frequency is thus

$$\delta\omega = \int_0^{\Delta t} \frac{d\omega}{dt}\,dt = \frac{e}{2m_{\mathrm{e}}c}\int_0^{\Delta t}\frac{dB}{dt}\,dt = \frac{e}{2m_{\mathrm{e}}c}\,B_0 , \qquad \text{(S-13.397)}$$

in agreement with (S-13.391). The increasing magnetic field corresponds to an increasing centripetal force, thus an increasing angular velocity is required in order to follow an orbit of constant radius. The induced azimuthal electric field provides the required accelaration of the electron, as shown by the equality of (S-13.391) and (S-13.397). In other words, at any time $t^* < \Delta t$ the magnetic field is $B(t^*)$ and the orbital angular frequency is

$$\omega(t^*) = \omega_0 + \frac{e}{2m_{\mathrm{e}}c}\int_0^{t^*}\frac{dB}{dt}\,dt = \omega_0 + \frac{e}{2m_{\mathrm{e}}c}\,B(t^*) = \omega_0 + \delta\omega(t^*) . \qquad \text{(S-13.398)}$$

Thus the centripetal force required for keeping the electron on a circular orbit of radius r at time t^* is

$$f_c(t^*) = m_e r \omega^2(t) \simeq m_e r \left[\omega_0^2 + \frac{\omega_0 e}{m_e c} B(t^*) \right] = m_e r \left[\omega_0^2 + 2\omega_0 \, \delta\omega(t^*) \right] ,$$

$$\text{(S-13.399)}$$

where we have neglected the quadratic term in $\delta\omega(t^*)$, consistently with the treatment that led us to (S-13.391). On the other hand the total force acting on the electron is

$$f_t(t^*) = \frac{e^2}{r^2} + \frac{e r \omega(t^*) B(t^*)}{c} = m_e \omega_0^2 r + \frac{e r B(t^*)}{c} \left[\omega_0 + \frac{e}{2 m_e c} B(t^*) \right]$$

$$\simeq m_e r \left[\omega_0^2 + \frac{\omega_0 e}{m_e c} B(t^*) \right] = m_e r \left[\omega_0^2 + 2\omega_0 \, \delta\omega(t^*) \right] , \qquad \text{(S-13.400)}$$

where, again, we have neglected the quadratic term in $\delta\omega(t^*)$. Thus, with our assumption of axial symmetry around the orbit axis, the shape of the orbit is unperturbed up to terms in $[\delta\omega(t^*)]^2$.

(c) In the absence of the magnetic field the canonical angular momentum simply equals the kinetic angular momentum

$$L_i^c = L_i^{kin} = \mathbf{r} \times (m\boldsymbol{\omega}_0 \times \mathbf{r}) = \hat{\mathbf{z}} r^2 m \omega_0 . \qquad \text{(S-13.401)}$$

After applying the field the kinetic angular momentum is, within our approximations,

$$L_f^{kin} = \hat{\mathbf{z}} r^2 m (\omega_0 + \delta\omega) . \qquad \text{(S-13.402)}$$

From this we obtain the final canonical angular momentum by adding the contribution of $\mathbf{r} \times (-e)\mathbf{A}/c$

$$L_f^c = \hat{\mathbf{z}} r^2 m (\omega_0 + \delta\omega) - \frac{e}{c} \mathbf{r} \times \mathbf{A} = \hat{\mathbf{z}} r^2 m (\omega_0 + \delta\omega) - \frac{e B_0}{2c} \mathbf{r} \times (-\hat{\mathbf{x}} r_y + \hat{\mathbf{y}} r_x)$$

$$= \hat{\mathbf{z}} \left[r^2 m (\omega_0 + \delta\omega) - r^2 \frac{e B_0}{2c} \right] = \hat{\mathbf{z}} r^2 m \omega_0 = L_i^c . \qquad \text{(S-13.403)}$$

S-13.32   Force-Free Magnetic Field

(a) The density of the magnetic force is $\mathbf{J} \times \mathbf{B}/c$. Thus, if $\mathbf{J}$ and $\mathbf{B}$ are both different from zero, $\mathbf{J}$ and $\mathbf{B}$ must be parallel (or antiparallel) to each other for the force to be zero.

(b) If $\mathbf{J}$ and $\mathbf{B}$ are parallel (or antiparallel) to each other the following relation holds

$$\mathbf{J}(r) = f(\mathbf{r}) \mathbf{B}(r) , \qquad \text{(S-13.404)}$$

where $f(\mathbf{r})$ is a generic scalar function of the coordinates, here r is the radial cylindrical coordinate while $\mathbf{r}$ is the position vector. We thus have

$$\mathbf{\nabla} \times \mathbf{B}(r) = \frac{4\pi}{c} \mathbf{J}(r) = \frac{4\pi}{c} f(\mathbf{r}) \mathbf{B}(r) \,. \tag{S-13.405}$$

The conditions $\mathbf{\nabla} \cdot \mathbf{B} = 0$ (always valid) and $\mathbf{\nabla} \cdot \mathbf{J} = 0$ (valid in stationary conditions) further imply

$$0 = \mathbf{\nabla} \cdot \mathbf{J} = \mathbf{\nabla} \cdot (f\mathbf{B}) = f \mathbf{\nabla} \cdot \mathbf{B} + \mathbf{B} \cdot \mathbf{\nabla} f = \mathbf{B} \cdot \mathbf{\nabla} f \,, \tag{S-13.406}$$

this means that $f(\mathbf{r})$ must be constant along the field lines of $\mathbf{B}$. Here we consider the special case

$$f(\mathbf{r}) = \text{const} = \alpha \frac{c}{4\pi} \,, \tag{S-13.407}$$

where α has the dimensions of the inverse of a length. Inserting (S-13.407) into (S-13.405) we obtain

$$\mathbf{\nabla} \times \mathbf{B}(r) = \alpha \mathbf{B}(r) \,. \tag{S-13.408}$$

(c) The curl of the left-hand side of (S-13.408) is

$$\mathbf{\nabla} \times (\mathbf{\nabla} \times \mathbf{B}) = \mathbf{\nabla}(\mathbf{\nabla} \cdot \mathbf{B}) - \nabla^2 \mathbf{B} = -\nabla^2 \mathbf{B} \,, \tag{S-13.409}$$

thus we have, equating the curls of both sides,

$$- \nabla^2 \mathbf{B}(r) = \alpha \mathbf{\nabla} \times \mathbf{B} = \alpha^2 \mathbf{B}(r) \,. \tag{S-13.410}$$

Now we consider the z component of (S-13.410), remembering that $\mathbf{B}$ depends on r only,

$$\nabla^2 B_z(r) = \frac{1}{r} \frac{\partial}{\partial r} \left[r \frac{\partial B_z(r)}{\partial r} \right] = -\alpha^2 B_z(r) \,, \tag{S-13.411}$$

and introduce the dimensionless variable $\rho = |\alpha| r$, obtaining

$$\frac{1}{\rho} \frac{\partial}{\partial \rho} \left[\rho \frac{\partial B_z(\rho)}{\partial \rho} \right] = -B_z(\rho) \,. \tag{S-13.412}$$

(d) Equation (S-13.412) can be rewritten as

$$\frac{\partial^2 B_z(\rho)}{\partial \rho^2} + \frac{1}{\rho} \frac{\partial B_z(\rho)}{\partial \rho} + B_z(\rho) = 0 \,, \tag{S-13.413}$$

which is the Bessel differential equation

$$x^2 \frac{d^2 y}{dx^2} + x \frac{dy}{dx} + \left(x^2 - n^2 \right) y = 0 \,, \tag{S-13.414}$$

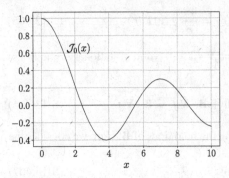

Fig. S-13.16

for $n = 0$. The solution of (S-13.413) which is nonsingular at the origin, as required by the condition $B_z(0) = 4\pi J_z(0)/(\alpha c) = 4\pi J_0/(\alpha c)$, is

$$B_z(\rho) = B_z(0)\,\mathcal{J}_0(\rho) \ . \quad \text{(S-13.415)}$$

where $\mathcal{J}_0(\rho)$ is the Bessel function of the first kind of order 0, plotted in Fig. S-13.16. From (S-13.404) and (S-13.407) we further have

$$J_z(\rho) = \frac{c\alpha}{4\pi} B_z(0)\mathcal{J}_0(0) \ , \quad \text{(S-13.416)}$$

thus $B_z(0) = 4\pi J_0/(c\alpha)$ since $\mathcal{J}_0(0) = 1$. The corresponding expression for $B_\phi(\rho)$ is obtained by considering the azimuthal components of (S-13.408)

$$\alpha B_\phi(r) = \left[\nabla \times \mathbf{B}(r)\right]_\phi = -\frac{\partial B_z(r)}{\partial r} \ . \quad \text{(S-13.417)}$$

Introducing the dimensionless variable ρ we have

$$B_\phi(\rho) = -(\text{sgn}\,\alpha)\frac{\partial B_z(\rho)}{\partial \rho} = (\text{sgn}\,\alpha)B_z(0)\,\mathcal{J}_1(\rho) = \frac{4\pi}{c\alpha}J_\phi(\rho), \quad \text{(S-13.418)}$$

where we used the relation $d\mathcal{J}_0(\rho)/d\rho = -\mathcal{J}_1(\rho)$. The condition $B_z(R) = 0$ implies that the product αR corresponds to a zero of the Bessel function $\mathcal{J}_0$. Known R, we have a discrete set of possible values for α. The first values are $\alpha R \simeq 2.40,\ 5.52,\ 8.63, \ldots$, see Fig. S-13.16.

A qualitative physical insight into the problem can be gained by remembering that parallel electric currents attract each other, while antiparallel currents repel each other. Now consider two small volume elements of the conducting medium located at (r, ϕ, z) and $(r, \phi + \pi, z)$, respectively. The J_z components of the current density in the two volume elements are parallel, while the J_ϕ components are antiparallel. We have evaluated the dependencies of J_z and J_ϕ on r such that the two contributions, summed over all the volume elements, cancel out.

S-13.33 Field Inside a Spherical Shell

(a) The similarity with Problem 3.4 suggests that the polarization charges of the shell give origin to a dipole-type contribution to the field outside the shell, justifying the ansatz (13.26). We assume $\mathbf{p}_0$, $\mathbf{p}_1$, $\mathbf{E}_1$ and $\mathbf{E}_2$ to be all parallel to $\mathbf{E}_0$, and we use a spherical coordinate system (r, θ, ϕ) with the origin located at the shell center and the polar axis parallel to $\mathbf{E}_0$. The boundary conditions are the continuities of the components $(\varepsilon_r \mathbf{E})_\perp$ and $\mathbf{E}_\parallel$ at the interfaces between dielectric and vacuum, i.e., at $r = b$ and $r = a$. The directions perpendicular ($\perp$) and ($\parallel$) to the interface surfaces coincide with the unit vectors $\hat{\mathbf{r}}$ and $\hat{\boldsymbol{\theta}}$, respectively. The electric field generated by an electric dipole is, in spherical coordinates,

$$\mathbf{E}_p = \frac{(3\mathbf{p} \cdot \hat{\mathbf{r}})\,\hat{\mathbf{r}} - \mathbf{p}}{r^3} = \hat{\mathbf{r}}\,\frac{2p\cos\theta}{r^3} + \hat{\boldsymbol{\theta}}\,\frac{p\sin\theta}{r^3} \,, \tag{S-13.419}$$

while we have for the external field

$$\mathbf{E}_0 = \hat{\mathbf{r}}\,E_0 \cos\theta - \hat{\boldsymbol{\theta}}\,E_0 \sin\theta \,, \tag{S-13.420}$$

the ϕ components vanishing because of the cylindrical symmetry around the polar axis. Thus the boundary conditions at $r = b$ and $r = a$ are written

$$E_0 + \frac{2p_0}{b^3} = \varepsilon_r \left(E_1 + \frac{2p_1}{b^3} \right) \,, \tag{S-13.421}$$

$$-E_0 + \frac{p_0}{b^3} = -E_1 + \frac{p_1}{b^3} \,, \tag{S-13.422}$$

$$\varepsilon_r \left(E_1 + \frac{2p_1}{a^3} \right) = E_2 \,, \tag{S-13.423}$$

$$-E_1 + \frac{p_1}{a^3} = -E_2 \,, \tag{S-13.424}$$

or equivalently, in matrix form,

$$\begin{pmatrix} 2/b^3 & -\varepsilon_r & -2\varepsilon_r/b^3 & 0 \\ 1/b^3 & 1 & -1/b^3 & 0 \\ 0 & \varepsilon_r & 2\varepsilon_r/a^3 & -1 \\ 0 & -1 & 1/a^3 & 1 \end{pmatrix} \begin{pmatrix} p_0 \\ E_1 \\ p_1 \\ E_2 \end{pmatrix} = \begin{pmatrix} -E_0 \\ E_0 \\ 0 \\ 0 \end{pmatrix} \,. \tag{S-13.425}$$

We can use Cramer's method to determine the four unknown values. The determinant of the system matrix is

$$\begin{aligned} D &= \frac{1}{b^3} \left(2 \begin{vmatrix} 1 & -1/b^3 & 0 \\ \varepsilon_r & 2\varepsilon_r/a^3 & -1 \\ -1 & 1/a^3 & 1 \end{vmatrix} - \begin{vmatrix} -\varepsilon_r & -2\varepsilon_r/b^3 & 0 \\ \varepsilon_r & 2\varepsilon_r/a^3 & -1 \\ -1 & 1/a^3 & 1 \end{vmatrix} \right) \\ &= \frac{1}{b^3} \left(2 \begin{vmatrix} 1 & -1/b^3 & 0 \\ \varepsilon_r - 1 & (2\varepsilon_r + 1)/a^3 & 0 \\ -1 & 1/a^3 & 1 \end{vmatrix} - \begin{vmatrix} -\varepsilon_r & -2\varepsilon_r/b^3 & 0 \\ \varepsilon_r - 1 & (2\varepsilon_r + 1)/a^3 & 0 \\ -1 & 1/a^3 & 1 \end{vmatrix} \right) \\ &= \frac{1}{b^3} \left(2 \begin{vmatrix} 1 & -1/b^3 \\ \varepsilon_r - 1 & (2\varepsilon_r + 1)/a^3 \end{vmatrix} - \begin{vmatrix} -\varepsilon_r & -2\varepsilon_r/b^3 \\ \varepsilon_r - 1 & (2\varepsilon_r + 1)/a^3 \end{vmatrix} \right) \end{aligned}$$

$$= \frac{1}{b^3}\left(2\frac{2\varepsilon_r+1}{a^3}+2\frac{\varepsilon_r-1}{a^3}+\varepsilon_r\frac{2\varepsilon_r+1}{a^3}-2\varepsilon_r\frac{\varepsilon_r-1}{b^3}\right)$$

$$= \frac{1}{b^3}\left(\frac{(2\varepsilon_r+1)(\varepsilon_r+2)}{a^3}-\frac{2(\varepsilon_r-1)^2}{b^3}\right)\,. \tag{S-13.426}$$

The parameter E_2, i.e., the electric field inside the shell cavity, is obtained by dividing the determinant obtained by replacing the fourth column by the column vector of the constant terms by D,

$$E_2 = \frac{E_0}{D}\begin{vmatrix} 2/b^3 & -\varepsilon_r & -2\varepsilon_r/b^3 & -1 \\ 1/b^3 & 1 & -1/b^3 & 1 \\ 0 & \varepsilon_r & 2\varepsilon_r/a^3 & 0 \\ 0 & -1 & 1/a^3 & 0 \end{vmatrix} = \frac{E_0}{D}\begin{vmatrix} 0 & -(\varepsilon_r+2) & 2(1-\varepsilon_r)/b^3 & -3 \\ 1/b^3 & 1 & -1/b^3 & 1 \\ 0 & \varepsilon_r & 2\varepsilon_r/a^3 & 0 \\ 0 & -1 & 1/a^3 & 0 \end{vmatrix}$$

$$= -\frac{E_0}{b^3 D}\begin{vmatrix} -(\varepsilon_r+2) & 2(1-\varepsilon_r)/b^3 & -3 \\ \varepsilon_r & 2\varepsilon_r/a^3 & 0 \\ \varepsilon_r-1 & (2\varepsilon_r+1)/a^3 & 0 \end{vmatrix} = \frac{3E_0}{b^3 D}\begin{vmatrix} \varepsilon_r & 2\varepsilon_r/a^3 \\ \varepsilon_r-1 & (2\varepsilon_r+1)/a^3 \end{vmatrix}$$

$$= \frac{3E_0}{a^3 b^3 D}((2\varepsilon_r+1)\varepsilon_r - 2\varepsilon_r(\varepsilon_r-1))$$

$$= \frac{9\varepsilon_r E_0}{a^3 b^3 D} = \frac{9\varepsilon_r b^3}{b^3(\varepsilon_r+2)(2\varepsilon_r+1)-a^3(\varepsilon_r-1)^2}\, E_0\,. \tag{S-13.427}$$

Quantities E_1, p_0 and p_1 are obtained analogously, by dividing the respective determinants by D.

Alternatively we can subtract (S-13.421) from (S-13.422) multiplied by two, and add (S-13.423) to (S-13.424), obtaining a system of two linear equations in the two unknowns E_1 and p_1

$$b^3(\varepsilon_r+2)E_1 + 2(\varepsilon_r-1)p_1 = 3b^3 E_0\,, \tag{S-13.428}$$

$$a^3(\varepsilon_r-1)E_1 + (2\varepsilon_r+1)p_1 = 0\,, \tag{S-13.429}$$

with solutions

$$E_1 = \frac{3b^3(2\varepsilon_r+1)}{b^3(\varepsilon_r+2)(2\varepsilon_r+1)-a^3(\varepsilon_r-1)^2}\, E_0\,, \tag{S-13.430}$$

$$p_1 = -\frac{3b^3 a^3(\varepsilon_r-1)}{b^3(\varepsilon_r+2)(2\varepsilon_r+1)-a^3(\varepsilon_r-1)^2}\, E_0\,. \tag{S-13.431}$$

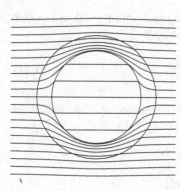

Fig. S-13.17

Knowing E_1 and p_1 we can easily obtain E_2 by substituting p_1 into (S-13.424)

$$E_2 = \frac{9\varepsilon_r b^3}{b^3(\varepsilon_r + 2)(2\varepsilon_r + 1) - a^3(\varepsilon_r - 1)^2} E_0 ,$$
$$(S\text{-}13.432)$$

in agreement with (S-13.427), and p_0 by substituting E_1 and p_1 into (S-13.422)

$$p_0 = \frac{b^6(2\varepsilon_r + 1)(\varepsilon_r - 1) - a^3 b^3(\varepsilon_r^2 + \varepsilon_r - 2)}{b^3(\varepsilon_r + 2)(2\varepsilon_r + 1) - a^3(\varepsilon_r - 1)^2} E_0 .$$
$$(S\text{-}13.433)$$

Note that the condition $E_2 < E_0$ reduces to $(\varepsilon_r - 1)^2 > 0$ which is valid for any value of ε_r: this means that the external field is screened by the shell.

The field lines of the electric field in the presence of a dielectric shell are shown in Fig. S-13.17 for the case of $\varepsilon_r = 10$ and $b/a = 4/3$.

We note that for $\varepsilon_r = 1$ we have $E_1 = E_2 = E_0$ and $p_0 = p_1 = 0$, as expected. As further checks we can evaluate the limit for $a \to 0$

$$\lim_{a \to 0} E_1 = \frac{3E_0}{\varepsilon_r + 2} , \qquad (S\text{-}13.434)$$

which is the field inside a dielectric sphere in the presence of an external uniform field E_0. The limit $b \to \infty$ corresponds to a cavity of radius a in an infinite dielectric medium with field

$$E_{\text{diel}} = \lim_{b \to \infty} E_1 = \frac{3E_0}{\varepsilon_r + 2} \qquad (S\text{-}13.435)$$

far from the cavity. The field inside the cavity is

$$E_{\text{cav}} = \lim_{b \to \infty} E_2 = \frac{9\varepsilon_r E_0}{(\varepsilon_r + 2)(2\varepsilon_r + 1)} = \frac{3\varepsilon_r E_{\text{diel}}}{2\varepsilon_r + 1} = \frac{3E_{\text{diel}}}{1/\varepsilon_r + 2} , \qquad (S\text{-}13.436)$$

in agreement with (S-3.32), as expected.

(b) Due to the analytical similarity of the equations of electrostatics and magnetostatics in the absence of free charges and currents, we can obtain the solution for the problem of the magnetic spherical shell in an external uniform magnetic field $\mathbf{B}_0$ from point (a), through the replacements $\mathbf{E} \to \mathbf{H}$, $\mathbf{D} \to \mathbf{B}$, and $\varepsilon_r \to \mu_r$. Noting that $E_0 = D_0$, $B_0 = H_0$, and $D_2 = E_2$ we immediately obtain

$$B_2 = \frac{9\mu_r b^3}{(2\mu_r + 1)(\mu_r + 2)b^3 - 2(\mu_r - 1)^2 a^3} B_0 . \qquad (S\text{-}13.437)$$

As in the dielectric case, $B_2 < B_0$, i.e., the external field is screened by the shell for any value of $\mu_r \neq 1$, including the case $\mu_r < 1$, i.e., diamagnetic materials. As a special case, $\mu_r = 0$ corresponds to a superconducting shell with $B_2 = 0$, as expected. In the case of a strongly paramagnetic material, with $\mu_r \gg 1$,

$$B_2 \simeq \frac{9}{2\mu_r(1 - a^3/b^3)} B_0 \,. \tag{S-13.438}$$

The effect of the highly paramagnetic material is to "concentrate" the field lines inside the shell, thus screening the inner cavity, as shown in Fig. S-13.17.

S-13.34   Resonances in a Spherical Shell

(a) To describe the response of the shell it is sufficient to generalize the results for a dielectric shell discussed in Problem 13.33 to the case of a time-dependent harmonic field, which corresponds to insert the metal permittivity in the expressions for the fields. For instance, taking the expression (S-13.427) for the uniform field inside the cavity we obtain

$$
\begin{aligned}
\mathbf{E}_c &= \frac{9\varepsilon_r b^3}{(2\varepsilon_r + 1)(\varepsilon_r + 2)b^3 - 2(\varepsilon_r - 1)^2 a^3} \mathbf{E}_0 \\
&= \frac{9(1 - \omega_p^2/\omega^2)}{(3 - 2\omega_p^2/\omega^2)(3 - \omega_p^2/\omega^2) - 2(a/b)^3 \omega_p^4/\omega^2} \mathbf{E}_0 \\
&\equiv \frac{9(1 - \omega_p^2/\omega^2)}{D(\omega)} \mathbf{E}_0 \,.
\end{aligned}
\tag{S-13.439}
$$

Resonance frequencies are identified as zeros of the denominator $D(\omega)$ in the above expressions (obviously the denominator is common also to the dipole moment and current density of the shell). Before calculating the full expression, it may be worth noting that in both the case of $a \to 0$ at fixed b (dielectric sphere) and the case of $b \to \infty$ at fixed a (spherical cavity in a dielectric medium) the denominator becomes

$$D(\omega) \simeq \left(3 - 2\frac{\omega_p^2}{\omega^2}\right)\left(3 - \frac{\omega_p^2}{\omega^2}\right)\,, \tag{S-13.440}$$

which vanishes for $\omega = \omega_p/\sqrt{3}$ and $\omega = \omega_p\sqrt{2/3}$. The first value corresponds to the case $\varepsilon_r + 2 = 0$, i.e., to the well-known Mie resonance frequency for a sphere (see Problem 1.5), the second value to the resonance of a spherical cavity which, according to (S-3.32) of Problem 3.4, occurs for $1 + 2\varepsilon_r = 0$.

In the general case, posing $D(\omega) = 0$ yields a quadratic equation in $(\omega/\omega_p)^2$,

$$\left(\frac{\omega}{\omega_p}\right)^4 - \left(\frac{\omega}{\omega_p}\right)^2 + \frac{2}{9}\left[1 - \left(\frac{a}{b}\right)^3\right] = 0 \,, \tag{S-13.441}$$

with the two solutions

$$\left(\frac{\omega_\pm}{\omega_p}\right)^2 = \frac{1}{2}\left\{1 \pm \sqrt{1 - \frac{8}{9}\left[1 - \left(\frac{a}{b}\right)^3\right]}\right\} . \qquad \text{(S-13.442)}$$

Both roots are positive, so that there are two resonant values for ω. At the $a/b \to 0$ limit we recover the frequencies of the above mentioned "cavity' and "sphere" oscillation modes.

At the limit $d = b - a \ll b$, we have $(a/b)^3 \simeq 1 - 3d/b$ at the first order in d/b, which leads to

$$\left(\frac{\omega_\pm}{\omega_p}\right)^2 \simeq \frac{1}{2} \pm \frac{1}{2}\sqrt{1 - \frac{8d}{3b}} \simeq \frac{1}{2} \pm \frac{1}{2}\left(1 - \frac{4d}{3b}\right)$$

$$= \begin{cases} 1 - \dfrac{2d}{3b}, \\[2mm] \dfrac{2d}{3b}. \end{cases} \qquad \text{(S-13.443)}$$

Thus we see that in a thin shell the resonant frequencies ω_+ and ω_- are slightly less and much less than the plasma frequency ω_p, respectively.

From a qualitative viewpoint, the two resonances can be viewed as a sort of superpositions of the above mentioned oscillation modes of the sphere and the cavity: these modes are independent for a thick shell ($a/b \to 0$) but become linearly coupled for a thin shell. When the two oscillations are in phase, the surface charge density has the same angular dependence on both the outer and inner surfaces of the shell. Thus, both charge distributions cooperate to enhance the field in the cavity ($\mathbf{E}_c$) and to largely cancel the field inside the metal layer ($\mathbf{E}_m$), as shown in Fig. S-13.18a. The electrostatic recoil force on electrons is thus weakened, reducing the oscillation frequency to values much smaller than ω_p as found for the ω_- mode. When the two oscillations are out of phase, the surface charge densities are opposite to each other, which leads

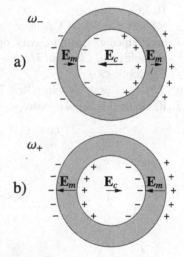

Fig. S-13.18

to a large cancellation of $\mathbf{E}_c$, as shown in Fig. S-13.18b. At the thin shell limit, the surface charge distribution produces a near-plane capacitor-like field in the metal layer, so the frequencies approaches that of planar plasma oscillations (Problem 1.4), i.e., $\omega_+ \lesssim \omega_p$.

(b) The laser source frequency is $\omega_L = 2\pi c/\lambda_L \simeq 2.5 \times 10^{15}$ Hz. Since ω_p is significantly larger than ω_L, the best option is to excite the shell resonance corresponding to the second of the two resonant frequencies (S-13.443), i.e., to achieve the condition

$$\omega_L = \sqrt{\frac{2d}{3b}} \, \omega_p \, , \qquad (S\text{-}13.444)$$

which is possible by choosing an appropiately small d/b ratio. Since $\omega_L/\omega_p \simeq 1/5.6$, we get $d/b \simeq 4.8 \times 10^{-2}$. Assuming $b = 80$ nm (we recall that $b \ll \lambda$ is required for the electrostatic modeling to be accurate) we obtain $d \simeq 3.8$ nm.

In laser surgery, nanoshells are inserted in those parts of tissues which must be removed. By tuning the resonant frequency to the laser source, energy is strongly absorbed by the nanoshells which heat up and burn the local tissue out. No significant damage is done to the surrounding tissue if the absorption is negligible at the used frequency, which is the case for infrared sources.

References

1. P.F. Cohadon, A. Heidmann, M. Pinard, Cooling of a mirror by radiation pressure. Phys. Rev. Lett. **83**, 3174 (1999)
2. J.D. Jackson, *Classical Electrodynamics*, 3rd edn. (Wiley, New York, 1998). 10.1.C
3. V.I. Veksler, Soviet J. Atomic Energy **2**, 525–528 (1957)
4. en.wikipedia.org/wiki/Cavity_optomechanics
5. H. Goldstein, *Classical Mechanics* (Addison-Wesley, Reading MA, Palo Alto, London, 1950)
6. T.W. Speiser, J. Geophys. Res. **70**, 4219 (1965)
7. R. Heller, M. Hippke, Astrophys. J. Lett. **835**, L32 (2017)

Appendix A
Mathematical Appendix

A.1 Gradient, Curl, Divergence and Laplacian

Vector equations are independent of the coordinate system used. Cartesian coordinates are used very often because they are the most convenient when the problem has no particular symmetry. However, in the case of particular symmetries, calculations can be greatly facilitated by a suitable choice of the coordinate system. Apart from the elliptical coordinates, used only in Problem 2.14, The only two special systems used in this book are the cylindrical and spherical coordinates.

A cylindrical coordinate system (r, ϕ, z) specifies a point position by the distance r from a chosen reference (*longitudinal*) axis z, the angle ϕ that r forms with a chosen reference plane $\phi = 0$ containing the z axis, and the distance, positive or negative, from a chosen reference plane perpendicular to the axis. The origin is the point where r and z are zero, for $r = 0$ the value of ϕ is irrelevant. Figure A.1 shows a cylindrical coordinate system, superposed to a Cartesian system sharing the same origin, with the z axes of the two systems are superposed, the xz plane corresponding to the $\phi = 0$ plane of the cylindrical system. We have the conversion relations

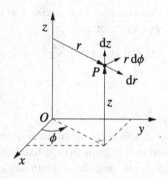

Fig. A.1

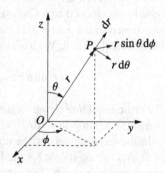

Fig. A.2

$$x = r \cos \phi, \quad y = r \sin \phi, \quad z = z. \quad \text{(A.1)}$$

The orthogonal line elements are dr, $r\,d\phi$ and dz, and the infinitesimal volume element is $r\,dr\,d\phi\,dz$.

© The Editor(s) (if applicable) and The Author(s), under exclusive license to Springer Nature Switzerland AG 2023
A. Macchi et al., *Problems in Classical Electromagnetism*,
https://doi.org/10.1007/978-3-031-22235-1

Table A.1 Gradient, curl, divergence and Laplacian in cylindrical and spherical coordinates

Cylindrical coordinates	Spherical coordinates
Components of the gradient of a scalar function V	
$\hat{\mathbf{r}}\,\frac{\partial V}{\partial r}$ $\hat{\boldsymbol{\phi}}\,\frac{1}{r}\frac{\partial V}{\partial \phi}$ $\hat{\mathbf{z}}\,\frac{\partial V}{\partial z}$	$\hat{\mathbf{r}}\,\frac{\partial V}{\partial r}$ $\hat{\boldsymbol{\theta}}\,\frac{1}{r}\frac{\partial V}{\partial \theta}$ $\hat{\boldsymbol{\phi}}\,\frac{1}{r\sin\theta}\frac{\partial V}{\partial \phi}$
Components of the curl of a vector function $\mathbf{A}$	
$\hat{\mathbf{r}}\left(\frac{1}{r}\frac{\partial A_z}{\partial \phi}-\frac{\partial A_\phi}{\partial z}\right)$ $\hat{\boldsymbol{\phi}}\left(\frac{\partial A_r}{\partial z}-\frac{\partial A_z}{\partial r}\right)$ $\hat{\mathbf{z}}\,\frac{1}{r}\left[\frac{\partial(rA_\phi)}{\partial r}-\frac{\partial A_r}{\partial \phi}\right]$	$\hat{\mathbf{r}}\,\frac{1}{r\sin\theta}\left[\frac{\partial}{\partial\theta}\left(A_\phi\sin\theta-\frac{\partial A_\theta}{\partial\phi}\right)\right]$ $\hat{\boldsymbol{\theta}}\left[\frac{1}{r\sin\theta}\frac{\partial A_r}{\partial\phi}-\frac{1}{r}\frac{\partial(rA_\phi)}{\partial r}\right]$ $\hat{\boldsymbol{\phi}}\,\frac{1}{r}\left[\frac{\partial(rA_\theta)}{\partial r}-\frac{\partial A_r}{\partial\theta}\right]$
Divergence of a vector function $\mathbf{A}$	
$\frac{1}{r}\frac{\partial(rA_r)}{\partial r}+\frac{1}{r}\frac{\partial A_\phi}{\partial\phi}+\frac{\partial A_z}{\partial z}$	$\frac{1}{r^2}\frac{\partial(r^2A_r)}{\partial r}+\frac{1}{r\sin\theta}\frac{\partial(A_\theta\sin\theta)}{\partial\theta}+\frac{1}{r\sin\theta}\frac{\partial A_\phi}{\partial\phi}$
Laplacian of a scalar function V	
$\frac{1}{r}\frac{\partial}{\partial r}\left(r\frac{\partial V}{\partial r}\right)+\frac{1}{r^2}\frac{\partial^2 V}{\partial\phi^2}+\frac{\partial^2 V}{\partial z^2}$	$\frac{1}{r^2}\frac{\partial}{\partial r}\left(r^2\frac{\partial V}{\partial r}\right)+\frac{1}{r^2\sin\theta}\frac{\partial}{\partial\theta}\left(\sin\theta\frac{\partial V}{\partial\theta}\right)+\frac{1}{r^2\sin^2\theta}\frac{\partial^2 V}{\partial\phi^2}$
Components of the Laplacian of a vector function $\mathbf{A}$	
$\hat{\mathbf{r}}\left(\nabla^2 A_r-\frac{A_r}{r^2}-\frac{2}{r^2}\frac{\partial A_\phi}{\partial\phi}\right)$ $\hat{\boldsymbol{\phi}}\left(\nabla^2 A_\phi-\frac{A_\phi}{r^2}+\frac{2}{r^2}\frac{\partial A_r}{\partial\phi}\right)$ $\hat{\mathbf{z}}\,\nabla^2 A_z$	$\hat{\mathbf{r}}\left[\nabla^2 A_r-\frac{2A_r}{r^2}-\frac{2}{r^2\sin\theta}\frac{\partial(A_\theta\sin\theta)}{\partial\theta}-\frac{2}{r^2\sin\theta}\frac{\partial A_\phi}{\partial\phi}\right]$ $\hat{\boldsymbol{\theta}}\left(\nabla^2 A_\theta-\frac{A_\theta}{r^2\sin\theta}+\frac{2}{r^2}\frac{\partial A_r}{\partial\theta}-\frac{2\cos\theta}{r^2\sin^2\theta}\frac{\partial A_\phi}{\partial\phi}\right)$ $\hat{\boldsymbol{\phi}}\left(\nabla^2 A_\phi-\frac{A_\phi}{r^2\sin\theta}+\frac{2}{r^2\sin^2\theta}\frac{\partial A_r}{\partial\phi}+\frac{2\cos\theta}{r^2\sin^2\theta}\frac{\partial A_\theta}{\partial\phi}\right)$

A spherical coordinate system (r,θ,ϕ) specifies a point position by the radial distance r from from a fixed origin, a polar angle θ measured from a fixed zenith direction, and the azimuth angle ϕ of the orthogonal projection of r on a reference plane that passes through the origin and is orthogonal to the zenith, measured from a fixed reference direction on that plane. Figure A.2 shows a spherical coordinate system, superposed to a Cartesian system sharing the same, origin, with the z axis superposed to the zenith axis, and the xz plane corresponding to the $\phi=0$ plane of the spherical system. We have the conversion relations

$$x = r\sin\theta\cos\phi, \quad y = r\sin\theta\sin\phi, \quad z = r\cos\theta. \tag{A.2}$$

The orthogonal line elements are dr, $r\,d\theta$, and $r\sin\theta d\phi$, and the infinitesimal volume element is $r^2\sin\theta\,d\theta\,d\phi\,dr$.

In the expressions for the components of the Laplacian of a vector function of Table A.1, $\nabla^2 A_i$ stands for the Laplacian of the vector component A_i treated as if it were a scalar function. The extra terms are due to the fact that the unit vectors $\hat{\mathbf{r}}$ and $\hat{\boldsymbol{\phi}}$ for the cylindrical coordinates, and $\hat{\mathbf{r}}$, $\hat{\boldsymbol{\theta}}$ and $\hat{\boldsymbol{\phi}}$ for the spherical coordinates, are not constants.

A.2　Vector Identities

Quantities $\mathbf{A}$, $\mathbf{B}$, and $\mathbf{C}$ are vectors or vector functions of the coordinates, f and g are scalar functions of the coordinates.

$$\mathbf{A} \cdot \mathbf{B} \times \mathbf{C} = \mathbf{A} \times \mathbf{B} \cdot \mathbf{C} = \mathbf{B} \cdot \mathbf{C} \times \mathbf{A} = \mathbf{C} \cdot \mathbf{A} \times \mathbf{B} \; ; \tag{A.3}$$

$$\mathbf{A} \times (\mathbf{B} \times \mathbf{C}) = (\mathbf{C} \times \mathbf{B}) \times \mathbf{A} = (\mathbf{A} \cdot \mathbf{C}) \mathbf{B} - (\mathbf{A} \cdot \mathbf{B}) \mathbf{C} \; ; \tag{A.4}$$

$$\nabla(fg) = f \nabla g + g \nabla f \; ; \tag{A.5}$$

$$\nabla \cdot (f\mathbf{A}) = f \nabla \cdot \mathbf{A} + \mathbf{A} \cdot \nabla f \; ; \tag{A.6}$$

$$\nabla \times (f\mathbf{A}) = f \nabla \times \mathbf{A} + \nabla f \times \mathbf{A} \; ; \tag{A.7}$$

$$\nabla \cdot (\mathbf{A} \times \mathbf{B}) = \mathbf{B} \cdot \nabla \times \mathbf{A} - \mathbf{A} \cdot \nabla \times \mathbf{B} \; ; \tag{A.8}$$

$$\nabla \times (\mathbf{A} \times \mathbf{B}) = \mathbf{A} (\nabla \cdot \mathbf{B}) - \mathbf{B} (\nabla \cdot \mathbf{A}) + (\mathbf{B} \cdot \nabla) \mathbf{A} - (\mathbf{A} \cdot \nabla) \mathbf{B} \; ; \tag{A.9}$$

$$\nabla(\mathbf{A} \cdot \mathbf{B}) = \mathbf{A} \times (\nabla \times \mathbf{B}) + \mathbf{B} \times (\nabla \times \mathbf{A}) + (\mathbf{A} \cdot \nabla)\mathbf{B} + (\mathbf{B} \cdot \nabla)\mathbf{A} \; ; \tag{A.10}$$

$$\nabla^2 \mathbf{A} = \nabla(\nabla \cdot \mathbf{A}) - \nabla \times \nabla \times \mathbf{A} \; , \tag{A.11}$$

$$\nabla \times (\nabla \times \mathbf{A}) = \nabla(\nabla \cdot \mathbf{A}) - \nabla^2 \mathbf{A} \; . \tag{A.12}$$

A.3　Hyperbolic Functions

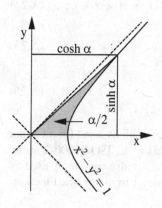

Fig. A.3

Hyperbolic functions are analogues of the ordinary trigonometric functions: as the points $(\cos \phi, \sin \phi)$ for $0 \le \phi < 2\pi$ form the unit circle $x^2 + y^2 = 1$, the points $(\cosh \alpha, \sinh \alpha)$ for $-\infty < \alpha < \infty$ form the right branch of the unit hyperbola $x^2 - y^2 = 1$, as shown in Fig. A.3. The argument α of the hyperbolic functions corresponds to twice the area of the hyperbolic sector shaded in Fig. A.3. Areas are considered negative for points on the hyperbola below the x axis. Mathematically, the functions *hyperbolic sine*, sinh, and *hyperbolic cosine*, cosh, are defined as

$$\sinh \alpha = \frac{e^\alpha - e^{-\alpha}}{2} \; , \tag{A.13}$$

$$\cosh \alpha = \frac{e^\alpha + e^{-\alpha}}{2} \; , \tag{A.14}$$

thus $\sinh \alpha$ and $\cosh \alpha$ are the odd and even parts of the exponential function e^α, respectively. From these we derive the hyperbolic *tangent, cotangent, secant* and *cosecant*

$$\tanh \alpha = \frac{\sinh \alpha}{\cosh \alpha} \,, \quad \coth \alpha = \frac{1}{\tanh \alpha} \,, \quad \mathrm{sech}\,\alpha = \frac{1}{\cosh \alpha} \,, \quad \mathrm{csch}\,\alpha = \frac{1}{\sinh \alpha} \,,$$
(A.15)

analogously to their trigonometric counterparts. The functions $\cosh \alpha$ and $\mathrm{sech}\,\alpha$ are even functions of α, all other hyperbolic functions are odd.

Some useful equalities are

$$\cosh \alpha + \sinh \alpha = \mathrm{e}^{\alpha} \,,$$
(A.16)

$$\cosh \alpha - \sinh \alpha = \mathrm{e}^{-\alpha} \,,$$
(A.17)

$$\cosh^2 \alpha - \sinh^2 \alpha = 1 \,,$$
(A.18)

$$\mathrm{sech}^2 \alpha = 1 - \tanh^2 \alpha \,,$$
(A.19)

$$\mathrm{csch}^2 \alpha = \coth^2 \alpha - 1 \,.$$
(A.20)

The inverse hyperbolic functions and their domains are

$$\mathrm{arsinh}\, x = \ln \left(x + \sqrt{x^2 + 1} \right) \,, \qquad\qquad -\infty < x < +\infty \,, \quad (A.21)$$

$$\mathrm{arcosh}\, x = \ln \left(x + \sqrt{x^2 - 1} \right) \,, \qquad\qquad 1 \le x < +\infty \,, \quad (A.22)$$

$$\mathrm{artanh}\, x = \frac{1}{2} \ln \left(\frac{1 + x}{1 - x} \right) \,, \qquad\qquad -1 < x < 1 \,, \quad (A.23)$$

$$\mathrm{arcoth}\, x = \frac{1}{2} \ln \left(\frac{x + 1}{x - 1} \right) \,, \qquad -\infty < x < -1 \text{ or } 1 < x < \infty \,, \quad (A.24)$$

$$\mathrm{arsech}\, x = \ln \left(\frac{1 + \sqrt{1 - x^2}}{x} \right) \,, \qquad\qquad 0 < x \le 1 \,, \quad (A.25)$$

$$\mathrm{arcsch}\, x = \ln \left(\frac{1}{x} + \sqrt{\frac{1}{x^2} + 1} \right) \,, \qquad\qquad x \ne 0 \,. \quad (A.26)$$

The letters *ar* in the symbols for the inverse hyperbolic functions stand for *area*, not for *arcus* as in the inverse trigonometric functions. This is the ISO 80000-2 norm, however, the prefix *arc* instead of *ar* (arcsinh, arccosh, …) is also commonly used by many authors in analogy with the nomenclature for the inverse trigonometric functions

Index

© The Editor(s) (if applicable) and The Author(s), under exclusive license to Springer
Nature Switzerland AG 2023
A. Macchi et al., *Problems in Classical Electromagnetism*,
https://doi.org/10.1007/978-3-031-22235-1

Printed in the United States
by Baker & Taylor Publisher Services